Longitudinal
Data Analysis

Chapman & Hall/CRC
Handbooks of Modern Statistical Methods

Series Editor

Garrett Fitzmaurice

Department of Biostatistics
Harvard School of Public Health
Boston, MA, U.S.A.

Aims and Scope

The objective of the series is to provide high-quality volumes covering the state-of-the-art in the theory and applications of statistical methodology. The books in the series are thoroughly edited and present comprehensive, coherent, and unified summaries of specific methodological topics from statistics. The chapters are written by the leading researchers in the field, and present a good balance of theory and application through a synthesis of the key methodological developments and examples and case studies using real data.

The scope of the series is wide, covering topics of statistical methodology that are well developed and find application in a range of scientific disciplines. The volumes are primarily of interest to researchers and graduate students from statistics and biostatistics, but also appeal to scientists from fields where the methodology is applied to real problems, including medical research, epidemiology and public health, engineering, biological science, environmental science, and the social sciences.

Published Titles

Longitudinal Data Analysis
Edited by Garrett Fitzmaurice, Marie Davidian,
Geert Verbeke, and Geert Molenberghs

Chapman & Hall/CRC

Handbooks of Modern Statistical Methods

Longitudinal Data Analysis

Edited by

Garrett Fitzmaurice
Marie Davidian
Geert Verbeke
Geert Molenberghs

CRC Press
Taylor & Francis Group
Boca Raton   London   New York

CRC Press is an imprint of the
Taylor & Francis Group, an **informa** business

A CHAPMAN & HALL BOOK

Chapman & Hall/CRC
Taylor & Francis Group
6000 Broken Sound Parkway NW, Suite 300
Boca Raton, FL 33487-2742

© 2009 by Taylor & Francis Group, LLC
Chapman & Hall/CRC is an imprint of Taylor & Francis Group, an Informa business

No claim to original U.S. Government works
Printed in the United States of America on acid-free paper
10 9 8 7 6 5 4 3 2 1

International Standard Book Number-13: 978-1-58488-658-7 (Hardcover)

Library of Congress Cataloging-in-Publication Data

Longitudinal data analysis / editors, Garrett Fitzmaurice ... [et al.].
      p. cm. -- (Chapman and Hall/CRC series of handbooks of modern statistical methods)
   Includes bibliographical references and index.
   ISBN 978-1-58488-658-7 (hardback : alk. paper)
   1. Longitudinal method. 2. Multivariate analysis. 3. Regression analysis. I. Fitzmaurice, Garrett M.,
1962- II. Title. III. Series.

QA278.L66 2008
519.5--dc22   2008020681

Visit the Taylor & Francis Web site at
http://www.taylorandfrancis.com

and the CRC Press Web site at
http://www.crcpress.com

Dedication

To Laura, Kieran, and Aidan
— G.F.

To Butch, Mom, and John
— M.D.

To Godewina, Lien, Noor, and Aart
— G.V.

To Conny, An, and Jasper
— G.M.

Contents

Preface

Longitudinal studies play a prominent role in the health, social, and behavioral sciences, as well as in public health, biological and agricultural sciences, education, economics, and marketing. They are indispensable to the study of change in an outcome over time. By measuring study participants repeatedly through time, longitudinal studies allow the direct study of temporal changes within individuals and the factors that influence change. Because the study of change is so fundamental to almost every discipline, there has been a steady growth in the number of studies using longitudinal designs. Moreover, the designs of many recent longitudinal studies have become increasingly complex.

There is a wide variety of challenges that arise in analyzing longitudinal data. By their very nature, the repeated measures arising from longitudinal studies are multivariate and have a complex random-error structure that must be appropriately accounted for in the analysis. Longitudinal studies also vary in the types of outcomes of interest. Although linear models have been the dominant approach for the analysis of longitudinal data when the outcome is continuous, in many applications the pattern of change is more faithfully characterized by a function that is non-linear in the parameters. In other settings, parametric models for longitudinal data are not sufficiently flexible to adequately capture the complex patterns of change in the outcome and their relationships to covariates; instead, more flexible functional forms are required. When the outcome of interest is discrete, there are broad classes of longitudinal models that may be suitable for analysis. However, there are distinctions among these models not only in approach, but in their relative targets of inference as well. As a result, greater care is required in the modeling of discrete longitudinal data. Another issue that complicates the analysis is the inclusion of time-varying covariates in models for longitudinal data. Longitudinal studies permit repeated measures not only of the outcome, but also of the covariates. The incorporation of covariates that change stochastically over time poses many intricate and complex analytic issues. Finally, longitudinal studies are also more prone to problems of missing data and attrition. The appropriate handling of missing data continues to pose one of the greatest challenges for the analysis of longitudinal data. These, and many other issues, increase the complexity of longitudinal data analysis.

The last 20 years have seen many remarkable advances in statistical methodology for analyzing longitudinal data. Although there are a number of books describing statistical models and methods for the analysis of longitudinal data, to date there is no volume that provides a comprehensive, coherent, unified, and up-to-date summary of the major advances. This has provided the main impetus for *Longitudinal Data Analysis*. This book constitutes a carefully edited collection of chapters that synthesize the state of the art in the theory and application of longitudinal data analysis. The book is comprised of 23 expository chapters, dealing with five broad themes. These chapters have been written by many of the world's leading experts in the field. Each chapter integrates and illustrates important research threads in the statistical literature, rather than focusing on a narrowly defined topic. Each part of the book begins with an introductory chapter that provides useful background material and a broad overview to set the stage for subsequent chapters. The book combines a good blend of theory and applications; many of the chapters include examples and case studies using data sets drawn from various disciplines. Many of the data sets used to illustrate methods can be downloaded from the Web site for the book (http://www.biostat.harvard.edu/~fitzmaur/lda), as can sample source code for fitting certain models.

Although our coverage of topics in the book is quite broad, it is certainly not complete. Our selection of topics required judicious choices to be made; we have decided to place greater emphasis on statistical models and methods that we think likely to endure. The book is intended to have a broad appeal. It should be of interest to all statisticians involved either in the development of methodology or the application of new and advanced methods to longitudinal research. We anticipate that the book will also be of interest to quantitatively oriented researchers from various disciplines.

Finally, the compilation of this book would not have been possible without the willingness, persistence, and dedication of each of the contributing authors; we thank them wholeheartedly for their tremendous efforts and the excellent quality of the chapters they have written. We would also like to thank the many friends and colleagues who have helped us produce this book. A special word of thanks to Butch Tsiatis and Nan Laird who reviewed several chapters and provided insightful feedback. Last, but not least, we thank Rob Calver, Aquiring Editor at Chapman & Hall/CRC Press of Taylor & Francis, for encouragement to undertake this project. The original seeds of this book arose from conversations Rob Calver had with a number of distinguished colleagues. We are grateful to all, most particularly to Rob, for his strong belief in the project and his enthusiasm and perseverance to see the project through from beginning to end.

Garrett Fitzmaurice
Boston, Massachusetts

Marie Davidian
Raleigh, North Carolina

Geert Verbeke
Leuven, Belgium

Geert Molenberghs
Diepenbeek, Belgium

Editors

Garrett Fitzmaurice is Associate Professor of Psychiatry (Biostatistics) at the Harvard Medical School, Associate Professor in the Department of Biostatistics at the Harvard School of Public Health, and Foreign Adjunct Professor of Biostatistics at the Karolinska Institute, Sweden. He is a Fellow of the American Statistical Association and a member of the International Statistical Institute. He has served as Associate Editor for *Biometrics*, the *Journal of the Royal Statistical Society, Series B*, and *Biostatistics*; currently, he is Statistics Editor for the journal *Nutrition*. His research and teaching interests are in methods for analyzing longitudinal and repeated measures data. A major focus of his methodological research has been on the development of statistical methods for analyzing repeated binary data and for handling the problem of attrition in longitudinal studies. Much of his collaborative research has concentrated on applications to mental health research, broadly defined. He has co-authored the textbook *Applied Longitudinal Analysis* (Wiley, 2004) and received the American Statistical Association's Excellence in Continuing Education Award for a short course on longitudinal analysis at the Joint Statistical Meetings in 2006.

Marie Davidian is William Neal Reynolds Distinguished Professor of Statistics at North Carolina State University and Adjunct Professor of Biostatistics and Bioinformatics at Duke University. She is a Fellow of the American Statistical Association, the Institute of Mathematical Statistics, and the American Association for the Advancement of Science. She has served as an Associate Editor for *Biometrics* and the *Journal of the American Statistical Association*, and was Coordinating Editor of *Biometrics* in 2000–2002. She is currently Executive Editor of *Biometrics*. Her research interests include the development of methods for analysis of longitudinal data arising in contexts such as pharmacokinetics, where non-linear, often mechanistically based models for individual behavior are used; and for joint modeling and analysis of longitudinal data and time-to-event outcomes.

Geert Verbeke is a Professor of Biostatistics at the Biostatistical Centre of the Katholieke Universiteit Leuven in Belgium. He has published a number of methodological articles on various aspects of models for longitudinal data analyses, with particular emphasis on mixed models. He held a visiting position in the Department of Biostatistics of the Johns Hopkins University in Baltimore, MD, as well as in the affiliated Institute of Gerontology. He is Past President of the Belgian Region of the International Biometric Society, International Program Chair for the International Biometric Conference in Montreal (2006), and Joint Editor of the *Journal of the Royal Statistical Society, Series A* (2005–2008). He has served as Associate Editor for several journals, including *Biometrics* and *Applied Statistics*. He is a Fellow of the American Statistical Association.

Geert Molenberghs is Professor of Biostatistics at the Universiteit Hasselt in Belgium. He received a B.S. degree in mathematics (1988) and a Ph.D. in biostatistics (1993) from the Universiteit Antwerpen. He published methodological work on surrogate markers in clinical trials, categorical data, longitudinal data analysis, and the analysis of non-response in clinical and epidemiological studies. He served as Joint Editor for *Applied Statistics* (2001–2004) and as Associate Editor for several journals, including *Biometrics* and *Biostatistics*. He was President of the International Biometric Society (2004–2005) and later Vice-President (2006). He was elected a Fellow of the American Statistical Association and received the Guy Medal in Bronze from the Royal Statistical Society. He is an elected member of the International Statistical Institute. He has held visiting positions at the Harvard School of Public Health (Boston). He has co-authored a book on surrogate marker evaluation in clinicial trials (Springer, 2005) and on incomplete data in clinical studies (Wiley, 2007).

Geert Molenberghs and Geert Verbeke have co-authored monographs on linear mixed models for longitudinal data (Springer, 2000) and on models for discrete longitudinal data (Springer, 2005). They received the American Statistical Association's Excellence in Continuing Education Award, based on short courses on longitudinal and incomplete data at the Joint Statistical Meetings of 2002, 2004, and 2005.

Contributors

Paul S. Albert Biometric Research Branch, National Cancer Institute
Rockville, Maryland

Tihomir Asparouhov Muthén & Muthén
Los Angeles, California

Babette A. Brumback Division of Biostatistics, University of Florida
Gainesville, Florida

Lyndia C. Brumback Department of Biostatistics, University of Washington
Seattle, Washington

James R. Carpenter Medical Statistics Unit, London School of Hygiene &
        Tropical Medicine
United Kingdom

Raymond Carroll Department of Statistics, Texas A&M University
College Station, Texas

Paul Catalano Department of Biostatistics, Dana Farber Cancer Institute
Boston, Massachusetts

Peter Diggle Department of Mathematics and Statistics, Lancaster
        University
United Kingdom

Christel Faes Center for Statistics, Hasselt University
Diepenbeek, Belgium

Steffen Fieuws Biostatistical Centre, Katholieke Universiteit
Leuven, Belgium

Dean A. Follmann Biostatistics Research Branch, National Institute of Allergy
        & Infectious Diseases
Bethesda, Maryland

Helena Geys Biometrics and Reporting, Johnson and Johnson
Beerse, Belgium

Robin Henderson School of Mathematics and Statistics, University of Newcastle
United Kingdom

Miguel A. Hernán Department of Epidemiology, Harvard School of Public Health
Boston, Massachusetts

Michael G. Kenward   Medical Statistics Unit, London School of Hygiene & Tropical
                          Medicine
                          United Kingdom

Xihong Lin   Department of Biostatistics, Harvard School of Public Health
                          Boston, Massachusetts

Mary J. Lindstrom   Department of Biostatistics & Medical Informatics,
                          University of Wisconsin
                          Madison, Wisconsin

Stuart Lipsitz   Division of General Medicine, Brigham and Women's Hospital
                          Boston, Massachusetts

Roderick Little   Department of Biostatistics, University of Michigan
                          Ann Arbor, Michigan

Hans-Georg Müller   Department of Statistics, University of California
                          Davis, California

Bengt Muthén   Graduate School of Education & Information Studies
                          University of California
                          Los Angeles, California

Peter Philipson   School of Mathematics and Statistics, University of Newcastle
                          United Kingdom

Sophia Rabe-Hesketh   Graduate School of Education, University of California
                          Berkeley, California

Stephen Raudenbush   Department of Sociology, University of Chicago
                          Chicago, Illinois

James M. Robins   Department of Epidemiology, Harvard School of Public Health
                          Boston, Massachusetts

Andrea Rotnitzky   Department of Economics, Di Tella University
                          Buenos Aires, Argentina

Anders Skrondal   Department of Statistics, London School of Economics
                          United Kingdom

S. J. Welham   Biomathematics and Bioinformatics Rothamsted Research
                          Harpenden, United Kingdom

PART I

Introduction and Historical Overview

CHAPTER 1

Advances in longitudinal data analysis: An historical perspective

Garrett Fitzmaurice and Geert Molenberghs

Contents

1.1 Introduction

There have been remarkable developments in statistical methodology for longitudinal data analysis in the past 25 to 30 years. Statisticians and empirical researchers now have access to an increasingly sophisticated toolbox of methods. As might be expected, there has been a lag between the recent developments that have appeared in the statistical journals and their widespread application to substantive problems. At least part of the reason why these advances have been somewhat slow to move into the mainstream is their limited implementation in widely available standard computer software. Recently, however, the introduction of new programs for analyzing multivariate and longitudinal data has made many of these methods far more accessible to statisticians and empirical researchers alike. Also, because statistical software is constantly evolving, we can anticipate that many of the more recent advances will soon be implemented. Thus, the outlook is bright that modern methods for longitudinal analysis will be applied more widely and across a broader spectrum of disciplines.

In this chapter, we take an historical perspective and review many of the key advances that have been made, especially in the past 30 years. Our review will be somewhat selective, and omissions are inevitable; our main goal is to highlight important and enduring developments in methodology. No attempt is made to assign priority to these methods. Our review will set the stage for the remaining chapters of the book, where the focus is on the current state of the art of longitudinal data analysis.

1.2 Early origins of linear models for longitudinal data analysis

The analysis of change is a fundamental component of so many research endeavors in almost every discipline. Many of the earliest statistical methods for the analysis of change were based on the analysis of variance (ANOVA) paradigm, as originally developed by

R. A. Fisher. One of the earliest methods proposed for analyzing longitudinal data was a mixed-effects ANOVA, with a single random subject effect. The inclusion of a random subject effect induced positive correlation among the repeated measurements on the same subject. Note that throughout this chapter we use the terms *subjects* and *individuals* interchangeably to refer to the participants in a longitudinal study. Interestingly, it was the British astronomer George Biddel Airy who laid the foundations for the linear mixed-model formulation (Airy, 1861), before it was put on a more formal theoretical footing in the seminal work of R. A. Fisher (see, for example, Fisher, 1918, 1925). Airy's work on a model for errors of observation in astronomy predated Fisher's more systematic study of related issues within the ANOVA paradigm (e.g., Fisher's [1921, 1925] writings on the intraclass correlation). Scheffé (1956) provides a fascinating discussion of the early contributions of 19th century astronomers to the development of the theory of random-effects models. As such, it can be argued that statistical methods for the analysis of longitudinal data, in common with classical linear regression and the method of least squares, have their earliest origins in the field of astronomy.

The mixed-effects ANOVA model has a long history of use for analyzing longitudinal data, where it is often referred to as the *univariate* repeated-measures ANOVA. Statisticians recognized that a longitudinal data structure, with N individuals and n repeated measurements, has striking similarities to data collected in a randomized block design, or the closely related split-plot design. So it seemed natural to apply ANOVA methods developed for these designs (e.g., Yates, 1935; Scheffé, 1959) to the repeated-measures data collected from longitudinal studies. In doing so, the individuals in the study are regarded as the blocks or main plots. The univariate repeated-measures ANOVA model can be written as

$$Y_{ij} = \boldsymbol{X}'_{ij}\boldsymbol{\beta} + b_i + e_{ij}, \quad i = 1, \ldots, N; j = 1, \ldots, n,$$

where Y_{ij} is the outcome of interest, $\boldsymbol{X}_{ij}$ is a design vector, $\boldsymbol{\beta}$ is a vector of regression parameters, $b_i \sim N(0, \sigma_b^2)$, and $e_{ij} \sim N(0, \sigma_e^2)$. In this model, the blocks or plot effects are regarded as random rather than fixed effects. The random effect, b_i, represents an aggregation of all the unobserved or unmeasured factors that make individuals respond differently. The consequence of including a single, individual-specific random effect is that it induces positive correlation among the repeated measurements, albeit with the following highly restrictive "compound symmetry" structure for the covariance: constant variance $\mathrm{Var}(Y_{ij}) = \sigma_b^2 + \sigma_e^2$ and constant covariance $\mathrm{Cov}(Y_{ij}, Y_{ik}) = \sigma_b^2$.

On the one hand, the univariate repeated-measures ANOVA model provided a natural generalization of Student's (1908) paired t-test to handle more than two repeated measurements, in addition to various between-subject factors. On the other hand, it can be argued that this model was a Procrustean bed for longitudinal data because the blocks or plots were random rather than fixed by design and there is no sense in which measurement occasions can ever be randomized. Importantly, it is only when the within-subject factor is randomly allocated to individuals that randomization arguments can be made to justify the "compound symmetry" structure for the covariance. There is no basis for this randomization argument in the case of longitudinal data, where the within-subject factor is the measurement occasions. Recognizing that the compound symmetry assumption is restrictive, and to accommodate more general covariance structures for the repeated measures, Greenhouse and Geisser (1959) suggested a correction to the numerator and denominator degrees of freedom of tests derived from the univariate repeated-measures ANOVA (see also Huynh and Feldt, 1976).

In spite of its restrictive assumptions, and many obvious shortcomings, the univariate repeated-measures ANOVA model can be considered a forerunner of more versatile regression models for longitudinal data. As we will discuss later, the notion of allowing effects to vary randomly from one individual to another is the basis of many modern regression models for longitudinal data analysis. Also, it must be remembered that the elegant computational formulae for balanced designs meant that the calculation of ANOVA tables was relatively straightforward, albeit somewhat laborious. For balanced data, estimates of variance components could be readily obtained in closed form by equating ANOVA mean squares to their expectations; sometime later, Henderson (1963) developed a related approach for unbalanced data. So, from an historical perspective, an undoubted appeal of the repeated-measures ANOVA was that it was one of the few models that could realistically be fit to longitudinal data at a time when computing was in its infancy. This explains why, in those days, the key issue perceived to arise with incomplete data was lack of balance.

A related approach for the analysis of longitudinal data with an equally long history, but requiring somewhat more advanced computations, is the repeated-measures *multivariate* analysis of variance (MANOVA). While the *univariate* repeated-measures ANOVA is conceptualized as a model for a *single* response variable, allowing for positive correlation among the repeated measures on the same individual via the inclusion of a random subject effect, MANOVA is a model for multivariable responses. As originally developed, MANOVA was intended for the simultaneous analysis of a single measure of a multivariate vector of substantively *distinct* response variables. In contrast, while longitudinal data are multivariate, the vector of responses are commensurate, being repeated measures of the same response variable over time. So, although MANOVA was developed for multiple, but distinct, response variables, statisticians recognized that such data share a common feature with longitudinal data, namely, that they are correlated. This led to the development of a very specific variant of MANOVA, known as repeated-measures analysis by MANOVA (or sometimes referred to as multivariate repeated-measures ANOVA).

A special case of the repeated-measures analysis by MANOVA is a general approach known as *profile analysis* (Box, 1950; see also Geisser and Greenhouse, 1958; Greenhouse and Geisser, 1959). It proceeds by constructing a set of derived variables, based on a linear combination of the original sequence of repeated measures, and using relevant subsets of these to address questions about longitudinal change and its relation to between-subject factors. These derived variables provide information about the mean level of the response, averaged over all measurement occasions, and also about change in the response over time. For the most part, the primary interest in a longitudinal analysis is in the analysis of the latter derived variables. The multiple derived variables representing the effects of measurement occasions are then analyzed by MANOVA.

Box (1950) provided one of the earliest descriptions of this approach, proposing the construction of derived variables that represent polynomial contrasts of the measurement occasions; closely related work can be found in Danford, Hughes, and McNee (1960), Geisser (1963), Potthoff and Roy (1964), Cole and Grizzle (1966), and Grizzle and Allen (1969). Alternative transformations can be used, as the MANOVA test statistics are invariant to how change over time is characterized in the transformation of the original repeated measures. Although the MANOVA approach is computationally more demanding than the univariate repeated-measures ANOVA, an appealing feature of the method is that it allows assumptions on the structure of the covariance among repeated measures to be relaxed. In standard applications of the method, no explicit structure is assumed for the covariance among repeated measures (other than homogeneity of covariance across different individuals).

There is a final related approach to longitudinal data analysis based on the ANOVA paradigm that has a long history and remains in widespread use. In this approach, the sequence of repeated measures for each individual is reduced to a single summary value (or, in certain cases, a set of summary values). The major motivation behind the use of this approach is that, if the sequence of repeated measures can be reduced to a single number summary, then ANOVA methods (or, alternatively, non-parametric methods) for the analysis of a univariate response can be applied. For example, the area under the curve (AUC) is one common measure that is frequently used to summarize the sequence of repeated measures on any individual. The AUC, usually approximated by the area of the trapezoids joining adjacent repeated measurements, can then be related to covariates (e.g., treatment or intervention groups) using ANOVA. Wishart (1938) provided one of the earliest descriptions of this approach in a paper with the almost unforgettable title "Growth-rate determinations in nutrition studies with the bacon pig, and their analysis"; closely related methods can be found in Box (1950) and Rao (1958).

Within a limited context, the three ANOVA-based approaches discussed thus far provided the basis for a longitudinal analysis. However, all of these methods had shortcomings that limited their usefulness in applications. The univariate repeated-measures ANOVA made very restrictive assumptions about the covariance structure for repeated measures on the same individual. The assumed compound symmetry form for the covariance is not appropriate for longitudinal data for at least two reasons. First, the constraint on the correlation among repeated measurements is somewhat unappealing for longitudinal data, where the correlations are expected to decay with increasing separation in time. Second, the assumption of constant variance across time is often unrealistic. In many longitudinal studies the variability of the response at the beginning of the study is discernibly different from the variability toward the completion of the study; this is especially the case when the first repeated measurement represents a "baseline" response. Finally, as originally conceived, the repeated-measures ANOVA model was developed for the analysis of data from designed experiments, where the repeated measures are obtained at a set of occasions common to all individuals, the covariates are discrete factors (e.g., representing treatment group and time), and the data are complete. As a result, early implementations of the repeated-measures ANOVA could not be readily applied to longitudinal data that were irregularly spaced or incomplete, or when it was of interest to include quantitative covariates in the analysis.

In contrast, the repeated-measures analysis by MANOVA did not make restrictive assumptions on the covariance among the longitudinal responses on the same individual. As a result, the correlations could assume any pattern and the variability could change over time. However, MANOVA had a number of features that also limited its usefulness. In particular, the MANOVA formulation forced the within-subject covariates to be the same for all individuals. There are at least two practical consequences of this constraint. First, repeated-measures MANOVA cannot be used when the design is unbalanced over time (i.e., when the vectors of repeated measures are of different lengths and/or obtained at different sequences of time). Second, the repeated-measures MANOVA (at least as implemented in existing statistical software packages) did not allow for general missing-data patterns to arise. Thus, if any individual has even a single missing response at any occasion, the entire data vector from that individual must be excluded from the analysis. This so-called "listwise" deletion of missing data from the analysis often results in dramatically reduced sample size and very inefficient use of the available data. Listwise deletion of missing data can also produce biased estimators of change in the mean response over time when the so-called "completers" (i.e., those with no missing data) are not a random sample from the target population. Furthermore, balance between treatment groups is destroyed, hence the early attraction of so-called imputation methods.

Finally, although the analysis of summary measures had a certain appeal due to the simplicity of the method, it too had a number of distinct drawbacks. By definition, it forces the data analyst to focus on only a single aspect of the repeated measures over time; when n repeated measures are replaced by a single-number summary, there must necessarily be some loss of information. Also, individuals with discernibly different response profiles can produce the same summary measure. A second potential drawback is that the covariates must be time-invariant; the method cannot be applied when covariates are time-varying. Furthermore, many of the simple summary measures are not so well defined when there are missing data or irregularly spaced repeated measures. Even in cases where the summary measure can be defined, the resulting analysis is not fully efficient. In particular, when some individuals have missing data or different numbers of repeated measures, the derived summary measures no longer have the same variance, thereby violating the fundamental assumption of homogeneity of variance for standard ANOVA models.

In summary, the origins of the statistical analysis of change can be traced back to the ANOVA paradigm. ANOVA methods have a long and extensive history of use in the analysis of longitudinal data. While ANOVA methods can provide a reasonable basis for a longitudinal analysis in cases where the study design is very simple, they have many shortcomings that have limited their usefulness in applications. In many longitudinal studies there is considerable variation among individuals in both the number and timing of measurements. The resulting data are highly unbalanced and not readily amenable to ANOVA methods developed for balanced designs. It was these features of longitudinal data that provided the impetus for statisticians to develop far more versatile techniques that can handle the commonly encountered problems of data that are unbalanced and incomplete, mistimed measurements, time-varying and time-invariant covariates, and responses that are discrete rather than continuous.

1.3 Linear mixed-effects model for longitudinal data

The linear mixed-effects model is probably the most widely used method for analyzing longitudinal data. Although the early development of mixed-effects models for hierarchical or clustered data can be traced back to the ANOVA paradigm (see, for example, Scheffé, 1959) and to the seminal paper by Harville (1977), their usefulness for analyzing longitudinal data, especially in the life sciences, was highlighted in the 1980s in a widely cited paper by Laird and Ware (1982). Goldstein (1979) is often seen as the counterpart for the humanities. The idea of allowing certain regression coefficients to vary randomly across individuals was also a recurring theme in the early contributions to growth curve analysis by Wishart (1938), Box (1950), Rao (1958), Potthoff and Roy (1964), and Grizzle and Allen (1969); these early contributions to growth curve modeling laid the foundation for the linear mixed-effects model. The idea of randomly varying regression coefficients was also a common thread in the so-called two-stage approach to analyzing longitudinal data. In the two-stage formulation, the repeated measurements on each individual are assumed to follow a regression model with distinct regression parameters for each individual. The distribution of these individual-specific regression parameters, or "random effects," is modeled in the second stage. A version of the two-stage formulation was popularized by biostatisticians working at the U.S. National Institutes of Health (NIH). They proposed a method for analyzing repeated-measures data where, in the first stage, subject-specific regression coefficients are estimated using ordinary least-squares regression. In the second stage, the estimated regression coefficients are then analyzed as summary measures using standard parametric (or non-parametric) methods. Interestingly, this method for analyzing repeated-measures data became known as the "NIH method." Although it is difficult to attribute the popularization of the NIH method to any single biostatistician at NIH, Sam Greenhouse, Max Halperin, and Jerry Cornfield introduced many biostatisticians to this technique. In the agricultural sciences, a similar approach was popularized in a highly cited paper by Rowell and Walters (1976). Rao (1965) put this two-stage approach on a more formal footing by specifying a parametric growth curve model that assumed normally distributed random growth curve parameters.

Although remarkably simple and useful, the two-stage formulation of the linear mixed-effects model introduced some unnecessary restrictions. Specifically, in the first stage, the covariates were restricted to be time-varying (with the exception of the column of 1s for the intercept); between-subject (or time-invariant) covariates could only be introduced in the second stage, where the individual-specific regression coefficients were modeled as a linear function of these covariates. The two-stage formulation placed unnecessary, and often inconvenient, constraints on the choice of the design matrix for the fixed effects. But, from an historical perspective, it provided motivation for the main ideas and concepts underlying linear mixed-effects models. The method can be viewed as based on summaries and consequently it shares the disadvantages with such methods.

In the early 1980s, Laird and Ware (1982), drawing upon a general class of mixed models introduced earlier by Harville (1977), proposed a flexible class of linear mixed-effects models for longitudinal data. These models could handle the complications of mistimed and incomplete measurements in a very natural way. The linear mixed-effects model is given by

$$Y_{ij} = \boldsymbol{X}'_{ij}\boldsymbol{\beta} + \boldsymbol{Z}'_{ij}\boldsymbol{b}_i + e_i$$

where $\boldsymbol{Z}_{ij}$ is a design vector for the random effects, $\boldsymbol{b}_i \sim N(\boldsymbol{0}, G)$, and $\boldsymbol{e}_i \sim N(\boldsymbol{0}, R_i)$. Commonly, it is assumed that $V_i = \sigma^2 I$, although additional correlation among the errors can be accommodated by allowing more general covariance structures for V_i (e.g., autoregressive). In addition, alternative distributions for the random effects can be entertained. The linear mixed-effects model proposed by Laird and Ware (1982) included the univariate repeated-measures ANOVA and growth curve models for longitudinal data as special cases. In addition, the Laird and Ware (1982) formulation of the model had two desirable features: first, there were fewer restrictions on the design matrices for the fixed and random effects; second, the model parameters could be estimated efficiently via likelihood-based methods. Previously, difficulties with estimation of mixed-effects models had held back their widespread application to longitudinal data. Laird and Ware (1982) showed how the expectation–maximization (EM) algorithm (Dempster, Laird, and Rubin, 1977) could be used to fit this general class of models for longitudinal data. Soon after, Jennrich and Schluchter (1986) proposed a variety of alternative algorithms, including Fisher scoring and Newton–Raphson. Currently, maximum likelihood and restricted maximum likelihood estimation, the latter devised to diminish the small-sample bias of maximum likelihood, are the most frequently employed routes for estimation and inference (Verbeke and Molenberghs 2000; Fitzmaurice, Laird, and Ware, 2004).

So, by the mid-1980s, a very general class of linear models for longitudinal data had been proposed that could handle issues of unbalanced data, due to either mistimed measurement or missing data, could handle both time-varying and time-invariant covariates, and provided a flexible, yet parsimonious, model for the covariance. Moreover, these developments appeared at a time when there were great advances in computing power. It was not too long before these methods were available at the desktop and were being applied to longitudinal data in a wide variety of disciplines. Nevertheless, many of the simple and simplifying procedures stuck, out of habit and/or because they have become part of standard operating procedures.

1.4 Models for non-Gaussian longitudinal data

The advances in methods for longitudinal data analysis discussed so far have been based on linear models for continuous responses that may be approximately normally distributed. Next, we consider some of the parallel developments when the response variable is discrete. The developments in methods for analyzing a continuous longitudinal response span more than a century, from the early work on simple random-effects models by the British astronomer Airy (1861) through the landmark paper on linear mixed-effects models for longitudinal data by Laird and Ware (1982). In contrast, many of the advances in methods for discrete longitudinal data have been concentrated in the last 25 to 30 years, harnessing the high-speed computing resources available at the desktop.

When the longitudinal response is discrete, linear models are no longer appropriate for relating changes in the mean response to covariates. Instead, statisticians have developed extensions of *generalized linear models* (Nelder and Wedderburn, 1972) for longitudinal data. Generalized linear models provide a unified class of models for regression analysis of independent observations of a discrete or continuous response. A characteristic feature of generalized linear models is that a suitable non-linear transformation of the mean response is assumed to be a linear function of the covariates. As we will discuss, this non-linearity raises some additional issues concerning the interpretation of the regression coefficients in models for longitudinal data. Statisticians have extended generalized linear models to handle longitudinal observations in a number of different ways; here we consider three broad, but quite distinct, classes of regression models for longitudinal data: (i) *marginal* or *population-averaged* models, (ii) *random-effects* or *subject-specific* models, and (iii) *transition* or *response conditional* models. These models differ not only in how the correlation among the repeated measures is accounted for, but also have regression parameters with discernibly different interpretations. These differences in interpretation reflect the different targets of inference of these models. Here we sketch some of the early developments of these models from an historical perspective; later chapters of this book will discuss many of these models in much greater detail. Because binary data are so common, we focus much of our review on models for longitudinal binary data. Most of the developments apply to, say, categorical data and counts equally well.

1.4.1 Marginal or population-averaged models

As mentioned above, the extensions of generalized linear models from the univariate to the multivariate response setting have followed a number of different research threads. In this section we consider an approach for extending generalized linear models to longitudinal data that leads to a class of regression models known as *marginal* or *population-averaged* models (see Chapter 3 of this volume). It must be admitted from the outset that the former term is potentially confusing; nonetheless it has endured *faute de mieux*. The term *marginal* in this context is used to emphasize that the model for the mean response at each occasion depends only on the covariates of interest, and not on any random effects or previous responses. This is in contrast to *mixed-effects* models, where the mean response depends not only on covariates but also on a vector of random effects, and to *transition* or generally conditional models (e.g., Markov models), where the mean response depends also on previous responses.

Marginal models provide a straightforward way to extend generalized linear models to longitudinal data. They directly model the mean response at each occasion, $E(Y_{ij}|\boldsymbol{X}_{ij})$, using an appropriate link function. Because the focus is on the marginal mean and its dependence on the covariates, marginal models do not necessarily require full distributional assumptions for the vector of repeated responses, only a regression model for the mean response. As we will discuss later, this can be advantageous, as there are few tractable likelihoods for marginal models for discrete longitudinal data.

Typically, a marginal model for longitudinal data has the following three-part specification:

1. The mean of each response, $E(Y_{ij}|\boldsymbol{X}_{ij}) = \mu_{ij}$, is assumed to depend on the covariates through a known link function

$$h^{-1}(\mu_{ij}) = \boldsymbol{X}'_{ij}\boldsymbol{\beta}.$$

2. The variance of each Y_{ij}, given the covariates, is assumed to depend on the mean according to

$$\mathrm{Var}(Y_{ij}|\boldsymbol{X}_{ij}) = \phi\, v(\mu_{ij}),$$

where $v(\mu_{ij})$ is a known *variance function* and ϕ is a scale parameter that may be known or may need to be estimated.

3. The conditional within-subject association among the vector of repeated responses, given the covariates, is assumed to be a function of an additional set of association parameters, $\boldsymbol{\alpha}$ (and may also depend upon the means, μ_{ij}).

Of the three, the first is the key component of a marginal model and specifies the model for the mean response at each occasion, $E(Y_{ij}|\boldsymbol{X}_{ij})$, and its dependence on the covariates. However, there is an implicit assumption in the first component that is often overlooked. Marginal models assume that the conditional mean of the jth response, given $\boldsymbol{X}_{i1}, \ldots, \boldsymbol{X}_{in}$, depends only on $\boldsymbol{X}_{ij}$, that is,

$$E(Y_{ij}|X_i) = E(Y_{ij}|\boldsymbol{X}_{i1}, \ldots, \boldsymbol{X}_{in}) = E(Y_{ij}|\boldsymbol{X}_{ij}),$$

where obviously $X_i = (\boldsymbol{X}_{i1}, \ldots, \boldsymbol{X}_{in})$; see Fitzmaurice, Laird, and Rotnitzky (1993) and Pepe and Anderson (1994) for a discussion of the implications of this assumption. With time-invariant covariates, this assumption necessarily holds. Also, with time-varying covariates that are fixed by design of the study (e.g., time since baseline, treatment group indicator in a crossover trial), the assumption also holds, as values of the covariates are determined *a priori* by study design and in a manner unrelated to the longitudinal response. However, when a time-varying covariate varies randomly over time, the assumption may no longer hold. As a result, somewhat greater care is required when fitting marginal models with time-varying covariates that are not fixed by design of the study. This problem has long been recognized by econometricians (see, for example, Engle, Hendry, and Richard, 1983), and there is now an extensive statistical literature on this topic (see, for example, Robins, Greenland, and Hu, 1999).

The second component specifies the marginal variance at each occasion, with the choice of variance function depending upon the type of response. For balanced longitudinal designs, a separate scale parameter, ϕ_j, can be specified at each occasion; alternatively, the scale parameter could depend on the times of measurement, with $\phi(t_{ij})$ being some parametric function of t_{ij}. Restriction to a single unknown parameter ϕ is especially limiting in the analysis of continuous responses where the variance of the repeated measurements is often not constant over the duration of the study.

The first two components of a marginal model specify the mean and variance of Y_{ij}, closely following the standard formulation of a generalized linear model. The only minor difference is that marginal models typically specify a common link function relating the vector of mean responses to the covariates. It is the third component that recognizes the characteristic lack of independence among longitudinal data by modeling the within-subject association among the repeated responses from the same individual. In describing this third component, we have been careful to avoid the use of the term *correlation* for two reasons. First, with a continuous response variable, the correlation is a natural measure of the linear dependence among the repeated responses and is variation independent of the mean response. However, this is not the case with discrete responses. With discrete responses, the correlations are constrained by the mean responses, and vice versa. The most extreme example of this arises when the response variable is binary. For binary responses, the correlations are heavily restricted to ranges that are determined by the means (or probabilities of success) of the responses. As a result, the correlation is not the most natural measure of within-subject association with discrete responses. For example, for two associated binary outcomes with probabilities of success equal to 0.2 and 0.8, the correlation can be no larger than 0.25.

Instead, the odds ratio is a preferable metric for association among pairs of binary responses. There are no restrictions for two outcomes, while they are mild for longer sequences of repeated measures. Second, for a continuous response that has a multivariate normal distribution, the correlations, along with the variances and the means, completely specify the joint distribution of the vector of longitudinal responses. This is not the case with discrete data. The vector of means and the covariance matrix do not, in general, completely specify the joint distribution of discrete longitudinal responses. Instead, the joint distribution requires specification of pairwise and higher-order associations among the responses.

This three-part specification of a marginal model makes transparent the extension of generalized linear models to longitudinal data. The first two parts of the marginal model correspond to the standard generalized linear model, albeit with no explicit distributional assumptions about the responses. It is the third component, the incorporation of a model for the within-subject association among the repeated responses from the same individual, that represents the main extension of generalized linear models to longitudinal data. A crucial aspect of marginal models is that the mean response and within-subject association are modeled separately. This separation of the modeling of the mean response and the association among responses has important implications for interpretation of the regression parameters $\boldsymbol{\beta}$. In particular, the regression parameters have *population-averaged* interpretations. They describe how the mean response in the population changes over time and how these changes are related to covariates. Note that the interpretation of $\boldsymbol{\beta}$ is not altered in any way by the assumptions made about the nature or magnitude of the within-subject association.

From an historical perspective, it is difficult to pinpoint the origins of marginal models. In the case of linear models, the earliest approaches based on the ANOVA paradigm fit squarely within the framework of marginal models. In a certain sense, the necessity to distinguish marginal models from other classes of models becomes critical only for discrete responses. The development of marginal models for discrete longitudinal data has its origins in likelihood-based approaches, where the three-part specification given above is extended by making full distributional assumptions about the $n \times 1$ vector of responses, $\boldsymbol{Y}_i = (Y_{i1}, \ldots, Y_{in})'$. Next, we trace some of these early developments and highlight many of the issues that have complicated the application of marginal models to discrete data, leading to the widespread use of alternative, semi-parametric methods.

At least three main research threads can be distinguished in the development of likelihood-based marginal models for discrete longitudinal data. Because binary data are so common, we focus much of this review on models for longitudinal binary data. One of the earliest likelihood-based approaches was proposed by Gumbel (1961), who posited a latent-variable model for multivariate binary data. In this approach, there is a vector of unobserved latent variables, say $L_{i1}, \ldots, L_{in}$, and each of these is related to the observed binary responses via

$$Y_{ij} = \begin{cases} 1 & L_{ij} \leq \boldsymbol{X}'_{ij}\boldsymbol{\beta}, \\ 0 & L_{ij} > \boldsymbol{X}'_{ij}\boldsymbol{\beta}. \end{cases}$$

Assuming a multivariate joint distribution for $L_{i1}, \ldots, L_{in}$ identifies the joint distribution for $Y_{i1}, \ldots, Y_{in}$, with

$$\Pr(Y_{i1} = 1, Y_{i2} = 1, \ldots, Y_{in} = 1) = \Pr(L_{i1} \leq \boldsymbol{X}'_{i1}\boldsymbol{\beta}, L_{i2} \leq \boldsymbol{X}'_{i2}\boldsymbol{\beta}, \ldots, L_{in} \leq \boldsymbol{X}'_{in}\boldsymbol{\beta})$$
$$= F(\boldsymbol{X}'_{i1}\boldsymbol{\beta}, \boldsymbol{X}'_{i2}\boldsymbol{\beta}, \ldots, \boldsymbol{X}'_{in}\boldsymbol{\beta}),$$

where $F(\cdot)$ denotes the joint cumulative distribution function of the latent variables. Furthermore, any dependence among the L_{ij} induces dependence among the Y_{ij}. For example, a bivariate logistic distribution for any L_{ij} and L_{ik} induces marginally a logistic regression model for Y_{ij} and Y_{ik},

$$E(Y_{ij}|\boldsymbol{X}_{ij}) = \frac{\exp(\boldsymbol{X}'_{ij}\boldsymbol{\beta})}{1 + \exp(\boldsymbol{X}'_{ij}\boldsymbol{\beta})},$$

with positive correlation between Y_{ij} and Y_{ik}.

Although Gumbel's (1961) model can accommodate more than two responses, the marginal covariance among the Y_{ij} becomes quite complicated. Other multivariate distributions for the latent variables, with arbitrary marginals (e.g., logistic or probit) for the Y_{ij} can be derived, but in general the joint distribution for $Y_{i1}, \ldots, Y_{in}$ is relatively complicated, as is the marginal covariance structure. As a result, these models were not widely adopted for the analysis of discrete longitudinal data. Closely related work, assuming a multivariate normal distribution for the latent variables, appeared in Ashford and Sowden (1970), Cox (1972), and Ochi and Prentice (1984). In the latter model, the Y_{ij} marginally follow a probit model,

$$E(Y_{ij}|\boldsymbol{X}_{ij}) = \Phi(\boldsymbol{X}'_{ij}\boldsymbol{\beta}),$$

where $\Phi(\cdot)$ denotes the normal cumulative distribution function, and the model allows both positive and negative correlation among the repeated binary responses, depending on the sign of the correlation among the underlying latent variables. This model is often referred to as the "multivariate probit model." Interestingly, the multivariate probit model can also be motivated through the introduction of random effects (see the discussion of generalized linear mixed models in Section 1.4.2).

One of the main drawbacks of the latent-variable model formulations that limited their application to longitudinal data is that they require n-dimensional integration over the joint distribution of the latent variables. In general, it can be computationally intensive to calculate or even approximate these integrals. In addition, the simple correlation structure assumed for the latent variables may be satisfactory for many types of clustered data but is somewhat less appealing for longitudinal data. In principle, however, a more complex covariance structure for the latent variables could be assumed.

At around the same time as Gumbel (1961) proposed his latent-variable formulation, a second approach to likelihood-based inferences was proposed by Bahadur (1961). Bahadur (1961) proposed an elegant expansion for an arbitrary probability mass function for a vector of responses $Y_{i1}, \ldots, Y_{in}$. The expansion for repeated binary responses is of the form

$$f(y_{i1}, \ldots, y_{in})$$

$$= \left\{ \prod_{j=1}^{n} (\pi_{ij})^{y_{ij}} (1 - \pi_{ij})^{1-y_{ij}} \right\}$$

$$\times \left\{ 1 + \sum_{j<k} \rho_{ijk} z_{ij} z_{ik} + \sum_{j<k<l} \rho_{ijkl} z_{ij} z_{ik} z_{il} + \cdots + \rho_{i1\ldots n_i} z_{i1} \ldots z_{in} \right\},$$

where

$$Z_{ij} = \frac{Y_{ij} - \pi_{ij}}{\sqrt{\pi_{ij}(1 - \pi_{ij})}},$$

$\pi_{ij} = E(Y_{ij})$, and $\rho_{ijk} = E(Z_{ij} Z_{ik}), \ldots, \rho_{i1\ldots n} = E(Z_{i1} \ldots Z_{in})$. Here, ρ_{ijk} is the pairwise or second-order correlation and the additional parameters relate to third- and higher-order correlations among the responses.

The Bahadur expansion has a particularly appealing property, shared with the multivariate probit model and many other marginal models, of being "reproducible" or "upwardly compatible" in the sense that the same model holds for any subset of the vector of responses. In addition, the multinomial probabilities for the vector of binary responses are relatively straightforward to obtain given the model parameters. Kupper and Haseman (1978) and Altham (1978) discussed applications of this model, albeit with very simple pairwise correlation structure and assuming higher-order terms are zero. The chief drawback of the Bahadur expansion that has limited its application to longitudinal data is its parameterization of the higher-order associations in terms of correlation parameters. As noted earlier, for discrete data there are severe restrictions on the correlations and dependence of the correlations on the means. Thus, for discrete data, the Bahadur model requires a complicated set of inequality constraints on the model parameters that make maximization of the likelihood very difficult. Except in very simple settings with a small number of repeated measures, the Bahadur model has not been widely applied to longitudinal data.

Because of the restrictions on the correlations, alternative multinomial models for the joint distribution of the vector of discrete responses have recently been proposed where the within-subject association is parameterized in terms of other metrics of association. For example, Dale (1984), McCullagh and Nelder (1989), Lipsitz, Laird, and Harrington (1990), Liang, Zeger, and Qaqish (1992), Becker and Balagtas (1993), Molenberghs and Lesaffre (1994), Lang and Agresti (1994), Glonek and McCullagh (1995), and others have proposed full likelihood approaches where the higher-order moments are parameterized in terms of marginal odds ratios. In closely related work, Ekholm (1991) parameterizes the association directly in terms of the higher-order marginal probabilities (see also Ekholm, Smith, and McDonald, 1995). An alternative approach is to parameterize the within-subject association in terms of conditional associations, leading to so-called "mixed-parameter" models (Fitzmaurice and Laird, 1993; Glonek, 1996; Molenberghs and Ritter, 1996). However, except in certain special cases (e.g., Markov models), these conditional association parameters have somewhat less appealing interpretations in the longitudinal setting; moreover, their interpretation is straightforward only in balanced longitudinal designs.

In virtually all of these later advances, the application of the methodology has been hampered by at least three main factors. First, unlike in the Bahadur model, there are no simple expressions for the joint probabilities in terms of the model parameters. This makes maximization of the likelihood somewhat difficult. Second, even with the current advances in computing, these models are difficult to fit except when the number of repeated measures is relatively small. Finally, many of these models are not robust to misspecification of the higher-order moments. That is, many of the likelihood-based methods require that the entire joint distribution be correctly specified. Thus, if the marginal model for the mean responses has been correctly specified but the model for any of the higher-order moments has not, then the maximum likelihood estimators of the marginal mean parameters will fail to converge in probability to the true mean parameters. The "mixed-parameter" models are an exception to the rule; however, even these models lose this robustness property when there are missing data.

A third approach to likelihood-based marginal models is to specify the entire multinomial distribution of the vector of repeated categorical responses and estimate the multinomial probabilities non-parametrically. This was the approach first proposed in Grizzle, Starmer, and Koch (1969). Specifically, they proposed a weighted least-squares (WLS) method for fitting a general family of models for categorical data; in recognition of its developers, the method is often referred to as the "GSK method." Koch and Reinfurt (1971) and Koch et al. (1977) later recognized how these models could be applied to discrete longitudinal data; Stanish, Gillings, and Koch (1978), Stanish and Koch (1984), and Woolson and Clarke (1984) further developed the methodology for longitudinal analysis.

The GSK method provides a very general family of models for repeated categorical data, allowing non-linear link functions to relate the marginal expectations to covariates. The GSK method stratifies individuals according to values of the covariates and fully specifies the multinomial distribution of the vector of repeated categorical responses within each stratum. This method, for example, allows the fitting of logistic regression models to repeated binary data, albeit with the restrictions that the longitudinal study design be balanced on time, all covariates must be categorical, and there are sufficient numbers of individuals within covariate strata to estimate the multinomial probabilities non-parametrically as the sample proportions. The method requires the estimation of the covariance among the repeated responses, within strata defined by covariate values; the covariance follows directly from the properties of the multinomial distribution. Asymptotically, the GSK method is equivalent to maximum likelihood estimation; thus, this approach was appealing for analyzing discrete longitudinal data when all of the conditions required for its use were met.

Although the GSK method was a landmark technique for the analysis of repeated categorical data, it had many restrictions that limited its usefulness. Specifically, it required that all covariates be categorical and sample sizes be of sufficient size to allow for stratification and separate estimation of the multinomial covariance in each covariate stratum. However, as the number of categorical covariates in the model increases, sparse data problems quickly arise due to Bellman's (1961) "curse of dimensionality." Furthermore, missing data are not easily handled by the GSK method because they require additional stratification by patterns of missingness. Thus, the GSK method was restricted to balanced designs with categorical covariates and relatively large sample sizes. It suffered from many of the same limitations as were noted for the repeated measures by MANOVA in Section 1.2.

Generally speaking, there have been a number of impediments to the application of likelihood-based marginal models for the analysis of discrete longitudinal data. The latent-variable model formulations, first proposed by Gumbel (1961), require high-dimensional integration over the joint distribution of the latent variables that is computationally too difficult. In contrast, methods that fully specify the multinomial probabilities for the response vector, such as the GSK method, are relatively straightforward to implement. However, the conditions for the use of the GSK method are typically not satisfied in many longitudinal settings. Alternative likelihood-based approaches that place more restrictions on the multinomial probabilities have proven to be substantially more difficult to implement. While the latter approaches do not require stratification and can incorporate a mixture of discrete and continuous covariates, they can be applied only in relatively simple cases.

For the most part, all of the likelihood-based approaches that have been proposed have been hampered by various combinations of the following factors. The first is the lack of a convenient joint distribution for discrete multivariate responses, with similar properties to the multivariate normal. Paradoxically, the joint distributions for discrete longitudinal data require specification of many higher-order moments despite the fact that there is, in some sense, substantially less information in the discrete than in the continuous data case. Second, with the exception of the Bahadur expansion, for many multinomial models there is no closed-form expression for the joint probabilities in terms of the model parameters. As such, there is no analog of the multivariate normal distribution for repeated categorical data that has simple and convenient properties. This makes maximization of the likelihood difficult. Third, all likelihood-based approaches face difficulties with sparseness of data once the number of repeated measures exceeds 5 or 6. Recall that a vector of repeated measures on a categorical response with C categories requires specification of $C^n - 1$ multinomial probabilities. For example, with a binary response measured at 10 occasions, there are $2^{10} - 1$ (or 1023) non-redundant multinomial probabilities. So, while data on 200 subjects may be more than adequate for estimation of the marginal probabilities at each occasion, they can be wholly inadequate for estimation of the joint probabilities of the vector of responses due to the curse of dimensionality. Finally, many of these difficulties are compounded by the fact that likelihood-based estimates of the interest parameters, $\boldsymbol{\beta}$, are quite sensitive to misspecification of the higher-order moments. Many of the proposed methods require that the entire joint distribution be correctly specified. For example, if the model for the mean response has been correctly specified but the model for the higher-order moments is incorrect, then the maximum likelihood estimator will fail to converge in probability to the true value of $\boldsymbol{\beta}$. Although some of the computational difficulties previously mentioned can be ameliorated by faster and more powerful computers, many of the other problems reflect the curse of dimensionality and cannot be easily handled with the typical amount of data collected in many longitudinal studies.

In the mid-1980s, remarkable advances in methodology for analyzing discrete longitudinal data were made when Liang and Zeger (1986) proposed the generalized estimating equations (GEE) approach. Because marginal models separately parameterize the model for the mean responses from the model for the within-subject association, Liang and Zeger (1986) recognized that it is possible to estimate the regression parameters in the former without making full distributional assumptions. The avoidance of distributional assumptions is potentially advantageous because, as we have discussed, there is no convenient and generally accepted specification of the joint multivariate distribution of Y_i for marginal models when the responses are discrete. The appeal of the GEE approach is that it only requires specification of that part of the probability mechanism that is of scientific interest, the marginal means. By avoiding full distributional assumptions for Y_i, the GEE approach provided a remarkably convenient alternative to maximum likelihood estimation of multinomial models for repeated categorical data, without many of the inherent complications of the latter. Chapter 3 provides a comprehensive account of the GEE methodology.

The GEE approach advocated in Liang and Zeger (1986) was a natural extension of the quasi-likelihood approach (Wedderburn, 1974) for generalized linear models to the multivariate response setting, where an additional set of nuisance parameters for the within-subject association must be incorporated. The foundation for the GEE approach relied on the theory of optimal estimating functions developed by Godambe (1960) and Durbin (1960). Liang and Zeger (1986) highlighted how the GEE provides a unified approach to the formulation and fitting of generalized linear models to longitudinal and clustered data. They demonstrated the versatility of the GEE method in handling unbalanced data, mixtures of discrete and continuous covariates, and arbitrary patterns of missingness. Until the publication of their landmark paper (Liang and Zeger, 1986), methods for the analysis of discrete longitudinal data had lagged behind corresponding methods for continuous responses. Soon after, marginal models were being widely applied to address substantive questions about longitudinal change across a broad spectrum of disciplines. Their work also generated much additional theoretical and applied research on the use of this methodology for analyzing longitudinal data. For example, to improve upon efficiency, Prentice (1988) proposed joint estimating equations for both the main regression parameters, $\boldsymbol{\beta}$, and the nuisance association parameters, $\boldsymbol{\alpha}$.

The essential idea behind the GEE approach is to extend quasi-likelihood methods, originally developed for a univariate response, by incorporating additional nuisance parameters for the covariance matrix of the vector of responses. For example, given a model for the pairwise correlations, the corresponding covariance matrix can be constructed as the product of the standard deviations and correlations

$$V_i = A_i^{1/2} \text{Corr}(\boldsymbol{Y}_i) A_i^{1/2},$$

where A_i is a diagonal matrix with $\text{Var}(Y_{ij}) = \phi\, v\,(\mu_{ij})$ along the diagonal, and $\text{Corr}(\boldsymbol{Y}_i)$ is a correlation matrix (here a function of $\boldsymbol{\alpha}$). In the GEE approach, V_i is referred to as a "working" covariance matrix to distinguish it from the true underlying covariance matrix of $\boldsymbol{Y}_i$. The term "working" in this context acknowledges uncertainty about the assumed model for the variances and within-subject associations. Because the GEE depend on both $\boldsymbol{\beta}$ and $\boldsymbol{\alpha}$, an iterative two-stage estimation procedure is required; this has been implemented in many widely available software packages. As noted by Crowder (1995), ambiguity concerning the definition of the working covariance matrix can, in certain cases, result in a breakdown of this estimation procedure.

In summary, the GEE approach has a number of appealing properties for estimation of the regression parameters in marginal models. First, in many longitudinal designs the GEE estimator of $\boldsymbol{\beta}$ is almost efficient when compared to the maximum likelihood estimator. For example, it can be shown that the GEE has a similar expression to the likelihood equations for $\boldsymbol{\beta}$ in a linear model for continuous responses that are assumed to have a multivariate normal distribution. The GEE also has an expression similar to the likelihood equations for $\boldsymbol{\beta}$ in certain models for discrete longitudinal data. As a result, for many longitudinal designs, there is relatively little loss of precision when the GEE approach is adopted as an alternative to maximum likelihood. Second, the GEE estimator has a very appealing robustness property, yielding a consistent estimator of $\boldsymbol{\beta}$ even if the within-subject associations among the repeated measures have been misspecified. It only requires that the model for the mean response be correct. This robustness property of GEE is important because the usual focus of a longitudinal study is on changes in the mean response. Although the GEE approach yields a consistent estimator of $\boldsymbol{\beta}$ under misspecification of the within-subject associations, the usual standard errors obtained under the misspecified model for the within-subject association are not valid. However, valid standard errors for the resulting estimator $\widehat{\boldsymbol{\beta}}$ can be obtained using the empirical or so-called *sandwich* estimator of $\text{Cov}(\widehat{\boldsymbol{\beta}})$. The sandwich estimator is also robust in the sense that, with sufficiently large samples, it provides valid standard errors when the assumed model for the covariances among the repeated measures is not correct.

1.4.2 Generalized linear mixed models

In the previous section, we discussed how marginal models can be considered an extension of generalized linear models that *directly* incorporate the within-subject association among the repeated measurements. In a certain sense, marginal models account for the consequences of the correlation among the repeated measures but do not provide any explanation for its potential source. An alternative approach for accounting for the within-subject association, and one that provides a source for the within-subject association, is via the introduction of random effects in the model for the mean response. Following the same basic ideas as in linear mixed-effects models, generalized linear models can be extended to longitudinal data by allowing a subset of the regression coefficients to vary randomly from one individual to another. These models are known as *generalized linear mixed (effects) models* (GLMMs), and they extend in a natural way the conceptual approach represented by the linear mixed-effects models discussed in Section 1.3; see Chapter 4 for a detailed overview. In GLMMs the model for the mean response is conditional upon both measured covariates and unobserved random effects; it is the inclusion of the latter that induces correlation among the repeated responses marginally, when averaged over the distribution of the random effects. However, as we discuss later, with non-linear link functions, the introduction of random effects has important ramifications for the interpretation of the "fixed-effects" regression parameters.

The generalized linear mixed model can be formulated using the following two-part specification:

1. Given a $q \times 1$ vector of random effects $\boldsymbol{b}_i$, the Y_{ij} are assumed to be conditionally independent and to have exponential family distributions with conditional mean depending upon both fixed and random effects,

$$h^{-1}\{E(Y_{ij}|\boldsymbol{b}_i)\} = \boldsymbol{X}'_{ij}\boldsymbol{\beta} + \boldsymbol{Z}'_{ij}\boldsymbol{b}_i,$$

for some known link function, $h^{-1}(\cdot)$. The conditional variance is assumed to depend on the conditional mean according to $\text{Var}(Y_{ij}|\boldsymbol{b}_i) = \phi\, v\{E(Y_{ij}|\boldsymbol{b}_i)\}$, where $v(\cdot)$ is a known variance function and ϕ is a scale parameter that may be known or may need to be estimated.

2. The random effects, $\boldsymbol{b}_i$, are assumed to be independent of the covariates, $\boldsymbol{X}_{ij}$, and to have a multivariate normal distribution, with zero mean and $q \times q$ covariance matrix G.

These two components completely specify a broad class of generalized linear mixed models. In principle, the conditional independence assumption in the first component is not necessary, but is commonly made. Similarly, any multivariate distribution can be assumed for the $\boldsymbol{b}_i$; in practice, however, it is common to assume that the $\boldsymbol{b}_i$ have a multivariate normal distribution.

Generalized linear mixed models have their foundation in simple random-effects models for binary and count data. The early literature on random-effects models for discrete data can be traced back to the development of random compounding models that introduced random effects on the response scale. For example, Greenwood and Yule (1920) introduced the negative binomial distribution as a compound Poisson distribution for count data, while Skellam (1948) provided an early discussion of the beta-binomial distribution for binary data. The beta-binomial model can be conceptualized as a two-stage model, where in the first stage the binary responses, $Y_{i1}, \ldots, Y_{in}$, are assumed to be conditionally independent with common success probability p_i, where $\Pr(Y_{ij} = 1|p_i) = E(Y_{ij}|p_i) = p_i$. In the second stage, the success probabilities, $p_1, \ldots, p_N$, are assumed to be independently distributed with a beta density. The mean of the success probabilities can be related to covariates via an appropriate link function, such as a logit or probit link function. Although the beta-binomial model accounts for overdispersion relative to the usual binomial variance, the model is somewhat more natural for clustered rather than longitudinal data. As a result, the model has been used in a wide variety of different clustered data applications (e.g., Chatfield and Goodhardt, 1970; Griffiths, 1973; Williams, 1975; Kupper and Haseman, 1978; Crowder, 1978, 1979; Otake and Prentice, 1984; Aerts et al., 2002).

The main feature of the beta-binomial model that has limited its usefulness for analyzing longitudinal data is that it produces the same marginal distribution at each measurement occasion. While this may not be so problematic in certain clustered data settings (e.g., in study designs where $\boldsymbol{X}_{i1} = \boldsymbol{X}_{i2} = \cdots = \boldsymbol{X}_{in}$), in a longitudinal study where interest is primarily in changes in the marginal means over time, this restriction on the marginal distributions is very unappealing. Nonetheless, the beta-binomial and other random compounding models motivated the later development of more versatile random-effects models. Recall that in the beta-binomial model, it is assumed that success probabilities vary randomly about a mean and the latter can be related to covariates via an appropriate link function, such as a logit link function. In contrast to this formulation, Pierce and Sands (1975) proposed an alternative model where the logit of p_i is assumed to vary about an expectation given by $\boldsymbol{X}'_{ij}\boldsymbol{\beta}$,

$$\text{logit}\{E(Y_{ij}|b_i)\} = \boldsymbol{X}'_{ij}\boldsymbol{\beta} + b_i,$$

where b_i has a normal distribution with zero mean and constant variance. The appealing feature of the model proposed by Pierce and Sands (1975) is that the fixed and random effects are combined together on the same logistic scale. This model is often referred to as the simple "logit-normal model" and is very similar in spirit to the random intercept model for continuous outcomes discussed in Section 1.2. Although this model was remarkably simple, it proved to be difficult to fit at the time because maximum likelihood estimation required maximization of the marginal likelihood, averaged over the distribution of the random effect. This required integration, and no analytic solutions were available. The fact that the integral cannot be evaluated in a closed form limited the application of this model.

In closely related work, Ashford and Sowden (1970) proposed a very similar model, except with probit rather than logit link function. Interestingly, Ashford and Sowden's (1970) model with random intercept and probit link function was equivalent to the equicorrelated latent-variable model discussed in Section 1.4.1, leading to identical inferences provided the correlation is positive. Despite the fact that maximum likelihood estimation for even the simple logit-normal model was computationally demanding with the computer resources available at the time, Korn and Whittemore (1979) proposed a far more ambitious version of the model, where

$$\text{logit}\{E(Y_{ij}|\boldsymbol{b}_i)\} = \boldsymbol{X}'_{ij}\boldsymbol{\beta} + \boldsymbol{Z}'_{ij}\boldsymbol{b}_i,$$

with $\boldsymbol{Z}_{ij} = \boldsymbol{X}_{ij}$. Although their model was very general and avoided some of the obvious drawbacks of the simple logit-normal model, it was difficult to fit and required a very long sequence of repeated measures on each subject.

From an historical perspective, the papers by Ashford and Sowden (1970), Pierce and Sands (1975), and Korn and Whittemore (1979) laid the conceptual foundations for generalized linear mixed models; much of the work that followed focused on the thorny problem of estimation. In GLMMs the marginal likelihood is used as the basis for inferences for the fixed-effects parameters, complemented with empirical Bayes estimation for the random effects. In general, evaluation and maximization of the marginal likelihood for GLMMs requires integration over the distribution of the random effects. While this is, strictly speaking, true for the linear mixed-effects model as well, there the integration can be done analytically, so effectively a closed form for the marginal likelihood function arises, in which case the application of maximum or restricted maximum likelihood is straightforward. In the absence of an analytical solution, and because high-dimensional numerical integration can be very trying, a variety of approaches has been suggested for tackling this problem.

Because no simple analytic solutions were available, Stiratelli, Laird, and Ware (1984) proposed an approximate method of estimation for the logit-normal model, based on empirical Bayes ideas, that circumvented the need for numerical integration. Specifically, they avoided the need for numerical integration by approximating the integrands with simple expansions whose integrals have closed forms. The paper by Stiratelli, Laird, and Ware (1984) led to the development of a general approach for fitting GLMMs, known as penalized quasi-likelihood (PQL). Various authors (e.g., Schall, 1991; Breslow and Clayton, 1993; Wolfinger, 1993) motivated PQL as a Laplace approximation to the marginal likelihood for GLMMs. Despite the generality of this method, and its implementation in a variety of commercially available software packages, the PQL method can often yield quite biased estimators of the variance components, which in turn leads to biased estimators of $\boldsymbol{\beta}$, especially for longitudinal binary data. This motivated research on bias corrections (e.g., Breslow and Lin, 1995) and on more accurate approximations based on higher-order Laplace approximations (e.g., Raudenbush, Yang, and Yosef, 2000). In general, the inclusion of higher-order terms for PQL has been shown to improve estimation. Breslow and Clayton (1993) also considered an alternative approach, related to PQL, known as marginal quasi-likelihood (MQL). MQL differs from PQL by being based on an expansion around the current estimates of the fixed effects and around $b_i = 0$. In general, MQL yields severely biased estimators of the variance components, providing a good approximation only when the variance of the random effects is relatively small.

There has also been much recent research on alternative methods, including approaches based on numerical integration (e.g., adaptive Gaussian quadrature) and Markov chain Monte Carlo algorithms. In particular, adaptive Gaussian quadrature, with the numerical integration centered around the empirical Bayes estimates of the random effects, permits maximization of the marginal likelihood with any desired degree of accuracy (e.g., Anderson and Aitkin, 1985; Hedeker and Gibbons, 1994, 1996). Adaptive Gaussian quadrature is especially appealing for longitudinal data where the dimension of the random effects is often relatively low. Monte Carlo approaches to integration, for example Monte Carlo EM

(McCulloch, 1997; Booth and Hobert, 1999) and Monte Carlo Newton–Raphson algorithms (Kuk and Cheng, 1997), have been proposed. The hierarchical formulation of GLMMs also makes Bayesian approaches quite appealing. For example, Zeger and Karim (1991) have proposed the use of Monte Carlo integration, via Gibbs sampling, to calculate the posterior distribution.

The normality assumption for the random effects in GLMMs leads, in general, to intractable likelihood functions, except in the case of the linear mixed model for continuous data. This is because the normal random-effects distribution is conjugate to the normal distribution for the outcome, conditional on the random effects. Lee and Nelder (1996, 2001, 2003) have extended this idea and propose using conjugate random-effects distributions in contexts other than the classical normal linear model.

Finally, it is worth emphasizing some differences between GLMMs and the marginal models discussed in Section 1.4.1. Although the introduction of random effects can simply be thought of as a means of accounting for and explaining the potential sources of the correlation among longitudinal responses, it has important implications for the interpretation of the regression coefficients in GLMMs. The fixed effects, $\boldsymbol{\beta}$, have somewhat different interpretations than the corresponding regression parameters in marginal models. In GLMMs the regression parameters have "subject-specific" interpretations. They represent the effects of covariates on changes in an individual's possibly transformed mean response per unit change in the covariate, while controlling for all other covariates *and* the random effects. This interpretation for $\boldsymbol{\beta}$ can be better appreciated by considering the following example of a simple logit-normal model given by

$$\log\left\{\frac{\Pr(Y_{ij}=1|\boldsymbol{X}_{ij},\boldsymbol{b}_i)}{\Pr(Y_{ij}=0|\boldsymbol{X}_{ij},\boldsymbol{b}_i)}\right\} = \boldsymbol{X}'_{ij}\boldsymbol{\beta} + b_i,$$

where b_i is assumed to have a univariate normal distribution with zero mean and constant variance. The interpretation of a component of $\boldsymbol{\beta}$, say β_k, is in terms of changes in any given *individual's* log odds of response for a unit change in the corresponding covariate, say X_{ijk}. Because β_k has interpretation that depends upon holding b_i fixed, it is referred to as a *subject-specific* effect. Note that this subject-specific interpretation of β_k is far more natural for a covariate that varies within an individual (i.e., a time-varying covariate). With a time-invariant covariate, problems of interpretation arise because a change in the value of the covariate also requires a change in the index i of X_{ijk} to, say, $X_{i'jk}$ (for $i \neq i'$). However, β_k then becomes confounded with differences between b_i and $b_{i'}$. One way around this is to think of the population, defined by all subjects sharing the same value of the random effect b_i. The effect of a covariate is then conditional on changing X_{ijk} within the fine population defined by b_i.

In summary, the distinction between the regression parameters in GLMMs and marginal models is best understood in terms of the targets of inference; a fuller discussion is given in Chapter 7. In GLMMs, the target of inference is the individual because the regression coefficients have interpretation in terms of contrasts of the transformed conditional means,

$$E(Y_{ij}|\boldsymbol{X}_{ij}, b_i).$$

In contrast, in marginal models the target of inference is the population because the regression parameters have interpretation in terms of contrasts of the transformed population means,

$$E(Y_{ij}|\boldsymbol{X}_{ij}).$$

For the special case of linear models, where an identity link function has been adopted, the fixed effects in the model for the conditional means,

$$E(Y_{ij}|\boldsymbol{X}_{ij}, \boldsymbol{b}_i) = \boldsymbol{X}'_{ij}\boldsymbol{\beta} + \boldsymbol{Z}'_{ij}\boldsymbol{b}_i,$$

also happen to have interpretation in terms of the population means because

$$E(Y_{ij}|\boldsymbol{X}_{ij}) = \boldsymbol{X}'_{ij}\boldsymbol{\beta}$$

when averaged over the distribution of the random effects. However, in general, for the non-linear link functions usually adopted for discrete data, this relationship no longer holds, and if

$$h^{-1}\{E(Y_{ij}|\boldsymbol{X}_{ij},\boldsymbol{b}_i)\} = \boldsymbol{X}'_{ij}\boldsymbol{\beta} + \boldsymbol{Z}'_{ij}\boldsymbol{b}_i,$$

then

$$h^{-1}\{E(Y_{ij}|\boldsymbol{X}_{ij})\} \neq \boldsymbol{X}'_{ij}\boldsymbol{\beta}$$

for any $\boldsymbol{\beta}$.

1.4.3 Conditional and transition models

There is a third way in which generalized linear models can be extended to handle longitudinal data. This is accomplished by modeling the mean and time dependence simultaneously via conditioning an outcome on other outcomes or on a subset of other outcomes (see, for example, Molenberghs and Verbeke, 2005, Part III). A particular case is given by so-called transition, or Markov, models. Transition models are appealing due to the sequential nature of longitudinal data. In transition models, the conditional distribution of each response is expressed as an explicit function of the past responses and the covariates. Transition models can be considered *conditional* models in the sense of modeling the conditional distribution of the response at any occasion given the previous responses and the covariates. The dependence among the repeated measures is thought of as arising due to past values of the response influencing the present observation. In transition models, it is assumed that

$$h^{-1}\left\{E(Y_{ij}|\boldsymbol{X}_{ij},\boldsymbol{H}_{ij})\right\} = \boldsymbol{X}'_{ij}\boldsymbol{\beta} + \sum_{r=1}^{s}\alpha_r f_r(\boldsymbol{H}_{ij}), \tag{1.1}$$

where $\boldsymbol{H}_{ij} = (Y_{i1},\dots,Y_{ij-1})$ denotes the history of the past responses at the jth occasion, and $f_r(\boldsymbol{H}_{ij})$ denote some known functions (often, but not necessarily, linear functions) of the history of the past responses. For example, a first-order autoregressive, AR(1), generalized linear model is obtained when

$$\sum_{r=1}^{s}\alpha_r f_r(\boldsymbol{H}_{ij}) = \alpha_1 f_1(\boldsymbol{H}_{ij}) = \alpha_1 Y_{ij-1}.$$

A more general autoregressive model of order s, say, AR(s), is obtained by incorporating the s previously generated values of the response. In general, models where the conditional distribution of the response at the jth occasion, given $\boldsymbol{H}_{ij}$, depends only on the s immediately prior responses are known as Markov models of order s. When the response variable is discrete, these models are referred to as Markov chain models. With discrete data and non-identity link functions, it may be necessary to transform the history of the past responses, $\boldsymbol{H}_{ij}$, in a manner similar to the transformation of the conditional mean, for example, $f_r(\boldsymbol{H}_{ij}) = h^{-1}(\boldsymbol{H}_{ij})$. Also, upon closer examination, the model given by (1.1) appears to make a strong assumption that the effects of the covariates are the same regardless of the actual history of past responses. However, this assumption can be relaxed by including interactions between relevant covariates and $\boldsymbol{H}_{ij}$.

There is an extensive history to the use of Markov chains to model equally spaced discrete longitudinal data with a finite number of states or categories (e.g., Anderson and Goodman, 1957; Cox, 1958; Billingsley, 1961). In the simplest of models for longitudinal data, a first-order Markov chain, the transition probabilities are assumed to be the same for each time interval. The resulting Markov chain can then be described in terms of the initial state and the set of transition probabilities. The transition probabilities are the conditional probabilities of going into each state, given the immediately preceding state. In a first-order Markov chain, there is dependence on the immediately preceding state but not on earlier outcomes. In the more general model given by (1.1), higher-order sequential dependence can be incorporated, with dependence on more than the immediately preceding state, and the transition probabilities can be allowed to vary over time. Moreover, the time dependence need not necessarily be a linear function of the history. Among others, Cox (1972), Korn and Whittemore (1979), Zeger, Liang, and Self (1985), and Ware, Lipsitz, and Speizer (1988) discuss transition models applicable to longitudinal data.

One appealing aspect of transition models is that the joint distribution of the vector of responses can be expressed as the product of a sequence of conditional distributions, that is,

$$f(y_{i1}, \ldots, y_{in}; \boldsymbol{\beta}, \boldsymbol{\alpha}) = \prod_{j=1}^{n} f(y_{ij}|y_{i1}, \ldots, y_{i,j-1}; \boldsymbol{\beta}, \boldsymbol{\alpha}).$$

Strictly speaking, for an sth-order Markov chain model, this is a conditional likelihood, given a set of s initial values. In the specification of the transition model given by (1.1), initial values of the responses are assumed to be incorporated into the covariates. In general, the unconditional distribution of the initial responses cannot be determined from the conditional distributions specified by (1.1). There are two ways to handle this initial value problem. The first is to treat the initial responses as a set of given constants rather than random variables and base estimation on the conditional likelihood, ignoring the contribution of the unconditional distribution of the initial responses. Maximization of the resulting likelihood is relatively straightforward; indeed, standard software for univariate generalized linear models can be used when $f_r(\boldsymbol{H}_{ij})$ does not depend on $\boldsymbol{\beta}$. Alternatively, the initial responses can be assigned the equilibrium distribution of the sequence of longitudinal responses. In general, the latter will yield more efficient estimates of the regression coefficient, $\boldsymbol{\beta}$.

Although Markov and autoregressive models have a long and extensive history of use for the analyses of time series data, their application to longitudinal data has been somewhat more limited. There are a number of features of transition models that limit their usefulness for the analysis of longitudinal data. In general, transition models have been developed for repeated measures that are equally separated in time; these models are more difficult to apply when there are missing data, mistimed measurements, and non-equidistant intervals between measurement occasions. In addition, estimation of the regression parameters $\boldsymbol{\beta}$ is very sensitive to assumptions concerning the time dependence; moreover, the interpretation of $\boldsymbol{\beta}$ changes with the order of the serial dependence. Finally, in many longitudinal studies $\boldsymbol{\beta}$ is not the usual target of inference because conditioning on the history of past responses may lead to attenuation of the effects of covariates of interest. That is, when a covariate is expected to influence the mean response at all occasions, its effect may be somewhat diminished if there is conditioning on the past history of the responses.

1.5 Concluding remarks

In the preceding sections, we traced the development of a very general and versatile class of linear mixed-effects models for longitudinal data when the response is continuous. These models can handle issues of unbalanced data, due to either mistimed measurement or missing data, time-varying and time-invariant covariates, and modeling of the covariance, in a flexible way. Linear mixed-effects models rely on assumptions of multivariate normality, and likelihood-based inferences for both the fixed and random effects are relatively straightforward. In contrast, when the longitudinal response is discrete, we have seen that there is more than one way to extend generalized linear models to the longitudinal setting. This has led to the development of "marginal" and "conditional" models for non-Gaussian longitudinal data; in the former, there is no conditioning on past responses or random effects, while in the latter there is conditioning on either the response history or a set of random effects. Although this classification is useful for pedagogical purposes, it should be recognized that this distinction between classes of models is somewhat artificial and is made, in part, to emphasize certain aspects of interpretation that arise when analyzing discrete longitudinal data. In contrast to the situation for linear models, conditioning on past responses or random effects has important implications for the regression parameters in models for discrete longitudinal data. While their interpretation obviously changes in all cases, also the parameter estimates are different because in fact the underlying estimands cannot be compared directly. However, it is possible to combine features of these models, thereby blurring the distinctions. For example, Conaway (1989, 1990) has suggested extending mixed-effects models to include lagged responses, while, more recently, Heagerty and Zeger (2000) have developed "conditionally specified" marginal model formulations.

In general, we have seen that likelihood-based approaches are somewhat more difficult to formulate in the non-Gaussian data setting than is the case with continuous responses. This has led to various avenues of research where more tractable approximations have been developed (e.g., PQL methods) and where likelihood-based approaches have been abandoned altogether in favor of semi-parametric methods (e.g., GEE approaches).

Our review of the developments of regression models for longitudinal data has focused exclusively on extensions of generalized linear models. Limitations of space have precluded a discussion of non-linear models (i.e., models where the relationship between the mean and covariates is non-linear in the regression parameters) for longitudinal data; see Chapter 5 of this volume and Davidian and Giltinan (1995) for a comprehensive and unified treatment of this topic. Perhaps not surprisingly, the development of non-linear regression models for longitudinal data has faced many of the challenges and issues that were discussed in Section 1.4.2.

We conclude this chapter by noting that in almost every discipline there is increased awareness of the importance of longitudinal studies for studying change over time and the factors that influence change. This has led to a steady growth in the availability of longitudinal data, often arising from relatively complex study designs. The analysis of longitudinal data continues to pose many interesting methodological challenges and is likely to do so for the foreseeable future. The goal of the remaining chapters in this book is to highlight the current state of the art of longitudinal data analysis and to provide a glimpse of future directions.

Acknowledgments

This work was supported by the following grants from the U.S. National Institutes of Health: R01GM029745 and R01MH054693.

References

Aerts, M., Geys, H., Molenberghs, G., and Ryan, L. (2002). *Topics in Modelling of Clustered Data.* Boca Raton, FL: Chapman & Hall/CRC.

Airy, G. B. (1861). *On the Algebraical and Numerical Theory of Errors of Observation and the Combination of Observations.* London: Macmillan.

Altham, P. M. E. (1978). Two generalizations of the binomial distribution. *Applied Statistics* **27**, 162–167.

Anderson, D. A. and Aitkin, M. (1985). Variance components models with binary response: Interviewer variability. *Journal of the Royal Statistical Society, Series B* **47**, 203–210.

Anderson, T. W. and Goodman, L. A. (1957). Statistical inference about Markov chains. *Annals of Mathematical Statistics* **28**, 89–110.

Ashford, J. R. and Sowden, R. R. (1970). Multivariate probit analysis. *Biometrics* **26**, 535–546.

Bahadur, R. R. (1961). A representation of the joint distribution of responses to n dichotomous items. In H. Solomon (ed.), *Studies in Item Analysis and Prediction*, pp. 158–168. Palo Alto, CA: Stanford University Press.

Becker, M. P. and Balagtas, C. C. (1993). Marginal modeling of binary cross-over data. *Biometrics* **49**, 997–1009.

Bellman, R. E. (1961). *Adaptive Control Processes*. Princeton, NJ: Princeton University Press.

Billingsley, P. (1961). Statistical methods in Markov chains. *Annals of Mathematical Statistics* **32**, 12–40.

Booth, J. G. and Hobert, J. P. (1999). Maximizing generalized linear mixed model likelihoods with an automated Monte Carlo EM algorithm. *Journal of the Royal Statistical Society, Series B* **61**, 265–285.

Box, G. E. P. (1950). Problems in the analysis of growth and wear data. *Biometrics* **6**, 362–389.

Breslow, N. E. and Clayton, D. G. (1993). Approximate inference in generalized linear mixed models. *Journal of the American Statistical Association* **88**, 9–25.

Breslow, N. E. and Lin, X. (1995). Bias correction in generalized linear models with a single component of dispersion. *Biometrika* **82**, 81–91.

Chatfield, C. and Goodhardt, G. J. (1970). The beta-binomial model for consumer purchasing behaviour. *Applied Statistics* **19**, 240–250.

Cole, J. W. L. and Grizzle, J. E. (1966). Applications of multivariate analysis of variance to repeated measurements experiments. *Biometrics* **22**, 810–828.

Conaway, M. (1989). Conditional likelihood methods for repeated categorical responses. *Journal of the American Statistical Association* **84**, 53–62.

Conaway, M. (1990). A random effects model for binary data. *Biometrics* **46**, 317–328.

Cox, D. R. (1958). The regression analysis of binary sequences (with discussion). *Journal of the Royal Statistical Society, Series B* **20**, 215–242

Cox, D. R. (1972). The analysis of multivariate binary data. *Applied Statistics* **21**, 113–120.

Crowder, M. J. (1978). Beta-binomial Anova for proportions. *Applied Statistics* **27**, 34–37.

Crowder, M. J. (1979). Inference about the intra-class correlation coefficient in the beta-binomial ANOVA for proportions. *Journal of the Royal Statistical Society, Series B* **41**, 230–234.

Crowder, M. J. (1995). On the use of a working correlation matrix in using generalised linear models for repeated measures. *Biometrika* **82**, 407–410.

Dale, J. R. (1984). Local versus global association for bivariate ordered responses. *Biometrika* **71**, 507–514.

Danford, M. B., Hughes, H. M., and McNee, R. C. (1960). On the analysis of repeated measurements experiments. *Biometrics* **16**, 547–565.

Davidian, M. and Giltinan, D. M. (1995). *Nonlinear Models for Repeated Measurement Data*. London: Chapman & Hall.

Dempster, A. P., Laird, N. M., and Rubin, D. B. (1977). Maximum likelihood from incomplete data via the EM algorithm (with discussion). *Journal of the Royal Statistical Society, Series B* **39**, 1–38.

Durbin, J. (1960). Estimation of parameters in time-series regression models. *Biometrika* **47**, 139–153.

Ekholm, A. (1991). Fitting regression models to a multivariate binary response. In G. Rosenqvist, K. Juselius, K. Nordström, and J. Palmgren (eds.), *A Spectrum of Statistical Thought: Essays in Statistical Theory, Economics, and Population Genetics in Honour of Johan Fellman*, pp. 19–32. Helsingfors: Swedish School of Economics and Business Administration.

Ekholm, A., Smith, P. W. F., and McDonald, J. W. (1995). Marginal regression analysis of a multivariate binary response. *Biometrika* **82**, 847–854.

Engle, R. F., Hendry, D. F., and Richard, J. F. (1983). Exogeneity. *Econometrica* **51**, 277–304.

Fisher, R. A. (1918). The correlation between relatives on the supposition of Mendelian inheritance. *Transactions of the Royal Society of Edinburgh* **52**, 399–433.

Fisher, R. A. (1921). On the probable error of a coefficient of correlation deduced from a small sample. *Metron* **1**, 3–32.

Fisher, R. A. (1925). *Statistical Methods for Research Workers*. Edinburgh: Oliver and Boyd.

Fitzmaurice, G. M. and Laird, N. M. (1993). A likelihood-based method for analysing longitudinal binary responses. *Biometrika* **80**, 141–151.

Fitzmaurice, G. M., Laird, N. M., and Rotnitzky, A. G. (1993). Regression models for discrete longitudinal responses (with discussion). *Statistical Science* **8**, 248–309.

Fitzmaurice, G. M., Laird, N. M., and Ware, J. H. (2004). *Applied Longitudinal Analysis*. Hoboken, NJ: Wiley.

Geisser, S. (1963). Multivariate analysis of variance for a special covariance case. *Journal of the American Statistical Association* **58**, 660–669.

Geisser, S. and Greenhouse, S. W. (1958). An extension of Box's results on the use of the F distribution in multivariate analysis. *Annals of Mathematical Statistics* **29**, 885–891.

Glonek, G. F. V. (1996). A class of regression models for multivariate categorical responses. *Biometrika* **83**, 15–28.

Glonek, G. F. V. and McCullagh, P. (1995). Multivariate logistic models. *Journal of the Royal Statistical Society, Series B* **57**, 533–546.

Godambe, V. P. (1960). An optimum property of regular maximum likelihood estimation. *Annals of Mathematical Statistics* **31**, 1208–1212.

Goldstein, H. (1979) *The Design and Analysis of Longitudinal Studies*. London: Academic Press.

Greenhouse, S. W. and Geisser, S. (1959). On methods in the analysis of profile data. *Psychometrika* **32**, 95–112.

Greenwood, M. and Yule, G. U. (1920). An enquiry into the nature of frequency distributions representative of multiple happenings with particular reference of multiple attacks of disease or of repeated accidents. *Journal of the Royal Statistical Society* **83**, 255–279.

Griffiths, D. A. (1973). Maximum likelihood estimation for the beta-binomial distribution and an application to the household distribution of the total number of cases of a disease. *Biometrics* **29**, 37–48.

Grizzle, J. E. and Allen, D. M. (1969). Analysis of growth and dose response curves. *Biometrics* **25**, 357–381.

Grizzle, J. E., Starmer, C. F., and Koch, G. G. (1969). Analysis of categorical data by linear models. *Biometrics* **15**, 489–504.

Gumbel, E. J. (1961). Bivariate logistic distributions. *Journal of the American Statistical Association* **56**, 335–349.

Harville, D. A. (1977). Maximum likelihood approaches to variance component estimation and to related problems. *Journal of the American Statistical Association* **72**, 320–338.

Heagerty, P. J. and Zeger, S. L. (2000), Marginalized multilevel models and likelihood inference (with discussion). *Statistical Science* **15**, 1–26.

Hedeker, D. and Gibbons, R. D. (1994). A random-effects ordinal regression model for multilevel analysis. *Biometrics* **50**, 933–944.

Hedeker, D. and Gibbons, R. D. (1996). MIXOR: A computer program for mixed-effects ordinal regression analysis. *Computer Methods and Programs in Biomedicine* **49**, 157–176.

Henderson, C. R. (1963). Selection index and expected genetic advance. In W. D. Hanson and H. F. Robinson (eds.), *Statistical Genetics and Plant Breeding*. Washington, D.C.: National Academy of Sciences–National Research Council.

Huynh, H. and Feldt, L. S. (1976). Estimation of the Box correction for degrees of freedom from sample data in the randomized block and split plot designs. *Journal of Educational Statistics* **1**, 69–82.

Jennrich, R. I. and Schluchter, M. D. (1986). Unbalanced repeated-measures models with structured covariance matrices. *Biometrics* **42**, 805–820.

Koch, G. G. and Reinfurt, D. W. (1971). The analysis of categorical data from mixed models. *Biometrics* **27**, 157–173.

Koch, G. G., Landis, J. R., Freeman, J. L., Freeman, D. H., and Lehnen, R. G. (1977). A general methodology for the analysis of experiments with repeated measurement of categorical data. *Biometrics* **33**, 133–158.

Korn, E. L. and Whittemore, A. S. (1979). Methods for analyzing panel studies of acute health effects of air pollution. *Biometrics* **35**, 795–802.

Kuk, A. Y. C. and Cheng, Y. W. (1997). The Monte Carlo Newton-Raphson algorithm. *Journal of Statistical Computation and Simulation* **59**, 233–250.

Kupper, L. L. and Haseman, J. K. (1978). The use of a correlated binomial model for the analysis of certain toxicological experiments. *Biometrics* **34**, 69–76.

Laird, N. M. and Ware, J. H. (1982). Random effects models for longitudinal data. *Biometrics* **38**, 963–974.

Lang, J. B. and Agresti, A. (1994). Simultaneous modeling joint and marginal distributions of multivariate categorical responses. *Journal of the American Statistical Association* **89**, 625–632.

Lee, Y. and Nelder, J. A. (1996). Hierarchical generalized linear models (with discussion). *Journal of the Royal Statistical Society, Series B* **58**, 619–678.

Lee, Y. and Nelder, J. A. (2001). Hierarchical generalized linear models: a synthesis of generalized linear models, random-effect models and structured dispersions. *Biometrika* **88**, 987–1006.

Lee, Y. and Nelder, J. A. (2003). Extended-REML estimators. *Journal of Applied Statistics* **30**, 845–856.

Liang, K.-Y. and Zeger, S. L. (1986). Longitudinal data analysis using generalized linear models. *Biometrika* **73**, 13–22.

Liang, K.-Y., Zeger, S. L., and Qaqish, B. (1992). Multivariate regression analyses for categorical data (with discussion). *Journal of the Royal Statistical Society, Series B* **54**, 2–24.

Lipsitz, S. R., Laird, N. M., and Harrington, D. P. (1990). Maximum likelihood regression methods for paired binary data. *Statistics in Medicine* **9**, 1417–1425.

McCullagh, P. and Nelder, J. (1989). *Generalized Linear Models*, 2nd ed. London: Chapman & Hall.

McCulloch, C. E. (1997). Maximum likelihood algorithms for generalized linear mixed models. *Journal of the American Statistical Association* **92**, 162–170.

Molenberghs, G. and Lesaffre, E. (1994). Marginal modeling of correlated ordinal data using a multivariate Plackett distribution. *Journal of the American Statistical Association* **89**, 633–644.

Molenberghs, G. and Ritter, L. (1996). Methods for analyzing multivariate binary data, with the association between outcomes of interest. *Biometrics* **52**, 1121–1133.

Molenberghs, G. and Verbeke, G. (2005). *Models for Discrete Longitudinal Data*. New York: Springer.

Nelder, J. A. and Wedderburn, R. W. M. (1972). Generalized linear models. *Journal of the Royal Statistical Society, Series A* **135**, 370–384.

Ochi, Y. and Prentice, R. L. (1984). Likelihood inference in a correlated probit regression model. *Biometrika* **71**, 531–543.

Otake, M. and Prentice, R. L. (1984). The analysis of chromosomally aberrant cells based on beta-binomial distribution. *Radiation Research* **98**, 456–470.

Pepe, M. S. and Anderson, G. L. (1994). A cautionary note on inference for marginal regression models with longitudinal data and general correlated response data. *Communications in Statistics – Simulation and Computation* **23**, 939–951.

Pierce, D. A. and Sands, B. R. (1975). Extra-Bernoulli variation in binary data. Technical Report 46, Department of Statistics, Oregon State University.

Potthoff, R. F. and Roy, S. N. (1964). A generalized multivariate analysis of variance model useful especially for growth curve problems. *Biometrika* **51**, 313–326.

Prentice, R. L. (1988). Correlated binary regression with covariates specific to each binary observation. *Biometrics* **44**, 1033–1068.

Rao, C. R. (1958). Some statistical methods for comparison of growth curves. *Biometrics* **14**, 1–17.

Rao, C. R. (1965). The theory of least squares when the parameters are stochastic and its application to the analysis of growth curves. *Biometrika* **52**, 447–458.

Raudenbush, S. W., Yang, H.-L., and Yosef, M. (2000). Maximum likelihood for generalized linear models with nested random effects via high-order, multivariate Laplace approximation. *Journal of Computational and Graphical Statistics* **9**, 141–157.

Robins, J. M., Greenland, S., and Hu, F.-C. (1999). Estimation of the causal effect of a time-varying exposure on the marginal mean of a repeated binary outcome (with discussion). *Journal of the American Statistical Association* **94**, 687–712.

Rowell, J. G. and Walters, D. E. (1976). Analysing data with repeated observations on each experimental unit. *Journal of Agricultural Science* **87**, 423–432.

Schall, R. (1991). Estimation in generalized linear models with random effects. *Biometrika* **78**, 719–727.

Scheffé, H. (1956) Alternative models for the analysis of variance. *Annals of Mathematical Statistics* **27**, 251–271.

Scheffé, H. (1959). *The Analysis of Variance.* New York: Wiley.

Skellam, J. G. (1948). A probability distribution derived from the binomial distribution by regarding the probability of success as variable between the sets of trials. *Journal of the Royal Statistical Society, Series B* **10**, 257–261.

Stanish, W. M., Gillings, D. B., and Koch, G. G. (1978). An application of multivariate ratio methods for the analysis of a longitudinal clinical trial with missing data. *Biometrics* **34**, 305–3117.

Stanish, W. M. and Koch, G. G. (1984). The use of CATMOD for repeated measurement analysis of categorical data. *Proceedings of the Ninth Annual SAS Users Group International Conference* **9**, 761–770.

Stiratelli, R., Laird, N. M., and Ware, J. H. (1984). Random effects models for serial observations with binary response. *Biometrics* **40**, 961–971.

"Student" (Gosset, W. S.) (1908). The probable error of the mean. *Biometrika* **6**, 1–25.

Verbeke, G. and Molenberghs, G. (2000) *Linear Mixed Models for Longitudinal Data.* New York: Springer.

Ware, J. H., Lipsitz, S. R., and Speizer, F. E. (1988). Issues in the analysis of repeated categorical outcomes. *Statistics in Medicine* **7**, 95–107.

Wedderburn, R. W. M. (1974). Quasi-likelihood functions, generalized linear models, and the Gauss-Newton method. *Biometrika* **61**, 439–447.

Williams, D. (1975). The analysis of binary responses from toxicological experiments involving reproduction and teratogenicity. *Biometrics* **31**, 949–952.

Wishart, J. (1938). Growth-rate determinations in nutrition studies with the bacon pig, and their analysis. *Biometrika* **30**, 16–28.

Wolfinger, R. (1993). Laplace's approximation for nonlinear mixed models. *Biometrika* **80**, 791–795.

Woolson, R. F. and Clarke, W. R. (1984). Analysis of categorical incomplete longitudinal data. *Journal of the Royal Statistical Society, Series A* **147**, 87–99.

Yates, F. (1935). Complex experiments (with discussion). *Supplement to the Journal of the Royal Statistical Society* **2**, 181–247.

Zeger, S. L. and Karim, M. R. (1991). Generalized linear models with random effects: A Gibbs sampling approach. *Journal of the American Statistical Association* **86**, 79–86.

Zeger, S. L., Liang, K.-Y., and Self, S. G. (1985). The analysis of binary longitudinal data with time-independent covariates. *Biometrika* **72**, 31–38.

PART II

Parametric Modeling of Longitudinal Data

CHAPTER 2

Parametric modeling of longitudinal data: Introduction and overview

Garrett Fitzmaurice and Geert Verbeke

Contents

2.1 Introduction

Chapter 1 traced the development of regression models for longitudinal observations, from the simple univariate repeated-measures ANOVA model through the versatile class of linear mixed-effect models for continuous responses. Generalizations of linear models for discrete longitudinal data have followed a number of different research directions. Two important extensions are marginal (or population-averaged) models and generalized linear mixed models (GLMMs). Marginal models and the generalized estimating equations approach to inference are the subject of Chapter 3. Generalized linear mixed models are discussed in detail in Chapter 4. Both of these broad classes of models can be considered extensions of generalized linear models for longitudinal data, albeit with somewhat different targets of inference. Chapter 7 highlights various aspects of interpretation of regression parameters in these two classes of models, emphasizing their different targets of inference and the distinct scientific questions that can be addressed by each class of models. Although these two classes of models have distinct targets of inference, they have proven to be very useful for analyzing longitudinal data across a wide spectrum of disciplines. In Section 2.2 we review marginal and generalized linear mixed models, emphasizing the main differences in approach.

The marginal and generalized linear mixed models for discrete longitudinal data discussed in Chapters 3 and 4 are *non-linear*, but in the very restricted sense that a suitable non-linear link function determines an appropriate scale on which the transformed mean response is linear in the regression parameters (and, possibly, also linear in the random effects). For example, in marginal models the population-averaged mean response is related to the covariates as follows:

$$h^{-1}\left\{E(Y_{ij}|\boldsymbol{X}_{ij})\right\} = \boldsymbol{X}'_{ij}\boldsymbol{\beta},$$

via a known, possibly non-linear link function $h^{-1}(\cdot)$. Similarly, in GLMMs the individual-specific means are related to covariates in the following manner:

$$h^{-1}\left\{E(Y_{ij}|\boldsymbol{X}_{ij},\boldsymbol{b}_i)\right\} = \boldsymbol{X}'_{ij}\boldsymbol{\beta} + \boldsymbol{Z}'_{ij}\boldsymbol{b}_i,$$

also via a known, possibly non-linear link function $h^{-1}(\cdot)$. In both cases the transformed mean response is linear in the regression parameters, although it can be non-linear in the covariates. For example, models with polynomial time trends of increasing order can often approximate relatively complex non-linear trends, albeit only within the observed range of the data. Thus, marginal and generalized linear mixed models are non-linear in a restricted way. In contrast, Chapter 5 considers regression models for continuous longitudinal data that are fundamentally non-linear. In these models, the mean response is non-linear in the regression parameters and the random effects,

$$E(Y_{ij}|\boldsymbol{X}_{ij},\boldsymbol{b}_i) = f(\boldsymbol{X}_{ij},\boldsymbol{\beta},\boldsymbol{b}_i),$$

where the regression function $f(\boldsymbol{X}_{ij},\boldsymbol{\beta},\boldsymbol{b}_i)$ depends on $\boldsymbol{\beta}$ and $\boldsymbol{b}_i$ in a non-linear fashion. The non-linear regression function, $f(\boldsymbol{X}_{ij},\boldsymbol{\beta},\boldsymbol{b}_i)$, may be chosen on purely empirical grounds or may be derived from theoretical considerations. For example, non-linear models for growth over time are often chosen on purely empirical grounds, incorporating monotonicity and an asymptote where growth is assumed to level off. On the other hand, non-linear pharmacokinetic models for drug concentration are based on well-understood mechanisms for drug absorption, distribution, and elimination.

A feature of non-linear models is that they typically require fewer parameters to describe within-individual change over time and their regression parameters often have useful and natural physical interpretations (e.g., describing acceleration in growth and asymptotes). Unlike linear models that incorporate non-linearity in the covariates via polynomial trends or non-linear transformations of the covariates, non-linear models also provide more reliable predictions outside the range of the observed data. Chapter 5 highlights how non-linear models provide great flexibility for modeling an individual's longitudinal response trajectory. Chapter 5 also underscores that complex modeling of the variability of the response (e.g., heterogeneous variance that depends on subject-specific regression coefficients) is also required in certain applications.

The non-linear regression models discussed in Chapter 5 can be considered a natural extension of the linear mixed-effects model to the case of a non-linear regression function; alternatively, they can be considered an extension of univariate non-linear models to the setting of correlated longitudinal data. In Section 2.3 we show how the formulation of non-linear mixed models is conceptually quite similar to linear mixed-effects models, and can be considered in two stages. In the first stage, within-subject variability is accounted for with a suitable non-linear regression model; in the second stage, between-subject variability is accounted for by allowing a subset of the non-linear regression parameters to vary randomly across individuals.

Interestingly, the adoption of a non-linear regression function raises many of the same issues concerning targets of inference that are discussed in detail in Chapter 7. In Section 2.3 we show that, similar to the case with GLMMs, the non-linear regression function makes marginalization over the distribution of the random effects intractable. As a result, the regression parameters in non-linear mixed-effects models for longitudinal data have *subject-specific* rather than *population-averaged* interpretations.

Finally, generalized (non-)linear mixed models have traditionally assumed a normal distribution for the random effects. In many cases, the normal distribution assumption has been made on grounds of convenience rather than with a strong scientific rationale. However, there is now increasing awareness that certain aspects of inference may not be robust to misspecification of the random-effects distribution. The focus of Chapter 6 is on regression models for longitudinal data with non-Gaussian random effects. In Section 2.4 we briefly review some established results on the sensitivity of inferences from the linear and generalized linear mixed-effects model to misspecification of the random-effects distribution.

2.2 Marginal and generalized linear mixed models

In this section we compare and contrast marginal and mixed-effects models for longitudinal data. There are a number of important distinctions between these two broad classes of models that go beyond simple differences in approaches to accounting for the within-subject association. In particular, we emphasize that these two classes of models have somewhat different targets of inference and therefore address subtly different questions regarding longitudinal change in the response. In this section we highlight the main distinctions and discuss the types of scientific questions addressed by each of the two classes of models.

A marginal model for the $n_i \times 1$ mean response vector, $\boldsymbol{\mu}_i = (\mu_{i1}, \ldots, \mu_{in_i})'$, is given by

$$h^{-1}(\boldsymbol{\mu}_i) = g\{E(\boldsymbol{Y}_i | X_i)\} = X_i\boldsymbol{\beta}, \tag{2.1}$$

where $h^{-1}(\cdot)$ is an appropriate non-linear vector-valued link function (e.g., logit for binary responses or log for count data) and X_i is an $n_i \times p$ matrix of covariates. The vector of regression parameters $\boldsymbol{\beta}$ in a marginal model has interpretation in terms of changes in the transformed mean response in the study population, and their relation to covariates. For example, when the components of the response vector $\boldsymbol{Y}_i$ are binary and a logit link function is adopted, with

$$\text{logit}(\boldsymbol{\mu}_i) = X_i\boldsymbol{\beta},$$

the regression parameters have interpretation in terms of changes in the log odds of success in the study population. For any known link function $h^{-1}(\cdot)$, the population means can be expressed in terms of the inverse link function, $h(\cdot)$,

$$h\{h^{-1}(\boldsymbol{\mu}_i)\} = \boldsymbol{\mu}_i = E(\boldsymbol{Y}_i | X_i) = h(X_i\boldsymbol{\beta}). \tag{2.2}$$

For example, when the components of $\boldsymbol{Y}_i$ are binary and a logit link function has been adopted, the corresponding model for $\boldsymbol{\mu}_i$ is

$$\boldsymbol{\mu}_i = \frac{\exp(X_i\boldsymbol{\beta})}{1 + \exp(X_i\boldsymbol{\beta})}.$$

Whether expressed as (2.1) or (2.2), the regression parameters $\boldsymbol{\beta}$ in a marginal model describe changes in the transformed population mean response vector, $\boldsymbol{\mu}_i$. Note also that $\boldsymbol{\mu}_i$ depends on the index i only via fixed and known covariate values.

Next, consider GLMMs where the conditional mean of $\boldsymbol{Y}_i$, given a $q \times 1$ vector of random effects $\boldsymbol{b}_i$, is

$$h^{-1}\{E(\boldsymbol{Y}_i | X_i, \boldsymbol{b}_i)\} = X_i\boldsymbol{\beta}^* + Z_i\boldsymbol{b}_i,$$

where Z_i is a $n_i \times q$ design matrix from the random effects. The random effects $\boldsymbol{b}_i$ are assumed to have a joint distribution with mean zero and $q \times q$ covariance matrix G. Here we denote the fixed effects by $\boldsymbol{\beta}^*$ to clearly distinguish them from the corresponding regression parameters in the marginal model in (2.1). The regression coefficients $\boldsymbol{\beta}^*$ have subject-specific interpretations in terms of changes in the transformed mean response for any individual. That is, to interpret any component of $\boldsymbol{\beta}^*$ we must consider a unit change in the corresponding covariate while holding $\boldsymbol{b}_i$ (and the remaining covariates) fixed. However, the most natural way to hold $\boldsymbol{b}_i$ fixed at a particular value is to focus on the conditional mean response vector of any given individual. This interpretation of the components of $\boldsymbol{\beta}^*$ is most natural when the covariate is time-varying.

Thus, unlike $\boldsymbol{\beta}$ in marginal models, $\boldsymbol{\beta}^*$ needs to be interpreted in terms of changes in the transformed mean response for any individual (or the notional comparison of individuals with the same values for $\boldsymbol{b}_i$). The regression parameters $\boldsymbol{\beta}^*$ do not describe changes in the transformed mean response in the study population. The implied model for the marginal means can only be obtained by averaging over the distribution of the random effects. This involves taking an expectation of a non-linear function of $\boldsymbol{b}_i$,

$$\boldsymbol{\mu}_i = E(\boldsymbol{Y}_i|X_i)$$

$$= E\{E(\boldsymbol{Y}_i|X_i, \boldsymbol{b}_i)\}$$

$$= E\{h(X_i\boldsymbol{\beta}^* + Z_i\boldsymbol{b}_i)\} \tag{2.3}$$

$$= \int_{-\infty}^{\infty} h(X_i\boldsymbol{\beta}^* + Z_i\boldsymbol{b}_i)f(\boldsymbol{b}_i)d\boldsymbol{b}_i,$$

where $f(\boldsymbol{b}_i)$ is the joint probability density function for $\boldsymbol{b}_i$. However, the marginal mean, $E(\boldsymbol{Y}_i|X_i)$, given by (2.3), does not, in general, have a closed-form expression and, moreover,

$$E(\boldsymbol{Y}_i|X_i) \neq h(X_i\boldsymbol{\beta}),$$

for any $\boldsymbol{\beta}$. That is, marginalization of a GLMM does not yield a generalized linear model for the marginal means. For example, consider the logistic regression model with a randomly varying intercept,

$$\text{logit}\{E(\boldsymbol{Y}_i|X_i, b_i)\} = X_i\boldsymbol{\beta}^* + b_i,$$

where $b_i \sim N(0, \sigma_b^2)$. The implied model for the marginal probability of success is

$$\boldsymbol{\mu}_i = E(\boldsymbol{Y}_i|X_i)$$

$$= E\{E(\boldsymbol{Y}_i|X_i, b_i)\}$$

$$= E\left\{\frac{\exp(X_i\boldsymbol{\beta}^* + b_i)}{1 + \exp(X_i\boldsymbol{\beta}^* + b_i)}\right\}$$

$$= \int_{-\infty}^{\infty} \frac{\exp(X_i\boldsymbol{\beta}^* + b_i)}{1 + \exp(X_i\boldsymbol{\beta}^* + b_i)} \frac{1}{\sqrt{2\pi\sigma_b^2}} \exp\left(-\frac{1}{2}\frac{b_i^2}{\sigma_b^2}\right) db_i.$$

This expression cannot be evaluated in closed form, and moreover is not of the logistic regression form,

$$\frac{\exp(X_i\boldsymbol{\beta})}{1 + \exp(X_i\boldsymbol{\beta})},$$

for any $\boldsymbol{\beta}$.

It is well known that for the special case of the logistic regression model with only a single randomly varying intercept (or subject effect),

$$\text{logit}\{E(\boldsymbol{Y}_i|X_i, b_i)\} = X_i\boldsymbol{\beta}^* + b_i,$$

where $b_i \sim N(0, \sigma_b^2)$, the following approximate relationship holds:

$$\text{logit}\{E(\boldsymbol{Y}_i|X_i)\} \approx \left(1 + k^2\sigma_b^2\right)^{-\frac{1}{2}} X_i\boldsymbol{\beta}^*,$$

where $k = \frac{16\sqrt{3}}{15\pi} = \sqrt{0.346}$ (Diggle et al., 2002, Section 7.4). This approximate relationship highlights how the logistic regression coefficients in the marginal model are attenuated relative to the corresponding fixed effects in the logistic regression model with a randomly varying intercept, with

$$\boldsymbol{\beta} \approx \frac{\boldsymbol{\beta}^*}{\sqrt{1 + 0.346\sigma_b^2}}.$$

Thus, when $\text{Var}(b_i) = \sigma_b^2 > 0$ the marginal logistic regression model parameters, $\boldsymbol{\beta}$, are smaller in absolute value than the fixed effects, $\boldsymbol{\beta}^*$, in the mixed-effects model. In addition, the discrepancy between $\boldsymbol{\beta}$ and $\boldsymbol{\beta}^*$ increases with increasing σ_b^2.

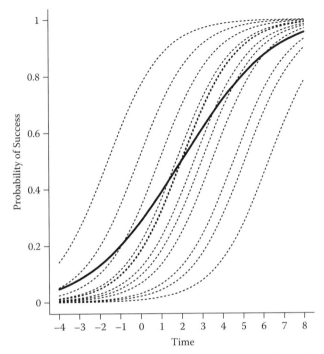

Figure 2.1 Comparison of conditional probabilities of success (dotted lines) and marginal probability of success (solid line), averaged over the distribution of the random effects.

The difference between $\boldsymbol{\beta}^*$ and $\boldsymbol{\beta}$ can also be illustrated graphically. Consider the following simple numerical illustration. Suppose that $\boldsymbol{Y}_i$ is a vector of binary responses and it is of interest to describe changes in the log odds of success over time. For simplicity, we assume that there are no covariates other than the times of measurement. A logistic regression model, with randomly varying intercepts, is given by

$$\text{logit}\{E(Y_{ij}|b_i)\} = \beta_0^* + \beta_1^* t_{ij} + b_i,$$

where b_i is assumed to have a normal distribution with zero mean and variance $\sigma_b^2 = \text{Var}(b_i)$. Figure 2.1 displays a plot of $E(Y_{ij}|b_i)$ versus t_{ij} for a random sample of b_i from a normal distribution with zero mean and variance $\sigma_b^2 = 4$ (with $\beta_0^* = -1.5$ and $\beta_1^* = 0.75$; $t_{ij} \in [-4, 8]$). Also displayed in Figure 2.1 is a plot of the marginal probability of success, integrated over the distribution of b_i. When the subject-specific logistic curves are compared to the population-averaged curve, it is apparent that the slopes of the former (determined by β_1^*) are steeper than the slope of the latter. Focusing on the range of probabilities from 0.3 to 0.7, where the logistic curves are approximately linear, the slopes for the subject-specific curves rise faster than the slope for the marginal success probabilities. This reinforces the notion that β^* does not characterize aspects of the population log odds of response, but instead describes changes in the log odds of success for an individual from the population.

Thus, in general, the fixed effects $\boldsymbol{\beta}^*$ in GLMMs are not comparable to the regression parameters $\boldsymbol{\beta}$ in marginal models. The lack of comparability reflects the distinct targets of inference associated with GLMMs and marginal models. That is, the fixed effects $\boldsymbol{\beta}^*$ describe the effects of covariates on changes in an individual's response over time, while the regression parameters $\boldsymbol{\beta}$ describe the effects of covariates on changes in the population mean response over time (see Chapter 7 for a comprehensive discussion of this topic).

One particular, well-known, exception is the special case of an identity link function (i.e., linear mixed-effects models). For an identity link function these issues concerning the interpretation of the regression parameters do not arise because $\boldsymbol{\beta}^* = \boldsymbol{\beta}$ and the regression parameters have the same interpretation in both classes of models. That is, marginalization of the linear mixed-effects model yields

$$\begin{aligned} E(\boldsymbol{Y}_i|X_i) &= E\{E(\boldsymbol{Y}_i|X_i, \boldsymbol{b}_i)\} \\ &= E(X_i\boldsymbol{\beta} + Z_i\boldsymbol{b}_i) \\ &= X_i\boldsymbol{\beta}. \end{aligned}$$

Simply stated, in the linear mixed-effects model $\boldsymbol{\beta}$ has a marginal interpretation because the average of the linear rates of change over time for individuals is the same as the linear rate of change over time in the population mean response. In contrast, for any *non-linear* function of $\boldsymbol{b}_i$, say $h(X_i\boldsymbol{\beta} + Z_i\boldsymbol{b}_i)$,

$$E\{h(X_i\boldsymbol{\beta} + Z_i\boldsymbol{b}_i)\} \neq h(X_i\boldsymbol{\beta}).$$

In summary, marginal and generalized linear mixed models represent natural extensions of generalized linear models to longitudinal data but differ in approaches and their respective targets of inference. The basic premise of marginal models is to make inferences about population means, albeit on a transformed scale (e.g., logit or log). The term "marginal" is used to emphasize that the mean response modeled is conditional only on the covariates and not on unobserved random effects (or on previous responses). A distinctive feature of marginal models is that the regression models for the mean response and the models for the within-subject association are specified separately. This separation of the model for the mean response from the model for the within-subject association ensures that the marginal model regression coefficients have interpretation that does not depend on the assumptions made about the within-subject association. Specifically, the regression coefficients in marginal models describe the effects of covariates on the population mean response.

In contrast, the basic premise of generalized linear mixed-effects models is that there is natural heterogeneity across individuals in the study population in a subset of the regression parameters. Generalized linear mixed models extend the conceptual approach of the linear mixed-effects model in a very natural way. The correlation among repeated measurements arises from their sharing of common random effects. Unlike the linear mixed-effects model, the regression parameters in GLMMs have subject-specific, rather than population-averaged, interpretations. That is, due to the non-linear link functions that are usually adopted for discrete responses, the fixed effects do not describe changes in the population-averaged mean response. Instead, they describe changes in an individual's mean response and the relation of these changes to covariates. As a result, GLMMs are most useful when the scientific objective is to make inferences about individuals rather than the study population.

Finally, the choice between marginal and generalized linear mixed models for longitudinal data can only be made on subject-matter grounds. For any given longitudinal study, different scientific questions will usually demand different analytic models. For example, a physician considering the potential benefits of a novel treatment for one of her patients might be more interested in the subject-specific effect of the treatment. On the other hand, public health researchers or health insurance assessors considering the potential reduction in morbidity or mortality in the population if patients were to receive the novel treatment would be more interested in the population-averaged effect of the treatment. When the answers to both of these questions are of interest, there is absolutely no contradiction in reporting estimates of both the subject-specific and population-averaged effects. Although the subject-specific and population-averaged effects are different, and address distinct scientific questions, they are not incompatible.

2.3 Non-linear mixed-effects models

In contrast to the marginal and generalized linear mixed models discussed previously, which allow for non-linearity in a restricted way, the regression models presented in Chapter 5 are fundamentally non-linear. Specifically, in non-linear mixed-effects models the mean response is assumed to be non-linear in the regression parameters and the random effects. It is insightful to develop these models in two stages: a model for intra-individual variability coupled with a model for inter-individual variability.

Stage 1: Intra-individual model. The first-stage model specifies the mean and covariance structure for a given individual. Here, for ease of exposition, we temporarily suppress dependence on covariates in the assumptions on the conditional mean and covariance. In the first stage we assume that the mean response for the ith individual at the jth occasion can be expressed in terms of a non-linear regression function and random error,

$$Y_{ij} = f(\boldsymbol{X}_{ij}, \boldsymbol{\beta}_i) + e_{ij},$$

where e_{ij} is a random error term with $E(e_{ij}|\boldsymbol{\beta}_i) = 0$. In this model the regression function depends in a non-linear way on a set of subject-specific regression parameters, $\boldsymbol{\beta}_i$. Although the functional form, $f(\cdot)$, is the same for all individuals, differences between individuals in their longitudinal response trajectories are accommodated by allowing for different $\boldsymbol{\beta}_i$ (as well as differences in the covariates, $\boldsymbol{X}_{ij}$). The random errors, e_{ij}, are commonly assumed to have a multivariate normal distribution, with $\boldsymbol{e}_i \sim N(\boldsymbol{0}, V_i)$. The covariance matrix V_i captures the variability (and covariation) of the repeated measures on the same individual. Note that V_i may depend upon i not only through its dimension but also through the subject-specific mean response given $\boldsymbol{\beta}_i$. Also, unlike standard GLMMs, the e_{ij} are not necessarily assumed to be conditionally independent, i.e., $\text{Cov}(\boldsymbol{Y}_i|\boldsymbol{\beta}_i) = V_i$ is not necessarily diagonal.

Stage 2: Inter-individual model. The first-stage model characterizes intra-individual variation in the response over time. In contrast, the second-stage model characterizes inter-individual variation in the regression parameters $\boldsymbol{\beta}_i$. For example, to account for inter-individual variation among the $\boldsymbol{\beta}_i$ it might be assumed that the $\boldsymbol{\beta}_i$ depend linearly on a set of covariates, say A_i,

$$\boldsymbol{\beta}_i = A_i \boldsymbol{\beta} + \boldsymbol{b}_i,$$

where the random effects, $\boldsymbol{b}_i$, are assumed to have zero mean and covariance matrix, G. Commonly, it is assumed that $\boldsymbol{b}_i \sim N(\boldsymbol{0}, G)$; however, see Chapter 6 for a detailed discussion of some of the ramifications of this choice for the distribution of the random effects. Note that in the above, a linear specification has been assumed. Alternatively, the inter-individual variation can be modeled as a non-linear function,

$$\boldsymbol{\beta}_i = d(A_i, \boldsymbol{\beta}, \boldsymbol{b}_i),$$

where $d(\cdot)$ is a known vector-valued function. That is, we can model $\boldsymbol{\beta}_i$ as a non-linear function of $\boldsymbol{\beta}$.

The two-stage specification of the non-linear mixed-effects model given above provides a very general and rich class of models for the analysis of longitudinal data. The first stage specifies the mean response for a given individual as a non-linear regression function. In addition, the conditional variance of Y_{ij}, given $\boldsymbol{b}_i$, is allowed to depend upon the subject-specific regression coefficients, $\boldsymbol{\beta}_i$. In this model, the $\boldsymbol{\beta}_i$ are subject-specific parameters, which are allowed to be non-linear functions of $\boldsymbol{\beta}$ (and $\boldsymbol{X}_{ij}$).

A characteristic of non-linear mixed-effects models is that inference usually focuses on features or mechanisms that underlie the subject-specific longitudinal response trajectories and how these vary across subjects in the population. Therefore, the interpretation of the fixed effects, $\boldsymbol{\beta}$, in a non-linear mixed-effects model is "subject-specific," in the same sense that the fixed effects are in GLMMs. By modeling the response at the individual level, the components of $\boldsymbol{\beta}$ describe the typical subject-specific effects of covariates on the mean response over time. In general, the marginal mean response (and covariance), when averaged over the distribution of the random effects, does not have a closed-form solution; moreover, it does not have the same functional form. That is,

$$E\left\{f(\boldsymbol{X}_{ij}, \boldsymbol{\beta}_i)\right\} \neq f(\boldsymbol{X}_{ij}, \boldsymbol{\beta}),$$

for any $\boldsymbol{\beta}$. This is precisely the same issue that arises in comparison of marginal models and GLMMs; thus, the discussion of the different targets of inference in Chapter 7 has great relevance here also.

The non-linear mixed-effects models have been introduced here from a two-stage modeling perspective. The same two-stage modeling perspective can be applied to both linear and generalized linear mixed models. For example, consider the logistic model with randomly varying intercepts, considered earlier. From a two-stage modeling perspective, stage 1 assumes that the intra-individual variation can be modeled as

$$Y_{ij} = \mu_{ij} + e_{ij},$$

for $\mu_{ij} = e^{(\boldsymbol{X}'_{ij}\boldsymbol{\beta}_i)}/\{1 + e^{(\boldsymbol{X}'_{ij}\boldsymbol{\beta}_i)}\}$. However, in contrast to the classical non-linear models, the error components e_{ij} are not assumed to be normally distributed. Instead, e_{ij} equals $1 - \mu_{ij}$ with probability μ_{ij}, and equals $0 - \mu_{ij}$ with probability $1 - \mu_{ij}$. The stage 2 model for inter-individual variation assumes that a subset of the regression parameters in $\boldsymbol{\beta}_i$ can be modeled as a known (non-)linear function of the covariates, while others are assumed to be fixed for all subjects.

Finally, it is worth pointing out that one can specify population-averaged non-linear models, e.g.,

$$Y_{ij} = f(\boldsymbol{X}_{ij}, \boldsymbol{\beta}) + \epsilon_{ij},$$

where the marginal expectation of Y_{ij} is specified as a non-linear function of $\boldsymbol{\beta}$ (and $\boldsymbol{X}_{ij}$),

$$E(Y_{ij}|\boldsymbol{X}_{ij}) = f(\boldsymbol{X}_{ij}, \boldsymbol{\beta}).$$

However, in many important areas of application (e.g., in studies of growth and in pharmacokinetic studies), the subject-specific approach embodied by non-linear mixed-effects models is somewhat more natural.

2.4 The random-effects distribution

A characteristic feature of all random-effects models, including GLMMs and non-linear mixed models, is that they specify a model for the response vector conditional on the random effects and covariates, combined with a model for the random effects. For ease of exposition, in the following we suppress the dependence on covariates. Let $f_i(\boldsymbol{y}_i|\boldsymbol{b}_i)$ denote the density function of the response vector $\boldsymbol{Y}_i$, and let $f(\boldsymbol{b}_i)$ denote the density for the random effects. In general, unless a fully Bayesian approach is adopted, model fitting and inference for random-effects models are based on the marginal model for $\boldsymbol{Y}_i$; the latter is obtained by integrating over the distribution of the random effects. Specifically, this marginal density function is given by

$$f_i(\boldsymbol{y}_i) = \int f_i(\boldsymbol{y}_i|\boldsymbol{b}_i)f(\boldsymbol{b}_i)d\boldsymbol{b}_i. \tag{2.4}$$

Although in practice one is usually primarily interested in estimating the fixed-effects parameters, it is often useful to obtain predictions of the random effects, $\boldsymbol{b}_i$, as well. Because the random effects reflect between-subject variability, this makes them particularly useful for detecting special or unusual response profiles (e.g., outlying individuals) or groups of individuals whose response profiles evolve differently over time. Also, estimates for the random effects are needed whenever there is interest in prediction of subject-specific evolutions. Inference for the random effects is usually based on their posterior distribution $f_i(\boldsymbol{b}_i|\boldsymbol{y}_i)$, given by

$$f_i(\boldsymbol{b}_i|\boldsymbol{y}_i) = \frac{f_i(\boldsymbol{y}_i|\boldsymbol{b}_i) \, f(\boldsymbol{b}_i)}{\int \, f_i(\boldsymbol{y}_i|\boldsymbol{b}_i) \, f(\boldsymbol{b}_i) \, d\boldsymbol{b}_i}. \tag{2.5}$$

The mean or mode of (2.5) can be used as point estimates for $\boldsymbol{b}_i$, yielding so-called empirical Bayes estimates.

It should be transparent that the random-effects distribution is crucial in the calculation of the marginal density (2.4), as well as the posterior density (2.5). One approach is to leave it completely unspecified and to use non-parametric maximum likelihood estimation, which maximizes the likelihood over all possible distributions for the random effects. The resulting estimate is then always discrete with finite support. Depending on the context, this may or may not be a realistic reflection of the true heterogeneity between individuals. Alternatively, the distribution of the random effects can be assumed to have a specific parametric form, thereby allowing classical likelihood-based inferential procedures to be applied. For example, in the two-stage model formulation presented in Section 2.3 it is common to assume that the random effects have a (multivariate) normal distribution. In the case of linear models this is mathematically convenient in the sense that the integral in (2.4) can be calculated analytically. The posterior density in (2.5) is then also multivariate normal. However, in generalized linear or in non-linear mixed models this is no longer the case and approximations to both (2.4) and (2.5) are needed.

An important issue related to the choice of random-effects distribution is the impact of its possible misspecification. For linear mixed-effects models with the classical normal assumption for the random effects, it has been shown (e.g., Verbeke and Lesaffre, 1997) that deviations from the normality assumption for the random effects have very little impact on the estimation of the fixed-effects parameters. However, standard errors require some correction in order to obtain valid inferences. This robustness property for estimation of the fixed effects is in strong contrast to the severe sensitivity that has been reported for the empirical Bayes estimates for the random effects (e.g., Verbeke and Lesaffre, 1996). Furthermore, for non-linear and generalized linear mixed models, misspecification of the random-effects distribution can lead to seriously biased estimates for the fixed-effects parameters. We refer to Neuhaus, Hauck, and Kalbfleisch (1992), Butler and Louis (1992), Pfeiffer et al. (2003), Heagerty and Zeger (2000), and Litiére, Alonso, and Molenberghs (2007) for more details on the effects of misspecification of the random-effects distribution in GLMMs.

Finally, it should be emphasized that empirical Bayes estimates obtained under specific parametric assumptions for the random effects cannot be used to assess the validity of these assumptions. This is because of the well-known result from Bayesian data analysis that prior assumptions often dominate the posterior density (2.5), except when many observations y_{ij} are available for each subject (i.e., except when n_i is very large). An illustration of this, in the context of the linear mixed model, can be found in Section 7.8 of Verbeke and Molenberghs (2000). Several authors have therefore proposed mixed models with relaxed distributional assumptions for the random effects. For example, Magder and Zeger (1996) and Verbeke and Lesaffre (1996) replaced the traditional normality assumption by mixtures of normals. Allowing some components in the mixture to have zero variance leads to latent-class models. Examples of the latter can be found in Fieuws, Spiessens, and Draney (2004) or in Muthén and Shedden (1999). Finally, Davidian and Gallant (1993), Zhang and Davidian (2001), and Ghidey, Lesaffre, and Eilers (2004) have applied smoothing techniques, yielding mixed models with semi-parametric, smooth random-effects distributions. Mixed models with non-Gaussian random effects will be discussed further in Chapter 6.

2.5 Concluding remarks

There is now a broad range of parametric regression models for longitudinal data available to statisticians. In this chapter we have given an overview of some general classes of approaches frequently used in practice. One class of models is based on the specification of the marginal distribution of the repeatedly measured responses. A second class of models starts by specifying a model for the responses conditional on a set of subject-specific random parameters. An appealing property of the latter models is that they clearly separate within-subject variability from between-subject variability. However, except in the special case of linear models with normal random effects, mixed-effects models lead, in general, to intractable likelihood functions. Because no simple analytic solutions are available, there has been much research effort focused on approximate methods of estimation.

One of the main differences between marginal and random-effects models is in the interpretation of their regression parameters. While a population-averaged interpretation is needed for the regression parameters in marginal models, the fixed-effects parameters in mixed models can only be interpreted at a subject-specific level. In general, the choice among these two broad classes of model should be made on substantive grounds, recognizing their different targets of inference and the distinct scientific questions that they address. To date, there has been a substantial amount of research focused on model formulation and model fitting; methods for model comparison and model diagnostics have received relatively less attention in the statistical literature.

Finally, it should be noted that missing data are a common problem in many longitudinal studies, i.e., the intended measurements on certain individuals are not obtained for a variety of known or (often) unknown reasons. Technically speaking, all of the models that have been mentioned in this chapter can handle the resulting unbalanced data structures. However, it is important to recognize that they may be based on somewhat different assumptions about possible relations between the reason(s) for missingness and the response variable of interest. This important topic will be discussed in much greater detail in Part 5 (see Chapters 17 through 22).

Acknowledgments

This work was supported by grants GM 29745 and MH 54693 from the U.S. National Institutes of Health.

References

Butler, S. M. and Louis, T. A. (1992). Random effects models with non-parametric priors. *Statistics in Medicine* **11**, 1981–2000.

Davidian, M. and Gallant, A. R. (1993). The nonlinear mixed effects model with a smooth random effects density. *Biometrika* **80**, 475–488.

Diggle P. J., Heagerty P. J., Liang K.-Y., and Zeger S. L. (2002). *Analysis of Longitudinal Data*, 2nd ed. Oxford: Oxford University Press.

Fieuws, S., Spiessens, B., and Draney, K. (2004). Mixture models. In P. De Boeck and M. Wilson (eds.), *Explanatory Item Response Models: A Generalized Linear and Nonlinear Approach*, pp. 317–340. New York: Springer.

Ghidey, W., Lesaffre, E., and Eilers, P. (2004). Smooth random effects distribution in a linear mixed model. *Biometrics* **60**, 945–953.

Heagerty, P. J. and Zeger, S. L. (2000). Marginalized multilevel models and likelihood inference. *Statistical Science* **15**, 1–26.

Litière, S., Alonso, A., and Molenberghs, G. (2007). The impact of a misspecified random-effects distribution on the estimation and the performance of inferential procedures in generalized linear mixed models. *Statistics in Medicine*. To appear.

Magder, L. S. and Zeger, S. L. (1996). A smooth nonparametric estimate of a mixing distribution using mixtures of Gaussians. *Journal of the American Statistical Association* **91**, 1141–1152.

Muthén, B. and Shedden, K. (1999). Finite mixture modeling with mixture outcomes using the EM-algorithm. *Biometrics* **55**, 463–469.

Neuhaus, J. M., Hauck, W. W., and Kalbfleisch, J. D. (1992). The effects of mixture distribution specification when fitting mixed-effects logistic models. *Biometrika* **79**, 755–762.

Pfeiffer, R. M., Hildesheim, A., Gail, M. H., Pee, D., Chen, C., Goldstein, A. M., and Diehl, S. R. (2003). Robustness of inference on measured covariates to misspecification of genetic random effects in family studies. *Genetic Epidemiology* **24**, 14–23.

Verbeke, G. and Lesaffre, E. (1996). A linear mixed-effects model with heterogeneity in the random-effects population. *Journal of the American Statistical Association* **91**, 217–221.

Verbeke, G. and Lesaffre, E. (1997). The effect of misspecifying the random effects distribution in linear mixed models for longitudinal data. *Computational Statistics and Data Analysis* **23**, 541–556.

Verbeke, G. and Molenberghs, G. (2000). *Linear Mixed Models for Longitudinal Data*. New York: Springer.

Zhang, D. and Davidian, M. (2001). Linear mixed models with flexible distributions of random effects for longitudinal data. *Biometrics* **57**, 795–802.

CHAPTER 3

Generalized estimating equations for longitudinal data analysis

Stuart Lipsitz and Garrett Fitzmaurice

Contents

3.1 Introduction

In this chapter we discuss the generalized estimating equations (GEE) approach for analyzing longitudinal data. Over the past 20 years, the GEE approach has proven to be an exceedingly useful method for the analysis of longitudinal data, especially when the response variable is discrete (e.g., binary, ordinal, or a count). When the longitudinal response is discrete, linear models (e.g., linear mixed-effects models) are not very appealing for relating changes in the mean response to covariates for at least two main reasons. First, with a discrete response there is intrinsic dependence of the variability on the mean. Second, the range of the mean response (e.g., a proportion or rate for a response that is binary or a count, respectively) is constrained. In the setting of regression modeling of a univariate response, both of these aspects of the response can be conveniently accommodated within generalized linear models via known variance and link functions.

However, a straightforward application of generalized linear models to longitudinal data is not appropriate, due to the lack of independence among repeated measures obtained on the same individual. There has been extensive statistical literature on extending generalized linear models to the longitudinal data setting. One approach for accounting for the within-subject association is via the introduction of random effects in generalized linear models. This leads to a class of models known as generalized linear mixed models (GLMMs); see

Chapter 4 for a very comprehensive and expository discussion of these models. Generalized linear mixed models represent one way to extend generalized linear models to longitudinal data; they extend in a natural way the conceptual approach represented by the linear mixed-effects models for continuous responses (e.g., Laird and Ware, 1982; see also Chapter 1). There is a second approach for extending generalized linear models to longitudinal data that leads to a class of regression models that are known as *marginal models*. The term *marginal* in this context is used to emphasize that the model for the mean response at each occasion depends only on the covariates of interest, and does not incorporate dependence on random effects or previous responses. This is in contrast to GLMMs, where the mean response is modeled not only as a function of covariates but is conditional also on random effects. The most salient feature of marginal models is a regression model, with appropriately specified link function, relating the mean response at each occasion to the covariates. For estimation of the regression model parameters, marginal models do not necessarily require distributional assumptions for the vector of longitudinal responses. When full distributional assumptions for the vector of responses are avoided, the marginal model is said to be *semi-parametric*, and this leads to a method of estimation known as *generalized estimating equations* (Liang and Zeger, 1986), the main focus of this chapter.

Thus, the GEE approach has its basis in one particular type of extension of generalized linear models to longitudinal, or more generally, cluster-correlated data. Before outlining the main features and properties of GEE methods in Section 3.2, it is useful to review some of its predecessors and consider the reasons why this method has been so widely adopted for longitudinal analyses.

Prior to the seminal companion papers on GEE by Liang and Zeger (1986) and Zeger and Liang (1986), statistical methods for the analysis of discrete longitudinal data had lagged somewhat behind the corresponding developments for continuous outcomes. The early foundations for statistical methods for estimation of marginal models for repeated categorical responses can be traced to a general approach developed by Grizzle, Starmer, and Koch (1969); this approach soon became known as the GSK method. Although the GSK method was developed originally as a very general method for the analysis of categorical data, Koch and Reinfurt (1971) and Koch et al. (1977) recognized that the method could be applied to the analysis of repeated measurements. The GSK method is founded on a weighted least-squares (WLS) approach that makes few assumptions about the within-subject association among the repeated categorical outcomes. Specifically, this WLS approach is based on a multinomial sampling model for the joint distribution of the vector of categorical outcomes and relies on the asymptotic normality of estimators of the corresponding multinomial probabilities. Recall that if a categorical outcome, with C levels, is measured repeatedly at n occasions, there are C^n possible response profiles; thus, the joint distribution of the vector of longitudinal responses is multinomial with C^n response probabilities ($C^n - 1$ non-redundant probabilities because the multinomial probabilities are constrained to sum to 1). In general, for a marginal model for the repeated categorical responses, the main interest is focused on the $n \times (C - 1)$ correlated marginal probabilities or transformations of these marginal probabilities (e.g., logit transformations). Moreover, these marginal probabilities can simply be expressed as linear transformations of the underlying multinomial probabilities. In the GSK method, the sample proportions (or means) at each occasion are obtained within each covariate stratum and grouped together to form a sample mean vector of length $n \times (C - 1)$; the sample covariance matrix is also estimated. A marginal model is specified by relating the sample means (or known functions of the sample means such as the empirical logits) at each occasion to a linear function of stratum covariates. Appealing to the multivariate central limit theorem, and the use of the so-called delta method, the observed sample means (or functions of the sample means) have approximate normal distributions, given sufficient sample sizes within each stratum. The WLS method, with the sample mean vectors as the outcome vectors and the inverse of the sample covariance matrices as the weight matrices, is then used to estimate the marginal model regression parameters.

At the time of its introduction, the GSK method provided a method for estimating a very general family of models for repeated categorical data, allowing non-linear link functions to relate the marginal expectations of the longitudinal responses to covariates. For example, the method allows the fitting of logistic regression models to repeated binary outcomes. At the heart of the GSK method are the notions of stratification of subjects on the basis of covariates and non-parametric estimation of the multinomial probabilities for the vector of repeated categorical responses within each stratum. Because the multinomial probabilities are estimated non-parametrically as the sample proportions within strata, asymptotically, the GSK method is equivalent to maximum likelihood (ML) estimation.

However, the GSK method has a number of restrictions that limit its usefulness for the analysis of longitudinal data. Because it relies on stratification, the method requires that all covariates be categorical or the reduction of quantitative covariates to categorical variables. Moreover, it also requires a relatively large number of subjects within each stratum to estimate the C^n multinomial probabilities; note that with $C = 4$ response levels for an outcome measured repeatedly at $n = 6$ occasions, there are $4^6 - 1$ or 4095 non-redundant multinomial probabilities. Implicitly, this means that the method can only be validly applied when (i) the number of repeated measures, n, is relatively small, thereby ensuring that the number of multinomial probabilities to be estimated does not grow exponentially; (ii) the number of covariate strata is relatively small; and (iii) the number of individuals is relatively large, ensuring a sufficient number of subjects within each stratum. Finally, the GSK method, as originally developed, also requires that the longitudinal study design be balanced on time in the sense that all subjects are measured at the same set of occasions; later refinements to the GSK method can handle balanced longitudinal designs with incompleteness due to missing observations.

As an alternative to GSK methods, full likelihood-based methods for estimating marginal models for repeated categorical data can, in principle, accommodate many of these common features of longitudinal studies (e.g., covariates that are both quantitative and categorical, missing data, and so on). However, in practice, ML fitting of marginal models for discrete longitudinal data has proven to be very challenging. Among the many challenges are: (i) it can be conceptually difficult to model higher-order associations in a flexible and interpretable manner that is consistent with the model for the marginal expectations; (ii) given a marginal model for the vector of repeated outcomes, the multinomial probabilities cannot, in general, be expressed in closed form as a function of the model parameters; and (iii) the number of multinomial probabilities grows exponentially with the number of repeated measures. As a result, ML estimation is feasible only for a relatively small number of repeated measures. Thus, although the development of likelihood-based methods for marginal models has been fertile ground for methodological research (e.g., Bahadur, 1961; McCullagh and Nelder, 1989; Zhao and Prentice, 1990; Lipsitz, Laird, and Harrington, 1990; Liang, Zeger, and Qaqish, 1992; Becker and Balagtas, 1993; Fitzmaurice, Laird, and Rotnitzky, 1993; Molenberghs and Lesaffre, 1994; Lang and Agresti, 1994; Glonek and McCullagh, 1995; Bergsma and Rudas, 2002; and many others), to date, ML methods that have been developed have also proven to be of only limited practical use.

Thus, by the end of the 1970s, the GSK method provided the first unified approach for extending generalized linear models to repeated categorical data, based on an application of non-iterative weighted least squares. The method was implemented in widely available statistical software (e.g., `proc catmod` in SAS). However, the method lacked versatility and could not accommodate many of the common features of longitudinal studies, namely mixtures of quantitative and categorical, time-invariant and time-varying, covariates and inherently unbalanced designs with both mistimed measurements and missing data. So, by the early 1980s, the time was ripe for new methods for the analysis of discrete longitudinal data that could handle all of these aspects of longitudinal data in a relatively flexible way.

Finally, we note that the WLS estimation at the heart of the GSK method is similar in spirit to non-iterative empirical logistic regression for binomial data (see, for example, Cox and Snell, 1989). Empirical logistic regression is also applicable only when individuals can be grouped into strata based on their covariates; at the time of its introduction, this greatly limited the usefulness of empirical logistic regression in practice. However, as computing advanced, empirical logistic (WLS) regression for binomial data was soon replaced by ML methods that were far more versatile. In a similar vein, the GSK method for discrete longitudinal data was replaced not by ML methods, for all the reasons mentioned earlier, but by the GEE approach developed by Liang and Zeger (1986). By and large, the GEE approach has overcome most of the limitations of its predecessors and has fundamentally changed the way empirical researchers can approach the analysis of discrete longitudinal data.

3.2 Generalized estimating equations (GEE) for longitudinal data

In this section we briefly review the main features of marginal models for longitudinal data and discuss the key properties of the GEE approach. Because the GEE approach can be considered a multivariate extension of quasi-likelihood estimation, we first discuss estimation of the regression parameters in generalized linear models for a *univariate* outcome, before considering estimation of the regression parameters in a marginal model for longitudinal outcomes.

3.2.1 Notation

Before we begin our discussion of marginal models and the GEE approach, we introduce some notation. We assume that N subjects are measured repeatedly over time. We let Y_{ij} denote the response variable for the ith subject on the jth measurement occasion ($i = 1, \ldots, N; j = 1, \ldots, n_i$). The response variable can be continuous or discrete (e.g., binary, ordinal, polytomous, or a count). The type of response variable (e.g., binary or count) will have implications for model specification. In general, we do not assume that subjects have the same number of repeated measures or that they are measured at a common set of occasions. To accommodate such *unbalanced* longitudinal data (i.e., repeated measurements that are not obtained at a common set of occasions), we assume that there are n_i repeated measurements of the response on the ith subject and that each Y_{ij} is observed at time t_{ij}. We can group the responses into an $n_i \times 1$ vector of responses denoted by $\boldsymbol{Y}_i$. Finally, associated with each response, Y_{ij}, there is a $p \times 1$ vector of covariates, $\boldsymbol{X}_{ij}$. We note that $\boldsymbol{X}_{ij}$ may include covariates whose values do not change throughout the duration of the study and covariates whose values change over time. The former are referred to as time-stationary or between-subject covariates (e.g., gender and fixed experimental treatments or interventions), whereas the latter are referred to as time-varying or within-subject covariates (e.g., time since baseline, current smoking status, and environmental exposures that can vary over time). We can group the vectors of covariates into an $n_i \times p$ matrix of covariates denoted by X_i.

3.2.2 Defining features of marginal models

As mentioned in Section 3.1, the defining feature of marginal models is a regression model relating the mean response at each occasion, via a suitable link function, to the covariates. With a marginal model, the main focus is on making inferences about population means.

As a result, marginal models for longitudinal data separately model the mean response and the within-subject association among the repeated responses. In a marginal model, the goal is to make inferences about the former, whereas the latter is regarded as a nuisance characteristic of the data that must be taken into account in order to make correct inferences about changes in the population mean response over time.

A marginal model for longitudinal data has the following three-part specification:

1. The conditional expectation of each response, $E(Y_{ij}|\boldsymbol{X}_{ij}) = \mu_{ij}$, is assumed to depend on the covariates through a known link function $h^{-1}(\cdot)$, e.g., logit(μ_{ij}) or log(μ_{ij}),

$$h^{-1}(\mu_{ij}) = \eta_{ij} = \boldsymbol{X}'_{ij}\boldsymbol{\beta},$$

   where $\boldsymbol{\beta}$ is a $p \times 1$ vector of marginal regression parameters.

2. The conditional variance of each Y_{ij}, given $\boldsymbol{X}_{ij}$, is assumed to depend on the mean according to

$$\mathrm{Var}(Y_{ij}) = \phi\, v(\mu_{ij}),$$

   where $v(\mu_{ij})$ is a known "variance function" (i.e., a known function of the mean, μ_{ij}) and ϕ is a scale parameter that may be fixed and known or may need to be estimated. Note that dependence of $\mathrm{Var}(Y_{ij})$ on the covariates $\boldsymbol{X}_{ij}$ is suppressed from notation.

3. The conditional within-subject association among the vector of repeated responses, given the covariates, is assumed to be a function of an additional vector of association parameters, say $\boldsymbol{\alpha}$ (and also depends upon the means, μ_{ij}). For example, the components of $\boldsymbol{\alpha}$ might represent the pairwise correlations or log odds ratios among the repeated responses.

This three-part specification of a marginal model makes the extension of generalized linear models to longitudinal data more transparent. The first two parts of the marginal model correspond to the standard generalized linear model, albeit with no explicit distributional assumptions about the responses. It is the third component, the incorporation of the within-subject association among the repeated responses from the same individual, that represents the main extension of generalized linear models to longitudinal data.

It is also worth emphasizing that this three-part specification of a marginal model can, in principle, be extended by making full distributional assumptions about the vector of responses. To do so would require that all two- and higher-way associations be specified in the third component of the model. However, for reasons mentioned in Section 3.1, ML fitting of marginal models for discrete longitudinal data has proven to be very challenging. Consequently, the third component of a marginal model is typically specified in terms of the two-way or pairwise association among the repeated responses from the same individual.

In summary, marginal models are a very natural way to extend generalized linear models to longitudinal responses. Marginal models specify a generalized linear model for the longitudinal responses at each occasion but also include a model for the within-subject association among the responses. A crucial aspect of marginal models is that the mean response and within-subject association are modeled separately. This separation of the modeling of the mean response and the association among responses has important implications for interpretation of the regression parameters in the model for the mean response. In particular, the regression parameters, $\boldsymbol{\beta}$, in the marginal model have so-called *population-averaged* interpretations. That is, they describe how the mean response in the population is related to the covariates. For example, regression parameters in a marginal model might have interpretation in terms of contrasts of the changes in the mean responses in subpopulations (e.g., different treatment, intervention, or exposure groups); see Chapter 7 for a detailed discussion of various aspects of interpretation of marginal model parameters.

3.2.3 Quasi-likelihood and generalized estimating equations

As noted in Section 3.2.2, the three-part marginal model specification does not require full distributional assumptions for the repeated responses, only a regression model for the mean response. The avoidance of distributional assumptions can be advantageous because, in general, there is no convenient specification of the joint multivariate distribution of Y_i for marginal models when the responses are discrete. When full distributional assumptions are avoided, the model is referred to as being *semi-parametric* because there is a parametric component β and a non-parametric component determined by the nuisance parameters for the second- and higher-order moments. The avoidance of distributional assumptions for Y_i leads to a method of estimation known as *generalized estimating equations*. Thus, the GEE approach can be thought of as providing a convenient alternative to ML estimation. The GEE approach proposed by Liang and Zeger (1986) is a multivariate generalization of the quasi-likelihood approach for generalized linear models introduced by Wedderburn (1974). To better understand this connection to quasi-likelihood estimation, in the following we briefly outline the quasi-likelihood approach for generalized linear models for a *univariate* response before discussing its extension to multivariate responses.

In the following, we now assume N independent observations of a *scalar* response variable, Y_i. Associated with the response, Y_i, there are p covariates, $X_{i1}, \ldots, X_{ip}$. We assume that primary interest is in relating the mean of Y_i, $\mu_i = E(Y_i | X_{i1}, \ldots, X_{ip})$, to the covariates. In generalized linear models, the distribution of the response is assumed to belong to the exponential family of distributions (e.g., normal, Bernoulli, binomial, and Poisson). As described in the first part of the specification of the marginal model in Section 3.2.2, in generalized linear models a transformation of the mean response, μ_i, is linearly related to the covariates via an appropriate link function,

$$h^{-1}(\mu_i) = \beta_0 + \beta_1 X_{i1} + \beta_2 X_{i2} + \cdots + \beta_p X_{ip},$$

where the link function $h^{-1}(\cdot)$ is a known function, such as $\log(\mu_i)$. The assumption that Y_i has an exponential family distribution has implications for the variance of Y_i. In particular, a feature of exponential family distributions is that the variance of Y_i can be expressed in terms of a known function of the mean and a scale parameter,

$$\text{Var}(Y_i) = \phi\, v(\mu_i),$$

where the scale parameter $\phi > 0$. The variance function, $v(\mu_i)$, describes how the variance of the response is functionally related to the mean of Y_i; the variance function was previously discussed in the second part of the specification of the marginal model in Section 3.2.2.

Next, we consider estimation of β. Assuming Y_i follows an exponential family density, with $\text{Var}(Y_i) = \phi\, v(\mu_i)$, the ML estimator of β is obtained as the solution to the likelihood score equations,

$$\sum_{i=1}^{N} \left(\frac{\partial \mu_i}{\partial \beta} \right)' \frac{1}{\phi\, v(\mu_i)} \{Y_i - \mu_i(\beta)\} = \mathbf{0},$$

where $\partial \mu_i / \partial \beta$ is the $1 \times p$ vector of derivatives, $\partial \mu_i / \partial \beta_k, k = 1, \ldots, p$. Interestingly, the likelihood equations for generalized linear models depend only on the mean and variance of the response (and the link function). Consequently, Wedderburn (1974) suggested using them as "estimating equations" for any choice of link or variance function, even when the particular choice of variance function does not correspond to an exponential family distribution. That is, Wedderburn (1974) proposed estimating β by solving the quasi-likelihood equations,

$$\sum_{i=1}^{N} \left(\frac{\partial \mu_i}{\partial \beta} \right)' V_i^{-1} \{Y_i - \mu_i(\beta)\} = \mathbf{0}. \tag{3.1}$$

Wedderburn (1974) showed that for any choice of weights, V_i, the quasi-likelihood estimator of β, say $\widehat{\beta}$, is consistent and asymptotically normal. The choice of weights $V_i = \text{Var}(Y_i)$ yields the estimator with smallest variance among all estimators in this class. Note that in generalized linear models, it is assumed that $V_i = \text{Var}(Y_i) = \phi\, v(\mu_i)$, and this assumption is sufficient to characterize the distribution within the exponential family. Thus, the optimal estimating equations (or quasi-likelihood equations) coincide with the score equations for the case where Y_i is assumed to have an exponential family distribution.

In summary, Wedderburn (1974) proposed estimators of β that do not require distributional assumptions on the response. This allows more flexible models for variability, e.g., incorporating overdispersion. Moreover, quasi-likelihood estimation only requires correct specification of the model for the mean to yield consistent and asymptotically normal estimators of β. That is, a key property of quasi-likelihood estimators is that they are consistent even when the variance of the response has been misspecified, that is, $V_i \neq \text{Var}(Y_i)$. Specifically, it can be shown that the asymptotic distribution of $\widehat{\beta}$, the estimator for β obtained from (3.1) with a particular choice of V_i, satisfies

$$\sqrt{N}(\widehat{\beta} - \beta) \to N(\mathbf{0}, C_\beta),$$

where

$$C_\beta = \lim_{N\to\infty} I_0^{-1} I_1 I_0^{-1},$$

$$I_0 = \frac{1}{N} \sum_{i=1}^{N} \left(\frac{\partial \mu_i}{\partial \beta}\right)' V_i^{-1} \left(\frac{\partial \mu_i}{\partial \beta}\right),$$

and

$$I_1 = \frac{1}{N} \sum_{i=1}^{N} \left(\frac{\partial \mu_i}{\partial \beta}\right)' V_i^{-1} \text{Var}(Y_i)\, V_i^{-1} \left(\frac{\partial \mu_i}{\partial \beta}\right).$$

Consistent estimators of the asymptotic covariance of the estimated regression parameters can be obtained using the empirical estimator of C_β first suggested by Cox (1961), and later proposed by Huber (1967), White (1982), and Royall (1986). The empirical variance estimator is obtained by evaluating $\partial \mu_i/\partial \beta$ at $\widehat{\beta}$ and substituting $(Y_i - \widehat{\mu}_i)^2$ for $\text{Var}(Y_i)$; this is widely known as the *sandwich* variance estimator. Moreover, it can be shown that the same asymptotic distribution holds when V_i is estimated rather than known, with V_i replaced by estimated weights, say $\widehat{V}_i$.

Next, we consider the multivariate extension of this quasi-likelihood approach to the setting of marginal models for longitudinal responses. In the following, $\boldsymbol{Y}_i$ denotes an $n_i \times 1$ vector of responses (similarly, $\boldsymbol{\mu}_i$ denotes an $n_i \times 1$ vector of means) and X_i is an $n_i \times p$ matrix of covariates. The fundamental idea underlying the GEE approach is to extend the quasi-likelihood equations (or estimating equations) to the multivariate setting by replacing Y_i and μ_i by their corresponding multivariate counterparts ($\boldsymbol{Y}_i$ and $\boldsymbol{\mu}_i$) and using a *matrix* of weights V_i. Thus, the estimating equations for β in a marginal model are given by

$$u_\beta(\boldsymbol{\beta}) = \sum_{i=1}^{N} D_i'\, V_i^{-1} \{\boldsymbol{Y}_i - \boldsymbol{\mu}_i(\boldsymbol{\beta})\} = \sum_{i=1}^{N} \left(\frac{\partial \boldsymbol{\mu_i}}{\partial \beta}\right)' V_i^{-1} \{\boldsymbol{Y}_i - \boldsymbol{\mu}_i(\boldsymbol{\beta})\} = \mathbf{0}, \qquad (3.2)$$

where $D_i = \partial \boldsymbol{\mu}_i/\partial \beta$ is now an $n_i \times p$ matrix of derivatives whose kth row is $\partial \mu_i/\partial \beta_k$. As with the quasi-likelihood approach, the optimal choice of V_i is to take $V_i = \text{Cov}(\boldsymbol{Y}_i)$, the $n_i \times n_i$ covariance matrix for $\boldsymbol{Y}_i$. Note that the GEE given by (3.2) is simply the multivariate analog of the quasi-likelihood equations given by (3.1). However, unlike the quasi-likelihood equations given by (3.1), the weight matrix, V_i, depends not only on β but also on the pairwise associations among the longitudinal responses.

3.2.4 Estimation: Generalized estimating equations

As mentioned in the previous section, GEE can be regarded as a multivariate extension of the quasi-likelihood estimating equations. Note that the scalar weight in (3.1) is replaced by the weight matrix, V_i, in (3.2). In general, the assumed covariance among the responses can be specified as

$$V_i = \phi A_i^{1/2} R_i(\boldsymbol{\alpha}) A_i^{1/2},$$

where $A_i = \text{diag}\{v(\mu_{ij})\}$ is a diagonal matrix with diagonal elements $v(\mu_{ij})$, which are specified entirely by the marginal means (i.e., by $\boldsymbol{\beta}$), $R_i(\boldsymbol{\alpha})$ is an $n_i \times n_i$ correlation matrix, and ϕ is a dispersion parameter. For the correlation matrix $R_i(\boldsymbol{\alpha})$, $\boldsymbol{\alpha}$ represents a vector of parameters associated with a specified model for $\text{Corr}(\boldsymbol{Y}_i)$, with typical element

$$\rho_{ist} = \rho_{ist}(\boldsymbol{\alpha}) = \text{Corr}(Y_{is}, Y_{it}; \boldsymbol{\alpha}), \quad s \neq t.$$

In the GEE approach, V_i is usually referred to as a "working covariance," where the term "working" is used to emphasize that V_i is only an approximation to the true covariance, say $\Sigma_i = \text{Cov}(\boldsymbol{Y}_i)$. Sometimes, $R_i(\boldsymbol{\alpha})$ is referred to as a "working correlation" matrix; however, this implicitly assumes that the marginal variance assumption, $\text{Var}(Y_{ij}) = \phi \, v(\mu_{ij})$, is correct. Since both $R_i(\boldsymbol{\alpha})$ and $\text{Var}(Y_{ij})$ can be incorrectly specified, we prefer the use of the term "working covariance." Note that if $R_i(\boldsymbol{\alpha}) = I$, the $n_i \times n_i$ identity matrix, then the GEE reduces to the quasi-likelihood estimating equations for a generalized linear model that assume the repeated measures are independent.

Liang and Zeger (1986), and also Prentice (1988), parameterized the within-subject correlations directly as a linear function of $\boldsymbol{\alpha}$ (typically $\rho_{ist}(\boldsymbol{\alpha}) = \alpha_{st}$, although $\rho_{ist}(\boldsymbol{\alpha})$ could, in principle, be allowed to depend on covariates). When $\boldsymbol{\alpha}$ is known, the only unknown quantity in (3.2) is $\boldsymbol{\beta}$ and the solution to (3.2) is a consistent estimator of $\boldsymbol{\beta}$. In the usual case where V_i, and specifically $\boldsymbol{\alpha}$, is unknown, then we must parameterize and estimate $\rho_{ist}(\boldsymbol{\alpha}) = \text{Corr}(Y_{is}, Y_{it})$. Some common examples of models for the correlation include: "exchangeable," in which $\rho_{ist} = \alpha$ for all $s < t$; first-order autoregressive (AR(1)),

$$\rho_{ist} = \alpha^{|t-s|},$$

where $0 < \alpha < 1$, and the correlation decreases as the time between measurements ($|t - s|$) increases; and "unstructured," in which $\rho_{ist} = \alpha_{st}$. For estimation of $\boldsymbol{\alpha}$, let

$$U_{ist}(\boldsymbol{\beta}) = \frac{(Y_{is} - \mu_{is})(Y_{it} - \mu_{it})}{\phi\{v(\mu_{is})v(\mu_{it})\}^{1/2}},$$

which has expected value ρ_{ist}; the U_{ist} can be grouped together to form the $n_i(n_i - 1)/2 \times 1$ vector $\boldsymbol{U}_i(\boldsymbol{\beta}) = (U_{i12}, U_{i13}, \ldots, U_{in_{i-1}n_i})'$. Also, let $\boldsymbol{\rho}_i(\boldsymbol{\alpha}) = E(\boldsymbol{U}_i; \boldsymbol{\alpha}) = (\rho_{i12}, \rho_{i13}, \ldots, \rho_{in_{i-1}n_i})'$. Then, a second set of (moment) estimating equations similar to (3.2) can be used to estimate $\boldsymbol{\alpha}$, given by

$$u_\alpha(\boldsymbol{\alpha}) = \sum_{i=1}^{N} E_i' W_i^{-1}\{\boldsymbol{U}_i(\boldsymbol{\beta}) - \boldsymbol{\rho}_i(\boldsymbol{\alpha})\} = \boldsymbol{0}, \tag{3.3}$$

where $W_i \approx \text{Cov}(\boldsymbol{U}_i)$ and $E_i = \partial\boldsymbol{\rho}_i(\boldsymbol{\alpha})/\partial\boldsymbol{\alpha}$. The working covariance matrix for $\boldsymbol{U}_i$ is typically specified as $\text{diag}\{\text{Var}(U_{ist})\}$. In their original paper on GEE, Liang and Zeger (1986) let W_i be the $(n_i \times n_i - 1)/2 \times (n_i \times n_i - 1)/2$ identity matrix, whereas Prentice (1988) suggested letting W_i be a diagonal matrix with the approximate variances of U_{ist} along the diagonal.

For the special case where the outcome is binary, an alternative to the correlation as a measure of association between pairs of binary responses is the odds ratio. The odds ratio has many desirable properties (e.g., see Bishop, Fienberg, and Holland, 1975, Chapter 11) and has a more straightforward interpretation. To estimate the odds ratio as a measure of association, a second set of estimating equations similar to (3.3) (Lipsitz, Laird, and Harrington, 1991) can be used in combination with (3.2) for the marginal regression parameters. A more efficient second set of estimating equations to estimate the odds ratio parameters, referred to as "alternating logistic regression," was later proposed by Carey, Zeger, and Diggle (1993), and is discussed in greater detail in the next section. We also note that Qu et al. (1992) proposed using tetrachoric correlations as measures of association between pairs of binary responses; the tetrachoric correlation is the correlation between underlying latent normal variables that are assumed to have been dichotomized to form the observed binary outcomes. Additional work extending the GEE approach to longitudinal ordinal and polytomous data has been developed by, for example, Miller, Davis, and Landis (1993), Lipsitz, Kim, and Zhao (1994), Kenward, Lesaffre, and Molenberghs (1994), Gange et al. (1995), Lumley (1996), and Heagerty and Zeger (1996). The main challenge with extending the GEE approach to ordinal and polytomous responses has been that the working covariance for ordinal and polytomous responses (other than "working independence"), in general, requires the specification and estimation of a large number of nuisance parameters.

Given any parameterization of the working covariance, V_i, in (3.2), and using Taylor series expansions similar to Prentice (1988), assuming that the regression model for the mean of $\boldsymbol{Y}_i$ has been correctly specified, $\widehat{\boldsymbol{\beta}}$ is consistent for $\boldsymbol{\beta}$, and also $\sqrt{N}(\widehat{\boldsymbol{\beta}} - \boldsymbol{\beta})$ has an asymptotic distribution which is multivariate normal with mean vector $\boldsymbol{0}$ and covariance matrix given by

$$C_\beta = \lim_{N \to \infty} N \left(\sum_{i=1}^N D_i' V_i^{-1} D_i \right)^{-1} \left\{ \sum_{i=1}^N D_i' V_i^{-1} \mathrm{Cov}(\boldsymbol{Y}_i) V_i^{-1} D_i \right\} \left(\sum_{i=1}^N D_i' V_i^{-1} D_i \right)^{-1},$$

(3.4)

and if $\mathrm{Cov}(\boldsymbol{Y}_i)$ is correctly specified, so that $V_i = \mathrm{Cov}(\boldsymbol{Y}_i)$, then (3.4) reduces to

$$\lim_{N \to \infty} N \left(\sum_{i=1}^N D_i' V_i^{-1} D_i \right)^{-1}.$$

(3.5)

Note that a key property of the GEE estimators of $\boldsymbol{\beta}$ is that they are consistent and asymptotically normal, for any choice of working covariance V_i, provided the regression model for the mean response has been correctly specified. Heuristically, using results from the method of moments, $\widehat{\boldsymbol{\beta}}$ is consistent for $\boldsymbol{\beta}$ regardless of whether V_i is the true covariance matrix of $\boldsymbol{Y}_i$ because

$$E\{u_\beta(\boldsymbol{\beta})\} = E \left[\sum_{i=1}^N D_i' V_i^{-1} \{\boldsymbol{Y}_i - \boldsymbol{\mu}_i(\boldsymbol{\beta})\} \right] = \sum_{i=1}^N D_i' V_i^{-1} E\{\boldsymbol{Y}_i - \boldsymbol{\mu}_i(\boldsymbol{\beta})\} = \boldsymbol{0},$$

(3.6)

and we are solving $u_\beta(\widehat{\boldsymbol{\beta}}) = \boldsymbol{0}$ for $\widehat{\boldsymbol{\beta}}$. In particular, for any positive definite and symmetric matrix V_i, the solution to (3.2) is a consistent estimator of $\boldsymbol{\beta}$. This is the key property of the GEE method and implies that V_i does not have to be correctly specified in order to obtain consistent estimators of $\boldsymbol{\beta}$.

Finally, C_β in (3.4) can be consistently estimated by $\widehat{C}_\beta$, which is obtained by replacing $\boldsymbol{\beta}$ and $\boldsymbol{\alpha}$ by their estimates, and also $\mathrm{Cov}(\boldsymbol{Y}_i)$ by $(\boldsymbol{Y}_i - \widehat{\boldsymbol{\mu}}_i)(\boldsymbol{Y}_i - \widehat{\boldsymbol{\mu}}_i)'$. The estimator $\widehat{C}_\beta$ is a sandwich variance estimator and has the same form as the sandwich variance estimator in our earlier discussion of quasi-likelihood estimation. Some authors refer to this sandwich variance estimator as the "robust" variance estimator because it is consistent for the asymptotic variance of $\widehat{\boldsymbol{\beta}}$ provided $E\{u_\beta(\widehat{\boldsymbol{\beta}})\} = \boldsymbol{0}$; that is, the sandwich variance estimator can be said to be robust to misspecification of the covariance among the repeated measures.

Obtaining GEE estimates requires an iterative algorithm. The structure of the GEE suggests the use of a specific iterative scheme, namely to iterate between estimating $\boldsymbol{\beta}$ (given the current estimate of $\boldsymbol{\alpha}$) as the solution to (3.2), and estimating $\boldsymbol{\alpha}$ (given the current estimate of $\boldsymbol{\beta}$) as the solution to (3.3) until convergence. In particular, the solution $(\widehat{\boldsymbol{\beta}}, \widehat{\boldsymbol{\alpha}})$ to (3.2) and (3.3) can be obtained by a Fisher scoring algorithm. Given a starting value for $\boldsymbol{\beta}$, say under the naive assumption of independence, the solution $(\widehat{\boldsymbol{\beta}}, \widehat{\boldsymbol{\alpha}})$ can be obtained by iterating between

$$\widehat{\boldsymbol{\beta}}^{(m+1)} = \widehat{\boldsymbol{\beta}}^{(m)} + \left[\sum_{i=1}^{N} D_i^{(m)'}\{V_i^{(m)}\}^{-1}D_i^{(m)}\right]^{-1}\sum_{i=1}^{N}D_i^{(m)'}\{V_i^{(m)}\}^{-1}\{\boldsymbol{Y}_i - \boldsymbol{\mu}_i(\widehat{\boldsymbol{\beta}}^{(m)})\},$$

(3.7)

and

$$\widehat{\boldsymbol{\alpha}}^{(m+1)} = \widehat{\boldsymbol{\alpha}}^{(m)} + \left[\sum_{i=1}^{N} E_i^{(m)'}\{W_i^{(m)}\}^{-1}E_i^{(m)}\right]^{-1}$$

$$\times \sum_{i=1}^{N} E_i^{(m)'}\{W_i^{(m)}\}^{-1}[\boldsymbol{U}_i\{\widehat{\boldsymbol{\beta}}^{(m)}\} - \boldsymbol{\rho}_i\{\widehat{\boldsymbol{\alpha}}^{(m)}\}],$$

until $\widehat{\boldsymbol{\beta}}^{(m+1)} = \widehat{\boldsymbol{\beta}}^{(m)}$ and $\widehat{\boldsymbol{\alpha}}^{(m+1)} = \widehat{\boldsymbol{\alpha}}^{(m)}$, where $D_i^{(m)} = D_i\{\widehat{\boldsymbol{\beta}}^{(m)}\}$, $V_i^{(m)} = V_i\{\widehat{\boldsymbol{\beta}}^{(m)}, \widehat{\boldsymbol{\alpha}}^{(m)}\}$, $E_i^{(m)} = E_i\{\widehat{\boldsymbol{\beta}}^{(m)}, \widehat{\boldsymbol{\alpha}}^{(m)}\}$, and $W_i^{(m)} = W_i\{\widehat{\boldsymbol{\beta}}^{(m)}, \widehat{\boldsymbol{\alpha}}^{(m)}\}$.

We note that, given the current estimate of $\boldsymbol{\beta}$, the estimator of $\boldsymbol{\alpha}$ proposed by Liang and Zeger (1986) is non-iterative. For example, suppose an "exchangeable" correlation pattern is assumed, in which $\rho_{ist} = \alpha$ for all $s < t$. Then, Liang and Zeger (1986) proposed estimating α, given the current estimate of $\boldsymbol{\beta}$, say $\widehat{\boldsymbol{\beta}}$, by

$$\widehat{\alpha} = \frac{1}{N^*}\sum_{i=1}^{N}\sum_{s<t}^{n_i}\frac{(Y_{is} - \widehat{\mu}_{is})(Y_{it} - \widehat{\mu}_{it})}{\widehat{\phi}\{v(\widehat{\mu}_{is})v(\widehat{\mu}_{it})\}^{1/2}}, \quad \text{where } N^* = \sum_{i=1}^{N}\frac{n_i(n_i-1)}{2};$$

alternatively, various degree-of-freedom corrections to account for the estimation of $\boldsymbol{\beta}$ have been suggested for the denominator N^*, e.g., $N^* - p$.

3.2.5 Properties of GEE estimators

In this section we consider properties of the GEE estimators. Before summarizing the key properties of the GEE approach, we note that there is an implicit assumption in the first component of a marginal model that is often overlooked. Recall that in a marginal model the conditional expectation of each response is assumed to depend on the covariates through a known link function

$$h^{-1}(\mu_{ij}) = \eta_{ij} = \boldsymbol{X}_{ij}'\boldsymbol{\beta}.$$

In marginal models, the primary interest lies in estimation of the regression parameters, $\boldsymbol{\beta}$, in the model for $E(Y_{ij}|\boldsymbol{X}_{ij})$. The key property for (asymptotically) unbiased estimators using GEE is given in (3.6), and can be rewritten as

$$E\left[D_i'V_i^{-1}\{\boldsymbol{Y}_i - \boldsymbol{\mu}_i(\boldsymbol{\beta})\}\right] = E_{x_i}\left[D_i'V_i^{-1}E_{y_i|x_i}\{\boldsymbol{Y}_i - \boldsymbol{\mu}_i(\boldsymbol{\beta})\}\right] = \boldsymbol{0}, \tag{3.8}$$

where $E_{x_i}(\cdot)$ denotes expectation with respect to the marginal distribution of X_i and $E_{y_i|x_i}(\cdot)$ denotes expectation with respect to the conditional distribution of $\boldsymbol{Y}_i$ given X_i. Note that the jth element of the vector $\boldsymbol{\mu}_i(\boldsymbol{\beta})$ is $\mu_{ij} = E(Y_{ij}|\boldsymbol{X}_{ij})$, but the elements of $E_{y_i|x_i}(\boldsymbol{Y}_i)$ in (3.8) are $E(Y_{ij}|\boldsymbol{X}_{i1}, \ldots, \boldsymbol{X}_{in_i})$. Thus, for (3.8) to hold, the conditional mean of the jth response, given $\boldsymbol{X}_{i1}, \ldots, \boldsymbol{X}_{in_i}$, must depend only on $\boldsymbol{X}_{ij}$, that is,

$$E(Y_{ij}|X_i) = E(Y_{ij}|\boldsymbol{X}_{i1}, \ldots, \boldsymbol{X}_{in_i}) = E(Y_{ij}|\boldsymbol{X}_{ij}); \tag{3.9}$$

see Fitzmaurice, Laird, and Rotnitzky (1993) and Pepe and Anderson (1994) for a more detailed discussion of this sufficient condition. With time-stationary covariates, this assumption poses no difficulties; it necessarily holds because $X_{ij} = X_{ik}$ for all occasions $k \neq j$. Also, with time-varying covariates that are fixed by design of the study (e.g., time since baseline, treatment group indicator in a crossover trial), the assumption also holds because values of the covariates at any occasion are determined a $priori$ by study design and in a manner completely unrelated to the longitudinal response. However, when a time-varying covariate varies randomly over time the assumption made in (3.9) may not hold. For example, the assumption will be violated when the current value of the response, say Y_{ij}, given the current covariates X_{ij}, predicts the subsequent value of $X_{i,j+1}$; see Chapter 23 for a more detailed discussion of stochastic time-varying covariates in longitudinal studies. When (3.9) is not satisfied, yet there is subject-matter interest in the dependence of Y_{ij} on X_{ij}, Pepe and Anderson (1994) recommend using GEE with a "working independence" assumption. Under a "working independence" assumption, the weight matrix is diagonal and the corresponding estimating equations simplify and are unbiased regardless of whether or not (3.9) is satisfied. In contrast, under alternative choices for the working covariance matrix, the estimating equations are not necessarily unbiased and may yield inconsistent estimators of the regression parameters.

Thus far, we have seen that the GEE approach provides a convenient alternative to ML estimation of the regression parameters in marginal models for longitudinal data, while also retaining a number of appealing properties. First, in many longitudinal designs the GEE estimator $\widehat{\beta}$ is almost as efficient as the ML estimator. For example, consider the case of linear models for continuous responses that are assumed to have a multivariate normal distribution. It can easily be shown that the generalized least-squares estimator of β in linear models can be considered a special case of the GEE approach. The GEE also has an expression similar to the likelihood equations for β in certain "mixed parameter" models for discrete longitudinal data (e.g., Fitzmaurice and Laird, 1993; Fitzmaurice, Laird, and Rotnitzky, 1993). As a result, for many longitudinal designs, there is little loss of efficiency when the GEE approach is adopted as an alternative to maximum likelihood. Moreover, because all GEE estimators of β are consistent and asymptotically normal, it is of interest to consider their efficiency under various working covariance assumptions. Indeed, it has been suggested in the literature that setting $R_i = I$, the identity matrix, leads to an estimator with nearly the same efficiency as the optimally weighted GEE (with $V_i = \Sigma_i$). Interestingly, for a balanced longitudinal design, with no missing data, there are cases where this claim has some justification. Specifically, there is relatively little loss of efficiency either for between-subject effects (where the covariate design $X_{ijk} = X_{ij'k}$ for all occasions $j \neq j'$) or for within-subject effects when the covariate design on time for the latter effects is the same for all subjects (i.e., $X_{ijk} \neq X_{ij'k}$ for some occasions $j \neq j'$, but $X_{ijk} = X_{i'jk}$ for all subjects $i \neq i'$). However, for the case of a within-subject effect when the covariate design on time is not the same for all subjects (i.e., $X_{ijk} \neq X_{ij'k}$ for some occasions $j \neq j'$ and $X_{ijk} \neq X_{i'jk}$ for some subjects $i \neq i'$), there can be a very discernible loss of efficiency under the non-optimal "working independence" assumption for the covariance, especially when the true correlations are moderately large (see, for example, Lipsitz et al., 1994). Heuristically, this result can be explained as follows. When using GEE methods, the estimated βs in a marginal model can be thought of as weighted averages of between- and within-subject contrasts. In the case of a between-subject effect (e.g., exposure or treatment group) or a within-subject effect when the covariate design on time is the same for all subjects (e.g., time since baseline), the respective between- or within-subject contrasts are weighted approximately equally, regardless of the assumed (or working) covariance. On the other hand, for a within-subject effect when the covariate design on time is not the same for all subjects, the GEE estimate is a weighted average of $both$ between-subject *and* within-subject contrasts that are weighted differently. Moreover, the weights for these different contrasts depend on the assumed covariance. So, for the latter case, the correlation is more important and determines the optimal weights for combining the contrasts. In addition, in the setting of unbalanced data, with $n_i \neq n$, each individual no longer contributes equally weighted components to even the between-subject effects, and thus we can expect a loss of efficiency for the GEE under "working independence" in that setting as well. So, in general, it can be advantageous to choose a working covariance that closely approximates the true covariance. The closer the working covariance matrix (V_i) approximates the true underlying covariance matrix (Σ_i), the greater the efficiency for estimation of β.

A second appealing property of the GEE estimator $\widehat{\beta}$ is its robustness, yielding a consistent estimator of β even if the within-subject associations among the repeated measures have been misspecified. It only requires that the model for the mean response be correct. This robustness property of GEE is important because the usual focus of a longitudinal study is on changes in the mean response. In general, there is usually far less interest in, and correspondingly less subject-matter knowledge of, the patterns of covariance among the repeated measures. Although the GEE approach yields a consistent estimator of β under misspecification of the within-subject associations, the usual standard errors obtained under the misspecified model for the within-subject association are not valid. Fortunately, in many cases, valid standard errors for $\widehat{\beta}$ can be obtained using the so-called sandwich variance estimator. A remarkable property of the sandwich estimator is that it is also robust in the sense that it provides valid standard errors when the assumed model for the covariances among the repeated measures is not correct. That is, with large sample sizes, the sandwich variance estimator yields correct standard errors. However, it is worth emphasizing that this robustness property of the sandwich variance estimator is an asymptotic property. In general, use of the sandwich variance estimator is best suited to balanced longitudinal designs where the number of subjects (N) is relatively large and the number of repeated measures (n) is relatively small. Moreover, the sandwich estimator is less appealing when the design is severely unbalanced and/or when there are few replications to estimate the true underlying covariance matrix. The use of the sandwich estimator implicitly relies on there being many replications of the vector of responses associated with each distinct set of covariate values.

Note that bias-corrected versions of the sandwich variance estimator have been proposed that have somewhat better finite-sample properties, including the jackknife (Paik, 1988; Mancl and DeRouen, 2001). Most of these bias-corrected versions involve a "degrees-of-freedom" correction. In cases where there are few, if any, replications of Y_i associated with each distinct set of covariate values, the use of the sandwich estimator can be problematic. In particular, standard errors based on the sandwich estimator tend to be biased downward. In addition, the sampling variability of the sandwich estimator of $\text{Cov}(\widehat{\beta})$ can be very large, resulting in an unstable estimator of variability. In these settings, it may be preferable to carefully model the covariances among the responses and use the "model-based" estimator of $\text{Cov}(\widehat{\beta})$ given by (3.5) and evaluated at $(\widehat{\beta}, \widehat{\alpha})$. This estimator of $\text{Cov}(\widehat{\beta})$ is referred to as a "model-based" estimator to remind us that it yields valid standard errors provided that the working covariance matrix, V_i, is a close approximation to the true underlying covariance matrix, Σ_i. In general, unlike the sandwich variance estimator, the "model-based" estimator does not require such a large number of replications; the sampling variability of the "model-based" estimator tends to be smaller than that of the sandwich variance estimator. The key to obtaining approximately unbiased variance estimators using the "model-based" estimator is to choose a model for the working covariance matrix that is close to the true covariance matrix.

Finally, for making inferences about $\boldsymbol{\beta}$, since the GEE approach is not likelihood-based, likelihood ratio tests are not available for hypotheses testing; in addition, this also has ramifications for inferences when there are missing data, a topic that will be discussed in Section 3.4. Instead, inferences typically rely on Wald test statistics based on quadratic forms. For example, if it is of interest to test whether or not the ℓth element of $\boldsymbol{\beta}$ is 0, $H_0 : \beta_\ell = 0$, then the Wald test statistic,

$$Z_\ell = \frac{\widehat{\beta}_\ell}{\sqrt{\widehat{\mathrm{Var}}(\widehat{\beta}_\ell)}} \sim N(0,1)$$

can be constructed, where $\widehat{\mathrm{Var}}(\widehat{\beta}_\ell)$ is the ℓth diagonal element of either the model-based or, more typically, the sandwich variance estimate. More generally, for a null hypothesis of the form $H_0 : L\boldsymbol{\beta} = 0$, versus the alternative $H_A : L\boldsymbol{\beta} \neq 0$, for an $r \times p$ matrix L of full rank, $r \leq p$, the Wald test statistic is

$$X^2 = (L\widehat{\boldsymbol{\beta}})' \{ L\widehat{\mathrm{Cov}}(\widehat{\boldsymbol{\beta}})L' \}^{-1} L\widehat{\boldsymbol{\beta}} \sim \chi_r^2,$$

where χ_r^2 denotes a chi-square distribution with r degrees of freedom.

Although this use of the Wald statistic allows for the comparison of nested models, it can be potentially problematic because Wald statistics are known to have less than optimal properties under the alternative in logistic regression models for univariate outcome data (Hauck and Donner, 1977). We conjecture that the same may be true for Wald statistics based on GEE estimates of marginal regression parameters for longitudinal binary data. As an alternative, and to circumvent this potential problem with the Wald statistic, one can base inferences on the score test statistic proposed by Rotnitzky and Jewell (1990) and Boos (1992). Recall the property of the GEE score vector $u_\beta(\boldsymbol{\beta})$ given in (3.6), namely $E\{u_\beta(\boldsymbol{\beta})\} = 0$. Furthermore, it can be shown that

$$V_u(\boldsymbol{\beta}) = \mathrm{Cov}\{u_\beta(\boldsymbol{\beta})\} = \{\mathrm{Cov}(\widehat{\boldsymbol{\beta}})\}^{-1}.$$

Finally, since the score vector is a sum of independent random variables, using the central limit theorem, it can be shown that

$$u_\beta(\boldsymbol{\beta}) \sim N\{0, V_u(\boldsymbol{\beta})\}.$$

Then, the "score" test statistic can be expressed as a quadratic form in the score vector,

$$u_\beta(\tilde{\boldsymbol{\beta}})' \{V_u(\tilde{\boldsymbol{\beta}})\}^{-1} u_\beta(\tilde{\boldsymbol{\beta}}) \sim \chi_r^2, \tag{3.10}$$

where $u_\beta(\tilde{\boldsymbol{\beta}})$ and $V_u(\tilde{\boldsymbol{\beta}})$ are the score vector and its variance under the alternative hypothesis, evaluated at $\tilde{\boldsymbol{\beta}}$, which is the estimate of $\boldsymbol{\beta}$ under the null hypothesis $H_0 : L\boldsymbol{\beta} = 0$.

3.3 Some extensions of GEE methods for longitudinal data

In this section we briefly review some extensions of the standard GEE methods proposed by Liang and Zeger (1986). In particular, we describe alternative estimators of the within-subject association, especially when the longitudinal responses are discrete. We also discuss joint estimation of the marginal mean and within-subject association parameters.

3.3.1 Alternative estimators of within-subject association parameters

As discussed in the previous section, when V_i is correctly specified in the standard GEE approach, the resulting estimator of $\boldsymbol{\beta}$ is semi-parametric efficient in the sense of having the smallest variance among all estimators in the class given by (3.2). However, the estimators of the correlation parameters $\boldsymbol{\alpha}$ proposed by Liang and Zeger (1986) are not efficient, in large part due to the use of a suboptimal weight matrix, W_i, in (3.3). This has led researchers to develop alternative estimating equations for $\boldsymbol{\alpha}$. Using the same set of estimating equations for $\boldsymbol{\beta}$ given in (3.2), we now consider two alternative estimating equations for $\boldsymbol{\alpha}$ (and thus for $V_i(\boldsymbol{\beta}, \boldsymbol{\alpha})$). We note that any set of estimating equations that use (3.2) to estimate $\boldsymbol{\beta}$ are often referred to as first-order GEE or "GEE1". Here, we consider two alternative GEE1s developed with the goal of providing more stable and/or efficient estimates of $\boldsymbol{\alpha}$ than the standard GEE of Liang and Zeger (1986).

One set of estimating equations for $\boldsymbol{\alpha}$ that has generated some interest in the statistical literature is that based on the notion of "Gaussian" estimation (see, for example, Lipsitz, Laird, and Harrington, 1992; Lee, Laird, and Johnston, 1999; Hall and Severini, 1998; Lipsitz et al., 2000; Fitzmaurice, Lipsitz, and Molenberghs, 2001; Wang and Carey, 2003). The key idea here is to base estimation of $\boldsymbol{\alpha}$ on the multivariate normal estimating equations for the correlations. In particular, let

$$\boldsymbol{\xi}_i = \boldsymbol{\xi}_i(\boldsymbol{\beta}) = \frac{1}{\sqrt{\phi}} A_i^{-1/2} (\boldsymbol{Y}_i - \boldsymbol{\mu}_i),$$

be the vector of standardized residuals, where A_i is a diagonal matrix with diagonal elements $v(\mu_{ij})$. Note that $\mathrm{Cov}(\boldsymbol{\xi}_i)$ is the correlation matrix of $\boldsymbol{Y}_i$, which we denoted earlier as $R_i = R_i(\boldsymbol{\alpha})$. Then, a second set of (moment) estimating equations can be obtained as the score equations for $\boldsymbol{\alpha}$ under the assumption that $\boldsymbol{\xi}_i \sim N(\boldsymbol{0}, R_i(\boldsymbol{\alpha}))$. Specifically, the estimating equations for $\boldsymbol{\alpha}$ are given by

$$u_\alpha(\boldsymbol{\alpha}) = \left[\frac{\partial}{\partial \boldsymbol{\alpha}} \left\{ \sum_{i=1}^N \log |R_i(\boldsymbol{\alpha})| - \sum_{i=1}^N \boldsymbol{\xi}_i'(\boldsymbol{\beta}) R_i^{-1}(\boldsymbol{\alpha}) \boldsymbol{\xi}_i(\boldsymbol{\beta}) \right\} \right] = \boldsymbol{0}. \tag{3.11}$$

The rth component of $\boldsymbol{u}_\alpha(\boldsymbol{\alpha})$ in (3.11) equals

$$\sum_{i=1}^N \mathrm{tr} \left[R_i^{-1}(\boldsymbol{\alpha}) \{ \boldsymbol{\xi}_i(\boldsymbol{\beta}) \boldsymbol{\xi}_i'(\boldsymbol{\beta}) - R_i(\boldsymbol{\alpha}) \} R_i^{-1}(\boldsymbol{\alpha}) \dot{\mathrm{R}}_{ir}(\boldsymbol{\alpha}) \right],$$

where $\dot{\mathrm{R}}_{ir}(\boldsymbol{\alpha}) = \partial R_i(\boldsymbol{\alpha}) / \partial \alpha_r$ and $\mathrm{tr}\{\cdot\}$ denotes the trace of a matrix. An appealing feature of the use of the multivariate normal estimating equations is that it ensures that the estimated correlation matrix, $R_i(\widehat{\boldsymbol{\alpha}})$, is non-negative definite when there are unbalanced data (i.e., $n_i \neq n$). In contrast, this is not the case for alternative GEE1 methods. Thus, in general, this leads to more stable estimation of $\boldsymbol{\alpha}$. As a slight modification to these multivariate normal estimating equations, Wang and Carey (2004) proposed estimating the correlation parameters by differentiating the Cholesky decomposition of the working correlation matrix; a similar approach was also used by Ye and Pan (2006). Furthermore, similar Gaussian or "quadratic" estimating equations have been proposed by Crowder (1995) and Qu, Lindsay, and Li (2000).

Motivated by the application of GEE methods to longitudinal binary outcomes, a second alternative set of estimating equations for $\boldsymbol{\alpha}$ were proposed by Carey, Zeger, and Diggle (1993) and Lipsitz and Fitzmaurice (1996). Specifically, they proposed a set of estimating equations based on the *conditional* residuals $\{Y_{it} - E(Y_{it}|Y_{is} = y_{is}, X_i)\}$, that is, deviations about *conditional* expectations. In contrast, note that the second set of estimating equations given by (3.3) are based on the *unconditional* residuals

$$\frac{(Y_{is} - \mu_{is})(Y_{it} - \mu_{it})}{\phi \{ v(\mu_{is}) v(\mu_{it}) \}^{1/2}} - \rho_{ist}.$$

The *conditional* expectations can be grouped together to form the $n_i(n_i - 1)/2 \times 1$ vector of conditional residuals, $(\boldsymbol{\mathcal{U}}_i - \boldsymbol{\eta}_i)$, where

$$\boldsymbol{\mathcal{U}}_i = \{\mathcal{U}_{i12}, \mathcal{U}_{i13}, \ldots, \mathcal{U}_{i,n-1,n}\}', \qquad \boldsymbol{\eta}_i = \{\eta_{i12}, \eta_{i13}, \ldots, \eta_{i,n-1,n}\}',$$

with $\mathcal{U}_{ist} = Y_{it}$ and $\eta_{ist} = E(Y_{it}|Y_{is} = y_{is}, X_i)$, for $s < t$. Note that for the special case where the Y_{it} are binary, the conditional mean equals

$$\eta_{ist}(\boldsymbol{\beta}, \boldsymbol{\alpha}) = E(Y_{it}|Y_{is} = y_{is}, \boldsymbol{\beta}, \boldsymbol{\alpha})$$

$$= y_{is}E(Y_{it}|Y_{is} = 1, \boldsymbol{\beta}, \boldsymbol{\alpha}) + (1 - y_{is})E(Y_{it}|Y_{is} = 0, \boldsymbol{\beta}, \boldsymbol{\alpha})$$

$$= y_{is}\Pr(Y_{it} = 1|Y_{is} = 1, \boldsymbol{\beta}, \boldsymbol{\alpha}) + (1 - y_{is})\Pr(Y_{it} = 1|Y_{is} = 0, \boldsymbol{\beta}, \boldsymbol{\alpha})$$

$$= y_{is}\left\{\Pr(Y_{is} = 1, Y_{it} = 1; \boldsymbol{\beta}, \boldsymbol{\alpha})/\Pr(Y_{is} = 1; \boldsymbol{\beta})\right\}$$

$$+(1 - y_{is})\left\{\Pr(Y_{is} = 0, Y_{it} = 1; \boldsymbol{\beta}, \boldsymbol{\alpha})/\Pr(Y_{is} = 0; \boldsymbol{\beta})\right\}$$

$$= y_{is}\left(\pi_{ist}/\mu_{is}\right) + (1 - y_{is})\left\{(\mu_{it} - \pi_{ist})/(1 - \mu_{is})\right\},$$

(3.12)

where $\pi_{ist} = \Pr(Y_{is} = 1, Y_{it} = 1; \boldsymbol{\beta}, \boldsymbol{\alpha})$. Then, a set of estimating equations for $\boldsymbol{\alpha}$ is given by

$$\boldsymbol{u}_\alpha(\boldsymbol{\alpha}) = \sum_{i=1}^{N} E_i' W_i^{-1}\{\mathcal{U}_i - \boldsymbol{\eta}_i(\boldsymbol{\beta}, \boldsymbol{\alpha})\} = \mathbf{0},$$

(3.13)

where $E_i = \partial\boldsymbol{\eta}_i/\partial\boldsymbol{\alpha}$ and $W_i = \mathrm{diag}\{\mathrm{Var}(Y_{it}|Y_{is} = y_{is})\}$. Note that for binary Y_{it}, $\mathrm{Var}(Y_{it}|Y_{is} = y_{is}) = \eta_{ist}(1 - \eta_{ist})$ since $E(Y_{it}^2|Y_{is} = y_{is}) = E(Y_{it}|Y_{is} = y_{is}) = \eta_{ist}$.

Although the *conditional* residuals $(\mathcal{U}_{ist} - \eta_{ist})$ are neither independent nor uncorrelated, Carey, Zeger, and Diggle (1993) argue that the correlations among the *conditional* residuals should be substantially smaller than the correlations among the *unconditional* residuals. Consequently, setting $W_i = \mathrm{diag}\{\eta_{ist}(1 - \eta_{ist})\}$ in (3.13) should provide a closer approximation to the optimal weight matrix. Kuk (2004) proposed a symmetrized version of these estimating equations that may be particularly useful for an exchangeable correlation structure.

We note that after some algebra, (3.12) can be shown to equal

$$E(Y_{it}|Y_{is} = y_{is}, \boldsymbol{\beta}, \boldsymbol{\alpha}) = \mu_{it} + (y_{is} - \mu_{is})\rho_{ist}\sqrt{\frac{\mathrm{Var}(Y_{it})}{\mathrm{Var}(Y_{is})}},$$

(3.14)

which is exactly the conditional expectation if (Y_{is}, Y_{it}) are assumed to be bivariate normal. Thus, even though these estimating equations were originally proposed for analysis of longitudinal binary data, the form of the conditional expectation in (3.14) suggests that they can be used also for non-binary longitudinal outcomes. The only additional issue that arises for non-binary outcomes is the specification of $\mathrm{Var}(Y_{it}|Y_{is} = y_{is})$. For binary outcome data, it is straightforward to specify this variance as $\mathrm{Var}(Y_{it}|Y_{is} = y_{is}) = \eta_{ist}(1 - \eta_{ist})$. For non-binary outcomes, one alternative is to use the conditional variance from the bivariate normal, with

$$\mathrm{Var}(Y_{it}|Y_{is} = y_{is}, \boldsymbol{\beta}, \boldsymbol{\alpha}) = \mathrm{Var}(Y_{it})\left(1 - \rho_{ist}^2\right).$$

As with the standard GEE1 discussed in Section 3.2.3, the two alternative approaches discussed in this section require an iterative algorithm. That is, both of these GEE1 methods require iterating between estimating $\boldsymbol{\beta}$ (given the current estimate of $\boldsymbol{\alpha}$) as the solution to (3.2), and estimating $\boldsymbol{\alpha}$ (given the current estimate of $\boldsymbol{\beta}$) as the solution to (3.11) or (3.13), until convergence.

3.3.2 *Second-order generalized estimating equations (GEE2)*

In the previous section, a number of alternative proposals for estimating $\boldsymbol{\alpha}$, and thus $V_i(\boldsymbol{\beta}, \boldsymbol{\alpha})$, were described. All of these approaches broadly fall within the framework of GEE1.

That is, all of the approaches discussed so far use estimating equations for $\boldsymbol{\beta}$ given by (3.2), and differ only in terms of how $\boldsymbol{\alpha}$, and hence $V_i(\boldsymbol{\beta}, \boldsymbol{\alpha})$, is estimated. In this section we discuss a second-order extension of generalized estimating equations, hereafter referred to as GEE2 (Zhao and Prentice, 1990; Liang, Zeger, and Qaqish, 1992). To do so, we need to introduce some additional notation. Let

$$\boldsymbol{\mathcal{Y}}_i = (Y_{i1}, \ldots, Y_{in_i}, Y_{i1}Y_{i2}, \ldots, Y_{i,n_i-1}Y_{in_i})' = (\boldsymbol{Y}_i', \boldsymbol{Z}_i')',$$

$\boldsymbol{\Pi}_i = E(\boldsymbol{\mathcal{Y}}_i | X_i, \boldsymbol{\beta}, \boldsymbol{\alpha})$ and, in an ever so slight departure from previous notation, $\mathcal{V}_i \approx \mathrm{Cov}(\boldsymbol{\mathcal{Y}}_i | X_i)$. Then, the GEE2 estimating equations for $\boldsymbol{\theta} = (\boldsymbol{\beta}', \boldsymbol{\alpha}')'$ are given by

$$u_2(\boldsymbol{\theta}) = \sum_{i=1}^{N} \mathcal{D}_i' \mathcal{V}_i^{-1} \{\boldsymbol{\mathcal{Y}}_i - \boldsymbol{\Pi}_i(\boldsymbol{\theta})\} = \mathbf{0}, \tag{3.15}$$

where $\mathcal{D}_i = \partial \boldsymbol{\Pi}_i(\boldsymbol{\theta}) / \partial \boldsymbol{\theta}$. Note that, unlike GEE1, $\mathcal{D}_i$ (and $\mathcal{V}_i$) in GEE2 is not block-diagonal and thus the estimating equations for $\boldsymbol{\beta}$ depend on $\boldsymbol{\mathcal{Y}}_i = (\boldsymbol{Y}_i', \boldsymbol{Z}_i')'$, and not simply on $\boldsymbol{Y}_i$. As a consequence, the GEE2 estimating equations for $\boldsymbol{\beta}$ given by (3.15) are quite different from the GEE1 estimating equations for $\boldsymbol{\beta}$ given by (3.2). In particular, for the GEE2 estimator of $\boldsymbol{\beta}$ to be consistent, the model for all elements of $\boldsymbol{\mathcal{Y}}_i$ must be correctly specified; that is, the first two moments of $\boldsymbol{Y}_i$ must be correctly specified. GEE2 can yield substantial bias in estimators of both $\boldsymbol{\beta}$ and $\boldsymbol{\alpha}$ (Fitzmaurice, Lipsitz, and Molenberghs, 2001) if assumptions about second moments are misspecified. This is in contrast to GEE1 where a consistent estimator of $\boldsymbol{\beta}$ is obtained provided only that the mean of $\boldsymbol{Y}_i$ has been correctly specified.

The main appeal of GEE2 is that it is almost fully efficient for both the marginal regression parameters, $\boldsymbol{\beta}$, and the within-subject associations, $\boldsymbol{\alpha}$. Note that the working covariance matrix $\mathcal{V}_i$ is usually obtained by making additional assumptions about the third and fourth moments of $\boldsymbol{Y}_i$. Some alternative specifications for the third and fourth moments are discussed in Zhao and Prentice (1990) and Liang, Zeger, and Qaqish (1992). Thus, a notable feature of GEE2 is that it makes additional assumptions about higher-order moments. Furthermore, specification of these higher-order moments can greatly increase the computational complexity and make implementation of the method somewhat more difficult; it is notable that none of the major statistical software packages currently have options for GEE2.

Finally, we note that some authors refer to the solution to (3.2) and (3.13) as GEE2; in contrast, we prefer to reserve the term GEE2 to refer to estimators that require the first two moments of $\boldsymbol{Y}_i$ to be correctly specified in order to yield consistent estimators of $\boldsymbol{\beta}$ and $\boldsymbol{\alpha}$. That is, in contrast to GEE1 estimators, GEE2 estimators of $\boldsymbol{\beta}$ are not robust to misspecification of the second moments.

3.4 GEE with missing data

Although most longitudinal studies are designed to collect complete data on all participants, missing data very commonly arise and must be properly accounted for in the analysis; otherwise, biased estimators of longitudinal change can result. When longitudinal data are missing, the data set is necessarily unbalanced over time because not all individuals have the same number of repeated measurements at a common set of occasions. This feature of missingness creates no difficulties for the standard GEE method as it can handle the unbalanced data without having to discard data on individuals with any missing data. That is, when some individuals' response vectors are only partially observed, the standard GEE approach circumvents the problem of missing data by simply basing inferences on the *observed* responses. However, the validity of this method of analysis will require that certain assumptions about the reasons for any missingness, often referred to as the *missing-data mechanism*, are tenable.

The missing-data mechanism can be thought of as a model that describes the probability that a response is observed or missing at any occasion. Here, we briefly review and distinguish two general types of missing-data mechanisms; see Chapter 17 for a more detailed description. The two missing-data mechanisms are referred to as *missing completely at random* (MCAR) and *missing at random* (MAR) (Rubin, 1976; Laird, 1988). These two mechanisms differ in terms of assumptions concerning whether or not missingness is related to responses that have been observed. The distinction between these two mechanisms determines the appropriateness of standard GEE methods. Specifically, the standard GEE method yields consistent marginal regression parameter estimators provided the responses are MCAR (see, for example, Laird, 1988).

Data are said to be MCAR when the probability that responses are missing is unrelated to either the specific values that, in principle, should have been obtained (the *missing responses*) or the set of observed responses. Thus, longitudinal data are MCAR when missingness in Y_i is simply the result of a chance mechanism that does not depend on either observed or unobserved components of Y_i. To better understand this missing-data process, consider the simple case where there are only two measurement occasions with the outcome fully observed on all subjects at the first occasion, and missing on some subjects at the second occasion. Because the outcome can be missing at the second occasion only, we can define the single indicator random variable R_{i2}, which equals 1 if Y_{i2} is observed and 0 if Y_{i2} is unobserved. The missing data are said to be MCAR if

$$\Pr(R_{i2} = 1 | Y_{i1}, Y_{i2}, X_i) = \Pr(R_{i2} = 1 | X_i);$$

that is, the probability that Y_{i2} is missing does not depend on the observed value of Y_{i1} or the possibly missing value of Y_{i2}. We note that the use of the term MCAR is sometimes restricted to the case where

$$\Pr(R_{i2} = 1 | Y_{i1}, Y_{i2}, X_i) = \Pr(R_{i2} = 1);$$

this distinction only becomes important when an analysis is based on a subset of the covariates in X_i that excludes a covariate that is predictive of R_{i2}. The essential feature of MCAR is that the observed data can by thought of as a random sample of the complete data. As a result, all of the moments of the observed data do not differ from the corresponding moments of the complete data. This property provides the validity for the standard GEE method that bases inferences on the observed responses.

In contrast to MCAR, data are said to be MAR when the probability that responses are missing depends on the set of observed responses, but is unrelated to the specific missing values that, in principle, should have been obtained. For the simple bivariate response example considered previously, the missing data are said to be MAR if

$$\Pr(R_{i2} = 1 | Y_{i1}, Y_{i2}, X_i) = \Pr(R_{i2} = 1 | Y_{i1}, X_i); \tag{3.16}$$

that is, the probability that Y_{i2} is missing depends on Y_{i1} (and X_i), but is conditionally independent of the possibly missing value of Y_{i2}. Because the missing-data mechanism now depends upon observed responses, the sample moments based on the available data are biased estimates of the corresponding moments in the target population. Consequently, when data are MAR, the standard GEE method that bases inferences on the observed responses can yield biased regression parameter estimates. Finally, if $\Pr(R_{i2} = 1 | Y_{i1}, Y_{i2}, X_i)$ depends in any way on the possibly missing outcome Y_{i2}, the missing data are said to be *not missing at random* (NMAR) or oftentimes referred to as being "non-ignorably" missing. The case of NMAR is beyond the scope of this chapter; see Chapters 20 and 22 for a more in-depth discussion of this topic.

When missing data are assumed to be MAR, there are two general approaches for handling this problem within the GEE framework: multiple imputation (e.g., Paik, 1997) and

weighted estimating equations (Robins, Rotnitzky, and Zhao, 1995). The idea behind multiple imputation is very simple: substitute or fill in the values that were not recorded with
imputed values. However, to reflect the uncertainty inherent in the imputation of the unobserved responses, the imputation process is repeated multiple times. The imputations
are typically based on some assumed model for the missing data given the observed data.
The attractive feature of imputation methods is that, once a filled-in data set has been
constructed, the standard GEE method for complete data can be applied. The multiple
filled-in data sets produce different sets of parameter estimates and their standard errors
that are then appropriately combined to provide a single estimate of the parameters of interest, together with standard errors that reflect the uncertainty inherent in the imputation of
the unobserved responses. There is an extensive literature on multiple imputation, and the
reader can find an excellent review of this topic in Chapter 21. We do not discuss multiple
imputation for GEE any further because, to date, there is no overwhelming concensus on
the best way to impute discrete longitudinal data; see Chapter 21 for more details.

The second general approach for handling data that are MAR is via weighted estimating
equations. In one of the simplest versions of the weighted estimating equations approach,
an individual's contribution to the standard GEE is weighted inversely by the probability
of being observed at the given times. The key idea behind weighting methods is that the underrepresentation of certain response profiles in the observed data is taken into account and
corrected. The weighted estimating equations approach is best suited to the case of monotone missing-data patterns as might arise when missingness is due to attrition or dropout.
A variety of different weighting methods that adjust for dropout have been proposed. These
approaches are often called propensity weighted or inverse probability weighted methods;
see Chapter 20 for an excellent discussion of these methods. In all of these methods the
underlying idea is to base estimation on the observed responses but weight them to account
for the probability of remaining in the study.

Inverse probability weighted methods were first proposed in the sample survey literature,
where the weights are known and based on the survey design (e.g., Horvitz and Thompson,
1952). In the weighted estimating equations approach, however, the weights are not known
but must be estimated based on an assumed model for dropout. The propensities for dropout
can be estimated as a function of the observed responses prior to dropout, and also as a
function of the covariates and any extraneous variables that are thought likely to predict
dropout.

Consider the simple example introduced earlier where there are only two measurement
occasions and any missingness is restricted to the second occasion. In this simple case, the
missing-data pattern is monotone. Furthermore, let us assume the data are MAR as in
(3.16) and let

$$\pi_i = \pi_i(\gamma) = \Pr(R_{i2} = 1 | Y_{i1}, X_i, \gamma),$$

a function of possibly unknown parameters γ. The basic idea underlying the weighted GEE
method is to weight each individual's contribution to the standard GEE by the inverse
probability of being observed, thereby accounting for those subjects with the same history
of responses and covariates (y_{i1}, X_i), but who were missing Y_{i2}. Specifically, the GEE in
(3.2) is weighted by the inverse probabilities to yield a simple "weighted GEE,"

$$\sum_{i=1}^{N} \left(\frac{R_{i2}}{\pi_i} + \frac{1 - R_{i2}}{1 - \pi_i} \right) \left(\frac{\partial \mu_i}{\partial \beta} \right)' V_i^{-1} \{ (Y_i - \mu_i(\beta)) \} = 0. \tag{3.17}$$

Note that when $R_{i2} = 1$, then $Y_i = (Y_{i1}, Y_{i2})$ and, similarly, $\mu_i = (\mu_{i1}, \mu_{i2})$; but when
$R_{i2} = 0$, then $Y_i = Y_{i1}$ and, similarly, $\mu_i = \mu_{i1}$.

The intuition behind this approach is that it reweights the observed data to mimic
what would likely be seen in a data set without missing data, thereby producing unbiased

estimating equations and consistent estimators for $\boldsymbol{\beta}$. In particular, under MAR as in (3.16),

$$\pi_i = \pi_i(\gamma) = \Pr(R_{i2} = 1 | Y_{i1}, Y_{i2}, X_i, \gamma) = \Pr(R_{i2} = 1 | Y_{i1}, X_i, \gamma),$$

or, equivalently,

$$E\left(\left.\frac{R_{i2}}{\pi_i}\right| Y_{i1}, Y_{i2}, X_i\right) = E\left(\left.\frac{R_{i2}}{\pi_i}\right| Y_{i1}, X_i\right) = 1.$$

Therefore, the estimating equations given by (3.17) are unbiased for $\mathbf{0}$ at the true $\boldsymbol{\beta}$,

$$E\left[\left(\frac{R_{i2}}{\pi_i} + \frac{1 - R_{i2}}{1 - \pi_i}\right)\left(\frac{\partial \boldsymbol{\mu}_i}{\partial \boldsymbol{\beta}}\right)' V_i^{-1}\{\mathbf{Y}_i - \boldsymbol{\mu}_i(\boldsymbol{\beta})\}\right]$$

$$= E\left[E\left\{\left(\frac{R_{i2}}{\pi_i} + \frac{1 - R_{i2}}{1 - \pi_i}\right)\left|\, Y_{i1}, Y_{i2}, X_i\right\}\left(\frac{\partial \boldsymbol{\mu}_i}{\partial \boldsymbol{\beta}}\right)' V_i^{-1}(\mathbf{Y}_i - \boldsymbol{\mu}_i(\boldsymbol{\beta}))\right]$$

$$= E\left[\left(\frac{\pi_i}{\pi_i} + \frac{1 - \pi_i}{1 - \pi_i}\right)\left(\frac{\partial \boldsymbol{\mu}_i}{\partial \boldsymbol{\beta}}\right)' V_i^{-1}\{\mathbf{Y}_i - \boldsymbol{\mu}_i(\boldsymbol{\beta})\}\right]$$

$$= 2E\left[\left(\frac{\partial \boldsymbol{\mu}_i}{\partial \boldsymbol{\beta}}\right)' V_i^{-1}\{\mathbf{Y}_i - \boldsymbol{\mu}_i(\boldsymbol{\beta})\}\right]$$

$$= \mathbf{0}.$$

As the estimating equations are unbiased for $\mathbf{0}$, using results from method of moments, the solution $\widehat{\boldsymbol{\beta}}$ to (3.17) defines a consistent estimator for $\boldsymbol{\beta}$.

Unlike in the survey sampling setting where π_i is known, here π_i is unknown and will need to be replaced in (3.17) with an estimate; this estimate can be obtained using, for example, a logistic regression model with "outcome" R_{i2} and "predictors" (y_{i1}, X_i). In general, the validity of most weighting methods requires that the model for the missingness probabilities, π_i, has been correctly specified. We note that the weighted GEE given above is a very simple special case of a general class of weighted estimators. In particular, Robins, Rotnitzky, and Zhao (1995) discuss more general weighted estimating equations for longitudinal data and the construction of "semi-parametric efficient" weighted estimating equations; see Chapter 20 for an excellent summary of these developments. Finally, Robins (2000) and Robins and Rotnitzky (2001) have also recently developed so-called "doubly robust" weighted estimators that relax the assumption that the model for the missingness probabilities, π_i, has been correctly specified, albeit requiring additional assumptions on the model for $\mathbf{Y}_i$ given X_i. Doubly robust methods require the specification of two models, one for the missingness probabilities and another for the distribution of the complete data. For doubly robust estimation, the weighted GEE is augmented by a function of the response. When this augmentation term is selected and modeled correctly according to the distribution of the complete data, the estimator of $\boldsymbol{\beta}$ is consistent even if the model for missingness is misspecified. On the other hand, if the model for missingness is correctly specified, the augmentation term does not need to be correctly specified to yield consistent estimators of $\boldsymbol{\beta}$ (Scharfstein, Rotnitzky, and Robins, 1999, see Section 3.2 of Rejoinder; Robins and Rotnitzky, 2001; van der Laan and Robins, 2003; Bang and Robins, 2005; see also Lipsitz, Ibrahim, and Zhao, 1999; Lunceford and Davidian, 2004). Thus, the appealing property of doubly robust methods is that they yield estimators that are consistent when either, but not necessarily both, the model for the missingness mechanism or the model for the distribution of the complete data has been correctly specified. These estimators are doubly robust in the sense of providing double protection against model misspecification. See Chapter 23 for a more detailed discussion of doubly robust estimators.

As noted previously, the weighted estimating equations approach is best suited to the case of monotone missing-data patterns. However, in many longitudinal studies an individual's response can be missing at one follow-up time, and be measured at the next follow-up time, resulting in a large class of distinct missingness patterns. This is often referred to as "intermittent" missingness or non-monotone missingness. In that setting, approximate methods have been proposed for handling missing data that are assumed to be MAR but not MCAR. For example, via simulations and asymptotic studies, Lipsitz et al. (2000) and Fitzmaurice, Lipsitz, and Molenberghs (2001) show that the GEE1 using the second set of "Gaussian" estimating equations described in Section 3.3.1 yields estimators of $\boldsymbol{\beta}$ with relatively small bias in many settings. However, this method does require that the within-subject association, usually considered a nuisance characteristic of the data, is correctly specified. The method does not, however, require estimation of the missingness probabilities, π_i.

3.5 Goodness-of-fit and model diagnostics

As is the case for any generalized linear model, model checking is an important aspect of the fitting of marginal models to longitudinal data. However, because GEE methods are not likelihood-based, many of the standard goodness-of-fit statistics and model diagnostics are not immediately available. In this section we very briefly review some recent literature on this topic; further research on model diagnostics is needed.

Recall that in linear regression with independent univariate outcome data, Cook's distance and so-called "DFBETAs" are widely used measures of influence (see, for example, Belsley, Kuh, and Welsch, 1980; Cook and Weisberg, 1982). Specifically, DFBETAs assess how each coefficient is changed by including the observation from the ith subject, thereby measuring the influence or effect that the ith subject has on the estimate of the regression parameters $\boldsymbol{\beta}$. It is calculated by deleting the ith observation, and recomputing the ordinary least-squares estimate of $\boldsymbol{\beta}$. Because ordinary least squares is non-iterative, fast computer algorithms have been developed to recompute the estimate of $\boldsymbol{\beta}$ for each subject. Pregibon (1981) extended DFBETAs to logistic regression using one-step estimators of $\boldsymbol{\beta}$. Preisser and Qaqish (1996) extended DFBETAs in a similar way within the GEE approach. Their one-step DFBETAs for GEE require deletion of the n_i repeated measures for the ith subject. That is, conditional on the estimate $\boldsymbol{\alpha}$, a "one-step" estimate of $\boldsymbol{\beta}$, say $\widehat{\boldsymbol{\beta}}_{-i}$, is obtained by excluding the n_i repeated measures for the ith subject and performing one step of the Fisher scoring algorithm described in Equation (3.7) with the full-data estimate $\widehat{\boldsymbol{\beta}}$ as the starting value,

$$\widehat{\boldsymbol{\beta}}_{-i} = \widehat{\boldsymbol{\beta}} + \left[\sum_{k=1, k \neq i}^{N} D_k'(\widehat{\boldsymbol{\beta}}) \{V_k(\widehat{\boldsymbol{\beta}}, \widehat{\boldsymbol{\alpha}})\}^{-1} D_k(\widehat{\boldsymbol{\beta}}) \right]^{-1} \sum_{k=1, k \neq i}^{N} D_k'(\widehat{\boldsymbol{\beta}}) V_k^{-1}(\widehat{\boldsymbol{\beta}}, \widehat{\boldsymbol{\alpha}}) \{ \boldsymbol{Y}_k - \boldsymbol{\mu}_k(\widehat{\boldsymbol{\beta}}) \}.$$

To obtain a unitless, composite measure of the influence of each subject on the entire set of regression coefficients, a statistic similar to Wilks's (1963) statistic for multivariate outliers,

$$w_i = (\widehat{\boldsymbol{\beta}}_{-i} - \widehat{\boldsymbol{\beta}})' \{ \widehat{\mathrm{Cov}}(\widehat{\boldsymbol{\beta}}) \}^{-1} (\widehat{\boldsymbol{\beta}}_{-i} - \widehat{\boldsymbol{\beta}}),$$

can be used. A large value of w_i (relative to the other w_is) indicates that the ith subject may have unusually high influence. Using Wilks's (1963) criterion as a very rough guide, $\{(N - p - 1)(N - 1)/(Np)\} w_i$ has an approximate F-distribution with p (the dimension of $\boldsymbol{\beta}$) and $N - p - 1$ degrees of freedom (for example, see Lipsitz, Laird, and Harrington, 1992).

Goodness-of-fit criteria for generalized estimating equations have also been developed. For example, Pan (2001) proposed an extension of Akaike's information criterion (AIC) to GEE. For the special case of longitudinal binary data, Barnhart and Williamson (1998) and Horton et al. (1999) proposed goodness-of-fit tests for GEE that are extensions of the Hosmer–Lemeshow goodness-of-fit test for logistic regression (Hosmer and Lemeshow, 1980). Here, we briefly review goodness-of-fit statistics for GEE. Suppose it is of interest to determine whether or not the marginal model

$$\mu_{ij} = \frac{\exp(\boldsymbol{X}'_{ij}\boldsymbol{\beta})}{1 + \exp(\boldsymbol{X}'_{ij}\boldsymbol{\beta})}$$

provides an adequate fit to the data. A standard approach is to fit a broader model (e.g., a model with interactions and/or polynomial and higher-order terms) and test whether the additional terms are significantly different from zero. Alternatively, a "global goodness-of-fit" statistic can be obtained by extending the Hosmer–Lemeshow statistic (Hosmer and Lemeshow, 1980; see also Tsiatis, 1980). Following the suggestion of Hosmer and Lemeshow for ordinary logistic regression, G (usually 10) groups can be formed based on combinations of the covariates $\boldsymbol{X}_{ij}$ in the logistic regression model. A test for goodness of fit is constructed by testing whether the additional regression coefficients for the $G - 1$ indicator variables differ from zero. Following Hosmer and Lemeshow, Horton et al. (1999) suggest forming groups based on deciles of the predicted probabilities from the given model,

$$\tilde{\mu}_{ij} = \frac{\exp(\boldsymbol{X}'_{ij}\widehat{\boldsymbol{\beta}})}{1 + \exp(\boldsymbol{X}'_{ij}\widehat{\boldsymbol{\beta}})}.$$

Note that each subject has n_i separate estimates of risk ($\tilde{\mu}_{ij}$s) and that there are $\sum_{i=1}^{N} n_i$ observations in total. Horton et al. (1999) suggest forming 10 groups of approximately equal size from the deciles of these predicted probabilities. For example, the first group contains the $\sum_{i=1}^{N} n_i/10$ ($Y_{ij}, \boldsymbol{X}_{ij}$)s with the smallest values of $\tilde{\mu}_{ij}$, and the last group contains the $\sum_{i=1}^{N} n_i/10$ ($Y_{ij}, \boldsymbol{X}_{ij}$)s with the largest values of $\tilde{\mu}_{ij}$. Finally, defining the $G - 1$ group indicators

$$I_{ijg} = \begin{cases} 1 \text{ if } \tilde{\mu}_{ij} \text{ is in group } g, \\ 0 \text{ otherwise}, \end{cases} \quad g = 1, \ldots, G - 1,$$

where the groups are based on "percentiles of risk," a test for goodness of fit is obtained by considering the alternative model

$$\text{logit}(\mu_{ij}) = \boldsymbol{X}'_{ij}\boldsymbol{\beta} + \gamma_1 I_{ij1} + \ldots + \gamma_{G-1} I_{ij,G-1}.$$

Specifically, the score statistic given in (3.10) can be used to test the null hypothesis

$$\text{H}_0 : \gamma_1 = \ldots = \gamma_{G-1} = 0.$$

Finally, graphical displays are very useful techniques for conveying information about the most salient features of longitudinal data. They can provide insights about patterns of change in the mean response over time (e.g., linearity or the lack thereof) and the choice of suitable functional forms for covariates. Graphical techniques, especially those based on residuals, are especially useful for assessing the adequacy of any postulated model for longitudinal data. They are also useful for identifying observations and individuals that are potential outliers. With appropriate transformations, residual diagnostics developed for standard linear regression can be extended to the longitudinal setting. However, an acknowledged difficulty with conventional residual diagnostics is that they are somewhat subjective in nature. What appears to be a random scatter to one individual might be considered evidence of systematic trend to another. That is, it can be very difficult to discern whether an apparent trend in a scatter plot of the residual reflects some aspect of model misspecification or is simply a reflection of natural variation. McCullagh and Nelder (1989, pp. 392–393) aptly summarize this problem when they state that "the practical problem is that any finite set of residuals can be made to yield some kind of pattern if we look hard enough, so that we have to guard against over-interpretation."

Recently, Lin, Wei, and Ying (2002) developed model-checking techniques based on "cumulative sums" and "moving sums" of residuals that help discern the "signal" from the "noise." The basic idea is to aggregate the residuals over certain coordinates. The coordinates typically used for these sums of residuals are the individual covariates (e.g., X_{ijk}, the kth covariate) and the fitted values, $h(\boldsymbol{X}_{ij}'\widehat{\boldsymbol{\beta}})$. These authors demonstrated that a key advantage of working with sums of residuals, rather than crude residuals, is that a reference distribution is available to ascertain their natural variation. That is, the *observed* sums of the residuals can be compared, both graphically and numerically, to a reference distribution under the assumption of a correctly specified marginal model for the mean. This allows for the determination of whether any apparent pattern is evidence of a systematic trend or simply due to natural variation. This model-checking technique developed by Lin, Wei, and Ying (2002) removes a large degree of subjectivity from the assessment of graphical displays of residuals and places residual diagnostics on a more objective footing.

Specifically, if the assumed model for the mean response is correct, then the cumulative sums of residuals are centered at zero. Moreover, the distribution of the cumulative sum can be approximated by that of a Gaussian process with zero mean whose realizations can be generated via computer simulation. It is relatively straightforward to generate realizations from the distribution of the cumulative sum, under the assumption that the model for the mean is correct (the details are omitted here and the interested reader is referred to Lin, Wei, and Ying, 2002). Thus, to assess whether any apparent trend in the *observed* cumulative sum of residuals reflects systematic trends rather than chance fluctuations, a number of realizations from the appropriate Gaussian process can be superimposed. To the extent that the curves generated from the null distribution tend to be closer to and intersect zero more often than the observed curve, this provides evidence of lack of fit. This assessment can be put on a more formal footing by comparing the maximum absolute value of the observed cumulative sum to a large number of realizations (say 10,000) from the null distribution.

An appealing feature of the graphical and numerical methods based on cumulative and moving sums of residuals is that they are valid regardless of the true joint distribution of the longitudinal response vector; in particular, they do not require correct specification of the covariance among the responses. As such, these graphical and numerical techniques for assessing the model for the mean response are relatively robust to assumptions about the distribution of the responses and assumptions about the covariance among the repeated measures.

3.6 Case study

In this section we present results of analyses of cardiovascular abnormalities from a longitudinal study of children infected with HIV-1 to illustrate some of the main ideas highlighted in earlier sections. Results from various cross-sectional and short-term longitudinal studies (Lipshultz et al., 1998) have suggested that children infected with HIV-1 might have higher risks of cardiovascular abnormalities. We consider this hypothesis using data from the Pediatric Pulmonary and Cardiac Complications (P2C2) of Vertically Transmitted HIV Infection Study (Lipshultz et al., 1998). This was a large, prospective longitudinal study designed to monitor heart disease and the progression of cardiac abnormalities in children born to HIV-infected women. In the P2C2 study, a birth cohort of 401 infants born to women infected with HIV-1 were scheduled to have their cardiovascular function measured approximately every year from birth to age 6, producing up to seven repeated measurements of cardiovascular function on each child. Of the 401 infants who participated in this study, 74 (18.8%) were HIV positive, and 319 (81.2%) were HIV negative.

Table 3.1 Data from 10 Randomly Selected Children from the P2C2 Study

Subject	HIV[b]	Mom Smoked[c]	Gest. Age (wks)	Low Birth Weight[d]	Heart Pumping Ability at Age[a] Birth	1	2	3	4	5	6
1	1	0	41	0	0	0	0	0	0	0	.
2	1	1	34	0	1	.	0	0	1	.	.
3	0	1	40	0	1	0	0	.	.	.	.
4	1	0	40	0	0	.	0	0	0	1	.
5	0	1	39	0	.	1	0	.	.	.	.
6	0	1	35	0	1	.	.	.	.	.	.
7	0	0	36	0	.	0	0	.	.	.	.
7	1	0	33	1	.	1	1	1	.	.	.
8	0	0	36	1	0	0	.	.	.	.	.
9	0	0	41	1	.	.	.	.	0	.	.
10	0	1	34	1	.	0	0	.	0	1	0

(. = missing)
[a] 1 = abnormal, 0 = normal.
[b] 1 = HIV positive, 0 = not HIV positive.
[c] 1 = mother smoked during pregnancy, 0 mother did not smoke.
[d] 1 = low birth weight for age, 0 = normal birth weight.

The question of main scientific interest is to determine if children infected with HIV-1 have worse heart function over time. In particular, we are interested in modeling the pumping ability of the heart (left ventricular fractional shortening) over time. The outcome variable for this analysis is a binary response denoting abnormally low left ventricular fractional shortening (1 = low fractional shortening, 0 = normal) at each occasion. For these data, we are interested in modeling the marginal means or, equivalently, the probabilities of abnormal heart function over time, and relating changes in these marginal probabilities to covariates. The main covariate is the indicator of HIV infection; it is of interest to estimate the effect of HIV infection on heart function over time. Previous results (Lipshultz et al., 1998, 2000, 2002) from the P2C2 study have shown that subclinical cardiac abnormalities develop early in children born with HIV, and that they are frequent, persistent, and often progressive.

Thus, for these analyses we are interested in assessing whether children with HIV have progressively worse heart function over time compared to non-HIV children. In the regression model for abnormal heart function, this translates into a test of the time-by-HIV status interaction. For these analyses, potential confounding variables that must be adjusted for include mother's smoking status during pregnancy (coded 1 = yes, 0 = no), gestational age (in weeks), and birth weight standardized for age (coded 1 = abnormal, 0 = normal). Data from 10 randomly selected children are displayed in Table 3.1.

We note that there is a substantial amount of missing data. Although each child was scheduled to have an echocardiogram every year for the first 6 years of life, including at birth, only 1 (0.25%) of the 401 study participants had outcomes measured at all seven occasions; see Table 3.2 for the frequency distribution of the number of repeated echocardiograms. Table 3.3 shows the number of subjects with echocardiogram measurements at each of the seven measurement occasions. From Table 3.2, we see that only 91 subjects (22.7%) were seen on more than three occasions. From Table 3.3, we see that 276 of the 401 children (68.8%) have baseline measurements; after birth, the percentage of children with echocardiogram measurements declines until only 7 (1.7%) of the 401 subjects have measurements at 6 years of age. For illustrative purposes, the analyses presented here assume that the missing responses are MCAR (Rubin, 1976; Laird, 1988). As discussed in Section 3.4, when the outcome data are MCAR, all GEE approaches yield consistent estimators of

Table 3.2 Frequency Distribution of the Number of Echocardiograms for Children in the P2C2 Study

Number of Echocardiograms	Number of Subjects	Percentage
1	148	36.91
2	104	25.94
3	58	14.46
4	50	12.47
5	30	7.48
6	10	2.49
7	1	0.25
Total	401	100.00

the regression parameters provided the model for the mean is correctly specified. However, we caution the reader that a substantive analysis of these data would require a far more careful treatment of the missing data (see the chapters in Part 5 for detailed discussions of this topic).

To examine the effect of HIV-1, we considered the following marginal logistic regression model for the probability of abnormal heart function at time t_j, denoted μ_{ij}:

$$\log\left(\frac{\mu_{ij}}{1-\mu_{ij}}\right) = \beta_0 + \beta_1 t_j + \beta_2 \mathrm{HIV}_i + \beta_3 t_j \mathrm{HIV}_i + \beta_4 \mathrm{smoke}_i + \beta_5 \mathrm{age}_i + \beta_6 \mathrm{wt}_i,$$

for $j = 1, \ldots, 7$, where $t_j = j - 1$, HIV_i equals 1 if the ith child is born with HIV-1 and equals 0 otherwise; smoke_i equals 1 if the mother smoked during pregnancy and 0 otherwise; age_i is the gestational age (in weeks); and wt_i equals 1 if the child's birth weight for gestational age was abnormal and 0 otherwise. Because there are so few children with echocardiogram measurements after age 4, there are insufficient data to fit an unstructured correlation matrix with $(7 \times 6)/2 = 21$ parameters,

$$\rho_{ist} = \mathrm{Corr}(Y_{is}, Y_{it}; \boldsymbol{\alpha}) = \rho_{st}.$$

Instead, to account for the within-subject association among the binary responses, we consider two possible "working correlation" patterns: "exchangeable," in which $\rho_{ist} = \alpha$ for all $s < t$, and AR(1), with

$$\rho_{ist} = \alpha^{|t-s|},$$

where $0 < \alpha < 1$.

Finally, to illustrate similarities and differences between various GEE estimation techniques, we compare the estimates of $(\boldsymbol{\beta}, \alpha)$ obtained using six approaches: (1) GEE under "working independence" (equivalent to ordinary logistic regression); (2) standard GEE1

Table 3.3 Frequency Distribution of the Number of Children with Echocardiograms at Each Occasion

Age at Visit (Years)	Number of Subjects	Percentage
Birth	276	68.83
1	267	66.58
2	154	38.40
3	123	30.67
4	83	20.70
5	37	9.23
6	7	1.75

with estimation of α based on *unconditional* residuals; (3) GEE1 with estimation of α based on *conditional* residuals; (4) GEE1 with Gaussian estimation of α; (5) GEE2 with $\mathcal{V}_i$ in (3.15) specified using the Bahadur distribution (Bahadur, 1961); and (6) maximum likelihood using the parametric Bahadur distribution, which is based on two- and higher-order correlations.

Because the Bahadur distribution is used in both the GEE2 and ML approaches, we provide a brief description here. We define the standardized binary variable S_{ij} as

$$S_{ij} = \frac{Y_{ij} - \mu_{ij}}{\{\mu_{ij}(1 - \mu_{ij})\}^{1/2}}.$$

The pairwise correlation between Y_{ij} and Y_{ik} is $\rho_{jk} = E(S_{ij}S_{ik})$, and the Rth-order correlation between the first R responses is defined as $\rho_{12...R} = E(S_{i1}S_{i2}...S_{iR})$. The Rth-order correlation between any R of the n repeated binary responses is defined similarly. The Bahadur representation of the $2^n - 1$ multinomial probabilities corresponding to the joint distribution of $(Y_{i1}, Y_{i2}, \ldots, Y_{in})$ is

$$\Pr\{Y_{i1} = y_1, Y_{i2} = y_2, \ldots, Y_{in} = y_n | X_i, \boldsymbol{\beta}, \boldsymbol{\alpha}\} = \left\{ \prod_{j=1}^{n} \mu_{ij}^{y_{ij}} (1 - \mu_{ij})^{1-y_{ij}} \right\}$$

$$\times \left\{ 1 + \sum_{j>k} \rho_{jk} s_{ij} s_{ik} + \sum_{j>k>l} \rho_{jkl} s_{ij} s_{ik} s_{il} + \cdots + \rho_{1...n} s_{i1} \cdots s_{in} \right\}.$$

For both GEE2 and ML approaches, we assumed all fifth- and higher-order correlations are 0 ($\rho_{jklmn} = \cdots = \rho_{1...n} = 0$). In addition, we assumed that all fourth-order correlations are the same, regardless of the sets of times ($\rho_{jklm} = \rho_{j'k'l'm'}$ for all $jklm \neq j'k'l'm'$), and that all third-order correlations are the same, regardless of the sets of times ($\rho_{jkl} = \rho_{j'k'l'}$ for all $jkl \neq j'k'l'$). The two models for the pairwise correlations ρ_{st} (exchangeable and autoregressive) are the same for all six estimation methods.

The logistic regression parameter estimates and standard errors are presented in Table 3.4 for the "working independence" GEE and GEE1 with exchangeable correlation estimated using *unconditional* residuals. Table 3.4 presents both model-based and empirical (or so-called sandwich) variance estimates. As might be expected, the most discernible differences between the model-based and empirical standard errors occur for the "working independence" GEE. This is because the naive assumption of independence among repeated binary responses obtained from the same child is "furthest" from the true underlying within-subject correlation. In general, when the true correlation among repeated measures is high, the differences between the model-based and empirical standard errors for the "working independence" GEE can be substantial; the former provide unrealistic estimates of the sampling variability. For the data from the P2C2 study, the estimated correlation is relatively modest (assuming an exchangeable correlation, the estimated correlation is 0.15); consequently, the differences between the model-based and empirical standard errors are not too extreme. For example, for the "working independence" GEE, the largest differences are in the estimated variance of the time-by-HIV infection interaction, with the empirical variance being almost 1.5 (or $[0.16/0.13]^2$) times larger than the model-based variance, and in the gestational age effect where the empirical variance is almost 1.4 (or $[0.36/0.31]^2$) times larger than the model-based variance. We note that for the "working independence" GEE the model-based variance is not necessarily always smaller than the empirical variance; their relative size depends on whether the parameter of interest is the effect of a time-varying (or within-subject) or time-stationary (or between-subject) covariate, the magnitude of the correlation, and the proportion of missing data over time (see Mountford et al., 2007, for a more detailed discussion). We also note that for the GEE1 with exchangeable correlation estimated using

Table 3.4 Regression Parameter Estimates with Model-Based and Empirical Standard Errors (SE) for Independence GEE and GEE1 with Exchangeable Correlation Estimated Using Unconditional Residuals

Effect	Method	Estimate	Model SE	SE	Empirical Wald Z	p-Value
Intercept	IND	2.284	1.236	1.508	1.51	0.130
	GEE1	1.763	1.416	1.468	1.20	0.230
Time	IND	−0.600	0.076	0.097	−6.16	<0.001
	GEE1	−0.630	0.078	0.096	−6.54	<0.001
HIV	IND	−0.069	0.252	0.266	−0.26	0.796
	GEE1	−0.081	0.255	0.266	−0.30	0.761
Time*HIV	IND	0.204	0.130	0.159	1.28	0.200
	GEE1	0.283	0.125	0.149	1.90	0.058
Mom Smoke	IND	−0.196	0.153	0.175	−1.12	0.263
	GEE1	−0.219	0.175	0.168	−1.31	0.191
Gest. Age	IND	−0.058	0.031	0.038	−1.52	0.130
	GEE1	−0.044	0.036	0.037	−1.19	0.233
Low Birth Wt.	IND	0.048	0.163	0.198	0.24	0.810
	GEE1	0.096	0.186	0.186	0.52	0.604

unconditional residuals, there are fewer differences between the model-based and empirical standard errors. However, the small number of discernible differences (e.g., the estimated variances of the time-by-HIV interaction) suggest that the model for the correlation could potentially be improved. That is, if the working model for the correlation has been correctly specified, in general, we would expect the model-based and empirical standard errors to be relatively similar.

Based on the regression parameter estimates and empirical standard errors from the GEE1 with exchangeable correlation, there is the suggestion that the pattern of change in risk of abnormal heart function differs by HIV-1 infection group ($Z = 1.90, p < 0.06$). Specifically, after adjustment for maternal smoking during pregnancy, gestational age, and birth weight, the risk decreases at a slower rate in the HIV-1 infected children. Note that the interpretation of the results is very similar to the interpretation of results from an ordinary logistic regression model for cross-sectional data. For example, children with HIV-1 infection have $\exp(\hat{\beta}_2 + \hat{\beta}_3 t_j) = \exp(-0.081 + 0.285 t_j)$ times the odds of having an abnormal pumping ability compared to children without HIV-1 infection at time t_j. Specifically, at birth ($t_1 = 0$), the ratio of their odds of having an abnormal pumping ability is approximately 1 ($\exp(-0.081) = 0.92$), while at year 6 the ratio of their odds is approximately 5.0 ($\exp(-0.081 + 6 \times 0.285) = 5.1$).

For illustrative purposes, Table 3.5 presents the parameter estimates (and empirical standard error estimates) obtained from all five methods for a "working exchangeable" correlation. Note that the regression parameter estimates and estimated standard errors from the three GEE1 methods are very similar to each other. In contrast, the regression parameter estimates from GEE2 are somewhat different. Recall that GEE2 requires the stronger assumption that *both* the first and second moments must be correctly specified. For these data, the empirical standard errors for GEE2 estimates are similar to those for GEE1; however, this result is not to be expected in general. Finally, as expected, ML yields the smallest estimated variances for the estimated regression parameters. The largest gain in efficiency is for the time-by-HIV infection interaction, the effect of greatest subject-matter interest.

Table 3.5 Comparison of Parameter Estimates (and Empirical Standard Errors) under an Exchangeable Correlation

Effect	Method	Estimate	SE	Wald Z	p-Value
Intercept	GEE1 (uncond)	1.763	1.468	1.20	0.230
	GEE1 (cond)	1.851	1.470	1.26	0.208
	GEE1 (Gaussian)	1.822	1.469	1.24	0.216
	GEE2	2.126	1.443	1.47	0.141
	ML	1.733	1.336	1.30	0.195
Time	GEE1 (uncond)	-0.630	0.096	-6.54	<0.001
	GEE1 (cond)	-0.626	0.096	-6.49	<0.001
	GEE1 (Gaussian)	-0.627	0.096	-6.51	<0.001
	GEE2	-0.583	0.095	-6.16	<0.001
	ML	-0.678	0.084	-8.10	<0.001
HIV	GEE1 (uncond)	-0.081	0.266	-0.30	0.761
	GEE1 (cond)	-0.079	0.265	-0.30	0.765
	GEE1 (Gaussian)	-0.080	0.265	-0.30	0.764
	GEE2	-0.047	0.274	-0.17	0.865
	ML	-0.037	0.255	-0.15	0.884
Time*HIV	GEE1 (uncond)	0.283	0.149	1.90	0.058
	GEE1 (cond)	0.271	0.151	1.80	0.072
	GEE1 (Gaussian)	0.275	0.150	1.83	0.068
	GEE2	0.264	0.156	1.70	0.089
	ML	0.287	0.122	2.35	0.019
Mom Smoke	GEE1 (uncond)	-0.219	0.168	-1.31	0.191
	GEE1 (cond)	-0.215	0.168	-1.28	0.201
	GEE1 (Gaussian)	-0.216	0.168	-1.29	0.198
	GEE2	-0.249	0.166	-1.50	0.134
	ML	-0.241	0.159	-1.51	0.131
Gest. Age	GEE1 (uncond)	-0.044	0.037	-1.19	0.233
	GEE1 (cond)	-0.047	0.037	-1.25	0.211
	GEE1 (Gaussian)	-0.046	0.037	-1.23	0.219
	GEE2	-0.055	0.036	-1.52	0.128
	ML	-0.043	0.034	-1.27	0.204
Low Birth Wt.	GEE1 (uncond)	0.096	0.186	0.52	0.604
	GEE1 (cond)	0.089	0.187	0.48	0.632
	GEE1 (Gaussian)	0.092	0.186	0.49	0.622
	GEE2	0.042	0.180	0.23	0.816
	ML	0.155	0.168	0.93	0.356
ρ	GEE1 (uncond)	0.153	0.120	1.28	0.201
	GEE1 (cond)	0.194	0.078	2.47	0.013
	GEE1 (Gaussian)	0.165	0.049	3.37	0.001
	GEE2	0.174	0.059	2.95	0.003
	ML	0.151	0.034	4.42	<0.001

The estimated relative efficiency for GEE1 with exchangeable correlation versus ML is $(0.12/0.15)^2 = 66\%$. However, a potential drawback of ML is that the full joint distribution of the binary outcomes must be correctly specified; thus, the assumptions made about all fifth- and higher-order correlations being zero must be correct for ML to yield asymptotically unbiased estimators of the regression parameters.

Although all GEE1 methods produce similar standard errors for the estimated regression parameters, this is not the case for the estimated correlation parameter. Specifically, the GEE1 based on conditional residuals greatly reduces the estimated variance of the correlation compared to GEE1 with unconditional residuals, while GEE1 based on Gaussian

Table 3.6 Comparison of Parameter Estimates (and Empirical Standard Errors) under an AR(1) Correlation

Effect	Method	Estimate	SE	Wald Z	p-Value
Intercept	GEE1 (uncond)	2.018	1.506	1.34	0.180
	GEE1 (cond)	1.817	1.508	1.20	0.228
	GEE1 (Gaussian)	1.978	1.506	1.31	0.190
	GEE2	2.118	1.429	1.48	0.138
	ML	1.763	1.352	1.30	0.193
Time	GEE1 (uncond)	−0.634	0.097	−6.50	<0.001
	GEE1 (cond)	−0.661	0.098	−6.72	<0.001
	GEE1 (Gaussian)	−0.639	0.098	−6.55	<0.001
	GEE2	−0.519	0.095	−5.48	<0.001
	ML	−0.637	0.088	−7.28	<0.001
HIV	GEE1 (uncond)	−0.072	0.264	−0.27	0.785
	GEE1 (cond)	−0.075	0.264	−0.28	0.777
	GEE1 (Gaussian)	−0.073	0.264	−0.28	0.783
	GEE2	−0.073	0.299	−0.24	0.808
	ML	−0.037	0.269	−0.14	0.891
Time*HIV	GEE1 (uncond)	0.242	0.157	1.54	0.123
	GEE1 (cond)	0.273	0.155	1.76	0.078
	GEE1 (Gaussian)	0.248	0.156	1.58	0.114
	GEE2	0.173	0.168	1.03	0.302
	ML	0.213	0.140	1.53	0.128
Mom Smoke	GEE1 (uncond)	−0.199	0.173	−1.15	0.250
	GEE1 (cond)	−0.200	0.172	−1.16	0.246
	GEE1 (Gaussian)	−0.199	0.173	−1.15	0.250
	GEE2	−0.266	0.174	−1.53	0.127
	ML	−0.206	0.172	−1.20	0.231
Gest. Age	GEE1 (uncond)	−0.050	0.038	−1.31	0.190
	GEE1 (cond)	−0.044	0.038	−1.15	0.249
	GEE1 (Gaussian)	−0.049	0.038	−1.28	0.202
	GEE2	−0.056	0.036	−1.55	0.122
	ML	−0.043	0.034	−1.26	0.207
Low Birth Wt.	GEE1 (uncond)	0.076	0.194	0.39	0.694
	GEE1 (cond)	0.100	0.192	0.52	0.603
	GEE1 (Gaussian)	0.081	0.193	0.42	0.677
	GEE2	0.040	0.188	0.21	0.830
	ML	0.096	0.173	0.55	0.581
ρ	GEE1 (uncond)	0.167	0.083	2.02	0.043
	GEE1 (cond)	0.289	0.063	4.63	<0.001
	GEE1 (Gaussian)	0.192	0.059	3.23	0.001
	GEE2	0.167	0.076	2.21	0.027
	ML	0.206	0.050	4.08	<0.001

estimation reduces the estimated variance even further. Note that GEE1 based on Gaussian estimation yields a smaller variance estimate than GEE2, although this result cannot be expected in general.

In Table 3.6, we consider estimates based on the five methods presented in Table 3.5 when the "working correlation" is assumed to be first-order autoregressive (AR(1)). In general, the pattern of results in Table 3.6 is similar to that in Table 3.5. With a modest correlation, it is perhaps not surprising that the estimates of the regression parameters under AR(1) or exchangeable should be so similar. The imbalance in the data due to missingness probably accounts for some of the small differences when the results in Table 3.5 and Table 3.6 are compared. Also, we remind the reader that all analyses have made the strong assumption that the missing data are MCAR.

In earlier sections, we remarked that for the GEE2 estimate of $\boldsymbol{\beta}$ to be consistent the model for the first two moments of $\boldsymbol{Y}_i$ must be correctly specified. This is in contrast to GEE1 where a consistent estimator of $\boldsymbol{\beta}$ is obtained provided only that the mean of $\boldsymbol{Y}_i$ has been correctly specified. To illustrate the sensitivity of GEE2 estimates of $\boldsymbol{\beta}$ to misspecification of the second moments, we fixed the exchangeable correlation at larger and smaller values than the estimate of $\widehat{\rho} = 0.174$ obtained in Table 3.5 and re-estimated $\boldsymbol{\beta}$. For example, when ρ is fixed at 0.7, the GEE2 estimate of the time-by-HIV status interaction, the parameter of main interest, is 0.390 (SE = 0.156), which is discernibly different from the estimate of 0.264 obtained in Table 3.5. Similarly, when ρ is fixed at 0.065, the GEE2 estimate of the time-by-HIV status interaction is 0.075 (SE = 0.192). This highlights how GEE2 estimates of $\boldsymbol{\beta}$ are sensitive to misspecification of the within-subject association.

Finally, in Section 3.2.4 we mentioned that for the special case where the outcome is binary, an alternative to the correlation as a measure of within-subject association is the odds ratio. The odds ratio has many desirable properties and, unlike the correlation, is not constrained by the marginal probabilities, μ_{ij}. For illustrative purposes, we replicated the GEE1 analyses in Table 3.5 and Table 3.6 where the within-subject association was parameterized in terms of log odds ratios rather than correlations. Recall that the joint distribution of Y_{is} and Y_{it} depends on both $\boldsymbol{\beta}$ and $\boldsymbol{\alpha}$, with

$$\pi_{ist} = E(Y_{is}Y_{it}|\boldsymbol{X}_{is}, \boldsymbol{X}_{it}, \boldsymbol{\beta}, \boldsymbol{\alpha}) = \Pr(Y_{is} = 1, Y_{it} = 1|\boldsymbol{X}_{is}, \boldsymbol{X}_{it}, \boldsymbol{\beta}, \boldsymbol{\alpha}).$$

This joint probability can be modeled in terms of either the marginal correlation or the marginal odds ratio. The marginal correlation between the responses at times s and t is

$$\rho_{ist} = \rho_{ist}(\boldsymbol{\alpha}) = \text{Corr}(Y_{is}, Y_{it}; \boldsymbol{\alpha}) = \frac{\pi_{ist} - \mu_{is}\mu_{it}}{\{\mu_{is}(1 - \mu_{is})\mu_{it}(1 - \mu_{it})\}^{1/2}}.$$

Note that the joint probability π_{ist} can be written in terms of the correlation coefficient as

$$\pi_{ist} = \mu_{is}\mu_{it} + \rho_{ist}\{\mu_{is}(1 - \mu_{is})\mu_{it}(1 - \mu_{it})\}^{1/2}.$$

Alternatively, instead of modeling the association between pairs of binary responses in terms of the marginal correlation, Lipsitz, Laird, and Harrington (1991), Liang, Zeger, and Qaqish (1992), and Carey, Zeger, and Diggle (1993) propose using the marginal odds ratio. The odds ratio between the responses at times s and t is

$$\psi_{ist} = \psi_{ist}(\boldsymbol{\alpha}) = \frac{\pi_{ist}(1 - \mu_{is} - \mu_{it} + \pi_{ist})}{(\mu_{is} - \pi_{ist})(\mu_{it} - \pi_{ist})}.$$

In terms of the odds ratio, the probability π_{ist} can then be written as

$$\pi_{ist} = \begin{cases} \dfrac{a_{ist} - \{a_{ist}^2 - 4\psi_{ist}(\psi_{ist} - 1)\mu_{is}\mu_{it}\}^{1/2}}{2(\psi_{ist} - 1)} & \text{if } \psi_{ist} \neq 1, \\[2mm] \mu_{is}\mu_{it} & \text{if } \psi_{ist} = 1, \end{cases}$$

where $a_{ist} = 1 - (1 - \psi_{ist})(\mu_{is} + \mu_{it})$. An "exchangeable" odds ratio model is given by $\psi_{ist} = \alpha$. As a binary data analog of the AR(1) model for correlation, Fitzmaurice and Lipsitz (1995) proposed the following serial pattern model for the odds ratio,

$$\psi_{ist} = \alpha^{1/|t-s|},$$

where $1 < \alpha < \infty$. Note that as $|t - s| \to 0$, $\psi_{ist} \to \alpha^\infty$ and there is perfect association. When observations are far apart in time, then as $|t - s| \to \infty$, $\psi_{ist} \to \alpha^0 = 1$ and the pairs of observations are independent. This serial odds ratio model mimics an AR(1) pattern with the strength of association declining with increasing time separation. Thus, we can express both the "exchangeable" and "serial" odds ratio models as linear models for the log odds ratio, with

$$\log(\psi_{ist}) = \left(\frac{1}{|t-s|}\right)\log(\alpha)$$

for the "serial" odds ratio model, and

$$\log(\psi_{ist}) = \log(\alpha)$$

for the "exchangeable" odds ratio model.

To estimate $\boldsymbol{\beta}$ and α, under an "exchangeable" or "serial" odds ratio pattern, we can use the estimating equations proposed by Lipsitz, Laird, and Harrington (1990) based on unconditional residuals or the estimating equations proposed by Carey, Zeger, and Diggle (1993) based on conditional residuals; in general, the latter yield more efficient estimators of $\boldsymbol{\alpha}$. The results of fitting the two models for the odds ratio pattern using both methods of estimation are presented in Table 3.7. The pattern of results in Table 3.7 is similar to

Table 3.7 Comparison of Parameter Estimates (and Empirical Standard Errors) Using Log Odds Ratio as a Measure of Within-Subject Association, Estimated Using Unconditional and Conditional Residuals

Effect	Method	Odds Ratio Model	Estimate	SE	Wald Z	p-Value
Intercept	GEE1 (uncond)	exc	1.870	1.491	1.25	0.210
	GEE1 (cond)	exc	1.797	1.492	1.20	0.228
	GEE1 (uncond)	serial	2.003	1.503	1.33	0.183
	GEE1 (cond)	serial	1.810	1.506	1.20	0.230
Time	GEE1 (uncond)	exc	−0.618	0.096	−6.46	<0.001
	GEE1 (cond)	exc	−0.623	0.096	−6.51	<0.001
	GEE1 (uncond)	serial	−0.618	0.097	−6.41	<0.001
	GEE1 (cond)	serial	−0.637	0.097	−6.58	<0.001
HIV	GEE1 (uncond)	exc	−0.075	0.265	−0.28	0.777
	GEE1 (cond)	exc	−0.078	0.265	−0.30	0.768
	GEE1 (uncond)	serial	−0.061	0.263	−0.23	0.816
	GEE1 (cond)	serial	−0.060	0.263	−0.23	0.819
Time*HIV	GEE1 (uncond)	exc	0.257	0.150	1.72	0.086
	GEE1 (cond)	exc	0.269	0.149	1.80	0.071
	GEE1 (uncond)	serial	0.232	0.154	1.51	0.132
	GEE1 (cond)	serial	0.256	0.152	1.68	0.093
Mom Smoke	GEE1 (uncond)	exc	−0.202	0.169	−1.20	0.232
	GEE1 (cond)	exc	−0.203	0.168	−1.21	0.228
	GEE1 (uncond)	serial	−0.194	0.172	−1.13	0.258
	GEE1 (cond)	serial	−0.194	0.171	−1.13	0.257
Gest. Age	GEE1 (uncond)	exc	−0.047	0.038	−1.24	0.214
	GEE1 (cond)	exc	−0.045	0.038	−1.19	0.234
	GEE1 (uncond)	serial	−0.050	0.038	−1.31	0.189
	GEE1 (cond)	serial	−0.044	0.038	−1.16	0.244
Low Birth Wt.	GEE1 (uncond)	exc	0.101	0.188	0.54	0.589
	GEE1 (cond)	exc	0.110	0.187	0.59	0.555
	GEE1 (uncond)	serial	0.087	0.192	0.45	0.652
	GEE1 (cond)	serial	0.111	0.190	0.58	0.559
Log(OR)	GEE1 (uncond)	exc	0.859	0.931	0.92	0.356
	GEE1 (cond)	exc	1.051	0.281	3.74	<0.001
	GEE1 (uncond)	serial	0.788	0.999	0.79	0.430
	GEE1 (cond)	serial	1.369	0.301	4.54	<0.001

those in Table 3.5 and Table 3.6, confirming that GEE1 methods are relatively insensitive to choice of models and methods of estimation for the within-subject association. Interestingly, the potential benefits of GEE1 based on conditional residuals for estimation of the within-subject association are highlighted in Table 3.7 where the standard error for the estimated log odds ratio is discernibly smaller than for GEE1 based on unconditional residuals. For the parameter of main interest, the time-by-HIV status interaction, the estimates of effects are similar to those found in Table 3.5 and Table 3.6.

3.7 Discussion and future directions

As discussed in the previous sections, Liang and Zeger (1986) developed generalized estimating equations to estimate the parameters of marginal models; the latter class of models represent a particular extension of generalized linear models to longitudinal data. Liang and Zeger (1986) introduced a class of moment-based estimating equations that yield consistent estimators of the regression parameters, and their variances, under relatively mild conditions. Over the past 20 years, the GEE approach has been the dominant method for estimation of marginal models for longitudinal data; so much so, indeed, that some authors confusingly refer to marginal models as "GEE models." However, it is worth emphasizing that GEE methods are but one approach for estimating marginal model parameters. Moreover, GEE methods should not be regarded as conjoined at the hip to marginal models; the estimating equation approach can be applied equally to alternative classes of models for longitudinal data, such as so-called transitional or response-conditional models (e.g., Markov models for longitudinal data) and GLMMs.

The GEE method is semi-parametric, in that the estimating equations are derived without fully specifying the joint distribution of the vector of repeated measures. This is a very appealing feature of the GEE approach, especially for the analysis of discrete longitudinal data, because for the latter case the total number of parameters in the saturated model for the joint distribution of the vector of responses grows exponentially with the number of repeated measures. In particular, GEE methods only require specification of the form of the first two moments of the outcome vector, and provide consistent estimators of the marginal regression parameters under the weak condition that only the first moment is correctly specified. That is, provided the marginal model for the mean response (the first moment) is correctly specified, the estimating equations yield estimators of the marginal regression parameters that are consistent and asymptotically normal, regardless of whether the second moments have been correctly specified.

Similar to the quasi-likelihood estimating equations originally proposed by Wedderburn (1974), the optimal choice of weights is to take $V_i = \text{Cov}(\boldsymbol{Y}_i)$. Naturally, there are certain trade-offs associated with the use of less than optimal weights. However, in general, if the correlation among repeated measures is not too high and/or the effects of primary interest are between-subject effects, then the efficiency of GEE estimators even under the naive assumption of independence is remarkably high (Liang and Zeger, 1986). However, when there is interest in the effects of non-stochastic time-varying covariates and/or the repeated measures are highly correlated, then suboptimal choices of weights (i.e., misspecification of the covariance) can result in a discernible loss of efficiency. Similarly, with inherently unbalanced longitudinal data (e.g., widely varying n_i), suboptimal choices of weights can also result in loss of efficiency. In general, more efficient estimators of the marginal regression parameters can be obtained by specifying and estimating the covariance among the repeated measures; however, the potential gains in efficiency will depend in a subtle way on both the magnitude of the correlation and the covariate design.

As mentioned earlier, the major impetus for the GEE approach was that it provided a convenient alternative to maximum likelihood estimation of marginal models. For the case of continuous outcomes, likelihood-based methods are widely used for the analysis of longitudinal data, for example, general linear models and linear mixed-effects models. In general, likelihood-based estimation of marginal models for discrete longitudinal data has proven to be very challenging, in large part because there is no simple unified joint likelihood for discrete longitudinal data. Unlike the multivariate normal distribution, which is completely specified by the first two moments of the outcome vector, for discrete longitudinal data complete specification of the joint distributions requires specifying third- and higher-order moments. Although various likelihood approaches have been proposed — for example, models based on second- and higher-order correlations (Bahadur, 1961; Zhao and Prentice, 1990) and models based on second- and higher-order odds ratios (McCullagh and Nelder, 1989; Lipsitz, Laird, and Harrington, 1990; Liang, Zeger, and Qaqish, 1992; Becker and Balagtas, 1993; Fitzmaurice, Laird, and Rotnitzky, 1993; Molenberghs and Lesaffre, 1994; Lang and Agresti, 1994; Glonek and McCullagh, 1995; Bergsma and Rudas, 2002) — none of these likelihood-based models have proven to be of real practical use except in relatively limited settings. As the number of repeated measures increases, the number of parameters that need to be specified and estimated proliferates rapidly for any of these joint distributions, and a solution to the likelihood equations quickly becomes intractable. Thus, in general, full likelihood approaches require the specification of too many nuisance parameters, are complicated algebraically, and ML estimation can be computationally prohibitive. Furthermore, to obtain unbiased estimators of the marginal regression parameters, the full joint distribution of the data must, in general, be correctly specified. In contrast, with GEE methods, only the first moment needs to be correctly specified in order to obtain unbiased estimators; moreover, the GEE approach is not computationally demanding.

We note that Heagerty (1999) and Heagerty and Zeger (2000) have recently developed a likelihood-based approach that combines the versatility of GLMMs for modeling the within-subject association with a marginal regression model for the mean response. They refer to these models as *marginalized* random-effects models; they are closely related to other approaches that formulate conditional models subject to marginal specification (e.g., Fitzmaurice and Laird, 1993; Azzalini, 1994; Heagerty, 2002; see also Wang and Louis, 2003). Recall that in the standard GLMM, the marginal means obtained by integrating over the random effects, in general, no longer follow a generalized linear model, due to the non-linearity of the link function typically adopted in regression models for discrete responses. In contrast, the *marginalized* random-effects model is specifically formulated such that the marginal mean follows a generalized linear model. Estimation for these *marginalized* random-effects models is as computationally demanding as for GLMMs, and ML estimation is, so far, limited to relatively low-dimensional random-effects distributions. However, these models appear to have the potential to overcome many of the limitations of previously proposed likelihood-based methods and to provide a likelihood-based alternative to GEE; further research is needed.

Finally, although an appealing feature of the GEE approach is its robustness to misspecification of the within-subject association, there are settings where it can be appealing to model the covariance. To date, the implementations of GEE in standard statistical software packages provide only very limited options for modeling the covariance. In particular, there are few choices of models for the "working covariance" when the data are highly unbalanced and irregularly spaced in time. This is in contrast to models for continuous responses (e.g., general linear models and linear mixed-effects models), where there are a broad class of models for the covariance. Future work is needed in both the formulation and implementation of flexible models for the working covariance in GEE methods.

Acknowledgments

This work was supported by grants AI 60373, GM 29745, and MH 54693 from the U.S. National Institutes of Health.

References

Azzalini, A. (1994). Logistic regression for autocorrelated data with application to repeated measures. *Biometrika* **81**, 767–775.

Bahadur, R. R. (1961). A representation of the joint distribution of responses to n dichotomous items. In H. Solomon (ed.), *Studies in Item Analysis and Prediction*, pp. 158–168. Palo Alto, CA: Stanford University Press.

Bang, H. and Robins, J. M. (2005). Doubly robust estimation in missing data and causal inference models. *Biometrics* **61**, 962–973.

Barnhart H. X. and Williamson J. M. (1998). Goodness-of-fit tests for GEE modeling with binary responses. *Biometrics* **54**, 720–729.

Becker, M. P. and Balagtas, C. C. (1993). Marginal modeling of binary cross-over data. *Biometrics* **49**, 997–1009.

Belsley, D. A., Kuh, E., and Welsch, R. E. (1980). *Regression Diagnostics*. New York: Wiley.

Bergsma, W. P. and Rudas, T. (2002). Marginal models for categorical data. *Annals of Statistics*, **30**, 140–159.

Bishop, Y. M. M., Fienberg, S. E., and Holland, P. W. (1975). *Discrete Multivariate Analysis: Theory and Practice*. Cambridge, MA: MIT Press.

Boos, D. D.(1992). On generalized score tests. *American Statistician* **46**, 327–333.

Carey, V., Zeger, S. L., and Diggle, P. J. (1993). Modelling multivariate binary data with alternating logistic regressions. *Biometrika* **80**, 517–526.

Cook, R. D. and Weisberg, S. (1982). *Residuals and Influence in Regression*. London: Chapman & Hall.

Cox, D. R. (1961). Tests of separate families of hypotheses. In J. Neyman (ed.), *Proceedings of the Fourth Berkeley Symposium on Mathematics, Probability and Statistics*, Vol. 1, pp. 105–123. Berkeley: University of California Press.

Cox, D. R. and Snell, E. J. (1989). *Analysis of Binary Data*, 2nd ed. London: Chapman & Hall.

Crowder, M. (1995). On the use of a working correlation matrix in using generalised linear models for repeated measures. *Biometrika* **82**, 407–410.

Fitzmaurice, G. M. and Laird, N. M. (1993). A likelihood-based method for analysing longitudinal binary responses. *Biometrika* **80**, 141–151.

Fitzmaurice, G. M., Laird, N. M., and Rotnitzky, A. G. (1993). Regression models for discrete longitudinal responses (with discussion). *Statistical Science* **8**, 248–309.

Fitzmaurice, G. M. and Lipsitz, S. R. (1995). A model for binary time series data with serial odds ratio patterns. *Applied Statistics* **44**, 51–61.

Fitzmaurice G. M., Lipsitz S. R., and Molenberghs G. (2001). Bias in estimating association parameters for longitudinal binary responses with drop-outs. *Biometrics* **57**, 15–21.

Gange S. J., Linton, K. L., Scott, A. J., DeMets, D. L., and Klein, R. (1995). A comparison of methods for correlated ordinal measures with ophthalmic applications. *Statistics in Medicine* **14**, 1961–1974.

Glonek, G. F. V. and McCullagh, P. (1995). Multivariate logistic models. *Journal of the Royal Statistical Society, Series B* **57**, 533–546.

Grizzle, J. E., Starmer, C. F., and Koch, G. G. (1969). Analysis of categorical data by linear models. *Biometrics* **15**, 489–504.

Hall, D. B. and Severini, T. A. (1998). Extended generalized estimating equations for clustered data. *Journal of the American Statistical Association* **93**, 1365–1375.

Hauck, W. W. and Donner, A. (1977). Wald's test as applied to hypotheses in logit analysis, *Journal of the American Statistical Association* **72**, 851–853.

Heagerty, P. J. (1999). Marginally specified logistic-normal models for longitudinal binary data. *Biometrics* **55**, 688–698.

Heagerty, P. J. (2002). Margininalized transition models and likelihood inference for longitudinal categorical data. *Biometrics* **58**, 342–351.

Heagerty, P. J. and Zeger, S. L. (1996). Marginal regression models for clustered ordinal measurements. *Journal of the American Statistical Association* **91**, 1024–1036.

Heagerty, P. J. and Zeger, S. L. (2000). Marginalized multilevel models and likelihood inference (with comments and a rejoinder by the authors). *Statistical Science* **15**, 1–26.

Horton, N. J., Bebchuk, J. D., Jones, C. L., Lipsitz, S. R., Catalano, P., and Fitzmaurice, G. M. (1999). Goodness of fit for GEE: An example with quality of life and prostate cancer. *Statistics in Medicine* **18**, 213–222.

Horvitz, D. G. and Thompson, D. J. (1952). A generalization of sampling without replacement from a finite universe. *Journal of the American Statistical Association* **47**, 663–685.

Hosmer, D. W. and Lemeshow, S. (2000). *Applied Logistic Regression*, 2nd ed. New York: Wiley.

Huber, P. J. (1967) The behavior of maximum likelihood estimates under nonstandard conditions. In L. LeCam and J. Neyman (eds.), *Proceedings of the Fifth Berkeley Symposium on Mathematical Statistics and Probability*, Vol. 1, pp. 221–233. Berkeley: University of California Press.

Kenward, M., Lesaffre, E., and Molenberghs, G. (1994). An application of maximum likelihood and generalized estimating equations to the analysis of ordinal data from a longitudinal study with cases missing at random. *Biometrics* **50**, 945–953.

Koch, G. G. and Reinfurt, D. W. (1971). The analysis of categorical data from mixed models. *Biometrics* **27**, 57–173.

Koch, G. G., Landis, J. R., Freeman, J. L., Freeman, D. H., and Lehnen, R. G. (1977). A general methodology for the analysis of experiments with repeated measurement of categorical data. *Biometrics* **33**, 133–158.

Kuk, A. Y. C. (2004). Permutation invariance of alternating logistic regression for multivariate binary data. *Biometrika*, **91**, 758–761.

Laird, N. M. (1988). Missing data in longitudinal studies. *Statistics in Medicine* **7**, 305–315.

Laird, N. M. and Ware, J. H. (1982). Random effects models for longitudinal data. *Biometrics* **38**, 963–974.

Lang, J. B. and Agresti, A. (1994). Simultaneous modeling joint and marginal distributions of multivariate categorical responses. *Journal of the American Statistical Association* **89**, 625–632.

Lee, H., Laird, N. M., and Johnston, G. (1999). Combining GEE and REML for estimation of generalized linear models with incomplete multivariate data. Unpublished manuscript.

Liang, K. Y. and Zeger, S. L. (1986). Longitudinal data analysis using generalized linear models. *Biometrika* **73**, 13–22.

Liang, K.-Y., Zeger, S. L., and Qaqish, B. (1992). Multivariate regression analyses for categorical data (with discussion). *Journal of the Royal Statistical Society, Series B* **54**, 3–40.

Lin, D. Y., Wei, L. J., and Ying, Z. (2002). Model-checking techniques based on cumulative residuals. *Biometrics* **58**, 1–12.

Lipshultz, S. E., Easley, K. A., Orav, E. J., Kaplan, S., Starc, T. J., Bricker, J. T., Lai, W.W., Moodie, D. S., McIntosh, K., Schluchter, M. D., and Colan, S. D. (1998). Left ventricular structure and function in children infected with human immunodeficiency virus: The Prospective P2C2 HIV Multicenter Study. Pediatric Pulmonary and Cardiac Complications of Vertically Transmitted HIV Infection (P2C2 HIV) Study Group. *Circulation* **97**, 1246–1256.

Lipshultz, S. E., Easley, K. A., Orav, E. J., Kaplan, S., Starc, T. J., Bricker, J. T., Lai, W.W., Moodie, D. S., Sopko, G., and Colan, S. D. (2000). Cardiac dysfunction and mortality in HIV-infected children: The Prospective P2C2 HIV Multicenter Study. Pediatric Pulmonary and Cardiac Complications of Vertically Transmitted HIV Infection (P2C2 HIV) Study Group. *Circulation* **102**, 1542–1548.

Lipshultz, S. E., Easley, K. A., Orav, E. J., Kaplan, S., Starc, T. J., Bricker, J. T., Lai, W.W., Moodie, D. S., Sopko, G., Schluchter, M. D., and Colan, S. D. (2002). Cardiovascular status of infants and children of women infected with HIV-1 (P^2C^2 HIV): A cohort study. *Lancet* **360**, 368–373.

Lipsitz, S. R. and Fitzmaurice, G. M. (1996). Estimating equations for measures of association between repeated binary responses. *Biometrics* **52**, 903–912.

Lipsitz, S. R., Ibrahim, J. G., and Zhao, L. P. (1999). A weighted estimating equation for missing covariate data with properties similar to maximum likelihood. *Journal of the American Statistical Association* **94**, 1147–1160.

Lipsitz, S. R., Kim, K., and Zhao, L. P. (1994). Analysis of repeated categorical data using generalized estimating equations. *Statistics in Medicine* **13**, 1149–1163.

Lipsitz, S. R., Laird, N. M., and Harrington, D. P. (1990). Maximum likelihood regression methods for paired binary data. *Statistics in Medicine* **9**, 1517–1525.

Lipsitz, S. R., Laird, N. M., and Harrington, D. P. (1991). Generalized estimating equations for correlated binary data: Using the odds ratio as a measure of association. *Biometrika* **78**, 153–160.

Lipsitz, S. R., Laird, N. M., and Harrington, D. P. (1992). A three-stage estimator for studies with repeated and possibly missing binary outcomes. *Applied Statistics* **41**, 203–213.

Lipsitz, S. R., Fitzmaurice, G. M., Orav, E. J., and Laird, N. M. (1994). Performance of generalized estimating equations in practical situations. *Biometrics* **50**, 270–278.

Lipsitz S. R., Molenberghs G., Fitzmaurice G. M., and Ibrahim J. G. (2000). GEE with Gaussian estimation of the correlations when data are incomplete. *Biometrics* **56**, 528–536.

Lumley, T. (1996). Generalized estimating equations for ordinal data: A note on working correlation structures. *Biometrics* **52**, 354–361.

Lunceford, J. K. and Davidian, M. (2004). Stratification and weighting via the propensity score in estimation of causal treatment effects: A comparative study. *Statistics in Medicine* **23**, 2937–2960.

Mancl, L. A. and DeRouen, T. A. (2001). A covariance estimator for GEE with improved small-sample properties. *Biometrics* **57**, 126–134.

McCullagh, P. and Nelder, J. A. (1989). *Generalized Linear Models*, 2nd ed. New York: Chapman & Hall.

Miller, M. E., Davis, C. S., and Landis, J. R. (1993). The analysis of longitudinal polytomous data: Generalized estimating equations and connections with weighted least squares. *Biometrics* **49**, 1033–1044.

Molenberghs, G. and Lesaffre, E. (1994). Marginal modeling of correlated ordinal data using a multivariate Plackett distribution. *Journal of the American Statistical Association* **89**, 633–644.

Mountford, W. K., Lipsitz, S. R., Lackland, D., Fitzmaurice, G. M., and Carter, R. E. (2007). Estimating the variance of estimated trends in proportions when there is no unique subject identifier. *Journal of the Royal Statistical Society, Series A* **170**, 185–193.

Paik, M. C. (1988). Repeated measurement analysis for nonnormal data in small samples. *Communications in Statistics — Simulation and Computation* **17**, 1155–1171.

Paik, M. C. (1997). The generalized estimating equation approach when data are not missing completely at random. *Journal of American Statistical Association* **92**, 1320–1329.

Pan, W. (2001). Akaike's information criterion in generalized estimating equations. *Biometrics* **57**, 120–125.

Pepe, M. S. and Anderson, G. L. (1994). A cautionary note on inference for marginal regression models with longitudinal data and general correlated response data. *Communications in Statistics — Simulation and Computation* **23**, 939–951.

Pregibon, D. (1981). Logistic regression diagnostics, *Annals of Statistics* **9**, 705–724.

Prentice, R. L. (1988). Correlated binary regression with covariates specific to each binary observation. *Biometrics* **44**, 1033–1048.

Preisser, J. S. and Qaqish, B. F. (1996). Deletion diagnostics for generalised estimating equations. *Biometrika* **83**, 551–562.

Qu, A., Lindsay, B. G., and Li, B. (2000) Improving generalised estimating equations using quadratic inference functions. *Biometrika* **87**, 823–836.

Qu, Y., Williams, G. W., Beck, G. J., and Medendorp, S. V. (1992). Latent variable models for clustered dichotomous data with multiple subclusters. *Biometrics* **48**, 1095–1102.

Robins, J. M. (2000). Robust estimation in sequentially ignorable missing data and causal inference models. In *Proceedings of the 1999 Joint Statistical Meetings.*

Robins, J. M., Rotnitzky, A., and Zhao, L. P. (1995). Analysis of semiparametric regression models for repeated outcomes in the presence of missing data. *Journal of the American Statistical Association* **90**, 106–121.

Robins, J. M. and Rotnitzky, A. (2001). Comment on "Inference for semiparametric models: Some questions and an answer," by P. J. Bickel and J. Kwon. *Statistica Sinica* **11**, 920–936.

Rotnitzky A. and Jewell N. P. (1990). Hypothesis testing of regression parameters in semiparametric generalized linear models for cluster correlated data. *Biometrika* **77**, 485–497.

Royall, R. M. (1986). Model robust confidence intervals using maximum likelihood estimators. *International Statistical Review* **54**, 221–226.

Rubin, D. B. (1976). Inference and missing data. *Biometrika* **63**, 581–592.

Scharfstein, D. O., Rotnitzky, A., and Robins, J. M. (1999). Adjusting for nonignorable drop-out using semiparametric nonresponse models. *Journal of the American Statistical Association* **94**, 1096–1120 (with rejoinder, 1135–1146).

Tsiatis, A. A. (1980). A note on a goodness-of-fit test for the logistic regression model. *Biometrika* **67**, 250–251.

van der Laan, M. J. and Robins, J. M. (2003). *Unified Methods for Censored Longitudinal Data and Causality.* New York: Springer.

Wang, Y. and Carey, V. (2003). Working correlation structure misspecification, estimation and covariate design: Implications for generalised estimating equations performance. *Biometrika* **90**, 29–41.

Wang, Y. and Carey, V. (2004). Unbiased estimating equations from working correlation models for irregularly timed repeated measures. *Journal of the American Statistical Association* **99**, 845–853.

Wang, Z. and Louis, T. A. (2003). Matching conditional and marginal shapes in binary mixed-effects models using a bridge distribution function. *Biometrika* **90**, 765–775.

Wedderburn, R. W. M. (1974). Quasi-likelihood functions, generalized linear models, and the Gauss-Newton method. *Biometrika* **61**, 439–447.

White, H. (1982). Maximum likelihood estimation under mis-specified models. *Econometrica* **50**, 1–26.

Wilks, S. S. (1963). Multivariate statistical outliers. *Sankhyā, Series B* **25**, 407–426.

Ye, H. and Pan, J. (2006). Modelling of covariance structures in generalised estimating equations for longitudinal data. *Biometrika* **93**, 927–941.

Zeger, S. L. and Liang, K.-Y. (1986). Longitudinal data analysis for discrete and continuous outcomes. *Biometrics* **42**, 121–130.

Zhao, L. P. and Prentice, R. L. (1990). Correlated binary regression using a quadratic exponential model. *Biometrika* **77**, 642–648.

CHAPTER 4

Generalized linear mixed-effects models

Sophia Rabe-Hesketh and Anders Skrondal

Contents

4.1 Introduction

Generalized linear mixed-effects models, more commonly known as generalized linear mixed models, are very popular in longitudinal data analysis. They are a natural combination of two modeling strands, linear mixed models and generalized linear models. Linear mixed models (e.g., Harville, 1977; Laird and Ware, 1982) are linear regression models that include normally distributed random effects in addition to fixed effects. A natural application is to longitudinal data where the random effects vary between subjects and induce within-subject dependence among repeated measurements after conditioning on observed covariates. Generalized linear models (Nelder and Wedderburn, 1972; Wedderburn, 1974) unify regression models for different response types such as linear models for continuous responses, logistic models for binary responses, and log-linear models for counts.

Generalized linear mixed models are generalized linear models that include multivariate normal random effects in the linear predictor. Early contributions explicitly discussing this idea are Wong and Mason (1985) and West (1985) in frequentist and Bayesian settings, respectively. The term "generalized linear mixed model" appears to have been coined by Gilmour, Anderson, and Rae (1985), although its widespread use is probably due to the highly cited paper by Breslow and Clayton (1993). In social statistics and other areas, generalized linear mixed models are also known as hierarchical, multilevel, or random-coefficient models.

The simplest cases of generalized linear mixed models are random-intercept models where there is a single normally distributed random effect. Random-intercept models are sometimes referred to as variance components, error components, or random-effects models. Early applications of such models include the random-intercept probit models discussed by Lawley (1943), Tucker (1948), and Lord (1952) in psychometrics. Similar models were later considered by Heckman and Willis (1976) and Chamberlain (1980) in econometrics. In statistics, Pierce and Sands (1975) and Anderson and Aitkin (1985) discussed random-intercept logit models, and Hinde (1982) and Brillinger and Preisler (1983) discussed random-intercept Poisson models.

Introducing random effects with conjugate distributions into special cases of generalized linear models has a very long history, including the negative binomial model for counts (Greenwood and Yule, 1920) and the beta-binomial model for proportions (Skellam, 1948). Although these models, which are strictly not generalized linear mixed models, are convenient from a computational point of view, a major limitation is that they cannot be extended to random-coefficient models where the effects of covariates vary randomly between subjects.

The plan of this chapter is as follows. We start by briefly introducing linear mixed models in Section 4.2 before describing generalized linear mixed models and different model formulations in Section 4.3. Specification of the linear predictor is treated in Section 4.4 where we delineate alternative formulations, discuss cross-sectional versus longitudinal effects and endogeneity, and consider the specification of the random part of the models. Parameter interpretation is discussed in Section 4.5 where different kinds of effects are distinguished for the fixed part and measures of heterogeneity and dependence are outlined for the random part. In Section 4.6 we survey several methods for estimating model parameters, focusing on the most commonly used approaches, before discussing different methods for assigning values to the random effects for individual subjects in Section 4.7. To illustrate the use of generalized linear mixed models, we apply them to data from a clinical trial with an ordinal response in Section 4.8. Finally, we close the chapter by considering some useful extensions to generalized linear mixed models.

4.2 Linear mixed models

It is useful to briefly introduce linear mixed models before discussing generalized linear mixed models in more detail. The linear mixed model can be written as

$$\boldsymbol{Y}_i = X_i\boldsymbol{\beta} + Z_i\boldsymbol{b}_i + \boldsymbol{e}_i, \qquad (4.1)$$

where $\boldsymbol{Y}_i$ is an n_i-dimensional vector of responses Y_{ij} for unit i, X_i an $n_i \times p$ matrix of covariates with fixed effects $\boldsymbol{\beta}$, Z_i an $n_i \times q$ matrix of covariates with random effects $\boldsymbol{b}_i$, and $\boldsymbol{e}_i$ a vector of residual errors. For a given occasion j, the linear mixed model can be expressed as

$$Y_{ij} = \boldsymbol{x}'_{ij}\boldsymbol{\beta} + \boldsymbol{z}'_{ij}\boldsymbol{b}_i + e_{ij},$$

where $\boldsymbol{x}'_{ij}$ and $\boldsymbol{z}'_{ij}$ are the jth rows of the corresponding matrices X_i and Z_i, respectively.

We allow the covariate vectors $\boldsymbol{x}_{ij}$ and $\boldsymbol{z}_{ij}$ to be random since it is usually unreasonable to treat them as fixed in non-experimental settings. It is assumed that the covariates are strictly exogenous (e.g., Chamberlain, 1984) in the sense that

$$E(e_{ij}|\boldsymbol{b}_i, X_i, Z_i) = E(e_{ij}|\boldsymbol{b}_i, \boldsymbol{x}_{ij}, \boldsymbol{z}_{ij}) = 0,$$

and

$$E(\boldsymbol{b}_i|X_i, Z_i) = E(\boldsymbol{b}_i) = \boldsymbol{0}.$$

It follows that the conditional expectation of the response, given the random effects and covariates, is

$$\mu_{ij} \equiv E(Y_{ij}|\boldsymbol{b}_i, X_i, Z_i) = E(Y_{ij}|\boldsymbol{b}_i, \boldsymbol{x}_{ij}, \boldsymbol{z}_{ij}) = \boldsymbol{x}_{ij}'\boldsymbol{\beta} + \boldsymbol{z}_{ij}'\boldsymbol{b}_i.$$

Thus, once we have conditioned on $\boldsymbol{b}_i$, $\boldsymbol{x}_{ij}$, and $\boldsymbol{z}_{ij}$, there are no effects of $\boldsymbol{x}_{ij'}$ and $\boldsymbol{z}_{ij'}$ on Y_{ij} for $j' \neq j$. Note that in a slight departure from the notation used in other chapters of this book, μ_{ij} denotes the *conditional* mean of Y_{ij} given $\boldsymbol{b}_i$, rather than the *marginal* mean of Y_{ij} (averaged over the distribution of $\boldsymbol{b}_i$).

The random effects are assumed to have multivariate normal distributions $\boldsymbol{b}_i \sim N(\boldsymbol{0}, G)$ and the residual errors multivariate normal distributions $\boldsymbol{e}_i \sim N(\boldsymbol{0}, V_i)$, both independent across subjects given the covariates. It is furthermore usually assumed that $V_i = \sigma_e^2 I_{n_i}$. In this case, the responses for subject i at different occasions j are conditionally independent, given the covariates and random effects, and have conditional normal distributions

$$f(y_{ij}|\boldsymbol{b}_i, \boldsymbol{x}_{ij}, \boldsymbol{z}_{ij}) = \frac{1}{\sigma_e\sqrt{2\pi}} \exp\left[-\frac{(y_{ij} - \mu_{ij})^2}{2\sigma_e^2}\right].$$

Linear mixed models and related models for longitudinal data are discussed in Verbeke and Molenberghs (2000) and Skrondal and Rabe-Hesketh (2008).

4.3 Generalized linear mixed models

4.3.1 Generalized linear model formulation

In the biostatistical and statistical literatures, the generalized linear mixed model is often written as

$$h^{-1}\{E(\boldsymbol{Y}_i|\boldsymbol{b}_i, X_i, Z_i)\} = X_i\boldsymbol{\beta} + Z_i\boldsymbol{b}_i,$$

where $h^{-1}(\cdot)$ is a vector-valued link function. As for linear mixed models, it is assumed that the random effects are multivariate normal and that the covariates are strictly exogenous. The responses are assumed to be conditionally independent, given the covariates and random effects, and have conditional distributions from the exponential family

$$f(y_{ij}|\boldsymbol{b}_i, \boldsymbol{x}_{ij}, \boldsymbol{z}_{ij}) = \exp[\phi^{-1}\{y_{ij}\theta_{ij} - \psi(\theta_{ij})\} + c(y_{ij}, \phi)].$$

Here ϕ is a dispersion parameter, θ_{ij} is the canonical or natural parameter which is a function of the linear predictor η_{ij},

$$\eta_{ij} = \boldsymbol{x}_{ij}'\boldsymbol{\beta} + \boldsymbol{z}_{ij}'\boldsymbol{b}_i, \tag{4.2}$$

and $\psi(\cdot)$ and $c(\cdot)$ are known functions. The conditional expectation becomes

$$\mu_{ij} = h(\eta_{ij}) = \psi'(\theta_{ij}) \equiv \frac{\partial\psi(\theta_{ij})}{\partial\theta_{ij}}$$

and the conditional variance is given by

$$\text{Var}(Y_{ij}|\eta_{ij}) = \phi\psi''(\theta_{ij}) = \phi V(\mu_{ij}),$$

where $V(\mu_{ij})$ is the variance function.

The following are the most common special cases of generalized linear mixed models:

- Linear mixed model for *continuous* responses:

$$\mu_{ij} = \eta_{ij}$$
$$Y_{ij}|\mu_{ij} \sim N\left(\mu_{ij}, \sigma_e^2\right). \tag{4.3}$$

- Mixed logit or probit model for *dichotomous* responses:

$$\text{logit}(\mu_{ij}) \equiv \log\left(\frac{\mu_{ij}}{1-\mu_{ij}}\right) = \eta_{ij} \quad \text{or} \quad \Phi^{-1}(\mu_{ij}) = \eta_{ij}$$
$$Y_{ij}|\mu_{ij} \sim \text{Bernoulli}(\mu_{ij}), \tag{4.4}$$

where $\Phi^{-1}(\cdot)$ is the inverse standard normal cumulative distribution function or probit link.

- Mixed logit or probit model for *ordinal* responses with categories $s = 0, \ldots, S-1$. Letting γ_{ij}^s denote the conditional cumulative probability

$$\Pr(Y_{ij} > s|\eta_{ij}) \equiv \gamma_{ij}^s,$$

the model can be specified as

$$\text{logit}(\gamma_{ij}^s) = \eta_{ij} - \kappa^{s+1} \quad \text{or} \quad \Phi^{-1}(\gamma_{ij}^s) = \eta_{ij} - \kappa^{s+1}$$
$$\pi_{ij}^s \equiv \Pr(Y_{ij} = s|\eta_{ij}) = \gamma_{ij}^{s-1} - \gamma_{ij}^s$$
$$Y_{ij}|\eta_{ij} \sim \text{Multinomial}\left(1, \left(\pi_{ij}^0, \ldots, \pi_{ij}^{S-1}\right)\right), \tag{4.5}$$

where κ^s, $s = 1, \ldots, S-1$, are model parameters and $\kappa^S = \infty$. When $S = 2$, the model becomes the binary logit or probit model shown in (4.4) with $\gamma_{ij}^0 = \pi_{ij}^1 = \mu_{ij}$ and $\kappa^1 = 0$. If the responses are frequencies of the outcomes of n_B "trials," μ_{ij} becomes the expected proportion (not the expected frequency) and the conditional distribution of the frequencies is Multinomial$(n_B, (\pi_{ij}^0, \ldots, \pi_{ij}^{S-1}))$.

- Mixed Poisson model for *counts*:

$$\log(\mu_{ij}) = \eta_{ij}$$
$$Y_{ij}|\mu_{ij} \sim \text{Poisson}(\mu_{ij}). \tag{4.6}$$

In all the models above except linear mixed models, the dispersion parameter ϕ is known *a priori*. Given the random effects and covariates, the variance of the responses is therefore a function of the conditional expectation only. For Poisson, binomial, or multinomial models (with $n_B > 1$), the model-implied variance function may be violated by the data. In this case, quasi-likelihood approaches (see Section 4.6.2) can be used to estimate the dispersion parameter. In likelihood or Bayesian methods, overdispersion is instead induced by including additional random effects varying over both occasions and subjects (see Section 4.4.3).

Generalized linear mixed models are often written as non-linear models with an error term (e.g., Goldstein, 2003), perhaps because the popular penalized quasi-likelihood algorithm described in Section 4.6.2 relies on such a formulation. For instance, mixed logit or probit models for binary responses are written as

$$Y_{ij} = \pi_{ij} + \epsilon_{ij} z_{ij}^{(1)}, \qquad \pi_{ij} = h(\eta_{ij}), \qquad z_{ij}^{(1)} = \sqrt{\pi_{ij}(1-\pi_{ij})}.$$

Constraining the variance of ϵ_{ij} to one, we obtain the required Bernoulli variance $\pi_{ij}(1-\pi_{ij})$. However, this formulation is awkward because it requires a very peculiar distribution of ϵ_{ij} to produce valid 0 or 1 responses. More importantly, the formulation gives the false impression that the variance of ϵ_{ij} could be estimated and should therefore be avoided (see Skrondal and Rabe-Hesketh, 2007).

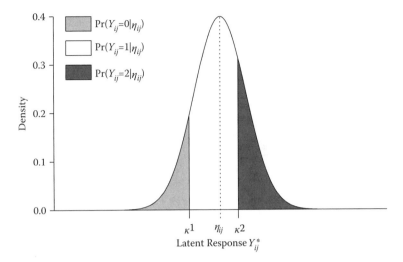

Figure 4.1 Latent-response formulation for ordinal response with $S=3$ categories.

4.3.2 Latent-response formulation

When the responses are dichotomous or ordinal, it is useful to think of an underlying or latent continuous response Y_{ij}^* that is partitioned or "categorized" into the observed discrete response using thresholds κ^s. In this case, Y_{ij} takes the value s $(s = 0, \ldots, S-1)$ if $\kappa^s < Y_{ij}^* \leq \kappa^{s+1}$, with $\kappa^0 = -\infty$ and $\kappa^S = \infty$. This latent-response formulation is illustrated for $S=3$ categories and $Y_{ij}^* \sim N(\eta_{ij}, \sigma_e^2)$ in Figure 4.1. The response probabilities are given by the areas under the normal density curve with mean η_{ij} and variance σ_e^2,

$$\Pr(Y_{ij} = s | \eta_{ij}) = \Phi\left(\frac{\kappa^{s+1} - \eta_{ij}}{\sigma_e}\right) - \Phi\left(\frac{\kappa^s - \eta_{ij}}{\sigma_e}\right),$$

where $\Phi(\cdot)$ is the standard normal cumulative distribution function.

It is evident that identification restrictions must be imposed because we can add a constant to the thresholds and add the same constant to the linear predictor without changing the probabilities. Furthermore, we can multiply the numerator and denominator by the same constant without changing the probabilities. To ensure identification, the constraints $\beta_0 = 0$ and $\sigma_e = 1$ are typically imposed. The cumulative probability that the response exceeds category s then becomes

$$\gamma_{ij}^s = \Phi(\eta_{ij} - \kappa^{s+1}),$$

giving the cumulative probit model for ordinal responses in (4.5).

An advantage of this latent-response formulation is that a *linear* mixed model can be specified for the latent continuous responses Y_{ij}^* underlying the observed responses Y_{ij},

$$Y_{ij}^* = \boldsymbol{x}_{ij}'\boldsymbol{\beta} + \boldsymbol{z}_{ij}'\boldsymbol{b}_i + e_{ij}.$$

Grouped or interval censored continuous responses can be modeled in the same way by constraining the threshold parameters to equal the limits of the censoring intervals. By allowing subject-specific right-censoring, this approach can be used for discrete time survival analysis (e.g., Rabe-Hesketh, Yang, and Pickles, 2001).

If a logistic distribution is specified for e_{ij}, we obtain a logit model. Using a Gumbel distribution produces the complementary log-log link which is sometimes used in discrete time survival analysis. Gumbel distributions can also be used to specify models for comparative responses such as discrete choices, rankings, or pairwise comparisons (e.g., Skrondal and Rabe-Hesketh, 2003).

While equivalent to the generalized linear formulation, the latent-response formulation is useful for interpretation (see Section 4.5), investigation of identification (Rabe-Hesketh and Skrondal, 2001), and for estimation (see Section 4.6.3).

4.4 Specification of the linear predictor

4.4.1 Reduced form and two-stage formulation

Here we discuss in more detail how the linear predictor in (4.2) is usually specified, either directly, or indirectly via a two-stage formulation. The direct approach is sometimes called the "reduced form," as will become apparent when the two-stage formulation is discussed.

Reduced form

The vector of covariates $\boldsymbol{x}_{ij}$, having fixed coefficients, often includes both subject-specific (or time-constant) and time-varying covariates. The simplest type of generalized linear mixed model is a random-intercept model where $\boldsymbol{z}_{ij} = 1$ with corresponding random intercept b_{i0}. The random intercept can be interpreted as the effect of all unobserved subject-specific variables on the linear predictor. So-called random-coefficient models also include random slopes of time-varying covariates, interpretable as interactions of unobserved subject-specific covariates with observed time-varying covariates. It is also possible to include a random coefficient of a time-constant covariate z_i to produce a heteroscedastic random intercept $b_{i0} + b_{i1}z_i$ (e.g., Snijders and Bosker, 1999). The variables in $\boldsymbol{z}_{ij}$ are in practice a subset of the variables in $\boldsymbol{x}_{ij}$ so that the corresponding subset of $\boldsymbol{\beta}$ represents the expectations (over subjects) of the effects of the $\boldsymbol{z}_{ij}$ on the scale of the linear predictor.

Two-stage formulation

Some researchers advocate a two-stage formulation. For instance, Raudenbush and Bryk (2002) first specify a within-subject or level-1 model, where the occasion-specific linear predictor is a function of time-varying covariates $\boldsymbol{v}_{ij}$,

$$\eta_{ij} = \boldsymbol{v}'_{ij}\boldsymbol{\delta}_i.$$

They then specify a between-subject or level-2 model for the coefficients $\boldsymbol{\delta}_i$,

$$\boldsymbol{\delta}_i = \Gamma \boldsymbol{w}_i + \boldsymbol{b}_i,$$

where $\boldsymbol{w}_i$ is a vector of time-constant covariates with corresponding regression coefficient matrix Γ. Usually, some elements of $\boldsymbol{\delta}_i$ are constant and some depend only on observed covariates. The reduced-form model is obtained by substituting the level-2 model into the level-1 model giving the mixed-effects model in (4.2).

We illustrate the two-stage formulation by considering a growth curve model with a random intercept, a random slope of time t_{ij}, fixed slopes for a time-varying covariate v_{ij}, and a time-constant covariate w_i. The within-subject model is

$$\eta_{ij} = \delta_{i0} + \delta_{i1}t_{ij} + \delta_{i2}v_{ij},$$

and the between-subject models are

$$\delta_{i0} = \gamma_{00} + \gamma_{01}w_i + b_{i0},$$

$$\delta_{i1} = \gamma_{10} + \gamma_{11}w_i + b_{i1},$$

$$\delta_{i2} = \gamma_{20}.$$

Substituting the level-2 models into the level-1 model produces the reduced form with a so-called "cross-level interaction" term $\gamma_{11}w_i t_{ij}$.

Although the two formulations (two-stage and reduced form) are equivalent, the choice of formulation can have an impact on the types of models being considered. For instance, the two-stage formulation encourages inclusion of many cross-level interactions and few same-level interactions. In contrast, interactions are more rarely included in the reduced-form formulation.

Due to the abundance of interactions in the two-stage formulation, there is a tradition of grand-mean centering all variables (except time) by subtracting the sample mean (across occasions and subjects) to make the coefficients more interpretable. In the example above, grand-mean centering of w_i makes γ_{10} interpretable as the mean effect of t_{ij} when w_i takes its mean value.

Note that the two-stage formulation described above differs from that of Laird and Ware (1982) where (4.1) is specified at stage 1 and only *random* between-subject variation in terms of the multivariate normal distribution of $\boldsymbol{b}_i$ is specified at stage 2.

4.4.2 Longitudinal effects, cross-sectional effects, and endogeneity

In longitudinal studies, it is often said that "subjects serve as their own controls" when considering the effects of time-varying covariates. This seems to imply that all subject-level observed and unobserved covariates have been controlled for. Unfortunately, this is not true in generalized linear mixed models since omitted subject-level covariates may correlate and hence be confounded with the time-varying variables of interest.

For simplicity, consider a linear random-intercept model with a single time-varying covariate z_{ij} and a single unobserved subject-specific covariate w_i. If the omitted covariate w_i is correlated with z_{ij}, the effect of z_{ij} is confounded and there will be omitted variable bias. In this case, z_{ij} is said to be endogenous or correlated with the random intercept b_{i0} since the random intercept represents the effects of all unobserved subject-level covariates including w_i.

Note that w_i cannot be correlated with z_{ij} if $\bar{z}_{i\cdot}$, the mean of z_{ij} over time for subject i, is the same for all subjects. We can therefore avoid omitted subject-level variable bias by subject-mean centering z_{ij}, that is, by subtracting the subject means from the original variable, forming the instrumental variable $z_{ij}^d = z_{ij} - \bar{z}_{i\cdot}$. The regression coefficient of z_{ij}^d can then be interpreted as a purely within-subject or longitudinal effect. The subject mean itself should also be included since it is unrealistic to assume that a covariate has a within-subject effect but no between-subject or cross-sectional effect,

$$Y_{ij} = \beta_0 + \beta_1(z_{ij} - \bar{z}_{i\cdot}) + \beta_2 \bar{z}_{i\cdot} + b_{i0} + e_{ij}$$
$$= \beta_0 + \beta_1 z_{ij} + (\beta_2 - \beta_1)\bar{z}_{i\cdot} + b_{i0} + e_{ij}.$$

A test of the null hypothesis $\beta_1 = \beta_2$ can be used to assess the exogeneity of z_{ij} (in the sense that $E(b_{0i}|z_{ij}) = 0$), which is equivalent to the Durbin–Wu–Hausman test (e.g., Hausman, 1978). Neuhaus and Kalbfleisch (1998) advocate using this decomposition in both linear and logistic mixed models to avoid inconsistency due to model misspecification. Diggle et al. (2002) consider a similar decomposition but use the covariate value at baseline z_{i1} instead of $\bar{z}_{i\cdot}$ to distinguish between longitudinal and cross-sectional effects.

In the two-stage formulation, it is also common practice to subtract the subject-specific mean from the time-varying covariates (known as "group-mean centering"). One motivation for this is to make the random intercept δ_{i0} in the level-1 model interpretable as the adjusted mean response (for subject i), instead of the mean response when the original covariates are zero. Unfortunately, the cluster mean itself is often not included in the level-2 model, leading to potential confounding with the effects of other subject-level variables.

If only longitudinal or within-subject effects are of interest, fixed subject-specific intercepts can be used instead of random intercepts. In models for dichotomous responses, estimating these intercepts along with the regression coefficients of interest leads to inconsistency for the latter, known as the incidental parameter problem (Neyman and Scott, 1948). For binary logistic models, this can be overcome by using conditional maximum likelihood estimation, conditioning on the sum of the responses for each subject (e.g., Rasch, 1960; Breslow and Day, 1980). Note that such fixed-effects models are not generalized linear mixed models.

4.4.3 Specification of the random part

The covariance matrix G of the random effects is not invariant to translation of the covariates in z_{ij}. Specifically, translating any of these variables changes the interpretation (and estimates) of the random-intercept variance and the covariances or correlations between the random intercept and other random coefficients. This is because the intercept refers to the value of the linear predictor when the covariates are zero.

The lack of translation invariance is illustrated in Figure 4.2 for a linear model with a random intercept and random slope of a single covariate z_{ij}. There is a positive correlation between the intercepts and slopes of the subject-specific regression lines (see dotted line). However, translating z_{ij} by subtracting 4 (as shown in the top axis) yields a model with a negative correlation (see dashed line). Also note that the magnitude of the random-intercept variance changes after translation of z_{ij}.

Translation of z_{ij} can be counteracted by appropriate changes in the covariance matrix G yielding an equivalent model. Importantly, this is no longer the case if a diagonal covariance matrix is specified, and G should therefore generally be left unrestricted.

In linear mixed models, the conditional within-subject covariance matrix V_i (given X_i, Z_i, and b_i) is usually specified as $V_i = \sigma_e^2 I_{n_i}$. However, in longitudinal settings, it sometimes makes sense to allow for serial dependence among the residuals, for instance using a first-order autoregressive structure. For probit models such a correlation structure can also be used for the residuals in the latent-response formulation. In particular, Heckman (1978, 1981) discusses a very general class of probit models with dependent residuals and current responses regressed on previous latent responses as well as observed responses.

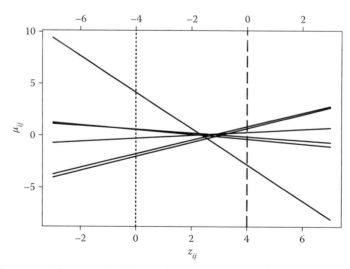

Figure 4.2 Subject-specific regression lines for linear random-coefficient model, illustrating lack of invariance under translation of explanatory variable. (Source: Skrondal and Rabe-Hesketh, 2004.)

In Poisson models for counts or binomial (or multinomial) models for proportions (with $n_B > 1$), a random intercept $u_{ij} \sim N(0, d)$ varying between both occasions and subjects is sometimes introduced in the linear predictor to produce overdispersion. For instance, for log-linear Poisson models, the conditional variance of the response given the subject-level random effects $\boldsymbol{b}_i$ (and the covariates) becomes

$$\text{Var}(Y_{ij}|\boldsymbol{b}_i, \boldsymbol{x}_{ij}, \boldsymbol{z}_{ij}) = \mu_{ij}\{1 + \mu_{ij}[\exp(d) - 1]\},$$

where $\mu_{ij} = E(Y_{ij}|\boldsymbol{b}_i, \boldsymbol{x}_{ij}, \boldsymbol{z}_{ij})$, not conditioning on u_{ij}. This variance is larger than the ordinary conditional Poisson variance except when $d = 0$.

4.5 Interpretation of model parameters

4.5.1 Fixed part: conditional and marginal relationships

As we have seen, generalized linear mixed models are expressed by combining a conditional response distribution $f(y_{ij}|\boldsymbol{b}_i, \boldsymbol{x}_{ij}, \boldsymbol{z}_{ij})$, given the random effects $\boldsymbol{b}_i$ and the covariates, with a multivariate normal random-effects distribution $\varphi(\boldsymbol{b}_i; \boldsymbol{0}, G)$ with mean zero and covariance matrix G. The covariates and random effects determine the person-specific or conditional mean μ_{ij} and the regression coefficients $\boldsymbol{\beta}$ can therefore be interpreted as subject-specific or conditional effects of covariates $\boldsymbol{x}_{ij}$, given the random effects $\boldsymbol{b}_i$. The marginal expectation for occasion j (not conditioning on $\boldsymbol{b}_i$ but still on $\boldsymbol{x}_{ij}$ and $\boldsymbol{z}_{ij}$) can be obtained as

$$E(Y_{ij}|\boldsymbol{x}_{ij}, \boldsymbol{z}_{ij}) = \int \varphi(\boldsymbol{b}_i; \boldsymbol{0}, G) \, h(\boldsymbol{x}'_{ij}\boldsymbol{\beta} + \boldsymbol{z}'_{ij}\boldsymbol{b}_i) \, d\boldsymbol{b}_i.$$

Conditional effects express comparisons holding the subject-specific random effects (and covariates) constant. For this reason, conditional effects are sometimes referred to as "subject-specific effects." In contrast, marginal effects express comparisons of entire subpopulations or population strata defined by covariate values and are sometimes referred to as "population-averaged effects."

In linear mixed models $h(\cdot)$ is the identity function so that the marginal expectation is simply $\boldsymbol{x}'_{ij}\boldsymbol{\beta}$. The parameters $\boldsymbol{\beta}$ can therefore be interpreted as either conditional or marginal effects. Under the assumptions stated in Section 4.2, the marginal distribution is multivariate normal with covariance matrix

$$\Sigma_i \equiv \text{Cov}(\boldsymbol{Y}_i|X_i, Z_i) = Z_i G Z'_i + \sigma_e^2 I. \tag{4.7}$$

Importantly, the conditional and marginal effects differ for other link functions. This is most easily seen for the probit link by using the latent-response formulation described in Section 4.3.2. The model for the latent response can be written as

$$Y^*_{ij} = \boldsymbol{x}'_{ij}\boldsymbol{\beta} + \xi_{ij}, \qquad \xi_{ij} = \boldsymbol{z}'_{ij}\boldsymbol{b}_i + e_{ij},$$

where ξ_{ij} is the "total residual" or random part (conditioning on $\boldsymbol{x}_{ij}$ and $\boldsymbol{z}_{ij}$) of the model with variance

$$\sigma_{i,jj} \equiv \boldsymbol{z}'_{ij} G \boldsymbol{z}_{ij} + \sigma_e^2.$$

The marginal cumulative response probabilities become

$$\Pr(Y_{ij} > s|\boldsymbol{x}_{ij}, \boldsymbol{z}_{ij}) = \Pr(Y^*_{ij} > \kappa^{s+1}|\boldsymbol{x}_{ij}, \boldsymbol{z}_{ij}) = \Pr(\boldsymbol{x}'_{ij}\boldsymbol{\beta} + \xi_{ij} > \kappa^{s+1}|\boldsymbol{x}_{ij}, \boldsymbol{z}_{ij})$$

$$= \Pr(-\xi_{ij} \leq \boldsymbol{x}'_{ij}\boldsymbol{\beta} - \kappa^{s+1}|\boldsymbol{x}_{ij}, \boldsymbol{z}_{ij})$$

$$= \Pr\left(\frac{\xi_{ij}}{\sqrt{\sigma_{i,jj}}} \leq \frac{\boldsymbol{x}'_{ij}\boldsymbol{\beta} - \kappa^{s+1}}{\sqrt{\sigma_{i,jj}}} \,\Bigg|\, \boldsymbol{x}_{ij}, \boldsymbol{z}_{ij}\right)$$

$$= \Phi\left(\frac{\boldsymbol{x}'_{ij}\boldsymbol{\beta} - \kappa^{s+1}}{\sqrt{\sigma_{i,jj}}}\right).$$

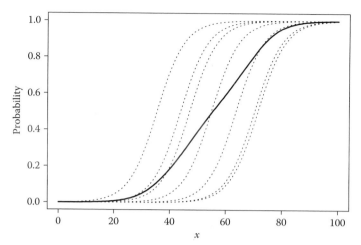

Figure 4.3 Conditional relationships (dotted curves) and marginal relationship (solid curve) for a random-intercept logistic model.

Here the probit link is preserved for the marginal probabilities but with different regression coefficients; the marginal effects $\boldsymbol{\beta}/\sqrt{\sigma_{i,jj}}$ are attenuated, or closer to zero, compared to the conditional effects $\boldsymbol{\beta}$.

For the log link, it can be shown that the link function is also preserved with

$$E(Y_{ij}|\boldsymbol{x}_{ij}, \boldsymbol{z}_{ij}) = \exp(\boldsymbol{x}'_{ij}\boldsymbol{\beta} + \boldsymbol{z}'_{ij}G\boldsymbol{z}_{ij}/2).$$

Here the covariates without random coefficients have the same marginal and conditional effects. In a random-intercept model, only the intercept changes after marginalization, increasing from β_0 to $\beta_0+g/2$.

Apart from the probit and log link, the link function for the marginal model is generally different from that for the conditional model and usually does not have a simple form. If the marginal relationship is approximated by the same link function (as for the conditional model), then the corresponding regression coefficients are attenuated compared to the conditional ones for the logit and complementary log-log links. This is illustrated for a random-intercept logistic regression model with a single covariate in Figure 4.3, where the average curve resembles a logistic curve with a smaller regression coefficient. See Ritz and Spiegelman (2004) and McCulloch (2008) for further discussion of marginal versus conditional relationships.

For linear mixed models, the maximum likelihood estimators of the regression coefficients are consistent under misspecification of the random part of the model. However, this is not the case for other generalized linear mixed models. Loosely speaking, this is because the specification of the random part determines the relationship between the conditional and marginal effects, and the latter are directly "fitted" to the data. This lack of robustness to the dependence structure is commonly used as a reason for estimating marginal effects directly using generalized estimating equations (see Chapter 3). Note that marginal effects can also be estimated directly using random-effects models. For instance, Heagerty and Zeger (2000) consider random-coefficient models with a standard link function for the marginal relationship (and a non-standard conditional relationship) so that the parameters have marginal interpretations. Wang and Louis (2004) suggest a non-standard random-effects distribution which produces matching marginal and conditional relationships.

For models that can be expressed using the latent-response formulation, the regression coefficients $\boldsymbol{\beta}$ can be directly interpreted as effects on the expected latent response. In random-intercept models, the magnitude of a coefficient can be assessed by dividing it by the estimated total residual standard deviation, $\sqrt{\widehat{g}+1}$ for probit models and $\sqrt{\widehat{g}+\pi^2/3}$ for logit models. The resulting ("half-standardized") estimates will be comparable between logit and probit models, as well as between marginal ($g=0$) and conditional models.

4.5.2 Random part

Measures of heterogeneity

An obvious way of interpreting estimated standard deviations of the random effects is to produce percentiles of the effects based on the normality assumption. For instance, for a covariate with an estimated fixed coefficient $\widehat{\beta}_k$ and random coefficient b_{ik}, form the approximate 2.5th and 97.5th percentiles using $\widehat{\beta}_k \pm 1.96\sqrt{\widehat{g}_{kk}}$ to express the range of effects that are likely to occur. In a random-intercept model, $\widehat{\beta}_0 \pm 1.96\sqrt{\widehat{g}_{00}}$ gives a range of intercepts which can be converted to a range of conditional expectations μ_{ij} (e.g., probabilities for dichotomous responses) by plugging particular covariate values into the linear predictor and applying the inverse link function. Duchateau and Janssen (2005) suggest plotting the entire density of μ_{ij} for given covariate values. The same approach can be used in random-coefficient models by using the standard deviation of the $z_{ij}'b_i$ at particular values of z_{ij}.

For logit models with normally distributed random intercepts, Larsen et al. (2000) and Larsen and Merlo (2005) suggest a useful measure of heterogeneity. They consider repeatedly sampling two subjects with the same covariate values and forming the odds ratio comparing the subject with the larger random intercept to the other subject. Larsen *et al.* then express heterogeneity as the median of these odds ratios. The median and other percentiles can be obtained from the following cumulative distribution function (for $a > 1$):

$$\widehat{\Pr}(\exp(|b_i - b_{i'}|) \le a) = \widehat{\Pr}\left(\frac{|b_i - b_{i'}|}{\sqrt{2\widehat{g}}} \le \frac{\ln(a)}{\sqrt{2\widehat{g}}}\right) = 2\Phi\left(\frac{\ln(a)}{\sqrt{2\widehat{g}}}\right) - 1.$$

Measures of dependence

Since unobserved between-subject heterogeneity induces within-subject dependence, as shown for a linear mixed model in (4.7), we can interpret the random part using measures of dependence. In the simple case of a linear random-intercept model, the residual correlation between the responses at any two occasions j and j' becomes

$$\rho \equiv \text{Corr}(Y_{ij}, Y_{ij'}|X_i, Z_i) = \frac{g}{g + \sigma_e^2},$$

also known as the (residual) intraclass correlation. This can also be interpreted as the proportion of the total residual variance that is due to unobserved between-subject heterogeneity. In linear random-coefficient models, the covariance matrix of the total residuals is a function of covariates, and there is no simple summary measure.

For generalized linear mixed models, the correlations $\text{Corr}(Y_{ij}, Y_{ij'}|X_i, Z_i)$ generally depend on covariates even if only a random intercept is included (e.g., Browne et al., 2005) and are therefore not useful measures of dependence. Fortunately, the correlations of the latent responses $\text{Corr}(Y_{ij}^*, Y_{ij'}^*|X_i, Z_i)$ in logit and probit random intercept models are constant and obtained by simply substituting either 1 (probit) or $\pi^2/3$ (logit) for σ_e^2 above.

4.6 Estimation

In this section we focus on the most commonly used estimation methods for generalized linear mixed models: maximum likelihood, penalized quasi-likelihood, and Markov chain Monte Carlo. These and other estimation methods are treated in more detail in Skrondal and Rabe-Hesketh (2004, Chapter 6).

4.6.1 Maximum likelihood estimation

The marginal log-likelihood for a two-level generalized linear mixed model can be written as

$$\ell(\boldsymbol{\beta}, G) = \ln \prod_{i=1}^{N} \int \varphi(\boldsymbol{b}_i; \boldsymbol{0}, G) \prod_{j=1}^{n_i} f(y_{ij}|\boldsymbol{b}_i)\, d\boldsymbol{b}_i,$$

where $\varphi(\boldsymbol{b}_i; \boldsymbol{0}, G)$ is the multivariate normal density with mean $\boldsymbol{0}$ and covariance matrix G. To simplify notation, we have suppressed explicit conditioning on $\boldsymbol{x}_{ij}$ and $\boldsymbol{z}_{ij}$ in the conditional response distributions $f(y_{ij}|\boldsymbol{b}_i)$.

It is useful to change the variables of integration to independent standard normally distributed random effects $\boldsymbol{v}_i$, using for instance the Cholesky decomposition Q of the covariance matrix G so that $\boldsymbol{b}_i = Q\boldsymbol{v}_i$. The log-likelihood can then be written as

$$\ell(\boldsymbol{\beta}, G) = \ln \prod_{i=1}^{N} \int_{-\infty}^{\infty} \phi(v_{iq}) \dots \left[\int_{-\infty}^{\infty} \phi(v_{i1}) \prod_{j=1}^{n_i} f(y_{ij}|\boldsymbol{v}_i) dv_{i1} \right] \dots dv_{iq},$$

where $\phi(\cdot)$ is the univariate standard normal density function. Each unidimensional integral can be approximated by Gauss–Hermite quadrature as

$$\int_{-\infty}^{\infty} \phi(v_{i1}) \prod_{j=1}^{n_i} f(y_{ij}|\boldsymbol{v}_i) dv_{i1} \approx \sum_{r=1}^{R} p_r \prod_{j=1}^{n_i} f(y_{ij}|(a_r, v_{i2}, \dots, v_{iq})), \qquad (4.8)$$

where $\sqrt{\pi} p_r$ and $a_r/\sqrt{2}$ are the weights and locations of the $(2R-1)$th-degree Gauss–Hermite quadrature rule (e.g., Stroud and Secrest, 1966). This is illustrated for a five-point or ninth-degree quadrature rule in the top panel of Figure 4.4. The approximation becomes more accurate as more quadrature points R are used. Unfortunately, using a large number of points can be computationally expensive since the number of terms required to evaluate each multivariate integral is R^q. Instead of using unidimensional rules for each dimension (so-called Cartesian product quadrature) as illustrated for $q=2$ in the top panel of Figure 4.5, multidimensional, so-called spherical quadrature rules can achieve higher accuracy with fewer points (e.g., Rabe-Hesketh, Skrondal, and Pickles, 2005).

Gaussian quadrature works well if the product of conditional response distributions in Equation (4.8) is well approximated by a low-degree polynomial in v_{i1}. However, in practice a large number of quadrature points is often required to approximate the likelihood. This will be the case for large n_i and/or large intraclass correlations and/or when the conditional response distributions are normal or Poisson (e.g., Lesaffre and Spiessens, 2001; Rabe-Hesketh, Skrondal, and Pickles, 2002). The integrand can then have a very sharp peak between adjacent quadrature points a_r and a_{r+1} as shown in the top panel of Figure 4.4.

Adaptive quadrature largely overcomes these problems. For simplicity we consider a random-intercept model with integrand

$$\phi(v_i) \prod_{j=1}^{n_i} f(y_{ij}|v_i)$$

which is proportional to the posterior density of v_i given $\boldsymbol{Y}_i$. The posterior can often be well approximated by a normal density $\varphi(v_j; \mu_j, \tau_j^2)$ with cluster-specific mean μ_j and variance τ_j^2. Instead of treating the random-effects density as the "weight function" when applying the quadrature rule as in (4.8), we therefore rewrite the integral as

$$\int_{-\infty}^{\infty} \phi(v_i) \prod_{j=1}^{n_i} f(y_{ij}|v_i) dv_i = \int_{-\infty}^{\infty} \varphi(v_i; \mu_i, \tau_i^2) \left[\frac{\phi(v_i) \prod_{j=1}^{n_i} f(y_{ij}|v_i)}{\varphi(v_i; \mu_i, \tau_i^2)} \right] dv_i$$

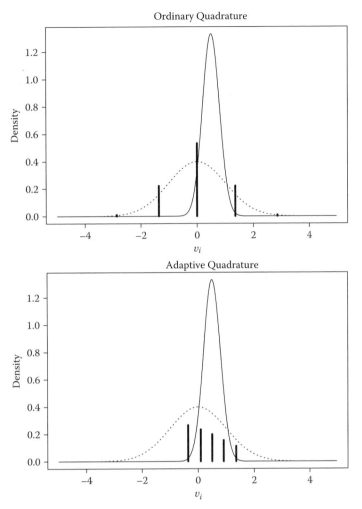

Figure 4.4 Prior (dotted curve) and posterior (solid curve) densities and quadrature weights (bars) for ordinary quadrature and adaptive quadrature in one dimension. Note that the integrand is proportional to the posterior density. (Source: Rabe-Hesketh, Skrondal, and Pickles, 2002.)

and treat the normal density approximating the posterior density as the weight function (Rabe-Hesketh, Skrondal, and Pickles, 2005). This leads to the approximation

$$\int_{-\infty}^{\infty} \phi(v_i) \prod_{j=1}^{n_i} f(y_{ij}|v_i)dv_i \approx \sum_{r=1}^{R} p_{ir} \prod_{j=1}^{n_i} f(y_{ij}|a_{ir}),$$

where the subject-specific locations and weights are given by

$$a_{ir} \equiv \tau_i a_r + \mu_i,$$

and

$$p_{ir} \equiv \sqrt{2\pi} \, \tau_i \, \exp\left(a_r^2/2\right) \, \phi(\tau_i a_r + \mu_i) \, p_r.$$

The superiority of adaptive quadrature can be seen in Figure 4.4 which illustrates for $R=5$ how adaptive quadrature translates and scales the locations so that they lie directly under the integrand. Figure 4.5 illustrates the two-dimensional adaptive quadrature transformation for a grid of quadrature points. Simulations reported in Rabe-Hesketh, Skrondal, and Pickles (2005) show that adaptive quadrature performs very well for a wide range of cluster sizes and intraclass correlations.

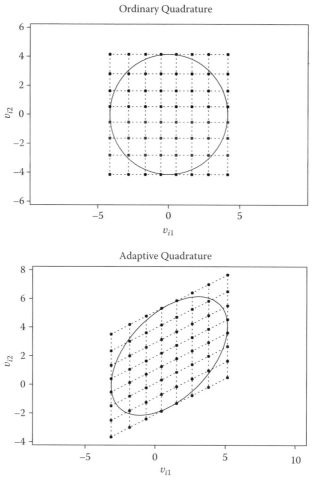

Figure 4.5 Locations for ordinary and adaptive quadrature using two-dimensional Cartesian rules, where $R=8$, $\mu_1=1$, $\mu_2=2$, $\tau_1=\tau_2=1$ and the posterior correlation is 0.5. (Source: Rabe-Hesketh, Skrondal, and Pickles, 2005.)

Different variants of Monte Carlo integration are also sometimes used (e.g., Meng and Schilling, 1996; McCulloch, 1997; Ng et al., 2006). For probit models, the Geweke–Hajivassiliou–Keane simulator (e.g., Hajivassiliou and Ruud, 1994) can be used to integrate over the latent responses.

To obtain maximum likelihood estimates, numerical or Monte Carlo integration is combined with optimization algorithms such as Newton–Raphson, Fisher scoring, or expectation–maximization (EM).

Software implementing maximum likelihood estimation of generalized linear mixed models using adaptive quadrature includes the Stata programs `xtlogit`, `xtmelogit`, etc., and `gllamm` (Rabe-Hesketh, Skrondal, and Pickles, 2004b; Rabe-Hesketh and Skrondal, 2008) and SAS `PROC NLMIXED` (Wolfinger, 1999; Molenberghs and Verbeke, 2005).

4.6.2 Quasi-likelihood methods

Marginal quasi-likelihood (MQL) and penalized quasi-likelihood (PQL) methods (e.g., Goldstein, 1991; Breslow and Clayton, 1993) are analogous to iteratively reweighted least squares for generalized linear models in that the model is linearized (here to a linear mixed model) in each iteration. Using estimates β^k, G^k, V^k, and $\boldsymbol{b}_i^k$ from the "current" iteration k, the model for y_{ij} is linearized by expanding $h(\eta_{ij})$ as a first-order Taylor series around

$$\eta_{ij}^k = \boldsymbol{x}_{ij}'\boldsymbol{\beta}^k + \boldsymbol{z}_{ij}'\boldsymbol{b}_i^k, \tag{4.9}$$

giving

$$Y_{ij} \approx h(\eta_{ij}^k) + \boldsymbol{x}_{ij}'[\boldsymbol{\beta} - \boldsymbol{\beta}^k]h'(\eta_{ij}^k) + \boldsymbol{z}_{ij}'[\boldsymbol{b}_i - \boldsymbol{b}_i^k]h'(\eta_{ij}^k) + \epsilon_{ij}. \tag{4.10}$$

Here, ϵ_{ij} is a heteroscedastic error term with variance $\phi V(\mu_{ij}^k)$. Note that this expression is linear in the unknown parameters $\boldsymbol{\beta}$. The sum of the terms involving known current values $\boldsymbol{\beta}^k$ and $\boldsymbol{b}_i^k$ is treated as an offset o_{ij},

$$o_{ij} = h(\eta_{ij}^k) - h'(\eta_{ij}^k)\boldsymbol{x}_{ij}'\boldsymbol{\beta}^k - h'(\eta_{ij}^k)\boldsymbol{z}_{ij}'\boldsymbol{b}_i^k,$$

and the terms involving random effects $\boldsymbol{b}_i$ contribute to the total residual ξ_{ij},

$$\xi_{ij} = h'(\eta_{ij}^k)\boldsymbol{z}_{ij}'\boldsymbol{b}_i + \epsilon_{ij},$$

giving

$$Y_{ij} = o_{ij} + h'(\eta_{ij}^k)\boldsymbol{x}_{ij}'\boldsymbol{\beta} + \xi_{ij}.$$

Multiplying $\boldsymbol{x}_{ij}$ by $h'(\eta_{ij}^k)$, $\boldsymbol{\beta}^{k+1}$ can be obtained by using generalized least squares based on the model-implied covariance matrix Σ^k of the total residuals ξ_{ij}. The parameters of the random part are updated by fitting the model-implied covariance matrix Σ^{k+1} to the sample covariance matrix of the estimated total residuals; see Goldstein (2003, Appendix 2.1) for details.

There are several variants of this algorithm (Goldstein, 1991, 2003; Longford, 1993, 1994). In MQL $\boldsymbol{b}_i^k$ is set to zero in (4.9) and (4.10), whereas the expansion is improved by setting the random effects equal to the posterior modes based on the linearized model in PQL. Hence, the difference between MQL and PQL is in the offset used. Since MQL sets the random effects to zero, the fixed-effects estimates are essentially marginal effects which are attenuated relative to the required conditional effects as discussed in Section 4.5.1.

The algorithms have been improved considerably by using a second-order Taylor expansion in the random effects (Goldstein and Rasbash, 1996); see Goldstein (2003, Appendix 4.1) for details. This method is referred to as PQL-2 in contrast to the first-order version PQL-1. PQL is computationally very efficient since numerical integration is avoided. Furthermore, the approach can be used for models with crossed random effects (e.g., Breslow and Clayton, 1993; Goldstein, 1987), and (spatially or temporally) autocorrelated random effects (e.g., Breslow and Clayton, 1993).

Quasi-likelihood methods work well when the conditional distribution of the responses given the random effects is close to normal, for example for a Poisson distribution with mean 5 or greater (Breslow, 2005) or 7 or greater (McCulloch and Searle, 2001) or if the responses are proportions with large binomial denominators n_B. The methods also work well if the conditional joint distributions of the responses belonging to each cluster are nearly normal or, equivalently, if the posterior distribution of the random effects is nearly normal. Even for dichotomous responses, this becomes increasingly the case as the cluster sizes increase. However, PQL performs poorly for dichotomous responses with small cluster sizes (e.g., Rodriguez and Goldman, 1995, 2001; Breslow and Lin, 1995; Lin and Breslow, 1996; Breslow, Leroux, and Platt, 1998; Goldstein and Rasbash, 1996; Browne and Draper, 2006; McCulloch and Searle, 2001; Breslow, 2005). In such situations, PQL-2 yields better results than PQL-1. However, Rodriquez and Goldman (2001) found that estimates of both fixed and random parameters were attenuated even for PQL-2. Moreover, PQL-2 is sometimes numerically unstable, a problem reported by Rodriquez and Goldman (2001) for one of their examples.

The estimated standard errors for $\widehat{\boldsymbol{\beta}}$ do not take into account the imprecision in the estimates of $\widehat{G}$ when using quasi-likelihood methods. This can result in large biases since the fixed-effects estimates are generally correlated with the variance estimates in generalized linear mixed models. Another drawback of MQL and PQL is that no likelihood is provided, precluding for instance the use of likelihood ratio testing.

Other approximate methods that avoid evaluating the full marginal likelihood include the h-likelihood (e.g., Lee, Nelder, and Pawitan, 2006), high-order Laplace approximations (e.g., Raudenbush, Yang, and Yosef, 2000), and limited information or pseudo-likelihood methods based on bivariate information for probit models (e.g., Muthén, 1984; Renard, Molenberghs, and Geys, 2004).

Software implementing PQL-1 includes SAS `proc glimmix` (SAS Institute, 2006), the S-Plus or R function `glmmPQL` (Venables and Ripley, 2002) and the stand-alone programs HLM (Raudenbush et al., 2004) and MLwiN (Rasbash et al., 2005). MLwiN also provides PQL-2.

4.6.3 Markov chain Monte Carlo

In the Bayesian approach, both the "fixed" coefficients $\boldsymbol{\beta}$ and the random coefficients $\boldsymbol{b}_i$ are random parameters with prior distributions (as are κ^s, σ_e^2, etc., if applicable). In addition, the prior distribution of $\boldsymbol{b}_i$ depends on so-called hyperparameters (the unique elements of G), which are also random and have hyperprior distributions. Due to these different "layers" of priors, Bayesians often refer to mixed models as hierarchical Bayesian models.

Bayesian inference is based on the posterior distribution of the parameters given the observed data. Although the full multivariate posterior distribution is theoretically of interest, posterior means are invariably reported as parameter estimates in practice, along with posterior standard deviations taking the place of standard errors. The posterior distribution of $\boldsymbol{\beta}$ and G, marginalized over the random effects $\boldsymbol{b}_i$, is proportional to the prior distributions of $\boldsymbol{\beta}$ and G times the marginal likelihood (see Section 4.6.1). Loosely speaking, the posterior distribution updates prior "knowledge" (represented by the prior distributions) with information in the observed data (represented by the likelihood).

True Bayesians would specify prior distributions to incorporate prior beliefs or knowledge regarding the parameters. An example would be a prior for a treatment effect in a clinical trial based on elicited expert opinion (e.g., Spiegelhalter, Friedman, and Parmar, 1994). However, in practice, priors are often specified only for the pragmatic reason that Markov chain Monte Carlo (MCMC) methods can then be used for estimating complex models. In this case, the priors are typically specified as "non-informative" (also denoted as flat, vague, or diffuse) to minimize their effect on statistical inference. The likelihood component of the posterior then dominates so that the posterior becomes nearly proportional to the likelihood.

Until the advent of MCMC, Bayesian inference was unfeasible for complex models because the posterior distribution (and its mean or mode) is usually analytically intractable. MCMC is a simulation-based approach where parameters are drawn from their posterior distribution so that features of the posterior can be estimated from the simulated data. The ingenious device is to use a first-order Markov chain in which the parameters are sequentially updated depending on their current values. The chain gradually "forgets" its initial state and eventually converges to the required posterior distribution as stationary distribution.

There are several ways of constructing such a Markov chain. The most popular method is the so-called Gibbs sampler, which utilizes the fact that the conditional distribution of each random variable, given all the other random variables, may be relatively simple in spite of a complicated joint distribution. In its basic form, the Gibbs sampler proceeds by sequentially drawing each parameter from its conditional distribution given the current values of the other parameters. Gibbs sampling has been used for generalized linear mixed models by Zeger and Karim (1991) and Clayton (1996), among others.

Here we sketch how a Gibbs sampler proceeds for the random-intercept binary probit model where implementation is relatively simple. In the latent-response formulation the model can be expressed as

$$Y_{ij}^* = \boldsymbol{x}_{ij}'\boldsymbol{\beta} + b_{i0} + e_{ij}, \qquad b_i \sim N(0, g) \quad e_{ij} \sim N(0, 1),$$

$$Y_{ij} = I(Y_{ij}^* > 0),$$

where $I(\cdot)$ is the indicator function. A multivariate normal prior is used for $\boldsymbol{\beta}$, an inverse-gamma prior for g, and g and $\boldsymbol{\beta}$ are assumed to be independent.

Let $\boldsymbol{Y}$, $\boldsymbol{Y}^*$, $\boldsymbol{X}$, and $\boldsymbol{b}$ refer to the collection of the corresponding subject-specific vectors and matrices. The posterior distribution can be expressed as

$$\Pr(\boldsymbol{Y}^*, \boldsymbol{b}, \boldsymbol{\beta}, g | \boldsymbol{Y}, \boldsymbol{X}) \propto \Pr(\boldsymbol{Y}, \boldsymbol{Y}^* | \boldsymbol{X}, \boldsymbol{b}, \boldsymbol{\beta}) \Pr(\boldsymbol{b}|g) \Pr(g) \Pr(\boldsymbol{\beta}),$$

where the first term on the right-hand side is given by

$$\prod_{i=1}^{N} \prod_{j=1}^{n_i} [I(Y_{ij}^* > 0)I(Y_{ij} = 1) + I(Y_{ij}^* \leq 0)I(Y_{ij} = 0)] \varphi(Y_{ij}^*; \ \boldsymbol{x}_{ij}'\boldsymbol{\beta} + b_{i0}, 1).$$

The Gibbs sampler is based on the following conditional distributions:

1. Independent truncated normal full conditionals of Y_{ij}^*,

$$\Pr(Y_{ij}^*|\boldsymbol{Y}_i, X_i, b_{i0}, \boldsymbol{\beta}) = \varphi^-(Y_{ij}^*; \ \boldsymbol{x}_{ij}'\boldsymbol{\beta} + b_{i0}, 1)^{y_{ij}} \varphi^+(Y_{ij}^*; \ \boldsymbol{x}_{ij}'\boldsymbol{\beta} + b_{i0}, 1)^{1-y_{ij}},$$

where $\varphi^-(\cdot)$ is a left-truncated normal density equal to 0 when $Y_{ij}^* \leq 0$ and $\varphi^+(\cdot)$ is a right-truncated normal density equal to 0 when $Y_{ij}^* > 0$. Importantly, having simulated y_{ij}^*, the full conditionals of the other random variables become independent of Y_{ij}.

2. Multivariate normal full conditional $\Pr(\boldsymbol{\beta}|\boldsymbol{Y}^*, \boldsymbol{X}, \boldsymbol{b})$.

3. Independent normal full conditionals $\Pr(b_i|\boldsymbol{Y}_i^*, X_i, \boldsymbol{\beta}, g)$. The mean and variance are given by expressions (4.12) and (4.13) (in a later section on empirical Bayes) with $\boldsymbol{Y}_i^*$ replacing $\boldsymbol{Y}_i$ and current $\boldsymbol{\beta}$ and g replacing $\widehat{\boldsymbol{\beta}}$ and $\widehat{g}$. Note that the full conditional would not be multivariate normal if we had not augmented the data with y_{ij}^*.

4. Inverse-gamma full conditional $\Pr(g|\boldsymbol{Y}^*, \boldsymbol{X}, \boldsymbol{b}, \boldsymbol{\beta})$.

The fairly straightforward application of the Gibbs sampler for the probit model relies on data augmentation with latent responses (e.g., Tanner and Wong, 1987). More complex procedures must be invoked in other cases such as the random-intercept logit model where Zeger and Karim (1991) use rejection sampling, Spiegelhalter et al. (1996) adaptive rejection sampling (Gilks and Wild, 1992), and Browne and Draper (2006) a hybrid Metropolis–Gibbs approach. When the conditional distributions are complicated, the Metropolis–Hastings algorithm (see Chib and Greenberg, 1985) is often used.

A definite merit of MCMC methods is that they can be used to estimate complex models for which other methods are either unfeasible or work poorly. Simulations reported by Browne and Draper (2006) suggest that MCMC methods work well for generalized linear mixed models. Challenges of using MCMC include choosing the burn-in (the number of initial iterations to discard) and deciding when to stop the chain to ensure acceptable estimates (e.g., Gelman and Rubin, 1996). Finally, the specification of non-informative priors for variance parameters in mixed-effects models is problematic (e.g., Natarajan and Kass, 2000; Hobert, 2000).

Software implementing Markov chain Monte Carlo includes the general program BUGS or WinBUGS (Spiegelhalter et al., 1996; see also Congdon, 2006) and the program MLwiN (Browne, 2005), which is custom-made for generalized linear mixed models.

4.7 Assigning values to the random effects for individual subjects

In Bayesian inference the random effects have a similar status to that of the model parameters, and if MCMC methods are used it is straightforward to make inferences regarding random effects based on their posterior distribution. In a non-Bayesian setting the standard approach is first to estimate the model parameters and subsequently to obtain predictions or estimates of the random effects for each subject i, treating the model parameters as known. The predominant approach to assigning values to random effects is empirical Bayes prediction, but it is useful to first consider maximum likelihood estimation.

4.7.1 Maximum likelihood estimation

For a given subject i, maximum likelihood estimates $\widehat{\boldsymbol{b}}_i$ of the random effects $\boldsymbol{b}_i$ can be obtained by maximizing the joint distribution of the responses given the random effects (with parameter estimates plugged in),

$$\prod_{j=1}^{n_i} f(y_{ij}|\boldsymbol{b}_i, \boldsymbol{x}_{ij}, \boldsymbol{z}_{ij}; \widehat{\boldsymbol{\beta}}, \widehat{G}),$$

with respect to the random effects $\boldsymbol{b}_i$.

For linear mixed models, the maximum likelihood estimator can be written in closed form as

$$\widehat{\boldsymbol{b}}_i = (Z_i'Z_i)^{-1}Z_i'(\boldsymbol{y}_i - X_i\widehat{\boldsymbol{\beta}}),$$

which is just the ordinary least-squares estimator treating the estimated total residuals as responses. For a linear random-intercept model, the maximum likelihood estimator becomes the mean estimated total residual for subject i,

$$\widehat{b}_{i0} = \frac{1}{n_i} \sum_{j=1}^{n_i} (y_{ij} - \boldsymbol{x}_{ij}'\widehat{\boldsymbol{\beta}}). \tag{4.11}$$

A problem with the maximum likelihood estimator is that it is inefficient for small cluster sizes n_i. For some subjects, no estimates can be obtained because some of the $\boldsymbol{z}_{ij}$ do not vary over occasions or, in the case of dichotomous responses, because all responses are the same (all 0 or all 1), producing "estimates" of $\pm\infty$. Finally, it seems strange to estimate the random effects as if they were fixed parameters.

4.7.2 Empirical Bayes prediction

As in full Bayesian inference, empirical Bayes is based on the posterior distribution of the random effects given the data. The crucial difference is that in empirical Bayes the model parameters $\boldsymbol{\beta}$ and G are treated as fixed and replaced by estimates $\widehat{\boldsymbol{\beta}}$ and $\widehat{G}$. The empirical Bayes prediction for subject i is the posterior expectation

$$\widetilde{\boldsymbol{b}}_i = \frac{\int \boldsymbol{b}_i\, \varphi(\boldsymbol{b}_i; \boldsymbol{0}, \widehat{G}) \prod_{j=1}^{n_i} f(y_{ij}|\boldsymbol{b}_i, \boldsymbol{x}_{ij}, \boldsymbol{z}_{ij}; \widehat{\boldsymbol{\beta}})\, d\boldsymbol{b}_i}{\int \varphi(\boldsymbol{b}_i; \boldsymbol{0}, \widehat{G}) \prod_{j=1}^{n_i} f(y_{ij}|\boldsymbol{b}_i, \boldsymbol{x}_{ij}, \boldsymbol{z}_{ij}; \widehat{\boldsymbol{\beta}})\, d\boldsymbol{b}_i}.$$

There are generally no closed-form expressions for the empirical Bayes predictor except for linear mixed models where it becomes

$$\widetilde{\boldsymbol{b}}_i = \widehat{G}Z_i'\widehat{\Sigma}_i^{-1}(\boldsymbol{y}_i - X_i\widehat{\boldsymbol{\beta}}). \tag{4.12}$$

This predictor can alternatively be motivated as the best linear unbiased predictor (BLUP) regardless of distributional assumptions (e.g., Robinson, 1991).

It is instructive to consider the simple case of a linear random-intercept model where the empirical Bayes predictor can be expressed as

$$\widetilde{b}_{i0} = R_i \widehat{b}_{i0}, \qquad R_i = \frac{\widehat{g}}{\widehat{g} + \widehat{\sigma}_e^2/n_i}.$$

Here R_i is the "reliability" (variance of "truth" divided by variance of estimator) of the maximum likelihood estimator $\widehat{b}_{i0}$ in (4.11). The maximum likelihood estimator is reliable for large cluster sizes n_i and/or small $\widehat{\sigma}_e^2$ and/or large $\widehat{g}$. In this case, the empirical Bayes predictor is close to the maximum likelihood estimator. In contrast, when the maximum likelihood estimator is unreliable the empirical Bayes predictor is pulled or shrunk toward 0, the "prior" mean of the random intercept.

It is natural to consider the posterior standard deviation as a measure of uncertainty regarding empirical Bayes prediction. In linear mixed models, the posterior covariance matrix

$$\text{Cov}(\boldsymbol{b}_i|\boldsymbol{y}_i, X_i, Z_i; \widehat{\boldsymbol{\beta}}, \widehat{G}) = \widehat{G} - \widehat{G}'Z_i'\Sigma^{-1}Z_i\widehat{G} \tag{4.13}$$

is equal to the sampling covariance matrix of the prediction errors $\text{Cov}(\widetilde{\boldsymbol{b}}_i - \boldsymbol{b}_i|X_i, Z_i; \widehat{\boldsymbol{\beta}}, \widehat{G})$ and the sampling covariance matrix of the empirical Bayes predictions is given by

$$\text{Cov}(\widetilde{\boldsymbol{b}}_i|X_i, Z_i; \widehat{\boldsymbol{\beta}}, \widehat{G}) = \widehat{G} - \text{Cov}(\boldsymbol{b}_i|\boldsymbol{y}_i, X_i, Z_i; \widehat{\boldsymbol{\beta}}, \widehat{G}).$$

The standard deviation of the prediction errors is sometimes referred to as the "comparative standard error" and the standard deviation of the empirical Bayes predictions as the "'diagnostic standard error" (e.g., Goldstein, 2003). Unfortunately, there are usually no closed-form expressions for the covariance matrices in generalized linear mixed models and the above equalities no longer hold. The posterior covariance matrix can be obtained, for instance, by numerical integration using adaptive quadrature.

Note that the model parameters are treated as known above so that the true sampling variances and the true (fully Bayesian) posterior variances are underestimated. We refer to Skrondal and Rabe-Hesketh (2004, Chapter 7) for a more detailed discussion of methods for assigning values to random effects.

4.8 Application

Koch et al. (1989) analyzed data from a clinical trial comparing two treatments for respiratory illness. Of the eligible patients, 54 were randomized to active treatment and 57 to placebo, where randomization was stratified by center. The patients' respiratory status was determined prior to randomization and during four visits after randomization. The dichotomized response was analyzed by Davis (1991) and Everitt and Pickles (1999), among others. Here we analyze the original ordinal measure of respiratory status (0 = terrible, 1 = poor, 2 = fair, 3 = good, 4 = excellent).

The explanatory variables are:

- drug, dummy variable for active treatment group (x_{i1});
- male, dummy variable for patient being male (x_{i2});
- age, age in years at baseline (x_{i3});
- bl, respiratory status at baseline (x_{i4});
- center, dummy variable for center 2 (x_{i5}).

As is often the case in clinical trials, all explanatory variables are time-invariant. The main purpose of the study was to estimate the effect of the drug on respiratory status, but the other covariate effects were also of interest.

We first estimated a proportional odds models with main effects of all explanatory variables listed previously and a random intercept for patient. For patient i at visit $j = 1, 2, 3, 4,$

Table 4.1 Maximum Likelihood Estimates and Approximate 95% Confidence Intervals for Proportional Odds Model (POM), Random-Intercept Proportional Odds Model (RI-POM), and Random-Intercept "Non-Proportional" Odds Model (RI-NPOM)

	POM		RI-POM		RI-NPOM	
	Est	(95% CI)	Est	(95% CI)	Est	(95% CI)
Fixed part: Odds ratios						
$\exp(\beta_1)$ [drug]	3.38	$(1.90, 6.01)$	7.70	$(2.93, 20.24)$	—	
$\exp(\beta_1^1)$					33.42	$(6.44, 173.55)$
$\exp(\beta_1^2)$					8.23	$(2.56, 26.45)$
$\exp(\beta_1^3)$					7.77	$(2.78, 21.71)$
$\exp(\beta_1^4)$					5.67	$(2.01, 16.02)$
$\exp(\beta_2)$ [male]	0.76	$(0.35, 1.63)$	0.55	$(0.16, 1.91)$	0.54	$(0.16, 1.82)$
$\exp(\beta_3)$ [age]	0.98	$(0.96, 1.00)$	0.97	$(0.93, 1.00)$	0.97	$(0.93, 1.00)$
$\exp(\beta_4)$ [bl]	2.46	$(1.76, 3.43)$	4.08	$(2.48, 6.71)$	4.11	$(2.52, 6.71)$
$\exp(\beta_5)$ [center]	1.48	$(0.76, 2.89)$	1.99	$(0.73, 5.42)$	1.95	$(0.73, 5.22)$
Fixed part: Thresholds						
κ^1	−0.66	$(-1.90, 0.57)$	−1.40	$(-3.44, 0.65)$	−1.00	$(-3.03, 1.03)$
κ^2	0.23	$(-0.97, 1.44)$	0.09	$(-1.93, 2.11)$	0.14	$(-1.86, 2.15)$
κ^3	1.84	$(0.58, 3.09)$	2.61	$(0.56, 4.65)$	2.63	$(0.61, 4.65)$
κ^4	3.00	$(1.67, 4.34)$	4.49	$(2.41, 6.57)$	4.34	$(2.28, 6.39)$
Random part: Variances						
g			4.71		4.49	
Log-likelihood		—		−519.75		−516.08

the (conditional) cumulative log odds were modeled as

$$\text{logit}\left(\gamma_{ij}^s\right) = \beta_1 x_{i1} + \beta_2 x_{i2} + \beta_3 x_{i3} + \beta_4 x_{i4} + \beta_5 x_{i5} + b_{i0} - \kappa^{s+1}, \qquad b_{i0} \sim N(0, g).$$

We then included the time-varying visit number and the interaction between visit and drug, but returned to the original model because the estimated effects and change in log-likelihood were negligible. Maximum likelihood estimates from gllamm (Rabe-Hesketh, Skrondal, and Pickles, 2004b; Rabe-Hesketh and Skrondal, 2008) using 20-point adaptive quadrature are given under RI-POM in Table 4.1. For comparison, estimates of the corresponding marginal model (without the random intercept) are given under POM in the same table with "robust" confidence intervals based on the sandwich estimator for clustered data.

The estimates for both models suggest that the treatment is beneficial, but apart from baseline respiratory status, the other explanatory variables do not have statistically significant effects at the 5% level. For the random-intercept model, the estimated odds ratio of 7.7 for drug can be interpreted as the odds ratio comparing a patient from the active treatment group with another patient from the placebo group, both having identical covariate and random-intercept values. Conditioning on the random intercept can be thought of as controlling for unobserved covariates (or unobserved heterogeneity), leading to an estimate of the "causal" effect of treatment, i.e., treatment effect for a given patient. This interpretation is natural here since the drug variable is exogenous due to randomization in the sense that it is uncorrelated with the random intercept (and the residual error in the latent-response formulation). To visualize the treatment effect, the top panel of Figure 4.6 shows the model-implied cumulative probabilities for the drug and placebo condition for given covariate values and a (realized) random intercept equal to zero.

The residual intraclass correlation of the latent responses is estimated as 0.59. For two randomly sampled patients with identical covariate values, the 25th, 50th, and 75th percentiles

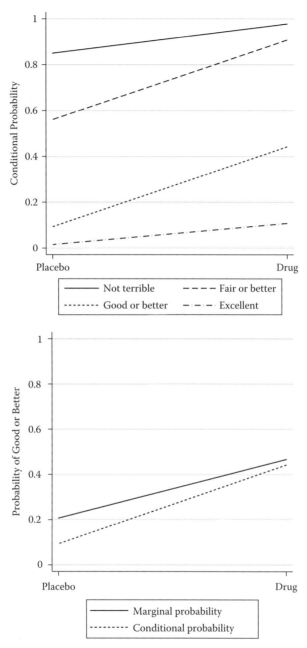

Figure 4.6 Predicted probabilities for 32-year-old females from center 1 with poor respiratory status at baseline by treatment group (assumes proportional odds for drug). Top: Conditional cumulative probabilities for $b_{i0} = 0$. Bottom: Conditional probabilities of good or better for $b_{i0} = 0$ and corresponding marginal probabilities.

of the odds ratios (comparing the patient with the larger random intercept to the other patient) are 2.7, 7.9, and 34.2, respectively.

Due to a substantial random-intercept variance, the odds-ratio estimates for the marginal model are attenuated relative to those for the conditional model. The attenuation of the treatment effect is evident from the bottom panel of Figure 4.6, which shows the model-based conditional (with $b_{i0} = 0$) and marginal probabilities that respiratory status is good or better for given covariate values. The marginal probabilities were obtained by integrating the conditional probabilities over the random-intercept distribution and can be interpreted as the predicted prevalences for the two population strata defined by the covariate values.

To illustrate the proportional odds assumption for treatment, the top panel of Figure 4.7 shows the same conditional cumulative probabilities as the top panel of Figure 4.6, but now using a log odds scale, leading to parallel curves. Since the treatment effect is of primary interest, we assessed the proportional odds assumption for drug using the model

$$\text{logit}(\gamma_{ij}^s) = \beta_1^{s+1}x_{i1} + \beta_2 x_{i2} + \beta_3 x_{i3} + \beta_4 x_{i4} + \beta_5 x_{i5} + b_{i0} - \kappa^{s+1}, \qquad b_{i0} \sim N(0, g),$$

with category-specific regression coefficients β^{s+1} for drug. The estimates given under NPOM-RI in Table 4.1 and the bottom panel of Figure 4.7 suggest that the proportional odds assumption may be violated, with the drug appearing to be particularly effective at increasing the odds of respiratory status not being terrible. However, according to a likelihood ratio test, there is no overwhelming evidence against the proportional odds assumption, with a p-value of 0.06.

In this application we have conditioned on respiratory status at baseline. However, the baseline response would be expected to be affected by the random intercept (representing unobserved heterogeneity) just like the subsequent responses. Conditioning on such an endogenous variable generally leads to inconsistent parameter estimates. Fortunately, due to randomization, the coefficient of the treatment dummy variable can still be estimated consistently.

4.9 Some extensions

In this chapter we have considered generalized linear mixed models with subject-specific random effects $\boldsymbol{b}_i$, but in longitudinal models occasion-specific random effects $\boldsymbol{a}_j$ are sometimes used as well. This makes sense if all subjects are influenced by the same unobserved, time-varying covariates, for instance weather conditions. Furthermore, there must be an adequate number of occasions and an adequate number of observations per occasion. Such crossed-effects or two-way error component models are common in econometrics (e.g., Baltagi, 2005). Models with nested random effects at different hierarchical "levels" are useful if subjects are nested in clusters such as hospitals, schools, geographic areas, households, or twin pairs. Estimation of models with nested random effects is not much more complex than estimation of standard generalized linear mixed models. In contrast, estimation of models with crossed random effects remains a challenge except in linear mixed models.

In generalized linear mixed models it is assumed that the random effects have a multivariate normal distribution. Approaches to relaxing this assumption include discrete distributions (e.g., Nagin and Land, 1993), multivariate t-distributions (e.g., Pinheiro, Liu, and Wu, 2001), finite mixtures of normal distributions (e.g., Uebersax, 1993; Magder and Zeger, 1996), truncated Hermite series expansions (e.g., Gallant and Nychka, 1987; Davidian and Gallant, 1992), non-parametric maximum likelihood estimation (e.g., Lindsay, 1995), and, in a Bayesian setting, semi-parametric mixtures of Dirichlet processes (e.g., Müller and Roeder, 1997).

The random part $Z_i \boldsymbol{b}_i$ in generalized linear mixed models is quite restrictive because each random effect merely multiplies an observed variable. Generalized linear latent and mixed models or GLLAMMs (Rabe-Hesketh, Skrondal, and Pickles, 2004a; Skrondal and Rabe-Hesketh, 2004) extend generalized linear mixed models by allowing the random effects to be multiplied by different parameters for different responses. Specifically, the random part is structured as

$$\sum_{l=2}^{L} \sum_{m=1}^{M_l} b_m^{(l)} Z_m^{(l)} \boldsymbol{\lambda}_m^{(l)},$$

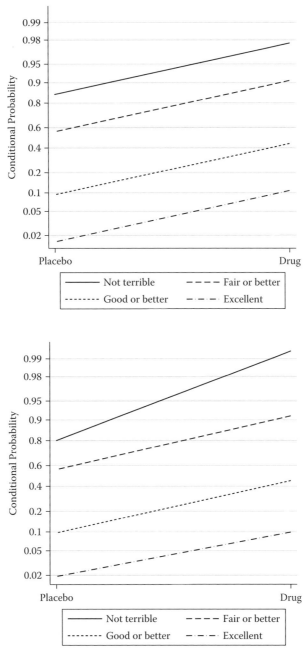

Figure 4.7 Predicted probabilities on log odds scale for 32-year-old females from center 1 with poor respiratory status at baseline with $b_{i0} = 0$ by treatment group. Top: Model imposing proportional odds for drug. Bottom: Model not imposing proportional odds for drug.

where l denotes the hierarchical level (e.g., in clustered longitudinal data subjects are at level 2 and clusters at level 3), $b_m^{(l)}$ is the mth random effect at level l, $Z_m^{(l)}$ is a corresponding covariate matrix, and $\boldsymbol{\lambda}_m^{(l)}$ is a parameter vector. Furthermore, the random effects can be regressed on observed covariates and other random effects (at the same or higher hierarchical levels).

Apart from generalized linear mixed models, useful special cases of GLLAMMs include factor or two-parameter item response models where the random effects (referred to as latent variables in that context) are multiplied by different parameters or "factor loadings" for different responses. This idea can also be used to model non-linear growth (Meredith and Tisak, 1990) or for "shared random-effects models" (see Part 4 and Chapter 19) where random effects induce dependence among different response processes, and "factor loadings" are therefore needed to scale the random effects differently for different responses. A less obvious example is covariate measurement error models (e.g., Rabe-Hesketh, Pickles, and Skrondal, 2003). Skrondal and Rabe-Hesketh (2004) explore a wide range of applications of GLLAMMs in different disciplines.

References

Anderson, D. A. and Aitkin, M. (1985). Variance component models with binary responses: Interviewer variability. *Journal of the Royal Statistical Society, Series B* **47**, 203–210.

Baltagi, B. H. (2005). *The Econometrics of Panel Data*, 3rd ed. Chichester: Wiley.

Breslow, N. E. (2005). Whither PQL? In D. Y. Lin and P. J. Heagerty (eds.), *Proceedings of the Second Seattle Symposium in Biostatistics: Analysis of Correlated Data*, pp. 1–22. New York: Springer.

Breslow, N. E. and Clayton, D. G. (1993). Approximate inference in generalized linear mixed models. *Journal of the American Statistical Association* **88**, 9–25.

Breslow, N. E. and Day, N. (1980). *Statistical Methods in Cancer Research. Vol I — The Analysis of Case-Control Studies*. Lyon: IARC.

Breslow, N. E., Leroux, B., and Platt, R. (1998). Approximate hierarchical modeling of discrete data in epidemiology. *Statistical Methods in Medical Research* **4**, 49–62.

Breslow, N. E. and Lin, X. (1995). Bias correction in generalised linear mixed models with a single component of dispersion. *Biometrika* **82**, 81–91.

Brillinger, D. R. and Preisler, H. K. (1983). Maximum likelihood estimation in a latent variable problem. In S. Karlin, T. Amemiya, and L. A. Goodman (eds.), *Studies in Econometrics, Time Series, and Multivariate Statistics*. New York: Academic Press.

Browne, W. J. (2005). *MCMC estimation in MLwiN*. University of Bristol, Bristol. Downloadable from http://www.cmm.bris.ac.uk/MLwiN/download/MCMC%20est_2005.pdf.

Browne, W. J. and Draper, D. (2006). A comparison of Bayesian and likelihood methods for fitting multilevel models. *Bayesian Analysis* **1**, 473–514.

Browne, W. J., Subramanian, S. V., Jones, K., and Goldstein, H. (2005). Variance partitioning in multilevel logistic models that exhibit overdispersion. *Journal of the Royal Statistical Society, Series A* **168**, 599–613.

Chamberlain, G. (1980). Analysis of covariance with qualitative data. *Review of Economic Studies* **47**, 225–238.

Chamberlain, G. (1984). Panel data. In Z. Griliches and M. D. Intriligator (eds.), *Handbook of Econometrics*, Volume II, pp. 1247–1318. Amsterdam: North-Holland.

Chib, S. and Greenberg, E. (1995). Understanding the Metropolis-Hastings algorithm. *American Statistician* **49**, 327–335.

Clayton, D. G. (1996). Generalized linear mixed models. In W. R. Gilks, S. Richardson, and D. J. Spiegelhalter (eds.), *Markov Chain Monte Carlo in Practice*, pp. 275–301. London: Chapman & Hall.

Congdon, P. (2006). *Bayesian Statistical Modelling*, 2nd ed. Chichester: Wiley.

Davidian, M. and Gallant, A. R. (1992). Smooth nonparametric maximum likelihood estimation for population pharmacokinetics, with application to quinidine. *Journal of Pharmacokinetics and Biopharmaceutics* **20**, 529–556.

Davis, C. S. (1991). Semi-parametric and non-parametric methods for the analysis of repeated measurements with applications to clinical trials. *Statistics in Medicine* **10**, 1959–1980.

Diggle, P. J., Heagerty, P. J., Liang, K.-Y., and Zeger, S. L. (2002). *Analysis of Longitudinal Data*, 2nd ed. Oxford: Oxford University Press.

Duchateau, L. and Janssen, P. (2005). Understanding heterogeneity in generalized mixed and frailty models. *American Statistician* **59**, 143–146.

Everitt, B. S. and Pickles, A. (1999). *Statistical Aspects of the Design and Analysis of Clinical Trials*. London: Imperial College Press.

Gallant, A. R. and Nychka, D. W. (1987). Semi-nonparametric maximum likelihood estimation. *Econometrica* **55**, 363–390.

Gelman, A. and Rubin, D. B. (1996). Markov chain Monte Carlo methods in biostatistics. *Statistical Methods in Medical Research* **5**, 339–355.

Gilks, W. R. and Wild, P. (1992). Adaptive rejection sampling for Gibbs sampling. *Applied Statistics* **41**, 337–348.

Gilmour, A. R., Anderson, R. D., and Rae, A. L. (1985). The analysis of binomial data by a generalized linear mixed model. *Biometrika* **72**, 593–599.

Goldstein, H. (1987). Multilevel covariance component models. *Biometrika* **74**, 430–431.

Goldstein, H. (1991). Nonlinear multilevel models, with an application to discrete response data. *Biometrika* **78**, 45–51.

Goldstein, H. (2003). *Multilevel Statistical Models*, 3rd ed. London: Arnold.

Goldstein, H. and Rasbash, J. (1996). Improved approximations for multilevel models with binary responses. *Journal of the Royal Statistical Society, Series A* **159**, 505–513.

Greenwood, M. and Yule, G. U. (1920). An inquiry into the nature of frequency distributions of multiple happenings, with particular reference to the occurence of multiple attacks of disease or repeated accidents. *Journal of the Royal Statistical Society, Series A* **83**, 255–279.

Hajivassiliou, V. A. and Ruud, P. A. (1994). Classical estimation methods for LDV models using simulation. In R. F. Engle and D. L. McFadden (eds.), *Handbook of Econometrics*, Volume IV, pp. 2383–2441. New York: Elsevier.

Harville, D. A. (1977). Maximum likelihood approaches to variance components estimation and related problems. *Journal of the American Statistical Association* **72**, 320–340.

Hausman, J. A. (1978). Specification tests in econometrics. *Econometrica* **46**, 1251–1271.

Heagerty, P. J. and Zeger, S. L. (2000). Marginalized multilevel models and likelihood inference. *Statistical Science* **15**, 1–26.

Heckman, J. J. (1978). Dummy endogenous variables in a simultaneous equation system. *Econometrica* **46**, 931–959.

Heckman, J. J. (1981). Statistical models for discrete panel data. In C. F. Manski and D. L. McFadden (eds.), *Structural Analysis of Discrete Data with Econometric Applications*, pp. 114–178. Cambridge, MA: MIT Press.

Heckman, J. J. and Willis, R. J. (1976). Estimation of a stochastic model of reproduction: An econometric approach. In N. Terleckyj (ed.), *Household Production and Consumption*. New York: National Bureau of Economic Research.

Hinde, J. (1982). Compound Poisson regression models. In R. Gilchrist (ed.), *GLIM 82: Proceedings of the International Conference on Generalised Linear Models*, pp. 109–121. New York: Springer.

Hobert, J. P. (2000). Hierarchical models: A current computational perspective. *Journal of the American Statistical Association* **95**, 1312–1316.

Koch, G. G., Carr, G. J., Amara, I. A., Stokes, M. E., and Uryniak, T. J. (1989). Categorical data analysis. In D. A. Berry (ed.), *Statistical Methodology in the Pharmaceutical Sciences*, pp. 389–473. New York: Marcel Dekker.

Laird, N. M. and Ware, J. H. (1982). Random effects models for longitudinal data. *Biometrics* **38**, 963–974.

Larsen, K. and Merlo, J. (2005). Appropriate assessment of neighborhood effects on individual health: Integrating random and fixed effects in multilevel logistic regression. *American Journal of Epidemiology* **161**, 81–88.

Larsen, K., Petersen, J. H., Budtz-Jørgensen, E., and Endahl, L. (2000). Interpreting parameters in the logistic regression model with random effects. *Biometrics* **56**, 909–914.

Lawley, D. N. (1943). On problems connected with item selection and test construction. In *Proceedings of the Royal Society of Edinburgh* **61**, 273–287.

Lee, Y., Nelder, J. A., and Pawitan, Y. (2006). *Generalized Linear Models with Random Effects: Unified Analysis via H-likelihood*. Boca Raton, FL: Chapman & Hall/CRC.

Lesaffre, E. and Spiessens, B. (2001). On the effect of the number of quadrature points in a logistic random-effects model: An example. *Applied Statistics* **50**, 325–335.

Lin, X. and Breslow, N. E. (1996). Bias correction in generalized linear mixed models with multiple components of dispersion. *Journal of the American Statistical Association* **91**, 1007–1016.

Lindsay, B. G. (1995). *Mixture Models: Theory, Geometry and Applications*. Hayward, CA: Institute of Mathematical Statistics.

Longford, N. T. (1993). *Random Coefficient Models*. Oxford: Oxford University Press.

Longford, N. T. (1994). Logistic regression with random coefficients. *Computational Statistics & Data Analysis* **17**, 1–15.

Lord, F. M. (1952). A theory of test scores. *Psychometric Monograph* **No. 7**. New York: Psychometric Society.

Magder, L. S. and Zeger, S. L. (1996). A smooth nonparametric estimate of a mixing distribution using mixtures of Gaussians. *Journal of the American Statistical Association* **11**, 86–94.

McCulloch, C. E. (1997). Maximum likelihood algorithms for generalized linear mixed models. *Journal of the American Statistical Association* **92**, 162–170.

McCulloch, C. E. (2008). Joint modeling of mixed outcome types using latent variables. *Statistical Methods in Medical Research* **17**, 53–73.

McCulloch, C. E. and Searle, S. R. (2001). *Generalized, Linear and Mixed Models*. New York: Wiley.

Meng, X.-L. and Schilling, S. G. (1996). Fitting full-information item factor models and an empirical investigation of bridge sampling. *Journal of the American Statistical Association* **91**, 1254–1267.

Meredith, W. and Tisak, J. (1990). Latent curve analysis. *Psychometrika* **55**, 107–122.

Molenberghs, G. and Verbeke, G. (2005). *Models for Discrete Longitudinal Data*. New York: Springer.

Müller, P. and Roeder, K. (1997). A Bayesian semiparametric model for case-control studies with errors in variables. *Biometrika* **84**, 523–537.

Muthén, B. O. (1984). A general structural equation model with dichotomous, ordered categorical and continuous latent indicators. *Psychometrika* **49**, 115–132.

Nagin, D. S. and Land, K. C. (1993). Age, criminal careers, and population heterogeneity: Specification and estimation of a nonparametric mixed Poisson model. *Criminology* **31**, 327–362.

Natarajan, R. and Kass, R. E. (2000). Reference Bayesian methods for generalized linear mixed models. *Journal of the American Statistical Association* **95**, 227–237.

Nelder, J. A. and Wedderburn, R. W. M. (1972). Generalised linear models. *Journal of the Royal Statistical Society, Series A* **135**, 370–384.

Neuhaus, J. M. and Kalbfleisch, J. D. (1998). Between- and within-cluster covariate effects in the analysis of clustered data. *Biometrics* **54**, 638–645.

Neyman, J. and Scott, E. L. (1948). Consistent estimates based on partially consistent observations. *Econometrica* **16**, 1–32.

Ng, E. S. W., Carpenter, J. R., Goldstein, H., and Rasbash, J. (2006). Estimation in generalised linear mixed models with binary outcomes by simulated maximum likelihood. *Statistical Modelling* **6**, 23–42.

Pierce, D. A. and Sands, B. R. (1975). Extra-binomial variation in binary data. Technical report, No. 46, Department of Statistics, Oregon State University.

Pinheiro, J. C., Liu, C., and Wu, Y. (2001). Efficient algorithms for robust estimation in linear mixed-effects models using the multivariate t-distribution. *Journal of Computational Graphics and Statistics* **10**, 249–276.

Rabe-Hesketh, S., Pickles, A., and Skrondal, A. (2003). Correcting for covariate measurement error in logistic regression using nonparametric maximum likelihood estimation. *Statistical Modelling* **3**, 215–232.

Rabe-Hesketh, S. and Skrondal, A. (2001). Parameterization of multivariate random effects models for categorical data. *Biometrics* **57**, 1256–1264.

Rabe-Hesketh, S. and Skrondal, A. (2008). *Multilevel and Longitudinal Modeling Using Stata*, 2nd ed. College Station, TX: Stata Press.

Rabe-Hesketh, S., Skrondal, A., and Pickles, A. (2002). Reliable estimation of generalized linear mixed models using adaptive quadrature. *Stata Journal* **2**, 1–21.

Rabe-Hesketh, S., Skrondal, A., and Pickles, A. (2004a). Generalized multilevel structural equation modeling. *Psychometrika* **69**, 167–190.

Rabe-Hesketh, S., Skrondal, A., and Pickles, A. (2004b). Gllamm manual. Technical Report 160, UC Berkeley Division of Biostatistics Working Paper Series. Downloadable from http://www.bepress.com/ucbbiostat/paper160/.

Rabe-Hesketh, S., Skrondal, A., and Pickles, A. (2005). Maximum likelihood estimation of limited and discrete dependent variable models with nested random effects. *Journal of Econometrics* **128**, 301–323.

Rabe-Hesketh, S., Yang, S., and Pickles, A. (2001). Multilevel models for censored and latent responses. *Statistical Methods in Medical Research* **10**, 409–427.

Rasbash, J., Steele, F., Browne, W. J., and Prosser, R. (2005). *A User's Guide to MLwiN Version 2.0*. University of Bristol, Bristol. Downloadable from http://www.cmm.bris.ac.uk/MLwiN/download/userman_2005.pdf.

Rasch, G. (1960). *Probabilistic Models for Some Intelligence and Attainment Tests*. Copenhagen: Danmarks Pædagogiske Institut.

Raudenbush, S. W. and Bryk, A. S. (2002). *Hierarchical Linear Models*. Thousand Oaks, CA: Sage.

Raudenbush, S. W., Yang, M., and Yosef, M. (2000). Maximum likelihood for generalized linear models with nested random effects via high-order, multivariate Laplace approximation. *Journal of Computational and Graphical Statistics* **9**, 141–157.

Raudenbush, S. W., Bryk, A. S., Cheong, Y. F., Congdon, R., and du Toit, M. (2004). *HLM 6: Hierarchical Linear and Nonlinear Modeling*. Lincolnwood, IL: Scientific Software International.

Renard, D., Molenberghs, G., and Geys, H. (2004). A pairwise likelihood approach to estimation in multilevel probit models. *Computational Statistics and Data Analysis* **44**, 649–667.

Ritz, J. and Spiegelman, D. (2004). A note about the equivalence of conditional and marginal regression models. *Statistical Methods in Medical Research* **13**, 309–323.

Robinson, G. K. (1991). That BLUP is a good thing: The estimation of random effects. *Statistical Science* **6**, 15–51.

Rodriguez, G. and Goldman, N. (1995). An assessment of estimation procedures for multilevel models with binary responses. *Journal of the Royal Statistical Society, Series A* **158**, 73–89.

Rodriguez, G. and Goldman, N. (2001). Improved estimation procedures for multilevel models with binary response: A case study. *Journal of the Royal Statistical Society, Series A* **164**, 339–355.

SAS Institute (2006). *The GLIMMIX Procedure, June 2006*. SAS Institute, Cary, NC. Downloadable from http://www2.sas.com/proceedings/sugi30/196-30.pdf.

Skellam, J. G. (1948). A probability distribution derived from the binomial distribution by regarding the probability of a success as variable between the sets of trials. *Journal of the Royal Statistical Society, Series B* **10**, 257–261.

Skrondal, A. and Rabe-Hesketh, S. (2003). Multilevel logistic regression for polytomous data and rankings. *Psychometrika* **68**, 267–287.

Skrondal, A. and Rabe-Hesketh, S. (2004). *Generalized Latent Variable Modeling: Multilevel, Longitudinal, and Structural Equation Models*. Boca Raton, FL: Chapman & Hall/CRC.

Skrondal, A. and Rabe-Hesketh, S. (2007). Redundant overdispersion parameters in multilevel models. *Journal of Educational and Behavioral Statistics* **32**, 419–430.

Skrondal, A. and Rabe-Hesketh, S. (2008). Multilevel and related models for longitudinal data. In J. de Leeuw and E. Meijer (eds.), *Handbook of Multilevel Analysis*, pp. 275–299. New York: Springer.

Snijders, T. A. B. and Bosker, R. J. (1999). *Multilevel Analysis*. London: Sage.

Spiegelhalter, D. J., Freedman, L. S., and Parmar, M. K. B. (1994). Bayesian approaches to randomized trials (with discussion). *Journal of the Royal Statistical Society, Series A* **157**, 357–416.

Spiegelhalter, D. J., Thomas, A., Best, N. G., and Gilks, W. R. (1996). *BUGS 0.5 Bayesian Analysis Using Gibbs Sampling. Manual (Version ii)*. MRC Biostatistics Unit, Cambridge. Downloadable from http://www.mrc-bsu.cam.ac.uk/bugs/documentation/Download/manual05.pdf.

Stroud, A. H. and Secrest, D. (1966). *Gaussian Quadrature Formulas*. Englewood Cliffs, NJ: Prentice Hall.

Tanner, M. A. and Wong, W. H. (1987). The calculation of posterior distributions by data augmentation. *Journal of the American Statistical Association* **82**, 528–540.

Tucker, L. R. (1948). A method for scaling ability test items taking item unreliablity into account. *American Psychologist* **3**, 309–310.

Uebersax, J. S. (1993). Statistical modeling of expert ratings on medical treatment appropriateness. *Journal of the American Statistical Association* **88**, 421–427.

Venables, W. N. and Ripley, B. D. (2002). *Modern Applied Statistics with S*, 4th ed. New York: Springer.

Verbeke, G. and Molenberghs, G. (2000). *Linear Mixed Models for Longitudinal Data*. New York: Springer.

Wang, Z. and Louis, T. A. (2004). Marginalized binary mixed-effects models with covariate-dependent random effects and likelihood inference. *Biometrics* **60**, 884–891.

Wedderburn, R. W. M. (1974). Quasi-likelihood functions, generalized linear models, and the Gauss-Newton method. *Biometrika* **61**, 439–447.

West, M. (1985). Generalized linear models: Scale parameters, outlier accommodation and prior distributions. In J. M. Bernardo, M. H. DeGroot, D. V. Lindley, and A. F. M. Smith (eds.), *Bayesian Statistics 2*, pp. 531–557. Amsterdam: Elsevier.

Wolfinger, R. D. (1999). Fitting non-linear mixed models with the new NLMIXED procedure. Technical report, SAS Institute, Cary, NC.

Wong, G. Y. and Mason, W. M. (1985). The hierarchical logistic regression model for multilevel analysis. *Journal of the American Statistical Association* **80**, 513–524.

Zeger, S. L. and Karim, M. R. (1991). Generalized linear models with random effects: A Gibbs sampling approach. *Journal of the American Statistical Association* **86**, 79–86.

CHAPTER 5

Non-linear mixed-effects models

Marie Davidian

Contents

5.1 Introduction

In many applications, particularly in the biological sciences, the time course of a continuous response for an individual may be characterized by a function that is non-linear in one or more parameters, where an "individual" may be a human subject, an animal, a plant, an agricultural plot, a laboratory sample, or other observational unit. The non-linear function may be chosen on empirical grounds for its ability to represent faithfully the apparent individual-specific response-time relationship. For example, the logistic growth model, discussed in Section 5.2.1, is often adopted for this reason to describe the S shape, typical of relationships between continuous measures of growth and time on a given plant or plot. Alternatively, a non-linear model for individual behavior may arise from theoretical, mechanistic considerations that are of central scientific interest. In pharmacokinetics, as we demonstrate in Section 5.2.2, non-linear models for drug concentrations achieved over time following administration of a drug are derived from representing the body as a system of "compartments," embodying assumptions on how processes of drug absorption, distribution, and elimination take place within a subject.

Whether or not an appropriate non-linear model for individual response is motivated by empirical or theoretical considerations, it is often the case that its parameters reflect directly phenomena of interest. Thus, the parameters in these individual-specific non-linear models have scientifically meaningful interpretations, and substantive questions may be formulated on their basis. This is especially true when the model is based on mechanistic assumptions. For example, the parameters in a standard pharmacokinetic compartment model are quantities characterizing explicitly, for a given subject, processes such as absorption and elimination that underlie observed drug concentrations. Pharmacokineticists wish to understand the "typical" behavior of these quantities in and the extent to which they vary across the population of subjects to assist them in the development of dosing recommendations.

In such settings, longitudinal continuous response data are collected on each individual in a sample drawn from a population of interest. From a statistical point of view, a primary goal is to make inferences on how the underlying features represented by the parameters in an appropriate individual-specific non-linear model behave across the population. The *non-linear mixed-effects model*, also referred to as the hierarchical non-linear model, is a natural framework within which to address this objective. In this chapter, we provide an introduction to this important class of statistical models for longitudinal data.

The class of non-linear mixed-effects models is similar to that of generalized linear mixed-effects models for longitudinal data involving individual-specific random effects, discussed in Chapter 4. However, because non-linear mixed-effects models are usually predicated on a non-linear function whose parameters enjoy explicit, substantive meaning, their applicability and interpretation can be somewhat different from those of generalized linear mixed-effects models, although the formulation and implementation of both models share common features. We concentrate here on situations where the focus of inference is on the behavior in the population of scientifically meaningful, individual-specific parameters in a non-linear model for individual trajectories arising from subject-matter considerations.

The remainder of this chapter is organized as follows. In Section 5.2, to exhibit more concretely the foregoing perspective, we discuss three major applications where the non-linear mixed-effects model is an appropriate inferential framework. Section 5.3 develops the conception and formulation of the basic model. In Section 5.4, we review the most popular approaches to inference under the model, and in Section 5.5 we demonstrate some of these, as incorporated in widely available software packages. Section 5.6 discusses extensions to the basic model and further developments.

5.2 Applications

5.2.1 Soybean growth

Studies of growth, such as those in agricultural research, are classical examples of a setting where the non-linear mixed-effects model is a natural basis for inference. Davidian and Giltinan (1995, Sections 1.1.3 and 11.2) discussed a study carried out over three growing seasons for which a central objective was to compare features of the patterns of growth of a commercial variety of soybean, denoted by F, to those of an experimental variety, referred to as P. In each season, eight plots were planted with F and eight with P, and, roughly weekly thereafter, six plants were taken from each plot at random. Leaves from the plants were aggregated, ground up, and weighed, and the resulting total weight was divided by 6

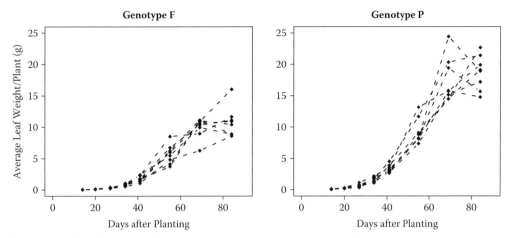

Figure 5.1 Growth profiles for eight plots planted with genotype F and eight plots planted with genotype P from the soybean growth study.

to yield a (continuous) measure of average leaf weight per plant (g) for the plot for that week.

Figure 5.1 shows the average leaf weights per plant over time for the eight plots planted with each genotype in the 1989 growing season, where longitudinal growth measures for each plot are connected by dashed lines. Subject to some intra-plot variation, each plot's growth profile follows roughly an S shape in which growth begins slowly and then shows a linear trend during the middle of the growing season followed by a "leveling off" at the end, a pattern typical of many growth studies. Such plot-specific behavior is fully expected from a biological point of view, and one main goal of the soybean study was to contrast the asymptotic behavior for the two genotypes, knowledge of which would be useful for informing harvesting and management practices. Although all plots within a genotype follow the S pattern, the apparent limiting value appears to differ among them, presumably due to natural biological plot-to-plot variation. Thus, more specifically, the researchers were interested in comparing the "typical" limiting growth values for the two genotypes and in understanding the extent of their inter-plot variation within each genotype.

Accordingly, a statistical model in which this objective may be formalized should acknowledge that the apparent growth trajectory for an individual plot approaches a limiting value. Notably, a linear, polynomial (quadratic) model would likely provide a poor approximation to this asymptotic behavior. Alternatively, a popular description of individual profiles that accommodates these features is the logistic growth model. Letting t denote time since planting and $m(t)$ the measure of growth at t, average leaf weight per plant here, one version of the model is

$$m(t) = \frac{\theta_1}{1 + \exp\{-\theta_3(t - \theta_2)\}}, \quad \theta_1, \theta_3 > 0, \tag{5.1}$$

where $\boldsymbol{\theta} = (\theta_1, \theta_2, \theta_3)'$ are parameters characterizing features of the trajectory. For example, taking $t \to \infty$, it follows that θ_1 explicitly characterizes the limiting growth value.

The logistic model (5.1) is often chosen as a convenient empirical representation without regard to theoretical assumptions on the growth process. However, (5.1) may be derived from a mechanistic point of view. Assuming that growth rate relative to current growth value declines linearly with increasing growth value, we have, writing $\dot{m}(t) = d/dt\{m(t)\}$,

$$\frac{\dot{m}(t)}{m(t)} = k\left\{1 - \frac{m(t)}{a}\right\}, \quad k, a > 0. \tag{5.2}$$

Integration of (5.2) yields (5.1), with $\theta_1 = a$, $\theta_3 = k$, and θ_2 such that $\theta_1/(1 + e^{\theta_2\theta_3})$ is the initial growth value at $t = 0$. The parameter θ_3 is a scaling factor for time and thereby influences the growth rate; larger values of θ_3 lead to steeper relationships. When $\theta_2 > 0$, $m(t)$ has a visible inflection point at θ_2, and θ_2 is the time at which the maximal growth rate occurs. Thus, from the perspective of (5.2), all elements of $\boldsymbol{\theta}$ describe meaningful aspects of the assumed growth mechanism. Other growth models arise from alternative theoretical considerations; Seber and Wild (1989, Chapter 7) gave an excellent review.

While the underlying trajectory for a given plot is well described by a model such as (5.1), as noted previously, features such as steepness and limiting growth value vary across plots. In the context of (5.1), a natural way to represent this is to assume that, while the underlying growth pattern for a particular plot follows (5.1), the value of $\boldsymbol{\theta}$ differs across plots. This is the view embodied in the non-linear mixed-effects model, formalized in Section 5.3.

5.2.2 Pharmacokinetics

Pharmacokinetics is often referred to as the study of "what the body does to a drug." Formally, pharmacokinetics attempts to describe the fate of a drug once it is introduced into the body by characterizing within-subject processes of absorption, distribution, and elimination that govern concentrations achieved over time. This knowledge is critical for establishing dosing guidelines seeking to maintain concentrations in a "therapeutic window" where clinical benefit is likely to be realized; see below. An outstanding overview of pharmacokinetics and the companion area of pharmacodynamics was given by Giltinan (2006).

As noted in Section 5.1, a standard approach is to represent the body by a series of hypothetical compartments that, under a specified route of administration of the drug into the body (intravenously, orally, etc.), leads to a system of differential equations in terms of parameters related to the pharmacokinetic processes. Analogous to (5.2), solution of these yields a mathematical model for concentration as a function of time depending on parameters embodying assumptions on how these processes take place. Gibaldi and Perrier (1982) provided a comprehensive account of compartmental pharmacokinetic modeling. Although usually vastly simplified characterizations of undoubtedly more complex mechanisms, these models can provide excellent descriptions of observed concentration–time profiles from human and animal subjects and have demonstrated great utility in informing effective dosing strategies. As for growth data, a fundamental principle is that, although the same compartmental model may describe adequately intra-individual pharmacokinetics for all subjects, the parameters that fully determine the model for any given subject vary across subjects.

Pharmacokinetic studies in humans are generally of two types. So-called "intensive studies" typically involve a small number of subjects, each of whom may be given a single dose of the agent and from whom frequent blood samples are drawn to obtain detailed concentration–time profiles. Intensive studies are often carried out early in drug development on healthy, homogeneous individuals, and the objectives are to identify an appropriate compartmental model and to obtain preliminary information on the "typical" values of the pharmacokinetic parameters. "Population studies" are usually conducted in later stages of drug development or when a drug is already in routine use. These studies involve a large number of subjects drawn from the heterogeneous patient population, from whom only sparse concentration data may be obtained over multiple dosing intervals along with extensive demographical and physiological information. A compartmental model ideally is adopted based on prior intensive studies, and a main goal is understanding inter-subject variation in the pharmacokinetic parameters and their associations with subject characteristics. Armed with the likely values of the parameters for subjects from a subpopulation having specific characteristics and the extent of their variation, pharmacokineticists can

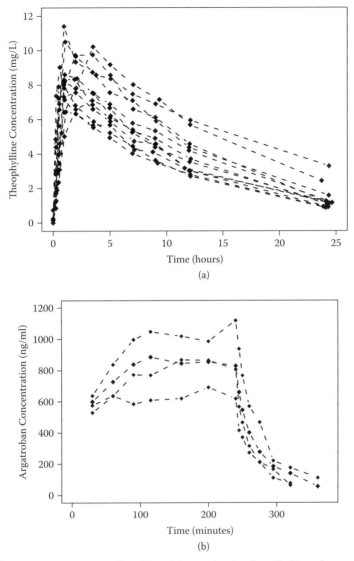

Figure 5.2 (a) Drug concentrations for 12 participants in the theophylline pharmacokinetic study. (b) Profiles for four subjects receiving an intravenous infusion of 4.5 μg/kg-min for 240 minutes (4 hours) in the argatroban pharmacokinetic study.

obtain from the model predictions of the concentrations that may be achieved under different dosing strategies and use these to design tailored dosing regimens to maintain concentrations for such subjects in a desired therapeutic range.

Figure 5.2 depicts data from two well-known intensive studies. Figure 5.2a shows concentration–time profiles for 12 subjects administered the same oral dose (on a per-weight basis, mg/kg) of the anti-asthmatic theophylline (Davidian and Giltinan, 1995, Section 5.5). The pattern, for any subject, of rapid rise to a peak concentration followed by apparent exponential decay may be conceptualized by a one-compartment model with first-order absorption and elimination and the associated system of differential equations and initial conditions. The model corresponds roughly to viewing the body as a single "blood compartment." Letting $A(t)$ denote the amount of theophylline present in the compartment at time t following oral dose D at $t = 0$, and writing $\dot{A}(t) = d/dt\{X(t)\}$ as before, the model is

$$\dot{A}(t) = k_a X_a(t) - k_e A(t), \qquad A(0) = 0$$
$$\dot{A}_a(t) = -k_a A_a(t), \qquad\qquad A_a(0) = FD,$$

(5.3)

where $A_a(t)$ is the amount of drug at t in an "absorption depot" (e.g., the gut) from which it is absorbed into the system at fractional rate k_a (units of 1/time). The drug is eliminated from the system at fractional rate $k_e = Cl/V$, expressed in terms of clearance Cl, the volume of blood "cleared" of drug per unit time, and a hypothetical "volume" of the compartment, V, known as the volume of distribution, roughly interpreted as the volume necessary to account for all drug in the system. The so-called bioavailability, F, is not estimable from data like those in Figure 5.2 and is often set equal to 1. Adopting this convention, solution of (5.3) for $A(t)$ and division by V yields the expression for the concentration $m(t)$ (amount/unit volume) of drug at time t,

$$m(t) = \frac{A(t)}{V} = \frac{k_a D}{V(k_a - Cl/V)}[\exp\{-(Cl/V)t\} - \exp(-k_a t)],$$

(5.4)

characterized in terms of the pharmacokinetic parameters $\boldsymbol{\theta} = (k_a, V, Cl)'$. Figure of Davidian and Giltinan (1995) shows individual regression fits of (5.4) to data for four of the subjects that demonstrate its relevance for describing the concentration–time relationship.

Figure 5.2b presents data from 4 of the 37 subjects in an intensive study of the anti-coagulant argatroban (Davidian and Giltinan, 1995, Section 9.5). Subjects were assigned to receive an intravenous infusion of argatroban for 4 hours at different rates; the figure shows concentration–time profiles for subjects infused at a rate of 4.5 μg/kg-min. Each subject shows a similar pattern, replicated at other infusion rates, of a gradual concentration increase approaching an apparent asymptote followed by an abrupt exponential decline upon termination of the infusion. Following Gibaldi and Perrier (1982, Chapter 1), a one-compartment model similar to (5.3), where absorption of the oral dose is replaced by continuous infusion at rate D, leads to a differential equation for the amount $A(t)$ whose solution divided by apparent volume V yields the expression for argatroban concentration $m(t)$ at time t,

$$m(t) = \frac{D}{Cl}\left[\exp\left\{-\frac{Cl}{V}(t - t_{\text{inf}})_+\right\} - \exp\left(-\frac{Cl}{V}t\right)\right], \quad \boldsymbol{\theta} = (Cl, V)',$$

(5.5)

under infusion of duration t_{inf}, where $x_+ = 0$ if $x \le 0$ and $x_+ = x$ if $x > 0$ (so $t_{\text{inf}} = 240\,\text{min}$ here). In (5.5), as in (5.4), Cl and V represent, respectively, clearance, summarizing the mechanism of elimination, and volume of distribution, related to how the drug distributes throughout the system.

Models with more than one compartment are more realistic but involve greater numbers of pharmacokinetic parameters and lead to more complex representations. Sufficiently complicated models may not admit a closed-form solution for concentration $m(t)$ as in (5.4) or (5.5); however, in principle, these may be solved numerically for $m(t)$ using ordinary differential equation solver routines. Even for simple compartment models for which analytical solutions are possible, under repeated dosing schemes, the expression for concentration at

Table 5.1 Data for a Subject in the Quinidine Study

Time (h)	Conc. (mg/L)	Dose (mg)	Age (years)	Weight (kg)	Creatinine (ml/min)	Glyco. (mg/dl)
0.00	–	166	75	108	> 50	69
6.00	–	166	75	108	> 50	69
11.00	–	166	75	108	> 50	69
17.00	–	166	75	108	> 50	69
23.00	–	166	75	108	> 50	69
27.67	0.7	–	75	108	> 50	69
29.00	–	166	75	108	> 50	94
35.00	–	166	75	108	> 50	94
41.00	–	166	75	108	> 50	94
47.00	–	166	75	108	> 50	94
53.00	–	166	75	108	> 50	94
65.00	–	166	75	108	> 50	94
71.00	–	166	75	108	> 50	94
77.00	0.4	–	75	108	> 50	94
161.00	–	166	75	108	> 50	88
168.75	0.6	–	75	108	> 50	88

Note: Height = 72 inches, Caucasian, smoker, no ethanol abuse, no CHF. Conc. = quinidine concentration, Glyco. = α_1-acid glycoprotein concentration, CHF = congestive heart failure.

time t may nonetheless be fairly complex, as it must account for the accumulation of drug in the system following multiple doses. This is generally the case for population studies, where subjects are observed haphazardly under chronic dosing conditions; that is, while they are receiving multiple doses at periodic intervals.

An example is provided by a population study of the anti-arrhythmic drug quinidine, which was reported by Verme et al. (1992) and which has been cited by numerous authors (e.g., Davidian and Giltinan, 1995, Sections 1.1.2 and 9.3; Wakefield, 1996). The 136 subjects in the study were hospitalized and undergoing routine treatment with oral quinidine for atrial fibrillation or ventricular arrhythmia. Table 5.1 shows abridged data for one subject, which typify the information collected. Demographical and physiological characteristics included age; weight; height; ethnicity/race (Caucasian/Black/Hispanic); smoking status (yes/no); ethanol abuse (no/current/previous); congestive heart failure (no or mild/moderate/severe); creatinine clearance, a measure of renal function (≤ 50 ml/min indicates renal impairment); and α_1-acid glycoprotein concentration, the level of a molecule that binds quinidine. In addition, dosing history (times, amounts) were recorded along with quinidine concentrations.

The one-compartment model with first-order absorption and elimination (5.3) was used by Verme et al. (1992) to describe quinidine concentration. In this study, subjects were observed over multiple dosing intervals. An assumption that is often reasonable in this setting is that pharmacokinetic behavior, and hence the relevant compartmental system used to represent it, is unchanged regardless of the number of doses given. Another is that achieved concentrations are governed by the so-called "principle of superposition" (e.g., Giltinan, 2006), which dictates that a new dose contributes in additive fashion to the amount of drug already present in the system due to previous doses. Moreover, under repeated dosing, drug accumulates in the system until a "steady state" is reached at which, roughly, rate of administration of drug is equal to the rate of elimination (e.g., Giltinan, 2006). Adopting (5.3) to describe pharmacokinetic behavior, then, these assumptions lead to rather complicated expressions for quinidine concentration at any time that involve additive contributions of terms of the form (5.4), which may be written in a recursive fashion, as follows. Denoting the ℓth (dose time, amount) by (s_ℓ, D_ℓ), the amount of quinidine in the absorption depot, $A_a(s_\ell)$, and the concentration of quinidine in the blood, $m(s_\ell)$, at dose time s_ℓ for a subject who has not yet achieved a steady state are given by

$$A_a(s_\ell) = A_a(s_{\ell-1}) \exp\{-k_a(s_\ell - s_{\ell-1})\} + D_\ell,$$

$$m(s_\ell) = m(s_{\ell-1}) \exp\{-k_e(s_\ell - s_{\ell-1})\} + A_a(s_{\ell-1}) \frac{k_a}{V(k_a - k_e)}$$
$$\times [\exp\{-k_e(s_\ell - s_{\ell-1})\} - \exp\{-k_a(s_\ell - s_{\ell-1})\}],$$

and the concentration of quinidine at time t in the next dosing interval $(s_\ell, s_{\ell+1})$ is

$$m(t) = m(s_\ell) \exp\{-k_e(t - s_\ell)\} + A_a(s_\ell) \frac{k_a}{V(k_a - k_e)}$$
$$\times [\exp\{-k_e(t - s_\ell)\} - \exp\{-k_a(t - s_\ell)\}], \quad s_\ell < t < s_{\ell+1}, \tag{5.6}$$

where $k_e = Cl/V$. Once a steady state has been reached, a further set of equations governs the values of $A_a(s_\ell)$ and $m(s_\ell)$ at dose times, which we exclude here for brevity; see Davidian and Giltinan (1995, Section 1.1.2) for details. Although these expressions turn out to be complex because of the multiple dosing, they arise from the compartment model (5.3) and hence embody plausible biological assumptions in terms of the meaningful parameters $\boldsymbol{\theta} = (k_a, V, Cl)'$ of genuine scientific interest.

As for the theophylline and argatroban studies, it is assumed that quinidine concentrations over time for all subjects may be represented by (5.6); however, the parameters $\boldsymbol{\theta}$ governing this relationship vary across subjects. Accordingly, as discussed above, a key objective is to identify systematic associations between the elements of $\boldsymbol{\theta}$ and demographical and physiological characteristics to be used in developing individualized dosing regimens tailored to specific demographical and physiological features. Of course, these characteristics cannot account for all variation in pharmacokinetics across subjects; hence, it is also of interest to quantify the extent of remaining "biological" variation "unexplained" by systematic associations. In Section 5.3 we demonstrate how the non-linear mixed-effects model provides a convenient framework in which to cast this problem.

Interest in pharmacokinetics is not limited to drugs. For example, toxicokinetics refers to the study of pharmacokinetics of environmental, chemical, or other agents in the context of assessment of their possible toxic effects. Toxicokinetic studies are often conducted in animal models and involve exposure of each animal to the agent and collection of frequent blood or other samples from which concentrations are ascertained. It is standard to entertain more detailed compartmental models than those used in human population studies that represent the body by physiologically identifiable compartments such as fatty tissues, poorly- and well-perfused tissues, the liver, and so on. Understanding of the parameters in the model, such as rates at which a compound is metabolized in the liver, can help guide public health recommendations. These so-called physiologically based pharmacokinetic (PBPK) models generally are complex systems of differential equations that do not admit closed-form solutions for observable concentrations, and so must be solved numerically. Moreover, they contain numerous parameters, many of which are not identifiable from longitudinal concentration data, a complication we discuss in subsequent sections. In principle, however, they involve the same considerations as the models discussed previously, and hence may be incorporated in the non-linear mixed-effects model to facilitate these objectives. Gelman, Bois, and Jiang (1996) and Mezzetti et al. (2003) offered accounts of analyses of such studies.

5.2.3 HIV dynamics

Human immunodeficiency virus type 1 (HIV) progressively destroys the body's ability to fight infection by killing or damaging cells in the immune system. Since the mid-1990s, when Ho et al. (1996) first demonstrated that the time course of HIV RNA (viral) load, a measure of the concentration of copies of viral RNA in the body, can be described mathematically, there has been considerable interest in developing models to represent hypothesized mechanisms governing the interplay between HIV and the immune system. These so-called HIV dynamic models have led to advances in understanding of plausible mechanisms underlying HIV pathogenesis and in developing antiretroviral (ARV) treatment strategies for HIV-infected individuals; see Adams et al. (2005), Huang, Liu, and Wu (2006), and numerous references therein.

As in pharmacokinetics, these models are predicated on representing processes involved in the virus–immune system interplay via hypothetical compartments, where, here, the compartments characterize different populations of virus, immune system cells targeted by the virus, and so on, that interact within a subject. As noted by Huang, Liu, and Wu (2006), early applications used simplified models admitting closed-form expressions for, for example, viral load over time; although useful for short-term insights, such models cannot represent the dynamics of the virus–immune system over longer periods. Accordingly, later work has focused on complex systems that, similar to PBPK models, involve several compartments. We illustrate with a model studied by Adams et al. (2007), involving compartments denoted by T_1, type 1 target cells, for example, CD4$^+$ cells (cells/μl); T_2, type 2 target cells, such as macrophages (cells/μl); V_I and V_{NI}, infectious and non-infectious free virus, respectively (RNA copies/ml); and E, cytotoxic T-lymphocytes (cells/μl). A superscript asterisk (*) denotes infected target cells; and, for example, with $T_1(t) = $ concentration of type 1 target cells at time t, the model is

$$\dot{T}_1 = \lambda_1 - d_1 T_1 - \{1 - \varepsilon_1 u(t)\}k_1 V_I T_1,$$
$$\dot{T}_2 = \lambda_2 - d_2 T_2 - \{1 - f\varepsilon_1 u(t)\}k_2 V_I T_2,$$
$$\dot{T}_1^* = \{1 - \bar{\varepsilon}_1(t)\}k_1 V_I T_1 - \delta T_1^* - m_2 E T_1^*,$$
$$\dot{T}_2^* = \{1 - f\varepsilon_1 u(t)\}k_2 V_I T_2 - \delta T_2^* - m_2 E T_2^*,$$
$$\dot{V}_I = \{1 - \varepsilon_2 u(t)\}10^3 N_T \delta(T_1^* + T_2^*) - cV_I - \{1 - \varepsilon_1 u(t)\}\rho_1 10^3 k_1 T_1 V_I \qquad (5.7)$$
$$\qquad -\{1 - f\varepsilon_1 u(t)\}\rho_2 10^3 k_2 T_2 V_I,$$
$$\dot{V}_{NI} = \varepsilon_2 u(t)10^3 N_T \delta(T_1^* + T_2^*) - cV_{NI},$$
$$\dot{E} = \lambda_E + \frac{b_E(T_1^* + T_2^*)}{(T_1^* + T_2^*) + K_b}E - \frac{d_E(T_1^* + T_2^*)}{(T_1^* + T_2^*) + K_d}E - \delta_E E,$$

with an initial condition vector $\{T_1(0), T_2(0), T_1^*(0), T_2^*(0), V_I(0), V_{NI}(0), E(0)\}'$, where we suppress most dependence on t for brevity, and the factors of 10^3 convert between μl and ml scales. The system (5.7) depends on numerous meaningful parameters $\boldsymbol{\theta}$ that are thought to vary across subjects; for example, c (1/day), the natural death rate of the virus; δ (1/day), the death rate of infected target cells; and λ_k, $k = 1, 2$, production rates (cells/μl-day) of type 1 and 2 target cells. The function $u(t)$, $0 \leq u(t) \leq 1$, represents time-dependent input of ARV therapy, with $u(t) = 0$ corresponding to fully off-treatment and $u(t) = 1$ to fully on-treatment. The parameters ε_k, $k = 1, 2$, $0 \leq \varepsilon_k \leq 1$, are efficacies of reverse transcriptase inhibitor treatments for blocking new infections and protease inhibitors for causing infected cells to produce non-infectious virus, respectively. See Adams et al. (2007) for a complete description.

In an HIV study, longitudinal data on combinations of one or more of these compartments are collected, and one objective is to learn about the "typical" values of at least some of the parameters in a model like (5.7) and how they vary across individuals in order to gain insight into viral mechanisms and their possible associations with subject characteristics. Usually, repeated measurements of total CD4$^+$ cells, $T_1 + T_1^*$, and total viral load, $V_i + V_{NI}$, are available on each subject. Needless to say, solution of (5.7) can only be carried out numerically. A further complication is that total viral load measurements may be left-censored by the lower limit of quantification of the assay. Finally, as with PBPK models, the available data fail to identify all the parameters. Here, then, multivariate, possibly censored responses are available on "part" of a mechanistic model from which analytical expressions for these responses are not possible and for which many of the parameters are not identifiable. Despite these difficulties, conceptually, this situation is similar to those in Section 5.2.1 and Section 5.2.2, suggesting again the relevance of the non-linear mixed-effects model.

5.2.4 Summary

There are numerous subject-matter areas, in addition to those reviewed in Section 5.2.1 through Section 5.2.3, where a scientifically relevant model for individual longitudinal trajectories of a continuous response is available and interest focuses on how the meaningful parameters in this model vary in the population and are associated with individual characteristics. A short list includes forestry, dairy science, cancer dynamics, population dynamics, and fisheries science; Davidian and Giltinan (2003) offered an extensive bibliography. We now formalize the non-linear mixed-effects model as the appropriate inferential framework for this setting.

5.3 Model formulation

5.3.1 Basic set-up

Consider a study involving N individuals drawn from a population of interest, indexed by i, where, on the ith individual, n_i measurements of a continuous, univariate response, such as average leaf weight per plant or drug concentration, are observed at times $t_{i1}, \ldots, t_{in_i}$. We discuss multivariate response in Section 5.6. More precisely, let Y_{ij} denote the response measurement on the ith individual at time t_{ij}, $i = 1, \ldots, N$, $j = 1, \ldots, n_i$.

Let $\boldsymbol{U}_i$ denote a vector of covariates representing conditions under which i is observed. For example, in the theophylline study in Section 5.2.2, each subject received a single oral dose D_i, say, at time zero, so that $\boldsymbol{U}_i = D_i$, while in the argatroban study, subject i received an infusion at rate D_i for the same duration t_{inf}, implying $\boldsymbol{U}_i = (D_i, t_{\text{inf}})'$. Each subject i in the population study of quinidine received $d_i \geq 1$ repeated doses; thus, $\boldsymbol{U}_i$ contains i's entire dosing history consisting of time–dose pairs $(s_{i\ell}, D_{i\ell})'$, $\ell = 1, \ldots, d_i$. There are no such conditions in the soybean study in Section 5.2.1, so $\boldsymbol{U}_i$ is null. In an HIV dynamic study, as in Section 5.2.3, one might have information on a subject's entire, continuous history of being fully on or off ARV therapy, so that $\boldsymbol{U}_i$ would be replaced by the continuous function $u_i(t)$ specifying subject i's known treatment status at any time during the observation period. The $\boldsymbol{U}_i$ may be regarded as "within-individual" covariates in the sense that they are required to describe the response–time relationship at the level of the individual.

Finally, assume that each individual i potentially has a vector of characteristics $\boldsymbol{A}_i$ that do not change over the observation period; for example, for the subject in the quinidine study in Table 5.1, height, age, ethnicity, smoking status, and so on, and, for a plot in the soybean study, an indicator of genotype. We defer the case where some characteristics change over time, as for α_1-acid glycoprotein concentration in Table 5.1, to Section 5.6. The $\boldsymbol{A}_i$ may be viewed as "between-individual" covariates because they are relevant only to how individuals differ but are not needed to characterize individual response–time relationships.

Letting $\boldsymbol{Y}_i = (Y_{i1}, \ldots, Y_{in_i})'$, and combining the within- and between-individual covariates into a vector $\boldsymbol{X}_i = (\boldsymbol{U}_i', \boldsymbol{A}_i')'$, we may summarize the observed data as independent pairs $(\boldsymbol{Y}_i, \boldsymbol{X}_i)$, $i = 1, \ldots, N$, although it is important to distinguish $\boldsymbol{U}_i$ and $\boldsymbol{A}_i$ in the following. The basic non-linear mixed-effects model may be expressed as a two-stage hierarchy:

$$\text{Stage 1: Individual-Level Model} \quad Y_{ij} = m(t_{ij}, \boldsymbol{U}_i, \boldsymbol{\theta}_i) + e_{ij}, \quad j = 1, \ldots, n_i \quad (5.8)$$

$$\text{Stage 2: Population Model} \quad \boldsymbol{\theta}_i = \boldsymbol{d}(\boldsymbol{A}_i, \boldsymbol{\beta}, \boldsymbol{b}_i), \quad i = 1, \ldots, N \quad (5.9)$$

In (5.8), $m(t, \cdot, \cdot)$ is a function of time, such as (5.1) or (5.4) through (5.6), depending on individual conditions $\boldsymbol{U}_i$ and an $(r \times 1)$ vector of parameters $\boldsymbol{\theta}_i$, specific to individual i, that describes the apparent individual-specific trajectory for i. For example, in (5.4),

$$\boldsymbol{\theta}_i = (k_{ai}, V_i, Cl_i)' = (\theta_{i1}, \theta_{i2}, \theta_{i3})',$$

and $\boldsymbol{U}_i = D_i$ as above. The intra-individual deviations $e_{ij} = Y_{ij} - m(t_{ij}, \boldsymbol{U}_i, \boldsymbol{\theta}_i)$ are assumed to be such that $E(e_{ij}|\boldsymbol{U}_i, \boldsymbol{\theta}_i) = 0$, so that

$$E(Y_{ij} \mid \boldsymbol{U}_i, \boldsymbol{\theta}_i) = m(t_{ij}, \boldsymbol{U}_i, \boldsymbol{\theta}_i) \quad \text{for each } j. \quad (5.10)$$

A standard assumption is that the e_{ij}, and hence Y_{ij}, are normally distributed, conditional on $\boldsymbol{U}_i$ and $\boldsymbol{\theta}_i$. Completing the individual-level model specification is discussed in Section 5.3.2.

In the population model (5.9), $\boldsymbol{d}$ is an r-dimensional function that describes the relationship between the elements of $\boldsymbol{\theta}_i$ and between-individual covariates $\boldsymbol{A}_i$ in terms of a $(p \times 1)$ fixed parameter $\boldsymbol{\beta}$, whose elements are referred to as fixed effects, and a $(q \times 1)$ vector $\boldsymbol{b}_i$ of random effects representing variation, "unexplained" once a systematic relationship between $\boldsymbol{\theta}_i$ and individual characteristics $\boldsymbol{A}_i$ is taken into account. The $\boldsymbol{b}_i$ are taken to be independent across i and to have a distribution that describes this variation with $E(\boldsymbol{b}_i|\boldsymbol{A}_i) = \boldsymbol{0}$. Commonly, this variation is assumed similar for all $\boldsymbol{A}_i$, represented by $\boldsymbol{b}_i$ independent of $\boldsymbol{A}_i$, so that

$$E(\boldsymbol{b}_i \mid \boldsymbol{A}_i) = E(\boldsymbol{b}_i) = \boldsymbol{0} \quad \text{and} \quad \text{Cov}(\boldsymbol{b}_i \mid \boldsymbol{A}_i) = \text{Cov}(\boldsymbol{b}_i) = G \quad (5.11)$$

for an unstructured covariance matrix G, and it is routine to take $\boldsymbol{b}_i \sim N(\boldsymbol{0}, G)$, so independent and identically distributed, similar to the standard assumptions for generalized linear mixed-effects models in Chapter 4. We adopt (5.11) in what follows, but the developments extend easily to situations where $\boldsymbol{b}_i$ and $\boldsymbol{A}_i$ are dependent. The $\boldsymbol{b}_i$ are also assumed independent of $\boldsymbol{U}_i$, which is natural in studies such as theophylline or argatroban where $\boldsymbol{U}_i$ (dose here) is dictated by design, and is likely also reasonable under systematic $\boldsymbol{U}_i$, such as chronic, regular dosing, as in the quinidine study.

The population model involves specification of the form of each element of $\boldsymbol{d}$. As an example of a typical population model, consider the quinidine study in Section 5.2.2, with $\boldsymbol{\theta}_i = (k_{ai}, V_i, Cl_i)'$ $(r = 3)$ and suppose that $\boldsymbol{A}_i = (w_i, \delta_i, g_i)'$, where w_i is weight; δ_i is an indicator of creatinine clearance, where $\delta_i = 1$ if > 50 ml/min; and g_i is α_1-acid glycoprotein concentration. The parameter's absorption rate, clearance, and volume in model (5.6) are known to be positive; moreover, it is widely accepted that they exhibit skewed distributions in the population. These considerations suggest a population model of the form similar to that considered by Davidian and Giltinan (1995, Section 9.3), given by

$$\begin{aligned}
k_{ai} &= \theta_{i1} = d_1(\boldsymbol{A}_i, \boldsymbol{\beta}, \boldsymbol{b}_i) = \exp(\beta_1 + b_{i1}), \\
V_i &= \theta_{i2} = d_2(\boldsymbol{A}_i, \boldsymbol{\beta}, \boldsymbol{b}_i) = \exp(\beta_2 + b_{i2}), \\
Cl_i &= \theta_{i3} = d_3(\boldsymbol{A}_i, \boldsymbol{\beta}, \boldsymbol{b}_i) = \exp(\beta_3 + \beta_4 w_i + \beta_5 \delta_i + \beta_6 g_i + b_{i3}),
\end{aligned} \quad (5.12)$$

where $\boldsymbol{b}_i = (b_{i1}, b_{i2}, b_{i3})'$ $(q = 3)$ and $\boldsymbol{\beta} = (\beta_1, \ldots, \beta_6)'$ $(p = 6)$. In (5.12), positivity of k_{ai}, V_i, and Cl_i is enforced automatically; moreover, if $\boldsymbol{b}_i$ is normal, then their distributions are lognormal (and hence skewed). The expression for Cl_i incorporates a systematic relationship with weight, creatinine clearance, and α_1-acid glycoprotein concentration that is linear on the log scale. The associated random effect b_{i3} accommodates biological variation left "unexplained" once this relationship is taken into account, allowing subjects sharing the same A_i value to still exhibit population variation in clearance. The assumption that b_i and A_i are independent implies that the extent of this variation is similar for all A_i; under (5.12), all three parameters exhibit constant coefficients of variation. Model (5.12) does not incorporate associations between either k_{ai} or V_i and the elements of A_i. Accordingly, an embedded assumption is that none of the variation in the population in either absorption rates or volumes may be attributed to systematic associations with weight, creatinine clearance, or α_1-acid glycoprotein concentration. Rather, population variation in these parameters is assumed to be entirely a consequence of "unexplained" biological variation.

In general, the chosen form of each element of d is based on knowledge of the situation, a combination of empirical and subject-matter considerations, and the inferential objectives. In the soybean study of Section 5.2.1, for which a goal was to compare "typical" asymptotic growth of genotypes F and P, $A_i = z_i$, where $z_i = 0$ or 1 if plot i is planted with F or P, respectively, determined by design. Here, one might begin by postulating a population model where the limiting growth value for plot i, θ_{i1}, satisfies

$$\text{(a)} \quad \theta_{i1} = \beta_1 + \beta_2 z_i + b_{i1} \quad \text{or} \quad \text{(b)} \quad \theta_{i1} = \exp(\beta_1 + \beta_2 z_i + b_{i1}), \tag{5.13}$$

and similarly for θ_{i2} and θ_{i3}. In (5.13)(a), with $E(b_{i1}) = 0$, "typical" value is interpreted as the mean of θ_{i1}, and β_2 represents the difference in mean limiting growth value between the two genotypes. In (5.13)(b), which enforces required positivity of θ_{i1}, with b_{i1} mean-zero normal, "typical" refers to median, as e^{β_1} is the median limiting growth for F, and e^{β_2} is the multiplicative factor by which this median value is altered for genotype P. Under (a) or (b), then, the desired comparison of "typical" limiting growth so defined may be addressed via inference on β_2; that is, determining whether the data contain adequate evidence to support $\beta_2 \neq 0$ versus a simpler model with $\beta_2 = 0$.

In contrast, in settings such as the quinidine study, where A_i is potentially high-dimensional and observational in nature, a main objective is to determine the form of d and the elements of A_i that should be incorporated in d to account for systematic associations between each component of θ_i and A_i. The functional form of each component of d may be dictated by positivity constraints and subject-matter knowledge (e.g., skewness of population distributions), as in (5.12), and the problem becomes one of "model building," as in ordinary regression, discussed further in Section 5.4.7. Once a final model like (5.12) is established, the "typical" values of meaningful parameters such as k_{ai}, Cl_i, and V_i for the subpopulations of subjects sharing the characteristics (elements of A_i) involved are estimated via inference on β.

In (5.12), each component of θ_i has an associated random effect representing unexplained population variation in that component, as is ordinarily the case in these models. In some circumstances, population models are postulated in which a component of θ_i has no such random effect; that is, in (5.12), the specification for Cl_i might instead be taken as

$$Cl_i = \exp(\beta_3 + \beta_4 w_i + \beta_5 \delta_i + \beta_6 g_i). \tag{5.14}$$

In (5.14), all variation in Cl_i values in the population is attributed to systematic associations with elements of A_i. In most biological contexts, it is implausible that there would be no additional, unexplained variation in a parameter like clearance among individuals sharing the same weight, renal status, and α_1-acid glycoprotein concentration. Thus, a specification like (5.14) is usually adopted as an approximation when the magnitude of unexplained variation in that component of θ_i is sufficiently small relative to that in the others that computational difficulties arise in fitting the model. Almost always, then, such a model should not be interpreted as embodying a belief in "perfect" associations in the population.

Model (5.13)(a) is an additive, linear model in elements of $\boldsymbol{\beta}$ and $\boldsymbol{b}_i$. Although (5.12) and (5.13)(b) demonstrate that more general, non-linear specifications are possible and may be appropriate in some subject-matter settings, some accounts of non-linear mixed-effects models (e.g., Lindstrom and Bates, 1990) restrict $\boldsymbol{d}$ to be of the traditional linear form

$$\boldsymbol{\theta}_i = A_i\boldsymbol{\beta} + B_i\boldsymbol{b}_i, \tag{5.15}$$

where A_i $(r \times p)$ is a "design matrix" whose rows depend on the elements of $\boldsymbol{A}_i$, and B_i $(r \times q)$ typically has elements that are zeros or ones, permitting some components of $\boldsymbol{\theta}_i$ to have no associated random effect when $r > q$; when $r = q$, $B_i = I_r$, an r-dimensional identity matrix. It is often possible to adopt a linear population model as in (5.15) by suitable reparameterization of the individual non-linear model $m(t, \cdot, \cdot)$ in (5.8). This is trivial for log-linear models like (5.12). Model (5.6) for quinidine concentration may be expressed alternatively in terms of the natural logarithms of absorption rate, volume, and clearance, so that $\boldsymbol{\theta}_i = (k_{ai}^*, V_i^*, Cl_i^*)'$, where $k_{ai}^* = \log(k_{ai})$, $V_i^* = \log(V_i)$, and $Cl_i^* = \log(Cl_i)$. Under this parameterization of (5.6), a population model achieving the same positivity and skewness requirements on the original scales of the pharmacokinetic parameters as does (5.12) is

$$\begin{aligned} k_{ai}^* &= \beta_1 + b_{i1}, \\ V_i^* &= \beta_2 + b_{i2}, \\ Cl_i^* &= \beta_3 + \beta_4 w_i + \beta_5 \delta_i + \beta_6 g_i + b_{i3}. \end{aligned} \tag{5.16}$$

It is straightforward to observe that (5.16) may be expressed in the linear form (5.15) with

$$A_i = \begin{pmatrix} 1 & 0 & 0 & 0 & 0 & 0 \\ 0 & 1 & 0 & 0 & 0 & 0 \\ 0 & 0 & 1 & w_i & \delta_i & g_i \end{pmatrix}, \quad B_i = I_3, \quad \boldsymbol{\beta} = (\beta_1, \ldots, \beta_6)'.$$

As observed by Davidian and Giltinan (1995, Section 4.2.3), in some areas, notably pharmacokinetics, the convention is to adopt non-linear population models as in (5.12), with the individual model $m(t, \cdot, \cdot)$ parameterized directly in terms of meaningful quantities, whereas statisticians traditionally have preferred expression via reparameterization of the individual model to achieve a linear second-stage specification, as in (5.16).

5.3.2 Within-individual considerations

As noted below expression (5.10), the individual-level model specification is completed by assumptions on the distribution of $\boldsymbol{Y}_i$, given $\boldsymbol{U}_i$ and $\boldsymbol{\theta}_i$. The considerations involved are those relevant in regression modeling of possibly serially correlated responses on a single individual, as we now describe, adopting a conceptual perspective laid out in Verbeke and Molenberghs (2000, Section 3.3), Diggle et al. (2002, Chapter 5), Davidian and Giltinan (2003, Section 2.2.2), and Fitzmaurice, Laird, and Ware (2004, Section 2.5).

Focusing on a single individual i, who is observed under conditions $\boldsymbol{U}_i$, we may think of the observations Y_{ij} at times t_{ij}, $j = 1, \ldots, n_i$, on i as arising from an individual-specific stochastic process

$$Y_i(t, \boldsymbol{U}_i) = m(t, \boldsymbol{U}_i, \boldsymbol{\theta}_i) + e_i(t, \boldsymbol{U}_i), \tag{5.17}$$

where $e_i(t, \boldsymbol{U}_i)$ is the "deviation" process with $E\{e_i(t, \boldsymbol{U}_i) \mid \boldsymbol{U}_i, \boldsymbol{\theta}_i\} = 0$, for all t, so that

$$E\{Y_i(t, \boldsymbol{U}_i) \mid \boldsymbol{U}_i, \boldsymbol{\theta}_i\} = m(t, \boldsymbol{U}_i, \boldsymbol{\theta}_i), \quad \text{for all } t. \tag{5.18}$$

Under (5.17), $Y_{ij} = Y_i(t_{ij}, \boldsymbol{U}_i)$ and $e_{ij} = e_i(t_{ij}, \boldsymbol{U}_i)$, and, following (5.18), (5.10) holds. Thus, the Y_{ij} at times t_{ij} may be viewed as intermittent observations on the process

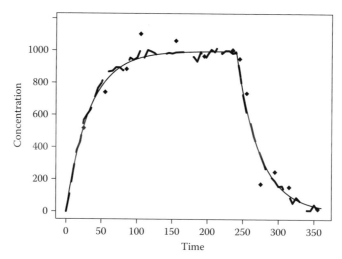

Figure 5.3 Conceptual depiction of intra-individual sources of variation. For individual i, the thin, solid line is $m(t, \boldsymbol{U}_i, \boldsymbol{\theta}_i)$, the "inherent trajectory" for i; the thick, dashed line is a realization of the response through continuous time; and the solid symbols are measurements of the realization at intermittent time points, contaminated by measurement error.

(5.17) for i, and the actual data collected on i are the result of a single realization of the process $Y_i(t, \boldsymbol{U}_i)$. The process $e_i(t, \boldsymbol{U}_i)$ is a convenient representation of all sources of variation acting within an individual at any time that cause a realization of the process $Y_i(t, \boldsymbol{U}_i)$ to deviate from the "smooth" behavior dictated by the individual-specific model $m(t, \boldsymbol{U}_i, \boldsymbol{\theta}_i)$.

Figure 5.3 gives a hypothetical depiction of such sources of variation in the context of an individual from the argatroban study. The dashed line represents an actual realization of the response, concentration here, if it could be observed continually, without measurement error. Thus, the figure takes the view that $m(t, \boldsymbol{U}_i, \boldsymbol{\theta}_i)$, which traces out the trajectory given by the solid line, is an "idealized" representation of any realization of the response. As noted by Davidian and Giltinan (2003, Section 2.2.2), in pharmacokinetics, actual, realized drug concentrations may not follow perfectly the smooth curve derived from a compartment model because the model is a simplification of the true physiology governing achieved concentrations and assumes, for example, that the drug is "well mixed" within the body. The solid symbols are actual, intermittent measurements recorded on the dashed realization; these deviate from the dashed line due to measurement error; for example, assay error.

Figure 5.3 thus identifies two intra-individual sources of variation that a model for the distribution of $\boldsymbol{Y}_i$, given $\boldsymbol{U}_i$ and $\boldsymbol{\theta}_i$ must acknowledge: "realization variation," due to the tendency for error-free realizations of the response to deviate from the "smooth" behavior dictated by $m(t, \cdot, \cdot)$; and measurement-error variation. From this point of view, (5.18) represents the "average" over all possible realizations of the longitudinal response trajectory (and over all possible measurement errors that could be committed) that could be observed on i over time. Accordingly, following Davidian and Giltinan (2003, Section 2.2.2), we may regard $m(t, \boldsymbol{U}_i, \boldsymbol{\theta}_i)$ as the "inherent tendency" for i's response to evolve over time under conditions $\boldsymbol{U}_i$, where $\boldsymbol{\theta}_i$ is an "inherent characteristic" of i that prescribes this tendency.

To formalize, we may view the deviation process $e_i(t, \boldsymbol{U}_i)$ as the sum of deviations due to these two sources, and write, following Davidian and Giltinan (2003, Section 2.2.2),

$$e_i(t, \boldsymbol{U}_i) = e_{R,i}(t, \boldsymbol{U}_i) + e_{M,i}(t, \boldsymbol{U}_i), \quad E\{e_{R,i}(t, \boldsymbol{U}_i)|\boldsymbol{U}_i, \boldsymbol{\theta}_i\} = E\{e_{M,i}(t, \boldsymbol{U}_i)|\boldsymbol{U}_i, \boldsymbol{\theta}_i\} = 0,$$

where $e_{R,i}(t, \boldsymbol{U}_i)$ represents that part of the overall deviation from $m(t, \boldsymbol{U}_i, \boldsymbol{\theta}_i)$ at t due to realization variation and $e_{M,i}(t, \boldsymbol{U}_i)$ that part due to measurement error. Alternative representations involving, for example, multiplicative deviations, are also possible; that given here serves to exemplify considerations the data analyst must make in adopting a model for intra-individual variation. Assumptions on these two components lead to a specification for the covariance matrix of $\boldsymbol{Y}_i$, given $\boldsymbol{U}_i$ and $\boldsymbol{\theta}_i$, as we now review.

Referring to Figure 5.3, it is natural to suppose that two "realization" deviations at times t and s, say, $e_{R,i}(t, \boldsymbol{U}_i)$ and $e_{R,i}(s, \boldsymbol{U}_i)$, would tend to be positively correlated for t and s close together in time, as realized deviations at nearby time points tend to deviate from the solid "inherent trajectory" $m(t, \boldsymbol{U}_i, \boldsymbol{\theta}_i)$ in the same direction. Deviations far apart in time seem as likely to deviate in the same or opposite direction. This suggests models for the autocorrelation function of the process $e_{R,i}(t, \boldsymbol{U}_i)$ for which correlation "dies out" as a function of the time interval. Ordinarily, models based on stationarity assumptions, such as the exponential correlation function given by

$$\text{Corr}\{e_{R,i}(t, \boldsymbol{U}_i), e_{R,i}(s, \boldsymbol{U}_i) \,|\, \boldsymbol{U}_i, \boldsymbol{\theta}_i\} = \exp(-\rho|t - s|), \quad \rho \geq 0, \tag{5.19}$$

are postulated; related models were described by Diggle et al. (2002, Chapter 5).

The nature of the response dictates assumptions that the analyst may be willing to make on $\text{Var}\{e_{R,i}(t, \boldsymbol{U}_i) \,|\, \boldsymbol{U}_i, \boldsymbol{\theta}_i\}$. If one believes that variation of realizations about the smooth trajectory are of similar magnitude over time and individuals, it is natural to assume $\text{Var}\{e_{R,i}(t, \boldsymbol{U}_i) \,|\, \boldsymbol{U}_i, \boldsymbol{\theta}_i\} = \sigma_R^2 \geq 0$, say; that is, constant "realization variance." Instead, if one believes that realization variance may depend on the magnitude of the "inherent" response, as might be expected if error-free responses have a skewed distribution, this variance can be taken to depend on $m(t, \boldsymbol{U}_i, \boldsymbol{\theta}_i)$ and possibly additional variance parameters.

Because well-maintained measuring devices tend to commit haphazard errors over repeated uses, it is conventional to assume that the measurement error deviation process $e_{M,i}(t, \boldsymbol{U}_i)$ exhibits no autocorrelation; that is, $\text{Corr}\{e_{M,i}(t, \boldsymbol{U}_i), e_{M,i}(s, \boldsymbol{U}_i) \,|\, \boldsymbol{U}_i, \boldsymbol{\theta}_i\} = 0$ for all $t > s$. An assumption about the variance $\text{Var}\{e_{M,i}(t, \boldsymbol{U}_i) \,|\, \boldsymbol{U}_i, \boldsymbol{\theta}_i\}$ should follow from knowledge of the measuring device. For many devices, such as weighing equipment, the magnitude of measurement errors is similar regardless of the size of the item measured, in which case it is natural to take $\text{Var}\{e_{M,i}(t, \boldsymbol{U}_i) \,|\, \boldsymbol{U}_i, \boldsymbol{\theta}_i\} = \sigma_M^2 \geq 0$, a constant. Alternatively, some assay procedures tend to involve measurement errors that increase in magnitude with increasing magnitude of the true response being measured. As discussed by Davidian and Giltinan (2003, p. 398), it is often reasonable to approximate this by taking $\text{Var}\{e_{M,i}(t, \boldsymbol{U}_i) \,|\, \boldsymbol{U}_i, \boldsymbol{\theta}_i\}$ to be a function of $m(t, \boldsymbol{U}_i, \boldsymbol{\theta}_i)$ and possibly additional variance parameters.

Combining all of the foregoing considerations, writing $e_{R,ij} = e_{R,i}(t_{ij}, \boldsymbol{U}_i)$ and $e_{M,ij} = e_{M,i}(t_{ij}, \boldsymbol{U}_i)$, so that $e_{ij} = e_{R,ij} + e_{M,ij}$, and letting $\boldsymbol{e}_i = (e_{i1}, \ldots, e_{in_i})' = \boldsymbol{e}_{R,i} + \boldsymbol{e}_{M,i}$ (in obvious notation), a model for $\text{Cov}(\boldsymbol{e}_i \,|\, \boldsymbol{U}_i, \boldsymbol{\theta}_i)$ and hence $\text{Cov}(\boldsymbol{Y}_i \,|\, \boldsymbol{U}_i, \boldsymbol{\theta}_i)$, may be deduced. It is common to assume that the processes $e_{R,i}(t, \boldsymbol{U}_i)$ and $e_{M,i}(t, \boldsymbol{U}_i)$ are independent, in which case $\text{Cov}(\boldsymbol{e}_i \,|\, \boldsymbol{U}_i, \boldsymbol{\theta}_i)$ is the sum of the covariance matrices of $\boldsymbol{e}_{R,i}$ and $\boldsymbol{e}_{M,i}$ dictated by the assumptions made on $e_{R,i}(t, \boldsymbol{U}_i)$ and $e_{M,i}(t, \boldsymbol{U}_i)$, which we write as

$$\text{Cov}(\boldsymbol{e}_i \,|\, \boldsymbol{U}_i, \boldsymbol{\theta}_i) = V_{R,i}(\boldsymbol{U}_i, \boldsymbol{\theta}_i, \boldsymbol{\alpha}_R) + V_{M,i}(\boldsymbol{U}_i, \boldsymbol{\theta}_i, \boldsymbol{\alpha}_M) = V_i(\boldsymbol{U}_i, \boldsymbol{\theta}_i, \boldsymbol{\alpha}) \quad (n_i \times n_i), \tag{5.20}$$

where $\boldsymbol{\alpha} = (\boldsymbol{\alpha}_R', \boldsymbol{\alpha}_M')'$. In (5.20), $V_{R,i}$ and $V_{M,i}$ are the assumed covariance matrices of $\boldsymbol{e}_{R,i}$ and $\boldsymbol{e}_{M,i}$, possibly depending on $\boldsymbol{U}_i$, $\boldsymbol{\theta}_i$, and parameters $\boldsymbol{\alpha}_R$ and $\boldsymbol{\alpha}_M$, say. Typically, $V_{R,i}$ would be a non-diagonal matrix, with off-diagonal elements embodying the assumption on autocorrelation of the realization process, while $V_{M,i}$ would be diagonal, reflecting the assumption of independent measurement errors. For example, assuming constant realization variance σ_R^2 as above with exponential autocorrelation function (5.19), $V_{R,i}(\boldsymbol{U}_i, \boldsymbol{\theta}_i, \boldsymbol{\alpha}_R)$ is the $(n_i \times n_i)$ matrix with (j, k)th element $\sigma_R^2 \exp(-\rho|t_{ij} - t_{ik}|)$, where $\boldsymbol{\alpha}_R = (\sigma_R^2, \rho)'$. Similarly, assuming constant measurement-error variance, $V_{M,i}(\boldsymbol{U}_i, \boldsymbol{\theta}_i, \boldsymbol{\alpha}_R) = \sigma_M^2 I_{n_i}$, and $\boldsymbol{\alpha}_M = \sigma_M^2$. Combining as in (5.20), we have $\boldsymbol{\alpha} = (\sigma_R^2, \rho, \sigma_M^2)'$.

Practical considerations usually suggest simplified versions of (5.20). Some responses may be ascertained with no or negligible measurement error; for example, the height of a child. In this case, $e_{M,i}(t, \boldsymbol{U}_i)$ would be deleted from the model, so that V_i in (5.20) would depend only on assumptions on the realization process. In many longitudinal studies in humans, the times t_{ij} at which responses are observed may be at widely spaced intervals over which autocorrelation of the realization process may be reasonably thought to have "died out" sufficiently so as to be negligible. In this case, the matrix $V_{R,i}$ would be taken to be diagonal, leading to adoption of the simple form

$$V_i(\boldsymbol{U}_i, \boldsymbol{\theta}_i, \boldsymbol{\alpha}) = \sigma_e^2 I_{n_i}, \quad \boldsymbol{\alpha} = \sigma_e^2, \tag{5.21}$$

so that $\mathrm{Var}(e_{ij} \,|\, \boldsymbol{U}_i, \theta_i) = \sigma_e^2$. The meaning of σ_e^2 in (5.21) would depend on the analyst's beliefs. If it were assumed that the realization process has constant variance σ_R^2 with negligible autocorrelation, and that measurement error has non-negligible, constant variance σ_M^2, then (5.20) implies (5.21) with $\sigma_e^2 = \sigma_R^2 + \sigma_M^2$. If measurement error were negligible, $\sigma_e^2 = \sigma_R^2$; on the other hand, if measurement error were thought to be the dominant source of intra-individual variation, $\sigma_e^2 = \sigma_M^2$. Although, ideally, (5.21) should be adopted only if the assumptions on intra-individual variation with which it is consistent are likely to hold, it is often used by default in practice with little thought given to the assumptions it represents. See Davidian and Giltinan (2003, Section 2.2.2) for more discussion of issues involved in specifying V_i.

Given a specification for $\mathrm{Cov}(\boldsymbol{Y}_i \,|\, \boldsymbol{U}_i, \boldsymbol{\theta}_i) = V_i(\boldsymbol{U}_i, \boldsymbol{\theta}_i, \boldsymbol{\alpha})$, as noted in Equation (5.10), it is routine to assume that the distribution of $\boldsymbol{Y}_i$, given $\boldsymbol{U}_i$ and $\boldsymbol{\theta}_i$, is multivariate normal with this covariance matrix and, from (5.10), mean vector $E(\boldsymbol{Y}_i \,|\, \boldsymbol{U}_i, \boldsymbol{\theta}_i) = \boldsymbol{m}_i(\boldsymbol{U}_i, \boldsymbol{\theta}_i)$, where

$$\boldsymbol{m}_i(\boldsymbol{U}_i, \boldsymbol{\theta}_i) = \{m(t_{i1}, \boldsymbol{U}_i, \boldsymbol{\theta}_i), \ldots, m(t_{in_i}, \boldsymbol{U}_i, \boldsymbol{\theta}_i)\}' \quad (n_i \times 1).$$

Another assumption sometimes made in pharmacokinetics is that each Y_{ij} is lognormally distributed, given $\boldsymbol{U}_i$ and $\boldsymbol{\theta}_i$. This is often incorporated by replacing Y_{ij} and $m(t_{ij}, \boldsymbol{U}_i, \boldsymbol{\theta}_i)$ in the foregoing development by $\log(Y_{ij})$ and $\log\{m(t_{ij}, \boldsymbol{U}_i, \boldsymbol{\theta}_i)\}$, respectively, and assuming $\log(Y_{ij})$, and hence the vector of $\log(Y_{ij})$, is conditionally normally distributed. In subsequent sections, we take the conditional distribution of $\boldsymbol{Y}_i$, given $\boldsymbol{U}_i$ and $\boldsymbol{\theta}_i$, to be multivariate normal, recognizing that this may be relevant on a transformed (log) scale.

5.3.3 Summary of the model

We summarize the non-linear mixed-effects model by writing the stage 1, individual-level model in (5.8) in vector form, incorporating the considerations in Section 5.3.2, and substituting the stage 2, population model in (5.9) for $\boldsymbol{\theta}_i$, so that conditioning is with respect to $\boldsymbol{X}_i$, the vector of all within- and between-individual covariates, and $\boldsymbol{b}_i$, the vector of random effects that dictates $\boldsymbol{\theta}_i$. We emphasize dependence of the conditional moments on $\boldsymbol{\beta}$ and $\boldsymbol{\alpha}$:

Stage 1: Individual-Level Model

$$E(\boldsymbol{Y}_i \,|\, \boldsymbol{X}_i, \boldsymbol{b}_i) = \boldsymbol{m}_i(\boldsymbol{U}_i, \boldsymbol{\theta}_i) = \boldsymbol{m}_i(\boldsymbol{X}_i, \boldsymbol{\beta}, \boldsymbol{b}_i),$$
$$\mathrm{Cov}(\boldsymbol{Y}_i \,|\, \boldsymbol{X}_i, \boldsymbol{b}_i) = V_i(\boldsymbol{U}_i, \boldsymbol{\theta}_i, \boldsymbol{\alpha}) = V_i(\boldsymbol{X}_i, \boldsymbol{\beta}, \boldsymbol{b}_i, \boldsymbol{\alpha}). \tag{5.22}$$

Stage 2: Population Model $\qquad \boldsymbol{\theta}_i = \boldsymbol{d}(\boldsymbol{A}_i, \boldsymbol{\beta}, \boldsymbol{b}_i), \quad \boldsymbol{b}_i \sim (\boldsymbol{0}, G). \tag{5.23}$

As indicated previously, the conventional assumption is that the distribution of $\boldsymbol{Y}_i$, given $\boldsymbol{X}_i$ and $\boldsymbol{b}_i$, is multivariate normal with moments given in (5.22); that $\boldsymbol{b}_i \sim N(\boldsymbol{0}, G)$, independently of $\boldsymbol{A}_i$ (and $\boldsymbol{U}_i$, and hence $\boldsymbol{X}_i$); and that $\boldsymbol{b}_i$ are independent across i. We do not explicitly adopt normality of the $\boldsymbol{b}_i$ in (5.23), as there are settings, discussed presently, where the analyst may wish to invoke a less restrictive assumption on the distribution of the $\boldsymbol{b}_i$.

5.3.4 Comparison to generalized linear mixed-effects models

As we have noted throughout, the non-linear mixed-effects model in (5.22) and (5.23) shares common features with the generalized linear mixed-effects model introduced in Chapter 4. We may now review the similarities and differences in more detail.

As discussed in Section 4.3.1, under a generalized linear mixed model, the elements of Y_i, Y_{ij}, which may be continuous or discrete, each have mean, conditional within- and between-individual covariates and random effects b_i; that is, a non-linear function of fixed effects β and b_i. However, as shown in (4.2), in contrast to the situation in this chapter, this non-linear function is taken to depend on β and b_i through a linear combination of these quantities, referred to as the "linear predictor." Thus, this function, denoted as $h(\cdot)$ in Chapter 4, has a single argument, in which it is usually monotone, and h^{-1} is referred to as the "link function." Instead of arising through mechanistic or other subject-matter considerations, h is generally chosen from a suite of standard link functions tied to the type of response (binary, count, continuous, etc.), as reviewed in Equation (4.3) through Equation (4.6). In contrast, non-linear mixed-effects models may have conditional mean of Y_{ij} that is much more general in form, as for the pharmacokinetic models in Equation (5.4) through Equation (5.6). Hence, we use the alternative notation $m(\cdot, \cdot, \cdot)$ to denote the conditional mean response to highlight that m often arises from theoretical rather than standard statistical modeling considerations.

Generalized linear mixed-effects models also conventionally take the elements of Y_i to be conditionally independent, so that the conditional covariance matrix V_i is diagonal by assumption, although, as discussed in Section 4.4.3, as with non-linear mixed-effects models, such conditional independence is not necessary. The conditional variances on the diagonal are a known function of the conditional mean times a dispersion parameter, which is dictated by the further assumption that, for each j, the conditional distribution of Y_{ij} has density from the scaled exponential family, which includes the normal density with constant variance as a special case. In non-linear mixed-effects models, this conditional distribution is almost always taken to be normal, but the conditional variance need not be constant and is specified according to beliefs about intra-individual sources of variation, as in Section 5.3.2.

In both models, it is usually assumed that $b_i \sim N(0, G)$, and both may be viewed as arising from a "two-stage" formulation, as in (5.22) and (5.23) for the non-linear mixed model and, as discussed in Section 4.4.1, for the generalized linear mixed model. Thus, both models are "subject-specific" in spirit. Given these and other similarities between the models, then, it is not surprising that standard methods for implementing non-linear mixed-effects models, discussed in Section 5.4, are similar to those for generalized linear mixed-effects models, reviewed in Section 4.6 and Section 4.7. However, it should be clear that the generalized linear mixed-effects model is much more of an empirically driven statistical framework than the non-linear mixed-effects model. The form of the latter is usually governed by subject-matter theory and considerations, which has implications for interpretation, discussed next.

5.3.5 Model interpretation and inferential objectives

Because most non-linear mixed-effects models embed a theoretical or mechanistic model for individual time trajectories in the two-stage statistical hierarchy (5.22) and (5.23), as we have pointed out repeatedly, the individual-specific parameters θ_i almost always represent real phenomena of scientific interest. Within the context of a particular population model d, the fixed-effects parameter β is clearly interpreted as defining the "typical value" of meaningful θ_i in the population and as quantifying the strength of systematic relationships between typical values and individual attributes A_i, as in the examples in (5.12) through (5.16). Accordingly, β characterizes the typical "subject-specific" behavior of these phenomena, and inferences on β are a central focus of any analysis. In contrast, as discussed by Davidian and Giltinan (2003, Section 2.4), there is generally no interest in so-called "population-averaged" effects, which have to do with describing the "typical response." Inferences on the average response are generally of little or no importance; rather, the response is of interest because it carries information on the underlying features, $\boldsymbol{\theta}_i$, and hence their typical behavior, represented by $\boldsymbol{\beta}$. See Chapter 7 for in-depth discussion of these considerations.

The covariance matrix G of the random effects provides information on the variation among subjects of the individual-specific parameters $\boldsymbol{\theta}_i$ above and beyond that explained by associations with available between-individual covariates. Likewise, it also has a "subject-specific" interpretation, and inferences on G are of considerable scientific importance as well. For example, if the magnitude of unexplained variation in pharmacokinetic parameters for subjects sharing common characteristics is relatively large, it may be difficult to develop broadly applicable dosing recommendations for such patients. If such circumstances are revealed via estimation of G, this may suggest attempting to identify additional individual attributes associated with the $\boldsymbol{\theta}_i$ that can further "explain" some of this apparent unexplained variation, hence rendering dosing guidelines for subpopulations feasible. Alternatively, this may indicate a need to base dosing on more "individualized" considerations.

"Individualized" dosing in pharmacokinetics exemplifies a setting in which the additional objective of inference on individual $\boldsymbol{\theta}_i$ is of interest. More generally, there may be interest in making individual-specific predictions of the (conditional) mean response at a future time t_0; for example, in the context of treatment of HIV, subject-specific predictions of $m(t_0, \boldsymbol{U}_i, \boldsymbol{\theta}_i)$ under a particular treatment strategy may be of value in informing patient management decisions. The hierarchical structure of (5.22) and (5.23) provides a natural framework for this enterprise that supports "borrowing" of information across individuals sharing common characteristics. Such "individual" inferences are discussed in Section 5.4.6.

5.4 Inferential approaches

In this section, we review the most popular methods used in practice for inference in non-linear mixed-effects models. Details were given by Davidian and Giltinan (1995, 2003) and in cited references. Accounts of further methodological developments are noted in Section 5.4.8.

5.4.1 Maximum likelihood estimation

A natural basis for inference in a model like (5.22) and (5.23) is maximum likelihood. The observed data are independent pairs $(\boldsymbol{Y}_i, \boldsymbol{X}_i)$, $i = 1, \ldots, N$, which we summarize as $(\boldsymbol{Y}, \boldsymbol{X})$, $\boldsymbol{Y} = (\boldsymbol{Y}_1', \ldots, \boldsymbol{Y}_N')'$ and $\boldsymbol{X} = (\boldsymbol{X}_1', \ldots, \boldsymbol{X}_N')'$, and the likelihood for the parameters in the model may be based on the joint density of $\boldsymbol{Y}$, given $\boldsymbol{X}$, which, by independence, factors into the product of N densities of $\boldsymbol{Y}_i$, given $\boldsymbol{X}_i$. We write the ith such density as

$$f_i(\boldsymbol{y}_i \mid \boldsymbol{x}_i; \boldsymbol{\gamma}, G), \quad \boldsymbol{\gamma} = (\boldsymbol{\beta}', \boldsymbol{\alpha}'). \tag{5.24}$$

Dependence of (5.24) on parameters $\boldsymbol{\gamma}$ and G follows from noting that, for each i, it may be represented as the integral with respect to $\boldsymbol{b}_i$ of the product of (i) the conditional density of $\boldsymbol{Y}_i$ given $\boldsymbol{X}_i$ and $\boldsymbol{b}_i$, $f_i(\boldsymbol{y}_i \mid \boldsymbol{x}_i, \boldsymbol{b}_i; \boldsymbol{\gamma})$, which from (5.22) depends on $\boldsymbol{\gamma}$ and is usually multivariate normal; and (ii) that of $\boldsymbol{b}_i$ given $\boldsymbol{X}_i$. Under the assumption that the $\boldsymbol{b}_i$ are independent of $\boldsymbol{X}_i$ for each i, density (ii) reduces to the marginal density of $\boldsymbol{b}_i$. When $\boldsymbol{b}_i$ is assumed to follow a specific distribution, as with the standard assumption that it is $N(\boldsymbol{0}, G)$, we emphasize dependence on parameters and write, for example, $f(\boldsymbol{b}_i; G)$. With these specifications, the log-likelihood for $(\boldsymbol{\gamma}, G)$ is

$$\ell(\boldsymbol{\gamma}, G) = \log \left\{ \prod_{i=1}^{N} f_i(\boldsymbol{y}_i \mid \boldsymbol{x}_i; \boldsymbol{\gamma}, G) \right\} = \log \left\{ \prod_{i=1}^{N} \int f_i(\boldsymbol{y}_i \mid \boldsymbol{x}_i, \boldsymbol{b}_i; \boldsymbol{\gamma}) \, f(\boldsymbol{b}_i; G) \, d\boldsymbol{b}_i \right\}. \tag{5.25}$$

The q-dimensional integrals in (5.25) are analytically intractable in all but the simplest situations; thus, algorithms to maximize $\ell(\gamma, G)$ in γ and the distinct elements of G require a means to evaluate these integrals. As numerical evaluation can be computationally intensive when required repeatedly in the context of an optimization algorithm when $q > 1$, historically, inference on γ and G has been advocated based on analytical approximations to (5.25). It is worth noting that pharmacokineticists were among the first to propose versions of such approximations for fitting non-linear mixed models (e.g., Beal and Sheiner, 1982; Steimer et al., 1984). Indeed, they pioneered the use of non-linear mixed-effects models for the types of inferential objectives discussed in this chapter; see Sheiner, Rosenberg, and Melmon (1972) and Sheiner, Rosenberg, and Marathe (1977).

We discuss analytical approximation methods in the next two sections and then return to numerical approximation methods in Section 5.4.4.

5.4.2 Inference based on individual estimates

When sufficient data are available on each individual, it may be possible to fit the individual-level "regression model" in (5.22) to these data alone to estimate $\boldsymbol{\theta}_i$ for individual i. This certainly requires that $n_i \geq r$, but, practically speaking, in most contexts, n_i would need to be much larger to achieve reliable estimates. When the intra-individual covariance matrix V_i is diagonal, such estimation of $\boldsymbol{\theta}_i$ along with parameters $\boldsymbol{\alpha}$ in V_i may be carried out via standard weighted regression methods for independent data, described in Davidian and Giltinan (1995, Chapter 2). In particular, when $V_i = \sigma_e^2 I_{n_i}$, ordinary non-linear least squares may be used. For non-diagonal V_i depending on intra-individual variance and correlation parameters, this is more involved; see Davidian and Giltinan (1995, Section 5.2). Because the parameters $\boldsymbol{\alpha}$ in V_i are usually assumed common across individuals, Davidian and Giltinan (1995, Section 5.2) proposed an approach based on "pooling" information from all N individuals to estimate $\boldsymbol{\alpha}$ while simultaneously estimating all of the $\boldsymbol{\theta}_i$, $i = 1, \ldots, N$.

Methods based on individual estimates $\widehat{\boldsymbol{\theta}}_i$, say, so obtained, use the $\widehat{\boldsymbol{\theta}}_i$, $i = 1, \ldots, N$, as "data" from which inference on the parameters $\boldsymbol{\beta}$ and G that dictate the population distribution of the true $\boldsymbol{\theta}_i$ may be drawn. However, appropriate account of the fact that the $\widehat{\boldsymbol{\theta}}_i$ are not the true $\boldsymbol{\theta}_i$ must be taken. This may be conceived as follows. For n_i "large," standard asymptotic theory for $\widehat{\boldsymbol{\theta}}_i$ may be regarded to hold conditional on $\boldsymbol{\theta}_i$ (and $\boldsymbol{U}_i$) for each i. That is, for large n_i, we have the approximation

$$\widehat{\boldsymbol{\theta}}_i \,|\, \boldsymbol{U}_i, \boldsymbol{\theta}_i \,\dot{\sim}\, N(\boldsymbol{\theta}_i, C_i), \tag{5.26}$$

where C_i depends on $\boldsymbol{\theta}_i$ and $\boldsymbol{\alpha}$ in general and may be estimated by $\widehat{C}_i$ obtained by substituting $\widehat{\boldsymbol{\theta}}_i$ and the estimate for $\boldsymbol{\alpha}$.

To demonstrate how (5.26) is used to facilitate inference on $\boldsymbol{\beta}$ and G, we consider a linear population model of the form (5.15); see Davidian and Giltinan (1995, Section 5.3.4) for the case of general population models $\boldsymbol{d}$. From (5.26), substituting $\widehat{C}_i$ for C_i, we have the further approximation $\widehat{\boldsymbol{\theta}}_i \,|\, \boldsymbol{U}_i, \boldsymbol{\theta}_i \,\dot{\sim}\, N(\boldsymbol{\theta}_i, \widehat{C}_i)$, which may be written equivalently as $\widehat{\boldsymbol{\theta}}_i \approx \boldsymbol{\theta}_i + \boldsymbol{e}_i^*$, where the "deviation" $\boldsymbol{e}_i^*$, given $\boldsymbol{U}_i$ and $\boldsymbol{\theta}_i$, is approximately $N(\boldsymbol{0}, \widehat{C}_i)$. Substituting the population model yields the approximate "linear mixed-effects model"

$$\widehat{\boldsymbol{\theta}}_i \approx A_i \boldsymbol{\beta} + B_i \boldsymbol{b}_i + \boldsymbol{e}_i^*, \quad \boldsymbol{b}_i \sim N(\boldsymbol{0}, G), \quad \boldsymbol{e}_i^* \,|\, \boldsymbol{U}_i, \boldsymbol{\theta}_i \,\dot{\sim}\, N(\boldsymbol{0}, \widehat{C}_i), \tag{5.27}$$

of the form in (4.1), with, treating the $\widehat{C}_i$ as fixed, known "residual" covariance matrices $\widehat{C}_i$. This suggests that standard methods for fitting these models may be used for inferences on $\boldsymbol{\beta}$ and G. Steimer et al. (1984) proposed using an EM algorithm, described by Davidian and Giltinan (1995, Section 5.3.2) for $r = q$ and $B_i = I_r$. Alternatively, one may use linear mixed-model software such as SAS `proc mixed` (Littell et al., 1996) or the S-Plus/R routine lme (Pinheiro and Bates, 2000). Davidian and Giltinan (2003, Section 3.2) showed that, by pre-multiplying both sides of (5.27) by a "square root" matrix $\widehat{C}_i^{-1/2}$ of $\widehat{C}_i^{-1}$, (5.27) becomes

$$\widetilde{\boldsymbol{Y}}_i \approx \widetilde{X}_i \boldsymbol{\beta} + \widetilde{Z}_i \boldsymbol{b}_i + \widetilde{\boldsymbol{e}}_i, \quad \widetilde{\boldsymbol{Y}}_i = \widehat{C}_i^{-1/2} \widehat{\boldsymbol{\theta}}_i, \quad \widetilde{X}_i = \widehat{C}_i^{-1/2} A_i, \quad \widetilde{Z}_i = \widehat{C}_i^{-1/2} B_i, \qquad (5.28)$$

where $\widetilde{\boldsymbol{e}}_i \sim N(\boldsymbol{0}, I_r)$. This version of the "linear mixed-effects model" has diagonal "residual" covariance matrix of the popular form $\sigma_{\widetilde{e}}^2 I_r$ (Section 4.2) with $\sigma_{\widetilde{e}}^2 = 1$, so (5.28) is easily implemented in this software by adopting this simple intra-individual covariance structure and constraining this "residual variance" to equal 1. In any case, approximate standard errors for the components of the resulting estimator for $\boldsymbol{\beta}$ are obtained from the usual theory for linear mixed-effects models treating (5.27) as exact and $\widehat{C}_i$ as fixed.

That this approach follows from analytical approximation of the integrals in (5.25) may not be readily apparent. As noted by Davidian and Giltinan (2003, Section 3.2), one may view the $\widehat{\boldsymbol{\theta}}_i$ as approximate "sufficient statistics" for the $\boldsymbol{\theta}_i$ and reason that, with a change of variables in the integrals, this method follows from replacing $f_i(\boldsymbol{y}_i \mid \boldsymbol{x}_i, \boldsymbol{b}_i; \boldsymbol{\gamma})$ by the (normal) density $f(\widehat{\boldsymbol{\theta}}_i \mid \boldsymbol{U}_i, \boldsymbol{\theta}_i; \boldsymbol{\beta})$ corresponding to the approximation (5.26) with $\widehat{C}_i$ in place of C_i.

5.4.3 Inference based on analytic approximation to the likelihood

In many settings, such as population studies of pharmacokinetics, rich data on each individual to facilitate individual fitting are not available. Under these conditions, and more generally regardless of n_i, an alternative approach is to approximate the integrals in (5.25) more directly by deriving an approximation to $f_i(\boldsymbol{y}_i \mid \boldsymbol{x}_i; \boldsymbol{\gamma}, G)$.

We discuss two classes of approaches to obtaining an approximation to $f_i(\boldsymbol{y}_i \mid \boldsymbol{x}_i; \boldsymbol{\gamma}, G)$, both of which may be motivated by considering a first-order Taylor series as follows. Write the model (5.22) and (5.23) under normality assumptions at both stages equivalently as

$$\boldsymbol{Y}_i = \boldsymbol{m}_i(\boldsymbol{X}_i, \boldsymbol{\beta}, \boldsymbol{b}_i) + V_i^{1/2}(\boldsymbol{X}_i, \boldsymbol{\beta}, \boldsymbol{b}_i, \boldsymbol{\alpha}) \, \boldsymbol{\varepsilon}_i, \quad \boldsymbol{b}_i \sim N(\boldsymbol{0}, G), \qquad (5.29)$$

where $V_i^{1/2}$ is a $(n_i \times n_i)$ matrix such that $V_i^{1/2}(V_i^{1/2})' = V_i$; and $\boldsymbol{\varepsilon}_i$ $(n_i \times 1)$ is $N(\boldsymbol{0}, I_{n_i})$, conditional on $\boldsymbol{X}_i$ and $\boldsymbol{b}_i$. Taking a linear Taylor series of (5.29) about $\boldsymbol{b}_i = \boldsymbol{b}_i^*$ "close" to $\boldsymbol{b}_i$, ignoring as negligible the resulting term depending on the cross-product $(\boldsymbol{b}_i - \boldsymbol{b}_i^*)\boldsymbol{\varepsilon}_i$, and letting $Z_i(\boldsymbol{X}_i, \boldsymbol{\beta}, \boldsymbol{b}_i^*) = \partial/\partial \boldsymbol{b}_i \{\boldsymbol{m}_i(\boldsymbol{X}_i, \boldsymbol{\beta}, \boldsymbol{b}_i)\}|_{\boldsymbol{b}_i = \boldsymbol{b}_i^*}$ yields

$$\boldsymbol{Y}_i = \boldsymbol{m}_i(\boldsymbol{X}_i, \boldsymbol{\beta}, \boldsymbol{b}_i^*) - Z_i(\boldsymbol{X}_i, \boldsymbol{\beta}, \boldsymbol{b}_i^*)\boldsymbol{b}_i^* + Z_i(\boldsymbol{X}_i, \boldsymbol{\beta}, \boldsymbol{b}_i^*)\boldsymbol{b}_i + V_i^{1/2}(\boldsymbol{X}_i, \boldsymbol{\beta}, \boldsymbol{b}_i^*, \boldsymbol{\alpha}) \, \boldsymbol{\varepsilon}_i. \quad (5.30)$$

That (5.30) is linear in $\boldsymbol{b}_i$ is the key feature leading to the two approaches.

The first approach we discuss is generally considered in the non-linear mixed-effects model context due to Beal and Sheiner (1982) and is analogous to the marginal quasi-likelihood method for generalized linear mixed-effects models reviewed in Section 4.6.2. Beal and Sheiner (1982) considered (5.30) with $\boldsymbol{b}_i^* = \boldsymbol{0}$, the assumed mean of the $\boldsymbol{b}_i$. Under this condition, it follows from (5.30) that the distribution of $\boldsymbol{Y}_i$ given $\boldsymbol{X}_i$ is approximately normal, depending on $\boldsymbol{\gamma}$ and G, with approximate mean and variance

$$E(\boldsymbol{Y}_i \mid \boldsymbol{X}_i) \approx \boldsymbol{m}_i(\boldsymbol{X}_i, \boldsymbol{\beta}, \boldsymbol{0}), \qquad (5.31)$$

$$\text{Var}(\boldsymbol{Y}_i \mid \boldsymbol{X}_i) \approx Z_i(\boldsymbol{X}_i, \boldsymbol{\beta}, \boldsymbol{0}) \, G \, Z_i'(\boldsymbol{X}_i, \boldsymbol{\beta}, \boldsymbol{0}) + V_i(\boldsymbol{X}_i, \boldsymbol{\beta}, \boldsymbol{0}, \boldsymbol{\alpha}). \qquad (5.32)$$

This suggests approximating $f_i(\boldsymbol{y}_i \mid \boldsymbol{x}_i; \boldsymbol{\gamma}, G)$ by a normal density with these moments. Accordingly, Beal and Sheiner (1982) proposed approximating the log-likelihood (5.25) by substituting this normal density for each term in the product, eliminating the integral, and maximizing in $\boldsymbol{\gamma}$ and G, a procedure they refer to as the "first-order method." This is implemented as the fo method in the software package NONMEM (Boeckmann, Sheiner, and

Beal, 1992; NONMEM, 2006), which is a suite of Fortran programs focused on pharmacokinetic analysis. SAS `proc nlmixed` (SAS Institute, 2006) also carries out this method via the `method=firo` option.

Recognizing that (5.31) and (5.32) define an approximate "population-averaged" model for the mean and covariance matrix of $\boldsymbol{Y}_i$ conditional on $\boldsymbol{X}_i$, other authors (e.g., Vonesh and Carter, 1992; Davidian and Giltinan, 1995, Section 6.2.3) proposed that, instead, inference be based on solving a set of generalized estimating equations (GEEs, see Chapter 3) for $\boldsymbol{\gamma}$ and the distinct elements of G based on these moments, where the equation for $\boldsymbol{\beta}$ is linear in $\boldsymbol{Y}_i$. As explained by Davidian and Giltinan (2003, Section 3.3.1), although (5.31) and (5.32) are not of the "usual" form for means and covariance matrices in most accounts of the use of GEEs, this approach is broadly applicable. Moreover, these approaches lead to different estimators from that in the previous paragraph. The SAS macro `nlinmix` (Littell et al., 1996, Chapter 12), available at `http://support.sas.com/ctx/samples/index.jsp?sid=539`, implements this type of approach under (5.31) and (5.32) with the `expand=zero` option.

A shortcoming of methods based on (5.31) and (5.32) is that (5.31) is clearly a poor approximation to the true $E(\boldsymbol{Y}_i \,|\, \boldsymbol{X}_i)$, which may result in biased (inconsistent) estimation of $\boldsymbol{\beta}$. A second class of approaches may be viewed as attempting to improve on matters by taking $\boldsymbol{b}_i^*$ in (5.30) to be something "closer" to $\boldsymbol{b}_i$. As demonstrated by Lindstrom and Bates (1990), Wolfinger (1993), Vonesh (1996), and Wolfinger and Lin (1997) from different points of view, a natural contender is $\widehat{\boldsymbol{b}}_i$, the mode of the posterior density for $\boldsymbol{b}_i$,

$$f(\boldsymbol{b}_i \,|\, \boldsymbol{y}_i, \boldsymbol{x}_i; \boldsymbol{\gamma}, G) = \frac{f_i(\boldsymbol{y}_i \,|\, \boldsymbol{x}_i, \boldsymbol{b}_i; \boldsymbol{\gamma}) \, f(\boldsymbol{b}_i; G)}{f_i(\boldsymbol{y}_i \,|\, \boldsymbol{x}_i; \boldsymbol{\gamma}, G)}. \tag{5.33}$$

This leads to the approximate moments

$$E(\boldsymbol{Y}_i \,|\, \boldsymbol{X}_i) \approx \boldsymbol{m}_i(\boldsymbol{X}_i, \boldsymbol{\beta}, \widehat{\boldsymbol{b}}_i) - Z_i(\boldsymbol{X}_i, \boldsymbol{\beta}, \widehat{\boldsymbol{b}}_i)\widehat{\boldsymbol{b}}_i, \tag{5.34}$$

$$\mathrm{Var}(\boldsymbol{Y}_i \,|\, \boldsymbol{X}_i) \approx Z_i(\boldsymbol{X}_i, \boldsymbol{\beta}, \widehat{\boldsymbol{b}}_i) \, G \, Z_i'(\boldsymbol{X}_i, \boldsymbol{\beta}, \widehat{\boldsymbol{b}}_i) + V_i(\boldsymbol{X}_i, \boldsymbol{\beta}, \widehat{\boldsymbol{b}}_i, \boldsymbol{\alpha}). \tag{5.35}$$

Various methods based on these developments have been proposed, all of which involve iteration between (i) updating $\widehat{\boldsymbol{b}}_i$ by maximizing (5.33) (or an approximation to it) with $\widehat{\boldsymbol{\gamma}}$ and $\widehat{G}$ substituted, and (ii) holding $\widehat{\boldsymbol{b}}_i$ fixed and updating estimation of $\boldsymbol{\gamma}$ and G by either (a) maximizing the approximate normal log-likelihood based on treating $\boldsymbol{Y}_i$, given $\boldsymbol{X}_i$, as normal with moments (5.34) and (5.35) or (b) by solving a corresponding set of GEEs. The NONMEM package with the `foce` option implements step (ii) via approach (a). The SAS macro `nlinmix` with `expand=eblup` (Littell et al., 1996, Chapter 12) and the S-Plus/R function `nlme` (Pinheiro and Bates, 2000) carry out step (ii) by approach (b) and differ somewhat in the way they approximate step (i) and operationalize step (ii). In the case where $m(\cdot, \cdot, \cdot)$ is defined as the solution to a system of differential equations that must be obtained numerically, software that combines this task with such iterative schemes is available, including NONMEM; `nlmem`, which builds on the `nlinmix` macro (Galecki, 1998); and the R package `nlmeODE` (Tornoe, 2006), which builds on `nlme`.

Davidian and Giltinan (2003, Section 3.3.2) mentioned additional software implementations of these so-called "first-order conditional" methods, which are similar in spirit to the penalized quasi-likelihood approach for generalized linear mixed models outlined in Section 4.6.2. Davidian and Giltinan (1995, Section 6.2.3), Vonesh and Chinchilli (1997, Section 8.2), and Davidian and Giltinan (2003, Sections 3.3.1, 3.3.2) offered further discussion on connections between all of the foregoing methods and GEEs, and the latter authors noted the motivation for these methods via approximation of the integrals in (5.25) using Laplace's method (Wolfinger, 1993; Vonesh, 1996; Wolfinger and Lin, 1997).

For both the "first-order" and "first-order conditional" approaches, standard errors for the estimator for $\boldsymbol{\beta}$ are obtained from the usual large-N formulae evaluated at the final estimates for the approximate method used (normal theory maximum likelihood or GEE), treating the moments in either (5.31) and (5.32) or (5.34) and (5.35) as exact.

5.4.4 Inference based on numerical approximation to the likelihood

The methods in the previous two sections are based on analytical manipulations to justify approximations to either the log-likelihood (5.25) or at least to the first two moments of $f_i(\boldsymbol{y}_i \,|\, \boldsymbol{x}_i; \boldsymbol{\gamma}, G)$ in (5.24) so as to obviate the need to "do" the integrals in (5.25). An alternative approach is to instead approximate these integrals by deterministic or stochastic numerical integration techniques.

Under the assumption that $f(\boldsymbol{b}_i; G)$ is a q-variate normal density, it is natural to use Gauss–Hermite quadrature to carry out the q-dimensional integrals in (5.25) numerically. A quadrature rule approximates an integral by a weighted average of the integrand over a q-dimensional grid of values, and its accuracy increases with the number of grid points. Simultaneously, however, the computational burden increases with q, but using a sparse grid can compromise accuracy, which in turn can lead to a poor approximation to the log-likelihood. Embedding repeatedly such numerical integration in an optimization, like the one maximizing (5.25), was computationally prohibitive in the early days of non-linear mixed modeling, but modern advances in computing have rendered this feasible.

In the context of the non-linear mixed model (5.22) and (5.23) and the log-likelihood (5.25), Davidian and Gallant (1993), Pinheiro and Bates (1995, Section 2.4), and Davidian and Giltinan (1995, Section 7.3) showed how to simplify the problem by transforming the relevant q-dimensional integrals into a series of one-dimensional integrals. Section 4.6.1 describes this in the similar context of generalized linear mixed models. SAS `proc nlmixed` (SAS Institute, 2006) implements Gauss–Hermite quadrature via the `method=gauss noad` option. Pinheiro and Bates (1995; 2000, Chapter 7) instead proposed the use of so-called adaptive Gaussian quadrature to approximate the integrals, which in effect "centers" and "scales" the grid about $\widehat{\boldsymbol{b}}_i$ maximizing (5.33). This tactic can greatly reduce the number of grid points required to achieve acceptable accuracy; see Section 4.6.1. Adaptive Gaussian quadrature is the default method implemented by `proc nlmixed` (SAS Institute, 2006). In all of these cases, standard errors for the components of the resulting estimators are obtained via standard likelihood theory.

Alternatively, various forms of Monte Carlo integration have been used. Davidian and Gallant (1993) suggest carrying this out by brute force, while Pinheiro and Bates (1995, Section 2.3) described the use of importance sampling, which is incorporated in SAS `proc nlmixed` (SAS Institute, 2006) with the `isamp` option. Walker (1996) proposed an EM algorithm that uses Monte Carlo integration at the "E-step" to maximize (5.25). Section 4.6.1 reviews parallel developments for generalized linear mixed-effects models.

The assumption of normally distributed $\boldsymbol{b}_i$ underlies much of the literature on mixed-effects models. In the non-linear mixed-effects model context, it is reasonable to question whether this is always a realistic model for natural population variation in biological or other phenomena, even when these are placed on a transformed scale as in (5.16). For example, a between-individual binary characteristic that is systematically associated with components of $\boldsymbol{\theta}_i$ may have been unmeasured, so that the apparent variation unexplained by observed covariates is best represented by a bimodal distribution. Several authors have considered relaxing the normality assumption in various ways; see, for example, Mallet (1986), Schumitzky (1991), Davidian and Gallant (1992, 1993), Mentré and Mallet (1994), Davidian and Giltinan (1995, Chapter 7), and Müller and Rosner (1997).

5.4.5 Inference based on a Bayesian framework

Hierarchical models such as the non-linear mixed-effects model (5.22) and (5.23) lend themselves well to Bayesian inference, implementation of which has been made accessible by modern computing power coupled with advances in Markov chain Monte Carlo (MCMC) methods, which facilitate the high-dimensional integration involved. We do not provide here a detailed account of the use of MCMC techniques to operationalize the non-linear mixed-effects model when placed in a Bayesian framework but rather only sketch the basic premise and refer the reader to Rosner and Müller (1994), Wakefield et al. (1994), Davidian and Giltinan (1995, Chapter 8), Bennett, Racine-Poon, and Wakefield (1996), and Wakefield (1996) for introductions and demonstrations.

As discussed at the outset of Section 4.6.3, in the Bayesian paradigm, the "fixed" parameters γ and G and the random-effects $\boldsymbol{b}_i$, $i = 1, \ldots, N$, are all treated as random parameters with prior distributions. It is customary to view the distribution of the $\boldsymbol{b}_i$ as a prior distribution, whose parameters (G in the case of normal $\boldsymbol{b}_i$) are themselves random. Thus, when placed in a Bayesian formulation, the two hierarchical stages of the non-linear mixed-effects model given in (5.22) and (5.23) are supplemented by a third, "hyperprior," stage:

$$\text{Stage 3: Hyperprior} \qquad (\boldsymbol{\beta}, \boldsymbol{\alpha}, G) \sim f(\boldsymbol{\beta}, \boldsymbol{\alpha}, G), \qquad (5.36)$$

where $f(\boldsymbol{\beta}, \boldsymbol{\alpha}, G)$ is the assumed joint hyperprior density; and we emphasize that γ is partitioned into fixed effects $\boldsymbol{\beta}$ and intra-individual covariance parameters $\boldsymbol{\alpha}$. Usually, (5.36) is taken to factor as $f(\boldsymbol{\beta})f(\boldsymbol{\alpha})f(G)$, and these components are ordinarily taken to have convenient forms reflecting vague prior knowledge, although in some settings they may be constructed to incorporate known constraints on and prior knowledge of the values of some model parameters; see Gelman, Bois, and Jiang (1996) for an example.

Bayesian inference on $\boldsymbol{\beta}$, $\boldsymbol{\alpha}$, G, and the $\boldsymbol{b}_i$ is based on their posterior distributions given $(\boldsymbol{Y}, \boldsymbol{X})$ (with other parameters integrated out, in contrast to [5.33], where γ and G are treated as fixed). The forms of the posteriors are implied by the three-stage hierarchy (5.22), (5.23), and (5.36). Thus, the Bayesian estimator for $\boldsymbol{\beta}$ is the mode of the posterior density $f(\boldsymbol{\beta} \mid \boldsymbol{y}, \boldsymbol{x})$, and that for $\boldsymbol{b}_i$ is the mode of $f(\boldsymbol{b}_i \mid \boldsymbol{y}, \boldsymbol{x})$. Uncertainty in these estimators is measured by the spread of these densities. Of course, the high-dimensional integration required to derive relevant posterior densities analytically is intractable. Accordingly, as demonstrated by Wakefield et al. (1994) and other authors cited above in this context, samples from the required posterior distributions can be obtained by MCMC techniques, from which desired inferences may be derived empirically.

The BUGS and WinBUGS software (Spiegelhalter et al., 1996) implements MCMC methods easily for general problems, but some intervention is required for the non-linear mixed-effects model, owing to the non-linearity of m in $\boldsymbol{b}_i$. The software package PKBugs (2004) is a WinBUGS interface that implements Bayesian analysis of population pharmacokinetic studies represented by the non-linear mixed-effects model, including a facility to handle recursive models like (5.6) that involve complex dosing histories.

5.4.6 Inference on individuals

As noted at the end of Section 5.3.5, inference on individuals may be of interest. If n_i is sufficiently large, an obvious approach is to use an appropriate regression technique based on individual i's data only, such as those outlined at the beginning of Section 5.4.2, to estimate $\boldsymbol{\theta}_i$, similar in spirit to the use of individual maximum likelihood, discussed in Section 4.7.1 for generalized linear mixed models. However, this method will suffer from low precision or may be infeasible when n_i is small. Moreover, it disregards the facts that individuals are assumed to be drawn at random from a population and that, hence, information from individuals sharing common characteristics may enhance inference on a randomly chosen individual i. This concept of "borrowing strength" across individuals is well known in related contexts, such as the linear mixed-effects model (e.g., Carlin and Louis, 2000, Section 3.3). It also follows naturally from a Bayesian perspective, as the posterior distributions of $\boldsymbol{b}_i$ or $\boldsymbol{\theta}_i$, given $(\boldsymbol{Y}, \boldsymbol{X})$, are obvious constructs on which to base such individual inference. These posterior distributions will be available following Bayesian analysis (e.g., empirically from an MCMC implementation), and the mode (or mean) of the posterior is a natural "estimate" of such individual random effects.

A similar approach may be pursued following a frequentist analysis using the methods in Section 5.4.2 through Section 5.4.4. This is accomplished by basing inference on the posterior distribution in (5.33), in which $\boldsymbol{\gamma}$ and G appear and are regarded as fixed. In the spirit of empirical Bayes inference (e.g., Carlin and Louis, 2000, Chapter 3), an obvious strategy is to substitute the estimates $\widehat{\boldsymbol{\gamma}}$ and $\widehat{G}$ in (5.33) as before and to obtain the posterior mode or mean $\widehat{\boldsymbol{b}}_i$ on its basis. "Estimation" of $\boldsymbol{\theta}_i$ then follows by substituting $\widehat{\boldsymbol{b}}_i$ and $\widehat{\boldsymbol{\beta}}$ in the population model to yield $\boldsymbol{\theta}_i = \boldsymbol{d}(\boldsymbol{A}_i, \widehat{\boldsymbol{\beta}}, \widehat{\boldsymbol{b}}_i)$. Once the empirical Bayes "estimate" of $\boldsymbol{\theta}_i$ is in hand, it may be used to facilitate, for example, prediction of individual concentration profiles that would be achieved by subject i under different dosing regimens in pharmacokinetic analysis. See Section 4.7.1 for parallel development for generalized linear mixed models.

5.4.7 Population model selection

The inferential objectives in Section 5.3.5 hinge critically on the relevance of the population model $\boldsymbol{d}$, which is taken as pre-specified in the previous sections. Several formal and informal procedures have been discussed for assisting the analyst in identifying the appropriate elements of $\boldsymbol{A}_i$ to include in the population model. When a (frequentist) analysis is based on the log-likelihood (5.25), either under the approximations in Section 5.4.3 or using numerical methods as in Section 5.4.4, a common strategy is to begin with a "full" model for $\boldsymbol{d}$ incorporating many of the covariates in $\boldsymbol{A}_i$ in each component and to carry out likelihood ratio tests comparing the "full" model to "reduced," nested models. Alternatively, inspection of information criteria such as Akaike's (AIC) or the Bayesian (BIC) information criterion from fits of models of different complexity has been advocated (e.g., Pinheiro and Bates, 2000). Clearly, these methods present challenges when the dimension of $\boldsymbol{A}_i$ is large.

Ad hoc, graphical methods to assist the analyst in postulating the forms of the components of $\boldsymbol{d}$ have also been proposed, which are based on empirical Bayes or fully Bayesian "estimates" of individual $\boldsymbol{b}_i$ or $\boldsymbol{\theta}_i$. Maitre et al. (1991) were among the first to suggest the following "bottom-up" procedure for "building" population models in population pharmacokinetic analysis. One first fits the non-linear mixed model with an initial population model including no between-individual covariates $\boldsymbol{A}_i$; for example, $\boldsymbol{\theta}_i = \boldsymbol{\beta} + \boldsymbol{b}_i$. The components of the "estimates" of the $\boldsymbol{b}_i$ are then plotted against each element of $\boldsymbol{A}_i$, and the plots examined for systematic associations that may suggest both the need to take account of certain elements of $\boldsymbol{A}_i$ in particular components of the model and the functional forms of the relationships. A new population model is formulated on this basis and fitted, and the plots made anew. If the model has taken adequate account of associations between elements of $\boldsymbol{\theta}_i$ and $\boldsymbol{A}_i$, the plots should show a haphazard scatter. If evidence of relationships persists, the population model may be refined and refitted. Examples of the use of this approach in population model building for the quinidine data are given by Davidian and Gallant (1992, 1993), Davidian and Giltinan (1995, Section 9.3), and Wakefield (1996).

Bayes or empirical Bayes predictors are known to exhibit "shrinkage" toward the mean when the density of the $\boldsymbol{b}_i$ is taken as normal, so there is some concern that this may obscure the apparent relationships. Davidian and Gallant (1992, 1993) argued that using such "estimates" derived from methods that relax the normality assumption and estimate the density of b_i non-parametrically along with γ may not be as "shrunken" and hence may highlight important associations. Mandema, Verotta, and Sheiner (1992) argued that "one-at-a-time" examination of the relationships, as above, may miss important features, and they advocate fitting generalized additive models in all covariates to each component of the "estimates" and using model selection procedures to deduce joint relationships. Similarly, other model selection methods for regression could be used.

5.4.8 Remarks

The first-order methods discussed at the beginning of Section 5.4.3, for which the approximation to the log-likelihood (5.25) is based on linearization of the model about $b_i = 0$, yielding (5.30), have the potential for biased inference on β. Nonetheless, this approach has been popular for pharmacokinetic analysis, where early simulation studies (e.g., Sheiner and Beal, 1980) demonstrated that the biases could be modest in some circumstances. The so-called first-order conditional methods, based on the approximation in (5.34) and (5.35), have proven the most popular in practice. Although the analogous penalized quasi-likelihood method for generalized linear mixed-effects models may perform poorly for binary responses with small n_i in that context (see Section 4.6.2), this approach usually yields reliable inferences for non-linear mixed-effects models, both when n_i is large and under "sparse" data conditions with small n_i, as in population pharmacokinetic studies.

Methods based on individual estimates have considerable intuitive appeal, and, in our experience, work well with "large" n_i and are easy to explain to non-statistician collaborators. Essential to their success is the large-n_i result (5.26); thus, they should only be used when the analyst is confident of the validity of this approximation. These methods can be shown to be equivalent in a large-N/large-n_i sense to first-order conditional methods (Vonesh et al., 2002; Demidenko, 2004, Section 8.10). When these methods exhibit problematic behavior (e.g., when the estimated individual covariance matrices $\widehat{C}_i$ are close to singular), it is also often the case that first-order conditional methods show unsatisfactory performance.

Direct maximization of the log-likelihood, with numerical approximation of the integrals, yields valid inferences when it "works"; however, this can be computationally challenging and, despite the availability of all-purpose implementations such as SAS `proc nlmixed` (SAS Institute, 2006), achieving convergence of the optimization of the log-likelihood can be difficult when $q > 1$. Similarly, implementation of the model in a Bayesian framework via MCMC techniques can also involve considerable computational burden, more so than for some standard statistical models, and confidence that convergence of the chain has been achieved can be low in complex situations such as population pharmacokinetic analysis, as for the quinidine study (Wakefield, personal communication). However, the first-order conditional methods, which "approximate away" the integrations involved in these more intensive approaches, nonetheless involve complicated optimization and other computational hurdles and hence can pose challenges in their own right.

The message for the user is that, as in any setting involving a complex statistical model, caution must be exercised when undertaking non-linear mixed-effects model analysis. Sensitivity to starting values is a key issue, and the user should be prepared to try several sets of starting values in optimization routines and should be wary of "false convergence." Results should always be examined for subject-matter plausibility. Despite these caveats, the non-linear mixed-effects model and associated methods for its implementation are an enormously useful framework for the types of applications discussed in Section 5.2, and analyses based on them have led to many substantive advances. As interest in placing mechanistic models in a statistical framework continues to grow in many substantive areas, non-linear mixed models will see increasing application using these and other methods; see Section 5.6.

5.5 Demonstration of implementation

We provide a brief demonstration of implementation of a non-linear mixed-effects model using some of the approaches, reviewed in Section 5.4, by considering fitting the following non-linear mixed-effects model to the data from the argatroban study, introduced in Section 5.2.2. We use several popular software packages to carry out the analyses. See Davidian and Giltinan (1995, Section 9.5) for a more substantial analysis.

Let Y_{ij} denote the jth argatroban concentration at time t_{ij}, $j = 1, \ldots, n_i$, under an infusion of length $t_{\inf} = 240$ min at rate D_i μg/kg-min initiated at time 0 for subject i, so $\boldsymbol{U}_i = (D_i, t_{\inf})'$, as before, for each of the $N = 37$ subjects. For each of nine infusion rates ranging from 1.0 to 5.0 μg/kg-min in increments of 0.5, four subjects were assigned at random; one additional subject received an infusion at rate 4.38 μg/kg-min. In this controlled study, between-subject characteristics $\boldsymbol{A}_i$ were not collected. Note that under standard assumptions on the pharmacokinetics, an individual's pharmacokinetic parameters are inherent, static characteristics of the subject at the occasion of observation that govern concentrations achieved at any infusion rate; thus, the parameters do not change with infusion rate, and it is not appropriate to consider the rate groupings as "between-individual" characteristics. As discussed by Davidian and Giltinan (1995, Section 9.5), infusion at different rates was intentional, as the goal was for subjects in the study to achieve a wide range of concentrations so as to facilitate a pharmacodynamic analysis of how the concentrations are related to a measure of (roughly) the time until a certain extent of coagulation occurs.

For our analyses here, we focus on inference on the "typical" pharmacokinetic parameters and their variation in the population of subjects based on the one-compartment model (5.5). We consider the stage 1 individual-level model

$$E(Y_{ij} \mid \boldsymbol{U}_i, \boldsymbol{\theta}_i) = m(t_{ij}, \boldsymbol{U}_i, \boldsymbol{\theta}_i)$$
$$= \frac{D_i}{e^{Cl_i^*}} \left[\exp\left\{ -\frac{e^{Cl_i^*}}{e^{V_i^*}} (t_{ij} - t_{\inf})_+ \right\} - \exp\left(-\frac{e^{Cl_i^*}}{e^{V_i^*}} t_{ij} \right) \right], \qquad (5.37)$$

where $\boldsymbol{\theta}_i = (Cl_i^*, V_i^*)'$. Note that (5.37) is parameterized in terms of the logarithms of clearance and volume, as in (5.16), anticipating that the population distributions of these parameters on the original scale are skewed. For the intra-individual covariance structure, we adopt a standard model used in pharmacokinetic analyses, given by

$$\text{Var}(\boldsymbol{Y}_i \mid \boldsymbol{U}_i, \boldsymbol{\theta}_i) = V_i(\boldsymbol{U}_i, \boldsymbol{\theta}_i, \boldsymbol{\alpha}) = \sigma_e^2 \, \text{diag}\{m^{2\lambda}(t_{i1}, \boldsymbol{U}_i, \boldsymbol{\theta}_i), \ldots, m^{2\lambda}(t_{in_i}, \boldsymbol{U}_i, \boldsymbol{\theta}_i)\}, \quad (5.38)$$

where $\boldsymbol{\alpha} = (\sigma_e^2, \lambda)'$. In (5.38), it is assumed that sampling times are sufficiently intermittent for correlation of the realization process to be taken as negligible, which is routine in this context. Although this is often a reasonable assumption in pharmacokinetic analysis, Karlsson, Beal, and Sheiner (1995) cautioned that it may not always be so and proposed using autoregressive-type models to accommodate correlation. This assumption should be critically examined by the analyst. When intensive data are available, so that individual fits of the model are possible, subject-specific residuals may form the basis for an informal assessment; see Davidian and Giltinan (1995, Sections 4.4, 5.2.4, 11.3), Verbeke and Molenberghs (2000, Chapter 9), and Diggle et al. (2002, Section 3.4) for further discussion. The intra-subject variance in (5.38) reflects a common feature of pharmacokinetic data, that of intra-individual heteroscedasticity, under which variance increases with the magnitude of the response. One justification is that measurement error is the dominant source of intra-subject variation, and assays used to ascertain concentrations are well known to lead to such heteroscedasticity. The "power variance" model in (5.38), where variance is taken to be proportional to an unknown power-of-mean response, is a standard way to approximate this pattern. Alternatively, the power model may be justified as an empirical representation of the aggregate pattern of intra-subject variation due to both the realization process and measurement error. See Davidian and Giltinan (2003, Section 2.2.2) for more.

At the second stage, we take the population model to be

$$\boldsymbol{\theta}_i = \boldsymbol{\beta} + \boldsymbol{b}_i, \quad \boldsymbol{\beta} = (\beta_1, \beta_2)', \quad \boldsymbol{b}_i \sim N(\boldsymbol{0}, G), \tag{5.39}$$

so that β_1 and β_2 represent the means of the logarithms of clearance and volume in the population or, equivalently, e^{β_1} and e^{β_2} represent the medians. The square roots of the diagonal elements G_{11} and G_{22} of G roughly quantify the coefficients of variation of clearance and volume in the population. The correlation of the logarithms of the pharmacokinetic parameters in the population may be obtained using the off-diagonal element G_{12}.

We fitted the non-linear mixed-effects model given by (5.37) through (5.39) by several methods:

1. As in Section 5.4.2, obtaining individual estimates of the $\boldsymbol{\theta}_i$ and their estimated covariance matrices $\widehat{C}_i$, $i = 1, \ldots, 37$, along with estimates of σ_e^2 and λ using the "pooled" weighted regression algorithm given by Davidian and Giltinan (1993; 1995, Section 5.2), and then estimating $\boldsymbol{\beta}$ and G based on the approximate "linear mixed model" (5.27) by using SAS proc mixed to fit (5.28).

2. As in Section 5.4.3, estimating $\boldsymbol{\beta}$, G, and σ_e^2 using the first-order method implemented in version 8.01 of the SAS macro nlinmix with the option expand=zero. The macro does not support estimation of a within-individual variance parameter like λ; hence, we set $\lambda = 0.22$ based on the result of the fit in 1 above.

3. As in Section 5.4.3, estimating $\boldsymbol{\beta}$, G, and σ_e^2 using the first-order conditional method implemented in version 8.01 of the SAS macro nlinmix with the option expand=eblup. The intra-subject variance parameter λ was handled as in 2.

4. As in Section 5.4.3, estimating $\boldsymbol{\beta}$, G, σ_e^2, and λ using the first-order conditional method implemented in the R function nlme.

5. As in Section 5.4.4, using SAS proc nlmixed. This procedure supports different default models for the density $f_i(\boldsymbol{y}_i \,|\, \boldsymbol{x}_i, \boldsymbol{b}_i; \boldsymbol{\gamma})$; the normal model only allows constant variance. It would be possible for the user to write his or her own specification of this density; however, to illustrate implementation using the defaults, taking $\lambda \approx 0.25$ based on 1 and 4 above, we approximate (5.37) through (5.38) by the "transform-both-sides" model

$$s(Y_{ij}, \delta) = s\{m(t_{ij}, \boldsymbol{U}_i, \boldsymbol{\theta}_i), \delta\} + e_{ij}, \quad \boldsymbol{e}_i \,|\, \boldsymbol{U}_i, \boldsymbol{b}_i \sim N(\boldsymbol{0}, \sigma_e^2 I_{n_i}),$$

where $s(y, \delta) = (y^\delta - 1)/\delta$, $\delta \neq 0$, with $\delta = 0.75$ (Carroll and Ruppert, 1988, Section 4.2; Davidian and Giltinan, 1995, Section 2.7). We use the default adaptive Gaussian quadrature.

Abridged code carrying out 1 through 5 above is given in Section 5.7. Full code and output are available at the web site for this book given in the Preface.

The results presented in Table 5.2 show that estimates and standard errors for β_1, and to a somewhat lesser extent for β_2, are fairly consistent, with those based on individual estimates and on first-order conditional methods the most similar. The first-order estimates, for which bias may be expected, depart most dramatically from the others. Estimates of $\boldsymbol{\alpha}$ and G are not in as good agreement, with the results from the first-order method showing the greatest discrepancy. Based on the results from methods 1 and 2 through 5 in the table, taking into account that argatroban concentrations were measured in units of ng/ml = 1000 μg/ml, median argatroban clearance in the population of subjects is estimated to be about 4.4 μg/ml/kg ($\approx \exp(-5.43) \times 1000$); similarly, median volume of distribution is estimated as approximately 145.1 ml/kg, which corresponds to approximately 10 liters for a 70-kg subject. Assuming that subject-specific clearances and volumes are approximately lognormally

Table 5.2 Parameter Estimates Obtained from Fitting the Models (5.38) and (5.39) to the Argatroban Data by Several Methods

Method	β_1	β_2	σ_e	λ	G_{11}	G_{12}	G_{22}
1. Individual est.	-5.433 (0.062)	-1.927 (0.026)	23.47	0.22	0.137	6.06	6.17
2. First-order using `nlinmix`	-5.490 (0.066)	-1.828 (0.034)	26.45	—	0.158	-3.08	16.76
3 First-order cond. using `nlinmix`	-5.432 (0.062)	-1.926 (0.026)	23.43	—	0.138	5.67	4.76
4. First-order cond. using `nlme`	-5.433 (0.063)	-1.918 (0.025)	20.42	0.24	0.138	6.73	4.56
5. Likelihood using `nlmixed`	-5.424 (0.063)	-1.924 (0.030)	13.88	—	0.141	6.56	6.01

Note: Estimated standard errors for the elements of $\boldsymbol{\beta}$ are given in parentheses. Values for G_{12} and G_{22} are multiplied by 10^3.

distributed in the population, the estimate of G_{11} corresponds to roughly a 37% coefficient of variation for clearance in the population ($\approx \sqrt{0.14} \times 100$), which is fairly typical for many drugs. The coefficient of variation for volume of distribution may be calculated similarly using the estimate of G_{22} from `nlmixed` as about 8%. These values provide the pharmacokineticists with valuable information about "typical" elimination and distribution characteristics in the population and the extent to which these vary, which may be used to establish dosing requirements for this population to achieve argatroban concentrations in a desired range.

Figure 5.4 shows individual profiles obtained by estimation of $\boldsymbol{\theta}_i$ based on i's data only from method 1 above and based on empirical Bayes "estimates" for $\boldsymbol{\theta}_i$ from method 3 for two subjects. These two approaches to individual inference yield virtually identical results, which is not unexpected given the rich information (n_i relatively "large") available on each subject, so that the empirical Bayes "shrinkage" is modest.

This example carries some important object lessons for non-linear mixed-model analysis. The argatroban data are of relatively high quality, with a high "signal-to-noise" for and rich information on each subject. Nonetheless, methods 1 through 5 use different approximations and implementations, leading to somewhat different results. Indeed, in practical settings with "messier" or sparse data, more substantial intra-individual variation, more complex individual-level models, and so on, the analyst may confront numerical challenges, including

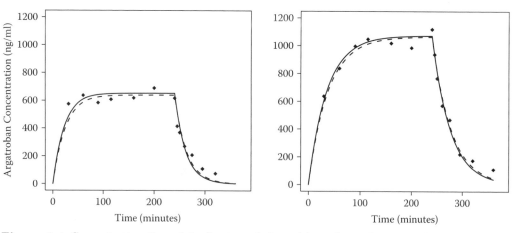

Figure 5.4 Concentration–time data for two of the subjects from the argatroban study, with individual weighted regression (dashed line) and empirical Bayes (solid line) fits of model (5.38) superimposed.

failure of some methods to converge; sensitivity to starting values and modeling assumptions; and more compelling disagreement in results. As with any complex statistical modeling exercise, it is the responsibility of the analyst to be aware of these issues and to exercise appropriate judgment in interpreting results obtained under these conditions.

5.6 Extensions and further developments

The 1990s were a period of vigorous development of the non-linear mixed-effects model, and methods for its implementation are now a routine fixture for data analysis and available in numerous commercial software packages, as we have noted. Subsequently, there have been numerous advances on several fronts, including new approaches to implementation (Vonesh et al., 2002; Lai and Shih, 2003); methods for handling missing and mismeasured data (Wang and Davidian, 1996; Wu and Wu, 2001, 2002; Li et al., 2004; Wu, 2004); methods for incorporating unspecified functions to be estimated non-parametrically into the model (Lindstrom, 1995; Ke and Wang, 2001; Li et al., 2002, 2004; Wu and Zhang, 2002; Lai, Shih, and Wong, 2006; Yang, Liu, and Wang, 2006); and extensions of the model to handle nested levels of random effects (Pinheiro and Bates, 2000, Section 7.1.2; Pinheiro and Chao, 2006), as might be the case in agricultural or forestry studies when plots are nested in blocks (Hall and Clutter, 2004) or in pharmacokinetics when the same individual is observed on multiple occasions, over which pharmacokinetic parameters may be expected to fluctuate within a subject (Karlsson and Sheiner, 1993). Because of the extent of activity, we listed only selected topics and references for brevity; see Davidian and Giltinan (2003) for more.

We highlighted here three additional issues. First, as noted in Section 5.3.1, there may be circumstances where components of the subject characteristics A_i change over the observation period on an individual, as for α_1-acid glycoprotein concentration in Table 5.1. Models for which the values of the individual-specific parameters differ across the observation times $t_{ij}, j = 1, \ldots, n_i$, for each subject, which we may write as θ_{ij}, say, do not pose any problems from a purely operational point of view. Taking the linear population model (5.15) as an example, analogous to linear and generalized linear mixed-effects models, one could write $\theta_{ij} = A_{ij}\beta + B_{ij}b_i$ (Pinheiro and Bates, 2000, Section 7.1) and use the inferential techniques we have described. The major issue is conceptual: from a subject-matter perspective, is such a model scientifically plausible? Models that arise from mechanistic considerations, such as (5.4) or (5.7), are founded on the assumption that the parameters involved are constants with respect to time. Thus, one must be prepared to offer scientific justification for such time-dependent parameters. One situation in which it is accepted that individual-specific parameters may change over observation periods (but not over each observation time j) is pharmacokinetics, as noted in the last paragraph. If a subject is observed over several treatment intervals, it is reasonable to expect that, although a basic compartment model with static parameters applies in any interval, fluctuations in the values of his or her pharmacokinetic parameters may occur that show association with other characteristics that also change. For the subject in Table 5.1, it is likely that α_1-acid glycoprotein concentration was measured intermittently at times 0, 29, and 161. Under the assumption that the pharmacokinetic parameters in (5.6) are constant over the intervals 0 to 29 hours, 29 to 77 hours, and after 161 hours (itself subject to scientific scrutiny), and indexing these intervals I_k by $k = 1, \ldots, a$ ($a = 3$), a standard modeling approach in the pharmacokinetic literature is to write a population model $\theta_{ij} = d(A_{ik}, \beta, b_i)$, where A_{ik} is the value of the subject characteristics for $t_{ij} \in I_k$. Within a specific subject, this assumes that "inter-interval" variation is entirely "explained" by the change in subject characteristics for that individual. Models that include nested random effects to relax this assumption are possible.

As the HIV example in Section 5.2.3 demonstrates, it is routine to collect more than one longitudinal response on each individual in many applications; see Hall and Clutter

(2004) for an example in forest science. Another important context is again pharmacokinetics, where both drug concentrations and responses, thought to reflect drug effect (e.g., a biomarker or other short-term effect measure), are collected over time on each subject; this was the case for the argatroban study. As noted at the beginning of Section 5.5, this was to support a pharmacokinetic/pharmacodynamic analysis, where both concentration–time and response–concentration relationships are studied together. A model for the response–concentration relationship is postulated in terms of subject-specific pharmacodynamic parameters and is linked to the pharmacokinetic model, and the entire model is fitted simultaneously; Davidian and Giltinan (1995, Section 9.5) demonstrated this for the argatroban study. For further discussion of the considerations involved in pharmacokinetic/pharmacodynamic modeling and analysis, see Davidian and Giltinan (1995, Section 9.2.2) and Giltinan (2006).

Finally, interest continues to heighten in the use of detailed dynamical systems such as the HIV model in (5.7) to describe the underlying mechanisms of multidimensional, complex biological, physical, and other phenomena. When realizations of the time course of such systems have been observed on each of several different individuals, such models, which may involve numerous individual-specific parameters, must be embedded in the non-linear mixed-effects model framework. As with the HIV example, multivariate responses may be available on only combinations of the compartments, and not all of the model parameters may be identifiable. Moreover, the system must be solved numerically. Surmounting the conceptual and computational challenges involved is a key area for continued development that may facilitate important substantive advances. For example, in the HIV context, design of time-dependent treatment strategies $u(t)$, for example, consisting of cycles of initiation and withdrawal of ARV therapy, that can maintain long-term viral suppression, are of great interest, as continuous therapy imposes serious patient burden and can lead to viral drug resistance. This may be assisted by the ability to test "virtually" such regimens by drawing "patients" (i.e., θ_i values) from the population, subjecting them to competing strategies, and simulating their ensuing long-term viral load trajectories to elucidate the effects of strategies on the population. The non-linear mixed-effects model provides a natural basis for this endeavor. If mechanistic models such as (5.7), embedded in a non-linear mixed-effects framework, can be fitted to multivariate, long-term longitudinal data on $CD4^+$ and viral load from multiple subjects, the results can serve to inform the design of realistic simulations.

As this last example illustrates, although non-linear mixed-effects models are now a standard tool in the data analyst's arsenal, there remain exciting new applications and opportunities for methodological development in this important class of statistical models.

5.7 Appendix

We provide abridged code for carrying out the analyses listed in Table 5.2. The data are in a plain text file `argconc.dat` with five columns: `obsno` (observation number), `indiv` (subject id), `dose` (infusion rate), `time` (time, min), and `conc` (argatroban concentration, ng/ml). The first three lines (for subject 1 with infusion rate 1 μg/kg-min) are

```
1 1 1 30 95.7
2 1 1 60 122
3 1 1 90 133
```

The data, full code, and output are available at the web site given in the Preface of this book.

1. *Estimation based on individual estimates.* We used customized R code to obtain the individual estimates $\widehat{\theta}_i$ and $\widehat{C}_i$ and to create a data set `argmix.dat` containing the information on $\widetilde{Y}_i$ (2×1), $\widetilde{X}_i$ (2×2), and $\widetilde{Z}_i$ (2×2) in the transformed "linear mixed model" (5.28). For each subject, there are two rows in `argmix.dat`, with columns `indiv` (subject

id), y (elements of $\widetilde{Y}_i$), x1 and x2 (columns of $\widetilde{X}_i$), and z1 and z2 (columns of $\widetilde{Z}_i$). The following SAS code takes these as input and calls `proc mixed` to fit (5.28):

```
data arg;    *  read in the transformed data;
  infile 'argmix.dat';
  input indiv y x1 x2 z1 z2;
run;
```

```
proc mixed data=arg method=ml;
  class id;
  model y = x1 x2 / noint solution chisq;
  random z1 z2 / subject=indiv type=un g gcorr gc;
  parms (0.14) (0.006) (0.006) (1) / eqcons=4;
run;
```

The `parms` statement allows the user to specify starting values for the elements of G and σ_e^2; the eqcons=4 statement constrains the fourth of these (σ_e^2) to be fixed at 1 in the optimization, as required. The starting values for G were suggested from an estimate of G obtained from the sample covariance matrix of the $\widehat{\theta}_i$.

2. *First-order method using the* `nlinmix` *macro.* Here, the original data in `argconc.dat` are read in and used directly by the macro.

```
%inc 'nlmm801.sas' / nosource; *  nlinmix macro;
```

```
data arg;
   infile 'argconc.dat';
   input obsno indiv dose time conc;
   tinf=240;
   t1=1;
   if time>tinf then t1=0;
   t2=tinf*(1-t1)+t1*time;
run;
```

```
%nlinmix(data=arg,
   model=%str(
     logcl=beta1+b1;
     logv=beta2+b2;
     cl=exp(logcl);
     v=exp(logv);
     predv=(dose/cl)*(1-exp(-cl*t2/v))*exp(-cl*(1-t1)*(time-tinf)/v);
   ),
   derivs=%str( wt=1/predv**(2*0.22); ),
   parms=%str(beta1=-6.0 beta2=-2.0),
   stmts=%str(
      class indiv;
      model pseudo_conc = d_beta1 d_beta2 / noint notest solution;
      random d_b1 d_b2 / subject=indiv type=un solution;
      weight wt;
    ),
   expand=zero,
   procopt=%str(maxiter=500 method=ml)
   )
run;
```

3. *First-order conditional method using the* `nlinmix` *macro.* The code is identical to that in 2 above except that `expand=zero` is replaced by `expand=eblup`.

4. *First-order conditional method using R function* `nlme`. The data are first read into the data frame `thedat`. A starting value of 0.5 is used for λ.

```
library(nlme)   #  access nlme()

thedat <- read.table("argconc.dat",col.names=c('obsno','indiv','dose',
   'time','conc'))

meanfunc <- function(x,b1,b2,dose){
   tinf <- 240
   cl <- exp(logcl)
   v <- exp(logv)
   t1 <- x<=tinf
   t2 <- tinf*(1-t1)+t1*x;
   f1 <- (dose/cl)*(1-exp(-cl*t2/v))*exp(-cl*(1-t1)*(x-tinf)/v)
   f1
}

arg.mlfit <- nlme(conc ~ meanfunc(time,logcl,logv,dose),
   fixed = list(logcl ~ 1,logv ~1),
   random = list(logcl ~ 1,logv ~ 1),
   groups = ~ indiv, data = thedat,
   start = list(fixed = c(-6.0,-2.0)),
   method="ML",
   verbose=T,
   weights=varPower(0.5))
```

5. *Maximum likelihood via adaptive Gaussian quadrature to perform the integrations using SAS* `proc nlmixed`. The data are read into a SAS data set in a manner identical to 2 and 3 above, with the additional line `conctrans=conc**0.75;` to construct the transformed response. The call to `proc nlmixed` looks like this:

```
proc nlmixed data=arg;
   parms beta1=-6.0 beta2=-2.0 s2b1=0.14 cb12=0.006 s2b2=0.006 s2=23.0;
   logcl=beta1+b1;
   logv=beta2+b2;
   cl=exp(logcl);
   v=exp(logv);
   pred=((dose/cl)*(1-exp(-cl*t2/v))*exp(-cl*(1-t1)*(time-tinf)/v))**0.75;
   model conctrans ~ normal(pred,s2);
   random b1 b2 ~ normal([0,0],[s2b1,cb12,s2b2]) subject=indiv;
run;
```

Acknowledgments

This work was supported by U.S. NIH grants R01-CA085848, R37-AI031789, R01-GM067299, and R01-AI071915. The author acknowledges the pioneering contributions of Lewis B. Sheiner (1940–2004) and Stuart L. Beal (1941–2006) to the formulation and widespread application of non-linear mixed-effects models in pharmacokinetic/pharmaco-dynamic analysis and more generally.

References

Adams, B. M., Banks, H. T., Davidian, M., Kwon, H. D., Tran, H. T., Wynne, S. N., and Rosenberg, E. S. (2005). HIV dynamics: Modeling, data analysis, and optimal treatment protocols. *Journal of Computational and Applied Mathematics* **184**, 10–49.

Adams, B. M., Banks, H. T., Davidian, M., and Rosenberg, E. S. (2007). Model fitting and prediction with HIV treatment interruption data. *Bulletin for Mathematical Biology* **69**, 563–584.

Beal, S. L. and Sheiner, L. B. (1982). Estimating population pharmacokinetics. *CRC Critical Reviews in Biomedical Engineering* **8**, 195–222.

Bennett, J. E., Racine-Poon, A., and Wakefield, J. C. (1996). MCMC for nonlinear hierarchical models. In W. R. Gilks, S. Richardson, and D. J. Spiegelhalter (eds.), *Markov Chain Monte Carlo in Practice*, pp. 339–357. London: Chapman & Hall.

Boeckmann, A. J., Sheiner, L. B., and Beal, S. L. (1992). *NONMEM User's Guide, Part V, Introductory Guide*. San Francisco: University of California.

Carlin, B. P. and Louis, T. A. (2000). *Bayes and Empirical Bayes Methods for Data Analysis*, 2nd ed. Boca Raton, FL: Chapman & Hall/CRC Press.

Carroll, R. J. and Ruppert, D. (1988). *Transformation and Weighting in Regression*. London: Chapman & Hall.

Davidian, M. and Gallant, A. R. (1992). Smooth nonparametric maximum likelihood estimation for population pharmacokinetics, with application to quinidine. *Journal of Pharmacokinetics and Biopharmaceutics* **20**, 529–556.

Davidian, M. and Gallant, A. R. (1993). The non-linear mixed effects model with a smooth random effects density. *Biometrika* **80**, 475–488.

Davidian, M. and Giltinan, D. M. (1993). Some simple methods for estimating intra-individual variability in non-linear mixed effects models. *Biometrics* **49**, 59–73.

Davidian, M. and Giltinan, D. M. (1995). *Non-linear Models for Repeated Measurement Data*. London: Chapman & Hall.

Davidian, M. and Giltinan, D. M. (2003). Non-linear models for repeated measurement data: An overview and update. *Journal of Agricultural, Biological, and Environmental Statistics* **8**, 387–419.

Demidenko, E. (2004). *Mixed Models*. Hoboken, NJ: Wiley.

Diggle, P. J., Heagerty, P., Liang, K.-Y., and Zeger, S. L. (2002). *Analysis of Longitudinal Data*, 2nd ed. Oxford: Oxford University Press.

Fitzmaurice, G. M., Laird, N. M., and Ware, J. H. (2004). *Applied Longitudinal Analysis*. Hoboken, NJ: Wiley.

Galecki, A. T. (1998). NLMEM: New SAS/IML macro for hierarchical non-linear models. *Computer Methods and Programs in Biomedicine* **55**, 207–216.

Gelman, A., Bois, F., and Jiang, L. M. (1996). Physiological pharmacokinetic analysis using population modeling and informative prior distributions. *Journal of the American Statistical Association* **91**, 1400–1412.

Gibaldi, M. and Perrier, D. (1982). *Pharmacokinetics*, 2nd ed. New York: Marcel Dekker.

Giltinan, D. M. (2006). Pharmacokinetics and pharmacodynamics. In P. Armitage and T. Colton (eds), *Encyclopedia of Biostatistics*, 2nd ed, pp. 4049–4062. Hoboken, NJ: Wiley.

Hall, D. B. and Clutter, M. (2004). Multivariate multilevel non-linear mixed-effects models for timber yield predictions. *Biometrics* **60**, 16–24.

Ho, D. D., Neumann, A. U., Perelson, A. S., Chen, W., Leonard, J. M., and Markowitz, M. (1996). Rapid turnover of plasma virions and CD4 lymphocytes in HIV-1 infection. *Nature* **373**, 123–126.

Huang, Y., Liu, D., and Wu, H. (2006). Hierarchical Bayesian models for estimation of parameters in a longitudinal HIV dynamic system. *Biometrics* **62**, 413–423.

Karlsson, M. O., Beal, S. L., and Sheiner, L. B. (1995). Three new residual error models for population PK/PD analyses. *Journal of Pharmacokinetics and Biopharmaceutics* **23**, 651–672.

Karlsson, M. O. and Sheiner, L. B. (1993). The importance of modeling inter-occasion variability in population pharmacokinetic analyses. *Journal of Pharmacokinetics and Biopharmaceutics* **21**, 735–750.

Ke, C. and Wang, Y. (2001). Semiparametric non-linear mixed models and their applications. *Journal of the American Statistical Association* **96**, 1272–1298.

Lai, T. L. and Shih, M. C. (2003). A hybrid estimator in non-linear and generalised linear mixed-effects models. *Biometrika* **90**, 859–879.

Lai, T. L., Shih, M. C., and Wong, S. P. (2006). A new approach to modeling covariate effects and individualization in population pharmacokinetics-pharmacodynamics. *Journal of Pharmacokinetics and Pharmacodynamics* **33**, 49–74.

Li, L., Brown, M. B., Lee, K. H., and Gupta, S. (2002). Estimation and inference for a spline-enhanced population pharmacokinetic model. *Biometrics* **58**, 601–611.

Li, L., Lin, X. H., Brown, M. B., Gupta, S., and Lee, K. H. (2004). A population pharmacokinetic model with time-dependent covariates measured with errors. *Biometrics* **60**, 451–460.

Lindstrom, M. J. (1995). Self-modeling with random shift and scale parameters and a free-knot spline shape function. *Statistics in Medicine* **14**, 2009–2021.

Lindstrom, M. J. and Bates, D. M. (1990). Non-linear mixed effects models for repeated measures data. *Biometrics* **46**, 673–687.

Littell, R. C., Milliken, G. A., Stroup, W. W., and Wolfinger, R. D. (1996). *SAS System for Mixed Models*. Cary, NC: SAS Institute, Inc.

Maitre, P. O., Buhrer, M., Thomson, D., and Stanski, D. R. (1991). A three-step approach combining Bayesian regression and NONMEM population analysis: Application to midazolam. *Journal of Pharmacokinetics and Biopharmaceutics* **19**, 377–384.

Mallet, A. (1986). A maximum likelihood estimation method for random coefficient regression models. *Biometrika* **73**, 645–656.

Mandema, J. W., Verotta, D., and Sheiner, L. B. (1992). Building population pharmacokinetic/pharmacodynamic models. *Journal of Pharmacokinetics and Biopharmaceutics* **20**, 645–656.

Mentré, F. and Mallet, A. (1994). Handling covariates in population pharmacokinetics. *International Journal of Biomedical Computing* **36**, 25–33.

Mezzetti, M., Ibrahim, J. G., Bois, F. Y., Ryan, L. M., Ngo, L., and Smith, T. J. (2003). A Bayesian compartmental model for the evaluation of 1,3-butadiene metabolism. *Applied Statistics* **52**, 291–305.

Müller, P. and Rosner, G. L. (1997). A Bayesian population model with hierarchical mixture priors applied to blood count data. *Journal of the American Statistical Association* **92**, 1279–1292.

NONMEM (2006). http://www.icondevsolutions.com/nonmem.htm.

Pinheiro, J. C. and Bates, D. M. (1995). Approximation to the log-likelihood function in the non-linear mixed-effects model. *Journal of Computational and Graphical Statistics* **4**, 12–35.

Pinheiro, J. C. and Bates, D. M. (2000). *Mixed-Effects Models in S and S-PLUS*. New York: Springer.

Pinheiro, J. C. and Chao, E. C. (2006). Efficient Laplacian and adaptive Gaussian quadrature algorithms for multilevel generalized linear mixed models. *Journal of Computational and Graphical Statistics* **15**, 58–81.

PKBugs (2004). PKBugs: An efficient interface for population PK/PD within WinBUGS. http://www.winbugs-development.org.uk/pkbugs/home.html.

Rosner, G. L. and Müller, P. (1994). Pharmacokinetic/pharmacodynamic analysis of hematologic profiles. *Journal of Pharmacokinetic and Biopharmaceutics* **22**, 499–524.

SAS Institute (2006). *SAS OnlineDoc, 9.1.3* Cary, NC: SAS Institute, Inc.

Schumitzky, A. (1991). Nonparametric EM algorithms for estimating prior distributions. *Applied Mathematics* **45**, 143–157.

Seber, G. A. F. and Wild, C. J. (1989). *Non-linear Regression*. New York: Wiley.

Sheiner, L. B. and Beal, S. L. (1980). Evaluation of methods for estimating population pharmacokinetic parameters. I. Michaelis Menten model: Routine clinical pharmacokinetic data. *Journal of Pharmacokinetics and Biopharmaceutics* **8**, 553–571.

Sheiner, L. B., Rosenberg, B., and Melmon, K. L. (1972). Modelling of individual pharmacokinetics for computer-aided drug dosage. *Computers and Biomedical Research* **5**, 441–459.

Sheiner, L. B., Rosenberg, B., and Marathe, V. V. (1977). Estimation of population characteristics of pharmacokinetic parameters from routine clinical data. *Journal of Pharmacokinetics and Biopharmaceutics* **5**, 445–479.

Spiegelhalter, D. J., Thomas, A., Best, N. G., and Gilks, W. R. (1996). *BUGS 0.5 Bayesian Inference using Gibbs Sampling Manual (version ii)*. Cambridge: MRC Biostatistics Unit. http://www.mrc-bsu.cam.ac.uk/bugs/documentation/contents.shtml.

Steimer, J. L., Mallet, A., Golmard, J. L., and Boisvieux, J. F. (1984). Alternative approaches to estimation of population pharmacokinetic parameters: Comparison with the non-linear mixed effect model. *Drug Metabolism Reviews* **15**, 265–292.

Tornoe, C. W. (2006). The `nlmeODE` package. Available at http://cran.r-project.org/src/contrib/Descriptions/nlmeODE.html.

Verbeke, G. and Molenberghs, G. (2000). *Linear Mixed Models for Longitudinal Data*. New York: Springer.

Verme, C. N., Ludden, T. M., Clementi, W. A., and Harris, S. C. (1992). Pharmacokinetics of quinidine in male patients: A population analysis. *Clinical Pharmacokinetics* **22**, 468–480.

Vonesh, E. F. (1996). A note on the use of Laplace's approximation for non-linear mixed-effects models. *Biometrika* **83**, 447–452.

Vonesh, E. F. and Carter, R. L. (1992). Mixed effects non-linear regression for unbalanced repeated measures. *Biometrics* **48**, 1–17.

Vonesh, E. F. and Chinchilli, V. M. (1997). *Linear and Nonlinear Models for the Analysis of Repeated Measurements*. New York: Marcel Dekker.

Vonesh, E. F., Wang, H., Nie, L., and Majumdar, D. (2002). Conditional second-order generalized estimating equations for generalized linear and non-linear mixed-effects models. *Journal of the American Statistical Association* **97**, 271–283.

Wakefield, J. (1996). The Bayesian analysis of population pharmacokinetic models. *Journal of the American Statistical Association* **91**, 271–283.

Wakefield, J. C., Smith, A. F. M., Racine-Poon, A., and Gelfand, A. E. (1994). Bayesian analysis of linear and non-linear population models by using the Gibbs sampler. *Applied Statistics* **43**, 201–221.

Walker, S. G. (1996). An EM algorithm for non-linear random effects models. *Biometrics* **52**, 934–944.

Wang, N. and Davidian, M. (1996). A note on covariate measurement error in nonlinear mixed effects models. *Biometrika* **83**, 801–812.

Wolfinger, R. D. (1993). Laplace's approximation for non-linear mixed models. *Biometrika* **80**, 791–795.

Wolfinger, R. D. and Lin, X. (1997). Two Taylor-series approximation methods for non-linear mixed models. *Computational Statistics and Data Analysis* **25**, 465–490.

Wu, H. L. and Wu, L. (2001). A multiple imputation method for missing covariates in non-linear mixed-effects models with application to HIV dynamics. *Statistics in Medicine* **20**, 1755–1769.

Wu, H. L. and Wu, L. (2002). Missing time-dependent covariates in human immunodeficiency virus dynamic models. *Applied Statistics* **51**, 297–318.

Wu, H. L. and Zhang, J. T. (2002). The study of long-term HIV dynamics using semi-parametric non-linear mixed-effects models. *Statistics in Medicine* **21**, 3655–3675.

Wu, L. (2004). Exact and approximate inferences for non-linear mixed-effects models with missing covariates. *Journal of the American Statistical Association* **99**, 700–709.

Yang, Y. C., Liu, A., and Wang, Y. D. (2006). Detecting pulsatile hormone secretions using non-linear mixed effects partial spline models. *Biometrics* **62**, 230–238.

CHAPTER 6

Growth mixture modeling: Analysis with non-Gaussian random effects

Bengt Muthén and Tihomir Asparouhov

Contents

6.1 Introduction

This chapter gives an overview of non-Gaussian random-effects modeling in the context of finite-mixture growth modeling developed in Muthén and Shedden (1999), Muthén (2001a, 2001b, 2004), and Muthén et al. (2002), and extended to cluster samples and cluster-level mixtures in Asparouhov and Muthén (2008). Growth mixture modeling represents unobserved heterogeneity between the subjects in their development using both random effects (e.g., Laird and Ware, 1982) and finite mixtures (e.g., McLachlan and Peel, 2000). This allows different sets of parameter values for mixture components corresponding to different unobserved subgroups of individuals, capturing latent trajectory classes with different growth curve shapes. This chapter discusses examples motivating modeling with such trajectory classes. A general latent-variable modeling framework is presented together with its maximum likelihood estimation. Examples from criminology, mental health, and education are analyzed. The choice of a normal or a non-parametric distribution for the random effects is discussed and investigated using a simulation study. The discussion will refer to growth mixture modeling techniques as implemented in the Mplus program (Muthén and Muthén, 1998–2007) and input scripts for the analyses are available at http://www.statmodel.com.

The outline of this chapter is as follows. Section 6.2 presents examples with substantive questions that motivate growth mixture analysis. Section 6.3 describes the general model. Section 6.4 discusses estimation and model assessment. Section 6.5 illustrates the modeling with a series of examples. Section 6.6 compares the parametric and non-parametric versions of the random-effect model. Section 6.7 concludes.

6.2 Examples

The following examples show the breadth of longitudinal studies that may be approached by growth mixture modeling.

6.2.1 Example 1: Clinical trials with placebo response

The first example concerns analysis of data from a double-blind 8-week randomized trial on depression medication (Leuchter et al., 2002). Of particular interest is how to assess medication effects in the presence of placebo response. Placebo response is an improvement in depression ratings that is unrelated to medication. The improvement is often seen as an early steep drop in depression, often followed by a later upswing. Figure 6.1 shows results for a two-class growth mixture model for the sample of 45 placebo group subjects using the Hamilton depression scale (Ham-D). The first two time points are before randomization and the next nine time points are after randomization. The responder class is shown in the left panel and the non-responder class in the right panel. The solid curve is the estimated mean curve, whereas the broken curves are observed individual trajectories for individuals classified as most likely belonging to this class.

Placebo response confounds the estimation of the true effect of medication and is an important phenomenon, given its high prevalence of 25 to 60%. Because placebo response is pervasive, the statistical modeling should account for this when estimating medication effects. This can be done by acknowledging the qualitative heterogeneity in trajectory shapes for responders and non-responders using growth mixture modeling. The estimation of medication effects using growth mixture modeling is described in Muthén et al. (2007). The medication effect is estimated in line with the approach of the next example.

6.2.2 Example 2: Randomized interventions with treatment effects varying across latent classes

The second example concerns a randomized preventive field trial conducted in Baltimore public schools (Dolan et al., 1993; Ialongo et al., 1999). The study applied a universal intervention aimed at reducing aggressive-disruptive behavior during first and second grade to improve reading and reduce aggression with outcomes assessed through middle school and beyond (Kellam et al., 1994). Children were followed from first to seventh grade with

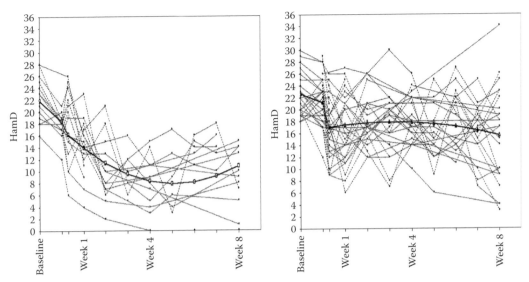

Figure 6.1 Two-class growth mixture model for depression in a placebo group.

respect to the course of aggressive behavior, and a follow-up to age 18 also allowed for the assessment of intervention impact on more distal events, such as the probability of juvenile delinquency as indicated by juvenile court records. The intervention was administered after one pre-intervention time point in the fall of first grade.

Key scientific questions addressed whether the intervention reduced the slope of the aggression trajectory across the grades, whether the intervention was different in impact for children who initially display higher levels of aggression, and whether the intervention impacted distal outcomes. Analyses of these hypotheses were presented in Muthén et al. (2002). Allowing for multiple trajectory classes in the growth model gave a flexible way to assess differential effects of the intervention. The analyses focused on boys and intervention status as defined by classroom assignment in the fall of first grade, resulting in a sample of 119 boys in the intervention group and 80 boys in the control group. Figure 6.2 shows results from a four-class growth mixture model for the 119 boys. For each combination of latent-trajectory class and intervention condition, the estimated mean growth curve is shown together with observed individual trajectories for individuals estimated to be most likely a member of the class. An intervention effect in terms of reducing aggressive behavior is seen for the high class and perhaps also for the low starting ("LS") class, whereas the other two classes show no effects.

6.2.3 Example 3: High school dropout predicted by failing math achievement development

The third example concerns growth mixture modeling of mathematics achievement development in U.S. schools. Muthén (2004) analyzed longitudinal math scores for students in grades 7 through 10 from the Longitudinal Study of American Youth (LSAY) and found a problematic trajectory class with an exceptionally low starting point in grade 7 as well as a low growth rate; see Figure 6.3. The class membership was strongly related to covariates such as grade 7 measures of having low schooling expectations and dropout thoughts. Taken together with the poor math development, this suggests that the class consists of students who are disengaged from school. Class membership was also highly predictive of dropping out by grade 12, a binary "distal" outcome. In a further analysis, Muthén (2004) carried out a growth mixture analysis where the clustering of students within schools was taken

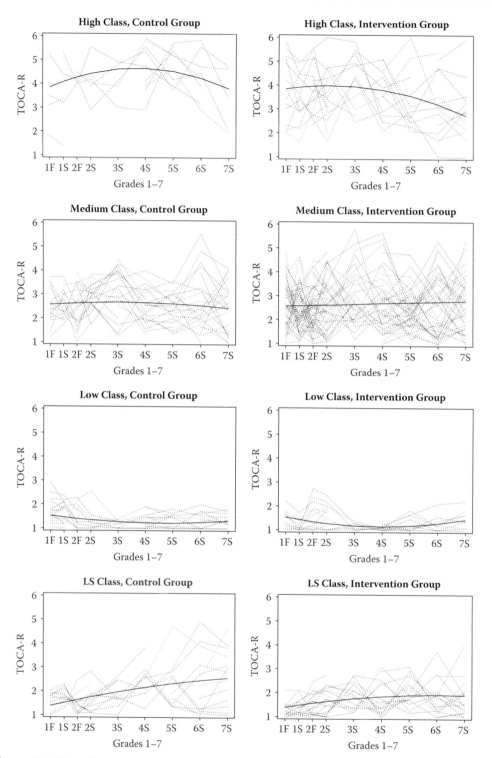

Figure 6.2 Four-class growth mixture model for aggressive behavior in control and intervention groups.

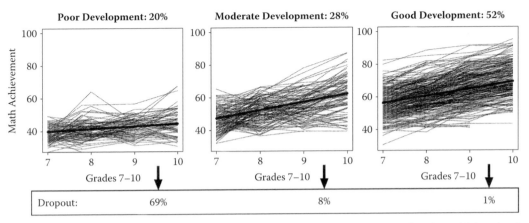

Figure 6.3 Three-class growth mixture model for math achievement related to high school dropout.

into account by allowing random-effect variation across schools. The school variation was represented in the random effects for the growth, the random intercept in the logistic regression for dropping out, and the random intercept in the multinomial regression predicting latent-class membership as a function of student-level covariates. Furthermore, school-level covariates corresponding to poverty of the school neighborhood and teaching quality in the school were used to predict across-school variation in the random coefficients.

6.2.4 Example 4: Age–crime curves

The fourth example concerns criminal activity of 13,160 males born in Philadelphia, Pennsylvania in 1958 (D'Unger et al., 1998; D'Unger, Land, and McCall, 2002; Loughran and Nagin, 2006). Annual counts of police contacts are available from age 4 to 26 of this birth cohort. The aggregate age–crime curve follows the well-known pattern of increasing annual convictions throughout the subjects' teenage years and decreasing annual convictions thereafter. The criminology literature has focused extensively on identifying groups of individuals with similar patterns or careers of delinquent and criminal offending. To quote D'Unger et al. (1998, p. 1595): "This question of how many latent classes of criminal careers are optimal, and why the number of categories itself is important, has gained salience for criminological theory in light of recent theoretical debates." The authors go on to mention Moffit (1993) as a key contributor to the notion of different trajectory classes, proposing a distinction between the trajectory of "life-course persistents" versus "adolescence limiteds" depending on the behaviors persisting over the life course or seen only during adolescence. The debate continues, as seen in Sampson and Laub (2005) discussing the "group-based" analysis approach of Nagin (1999, 2005), Nagin and Land (1993), and Roeder, Lynch, and Nagin (1999). The Philadelphia crime data will be analyzed in a new way in this chapter. The analyses to be presented have two special features. First, the outcome variable is a count variable that is very skewed, with a large number of zeros at each point in time. Second, it is of interest to contrast the "group-based" approach with random-effects models.

6.2.5 Example 5: Classification of schools based on achievement development

The fifth example extends the achievement analyses discussed in Example 3 by using a school-level latent-class variable, enabling a classification of schools as more or less successful. The LSAY data discussed in Example 3 are from a limited number of schools and analyses are instead performed on data from grades 8, 10, and 12 of the National Education Longitudinal Study (NELS). NELS surveyed 913 schools and a total of 14,217 students. In the analyses to be presented, student growth rate is regressed on the growth intercept in grade 8 and allows this relationship to vary across the school-level latent classes. It has been argued in the education literature that a weak relationship is an indicator of a school being egalitarian (e.g., Choi and Seltzer, 2006). The means of the random intercept and the intercept of the random growth rate are also allowed to vary across the school-level latent classes. Both types of school-level latent-class features are useful for determining school quality.

6.2.6 Other applications

Other applications of growth mixture modeling found in the literature include Verbeke and Lesaffre (1996) (see also Pearson et al., 1994), who considered different groups of males with linear or exponential growth in prostate-specific antigen (PSA); Muthén and Shedden (1999) and Muthén and Muthén (2000), with application to the development of heavy drinking and alcohol dependence; Lin et al. (2002), with application to PSA and prostate cancer, combining growth mixture modeling with survival analysis; Croudace et al. (2003), with application to bladder control; Muthén et al. (2003), with application to reading failure, including the modeling of a kindergarten process for phonemic awareness linked to a later process of word recognition; and Muthén and Masyn (2005), with application to aggressive behavior and juvenile delinquency, combining growth mixture modeling and discrete time survival analysis. Related applications to latent-class membership representing non-participation (non-compliance) and complier-average causal effect estimation in intervention studies (Angrist, Imbens, and Rubin, 1996) are given in Jo (2002) and Jo and Muthén (2003), and are also generalizable to longitudinal studies (see Yau and Little, 2001; Dunn et al., 2003; Muthén, Jo, and Brown, 2003), including time-varying compliance (Lin et al., 2006).

6.3 Growth mixture modeling

This section describes the general growth mixture modeling framework (see also Asparouhov and Muthén, 2008). The description is closely related to the implementation in the Mplus software version 4.2 and higher (Muthén and Muthén, 1998–2007). To familiarize readers with the general Mplus modeling framework, the section starts with a simple growth example put into the conventional linear mixed-effects model as well as the Mplus modeling framework.

6.3.1 Specification of a simple growth model

Consider a single growth process with no latent-trajectory classes, no clustering, linear growth for a continuous outcome y, a time-invariant covariate x, and a time-varying covariate w,

$$Y_{ij} = \eta_{0i} + \eta_{1i}a_{ij} + \kappa_i w_{ij} + \epsilon_{ij}, \tag{6.1}$$

where a_{ij} are time scores ($j = 1, 2, \ldots, T$), the random intercept η_{0i} and the random slope η_{1i} represent the growth process, κ_i is a random slope, and ϵ is a normally distributed residual. The random intercepts and slopes are expressed as

$$\eta_{0i} = \alpha_0 + \gamma_0 x_i + \zeta_{0i} \tag{6.2}$$

$$\eta_{1i} = \alpha_1 + \gamma_1 x_i + \zeta_{1i} \tag{6.3}$$

$$\kappa_i = \alpha_2 + \gamma_2 w_{ij} + \zeta_{2i} \tag{6.4}$$

where the αs and γs are parameters and the ζs are normally distributed residuals. In multi-level terms, Equation (6.1) represents level-1 variation across time and (6.2) through (6.4) represent level-2 variation across individuals. Consider the mixed linear model formulation for $Y_i = (Y_{i1}, \ldots, Y_{iT})'$,

$$\boldsymbol{Y}_i = X_i\boldsymbol{\beta} + Z_i\boldsymbol{b}_i + \boldsymbol{e}_i, \tag{6.5}$$

where some individuals may not be observed at all occasions T, leading to missing data. In this example, let

$$\Lambda_i = \begin{pmatrix} 1 & a_{i1} \\ 1 & a_{i2} \\ 1 & a_{i3} \\ \vdots & \vdots \\ 1 & a_{iT} \end{pmatrix},$$

so that in (6.5) we have

$$X_i = \begin{pmatrix} \Lambda_i \, \boldsymbol{w}_i \, \Lambda_i x_i \, \boldsymbol{w}_i x_i \end{pmatrix},$$
$$\beta = (\alpha_0, \alpha_1, \, \alpha_2, \, \gamma_0, \gamma_1, \, \gamma_2)',$$
$$Z_i = \begin{pmatrix} \Lambda_i \, \boldsymbol{w}_i \end{pmatrix},$$
$$\boldsymbol{b}_i = (\zeta_{0i}, \zeta_{1i}, \, \zeta_{2i})',$$
$$\boldsymbol{e}_i = (\epsilon_{i1}, \ldots, \epsilon_{iT})'.$$

The Mplus framework uses the general model expression for observed vectors $\boldsymbol{Y}_i$ and $\boldsymbol{X}_i$,

$$\boldsymbol{Y}_i = \boldsymbol{\nu} + \Lambda\boldsymbol{\eta}_i + K\boldsymbol{X}_i + \boldsymbol{\epsilon}_i, \tag{6.6}$$
$$\boldsymbol{\eta}_i = \boldsymbol{\alpha} + B\boldsymbol{\eta}_i + \Gamma\boldsymbol{X}_i + \boldsymbol{\zeta}_i, \tag{6.7}$$

implying

$$\boldsymbol{Y}_i = \boldsymbol{\nu} + \Lambda\,(I - B)^{-1}\,\boldsymbol{\alpha} + \Lambda\,(I - B)^{-1}\,\Gamma\,\boldsymbol{X}_i + K\,\boldsymbol{X}_i + \Lambda\,(I - B)^{-1}\,\boldsymbol{\zeta}_i + \boldsymbol{\epsilon}_i,$$

where the first four terms refer to fixed effects and the last two terms to random effects. The regression parameter arrays Λ, K, B, and Γ are allowed to vary across i as a function of observed variables or they can be unobserved random slopes. The model equations (6.6) and (6.7) capture the level-1 and level-2 expressions for the linear growth example in (6.1) and (6.2) through (6.4). The notation of (6.6) and (6.7) follows that of the linear growth example with $B = 0$ and with the vector $\boldsymbol{X}_i$ containing both the time-varying covariate w_{ij} and the time-invariant covariate x_i. The model of (6.6) and (6.7) includes the mixed linear model of (6.5) as a special case. In latent-variable modeling terms, (6.6) is referred to as the measurement part of the model, where the latent-variable vector $\boldsymbol{\eta}_i$ is measured by the indicators $\boldsymbol{Y}_i$. Here, Λ may contain parameters. A frequent example is when $a_{it} = a_t$, so that a_t can be treated as parameters, for example capturing deviations from linear growth shape (fixing two a_t values for identification, typically $a_1 = 0, a_2 = 1$). Another example is where multiple indicators of a factor are available at each time point, where different indicators have different factor loadings λ. With $\Lambda_i = \Lambda$, (6.6) also covers factor analysis with covariates. Furthermore, (6.7) is referred to as the structural part, containing regressions among the latent variables. The regression matrix B has zero diagonal elements, but the off-diagonal elements may be used to regress random effects on each other. For example, the growth slope (growth rate) η_{1i} or the random slope κ_i may be expressed as a function of the intercept (initial status) η_{0i}. More generally, (6.7) also covers structural equation modeling. In this way, the extensions of (6.6) and (6.7) to finite mixtures and cluster samples presented in this chapter pertain to not only growth models but also factor analysis and structural equation models, as well as combinations of such models and growth models (Muthén, 2002).

6.3.2 A general multilevel mixture model

Let Y_{kij} be the jth observed dependent variable for individual i in cluster k. Three types of variables are considered in the analyses to be presented: binary and ordered categorical variables, continuous normally distributed variables, and counts following the Poisson or zero-inflated Poisson distribution. Let C_{ki} be a latent categorical variable for individual i in cluster k which takes values $1, \ldots, L$. Let D_k be a cluster-level latent categorical variable for cluster k which takes values $1, \ldots, M$. The choice of L and M will be discussed in Section 6.4.1. To construct a model for observed binary and ordered categorical variables we proceed in line with Muthén (1984) by defining an underlying continuous, normally distributed latent variable Y_{kij}^* such that, for a set of threshold parameters τ_{cdsj},

$$Y_{kij} = s|_{C_{ki}=c,D_k=d} \Leftrightarrow \tau_{cdsj} < Y_{kij}^* < \tau_{cd,s+1,j}.$$

For continuous normally distributed variables we define $Y_{kij}^* = Y_{kij}$. For counts $Y_{kij}^* = \log(\lambda_{kij})$, where λ_{kij} is the rate of the Poisson distribution. Let $\boldsymbol{Y}_{ki}^*$ be the J-dimensional vector of all dependent variables and $\boldsymbol{X}_{ki}$ be the Q-dimensional vector of all individual-level covariates. Using latent-variable terms, the "measurement" part of the model is defined by

$$\boldsymbol{Y}_{ki}^*|_{C_{ki}=c,D_k=d} = \boldsymbol{\nu}_{cdk} + \Lambda_{cdk}\,\boldsymbol{\eta}_{ki} + K_{cdk}\,\boldsymbol{X}_{ki} + \boldsymbol{\epsilon}_{ki}, \qquad (6.8)$$

where $\boldsymbol{\nu}_{cdk}$ is a J-dimensional vector of intercepts, Λ_{cdk} is a $J \times m$ slope matrix for the m-dimensional random-effect vector $\boldsymbol{\eta}_{ki}$, K_{cdk} is a $J \times Q$ slope matrix for the covariates, and $\boldsymbol{\epsilon}_{ki}$ is a J-dimensional vector of residuals with mean zero and covariance matrix Θ_{cd}. For a categorical variable Y_{kij} a normality assumption for ϵ_{kij} is thus equivalent to a probit regression for Y_{kij} on η_{kij} and X_{kij}. Alternatively, ϵ_{kij} can have a logistic distribution, resulting in a logistic regression. For a count variable Y_{kij} the residual ϵ_{kij} is assumed to be zero. For normally distributed continuous variables Y_{kij} the residual variable ϵ_{kij} is assumed normally distributed.

The "structural" part of the model is defined by

$$\boldsymbol{\eta}_{ki}|_{C_{ki}=c,D_k=d} = \boldsymbol{\alpha}_{cdk} + B_{cdk}\boldsymbol{\eta}_{ki} + \Gamma_{cdk}\boldsymbol{X}_{ki} + \boldsymbol{\zeta}_{ki}, \qquad (6.9)$$

where $\boldsymbol{\alpha}_{cdk}$ is an m-dimensional vector of intercepts, B_{cdk} is an $m \times m$ structural regression parameter matrix, Γ_{cdk} is an $m \times Q$ slope parameter matrix, and $\boldsymbol{\zeta}_{ki}$ is an m-dimensional vector of normally distributed residuals with covariance matrix Ψ_{cd}. The model for the latent categorical variable C_{ki} is a multinomial logit model

$$\Pr(C_{ki} = c|D_k = d) = \frac{\exp(a_{cdk} + \boldsymbol{b}_{cdk}'\boldsymbol{X}_{ki})}{\sum_s \exp(a_{sdk} + \boldsymbol{b}_{sdk}'\boldsymbol{X}_{ki})}. \qquad (6.10)$$

Some parameters have to be restricted for identification purposes. For example, the variance of ϵ_{kij} should be 1 for categorical variables Y_{kij} under probit and $\pi^2/\sqrt{3}$ under logit. Also $a_{Ldk} = b_{Ldk} = 0$.

The multilevel part of the model is introduced as follows. Each of the intercepts, slopes, or loading parameters in Equation (6.8) through Equation (6.10) can be either a fixed coefficient or a random effect that varies across clusters k. Let $\boldsymbol{\eta}_k$ be the vector of all such random effects and let X_k be the vector of all cluster-level covariates. The between-level model for $\boldsymbol{\eta}_k$ is then

$$\boldsymbol{\eta}_k|_{D_k=d} = \boldsymbol{\mu}_d + B_d\boldsymbol{\eta}_k + \Gamma_d\boldsymbol{X}_k + \boldsymbol{\zeta}_k, \qquad (6.11)$$

where $\boldsymbol{\mu}_d$, B_d, and Γ_d are fixed parameters and $\boldsymbol{\zeta}_k$ is a normally distributed residual with covariance Ψ_d. The model for the between level categorical variable D is also a multinomial logit regression

$$\Pr(D_k = d) = \frac{\exp(a_d + \boldsymbol{b}_d'\boldsymbol{X}_k)}{\sum_s \exp(a_s + \boldsymbol{b}_s'\boldsymbol{X}_k)}. \qquad (6.12)$$

Equation (6.8) through Equation (6.12) comprise the definition of a multilevel latent-variable mixture model. There are many extensions of this model that are possible in the Mplus framework. For example, observed dependent variables can be incorporated on the between level. Other extensions arise from the fact that a regression equation can be constructed between any two variables in the model. Such equations can be fixed- or random-effect regressions. The model can also accommodate multiple latent-class variables on the within and the between level. Other types of dependent variables can also be incorporated in this model such as censored, nominal, semi-continuous, and time-to-event survival variables; see Olsen and Schafer (2001) and Asparouhov, Masyn, and Muthén (2006).

6.4 Estimation

The above model is estimated by the maximum likelihood estimator using the EM algorithm where the latent variables C_{ki}, $\boldsymbol{\eta}_{ki}$, D_k, and $\boldsymbol{\eta}_k$ are treated as missing data. The observed-data likelihood is given by

$$\prod_k \sum_d \Pr(D_k = d) \int \psi_k(\boldsymbol{\eta}_k) \prod_i \left(\sum_c \Pr(C_{ki} = c) \int f_{ki}(\boldsymbol{Y}_{ki}) \psi_{ki}(\boldsymbol{\eta}_{ki}) d\boldsymbol{\eta}_{ki} \right) d\boldsymbol{\eta}_k, \quad (6.13)$$

where f_{ki}, ψ_{ki}, and ψ_k are the likelihood functions for $\boldsymbol{Y}_{ki}$, $\boldsymbol{\eta}_{ki}$, and $\boldsymbol{\eta}_k$, respectively. Numerical integration is utilized in the evaluation of the above likelihood using both adaptive and non-adaptive quadrature (see Schilling and Bock, 2005). The method can be described as follows. Suppose that η is a continuously distributed random-effect variable with density function ψ. Then

$$\int f(\eta)\psi(\eta)d\eta \approx \sum_{q=1}^{Q} w_q f(n_q), \quad (6.14)$$

where n_q are the nodes of the numerical integration and w_q are the weights. The weights are computed as $w_q = \psi(n_q)/\sum_{i=1}^{Q} \psi(n_i)$. The numerical integration method approximates the continuous distribution for η with a categorical distribution, that is, we can assume that the variable η takes the values n_q with probabilities w_q. Using this method the likelihood (6.13) is approximated by

$$\prod_k \sum_d \Pr(D_k = d) \sum_q \Pr(\eta_k = n_{qk}) \prod_i \left(\sum_c \Pr(C_{ki} = c) \sum_r \Pr(\eta_{ki} = n_{rki}) f_{ki}(\boldsymbol{Y}_{ki}) \right)$$

$$= \prod_k \sum_{d,q} \Pr(D_k = d, \eta_k = n_{qk}) \prod_i \left(\sum_{c,r} \Pr(C_{ki} = c, \eta_{ki} = n_{rki}) f_{ki}(\boldsymbol{Y}_{ki}) \right), \quad (6.15)$$

where n_{qk} and n_{rki} are the nodes of the numerical integration.

The EM algorithm is as follows. First compute the posterior distribution for the latent variables. The posterior joint distribution for D_k and $\boldsymbol{\eta}_k$ is computed as follows:

$$p_{dqk} = \Pr(D_k = d, \boldsymbol{\eta}_k = n_{qk} | *)$$

$$= \frac{\Pr(D_k = d, \boldsymbol{\eta}_k = n_{qk}) \prod_i \left(\sum_{c,r} \Pr(C_{ki} = c, \boldsymbol{\eta}_{ki} = n_{rki}) f_{ki}(\boldsymbol{Y}_{ki}) \right)}{\sum_{d,q} \Pr(D_k = d, \boldsymbol{\eta}_k = n_{qk}) \prod_i \left(\sum_{c,r} \Pr(C_{ki} = c, \boldsymbol{\eta}_{ki} = n_{rki}) f_{ki}(\boldsymbol{Y}_{ki}) \right)}.$$

The posterior conditional joint distribution for C_{ki} and $\boldsymbol{\eta}_{ki}$ is computed as follows:

$$p_{crki|dq} = \Pr(C_{ki} = c, \boldsymbol{\eta}_{ki} = n_{rki} | *, D_k = d, \boldsymbol{\eta}_k = n_{qk})$$

$$= \frac{\Pr(C_{ki} = c, \boldsymbol{\eta}_{ki} = n_{rki}) f_{ki}(Y_{ki})}{\sum_{c,r} \Pr(C_{ki} = c, \boldsymbol{\eta}_{ki} = n_{rki}) f_{ki}(Y_{ki})}.$$

The expected complete-data log-likelihood is now given by

$$\sum_{dqk} p_{dqk} \log(\Pr(D_k = d, \boldsymbol{\eta}_k = n_{qk})) + \sum_{dcqrki} p_{dqk}\, p_{crki|dq} \log(\Pr(C_{ki} = c, \boldsymbol{\eta}_{ki} = n_{rki}))$$

$$+ \sum_{dcqrki} p_{dqk}\, p_{crki|dq} \log(f_{ki}(Y_{ki})),$$

which is maximized with respect to the model parameters.

An alternative algorithm for obtaining the maximum likelihood estimates can be constructed by directly optimizing (6.15) with standard maximization algorithms such as the Fisher scoring and the quasi-Newton algorithms. Such alternative algorithms can be used in combination with the EM algorithm to achieve faster convergence, an approach known as the accelerated EM algorithm (AEM). The AEM algorithm is implemented in Mplus.

A number of different integration methods can be used in (6.14). Mplus implements three different integration methods: rectangular, Gauss–Hermite, and Monte Carlo integration. In addition, adaptive integration can be used. With this method, the integration nodes are concentrated in the area where the posterior distribution of the random effects is non-zero.

The estimation implemented in Mplus allows missing at random data for all dependent variables (Little and Rubin, 2002). Non-ignorable missing data is discussed in Muthén et al. (2003). It should be noted that mixture models in general are prone to have multiple local maxima of the likelihood and the use of many different sets of starting values in the interactive maximization procedure is strongly recommended. An automatic random starts procedure is implemented in the Mplus program, where starting values given by the user or produced automatically by the program are randomly perturbed.

6.4.1 Model assessment

For comparison of fit of models that have the same number of classes and are nested, the usual likelihood ratio chi-square difference test can be used, as long as the requirement of not having parameters on the border of the admissible parameter space in the more restricted model is fulfilled. Comparison of models with different numbers of classes violates this requirement with zero probability parameters. Deciding on the number of classes is instead typically accomplished by a Bayesian information criterion (BIC) (Schwarz, 1978; Kass and Raftery, 1993),

$$BIC = -2 \log L + r \log n,$$

where r is the number of free parameters in the model and n is the sample size. The lower the BIC value, the better the model. The number of classes is increased until a BIC minimum is found. Although not chi-square distributed, the usual likelihood ratio statistic for comparing models with different number of classes can still be used, assessing the distribution of the statistic by bootstrap techniques. McLachlan and Peel (2000, Chapter 6) discuss a parametric bootstrapped likelihood ratio approach proposed by Aitkin, Anderson, and Hinde (1981). Although computationally intensive, it has been found to perform well in simulation studies using latent-class and growth mixture models, outperforming BIC in some instances (Nylund, Asparouhov, and Muthén, 2007).

The fit of the model to data for continuous variables can be studied by comparing for each class estimated moments with moments created by weighting the individual data by the estimated conditional probabilities (Roeder, Lynch and Nagin, 1999). To check how closely the estimated average curve within each class matches the data, it is also useful to randomly assign individuals to classes based on individual estimated conditional class probabilities. Plots of the observed individual trajectories together with the model-estimated average trajectory can be used to check assumptions using class membership determined by "pseudo-class" draws (Bandeen-Roche et al., 1997). Wang, Brown, and Bandeen-Roche (2005) present methods for residual checking based on these ideas. With categorical and count outcomes, model fit may be investigated with respect to univariate and bivariate frequency tables, as well as frequencies for response patterns that do not have too small expected counts. Finally, it is important to note that the need for latent classes may be due to non-normality of the outcomes rather than substantively meaningful subgroups (see McLachlan and Peel, 2000, pp. 14–17; Bauer and Curran, 2003). To support a substantive interpretation of the latent classes, the researcher should consider not only the outcome variable in question, but also antecedents (covariates predicting latent-class membership), concurrent outcomes, and distal outcomes (predictive validity); see also related arguments in Muthén (2004).

6.5 Analysis of Examples

This section presents analyses of the crime data of Example 4, the aggressive behavior data of Example 2, and the math achievement data of Example 5. The Example 4 analysis uses a growth mixture model for crime counts. Examples 2 and 5 consider multilevel growth mixture modeling of cluster data. Example 2 examines intervention effects that vary across both student-level and classroom-level latent-class variables. Example 5 considers students within school where student growth characteristics vary across a school-level latent-class variable.

6.5.1 Analysis of Example 4: Age–crime curves

The analysis of the Philadelphia data with counts of criminal activity for 13,160 males aged 4 through 26 will compare two different approaches: a "group-based" approach and growth mixture analysis (for more extended comparisons, see Kreuter and Muthén, 2007, 2008). The group-based analysis is associated with the work of Nagin and Land (1993), Nagin (1999, 2005), Roeder, Lynch, and Nagin (1999), and Jones, Nagin, and Roeder (2001). This approach is commonly seen in the criminology literature and was used by D'Unger et al. (1998), D'Unger, Land, and McCall (2002), and Loughran and Nagin (2006) for these data. The group-based analysis does not cover cluster sampling and has the further restrictions of zero within-class variances $\Psi_c = 0$, as well as $\Theta_c = \theta I$. The group-based approach is further discussed in Muthén (2004) where it is referred to as latent-class growth analysis (LCGA), given its similarity to latent-class analysis (LCA). Both LCGA and LCA search for classes of individuals defined by conditional independence of the repeated measures given class. In contrast, a growth mixture model (GMM) allows for within-class correlations between repeated measures. Such correlation may, for example, be due to omitted time-varying covariates. If within-class correlation is ignored, a distorted class formation is obtained. Within-class correlation is obtained in GMMs by allowing for random effects with non-zero within-class variances. Both LCGA and GMMs use a zero-inflated Poisson model in line with Roeder, Lynch, and Nagin (1999). For time point j, individual i, and cluster k,

$$Y_{kij}|_{C_{ki}=c} = \begin{cases} 0 & \text{with probability } \pi_{kij}, \\ \text{Poisson}(\lambda_{ckij}) & \text{with probability } 1 - \pi_{kij} \end{cases}$$

where λ is the Poisson rate. In line with previous modeling of the Philadelphia data, a quadratic growth curve is used. Drawing on (6.8) and (6.9) of the general model in Section 6.3.2, the growth mixture zero-inflated Poisson model for these data is expressed in terms of the log rate as

$$\log \lambda_{ij|C_i=c} = \eta_{0i} + \eta_{1i}\, a_{ij} + \eta_{2i}\, a_{ij}^2,$$
$$\eta_{0i|C_i=c} = \alpha_{0c} + \zeta_{0i},$$
$$\eta_{1i|C_i=c} = \alpha_{1c} + \zeta_{1i},$$
$$\eta_{2i|C_i=c} = \alpha_{2c} + \zeta_{2i}.$$

To make analysis results comparable to the LCGA of Loughran and Nagin (2006), a minority of individuals with more than 10 criminal offenses in any given year are deleted, reducing the sample size only from 13,160 to 13,126, and combining the data into two-year intervals. Loughran and Nagin (2006) settled on a four-class solution: non-offenders, adolescent-limited, and high and low chronic (persisting criminal activity at age 26). D'Unger et al. (1998) and D'Unger, Land, and McCall (2002) used a random subset ($n = 1000$) of the data and concluded based on BIC that a five-class LCGA solution was preferred. Their five classes were labeled: non-offenders, high and low adolescent-peaked, and high and low chronic.

Table 6.1 gives results for 1 through 4 classes of GMM and 4–8 classes for LCGA. In addition to log-likelihood values, number of parameters, and BIC, the table shows fit to the data in terms of the number of standardized residuals that are significant at the 5% level for the 10 most frequent response patterns across time (comprising 78% of the data and eliminating only patterns with observed frequency less than 100). The one-class GMM is the conventional random-effects model. Here, 5 of the 10 residuals show significant misfit, illustrating the need for a more flexible model. The two- and the three-class GMMs obtain considerably improved BIC values. The three-class GMM reduces the number of significant residuals from 5 to 1, indicating the appropriateness of the mixture modeling. The four-class GMM adds relatively little improvement. The three-class GMM displays the three themes of non-offenders, adolescent-limited, and chronic. Figure 6.4 shows the mean trajectories for the three-class GMM. The four-class GMM splits the adolescent-limited class into two, where the total percentage for those two classes is about the same as for the adolescent-limited class of the three-class GMM.

The four-class LCGA is the same as presented in Loughran and Nagin (2006) and the five-class LCGA shows the same types of trajectory classes as in D'Unger et al.'s analysis. Neither of these two models fit the data well. An eight-class LCGA is needed to get a reduction to one significant residual. In contrast, the three-class GMM has only one significant residual and the four-class GMM has none. With three classes the GMM gives a better BIC value than any of the LCGA models shown in Table 6.1. The BIC values for the four-class LCGA

Table 6.1 Age–Crime Curves: Log-Likelihood and BIC Comparisons for GMM and LCGA

Model	Log-Likelihood	# Parameters	BIC	# Significant Residuals
1-class GMM	−40,606	17	81,373	5
2-class GMM	−40,422	21	81,044	4
3-class GMM	−40,283	25	80,803	1
4-class GMM	−40,237	29	80,748	0
4-class LCGA	−40,643	23	81,503	4
5-class LCGA	−40,483	27	81,222	3
6-class LCGA	−40,410	31	81,114	3
7-class LCGA	−40,335	35	81,003	2
8-class LCGA	−40,263	39	80,896	1

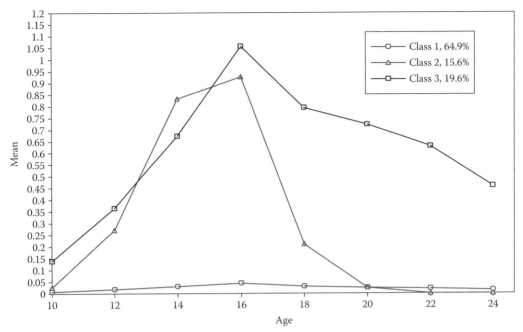

Figure 6.4 Estimated mean trajectories from a three-class growth mixture model for criminal activities.

used in Loughran and Nagin (2006) and the five-class model used in D'Unger et al. (1998) and D'Unger, Land, and McCall (2002) are considerably worse than the BIC value for the three-class GMM. Furthermore, the three-class GMM uses two parameters less than the five-class LCGA, but has a better log-likelihood by 200 points. This illustrates the importance of using random effects to allow for variations on the themes of the trajectory shapes of the classes. The LCGA approach leads to a proliferation of classes, all of which may not have substantive salience.

6.5.2 Analysis of Example 2: Varying intervention effects on classroom aggressive behavior

The Baltimore randomized field trial discussed in Section 6.2.2 was repeated for several cohorts of students. The Section 6.2.2 analysis considered cohort 1 data, whereas data from cohort 3 (Ialongo et al., 1999) are analyzed here. A total of 362 boys in 27 classrooms are considered over four time points: fall of first grade, spring of first grade, spring of second grade, and spring of third grade. The average number of boys per classroom is 13.4. It is of interest to study if teachers in classrooms with higher aggressiveness levels have a more difficult time successfully implementing the intervention aimed at reducing aggressive-disruptive behavior. For the first grade, there is substantial variation across classrooms in the aggressiveness scores as evidenced by the intraclass correlations at the four time points: $0.11, 0.16, 0.04, 0.00$. In addition to student-level trajectory classes, the use of latent classes on the classroom level makes it possible to more fully explore variation in intervention effects.

Drawing on the Section 6.3.2 general model, the two-level GMM is expressed as follows using a quadratic curve shape,

$$Y_{kij}|_{C_{ki}=c, D_k=d} = \eta_{0ki} + \eta_{1ki}\, a_{ij} + \eta_{2ki}\, a_{ij}^2 + \epsilon_{kij},$$

for $j = 1, 2, 3, 4$, with variation across students within classrooms expressed as

$$\eta_{0ki}|_{C_{ki}=c,D_k=d} = \alpha_{cdk0} + \zeta_{0ki},$$

$$\eta_{1ki}|_{C_{ki}=c,D_k=d} = \alpha_{cdk1} + \zeta_{1ki},$$

$$\eta_{2ki}|_{C_{ki}=c,D_k=d} = \alpha_{cdk2} + \zeta_{2ki},$$

and variation across classrooms expressed as

$$\alpha_{cdk0}|_{C_{ki}=c,D_k=d} = \alpha_{cd0} + \zeta_{30k},$$

$$\alpha_{cdk1}|_{C_{ki}=c,D_k=d} = \alpha_{cd1} + \gamma_{cd1} Z_k + \zeta_{31k},$$

$$\alpha_{cdk2}|_{C_{ki}=c,D_k=d} = \alpha_{cd2} + \gamma_{cd2} Z_k + \zeta_{32k}.$$

Here, $a_{i1} = 0$ to center the intercept η_0 at the pre-intervention time point. Z is a treatment-control dummy variable on the classroom level. For reasons of parsimony, the student-level latent-class variable C and the classroom-level latent-class variable D are taken to have an additive effect on the means α_{cd0}, α_{cd1}, and α_{cd2}. The γ intervention effects are, however, allowed to vary across combinations of C and D classes. The linear and quadratic slopes were found to have zero variance across classrooms. The intraclass correlation is captured by the classroom variation in the random intercept of the growth model, α_{cdk0}.

The latent categorical variable C_{ki} follows the multinomial logistic regression

$$\Pr(C_{ki} = c|D_k = d) = \frac{\exp(a_{cdk})}{\sum_s \exp(a_{sdk})},$$

where in this application

$$a_{cdk}|_{D_k=d} = a_c + \zeta_{ck}. \tag{6.16}$$

The analyses indicate that $V(\zeta_{ck}) = 0$, that is, the random intercepts for the latent-class variable C do not vary across classrooms. In other applications, however, this variance can be substantial.

As a first step, a model without the classroom-level latent-class variable D was explored. As judged by BIC, the conventional single-class random-effects growth model is clearly outperformed by growth mixture modeling, with a three-class model giving the lowest BIC. The log-likelihood for the conventional model is -4157.98 with 14 parameters and a BIC of 8398, while the three-class GMM has a log-likelihood of -4048.84 with 26 parameters and a BIC of 8251. The three-class model has a significant classroom variance for the random intercept. Second, two latent classes for D were added to the model resulting in latent classes with low versus high classroom-level aggression (51% versus 49%). The log-likelihood is -4041.42 with 34 parameters and a BIC of 8283. This BIC is not as good as for the previous model with no classroom-level latent-class variable, but it is not known how BIC performs in settings with multilevel latent-class variables. The three student-level latent-trajectory classes show a low-increasing class of 68%, a medium-increasing class of 19%, and a high-decreasing class of 12%. The mean curves for these three latent classes are shown in Figure 6.5 as pairs of control and intervention curves. Results for the latent class consisting of classrooms with low aggression level are given in the left plot and results for the latent class consisting of classrooms with high aggression level are given in the right plot. The plots suggest that in classrooms with a low level of aggression, students who are in the two highest trajectory classes benefit from the intervention. In classrooms with a high level of aggression, however, only students who are in the lowest trajectory class benefit from the intervention. This suggests that the intervention may be harder for teachers to implement well in high-aggressive classrooms. The results should be interpreted with caution, however, given the sample of only 27 classrooms and other competing models. An alternative model lets the C and D latent-class variables have an interactive effect on the

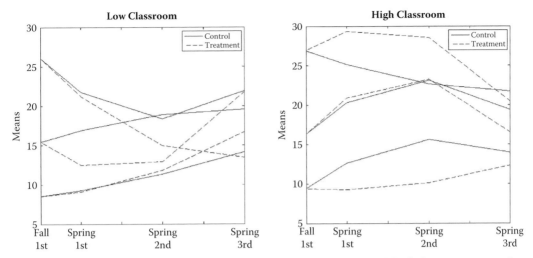

Figure 6.5 Estimated mean trajectories from a growth mixture model of classroom aggressive behavior.

random-effect growth means and lets the random-effect means of a_{cdk} in (6.16) be influenced by the latent classes of D. This significantly improves the log-likelihood, but the increased number of parameters on the classroom level results in a less stable solution. The resulting split into 7 and 20 classrooms for the latent classes of D causes estimated outcome mean differences with high variability.

6.5.3 Analysis of Example 5: Classification of schools based on achievement development

The NELS math achievement data from grades 8, 10, and 12 discussed in Section 6.2.5 are analyzed here. NELS surveyed 913 schools and a total of 14,217 students. The NELS analysis illustrates two features of the Section 6.3.2 model, taking into account the school clusters and using a school-level latent-class variable.

In the NELS analysis, student growth rate is regressed on the growth intercept in grade 8 using a random slope that varies across schools. This random slope and the means of the random intercept and the intercept of the random growth rate are allowed to vary across the school-level latent classes. Letting school-level latent classes influence student-level relations helps identify the school-level latent classes. Extending the example of Section 6.2.3 to clusters k and a cluster-level latent-class variable D_k, variation across grades is expressed as

$$Y_{kij|D_k=d} = \eta_{0ki} + \eta_{1ki} \, a_{ij} + \epsilon_{kij},$$

for $j = 1, 2, 3$, with variation across students expressed as

$$\eta_{0ki|D_k=d} = \alpha_{d0} + \zeta_{0ki},$$
$$\eta_{1ki|D_k=d} = \alpha_{d1} + \beta_{dk} \, \eta_{0ki} + \zeta_{1ki},$$

where the variation across schools is accomplished by the variation of α_{d0}, α_{d1}, and β_{dk} across the classes of D. A single-class growth model, that is, a conventional three-level analysis, obtains a log-likelihood of $-31{,}791$ with 10 parameters and a BIC of 63,678. A two-class GMM obtains a log-likelihood of $-31{,}545$ with 16 parameters and a BIC of 63,243. A three-class GMM obtains a log-likelihood of $-31{,}434$ with 22 parameters and a BIC of 63,079. A four-class model does not improve the log-likelihood further. The three-class model shows that the growth rate is significantly positively related to the growth intercept defined at grade 8 only for a class of 52% of the schools who have average growth over grades 8–12. A higher developing class of 25% and a lower developing class of 23% have small and insignificant relationships. This illustrates the possibility of finding clusters of schools with different achievement profiles. School-level covariates predicting school class membership can give further understanding of the school classes.

6.6 Parametric versus non-parametric random-effect models

Titterington, Smith, and Makov (1985) make a distinction in the use of finite-mixture modeling in terms of direct and indirect applications. A direct application uses mixtures to represent "the underlying physical phenomenon," whereas with the indirect application the mixture components "do not necessarily have a direct physical interpretation." The examples discussed so far can be seen as attempts at direct application, where trajectory classes are given substantive interpretation and results are presented for each mixture component rather than mixing over the classes. Examples of indirect applications include outlier detection and representation of non-normal distributions. Mixture modeling of non-normal distributions is the focus of this section. A growth model with a non-parametric representation of the random-effects distribution is presented and a simulation study compares the use of such a model to the conventional random-effect growth model assuming normality.

It has been argued that with categorical and count outcomes, the typical normality assumption for random effects in repeated-measurement modeling may be less well supported by data (see also Aitkin, 1999). Deviations from normality may strongly affect the results. With categorical and count outcomes, maximum likelihood leads to the use of numerical integration which is computationally heavy and intractable when the number of random effects is large. Numerical integration uses fixed quadrature points and weights according to a normal distribution. A non-parametric approach instead considers a discretized distribution, estimating the points and the weights using a finite-mixture model. The latent class means are the points and the class probabilities are the weights. In this way, the non-parametric approach both avoids the normality specification and is computationally less demanding.

In this section we describe and compare the general parametric and non-parametric random-effect models. Both of these models are special cases of the general model described in Equation (6.8) through Equation (6.12). Both of these modeling alternatives attempt to capture cluster-specific effects. The difference between the two models is the underlying assumption for the distributions of the cluster-specific random effects. In the parametric model the random effects are assumed to have conditionally normal distribution, that is, the conditional distribution of the random effects, given all covariates, is assumed to be normal. In the non-parametric model the random effects are assumed to have a non-parametric conditional distribution.

The parametric random-effect model is well established and frequently used in practice. Butler and Louis (1992) show that the normality assumption in the parametric model does not affect the fixed slopes in the model. Verbeke and Lesaffre (1996) show that more accurate estimates can be obtained for the random effects if a non-normal distribution is estimated. Aitkin (1999) gives the general modeling approach to the non-parametric random-effect models that we follow here. First we give the complete description of the two modeling alternatives and show how they fit in the general modeling framework (6.8) through (6.12).

6.6.1 Parametric random-effect model

This model is a special case of model (6.8) through (6.12) for the case of no categorical latent variables. The within-level model is given by

$$Y_{ki}^* = \nu_k + \Lambda_k \eta_{ki} + K_k X_{ki} + \epsilon_{ki}, \tag{6.17}$$

$$\eta_{ki} = \alpha + B_k \eta_{ki} + \Gamma_k X_{ki} + \zeta_{ki}. \tag{6.18}$$

The coefficients $\boldsymbol{\nu}_k$, Λ_k, K_k, B_k, and Γ_k can be either fixed coefficients that are the same across cluster or random effects that vary across cluster. Let $\boldsymbol{\eta}_k$ represent the vector of all such random effects. The between-level model is described by

$$\boldsymbol{\eta}_k = \boldsymbol{\mu} + B\boldsymbol{\eta}_k + \Gamma\,\boldsymbol{X}_k + \boldsymbol{\zeta}_k. \tag{6.19}$$

The random-effect residuals $\boldsymbol{\zeta}_{ki}$ and $\boldsymbol{\zeta}_k$ are assumed normally distributed. This assumption is the difference between the parametric and the non-parametric model. Note that the distributional assumption for $\boldsymbol{\epsilon}_{ki}$ is determined by the type of observed variable we model.

6.6.2 Non-parametric random-effect model

This model is a special case of model (6.8) through (6.12) where the random effects $\boldsymbol{\eta}_{ki}$ and $\boldsymbol{\eta}_k$ do not have normally distributed residuals $\boldsymbol{\zeta}_{ki}$ and $\boldsymbol{\zeta}_k$. The within-level model is given by

$$\boldsymbol{Y}_{ki}^* = \boldsymbol{\nu}_k + \Lambda_k \boldsymbol{\eta}_{ki} + K_k \boldsymbol{X}_{ki} + \boldsymbol{\epsilon}_{ki},$$

$$\boldsymbol{\eta}_{ki}|_{C_{ki}=c} = \boldsymbol{\alpha}_c + B_k \boldsymbol{\eta}_{ki} + \Gamma_k \boldsymbol{X}_{ki}, \tag{6.20}$$

$$\Pr(C_{ki} = c) = p_c, \tag{6.21}$$

where p_c are parameters to be estimated. The coefficients $\boldsymbol{\nu}_k$, Λ_k, K_k, B_k, and Γ_k can again be either fixed coefficients or random effects. Let $\boldsymbol{\eta}_k$ represent the vector of all random effects. The between-level model is given by

$$\boldsymbol{\eta}_k|_{D_k=d} = \boldsymbol{\mu}_d + B\boldsymbol{\eta}_k + \Gamma\boldsymbol{X}_k, \tag{6.22}$$

$$\Pr(D_k = d) = q_d, \tag{6.23}$$

where q_d are parameters to be estimated. The random-effect model (6.20) through (6.23) can alternatively be presented as in Equation (6.18) and Equation (6.19), considering the mixture across classes

$$\boldsymbol{\eta}_{ki} = \boldsymbol{\alpha} + B_k \boldsymbol{\eta}_{ki} + \Gamma_k \boldsymbol{X}_{ki} + \boldsymbol{\zeta}_{ki},$$

$$\boldsymbol{\eta}_k = \boldsymbol{\mu} + B\boldsymbol{\eta}_k + \Gamma\boldsymbol{X}_k + \boldsymbol{\zeta}_k,$$

where $\boldsymbol{\alpha} = \sum_c \boldsymbol{\alpha}_c p_c$, $\boldsymbol{\mu} = \sum_d \boldsymbol{\mu}_d q_d$, and $\boldsymbol{\zeta}_{ki}$ and $\boldsymbol{\zeta}_k$ are non-parametric zero-mean residuals that are freely estimated. The residual $\boldsymbol{\zeta}_{ki}$ takes the values $\boldsymbol{\alpha}_c - \boldsymbol{\alpha}$ with probability p_c and the residual $\boldsymbol{\zeta}_k$ takes values $\boldsymbol{\mu}_d - \boldsymbol{\mu}$ with probability q_d. The variance and covariance for the non-parametric effects can also be computed; for example, the variance of $\boldsymbol{\zeta}_{ki}$ is $\sum_c p_c(\boldsymbol{\alpha}_c - \boldsymbol{\alpha})(\boldsymbol{\alpha}_c - \boldsymbol{\alpha})'$.

6.6.3 Simulation study

A simulation study is conducted to compare the performance of the parametric and non-parametric random-effect models for data generated with non-normal random effects. Consider a logistic growth model with 10 binary items $U_1, \ldots, U_{10}$,

$$\log\left[\frac{\Pr(U_{ij} = 1)}{\Pr(U_{ij} = 0)}\right] = \eta_{0i} + \eta_{1i}\,a_{ij}, \tag{6.24}$$

where the time scores $a_{ij} = (j - 1)/2$, and η_0 and η_1 are non-normal random effects. Generation of η_0 and η_1 used the following finite mixture of normal distributions:

$$0.67 \cdot N(\mu_1, \sigma^2) + 0.09 \cdot N(\mu_2, \sigma^2) + 0.24 \cdot N(\mu_3, \sigma^3).$$

To generate η_0, the following parameters were used: $\mu_1 = 2$, $\mu_2 = 1$, $\mu_3 = 0$, and $\sigma = 0.4$. To generate η_1, the following parameters were used: $\mu_1 = -0.3$, $\mu_2 = -0.4$, $\mu_3 = -1$, and $\sigma = 0.1$. From these values, 100 samples of size 2000 were generated according to the model (6.24). The data were analyzed using the parametric linear model (PM) and the

Table 6.2 Comparing the Parametric (PM) and Non-Parametric (NPM) Random-Effect Models

Parameter	True Value	PM Bias	NPM Bias	PM MSE	NPM MSE
m_0	1.421	0.004	-0.010	0.0018	0.0018
m_1	-0.480	0.073	0.004	0.0056	0.0003
v_0	0.733	0.563	0.080	0.3302	0.0122
v_1	0.088	-0.033	-0.004	0.0012	0.0001
ρ	0.247	-0.006	0.003	0.0008	0.0002

non-parametric linear model (NPM). PM is a conventional single-class growth model with random normal effects as in (6.17) and (6.18). Drawing on (6.20) and (6.22), the NPM is expressed as

$$Y_{ij}^* = \eta_{0i} + \eta_{1i}\, a_{ij} + \epsilon_{ij},$$

$$\eta_{0i}|_{C_i=c} = \alpha_{0c},$$
$$\eta_{1i}|_{C_i=c} = \alpha_{1c},$$

so that the random effects are represented by a mixture distribution. A more general form would allow within-class variation for residuals ζ as in (6.9).

The parameter estimates are summarized in Table 6.2. The means of η_0 and η_1 are denoted by m_0 and m_1 and the variances by v_0 and v_1. The covariance of η_0 and η_1 is denoted by ρ. The results are presented for the non-parametric model with three nodes, since three nodes were determined to be sufficient for most replications using the McLachlan and Peel (2000) parametric likelihood ratio test. The estimates on which Table 6.2 is based are computed for the mixture over the three classes in line with (6.20) and (6.22). The results in Table 6.2 clearly indicate the advantages of the NPM method. The NPM parameter estimates have substantially smaller bias and smaller mean squared error (MSE) for several parameters.

In general it is difficult to evaluate model fit for random-effect models. There is no general unrestricted model that can be used for comparison. In this simulated example, however, there is such a model, namely, the completely unrestricted contingency table for the binary items. In addition, the Pearson chi-square test can be used to test the fit of the model. The data were generated according to a linear growth model with non-normal random effects; that is, the true model is a linear random-effect growth model. Both the PM and NPM models are linear random-effect growth models but are based on different assumptions on the distribution of the random effects; neither assumption specifies the true random-effect distribution. This situation is typical in practical applications, where the true random-effect distribution is unknown and the modeling assumptions are likely to deviate from the true distribution to some extent. It is assumed that the distributional misspecification will not interfere with the basic structure of the model and that the estimated model will provide a good fit for the data despite the distributional misspecifications. The Pearson test of fit can be used to directly compare the sensitivity of the PM and NPM models. If the Pearson test rejects the model, one concludes that the model fit is poor. In practical applications the lack of model fit could incorrectly be interpreted as evidence for deficiency in the linear growth structure of the model rather than as possible misspecification in the random-effect distribution.

In the current simulated example one wants the Pearson test to reject the model no more than the nominal 5% of the time. Table 6.3 contains the Pearson test of fit results for the PM linear growth model as well as the NPM linear growth model using 3, 4, 5, and 6 nodes. The rejection rate in Table 6.3 is the percentage of times the linear growth model was rejected incorrectly. Also presented are the average test statistic value and the degrees of freedom. These two values should generally be close because the expected value of the chi-square distribution is equal to the degrees of freedom. The parametric approach

Table 6.3 Pearson Test of Fit for the Binary Linear Growth Model

Model	PM	NPM(3)	NPM(4)	NPM(5)	NPM(6)
Rejection Rate	100%	22%	15%	8%	4%
Average Test Statistic	1252	1038	1025	1007	997
Degrees of Freedom	1018	1015	1012	1009	1006

rejected the correct model 100% of the time. There is a large difference between the degrees of freedom and the average test value for the parametric model. On the other hand, the non-parametric approach leads to the correct conclusion. The Pearson rejection rate converges to the nominal 5% value as the number of nodes is increased from 3 to 6. The average test statistic value also converges toward its expected value.

As yet another test of model fit, the estimated parametric and non-parametric models are used to compute response pattern frequencies. Table 6.4 shows the top five most frequent patterns in one of the generated data sets as well as the predicted frequency for these patterns, using both the parametric and the non-parametric random-effect models. For the non-parametric model three-node estimation is used. In this comparison as well one can clearly see the advantage of the non-parametric approach. The non-parametric pattern frequencies are more closely matching the observed values than the parametric pattern frequencies.

The simulation study clearly demonstrates the fact that model estimation with normal random effects is not robust to the distributional assumptions. If the random effects are truly non-normal, the parametric model can actually have a poor fit to the data. This phenomenon is exacerbated by the fact that in general random-effect models have no test of model fit. Thus, researchers can easily be mislead into believing that a random-effect model is a useful model. Violations of the normality assumptions can lead to poor estimates and inference. In addition, a comparison between the non-parametric model and the parametric model that is limited to the fixed-effect parameters can be misleading because even when the fixed parameter estimates as well as the random-effect parameter estimates are identical the models implied by the two approaches are quite different. Even when the parametric model fits the first- and second-order moments of the data, the model could have a much worse fit to the data than the non-parametric model simply because it will not fit higher-order moments. The non-parametric model described in this section overcomes these shortcomings and should be used more frequently in practice.

6.7 Conclusions

The range of growth mixture analyses described in this chapter shows the usefulness of extending the conventional random-effects repeated-measures model (mixed linear model) to include different latent classes of development. Growth mixture modeling is a powerful analytic tool when applied to randomized trials as well as to non-experimental research. The idea of detecting different treatment effects for individuals belonging to different trajectory classes has important implications for designing future intervention studies. It is possible to

Table 6.4 Observed and Estimated Pattern Frequencies for the Top Five Most Frequent Patterns

Pattern	Observed	PM-Estimated	NPM-Estimated
1111111111	141	246	135
0000000000	83	66	74
1000000000	61	39	69
1111111110	55	36	54
1111111101	49	33	47

select different interventions for individuals belonging to different trajectory classes using longitudinal screening procedures. One may attempt to classify individuals into their most likely trajectory class based on a set of initial repeated measurements before the intervention starts. Alternatively, one may administer a universal intervention and follow up with a targeted intervention for individuals who show little or no intervention effect.

In non-experimental research, the exploration of different trajectory classes provides a more nuanced picture of development, where different classes can have different antecedents and consequences. As a form of cluster analysis, both the use of individual-level and cluster-level latent-class variables defined from longitudinal data have great promise.

This chapter also studied non-parametric representation of random-effect distributions in growth models. It was shown that mixture modeling with latent classes can provide a good representation of non-normal effects in cases where the conventional approach drawing on normality gives misleading results. In this connection, it is interesting to consider the age–crime analysis of Example 4 in Section 6.5.1. Here, three substantively meaningful trajectory classes were found using growth mixture analysis, each showing within-class variation on the theme of the trajectory shape. The variation was captured by normally distributed random effects. The latent-class growth analysis that was presented may be seen as representing the random effects non-parametrically. In this example, the non-parametric approach leads to a proliferation of latent classes. An alternative is to postulate three substantive classes and allow for within-class variation by a non-parametric approach. The estimates would be mixed together across the non-parametric classes, but not across the substantive classes.

The statistical developments behind growth mixture modeling are still relatively recent and much statistical research remains to be done. For example, the quality of parameter recovery under different conditions needs to be more fully understood. The Monte Carlo simulation facility in Mplus is useful in this regard. Tests need to be developed further to compare models differing not only in the number of latent classes, but also in their random-effect specification. Residual checking of model fit needs to be further developed. Furthermore, much more practical experience is needed with these analyses to create guidelines for how to build models. Hopefully this chapter will stimulate such statistical work.

Acknowledgments

The research of the first author was supported under grant K02 AA 00230-01 from NIAAA, by grant 1 R21 AA10948-01A1 from NIAAA, and by NIMH under grant no. MH40859. The work has benefited from discussions in the Prevention Science and Methodology Group, directed by Hendricks Brown. The e-mail address of the first author is bmuthen@ucla.edu.

References

Aitkin, M. (1999). A general maximum likelihood analysis of variance components in generalized linear models. *Biometrics* **55**, 117–128.

Aitkin, M., Anderson, D., and Hinde, J. (1981). Statistical modelling of data on teaching styles (with discussion). *Journal of the Royal Statistical Society B* **144**, 419–461.

Angrist, J. D., Imbens, G. W., and Rubin, D. B. (1996). Identification of causal effects using instrumental variables. *Journal of the American Statistical Association* **91**, 444–455.

Asparouhov, T. and Muthén, B. (2008). Multilevel mixture models. In G. R. Hancock and K. M. Samuelson (eds.), *Advances in Latent Variable Mixture Models*, pp. 27–51. Charlotte, NC: Information Age Publishing.

Asparouhov, T., Masyn, K., and Muthén, B. (2006). Continuous time survival in latent variable models. In *2006 Proceedings of the American Statistical Association, Biometrics Section*, pp. 180–187. Alexandria, VA: American Statistical Association.

Bandeen-Roche, K., Miglioretti, D. L., Zeger, S. L., and Rathouz, P. J. (1997). Latent variable regression for multiple discrete outcomes. *Journal of the American Statistical Association* **92**, 1375–1386.

Bauer, D. J. and Curran, P. J. (2003). Distributional assumptions of growth mixture models: Implications for overextraction of latent trajectory classes. *Psychological Methods* **8**, 338–363.

Bock, R. D. and Aitkin, M. (1981). Marginal maximum likelihood estimation of item parameters: Application of an EM algorithm. *Psychometrika* **46**, 443–459.

Butler, S. M. and Louis, T. A. (1992). Random effects models with non-parametric priors. *Statistics in Medicine* **11**, 1981–2000.

Choi, K., and Seltzer, M. (2006). Modeling heterogeneity in relationships between initial status and rates of change: Treating latent variable regression coefficients as random coefficients in a three-level hierarchical model. Accepted for publication in *Journal of Educational and Behavioral Statistics*.

Croudace, T. J., Jarvelin, M. R., Wadsworth, M. E., and Jones, P. B. (2003). Developmental typology of trajectories to nighttime bladder control: Epidemiologic application of longitudinal latent class analysis. *American Journal of Epidemiology* **157**, 834–842.

Dolan, L., Kellam, S. G., Brown, C. H., Werthamer-Larsson, L., Rebok, G. W., Mayer, L. S., Laudolff, J., Turkkan, J. S., Ford, C., and Wheeler, L. (1993). The short-term impact of two classroom based preventive intervention trials on aggressive and shy behaviors and poor achievement. *Journal of Applied Developmental Psychology* **14**, 317–345.

D'Unger, A. V., Land, K. C., and McCall, P. L. (2002). Sex differences in age patterns of delinquent/criminal careers: Results from Poisson latent class analyses of the Philadelphia cohort study. *Journal of Quantitative Criminology* **18**, 349–375.

D'Unger, A. V., Land, K. C., McCall, P. L., and Nagin, D. S. (1998). How many latent classes of delinquent/criminal careers? Results from mixed Poisson regression analysis. *American Journal of Sociology* **103**, 1593–1630.

Dunn, G., Maracy, M., Dowrick, C., Ayuso-Mateos, J. L., Dalgard, O. S., Page, H., Lehtinen, V., Casey, P., Wilkinson, C., Vasquez-Barquero, J. L., and Wilkinson, G. (2003). Estimating psychological treatment effects from a randomized controlled trial with both non-compliance and loss to follow-up. *British Journal of Psychiatry* **183**, 323–331.

Ialongo, L. N. , Werthamer, L., Kellam, S. G., Brown, C. H., Wang, S., and Lin, Y. (1999). Proximal impact of two first-grade preventive interventions on the early risk behaviors for later substance abuse, depression and antisocial behavior. *American Journal of Community Psychology* **27**, 599–641.

Jo, B. (2002). Estimation of intervention effects with noncompliance: Alternative model specifications. *Journal of Educational and Behavioral Statistics* **27**, 385–409.

Jo, B. and Muthén, B. (2003). Longitudinal studies with intervention and noncompliance: Estimation of causal effects in growth mixture modeling. In S. P. Reise and N. Duan (eds.), *Multilevel Modeling: Methodological Advances, Issues, and Applications*, pp. 112–139. Mahwah, NJ: Lawrence Erlbaum Associates.

Jones, B. L., Nagin, D. S., and Roeder, K. (2001). A SAS procedure based on mixture models for estimating developmental trajectories. *Sociological Methods & Research* **29**, 374–393.

Kass R. E. and Raftery, A. E. (1993). Bayes factors. *Journal of the American Statistical Association* **90**, 773–795.

Kellam, S. G., Rebok, G. W., Ialongo, N., and Mayer, L. S. (1994). The course and malleability of aggressive behavior from early first grade into middle school: Results of a developmental epidemiologically-based preventive trial. *Journal of Child Psychology and Psychiatry* **35**, 359–382.

Kreuter, F. and Muthén, B. (2007). Analyzing criminal trajectory profiles: Bridging multilevel and group-based approaches using growth mixture modeling. *Journal of Quantitative Criminology*. To appear.

Kreuter, F. and Muthén, B. (2008). Longitudinal modeling of population heterogeneity: Methodological challenges to the analysis of empirically derived criminal trajectory profiles. In G. R.

Hancock and K. M. Samuelson (eds.), *Advances in Latent Variable Mixture Models*. Charlotte, NC: Information Age Publishing.

Laird, N. M. and Ware, J. H. (1982). Random-effects models for longitudinal data. *Biometrics* **38**, 963–974.

Leuchter, A. F., Cook, I. A., Witte, E. A., Morgan, M., and Abrams, M. (2002). Changes in brain function of depressed subjects during treatment with placebo. *American Journal of Psychiatry* **159**, 122–129.

Lin, J. Y., Ten Have, T. R., and Elliott, M. R. (2006). Longitudinal nested compliance class model in the presence of time-varying noncompliance. *Journal of the American Statistical Association*. To appear.

Lin, H., Turnbull, B. W., McCulloch, C. E., and Slate, E. (2002). Latent class models for joint analysis of longitudinal biomarker and event process data: application to longitudinal prostate-specific antigen readings and prostate cancer. *Journal of the American Statistical Association* **97**, 53–65.

Little, R. J. and Rubin, D. B. (2002). *Statistical Analysis with Missing Data*, 2nd ed. New York: Wiley.

Loughran, T. and Nagin, D. S. (2006). Finite sample effects in group-based trajectory models. *Sociological Methods & Research* **35**, 250–278.

McLachlan, G. J. and Peel, D. (2000). *Finite Mixture Models*. New York: Wiley.

Moffitt, T. E. (1993). Adolescence-limited and life-course persistent antisocial behavior: A developmental taxonomy. *Psychological Review* **100**, 674–701.

Muthén, B. (1984). A general structural equation model with dichotomous, ordered categorical, and continuous latent variable indicators. *Psychometrika* **49**, 115–132.

Muthén, B. (2001a). Second-generation structural equation modeling with a combination of categorical and continuous latent variables: New opportunities for latent class/latent growth modeling. In L. M. Collins and A. Sayer (eds.), *New Methods for the Analysis of Change*, pp. 291–322. Washington, DC: APA.

Muthén, B. (2001b). Latent variable mixture modeling. In G. A. Marcoulides and R. E. Schumacker (eds.), *New Developments and Techniques in Structural Equation Modeling*, pp. 1–33. Mahwah, NJ: Lawrence Erlbaum Associates.

Muthén, B. (2002). Beyond SEM: General latent variable modeling. *Behaviormetrika*, **29**, 81–117.

Muthén, B. (2004). Latent variable analysis: Growth mixture modeling and related techniques for longitudinal data. In D. Kaplan (ed.), *Handbook of Quantitative Methodology for the Social Sciences*, pp. 345–368. Newbury Park, CA: Sage.

Muthén, B., Jo, B., and Brown, C. H. (2003). Comment on the Barnard, Frangakis, Hill and Rubin article, Principal stratification approach to broken randomized experiments: A case study of school choice vouchers in New York City. *Journal of the American Statistical Association* **98**, 311–314.

Muthén, B. and Masyn, K. (2005). Discrete-time survival mixture analysis. *Journal of Educational and Behavioral Statistics* **30**, 27–58.

Muthén, B. and Muthén, L. (2000). Integrating person-centered and variable-centered analysis: Growth mixture modeling with latent trajectory classes. *Alcoholism: Clinical and Experimental Research* **24**, 882–891.

Muthén, B. and Muthén, L. (1998–2007). *Mplus User's Guide*. Los Angeles: Muthén and Muthén.

Muthén, B. and Shedden, K. (1999). Finite mixture modeling with mixture outcomes using the EM algorithm. *Biometrics* **55**, 463–469.

Muthén, B., Brown, C. H., Masyn, K., Jo, B., Khoo, S. T., Yang, C. C., Wang, C. P., Kellam, S., Carlin, J., and Liao, J. (2002). General growth mixture modeling for randomized preventive interventions. *Biostatistics* **3**, 459–475.

Muthén, B., Khoo, S. T., Francis, D., and Boscardin, K. (2003). Analysis of reading skills development from kindergarten through first grade: An application of growth mixture modeling to sequential processes. In S. P. Reise and N. Duan (eds.), *Multilevel Modeling: Methodological Advances, Issues, and Applications* (2003). Mahwah, NJ: Lawrence Erlbaum Associates.

Muthén, B., Brown, C. H., Leuchter, A., and Hunter, A. (2007). General approaches to analysis of course: Applying growth mixture modeling to randomized trials of depression medication. In P. E. Shrout (ed.), *Causality and Psychopathology: Finding the Determinants of Disorders and their Cures*. Washington, DC: American Psychiatric Publishing.

Nagin, D. S. (1999). Analyzing developmental trajectories: A semi-parametric, group-based approach. *Psychological Methods* **4**, 139–157.

Nagin, D. S. (2005). *Group-Based Modeling of Development*. Cambridge, MA: Harvard University Press.

Nagin, D. S. and Land, K. C. (1993). Age, criminal careers, and population heterogeneity: Specification and estimation of a nonparametric, mixed Poisson model. *Criminology* **31**, 327–362.

Nylund, K. L., Asparouhov, T., and Muthén, B. (2007). Deciding on the number of classes in latent class analysis and growth mixture modeling. A Monte Carlo simulation study. *Structural Equation Modeling*, **14**, 535–569.

Olsen, M. K. and Schafer, J. L. (2001). A two-part random-effects model for semicontinuous longitudinal data. *Journal of the American Statistical Association* **96**, 730–745.

Pearson, J. D., Morrell, C. H., Landis, P. K., Carter, H. B., and Brant, L. J. (1994). Mixed-effect regression models for studying the natural history of prostate disease. *Statistics in Medicine* **13**, 587–601.

Roeder, K., Lynch, K. G., and Nagin, D. S. (1999). Modeling uncertainty in latent class membership: A case study in criminology. *Journal of the American Statistical Association* **94**, 766–776.

Sampson, R. J. and Laub, J. H. (2005). Seductions of methods: Rejoinder to Nagin and Tremblay's "Developmental trajectory groups: Fact or fiction?" *Criminology* **43**, 905–913.

Schilling, S. and Bock, R. D. (2005). High-dimensional maximum marginal likelihood item factor analysis by adaptive quadrature. *Psychometrika* **70**, 533–555.

Schwarz, G. (1978). Estimating the dimension of a model. *Annals of Statistics* **6**, 461–464.

Titterington, D. M., Smith, A. F. M., and Makov, U. E. (1985). *Statistical Analysis of Finite Mixture Distributions*. New York: Wiley.

Verbeke, G. and Lesaffre, E. (1996). A linear mixed-effects model with heterogeneity in the random-effects population. *Journal of the American Statistical Association* **91**, 217–221.

Werthamer-Larsson, L., Kellam, S. G., and Wheeler, L. (1991). Effect of first-grade classroom environment on child shy behavior, aggressive behavior, and concentration problems. *American Journal of Community Psychology* **19**, 585–602.

Wang, C. P., Brown, C. H., and Bandeen-Roche, K. (2005). Residual diagnostics for growth mixture models: Examining the impact of a preventive intervention on multiple trajectories of aggressive behavior. *Journal of the American Statistical Association* **100**, 1054–1076.

Yau, L. H. Y. and Little, R.J. (2001). Inference for the complier-average causal effect from longitudinal data subject to noncompliance and missing data, with application to a job training assessment for the unemployed. *Journal of the American Statistical Association* **96**, 1232–1244.

CHAPTER 7

Targets of inference in hierarchical models for longitudinal data

Stephen W. Raudenbush

Contents

7.1 Introduction

7.1.1 Overview

The generalized linear model is widely applicable for studying outcomes of many types, including dichotomies, counts, polytomies, and continuous outcomes (McCullagh and Nelder, 1989). In conventional applications, a monotonic function of the mean known as a "link function" is regarded as a linear combination of known explanatory variables; conditional on these, outcomes are assumed statistically independent. However, in many interesting applications, this independence assumption is untenable. For example, in a cross-sectional study of students nested within schools, one can generally assume that children attending the same school are more similar than are children attending different schools. And in longitudinal studies, the focus of this chapter, outcomes collected on the same subject must generally be regarded as correlated.

A great deal of interest has focused on how to extend the generalized linear model to longitudinal data. Two prominent approaches have emerged, aptly labeled the "subject-specific" model and the "population-averaged" model (Zeger, Liang, and Albert, 1988). The availability of these two approaches, while extremely useful, has generated some confusion among applied researchers. For which research questions should the subject-specific approach be adopted? When is the population-averaged approach more appropriate?

The purpose of this chapter is to suggest a framework for answering these questions. My central arguments are that we can make these judgments intelligently when we identify three key "targets of inference" implied by the research questions that drive a particular study; and that these targets become clear when we construct the model in a hierarchical fashion as in Lindley and Smith (1972). The focus will be on longitudinal studies, though extensions to cross-sectional clustered data are natural.

7.1.2 Background

Ideally, research goals would drive the choice between the subject-specific and population-averaged models. In retrospect, however, it seems clear that a variety of secondary concerns have often driven this choice.

First, the two approaches have been associated with quite different methods of estimation. Analysts have tended to use maximum likelihood (or approximations to it) for subject-specific models (Stiratelli, Laird, and Ware, 1984; Breslow and Clayton, 1993; Raudenbush, Yang, and Yosef, 2000; Pinheiro and Bates, 2000). Maximum likelihood provides efficient estimates under parametric assumptions and facilitates comparatively weak assumptions about missing data. In contrast, analysts have used generalized estimating equations when estimating the population-averaged models (Zeger and Liang, 1986; Zeger, Liang, and Albert, 1988). These methods require fewer assumptions about the data distribution but are not efficient and may require stronger assumptions about missing data. Preferences about methods of estimation have therefore generated preferences for one model over the other. Thanks to Heagerty and Zeger (2000), however, we can now estimate the parameters of the population-averaged model as well as the subject-specific model by means of likelihood-based methods. This advance puts model choice on a more principled footing because model choice (subject-specific versus population-averaged) is no longer confounded with method of estimation.

Second, those who favor multilevel models for hierarchical data (Goldstein, 2003; Raudenbush and Bryk, 2002) have been inclined to favor subject-specific models. Such models differentiate variation within and between subjects and support empirical Bayes or Bayes inference about the personal trajectory of each subject. These methods are particularly valuable when information about some or all of the subjects is comparatively sparse yet the aim is to draw inferences about specific subjects and their development. Using the subject-specific model, methods are also well established for evaluating the fit of the model conditional on estimates of person-specific random effects. This helps to identify unusual cases and to probe the adequacy of the model at the level of the individual who is changing over time. Once again, however, Heagerty and Zeger's (2000) work is helpful. They have shown how to estimate subject-specific effects within the framework of the population-averaged model. This means that variation can be partitioned into between- and within-subject components using that approach while also enabling assessments of model fit at the subject level. As a result, the desire to conceptualize variation at multiple levels no longer precludes the use of the population-averaged model.

Finally, inference for subject-specific regression coefficients can be quite sensitive to the analyst's beliefs about variance components. Fitted results for population-averaged models are, in contrast, quite stable in the face of alternative specifications of the variance structure (Heagerty and Zeger, 2000). Analysts might therefore select the population-averaged model on grounds of their robustness. As I have argued earlier, however, fitted results based on the population-averaged model, while robust, may be incorrect or inappropriate if subject-specific questions drive the study (Raudenbush, 2000). This argument challenges us to be more explicit about how research questions logically affect the choice between the two approaches. That is the aim of this chapter.

7.1.3 Organization

Section 7.2 describes the two models of interest, showing when they are different. Section 7.3 defines the three targets of inference for these models. Section 7.4 illustrates the three targets of inference in the comparatively simple case using a continuous outcome and an identity link function, a case when the two models are equivalent. Section 7.5 and Section 7.6 illustrate the targets of inference in the case of a binary outcome using a logit link function. Section 7.5 illustrates application of the subject-specific model; Section 7.6 applies the population-averaged approach. Section 7.7 provides generalizations and concluding comments.

7.2 Two general approaches

Both modeling approaches are extensions of the standard generalized linear model

$$E(y_i|\boldsymbol{x}_i) = h^{-1}(\boldsymbol{x}_i'\boldsymbol{\beta}), \quad \text{Var}(y_i|\boldsymbol{x}_i) = v_i(\boldsymbol{\alpha}), \tag{7.1}$$

where y_i is a scalar outcome for person $i = 1, \ldots, N$; $\boldsymbol{x}_i$ is a column vector of known predictors; $\boldsymbol{\beta}$ is a column vector of unknown fixed parameters; $h(\cdot)$ is a link function, that is, a function of the mean that is equated to a linear model; and $v_i(\boldsymbol{\alpha})$ is the variance of y_i, given the covariates, that may depend on a parameter vector $\boldsymbol{\alpha}$. Note that, for ease of notation, in the remainder of this chapter we omit the explicit conditioning on the covariates. In the standard model, the outcomes y_i and $y_{i'}$ for subjects i and i' are independent. The project of interest in this chapter is to determine how to choose among models that relax this independence assumption. Note that the familiar ordinary least-squares regression model is a special case of (7.1) with $E(y_i) = h^{-1}(\boldsymbol{x}_i'\boldsymbol{\beta}) = \boldsymbol{x}_i'\boldsymbol{\beta}$ and $v_i(\alpha) = v_i(\sigma^2) = \sigma^2$, wherein h is the *identity link function* and the variance of y is the constant σ^2. In the case of binary data with logit link, we have $E(y_i) = h^{-1}(\boldsymbol{x}_i'\boldsymbol{\beta}) = [1 + \exp(-\boldsymbol{x}_i'\boldsymbol{\beta})]^{-1} = \phi_i$, and $v_i = \phi_i(1 - \phi_i)$; this is the well-known logistic regression model. The general approach applies to other outcomes and link functions, for example, count data with log link, ordinal data with cumulative logit link, multinomial data with multinomial logit link.

We shall consider two approaches to relaxing the independence assumption. The first is the subject-specific model. The model introduces a subject-specific random effect vector, $\boldsymbol{b}_i$, linearly within the link function, to capture the dependence of observations collected on the same subject:

$$E(\boldsymbol{Y}_i|\boldsymbol{b}_i) = h^{-1}(X_i\boldsymbol{\beta}^c + Z_i\boldsymbol{b}_i), \quad \text{Var}(\boldsymbol{Y}_i|\boldsymbol{b}_i) = V_i(\boldsymbol{\alpha}), \quad \text{Var}(\boldsymbol{b}_i) = G(\boldsymbol{\alpha}), \tag{7.2}$$

where $\boldsymbol{Y}_i$ is a vector of repeated measures having elements y_{ij}, $j = 1, \ldots, n_i$; X_i is a matrix having rows $\boldsymbol{x}'_i$; Z_i is a conformable matrix of known covariates; and $\boldsymbol{b}_i$ is a vector of subject-specific random effects. These random effects carry information about how subjects vary in their longitudinal trajectories; the elements of the outcome vector $\boldsymbol{Y}_i$ share these random effects and are therefore mutually dependent. Here $V_i(\boldsymbol{\alpha})$ is a diagonal matrix having elements $v_{ij}(\boldsymbol{\alpha})$ as in (7.1) while $G(\boldsymbol{\alpha})$ is typically a full variance–covariance matrix having dimension equal to the dimension of $\boldsymbol{b}_i$. While often labeled the subject-specific model, (7.2) is also known as the "conditional model" because it represents the transformed conditional mean of the response $\boldsymbol{Y}_i$, given the subject-specific random effect $\boldsymbol{b}_i$, as a linear function of the covariates. For that reason, we label its regression coefficients $\boldsymbol{\beta}^c$. Note than in (7.2), $\boldsymbol{\beta}^c$ quantifies the expected difference in the transformed conditional mean of $\boldsymbol{Y}$, that is $h[E(\boldsymbol{Y}|\boldsymbol{b})]$, for each unit difference in an element of X holding constant the random effects $\boldsymbol{b}$ and the remaining elements of X. A consequence of choosing a non-linear link function is that the marginal mean, $E(\boldsymbol{Y})$, will generally be an intractable non-linear function

$$E(\boldsymbol{Y}_i) = E[E(\boldsymbol{Y}_i|\boldsymbol{b}_i)] = \int h^{-1}(X_i\boldsymbol{\beta}^c + Z_i\boldsymbol{b}_i)p(\boldsymbol{b}_i)d\boldsymbol{b}_i, \tag{7.3}$$

where $p(\boldsymbol{b}_i)$ is the marginal distribution of the random effects, conventionally, but not necessarily, chosen to be $N[\boldsymbol{0}, G(\boldsymbol{\alpha})]$.

In contrast, the population-averaged model represents the transformed marginal mean of $\boldsymbol{Y}$ as a linear function of covariates. It is therefore often called the "marginal model." We then have

$$E(\boldsymbol{Y}_i) = h^{-1}(X'_i\boldsymbol{\beta}^m), \quad \text{Var}(\boldsymbol{Y}_i) = \Sigma_i[G(\boldsymbol{\alpha}), V_i(\boldsymbol{\alpha})]. \tag{7.4}$$

In this case, $\boldsymbol{\beta}^m$ quantifies the expected difference in the transformed marginal mean of $\boldsymbol{Y}$, that is $h[E(\boldsymbol{Y})]$, for each unit difference in an element of X averaging over person-specific random effects that are implicitly in the model. Heagerty and Zeger (2000) showed that these implicit random effects can be modeled as having variance–covariance structure $G(\boldsymbol{\alpha})$, as in the subject-specific model. They defined the conditional mean of $\boldsymbol{Y}$ given the random effects as $h^{-1}[\boldsymbol{\Delta}_i(\boldsymbol{\beta}^m, \boldsymbol{\alpha}) + \boldsymbol{b}_i]$, which can be found by solving the integral equation

$$E(\boldsymbol{Y}_i) = h^{-1}(X_i\boldsymbol{\beta}^m) = E[E(\boldsymbol{Y}_i|\boldsymbol{b}_i)] = \int h^{-1}[\boldsymbol{\Delta}_i(\boldsymbol{\beta}^m, \boldsymbol{\alpha}) + \boldsymbol{b}_i]p(\boldsymbol{b}_i)d\boldsymbol{b}_i \tag{7.5}$$

for $\boldsymbol{\Delta}_i(\boldsymbol{\beta}^m, \boldsymbol{\alpha})$. Recall that in the subject-specific model, the integrand $E(\boldsymbol{Y}_i|\boldsymbol{b}_i) = h^{-1}(X_i\boldsymbol{\beta}^c + Z_i\boldsymbol{b}_i)$ was tractable (see [7.3]) while the integral $E(\boldsymbol{Y}_i)$ was generally not tractable. In the case of the population-averaged model, the integrand $h^{-1}[\boldsymbol{\Delta}_i(\boldsymbol{\beta}^m, \boldsymbol{\alpha}) + \boldsymbol{b}_i]$ is generally intractable while the integral in (7.5), that is $E(\boldsymbol{Y}_i)$, is tractable. This fact has important implications for choice of model because it tends to be difficult to draw clear interpretations from intractable functions.

In light of (7.5), the marginal covariance structure of $\boldsymbol{Y}_i$ may be modeled as a full covariance matrix $\Sigma_i[G(\boldsymbol{\alpha}), V_i(\boldsymbol{\alpha})]$ that depends on $\text{Var}(\boldsymbol{b}_i) = G(\boldsymbol{\alpha})$ as well as on $\text{Var}(\boldsymbol{Y}_i|\boldsymbol{b}_i) = V_i(\boldsymbol{\alpha})$. Three comments clarify the relationships among the subject-specific and population-averaged models:

(i) As $\text{Var}(\boldsymbol{b}_i) = G(\boldsymbol{\alpha})$ approaches the null matrix, the subject-specific model (7.2) and the population-averaged model (7.4) converge to the standard generalized linear model (7.1) with appropriate change in notation: $E(y_{ij}) = h^{-1}(\boldsymbol{x}'_{ij}\boldsymbol{\beta})$, $\text{Var}(y_{ij}) = v_{ij}(\boldsymbol{\alpha})$, with $\boldsymbol{\beta}^c \approx \boldsymbol{\beta}^m \approx \boldsymbol{\beta}$. This means that when the random effects are uniformly small, estimation of the two models yields similar results so choice between them is of little practical significance despite the different definitions of the regression coefficients.

(ii) In the case of the identity link function, the subject-specific and population-averaged models are equivalent to each other and to the better-known linear mixed-effects model

(Verbeke and Molenberghs, 2000), also known as the hierarchical linear model (Rauden-bush and Bryk, 2002) or the multilevel linear model (Goldstein, 2003). For longitudinal data, this model was first discussed in detail in Laird and Ware (1982). Thus, we have $E(\boldsymbol{Y}_i) = E[E(\boldsymbol{Y}_i|\boldsymbol{b}_i)] = E[X_i'\boldsymbol{\beta}^c + Z_i'\boldsymbol{b}_i] = X_i'\boldsymbol{\beta}^c = X_i'\boldsymbol{\beta}^m$.

(iii) However, the two models are incompatible unless $G(\boldsymbol{\alpha})$ is small or the link is a linear transformation of the regression parameters. That is, the subject-specific and population-averaged models cannot both be correct: if the transformed conditional mean is linear in the regression parameters, the marginal mean cannot be (the solution of [7.3] for $h[E(\boldsymbol{Y})]$ is not linear in $\boldsymbol{\beta}^c$); and similarly, if the transformed marginal mean is linear in the regression parameters, the conditional mean will not be (the solution of [7.5] for $\boldsymbol{\Delta}(\boldsymbol{\beta}^m, \boldsymbol{\alpha})$ is not linear in $\boldsymbol{\beta}^m$). The key implication is that under these general conditions, the analyst must make a choice between the two models. How to make such a choice is the subject of this chapter.

7.3 Targets of inference

In studies collecting repeated measures on each subject, it is natural to require that a model explicitly describes change over time for each subject. Such a model should capture the effects of time-varying predictors (including, for example, age, time since the onset of a disease, or cumulative or intermittent exposure to time-varying treatments). In principle, the coefficients associated with these time-varying predictors can vary randomly across the population of persons, partly as a function of time-invariant subject characteristics. This between-subject variation in the association between time-varying predictors and outcomes reflects the fundamental idea that in most studies of human behavior, we can expect population heterogeneity in trajectories of change, in part because people may respond differently to interventions. This conceptual framework encourages us to formulate our model as a hierarchical model having two levels. Reconceptualizing longitudinal models in this way helps clarify potential targets of inference and therefore creates a framework for choosing between the subject-specific and population-averaged models.

Let us denote as "level-1 units" the time-series observations nested within "level-2 units," the subjects. The level-1 model describes the association between time-varying predictors and the time-varying outcome within person i. Fixed regression coefficients $\boldsymbol{\beta}$ and random effects $\boldsymbol{b}_i$ capture the person-specific contributions of the time-varying predictors to that person's development. I find it useful to define intermediate variables $\boldsymbol{\pi}_i = \boldsymbol{\pi}_i(\boldsymbol{\beta}, \boldsymbol{b}_i)$, typically linear functions of the regression coefficients $\boldsymbol{\beta}$ and the random effects $\boldsymbol{b}_i$ that characterize each subject's development and may be regarded as subject-specific growth parameters or, in the Bayesian literature, simply parameters or "micro-parameters" (Lindley and Smith, 1972; Dempster, Rubin, and Tsutakawa, 1981; Mason, Wong, and Entwistle, 1984). We then model the outcome as a function of these.

Level 1:

$$f[\boldsymbol{Y}_i|\boldsymbol{\pi}(\boldsymbol{\alpha}, \boldsymbol{\beta}, \boldsymbol{b}_i)].$$

The level-2 model describes the variation in these subject-specific growth parameters over a population of subjects.

Level 2:

$$p[\boldsymbol{\pi}_i(\boldsymbol{\beta}, \boldsymbol{b}_i)|\boldsymbol{\alpha}, \boldsymbol{\beta}].$$

In many cases, interest will focus on the marginal distribution of the outcomes, which can be obtained by integrating over the random effects.

Marginalization:

$$g(\boldsymbol{Y}_i|\boldsymbol{\alpha}, \boldsymbol{\beta}) = \int f[\boldsymbol{Y}_i|\boldsymbol{\pi}(\boldsymbol{\alpha}, \boldsymbol{\beta}, \boldsymbol{b}_i)]p[\boldsymbol{\pi}_i(\boldsymbol{\beta}, \boldsymbol{b}_i)|\boldsymbol{\alpha}, \boldsymbol{\beta}]\, d\boldsymbol{b}_i.$$

The three targets of inference will be:

(i) the associations between the predictors X and the outcomes $\mathbf{Y}$. These can be interpreted within the subject-specific model as associations conditional on the random effect or in the population-averaged model as the marginal association, averaging over the random effect.

(ii) the parameters of the level-1 model or functions thereof; more particularly, the latent subject-specific growth parameters $\boldsymbol{\pi}_i$; a special case arises when $\boldsymbol{\pi}_i = \boldsymbol{b}_i$. In practice, the published literature has virtually always studied these subject-specific growth parameters within the context of the subject-specific model. Heagerty and Zeger (2000) have shown how to study the subject-specific random effects, $\boldsymbol{b}_i$, within the population-averaged model.

(iii) the level-2 model that describes the population distribution of the subject-specific growth parameters. In the subject-specific model, this level-2 model is a linear model; that is, these latent variables $\boldsymbol{\pi}_i$ are linear functions of $\boldsymbol{\beta}^c$ and $\boldsymbol{b}_i$. In the population-averaged model, they are non-linear functions, often intractable, if they exist, of $\boldsymbol{\beta}^m$ and $\boldsymbol{b}_i$. This inclines one to adopt the readily interpretable subject-specific model for this third target of inference.

In the next section, I illustrate the differences between these targets of inference in a simple example of a linear model using an identity link function for normal-theory data.

7.4 An example with normal data and identity link

7.4.1 Data

We are interested in the growth of mathematics skill, denoted y_{ij}, of each child i across observation occasions $j = 1, \ldots, 5$ using the Sustaining Effects Study data (Carter, 1984; Bryk and Raudenbush, 1988). The first observation occurs in the spring of grade 1. The second and third observations occur, respectively, in the fall and spring of grade 2. The fourth and fifth observations occur, respectively, in the fall and spring of grade 3. We are interested in learning during the academic year (between fall and spring each year) and during the summer (between spring and fall). This interest generates two time-varying predictors. The covariate $(\text{Ayear})_{ij}$ measures the cumulative number of academic years of instruction since the first observation received by child i at observation time j. The covariate $(\text{Sum})_{ij}$ measures the cumulative number of summer sessions experienced by child i at observation time j since the first observation. The coding of these two "time-varying covariates" is given in the table below:

Occasion	Spring Grade 1	Fall Grade 2	Spring Grade 2	Fall Grade 3	Spring Grade 3
j	1	2	3	4	5
Ayear	0	0	1	1	2
Sum	0	1	1	2	2

7.4.2 Analytic model

The "level-1 model" describes the growth trajectory of child i:

$$y_{ij} = \pi_{0i} + \pi_{1i}(\text{Ayear})_{ij} + \pi_{2i}(\text{Sum})_{ij} + e_{ij}, \quad e_{ij} \sim N(0, \sigma^2)$$

$$= \begin{bmatrix} 1 & (\text{Ayear})_{ij} & (\text{Sum})_{ij} \end{bmatrix} \begin{bmatrix} \pi_{0i} \\ \pi_{1i} \\ \pi_{2i} \end{bmatrix} + e_{ij} = \boldsymbol{a}'_{ij}\boldsymbol{\pi}_i + e_{ij},$$

where $a_{ij} = [1 \quad (\text{Ayear})_{ij} \quad (\text{Sum})_{ij}]'$. Equivalently, we may define the vector $Y_i = (Y_{i1}, \ldots, Y_{in_i})'$ as

$$Y_i = A_i \pi_i + e_i,$$

where $Y_i = (y_{i1}, \ldots, y_{in_i})'$, $A_i = (a_{i1}, \ldots, a_{in_i})'$, $e_i = (e_{i1}, \ldots, e_{in_i})' \sim N(0, \sigma^2 I)$.

In general A_i is the $n_i \times p$ matrix of time-varying predictors, with $p = 3$ in this case; π_i is a vector of interesting latent subject-specific growth parameters that govern child i's development. The subject-specific latent-growth parameters of interest in our case are π_{0i}, that child's status at spring of grade 1, the outset of the study; π_{1i}, that child's rate of growth in mathematics skill per academic year; and π_{2i}, that child's rate of growth in mathematics skill per summer session.

Now looking across the population of children, we formulate the "level-2" model which describes how children vary in their growth trajectories. In this simple illustration, we allow each aspect of the trajectory to depend on Pov $= 1$ if the child is "poor," that is, from a low-income family and therefore eligible for a federal free lunch subsidy, and Pov $= 0$ if the child is "non-poor." Thus, we have the level-2 model

$$\pi_{0i} = \beta_{00} + \beta_{01}(\text{Pov})_i + b_{0i},$$
$$\pi_{1i} = \beta_{10} + \beta_{11}(\text{Pov})_i + b_{1i},$$
$$\pi_{2i} = \beta_{20} + \beta_{21}(\text{Pov})_i + b_{2i},$$

or, in matrix notation,

$$
\begin{bmatrix} \pi_{0i} \\ \pi_{1i} \\ \pi_{2i} \end{bmatrix} =
\begin{bmatrix} 1 & (\text{Pov})_i & 0 & 0 & 0 & 0 \\ 0 & 0 & 1 & (\text{Pov})_i & 0 & 0 \\ 0 & 0 & 0 & 0 & 1 & (\text{Pov})_i \end{bmatrix}
\begin{bmatrix} \beta_{00} \\ \beta_{01} \\ \beta_{10} \\ \beta_{11} \\ \beta_{20} \\ \beta_{21} \end{bmatrix} +
\begin{bmatrix} b_{0i} \\ b_{1i} \\ b_{2i} \end{bmatrix},
$$

that is,

$$\pi_i = W_i \beta + b_i, \quad b_i \sim N[0, G(\alpha)] \tag{7.6}$$

with

$$
G(\alpha) = \begin{bmatrix} g_{00} & g_{01} & g_{02} \\ g_{10} & g_{11} & g_{12} \\ g_{20} & g_{21} & g_{22} \end{bmatrix}.
$$

Substituting the level-1 model into the level-2 model, we have

$$Y_i = A_i W_i \beta + A_i b_i + e_i,$$

a special case of the mixed linear model

$$Y_i = X_i \beta + Z_i b_i + e_i; \quad \text{with} \quad X_i = A_i W_i \quad \text{and} \quad Z_i = A_i. \tag{7.7}$$

The mixed linear model is itself a special case of the conditional ("subject-specific") model (7.2) with

$$E(Y_i | \beta^c, b_i) = h^{-1}(X_i \beta^c + Z_i b_i) = X_i \beta + Z_i b_i,$$
$$\text{Var}(Y_i | \beta^c, b_i) = V_i(\alpha) = \sigma^2 I,$$
$$\text{Var}(b_i) = G(\alpha),$$
$$\alpha = (\sigma^2, g_{00}, g_{01}, g_{02}, g_{11}, g_{12}, g_{22})'.$$

In this case, with $h(\cdot)$ defined as the identity link function, (7.7) is also a special case of the marginal ("population-averaged") model (7.4),

$$E(\boldsymbol{Y}_i|\boldsymbol{\beta}^m) = h^{-1}(X_i\boldsymbol{\beta}^m) = X_i\boldsymbol{\beta},$$
$$\mathrm{Var}(\boldsymbol{Y}_i|\boldsymbol{\beta}^m) = \Sigma_i[G(\boldsymbol{\alpha}), V_i(\boldsymbol{\alpha})] = G(\boldsymbol{\alpha}) + V_i(\boldsymbol{\alpha}) = G(\boldsymbol{\alpha}) + \sigma^2 I.$$

Clearly, in this case, with $h(\cdot)$ defined as the identity link, both the conditional mean and the marginal mean are linear so that $\boldsymbol{\beta}^c = \boldsymbol{\beta}^m = \boldsymbol{\beta}$ and the marginal variance–covariance matrix of $\boldsymbol{Y}$ is an additive function of the conditional covariance matrix of $\boldsymbol{Y}$ given $\boldsymbol{\beta}$, and the covariance matrix of $\boldsymbol{b}_i$ itself.

7.4.3 Targets of inference

Three targets of inference are available. The first involves associations between predictors X and outcomes $\boldsymbol{Y}$. These are represented by the regression coefficients, $\boldsymbol{\beta}$. Because $\boldsymbol{\beta}^c = \boldsymbol{\beta}^m = \boldsymbol{\beta}$, the analyst need not distinguish between the influence of X on the marginal mean and the influence of X on the conditional mean. That is, our target can be defined as either $f(\boldsymbol{Y})$ or $f(\boldsymbol{Y}|\boldsymbol{b})$. The mixed model in detail is

$$\begin{aligned}
Y_{ij} = {}& \beta_{00} + \beta_{01}(\mathrm{Pov})_i + \beta_{10}(\mathrm{Ayear})_{ij} + \beta_{11}(\mathrm{Pov})_i \times (\mathrm{Ayear})_{ij} \\
& + \beta_{20}(\mathrm{Sum})_{ij} + \beta_{21}(\mathrm{Pov})_i \times (\mathrm{Sum})_{ij} + b_{0i} + b_{1i}(\mathrm{Ayear})_{ij} + b_{2i}(\mathrm{Sum})_{ij} + e_{ij}.
\end{aligned} \tag{7.8}$$

The second target of inference is the subject-specific growth parameter vector $\boldsymbol{\pi}_i = (\pi_{0i}, \pi_{1i}, \pi_{2i})'$, defined by the level-1 model. We want to know about the status and growth rate of a particular child, perhaps so that we can identify children at risk of slow cognitive growth or, perhaps, children who show exceptionally promising growth. We can make this inference using empirical Bayes methods to estimate $p(\boldsymbol{\pi}|\boldsymbol{Y}, \boldsymbol{\beta} = \widehat{\boldsymbol{\beta}}, \boldsymbol{\alpha} = \widehat{\boldsymbol{\alpha}})$, where $\widehat{\boldsymbol{\beta}}, \widehat{\boldsymbol{\alpha}}$ are consistent (typically maximum likelihood) estimates; or by means of fully Bayesian methods to estimate $p(\boldsymbol{\pi}|\boldsymbol{Y})$.

The third target of inference is the distribution of the latent subject-specific growth parameter $\boldsymbol{\pi}$, that is, $p(\boldsymbol{\pi}|\boldsymbol{\beta}, \boldsymbol{\alpha})$. For this purpose, we may regard the level-2 model as the *structural model* of interest while the level-1 model serves strictly as a *measurement model*. For example, in the model

$$E(\pi_{1i}) = \beta_{10} + \beta_{11}(\mathrm{Pov})_i \tag{7.9}$$

we are particularly interested in β_{11}, the mean difference in academic-year learning between poor and non-poor children. Note that $\beta_{10} + \beta_{11}$ can be interpreted equivalently as (a) the expected change in y for a unit increase in Ayear for a poor child or (b) the expected value of π_1 for a poor child (that is, a poor child's expected academic year growth rate). Thus, the $\boldsymbol{\beta}$ coefficients can be regarded equivalently as parameters of a structural model for the outcome $\boldsymbol{Y}$ (as in the mixed model [7.8]) or as parameters of a structural model for the latent-growth parameters (as in the level-2 model [7.9]). In the next section, we will see that this equivalence will hold generally in subject-specific models but not in population-averaged models.

We also might be especially interested in

$$\mathrm{Var}(\pi_1) = g_{11}, \tag{7.10}$$

the variance of the growth rates conditional on poverty status. We might similarly be interested in the distribution of summer growth rates.

We see then that target 3 focuses our interest on inferences about the population distribution of the latent-growth parameters, $\boldsymbol{\pi}$, including their mean (7.9) and their variance (7.10).

Table 7.1 Math Achievement by Poverty Status and Occasion in the Sustaining Effects Study

		Non-Poor			Poor		
Occasion		N	Mean	SD	N	Mean	SD
1.	Spring Grade 1	505	406	41	83	382	38
2.	Fall Grade 2	526	411	42	92	385	41
3.	Spring Grade 2	526	466	47	92	444	43
4.	Fall Grade 3	526	461	44	92	440	41
5.	Spring Grade 3	526	524	50	92	490	49

7.4.4 Results

The data are described in Table 7.1. The sample includes 618 participants, 14.9% of whom are poor. For both groups, the mean of Y appears to increase significantly with each academic year but not during summer. Poor children have lower means than do non-poor children.

Target 1: Association between predictors X *and outcomes* Y

Maximum likelihood point estimates for our first target of inference are given by

$$\widehat{E}(\boldsymbol{Y}) = 407.57 - 22.47 \times (\text{Pov})_i + 57.17 \times (\text{Ayear})_{ij} - 3.83 \times (\text{Pov})_i \times (\text{Ayear})_{ij}$$
$$+ 0.14 \times (\text{Sum})_{ij} + 0.49 \times (\text{Pov})_i \times (\text{Sum})_{ij}.$$

Associated standard errors (SEs) and approximate t-ratios are found in Table 7.2. Growth is comparatively rapid during the academic year for non-poor children, $\widehat{\beta}_{10} = 57.17$, SE $= 1.11$, $t = 51.31$, while growth during the summer for those children is essentially nil, $\widehat{\beta}_{20} = 0.14$, SE $= 1.04$, $t = 0.14$. The trajectory for poor children is similar in that their academic-year growth differs little from that of non-poor children, $\widehat{\beta}_{11} = -3.83$, SE $= 2.89$, $t = -1.32$; nor does their summer growth differ much from that of non-poor children, $\widehat{\beta}_{21} = 0.49$, SE $= 2.73$ $t = 0.18$. The main difference between the two groups is that the poor children have lower initial status (that is, status at spring grade 1), $\widehat{\beta}_{01} = -22.47$, SE $= 4.56$, $t = -4.93$.

Table 7.2 Fitted Model, Continuous Outcome, Identify Link

(a) Fixed Effects

Predictor	Coefficient	Estimate	SE	t
Intercept	β_{00}	407.57	1.75	233.40
Pov	β_{01}	-22.47	4.56	-4.39
Ayear	β_{10}	57.17	1.11	51.31
Pov $\times$ Ayear	β_{11}	-3.83	2.89	-1.32
Sum	β_{20}	0.14	1.04	0.14
Pov $\times$ Sum	β_{21}	0.49	2.73	0.18

(b) Covariance Components[a]

$$\widehat{\sigma}^2 = 600.07$$

$$\widehat{G}(\alpha) = \begin{bmatrix} 1146.18 & 53.64 & 11.30 \\ (0.16) & 92.07 & 6.60 \\ (0.20) & (0.42) & 2.74 \end{bmatrix}$$

[a] Lower diagonal entries of the matrix are correlations.

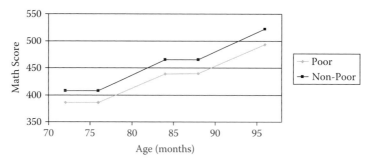

Figure 7.1 Math score as a function of student poverty.

In essence, the story is one of main effects of time of year (academic versus summer) and of child poverty, with little evidence of statistical interaction between poverty and time of year. This is vividly clear from a graph of the expected trajectories of the two groups, shown in Figure 7.1. The two mean trajectories are remarkably similar except for initial differences between the two groups; these initial differences do not diminish much over time, nor do they increase much.

Target 2: Subject-specific growth parameters

A practitioner may wish to identify children who are at risk in various ways, for example, children who do not seem to be benefiting much from academic instruction. These are children with unusually low values of π_1. Prediction intervals for subject-specific values of π_{1i} may be estimated from their posterior distribution using empirical Bayes (Dempster, Rubin, and Tsutakawa, 1981; Laird and Ware, 1982; see review and detailed illustrations by Raudenbush and Bryk, 2002). We are currently interested in

$$\pi_{1i} = \beta_{10} + \beta_{11}(\text{Pov})_i + b_{1i},$$

for $i = 1, \ldots, 618$. These can be estimated by $E(\pi_{1i}|\boldsymbol{Y}, \widehat{\beta}_{10}, \widehat{\beta}_{11}, \widehat{\boldsymbol{\alpha}})$, $\text{Var}(\pi_{1i}|\boldsymbol{Y}, \widehat{\beta}_{10}, \widehat{\beta}_{11}, \widehat{\boldsymbol{\alpha}})$ under full maximum likelihood or by $E(\pi_{1i}|Y, \widehat{\boldsymbol{\alpha}})$, $\text{Var}(\pi_{1i}|Y, \widehat{\boldsymbol{\alpha}})$ in the case of restricted maximum likelihood. Some would prefer a fully Bayesian solution, basing the posterior prediction intervals on $E(\pi_{1i}|\boldsymbol{Y})$, $\text{Var}(\pi_{1i}|\boldsymbol{Y})$. We use restricted maximum likelihood here. However, with $N = 618$ subjects, the full maximum likelihood and fully Bayes results will be nearly identical.

Figure 7.2 displays the prediction intervals, rank-ordered by the posterior mean and color-coded by poverty status. We see considerable overlap in the subject-specific intervals. However, poorer children (dark-colored intervals) tend to exhibit lower academic-year growth than do non-poor children (light-colored intervals). A small but non-trivial subset of children have posterior mean growth less than 40 and all of these children are poor. Similarly, a small set of children have very high growth rates and all of these children are non-poor.

However, we caution that subject-specific empirical Bayes or Bayes estimates based on shrinkage tend to exaggerate the association between covariates and components of $\boldsymbol{\pi}$. To the extent the data for a specific child are unreliable, the Bayes or empirical Bayes estimates will "borrow strength" from information about covariates gleaned from the whole sample. Getting a better picture of the association between covariates and components of $\boldsymbol{\pi}$ requires that we move to target 3.

Target 3: Distribution of the latent subject-specific growth parameters

The subject-specific Bayes or empirical Bayes estimates provide good approaches for estimating subject-specific growth parameters. But they do not accurately characterize the

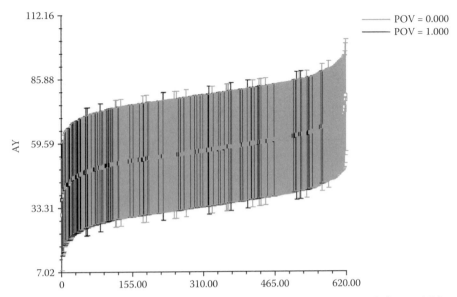

Figure 7.2 Subject-specific 95% posterior intervals for academic-year growth (poor children have dark shading, non-poor children have light shading).

distribution of these in the population. As discussed previously, the collection of empirical Bayes estimates tends to exaggerate associations between covariates and $\boldsymbol{\pi}$. Moreover, this collection tends to understate population heterogeneity. Thus, we find that the sample standard deviation of the empirical Bayes estimates of the academic year growth rates, that is, the standard deviation of the subject-specific point estimates of π_1 shown in Figure 7.2, is 5.94, which substantially undershoots the maximum likelihood estimate, $\sqrt{\widehat{\mathrm{Var}}(\pi_1)} = \sqrt{\hat{g}_{11}} = \sqrt{92.07} = 9.59$. The discrepancy is explainable by the "shrinkage" on which the Bayes and empirical Bayes methods depend for subject-specific accuracy. We also know that person-specific estimates based on ordinary least squares (OLS) will exhibit much larger dispersion than given by $\hat{g}_{11}$ estimates because the sampling variance of the OLS estimates will contribute to their dispersion. In our case, the standard deviation of these subject-specific OLS growth rate estimates is 24.06, more than twice the maximum likelihood estimate of the standard deviation of the true growth rates. We can thus conclude that most of the variation in the subject-specific OLS estimates is noise.

As an example of "target 3" inference, we might want to know the distribution of academic-year growth for non-poor kids and poor kids. Thus, we wish to estimate the parameters of the "level-2 model" (7.6). In this case, we are particularly interested in estimating $N[\beta_{10} + \beta_{11}(\mathrm{Pov}), \ g_{11}]$. The results in Table 7.2 tell us that the restricted maximum likelihood estimates are, for non-poor kids $N[57.17, \ 92.07]$ and for poor kids $N[53.35, \ 92.07]$. Under the normality assumption, we would expect to see about 95% of the growth rates between $57.17 \pm 1.96\sqrt{92.07} = (38.63, 75.98)$ for non-poor children and $53.35 \pm 1.96\sqrt{92.07} = (34.54, 72.16)$ for poor children. This inference is based on the assumption of homogeneity of variance as a function of poverty status, which can be checked. These results suggest a weaker association between child poverty and growth rates than was suggested by looking at the collection of empirical Bayes estimates (see discussion under target 2).

Target 3 might involve the distributions of any of the growth parameters $\boldsymbol{\pi} = (\pi_0, \pi_1, \pi_2)$ or combinations thereof. In the same way, target 2 might focus on subject-specific realizations $\boldsymbol{\pi}_i = (\pi_{0i}, \pi_{1i}, \pi_{2i})$, $i = 1, \ldots, N$, or combinations thereof.

Table 7.3 Proportion Attaining Basic
Skills, by Poverty Status and Occasion

	Occasion	Non-Poor	Poor
1.	Spring Grade 1	0.12	0.02
2.	Fall Grade 2	0.14	0.04
3.	Spring Grade 2	0.60	0.40
4.	Fall Grade 3	0.57	0.30
5.	Spring Grade 3	0.93	0.78

With normal-theory data and an identity link function, as mentioned, there is no distinction between the subject-specific and population-averaged models. In the next two sections we consider an example where that distinction becomes meaningful and where choice of model (subject-specific versus population-averaged) will depend on the target of inference.

7.5 An example with binary data, logit link, and a linear conditional mean

7.5.1 Data

Suppose that we have essentially the same scenario as in the previous section except that now mathematics skill is the binary outcome $y_{ij} = 1$ if child i has mastered "basic skill" in arithmetic operations (adding, subtracting, multiplying, and dividing) and $y_{ij} = 0$ if not. (We define this "basic skill" as 1 if a child exceeds a score of 450 on the test, the Iowa Test of Basic Skills). We expect that some children will exhibit this basic skill even at the outset of the study, in the spring of grade 1. Other children may not have achieved this level even by spring of grade 3. Table 7.3 shows the distribution of y as a function of time of year and poverty status.

7.5.2 Analytic model

We might use a duration time model for these data. However, measurement error in the outcome implies that some children will shift from 0 to 1 or from 1 to 0 by chance. Moreover, these data are not characterized by censoring. So a two-level hierarchical model, analogous to that used in the previous section, seems reasonable. We begin with the subject-specific model and consider the population-averaged model in the next section.

The "level-1 model" describes the growth trajectory of child i. First, we select a Bernoulli sampling model $\Pr(y_{ij} = 1|\boldsymbol{\pi}_i) = \Pr(y_{ij} = 1|\boldsymbol{\beta}^c, \boldsymbol{b}_i) = \phi_{ij}^c$ and a logit link function

$$h(\phi_{ij}^c) = \ln\left(\frac{\phi_{ij}^c}{1 - \phi_{ij}^c}\right) = \pi_{0i} + \pi_{1i}(\text{Ayear})_{ij} + \pi_{2i}(\text{Sum})_{ij}$$

$$= \begin{bmatrix} 1 & (\text{Ayear})_{ij} & (\text{Sum})_{ij} \end{bmatrix} \begin{bmatrix} \pi_{0i} \\ \pi_{1i} \\ \pi_{2i} \end{bmatrix} = \boldsymbol{a}_{ij}'\boldsymbol{\pi}_i.$$

Equivalently, we model the $n_i \times 1$ vector $h(\boldsymbol{\phi}_i^c) = [h(\phi_{i1}^c), \ldots, h(\phi_{in_i}^c)]'$ as

$$h(\boldsymbol{\phi}_i^c) = A_i \boldsymbol{\pi}_i.$$

The level-2 model is similar to that used in the previous section with continuous data and identity link function (7.6). The exception is that we now simplify the covariance structure of the random effects. Based on preliminary analysis, we find that the point estimates of g_{11} (the variance of the academic-year growth rate π_1) and g_{22} (the variance of the summer growth rate π_2) are zero. This may reflect the loss of information associated with dichotomizing the outcome. Thus, we constrain these as follows:

$$\pi_{0i} = \beta_{00}^c + \beta_{01}^c (\text{Pov})_i + b_{0i},$$
$$\pi_{1i} = \beta_{10}^c + \beta_{11}^c (\text{Pov})_i,$$
$$\pi_{2i} = \beta_{20}^c + \beta_{21}^c (\text{Pov})_i,$$

or, in matrix notation,

$$
\begin{bmatrix} \pi_{0i} \\ \pi_{1i} \\ \pi_{2i} \end{bmatrix} =
\begin{bmatrix} 1 & (\text{Pov})_i & 0 & 0 & 0 & 0 \\ 0 & 0 & 1 & (\text{Pov})_i & 0 & 0 \\ 0 & 0 & 0 & 0 & 1 & (\text{Pov})_i \end{bmatrix}
\begin{bmatrix} \beta_{00}^c \\ \beta_{01}^c \\ \beta_{10}^c \\ \beta_{11}^c \\ \beta_{20}^c \\ \beta_{21}^c \end{bmatrix} +
\begin{bmatrix} 1 \\ 0 \\ 0 \end{bmatrix} b_{0i},
$$

that is,

$$\boldsymbol{\pi}_i = W_i \boldsymbol{\beta}^c + R b_i, \quad b_i \sim N(0, G) \tag{7.11}$$

with $G = g_{00}$.

Substituting the level-1 model into the level-2 model, we have

$$h(\boldsymbol{\phi}_i^c) = A_i W_i \boldsymbol{\beta}^c + A_i R b_i,$$

a special case of the subject-specific model (sometimes called the "generalized linear mixed model" or "hierarchical generalized linear model")

$$h(\boldsymbol{\phi}_i^c) = X_i \boldsymbol{\beta}^c + Z_i b_i \quad \text{with } X_i = A_i W_i \quad \text{and} \quad Z_i = A_i R.$$

Under this model, the first two moments of interest are

$$E(\boldsymbol{Y}_i | \boldsymbol{\beta}^c, b_i) = h^{-1}(X_i \boldsymbol{\beta}^c + Z_i b_i) = \boldsymbol{\phi}_i^c,$$
$$\text{Var}(\boldsymbol{Y}_i | \boldsymbol{\beta}^c, b_i) = V_i(\boldsymbol{\alpha}) = \text{diag}\{\phi_{ij}^c (1 - \phi_{ij}^c)\},$$
$$\text{Var}(b_i) = G(\alpha),$$
$$\alpha = (g_{00}).$$

7.5.3 Targets of inference

Three targets of inference are again available. The first involves associations between predictors X and outcomes $\boldsymbol{Y}$. In our subject-specific model, these are represented by the conditional regression coefficients, $\boldsymbol{\beta}^c$. The mixed model in detail is

$$
\begin{aligned}
h[E(Y_{ij} | b_i)] = h(\phi_{ij}^c) &= \ln\left(\frac{\phi_{ij}^c}{1 - \phi_{ij}^c}\right) \\
&= \beta_{00}^c + \beta_{01}^c (\text{Pov})_i + \beta_{10}^c (\text{Ayear})_{ij} + \beta_{11}^c (\text{Pov})_i \times (\text{Ayear})_{ij} + \beta_{20}^c (\text{Sum})_{ij} \\
&\quad + \beta_{21}^c (\text{Pov})_i \times (\text{Sum})_{ij} + b_{0i}.
\end{aligned}
$$

The second and third targets are perfectly analogous to those in the previous section, involving inferences about subject-specific random coefficients, π_i, and their distribution (that is, the level-2 model [7.11]).

7.5.4 Results

Target 1: Association between predictors X and outcomes Y

Approximate maximum likelihood point estimates for our first target of inference are given by

$$
\begin{aligned}
\widehat{h}(\phi_{ij}^c) = {}&-2.57 - 1.11 \times (\text{Pov})_i + 3.07 \times (\text{Ayear})_{ij} + 0.03 \times (\text{Pov})_i \times (\text{Ayear})_{ij} \\
&- 0.02 \times (\text{Sum})_{ij} - 0.25 \times (\text{Pov})_i \times (\text{Sum})_{ij}.
\end{aligned}
$$

Table 7.4 Fitted Model, Binary Outcome, Logit Link, Subject-Specific Model

(a) Fixed Effects

Predictor	Coefficient	Estimate	SE	t
Intercept	β_{00}^c	−2.57	0.13	−19.62
Pov	β_{01}^c	−1.11	0.37	−2.97
Ayear	β_{10}^c	3.07	0.13	24.25
Pov × Ayear	β_{11}^c	0.03	0.33	0.93
Sum	β_{20}^c	−0.02	0.10	−0.26
Pov × Sum	β_{21}^c	−0.25	0.28	−0.90

(b) Covariance Components

$$\widehat{G}(\alpha) = \widehat{g}_{00} = 3.35$$

Associated SEs and approximate t-ratios are found in Table 7.4. The main outline of the results is similar to those when we used the continuous outcome: we see rapid increase in the log odds of achieving the basic skill level during the academic year and little increase during the summer. Child poverty is negatively related to basic skill acquisition at all time points but does not interact with time of year.

However, given that we have selected the subject-specific model, and given evidence of a large between-subject variance ($\widehat{g}_{00} = 3.35$), the predicted mean outcomes vary substantially as a function of the random effect, b_{0i}.

Figure 7.3 graphs the conditional expected trajectories for three hypothetical children. The first (Figure 7.3a) is at the 97.5th percentile of the random effect under the normal-theory model $b_{0i} \sim N(0, g_{00})$ and therefore has an estimated random effect of $b_{0(0.95)} = 1.96\sqrt{3.35} = 3.59$. For this child, the expected trajectory depends somewhat on child poverty at the first two time points, though the expected probability of attaining basic skill for this "high b_{0i}" group exceeds 0.90 for poor as well as non-poor children by the spring of grade 2 when these children are, on average, about 85 months old.

The second type of child (Figure 7.3b) is a "typical child" with a median random effect $b_{0(0.05)} = 0$. Few such children attain basic skills in spring of grade 1 or fall of grade 2, but nearly all do by the final time point, that is, spring of grade 3.

The third type of child is an at-risk child (Figure 7.3c), at the 2.5th percentile, having $b_{0(0.05)} = -1.96\sqrt{3.35} = -3.59$. For non-poor children in this group, only at the last time point does the expected probability of basic skill proficiency approach 0.50. For poor children in this group, the expected probability of reaching the basic skill level never exceeds 0.20, even at the last time point. It is these children — poor children with low random effects — who are most at risk of failing to achieve the basic skill level by the end of third grade.

Target 2: Subject-specific growth parameters

Inference closely follows the logic used in the previous example. To illustrate, Figure 7.4 plots 95% posterior intervals for $\pi_{0i} = \beta_{00}^c + \beta_{01}^c(\text{Pov})_i + b_{0i}$ ranked by estimated posterior mean and color-coded as a function of poverty status.

Target 3: Distribution of the latent-growth parameters

As an example of "target 3" inference, we might want to know the distribution of the probability of basic skill proficiency at the outset of the study — spring of grade 1 — for

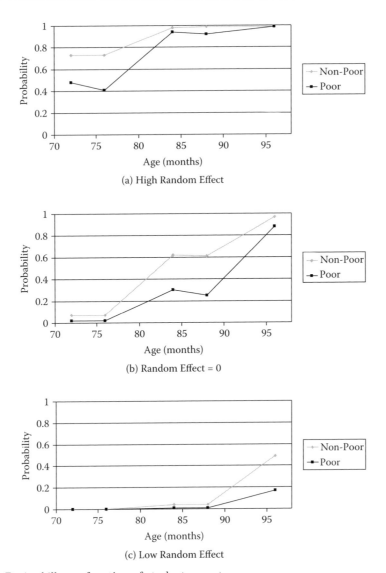

Figure 7.3 Basic skill as a function of student poverty.

non-poor and poor kids. Thus, we wish to estimate the parameters of the "level-2 model" (7.11). In this case, we are particularly interested in estimating the mean and variance of π_0. The results in Table 7.4 tell us that the approximate maximum likelihood estimates are, for non-poor kids $\widehat{E}(\pi_0|\text{Pov}=0) = -2.57$, $\widehat{\text{Var}}(\pi_0|\text{Pov}=0) = 3.35$, and for poor kids $\widehat{E}(\pi_0|\text{Pov}=1) = -3.68$, $\widehat{\text{Var}}(\pi_0|\text{Pov}=1) = 3.35$. Assuming π_0 to be normal in distribution, we can compute 95% plausible value intervals for π_0, that is, estimates of $\beta_{00}^c + \beta_{01}^c(\text{Pov})_i \pm 1.96\sqrt{g_{00}}$. Substituting estimates gives plausible values of $(-6.16, 1.07)$ for non-poor kids and $(-7.27, -0.09)$ for poor kids. Transforming these to probabilities yields $(0.002, 0.74)$ and $(0.0001, 0.52)$. There is tremendous variation within both groups, perhaps signaling the challenge facing schools in affording appropriate instruction for such a diverse population.

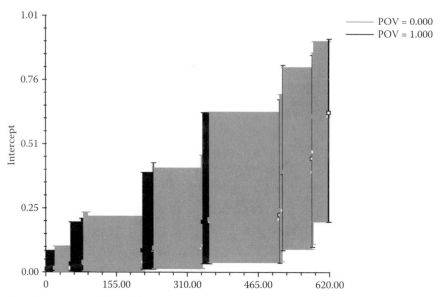

Figure 7.4 Subject-specific 95% posterior intervals for probability of basic skill proficiency at spring of grade 1 (poor children have dark shading, non-poor children have light shading).

7.6 An example with binary data, logit link, and a linear marginal mean

7.6.1 Data

We shall use the same data as in the previous example. Recall that Table 7.3 shows the distribution of y as a function of time and poverty status.

7.6.2 Analytic model

In the previous examples, we built the model "from the ground up" starting with a "level-1 model" for the change trajectory of subject i. This model defined certain subject-specific growth parameters that varied between subjects according to a "level-2 model." Combining the level-1 and level-2 model generated the mixed model for the conditional mean of $\boldsymbol{Y}$ given the random effect b.

Now that we are applying the population-averaged model, it makes more sense to begin with the mixed model for the marginal mean of Y and then to work backward, deriving the subject-specific level-1 model and the between-subject level-2 model in the spirit of Heagerty and Zeger (2000). Thus, we have the marginal mean

$$E(y_{ij}) = \Pr(y_{ij} = 1) = \phi_{ij}^m$$

and the logit link function

$$h\left(\phi_{ij}^m\right) = \ln\left(\frac{\phi_{ij}^m}{1 - \phi_{ij}^m}\right) = \beta_{00}^m + \beta_{01}^m(\text{Pov})_i + \beta_{10}^m(\text{Ayear})_{ij} + \beta_{11}^m(\text{Pov})_i \times (\text{Ayear})_{ij}$$

$$+ \beta_{20}^m(\text{Sum})_{ij} + \beta_{21}^m(\text{Pov})_i \times (\text{Sum})_{ij}$$

$$= \boldsymbol{x}_{ij}'\boldsymbol{\beta}^m,$$

with $\boldsymbol{x}'_{ij} = [1\,,(\text{Pov})_i,(\text{Ayear})_{ij},(\text{Pov}) \times (\text{Ayear})_{ij},(\text{Sum})_{ij},(\text{Pov})_i \times (\text{Sum})_{ij}]$ and with $\boldsymbol{\beta}^{m'} = [\beta^m_{00}\,,\ \beta^m_{01},\ \beta^m_{10},\ \beta^m_{11},\ \beta^m_{20},\ \beta^m_{21}\ ]$. Equivalently, we model the $n_i \times 1$ vector $h(\boldsymbol{\phi}^m_i) = [h(\phi^m_{i1}),\ldots,h(\phi^m_{in_i})]'$ as

$$h(\boldsymbol{\phi}^m_i) = X_i \boldsymbol{\beta}^m,$$

or

$$E(\boldsymbol{Y}_i) = \boldsymbol{\phi}^m_i = h^{-1}(X_i \boldsymbol{\beta}^m).$$

Suppose that we now postulate a subject-specific random effect b_i having a density $f(b|\alpha)$, for example, $b_i \sim N[0, G(\alpha)]$. We can then derive the conditional mean

$$E(\boldsymbol{Y}_i | \alpha, \boldsymbol{\beta}^m, b_i) = \boldsymbol{\Delta}_i(\alpha, \boldsymbol{\beta}^m) + b_i$$

by solving the integral equation

$$E(\boldsymbol{Y}_i) = h^{-1}(X_i \boldsymbol{\beta}^m) = \int h^{-1}[\boldsymbol{\Delta}_i(\alpha, \boldsymbol{\beta}^m) + b_i] p(b_i | \alpha)\, db_i$$

for $\boldsymbol{\Delta}_i(\alpha, \boldsymbol{\beta}^m) + b_i$.

This leads to three potential targets of inference. Target 1 involves the association between X and $\boldsymbol{Y}$, characterized now by the marginal linear model coefficients $\boldsymbol{\beta}^m$. Target 2 involves the person-specific random effects, b_i, or functions thereof, including $\boldsymbol{\Delta}_i(\alpha, \boldsymbol{\beta}^m) + b_i$ or possibly $\boldsymbol{\pi}[\boldsymbol{\Delta}_i(\alpha, \boldsymbol{\beta}^m) + b_i]$. Target 3 involves the distribution of the random effects, $f(b|\alpha)$, or the distribution of functions thereof.

In our example, we focus on target 1. It would also be of interest to compare results for target 2 with those based on the subject-specific model (Table 7.4). Assuming, however, that the results are similar, we would tend to prefer the subject-specific approach on grounds of tractability. In the subject-specific model, the transformed conditional mean has a linear structure in $\boldsymbol{\beta}^c$, b_i. This makes tractable interesting subject-specific growth parameters $\boldsymbol{\pi}(\boldsymbol{\beta}^c, b_i)$, such as academic-year learning rates or summer learning rates, which are linear in $\boldsymbol{\beta}^c$, b_i and can be modeled as having tractable, well-behaved distributions. In contrast, in the population-averaged model, the conditional mean is the generally intractable function $\boldsymbol{\Delta}_i(\alpha, \boldsymbol{\beta}^m)\ +\ b_i$ and it appears difficult to find readily interpretable expressions for interesting functions $\boldsymbol{\pi}(\boldsymbol{\beta}^c, b_i)$ and their distributions.

7.6.3 Results

Based on the reasoning described above, we focus on target 1: the association between predictors X and outcomes $\boldsymbol{Y}$ as characterized by the regression coefficients β^m. Using the results of Zeger, Liang, and Albert (1988), we can estimate the population-averaged model via the generalized estimating equation approach using as input the results from approximate maximum likelihood estimates of the subject-specific model found in Section 7.5. The point estimates for our first target of inference are given by

$$\widehat{h}(\phi^m_{ij}) = -1.89 - 0.88 \times (\text{Pov})_i + 2.28 \times (\text{Ayear})_{ij} + 0.04 \times (\text{Pov})_i \times (\text{Ayear})_{ij}$$
$$- 0.03 \times (\text{Sum})_{ij} - 0.16 \times (\text{Pov})_i \times (\text{Sum})_{ij}.$$

These results (see Table 7.5 for SEs) give estimates of the average probability of basic skill acquisition in two subpopulations (non-poor and poor children) at each of five time points. For non-poor children, the point estimates of the average probability of basic skills are $(0.13, 0.13, 0.59, 0.58, 0.93)$. For poor children, the corresponding average probabilities are $(0.06, 0.05, 0.35, 0.30, 0.82)$.

These average probabilities are plotted in Figure 7.5. These results are useful to the policy-maker who needs to know, at each time point, the fraction of poor and non-poor children who have mastered basic skills. Resources can then be concentrated on those who have

Table 7.5 Fitted Model, Binary Outcome, Logit Link,
Population-Averaged Model

Fixed Effects[a]

Predictor	Coefficient	Estimate	SE	t
Intercept	β_{00}^m	−1.89	0.10	−18.82
Pov	β_{01}^m	−0.88	0.27	−3.21
Ayear	β_{10}^m	2.78	0.10	23.36
Pov × Ayear	β_{11}^m	0.04	0.26	0.16
Sum	β_{20}^m	−0.03	0.07	−0.45
Pov × Sum	β_{21}^m	−0.16	0.20	−0.82

[a]Robust SEs are reported in this application.

not mastered those skills. Such results are not readily discernible from the subject-specific analysis (Figure 7.3).

We cannot view the trajectories plotted in Figure 7.5, however, as reflecting the expected trajectory of a hypothetical average or typical child, or, for that matter, of any specific child, real or hypothetical. It is true that there is a rough similarity in shape between the population-averaged results in Figure 7.5 and the estimated trajectory of a "typical child" (that is, a child with random effect $b_{0i} = 0$) as estimated under the subject-specific model and plotted in Figure 7.3b.

7.7 Conclusions

I have considered two approaches to inference for the generalized linear mixed model for longitudinal data. The first is the subject-specific model. Using this approach, one can regard the transformed conditional mean of the response, given the person-specific random effect, to be linear; this model is therefore also called the "conditional model" for short. The second is the population-averaged model. Using this approach, one regards the transformed marginal mean, averaging over the random effects, to be linear; thus, this model is also called the "marginal model." The two models can be constructed in a hierarchical fashion, revealing three targets of inference. However, the construction proceeds quite differently in the two cases, with consequences for model choice.

Using the subject-specific model, one begins with a model that describes how each subject's outcome changes over time. Specifically, at the first level, for each subject, the expected mean outcome given subject-specific random effects b_i is transformed using a link function, and this link function is equated to a linear model. The coefficients of this linear model

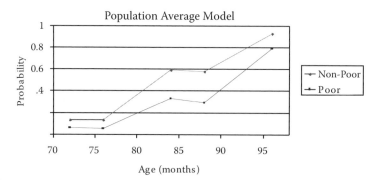

Figure 7.5 Basic skill as a function of student poverty.

are subject-specific growth parameters, in this chapter labeled π_i, and the level-1 "within-subject" variance is V_i. The second-level model characterizes population heterogeneity in growth by the population distribution of these values of $\pi_i = \pi(\beta^c, b_i)$ by a linear model having coefficients β^c and "between-subject" covariance structure $\text{Var}(b_i) = G$. The level-1 and level-2 are each linear models.

The strength of the subject-specific approach is that it nicely supports inferences about the level-1 model for each subject, that is, the person-specific growth parameter vector π_i itself as well as the within-subject variance V_i; it also supports inferences about the parameters of the level-2 model, characterized by between-subject coefficients β^c and the between-subject variance G.

Using the subject-specific model, one can also characterize the marginal distribution of Y_i as a function of the regression coefficients β^c. This is achieved by integration of the joint distribution of Y_i and the random effects b_i with respect to the random effects b_i. Unfortunately, this integration leads to a generally intractable form, making interpretation of the marginal association between X_i and Y_i, the third target of inference, difficult. Thus, the subject-specific model is not generally convenient for pursuing this association. Rather, the subject-specific model characterizes the X_i, Y_i association in terms of the conditional association, that is, the association between X_i and Y_i holding constant the random effect b_i.

The population-averaged model can also be constructed as a hierarchical model, but the order is reversed. The population-averaged model does not begin by characterizing the growth of a particular subject in some tractable way. Rather, it begins with a tractable form for the marginal association between X_i and Y_i as characterized by regression coefficients β^m. If we are willing to assume the existence of subject-specific random effects b_i, an analog to the level-2 model described above, we can also identify the trajectory for a particular subject as the conditional association between X_i and Y_i given b_i. However, such a characterization of the subject-specific association will not generally be tractable. Thus, the population-averaged model is not especially convenient for characterizing the trajectories of change for particular types of subjects.

There are important exceptions to these general conclusions. First, if the link function is a linear transformation of the mean, such as the identity link commonly used for normal-theory outcomes, there is no distinction between β^c and β^m. Second, as the between-subject variance matrix G approaches the null matrix, the two models converge to the standard generalized linear model, in which case there is no distinction between β^c and β^m. Finally, if the level-2 model is the conjugate of the level-1 model, the marginal model will also generally be tractable and the distinction between the subject-specific and population-averaged models will diminish as well. For example, if the first level of the model is Poisson and the random effect follows a gamma distribution, the marginal distribution will be the negative binomial. A serious limitation in this conjugate approach is that it is limited in characterizing the joint distribution of multiple random effects.

How, then, should we approach the task of choosing between the subject-specific and the population-averaged models? I have approached that question by posing three targets of inference. The conclusions I draw are the following.

Target 1 is the association between X_i and Y_i. If one is interested in the conditional association, that is, the association between X_i and Y_i holding constant the random effect, one should choose the subject-specific model. This was illustrated in Figure 7.3 where expected trajectories of change in basic math skills were plotted as a function of time of year and child poverty for three types of children: children high, medium, or low on the random effect. In contrast, if one is interested in how the average level of basic skills changes as a function of time of year for subpopulations defined by poverty status, one should choose the population-averaged model as illustrated in Figure 7.5.

Target 2 is an inference about the development of a single child. If obtaining a good picture of development for a specific child is of interest, one can employ either the subject-specific approach, as illustrated here in Figure 7.4, or the population-averaged approach as developed by Heagerty and Zeger (2000). One cannot say in advance which approach is preferable. If the linear model for the development of each child (that is, the subject-specific "level-1 model") is a good one, then the subject-specific approach would seem preferable. In contrast, if the linear model for the marginal association between X_i and $\mathbf{Y}_i$ is correct, the population-averaged model would be preferable. In the general case (non-linear link function, $G > 0$, non-conjugate models), the subject-specific and population-averaged models will not simultaneously be true: linearity in the subject-specific model implies non-linearity in the population-averaged model, while linearity in the population-averaged model implies non-linearity in the subject-specific model. Nor can one appeal to the population-averaged on robustness grounds. Inferences about $\boldsymbol{\beta}^m$ are quite robust to misspecification of the random-effects variance structure in the population-averaged model. However, inferences about the random effects using this approach will not be. While the two approaches are generally incompatible for this target of inference, the practical results of using the two may in fact be small. More study is needed of whether and when the two approaches would produce discrepant results for this target.

Target 3 is the population distribution of person-specific growth parameters $\pi_i = \pi(\boldsymbol{\beta}^c, b_i)$. Here the subject-specific approach seems the only viable one. Using that approach, the model for π_i is linear in $(\boldsymbol{\beta}^c, \boldsymbol{b}_i)$, and $\boldsymbol{\pi}_i$, in turn, has a well-defined role in shaping the development of each subject. The distribution of such subject-specific growth parameters does not appear well defined within the framework of the population-averaged model.

Acknowledgments

The research reported here was supported by a grant from the Spencer Foundation. I wish to thank Guanglei Hong for her thoughtful comments on an earlier draft.

References

Breslow, N. E. and Clayton, D. G. (1993). Approximate inference in generalized linear mixed models. *Journal of the American Statistical Association* **88**, 9–25.

Bryk, A. S. and Raudenbush, S. W. (1988). Toward a more appropriate conceptualization of research on school effects: A three-level hierarchical linear model. *American Journal of Education* **97**, 65–108.

Carter, L. F. (1984). The sustaining effects study of compensatory and elementary education. *Educational Researcher* **13**, 4–13.

Dempster, A. P., Rubin, D. B., and Tsutakawa, R. D. (1981). Estimation in covariance components models. *Journal of the American Statistical Association* **76**, 341–353.

Goldstein, H. (2003). *Multilevel Statistical Models*, 3rd ed. London: Edward Arnold.

Heagerty, P. J. and Zeger, S. L. (2000). Marginalized multilevel models and likelihood inference. *Statistical Science* **15**, 1–26.

Laird, N. M. and Ware, J. H. (1982). Random-effects models for longitudinal data. *Biometrics* **38**, 963–974.

Lindley, D. V. and Smith, A. F. M. (1972). Bayes estimates for the linear model. *Journal of the Royal Statistical Society, Series B* **34**, 1–41.

Mason, W. M., Wong, G. Y., and Entwistle, B. (1984). Contextual analysis through the multi-level linear model. In S. Leinhardt (ed.), *Sociological Methodology 1983–1984*, pp. 72-103. San Francisco: Jossey-Bass.

McCullagh, P. and Nelder, J. (1989). *Generalized Linear Models*. London: Chapman & Hall.

Pinheiro, J. C. and Bates, D. M. (2000). *Mixed-Effects Models in S and S-PLUS*. New York: Springer.

Raudenbush, S. W. (2000). Comment on Heagerty, P. J. and Zeger, S. L., Marginal multilevel models and likelihood inference. *Statistical Science* **15**(1), 22–24.

Raudenbush, S. W. and Bryk, A. S. (2002). *Hierarchical Linear Models*, 2nd ed. Thousand Oaks, CA: Sage.

Raudenbush, S. W., Yang, H.-L., and Yosef, M. (2000). Maximum likelihood for generalized linear models with nested random effects via high-order, multivariate Laplace approximation. *Journal of Computational and Graphical Statistics* **9**, 141–157.

Stiratelli, R., Laird, N., and Ware, J. (1984). Random effects models for serial observations with binary responses. *Biometrics* **40**, 961–970.

Verbeke, G. and Molenberghs, G. (1997). *Linear Mixed Models for Longitudinal Data*. New York: Springer.

Zeger, S. L. and Liang, K. Y. (1986). Longitudinal data analysis for discrete and continuous outcomes. *Biometrics* **42**, 121–130.

Zeger S. L., Liang, K. Y., and Albert, P. S. (1988). Models for longitudinal data: A generalized estimating equation approach. *Biometrics* **44**, 1049–1060.

PART III

Non-Parametric and Semi-Parametric Methods for Longitudinal Data

CHAPTER 8

Non-parametric and semi-parametric regression methods: Introduction and overview

Xihong Lin and Raymond J. Carroll

Contents

8.1 Introduction and overview

Parametric regression methods for longitudinal data have been well developed in the last 20 years. Such methods can be classified broadly as estimating equation based methods, such as generalized estimating equations (Liang and Zeger, 1986), and their extensions (Chapter 3), and mixed-effects models (Laird and Ware, 1982; Breslow and Clayton, 1993; see also Chapter 4). Diggle et al. (2002) provide an excellent overview of these parametric regression methods. For recent developments, see Chapter 3 through Chapter 6.

A major limitation of these methods is that the relationship of the mean of a longitudinal response to covariates is assumed fully parametric. Although such parametric mean models enjoy simplicity, they have suffered from inflexibility in modeling complicated relationships between the response and covariates in various longitudinal studies. Examples include hormone profiles in a menstrual cycle in reproductive health (Brumback and Rice, 1998; Zhang et al., 1998); longitudinal CD4 trajectories in AIDS research (Zeger and Diggle, 1994; Lin and Ying, 2001); age effects on childhood respiratory disease (Diggle et al., 2002; Lin and Zhang, 1999); time trajectories in speech research and growth curves (Brumback and Lindstrom, 2004; Gasser et al., 1984); time-varying treatment/exposure effects (Hogan, Lin, and Herman, 2004; Huang, Wu, and Zhou, 2002); and time course analysis of microarray gene expressions (Luan and Li, 2003; Storey et al., 2005). These practical applications have placed a strong demand in the last 10 years on developing non-parametric and semi-parametric regression methods for longitudinal data, where flexible functional forms can be estimated from the data to capture possibly complicated relationships between longitudinal outcomes and covariates.

Non-parametric and semi-parametric regression methods for independent data have been well developed in the last two decades. Non-parametric regression methods can be broadly classified into kernel methods (Wand and Jones, 1995), which are often based on local likelihoods (Fan and Gijbels, 1996), and splines, which include smoothing splines (Green and Silverman, 1994; Wahba, 1990), penalized splines (Eilers and Marx, 1996; Ruppert, Wand, and Carroll, 2003), and regression splines (Stone et al., 1997). Both smoothing splines and penalized splines are based on penalized likelihoods. Silverman (1984) demonstrated a close connection between kernel smoothing and smoothing spline smoothing, and showed that kernels and smoothing splines are asymptotically equivalent for independent data and that splines are higher-order kernels.

Semi-parametric regression methods for independent data have been equally well developed (Härdle, Liang, and Gao, 1999; Green and Silverman, 1994, Chapter 4). Such models are sometimes referred to as (generalized) partial linear models, where the mean or the transformed mean (by a parametric link function) of an outcome variable is modeled in terms of parametric functions of a subset of the covariates and non-parametric functions of other covariates. Profile-kernel and profile-spline methods have been proposed for estimation in such partial linear models (Heckman, 1984; Speckman, 1988; Carroll et al., 1997).

Non-parametric and semi-parametric regression methods for longitudinal data using kernel and spline methods have enjoyed substantial developments in the last 10 years. Chapter 9 through Chapter 12 provide reviews of these methods. To help the reader understand these developments for longitudinal data, in the next section we provide an overview of non-parametric and semi-parametric regression methods using kernels and splines for independent data.

8.2 Brief review of non-parametric and semi-parametric regression methods for independent data

8.2.1 Local polynomial kernels

Traditional kernel regression estimates a non-parametric regression function at a target point using local weighted averages; for example, the Nadaraya–Watson estimator. The most popular kernel regression method is local polynomial regression (Wand and Jones, 1994; Fan and Gijbels, 1996). Consider the simplest non-parametric regression model,

$$Y_i = \theta(Z_i) + \epsilon_i, \tag{8.1}$$

where Y_i is a scalar continuous outcome, Z_i is a scalar covariate, $\theta(z)$ is an unknown smooth function, and $\epsilon_i \sim N(0, \sigma^2)$ and is independent and identically distributed ($i = 1, \ldots, N$). The idea of the local dth-order polynomial regression estimator of $\theta(z)$ is to approximate $\theta(Z_i)$ locally around any arbitrary point z by a dth-order polynomial as $\theta(Z_i) \approx \alpha_0 + \cdots + \alpha_d(Z_i - z)^d = \boldsymbol{Z}_i(z)'\boldsymbol{\alpha}$, where $\boldsymbol{Z}_i(z) = \{1, \ldots, (Z_i - z)^d\}$ and $\boldsymbol{\alpha} = (\alpha_0, \ldots, \alpha_d)'$, and to estimate $\boldsymbol{\alpha}$ by maximizing the local log-likelihood, apart from a constant, defined as

$$-\frac{1}{2\sigma^2} \sum_{i=1}^{N} K_h(Z_i - z)\{Y_i - \boldsymbol{Z}_i(z)'\boldsymbol{\alpha}\}^2,$$

where $K_h(s) = h^{-1}K(s/h)$, h is a bandwidth, and $K(\cdot)$ is a kernel function, which is often chosen as a symmetric density function with mean 0. Commonly used kernel functions include the Gaussian, uniform, and Epanechnikov kernels, the latter being $K(s) = \frac{3}{4}(1 - s^2)_+$, where $a_+ = a$ if $a > 0$ and 0 otherwise. The resulting kernel estimating equation is

$$\sum_{i=1}^{N} \boldsymbol{Z}_i(z) K_h(Z_i - z)\{Y_i - \boldsymbol{Z}_i(z)'\boldsymbol{\alpha}\} = 0. \tag{8.2}$$

The dth-order kernel estimator at the target point z is $\widehat{\theta}(z) = \widehat{\alpha}_0$. If $d = 0$, we have the traditional local average kernel estimator, which corresponds to the Nadaraya–Watson estimator,

$$\widehat{\theta}(z) = \frac{\sum_{i=1}^{N} K_h(Z_i - z) Y_i}{\sum_{i=1}^{N} K_h(Z_i - z)}. \tag{8.3}$$

The local linear kernel estimator ($d = 1$) has been commonly used because of its better bias properties. Bandwidth selection is important in kernel smoothing. The bandwidth h could be selected using cross-validation. Other approaches include plug-in estimators (Wand and Jones, 1994; Fan and Gijbels, 1996) and empirical bias bandwidth selection (Ruppert, 1997), among others.

A key feature of kernel smoothing for independent data is that it is local in the sense that $\widehat{\theta}(z)$ places more weight on the observations when Z_i values are in the neighborhood of z, and downweights those observations that are far from z. This can be seen from (8.3). Specifically, as the bandwidth $h \to 0$ and the sample size $N \to \infty$, only the observations in the shrinking neighborhood of z contribute to the estimation of $\theta(z)$. As we will see, this locality property is no longer true for superior kernel and spline smoothing for longitudinal data (Chapter 9; see also Lin et al., 2004; Welsh, Lin, and Carroll, 2002).

Fan (1993) showed that the local polynomial kernel estimator enjoys minimax efficiency among the class of all linear smoothers. The Epanechnikov kernel is optimal in the sense that it minimizes the mean squared error of the local polynomial kernel estimator. The local linear polynomial kernel estimator (8.2) can be extended easily to non-parametric regression for non-normal outcomes within the generalized linear model framework (Fan and Gijbels, 1996, Chapter 5).

8.2.2 Smoothing splines

A smoothing spline estimates the non-parametric regression function $\theta(z)$ using a piece-wise polynomial function with *all* the observed covariate values $\{Z_i\}$ used as knots, where smoothness constraints are assumed at the knots (Wahba, 1990; Green and Silverman, 1994). The most commonly used smoothing spline is the natural cubic smoothing spline, which assumes $\theta(z)$ is a piecewise cubic function, is linear outside of $\min(Z_i)$ and $\max(Z_i)$, and is continuous and twice differentiable with a step function third derivative at the knots $\{Z_i\}$. The natural cubic smoothing spline estimator can be obtained by maximizing a penalized log-likelihood as follows. Under the simple non-parametric model (8.1), the penalized log-likelihood can be written as

$$\frac{1}{2\sigma^2} \left[-\sum_{i=1}^{N} \{Y_i - \theta(Z_i)\}^2 - \lambda \int \{\theta^{(2)}(z)\}^2 dz \right] = \frac{1}{2\sigma^2} \left[-\sum_{i=1}^{N} \{Y_i - \theta(Z_i)\}^2 - \lambda \boldsymbol{\theta}' \Psi \boldsymbol{\theta} \right],$$

where λ is a smoothing parameter, Ψ is the cubic smoothing spline penalty matrix (Green and Silverman, 1994, Equation 2.3), $\boldsymbol{\theta} = \{\theta(Z_1), \ldots, \theta(Z_n)\}'$, and $\theta^{(2)}(z)$ denotes the second derivative of $\theta(z)$. The smoothing parameter λ controls the goodness of fit and the smoothness of the curve. The smoothing spline estimator interpolates the data if $\lambda = 0$, and assumes $\theta(z)$ to be linear if $\lambda \to \infty$.

The resulting cubic smoothing spline estimator takes the form of a ridge regression estimator,

$$\widehat{\boldsymbol{\theta}} = (I + \lambda \Psi)^{-1} \boldsymbol{Y}, \tag{8.4}$$

where $\boldsymbol{Y} = (Y_1, \ldots, Y_n)'$. Efficient algorithms, such as the Reinsch algorithm (Green and Silverman, 1994), can be used to calculate $\widehat{\boldsymbol{\theta}}$ in $O(N)$ arithmetic operations. The smoothing parameter λ can be estimated using cross-validation, generalized cross-validation (Wahba, 1990), and general maximum likelihood (GML) (Wahba, 1985).

There is a close connection between a smoothing spline estimator and a linear mixed model. Specifically, the GML esimator corresponds to the restricted maximum likelihood estimator in the corresponding mixed model. Such a connection, as well as the Bayesian formulation of the smoothing spline, will be discussed in more detail in Chapters 9, 11, and 12. Silverman (1984) showed that the smoothing spline estimator is asymptotically equivalent to the local average kernel estimator. Using Silverman's (1984) results, Nychka (1995) established the asymptotic properties of the smoothing spline estimator (8.4) by deriving its asymptotic bias and variance. Smoothing spline estimation has been extended to generalized linear models (Green and Silverman, 1994) and generalized additive models (Hastie and Tibshirani, 1990). Bayesian spline estimation can be found in Hastie and Tibshirani (2000). More discussions of the use of smoothing splines in longitudinal data can be found in Chapter 9 and Chapter 11.

8.2.3 Regression splines and penalized splines (P-splines)

A key advantage of a smoothing spline is that all the observed design points are used as knots. Hence, one does not need to choose knots. However, when the sample size is large, computational demands are significantly increased and make it difficult to compute.

Regression splines (Stone et al., 1997) are a basis function-based non-parametric regression method, which uses a small number of knots and proceeds with a parametric regression using the bases. Denote by $\{s_1, \ldots, s_L\}$ a set of L knots, where L is often small (e.g., 5 or 6), and by $\{B_1(z), \ldots, B_L(z)\}$ a set of basis functions (e.g., B-spline basis or plus-function basis). For the simple non-parametric regression (8.1), one approximates $\theta(z)$ by

$$\theta(z) \approx \sum_{l=1}^{L} B_l(z)\alpha_l. \tag{8.5}$$

Then one estimates $\boldsymbol{\alpha} = (\alpha_1, \ldots, \alpha_L)'$ by fitting the parametric model

$$Y_i = \sum_{l=1}^{L} B_l(z)\alpha_l + \epsilon_i, \tag{8.6}$$

via standard least squares. The resulting non-parametric regression spline estimator of $\theta(z)$ is $\widehat{\theta}(z) = \sum_{l=1}^{L} B_l(z)\widehat{\alpha}_l$, where $\widehat{\boldsymbol{\alpha}}$ is the maximum likelihood estimate under (8.6). A key advantage of the regression spline is its computational simplicity, since one only needs to fit a parametric model. However, choices of the number of knots and the locations of the knots are critical. Estimation of $\theta(z)$ could be sensitive to these choices. Adaptive knot allocation strategies have been recommended (Stone et al., 1997).

Penalized splines (P-splines) are a hybrid of regression splines and smoothing splines (Eilers and Marx, 1996; Ruppert, Wand, and Carroll, 2003). One approximates $\theta(z)$ using the basis expansion (8.5) with a large number of knots L, where L is often much smaller than the sample size N but much larger than the number of knots often used in regression splines (e.g., $L = 20$ to 30). P-spline estimation proceeds by fitting (8.6) with a quadratic penalty on $\{\alpha_l\}$. For example, if the $\{B_l(z)\}$ are plus basis functions and L is the number of interior knots, the dth-order P-spline model is

$$\theta(z; \boldsymbol{\alpha}) = \alpha_0 + \alpha_1 z + \cdots + \alpha_d z^d + \sum_{l=1}^{L} \alpha_{l+d}(z - s_l)_+^d.$$

One estimates $\theta(z; \boldsymbol{\alpha})$ by maximizing the penalized log-likelihood, apart from a constant,

$$\frac{1}{2\sigma^2} \left\{ -\sum_{i=1}^{N} \{Y_i - \theta(Z_i; \boldsymbol{\alpha})\}^2 - \lambda \sum_{l=1}^{L} \alpha_l^2 \right\}.$$

If $\{B_l(z)\}$ are B-spline basis functions, a second-order difference penalty of α can be used (Fahrmeir, Kneib, and Lang, 2004; Lang and Brezger, 2004). A key advantage of P-splines is that they reduce the computational burden of smoothing splines when the sample size is large, and are less sensitive to the allocation of the knots compared to regression splines. The smoothing parameter can be treated as a variance component using the connection between P-splines and mixed models and can be estimated using restricted maximum likelihood. Several recent attempts have been made to understand the theoretical properties of P-splines in special situations (Hall and Opsomer, 2005). More details about the use of regression splines and P-splines in longitudinal data can be found in Chapter 12.

8.3 Overview of non-parametric and semi-parametric regression for longitudinal data

Although non-parametric and semi-parametric regression methods have been well developed for independent data, their developments for longitudinal data have only occurred in recent years. A major difficulty in the analysis of longitudinal data is that the data are subject to within-subject correlation among repeated measures over time. This correlation presents significant challenges in the development of kernel and spline smoothing methods for longitudinal data; in particular, a need for developing non-conventional smoothing methods and a better understanding of their properties. Specifically, traditional local likelihood based kernel methods are not able to effectively account for the within-subject correlation (Lin and Carroll, 2000). A consistent and efficient non-parametric estimator for longitudinal data needs to be non-local (Welsh, Lin, and Carroll, 2002; Lin et al., 2004). Standard functional data analysis techniques are not directly applicable to longitudinal data, as repeated measures are often obtained at irregular sparse time points and are often more noisy (Yao, Müller, and Wang, 2005).

Chapter 9 provides an overview of both estimating equation based methods and likelihood based methods for non-parametric and semi-parametric regression using kernel and spline smoothing for longitudinal data. Chapter 10 surveys the use of functional data analysis methods for non-parametric regression in longitudinal data by treating data as samples of random curves. Chapter 11 reviews smoothing spline methods for longitudinal data, while Chapter 12 reviews penalized spline methods. One can find in these chapters detailed discussions of the attractive connection between spline estimation and mixed models (Brumback and Rice, 1998; Wang, 1998; Zhang et al., 1998; Lin and Zhang, 1999; Verbyla et al., 1999).

References

Breslow, N. E. and Clayton, D. G. (1993). Approximate inference in generalized linear mixed models. *Journal of the American Statistical Association* **88**, 9–25.

Brumback, B. and Rice, J. A. (1998). Smoothing spline models for the analysis of nested and crossed samples of curves (with discussion). *Journal of the American Statistical Association* **93**, 961–1006.

Brumback, L. C. and Lindstrom, M. J. (2004). Self modeling with flexible, random time transformations. *Biometrics* **60**, 461–470.

Carroll, R. J., Fan, J., Gijbels, I., and Wand, M. P. (1997). Generalized partially linear single-index models. *Journal of the American Statistical Association* **92**, 477–489.

Diggle, P. J., Heagerty, P. J., Liang, K. Y., and Zeger, S. L. (2002). *Analysis of Longitudinal Data*. Oxford: Oxford University Press.

Eilers, P. H. and Marx, B. D. (1996). Flexible smoothing with B-splines and penalities (with discussion). *Statistical Science* **11**, 89–121.

Fahrmeir, L., Kneib, T., and Lang, S. (2004). Penalized structured additive regression for space-time data: A Bayesian perspective. *Statistica Sinica* **14**, 731–761.

Fan, J. (1993). Local linear regression smoothers and their minimax efficiencies. *Annals of Statistics* **21**, 196–216.

Fan, J. and Gijbels, I. (1996). *Local Polynomial Modelling and Its Applications*. London: Chapman & Hall.

Gasser, T., Müller, H. G., Köhler, W., Molinari, L., and Prader, A. (1984). Nonparametric regression analysis of growth curves. *Annals of Statistics* **12**, 210–229.

Green, P. J. and Silverman, B. W. (1994). *Nonparametric Regression and Generalized Linear Models: A Roughness Penalty Approach*. London: Chapman & Hall.

Härdle, W., Liang, H., and Gao, J. (1999). *Partially Linear Models*. New York: Springer-Verlag.

Hastie, T. and Tibshirani, R. (1990). *Generalized Additive Models*. London: Chapman & Hall.

Hastie, T. and Tibshirani, R. (2000). Bayesian backfitting. *Statistical Science* **15**, 193–223.

Hall, P. and Opsomer, J. (2005). Theory for penalised spline regression. *Biometrika* **92**, 105–118.

Heckman, N. (1984). Spline smoothing in partial linear models. *Journal of the American Statistical Association* **48**, 244–248.

Hogan, J. W., Lin, X., and Herman, B. (2004) Mixtures of varying coefficient models for longitudinal data with discrete or continuous nonignorable dropout. *Biometrics* **60**, 854–864.

Huang, J., Wu, C., and Zhou, L. (2002). Varying-coefficient models and basis function approximation for the analysis of repeated measures. *Biometrika* **89**, 111–128.

Laird, N. M. and Ware, J. H. (1982). Random-effects models for longitudinal data. *Biometrics* **38**, 963–974.

Lang, S. and Brezger, A. (2004). Bayesian P-splines. *Journal of Computational and Graphical Statistics* **13**, 183–212.

Liang, K. Y. and Zeger, S. L. (1986). Longitudinal data analysis using generalized linear models. *Biometrika* **73**, 13–22.

Lin, D. and Ying, Z. (2001). Semiparametric and nonparametric regression analysis of longitudinal data. *Journal of the American Statistical Association* **96**, 103–126.

Lin, X. and Carroll, R. J. (2000). Nonparametric function estimation for clustered data when the predictor is measured without/with error. *Journal of the American Statistical Association* **95**, 520–534.

Lin, X. and Zhang, D. (1999). Inference in generalized additive mixed model using smoothing splines. *Journal of the Royal Statistical Society, Series B* **61**, 381–400.

Lin, X., Wang, N., Welsh, A., and Carroll, R. J. (2004). Equivalent kernels of smoothing splines in nonparametric regression for clustered data. *Biometrika* **91**, 177–193.

Luan, Y. and Li, H. (2003). Clustering of time-course gene expression data using a mixed-effects model with B-splines. *Bioinformatics* **19**, 474–482.

Nychka, D. (1995). Splines as local smoothers. *Annals of Statistics* **23**, 1175–1197.

Ruppert, D. (1997). Empirical-bias bandwidths for local polynomial nonparametric regression and density estimation. *Journal of the American Statistical Association* **92**, 1049–1062.

Ruppert, D., Wand, M. P., and Carroll, R. J. (2003). *Semiparametric Regression*. Cambridge: Cambridge University Press.

Silverman, B. (1984). Spline smoothing: the equivalent variable kernel method. *Annals of Statistics* **12**, 898–916.

Speckman, P. (1988). Kernel smoothing in partial linear models. *Journal of the Royal Statistical Society, Series B* **50**, 413–436.

Stone, C.J., Hansen, M., Kooperberg, C., and Truong, Y. K. (1997). Polynomial splines and their tensor products in extended linear modeling (with discussion). *Annals of Statistics* **25**, 1371–1470.

Storey, J. D., Xiao, W., Leek, J. T., Tompkins, R. G., and Davis, R. W. (2005). Significance analysis of time course microarray experiments. *Proceedings of the National Academy of Sciences* **102**, 12837–12842.

Verbyla, A. P., Cullis, B. R., Kenward, M. G., and Welham, S. J. (1999). The analysis of designed experiments and longitudinal data using smoothing splines. *Applied Statistics* **48**, 269–311.

Wahba, G. (1985). A comparison of GCV and GML for choosing the smoothing parameter in the generalized spline problem. *Annals of Statistics* **13**, 1378–1402.

Wahba, G. (1990). *Spline Models for Observational Data.* Philadephia: SIAM.

Wand, M. P. and Jones, M. C. (1995). *Kernel Smoothing.* London: Chapman & Hall.

Wang, Y. (1998). Mixed effects smoothing spline analysis of variance. *Journal of the Royal Statistical Society, Series B* **60**, 159–174.

Welsh, A. H., Lin, X., and Carroll, R. J. (2002). Marginal longitudinal nonparametric regression: Locality and efficiency of spline and kernel methods. *Journal of the American Statistical Association* **97**, 482–493.

Zeger, S. L. and Diggle, P. J. (1994). Semiparametric models for longitudinal data with application to CD4 cell numbers in HIV seroconverters. *Biometrics* **50**, 689–699.

Zhang, D., Lin, X., Raz, J., and Sowers, M. (1998). Semiparametric stochastic mixed models for longitudinal data. *Journal of the American Statistical Association* **93**, 710–719.

CHAPTER 9

Non-parametric and semi-parametric regression methods for longitudinal data

Xihong Lin and Raymond J. Carroll

Contents

9.1 Introduction

Significant developments in non-parametric and semi-parametric regression methods for longitudinal data have taken place in the last 10 years. The presence of the within-subject correlation among repeated measures over time presents major challenges in developing kernel and spline smoothing methods for longitudinal data. Classical local likelihood based kernel methods and their natural local estimating equation extensions fail to effectively account for the within-subject correlation. This difficulty has called for the development of a new class of non-local kernel estimators. Extension of spline smoothing to longitudinal data requires explicitly accounting for the within-subject correlation in constructing the penalized likelihood function. Statistical theory has recently been developed to understand the properties of these advanced kernel and spline methods for longitudinal data. In this chapter we provide a review of both estimating equation based methods and likelihood based methods for non-parametric and semi-parametric regression using kernel and spline smoothing for longitudinal data. Connections between splines (smoothing splines and P-splines) and mixed models are discussed. More detailed discussions about this connection can also be found in Chapter 11 and Chapter 12.

This chapter is organized as follows. Section 9.2 and Section 9.3 describe kernel and spline methods, and their relationship for longitudinal non-parametric regression with a single co-variate. The close connection between splines and mixed models is described. Section 9.4 discusses several extensions of single covariate non-parametric models, including generalized additive mixed models, self-modeling mixed models, and varying-coefficient models. Section 9.5 illustrates the non-parametric smoothing methods using data from a longitudinal study of the time course of progesterone levels. Section 9.6 describes semi-parametric regression models for longitudinal data in marginal models using estimating equations and mixed models using likelihood functions. Profile-kernel methods and joint maximization methods are described. Section 9.7 provides an illustration of these models using data from a longitudinal study of CD4 counts of HIV seroconverters. Section 9.8 provides a concluding discussion.

9.2 Single-covariate non-parametric regression for longitudinal data using kernel methods

9.2.1 The single-covariate generalized non-parametric model for longitudinal data

In this section we consider generalized marginal non-parametric models for longitudinal data. Let (Z_{ij}, Y_{ij}) be the covariate and the outcome variable for subject i $(i = 1, \ldots, N)$ measured at the jth time point $(j = 1, \ldots, n_i)$. The outcome variable Y_{ij} can be continuous or discrete, and might be correlated within the same subject. Suppose Y_{ij} marginally has mean $\mu_{ij} = E(Y_{ij}|Z_{ij})$ and variance $\text{Var}(Y_{ij}|Z_{ij}) = \phi^{-1}v(\mu_{ij})$, where $v(\cdot)$ is a variance function and ϕ is a scale parameter. Following Pepe and Anderson (1994), we assume the mean of Y_{ij} satisfies $E(Y_{ij}|Z_{i1}, \ldots, Z_{in_i}) = E(Y_{ij}|Z_{ij})$. The marginal mean μ_{ij} depends on Z_{ij} through

$$g(\mu_{ij}) = \theta(Z_{ij}), \tag{9.1}$$

where $\theta(\cdot)$ is an unknown smooth function and $g(\mu)$ is a known monotonic link function. Common link functions include the identity link $g(\mu) = \mu$ for normal outcomes; the logit link $g(\mu) = \log\{\mu/(1-\mu)\}$ or the probit link $g(\mu) = \Phi^{-1}(\mu)$ for binary outcomes; and the log link $g(\mu) = \log(\mu)$ for Poisson outcomes.

In this section we discuss several kernel smoothing methods for estimating $\theta(z)$ in the marginal model (9.1) for longitudinal data. We start in Section 9.2.2 with the kernel generalized estimating equations (GEE) estimator, which is an extension of the conventional local polynomial kernel estimator discussed in Section 8.2.1 to longitudinal data via the GEE framework. Unlike parametric GEE, this kernel GEE estimator is shown to fail to account for the within-subject correlation and loses efficiency if the true covariance matrix Σ_i of $\boldsymbol{Y}_i = (Y_{i1}, \ldots, Y_{in_i})'$ is used. In Section 9.2.3 we describe the seemingly unrelated (SUR) kernel estimator, which is generally obtained through solving iterative kernel estimating equations and effectively accounts for the within-subject correlation. Section 9.2.5 discusses extensions of the SUR kernel estimator to likelihood settings.

9.2.2 Local polynomial kernel GEEs

The kernel GEE estimator (Lin and Carroll, 2000) extends the generalized linear model version of the simple local polynomial kernel estimating equation (8.2) to longitudinal data by introducing a working covariance matrix, in a similar spirit to the GEE method for parametric regression (Liang and Zeger, 1986; see also Chapter 3). Specifically, at any given target point z, we once again approximate $\theta(Z_{ij})$ locally by a dth-order polynomial as $\theta(Z_{ij}) \approx \alpha_0 + \cdots + \alpha_d(Z_{ij} - z)^d = \mathbf{Z}_{ij}(z)'\boldsymbol{\alpha}$, where $\mathbf{Z}_{ij} = \{1, \ldots, (Z_{ij} - z)^d\}$. In what follows $f^{(d)}(z)$ denotes the dth-order derivative of $f(z)$ for any arbitrary function $f(z)$. The symmetric local polynomial kernel GEE estimating equation can be written as

$$\sum_{i=1}^{N} \mathbf{Z}_i(z)'\boldsymbol{\Delta}_i(z)\mathbf{K}_{ih}^{1/2}(z)\mathbf{V}_i^{-1}(z)\mathbf{K}_{ih}^{1/2}(z)\{\mathbf{Y}_i - \boldsymbol{\mu}_i(z)\} = 0, \qquad (9.2)$$

where $\mathbf{K}_{ih}(z) = \text{diag}\{K_h(Z_{ij} - z)\}$, $\boldsymbol{\mu}_i(z) = \{\mu_{i1}(z), \ldots, \mu_{in_i}(z)\}'$ with $\mu_{ij}(z) = g^{-1}\{\mathbf{Z}_{ij}(z)'\boldsymbol{\alpha}\}$, $\boldsymbol{\Delta}_i = \text{diag}\{\delta_{ij}(z)\}$ with $\delta_{ij}(z) = 1/g^{(1)}\{\mu_{ij}(z)\}$, $\mathbf{V}_i = \mathbf{S}_i^{1/2}\mathbf{R}_i(\boldsymbol{\gamma})\mathbf{S}_i^{1/2}$, $\mathbf{S}_i = \text{diag}[\phi^{-1}v\{\mu_{ij}(z)\}]$, and $\mathbf{R}_i$ is an invertible working correlation matrix, possibly depending on a parameter vector $\boldsymbol{\gamma}$, which can be estimated using the method of moments. Denote by $\widehat{\boldsymbol{\alpha}}$ the solution of (9.2). The kernel GEE estimator of $\theta(z)$ is $\widehat{\theta}_K(z) = \widehat{\alpha}_0$. Lin and Carroll (2000) also considered a non-symmetric kernel GEE estimating equation by replacing $\mathbf{K}_{ih}^{1/2}(z)\mathbf{V}_i^{-1}(z)\mathbf{K}_{ih}^{1/2}(z)$ by $\mathbf{V}_i^{-1}(z)\mathbf{K}_{ih}(z)$. The asymptotic performance of the kernel estimator from the non-symmetric kernel GEE estimating equation is similar.

One can solve the kernel GEE estimating equations (9.2) easily using the Fisher scoring algorithm via iteratively reweighted least squares. The covariance of $\widehat{\theta}(z)$ can be estimated using a sandwich estimator as $\text{Cov}\{\widehat{\theta}_K(z)\} = \mathbf{e}'\boldsymbol{\Omega}_1^{-1}\boldsymbol{\Omega}_2\boldsymbol{\Omega}_1^{-1}\mathbf{e}$, where $\mathbf{e} = (1, 0, \ldots, 0)'$, and

$$\boldsymbol{\Omega}_1 = \sum_{i=1}^{N} \mathbf{Z}_i(z)'\boldsymbol{\Delta}_i(z)\mathbf{K}_{ih}^{1/2}(z)\mathbf{V}_i^{-1}(z)\mathbf{K}_{ih}^{1/2}(z)\boldsymbol{\Delta}_i(z)\mathbf{Z}_i(z),$$

$$\boldsymbol{\Omega}_2 = \sum_{i=1}^{N} \mathbf{Z}_i(x)'\boldsymbol{\Delta}_i(z)\mathbf{K}_{ih}^{1/2}(z)\mathbf{V}_i^{-1}(z)\mathbf{K}_{ih}^{1/2}(z)\{\mathbf{Y}_i - \boldsymbol{\mu}_i(z)\}\{\mathbf{Y}_i - \boldsymbol{\mu}_i(z)\}' \otimes$$
$$\mathbf{K}_{ih}^{1/2}(z)\mathbf{V}_i^{-1}(z)\mathbf{K}_{ih}^{1/2}(z)\boldsymbol{\Delta}_i(z)\mathbf{Z}_i(z).$$

When the working correlation matrix $\mathbf{R}_i = \mathbf{I}_i$, where $\mathbf{I}_i$ is an identity matrix of dimension n_i, the kernel GEE estimator ignores the within-subject correlation. Such a working independence kernel GEE estimator reduces to the conventional local polynomial kernel estimator discussed in Section 8.2.1 and has been considered by many authors (Zeger and Diggle, 1994; Hoover et al., 1998; Fan and Zhang, 2000).

For parametric regression using GEEs, Liang and Zeger (1986) showed that the regression coefficients are consistent and are robust to misspecification of the working correlation matrix $\mathbf{R}_i(\boldsymbol{\gamma})$. When $\mathbf{R}_i$ is correctly specified, that is, $\mathbf{V}_i = \boldsymbol{\Sigma}_i$, where $\boldsymbol{\Sigma}_i = \text{Cov}(\mathbf{Y}_i)$, the GEE parametric regression coefficient estimators are most efficient.

Lin and Carroll (2000) showed that this result does not hold any more for the generalized non-parametric regression model (9.1) using the kernel GEE method. Specifically, we focus on the linear kernel estimator $d = 1$ for simplicity. Assuming the number of repeated measures per person n_i is finite as the number of subjects $N \to \infty$, the kernel GEE estimator $\widehat{\theta}_K(z)$ obtained from the estimating equation (9.2) has the following asymptotic properties:

1. The kernel GEE estimator $\widehat{\theta}_K(z)$ is consistent for any arbitrary working correlation matrix R_i. Its asymptotic bias is

$$E\{\widehat{\theta}_K(z)\} - \theta(z) = \frac{1}{2}h^2\theta^{(2)}(z) + o(h^2).$$

2. The asymptotic variance of $\widehat{\theta}_K(z)$ is

$$\text{Var}\{\widehat{\theta}_K(z)\} = \frac{1}{Nh} C_K(z; \boldsymbol{V}, \boldsymbol{\Sigma}) + o\{(Nh)^{-1}\},$$

where $C_K(z; \boldsymbol{V}, \boldsymbol{\Sigma})$ is some constant and is minimized when $\boldsymbol{V} = \boldsymbol{I}$, that is, the most efficient kernel GEE estimator is obtained by completely ignoring the within-subject correlation.

This result is in contrast to the parametric GEE result. A heuristic explanation of the kernel GEE result is that the kernel GEE estimating equation is local likelihood based. Hence, the resulting kernel GEE estimator $\widehat{\theta}_K(z)$ is local. In other words, at any given target point z, only the observations in the neighborhood of z affect estimation of $\widehat{\theta}_K(z)$. Given that the number of repeated measures n_i is finite ($n_i < \infty$), the probability that the observations in the neighborhood of z come from different subjects is asymptotically equal to 1. Hence, asymptotically the most optimal strategy is to ignore the within-subject correlation. Accounting for within-cluster correlation would result in a less efficient kernel estimator.

Chen and Jin (2005) proposed an improved local polynomial kernel GEE estimator, which uses a degenerate local working covariance matrix. If within-cluster correlation is taken into account, unlike the kernel GEE estimator, Chen and Jin (2005) showed that this modified kernel GEE estimator does not lose efficiency. In other words, the efficiency of this kernel estimator is the same no matter whether one ignores correlation or accounts for correlation, and is hence asymptotically the same as if the correlation is ignored.

9.2.3 The seemingly unrelated kernel estimator

The results of the kernel GEE estimator indicate that the traditional local likelihood principle fails to effectively account for the within-cluster correlation for non-parametric regression in longitudinal data. In order to account for the within-cluster correlation, one needs to move away from the traditional local likelihood principle for constructing the kernel estimator. The seemingly unrelated kernel estimator (Wang, 2003) provides such a vehicle.

Consider a dth-order polynomial kernel estimator. The SUR kernel estimator $\widehat{\theta}_K^*(z)$ solves the kernel estimating equation

$$\sum_{i=1}^{N} \sum_{j=1}^{n_i} K_h(Z_{ij} - z) \boldsymbol{Z}_{ij}(z)' \boldsymbol{V}_i^{-1} \{\boldsymbol{Y}_i - \boldsymbol{\mu}_{i(j)}(z)\} = 0, \tag{9.3}$$

where $\boldsymbol{V}_i$ is the working covariance matrix, $\boldsymbol{Z}_{ij}(z)$ is an $n_i \times (d+1)$ matrix of zeros except that the jth row is $\{1, (Z_{ij} - z), \ldots, (Z_{ij} - z)^d\}'$, and

$$\boldsymbol{\mu}_{i(j)}(z) = \left\{ \widehat{\theta}_K^*(Z_{i1}), \ldots, \widehat{\theta}_K^*(Z_{i,j-1}), \sum_{l=0}^{d} \widehat{\theta}_K^{(l)*}(z), \widehat{\theta}_K^*(Z_{i,j+1}), \ldots, \widehat{\theta}_K^*(Z_{in_i}) \right\}'.$$

An iterative algorithm can be used to solve (9.3). Specifically, denote the SUR kernel estimator at the lth iteration by $\widehat{\theta}_K^{[l]}(z)$; then the updated estimator of SUR kernel $\theta(z)$ at the $(l+1)$th iteration is $\widehat{\theta}_K^{[l+1]}(z) = \widehat{\alpha}_0$, where $\widehat{\boldsymbol{\alpha}} = (\alpha_0, \alpha_1, \ldots, \alpha_d)'$ solves

$$\sum_{i=1}^{N} \sum_{j=1}^{n_i} K_h(Z_{ij} - z) \boldsymbol{Z}_{ij}(z)' \boldsymbol{V}_i^{-1} \{\boldsymbol{Y}_i - \boldsymbol{\mu}_{i(j)}(z; \boldsymbol{\alpha})\} = 0, \tag{9.4}$$

where

$$\boldsymbol{\mu}_{i(j)}(z; \boldsymbol{\alpha}) = \left\{ \widehat{\theta}_K^{[l]}(Z_{i1}), \ldots, \widehat{\theta}_K^{[l]}(Z_{i,j-1}), \sum_{k=0}^{d} (Z_{ij} - z)^k \alpha_k, \widehat{\theta}_K^{[l]}(Z_{i,j+1}), \ldots, \widehat{\theta}_K^{[l]}(Z_{in_i}) \right\}'.$$

At convergence, it gives the SUR kernel estimator $\widehat{\theta}_K^*(z) = \widehat{\alpha}_0$. The Fisher scoring algorithm can be used to iteratively solve (9.4).

The SUR kernel estimator $\widehat{\theta}_K^*(z)$ is consistent. It effectively accounts for the within-cluster correlation. Specifically, for the local linear kernel ($d = 1$), the asymptotic bias and variance of $\widehat{\theta}_K^*(z)$ are

$$E\{\widehat{\theta}_K^*(z)\} = h^2 b_K^*(z) + o(h^2),$$

$$\mathrm{Var}\{\widehat{\theta}_K^*(z)\} = (Nh)^{-1} C_K^*(z; \boldsymbol{V}, \boldsymbol{\Sigma}) + o\{(Nh)^{-1}\},$$

where $b_K^*(z)$ depends on the density of Z and satisfies the Fredholm integral equation of the second kind (see Wang, 2003, Equation [14]; Lin et al., 2004, Equation [5]), and the constant $C_K^*(z; \boldsymbol{V}, \boldsymbol{\Sigma})$ is minimized when the working covariance matrix $\boldsymbol{V}$ equals the true covariance matrix $\boldsymbol{\Sigma}$ of $\boldsymbol{Y}$. It follows that the SUR kernel estimator is more efficient than the working independence kernel estimator in terms of variance comparison. The fact that $b_K^*(z)$ depends on the density of $\boldsymbol{Z}$ makes it difficult to examine whether $\boldsymbol{V} = \boldsymbol{\Sigma}$ minimizes the mean square of $\boldsymbol{\theta}_K^*(z)$. A two-stage method can be used to correct for this (Carroll et al., 2004; Lin and Carroll, 2006). Simulation results show that the SUR kernel estimator outperforms the working independence kernel GEE estimator in mean squared error (MSE) terms (Wang, 2003).

9.2.4 Comparison of the kernel GEE estimator and the SUR kernel estimator

To compare the kernel GEE estimator $\widehat{\theta}_K(z)$ and the SUR kernel estimator $\widehat{\theta}_K^*(z)$, we focus on the special case with an identity link $g(\mu) = \mu$, which is commonly used for normal outcomes, and the average kernel estimator ($d = 0$). This comparison provides us with some useful insight on the performance of these estimators. Specifically, consider the model

$$Y_{ij} = \theta(Z_{ij}) + \epsilon_{ij}, \tag{9.5}$$

where $\epsilon_i = (\epsilon_{i1}, \ldots, \epsilon_{in_i})'$ has mean 0 and true covariance $\boldsymbol{\Sigma}$.

Using the kernel GEE estimating equation (9.2), one can easily show that under (9.5) the average kernel GEE estimator assuming a working covariance matrix V_i has a closed form

$$\widehat{\theta}_K(z) = \frac{\sum_{i=1}^N \mathbf{1}_i' \boldsymbol{K}_i^{1/2}(z) \boldsymbol{V}_i^{-1} \boldsymbol{K}_i^{1/2}(z) \boldsymbol{Y}_i}{\sum_{i=1}^N \mathbf{1}_i' \boldsymbol{K}_i^{1/2}(z) \boldsymbol{V}_i^{-1} \boldsymbol{K}_i^{1/2}(z) \mathbf{1}_i}. \tag{9.6}$$

Because the kernel weights are only evaluated at the target point z, only the Y observations in the neighborhood of z contribute to the GEE kernel estimator $\widehat{\theta}_K(z)$.

Although the SUR kernel estimator $\widehat{\theta}_K^*(z)$ often needs to be obtained by iteratively solving (9.4), a not very difficult task, under the identity link model (9.5), a closed-form solution is available (Lin et al., 2004), namely

$$\widehat{\theta}_K^*(z) = \boldsymbol{K}_{wh}'(z)\{\boldsymbol{I} + (\widetilde{\boldsymbol{V}}^{-1} - \boldsymbol{V}^d)\boldsymbol{Z}_w\}^{-1} \widetilde{\boldsymbol{V}}^{-1} Y, \tag{9.7}$$

where

$$\boldsymbol{K}_{wh}(z) = \{\sum_{i=1}^N \sum_{j=1}^{n_i} K_h(Z_{ij} - z) v_i^{jj}\}^{-1} \{K_h(Z_{11} - z), \ldots, K_h(Z_{Nn_N} - z)\}'$$

denotes an $N \times 1$ vector, $N^* = \sum_{i=1}^N n_i$ is the total number of observations, $\boldsymbol{K}_w = \{\boldsymbol{K}_{wh}(Z_{11}), \ldots, \boldsymbol{K}_{wh}(Z_{n_i n_i})\}'$ is an $N^* \times N^*$ matrix, $\widetilde{\boldsymbol{V}} = \mathrm{diag}(\boldsymbol{V}_1, \ldots, \boldsymbol{V}_n)$, $\widetilde{\boldsymbol{V}}^d = \mathrm{diag}(\boldsymbol{V}_1^d, \ldots, \boldsymbol{V}_n^d)$, $\boldsymbol{V}_i^d = \mathrm{diag}(\boldsymbol{V}_i^{-1}) = \mathrm{diag}(v_i^{jj})$, and $\boldsymbol{Y} = (\boldsymbol{Y}_1', \ldots, \boldsymbol{Y}_N')'$. Write $\widehat{\theta}_K(z) = \boldsymbol{W}(z)'\boldsymbol{Y}$ and $\widehat{\theta}_K^*(z) = \boldsymbol{W}^*(z)\boldsymbol{Y}$. A graphical contrast of the weight functions $\boldsymbol{W}(z)$ and $\boldsymbol{W}^{\widehat{}*}(z)$ for cluster size $n = 3$ can be found in Figure 1(a) of Lin et al. (2004). It shows that the kernel GEE estimator is local, while the SUR kernel is not local and places large weights on the other observations within the same cluster.

9.2.5 SUR kernel estimation in likelihood settings

Lin and Carroll (2006) extended the SUR kernel method to likelihood settings. Suppose the data (Y_{ij}, Z_{ij}) have a likelihood function that can be written as $\ell\{\boldsymbol{Y}_i, \theta(Z_{i1}), \ldots, \theta(Z_{in_i})\}$, where $\theta(z)$ is a scalar smooth unknown function that is evaluated at n_i values of the co-variate Z, $\{Z_{i1}, \ldots, Z_{in_i}\}$. An example of such a likelihood framework is the generalized non-parametric mixed model. Specifically, given random effects $\boldsymbol{b}_i$, suppose the observations Y_{ij} are independent and follow the generalized non-parametric model

$$g(\mu_{ij}) = \theta(Z_{ij}) + \boldsymbol{U}_i'\boldsymbol{b}_i, \tag{9.8}$$

where $g(\cdot)$ is a monotonic link function, $\boldsymbol{U}_i$ is a covariate vector associated with random effects $\boldsymbol{b}_i$, and the random effects are assumed to follow $N\{\boldsymbol{0}, \boldsymbol{D}(\boldsymbol{\gamma})\}$, $\boldsymbol{\gamma}$ is a vector of variance components. This model is an extension of the generalized linear mixed model (Breslow and Clayton, 1993; see also Chapter 4) to model the covariate Z effect non-parametrically.

Assuming that $\boldsymbol{\gamma}$ is known, the log-likelihood for $\theta(\cdot)$ for the ith subject is

$$\ell\{\boldsymbol{Y}_i, \theta(Z_{i1}), \ldots, \theta(Z_{in_i})\} = \log \int \left\{ \exp \sum_{j=1}^{n_i} \ell\{Y_{ij}|\boldsymbol{b}_i; \theta(Z_{ij})\} + \ell(\boldsymbol{b}_i) \right\} d\boldsymbol{b}_i, \tag{9.9}$$

where $\ell(Y_{ij}|\boldsymbol{b}_i)$ is assumed to follow the exponential family (McCullagh and Nelder, 1989).

Consider the linear SUR kernel estimator $(d = 1)$ and assume $n_i = n$. The SUR kernel estimator in such a general likelihood setting can be constructed by solving the following SUR kernel estimating equation:

$$0 = \sum_{i=1}^{N} \sum_{j=1}^{n} K_h(Z_{ij} - z)\boldsymbol{Z}_{ij}(z)$$

$$\times \ell_{j\theta}\{\boldsymbol{Y}_i, Z_i, \widehat{\theta}_K^*(Z_{i1}), \ldots, \widehat{\theta}_K^*(z) + \widehat{\theta}_K^{(1)*}(z)(Z_{ij} - z), \ldots, \widehat{\theta}_K^*(Z_{im})\}, \tag{9.10}$$

where $\widehat{\theta}_K^{(1)*}(z)$ denotes the first derivative of $\widehat{\theta}(z)$, and $\ell_{j\theta}(\cdot) = \partial \ell(\boldsymbol{Y}, Z, \eta_1, \ldots, \eta_n)/\partial \eta_j$. One can solve Equation (9.10) for $\widehat{\theta}_K^*(z)$ in the following iterative fashion. Suppose that the current estimator of $\theta(\cdot)$ at the ℓth step is $\widehat{\theta}_K^{[\ell]}(\bullet)$. Then $\widehat{\theta}_K^{[\ell+1]}(z) = \widehat{\alpha}_0$, where $(\widehat{\alpha}_0, \widehat{\alpha}_1)$ solve

$$0 = \sum_{i=1}^{N} \sum_{j=1}^{n} K_h(Z_{ij} - z)\boldsymbol{Z}_{ij}(z)\ell_{j\theta}\{\boldsymbol{Y}_i, Z_i, \widehat{\theta}_{[\ell]}(Z_{i1}), \ldots, \alpha_0 + \alpha_1(Z_{ij} - z)/h, \ldots, \widehat{\theta}_{[\ell]}(Z_{in})\}.$$

At convergence, $\widehat{\theta}_K^*(z)$ solves the SUR kernel estimating equation (9.10). For the theoretical properties of $\widehat{\theta}_K^*(z)$, see Lin and Carroll (2006).

9.3 Single-covariate non-parametric spline methods for longitudinal data

9.3.1 The generalized smoothing spline estimator

To highlight the key features of the generalized smoothing spline estimator for longitudinal data, we focus here on normal outcomes under model (9.5). Assuming a working covariance matrix $\boldsymbol{V}_i$, the rth-order smoothing spline estimator minimizes

$$-\frac{1}{2N} \sum_{i=1}^{N} \{\boldsymbol{Y}_i - \boldsymbol{\theta}(Z_i)\}'\boldsymbol{V}_i^{-1}\{\boldsymbol{Y}_i - \boldsymbol{\theta}(Z_i)\} - \frac{1}{2}\lambda \int \{\theta^{(r)}(z)\}^2 dz$$

$$= -\frac{1}{2N} \sum_{i=1}^{N} \{\boldsymbol{Y}_i - \boldsymbol{\theta}(Z_i)\}'\boldsymbol{V}_i^{-1}\{\boldsymbol{Y}_i - \boldsymbol{\theta}(Z_i)\} - \frac{1}{2}\lambda\boldsymbol{\theta}'\boldsymbol{\Psi}\boldsymbol{\theta},$$

where $\boldsymbol{\theta}(Z_i) = \{\theta(Z_{i1}), \ldots, \theta(Z_{in_i})\}'$, λ is a smoothing parameter controlling for the trade-off between the goodness of fit and the smoothness of the curve, and $\boldsymbol{\Psi}$ is the smoothing matrix (Green and Silverman, 1994, Equation 2.3). For $r = 2$, we have the commonly used cubic smoothing spline estimator. It follows that the rth-order smoothing spline estimator of $\theta(t)$ is

$$\widehat{\boldsymbol{\theta}}_S = (\widetilde{\boldsymbol{V}}^{-1} + N\lambda\boldsymbol{\Psi})^{-1}\widetilde{\boldsymbol{V}}^{-1}\boldsymbol{Y}, \tag{9.11}$$

where $\widetilde{\boldsymbol{V}} = \text{diag}(\boldsymbol{V}_1, \ldots, \boldsymbol{V}_N)$ and $\boldsymbol{Y} = (\boldsymbol{Y}_1', \ldots, \boldsymbol{Y}_N')'$.

9.3.2 Connection between smoothing spline estimation and linear mixed models

An attractive feature of the smoothing spline estimator is that it has a close connection with linear mixed models (Brumback and Rice, 1998; Wang, 1998; Zhang et al., 1998; Verbyla et al., 1999). Specifically, consider the cubic smoothing spline estimator ($r = 2$), so that $\boldsymbol{\theta}$ can be reparameterized as

$$\boldsymbol{\theta} = \mathbf{1}\delta_0 + \boldsymbol{Z}\delta_1 + \boldsymbol{A}\boldsymbol{a}, \tag{9.12}$$

where, supposing there are no ties, $\boldsymbol{A} = \boldsymbol{L}(\boldsymbol{L}'\boldsymbol{L})^{-1}$, $\boldsymbol{L}$ is an $N^* \times (N^* - 2)$ full-rank matrix satisfying $\boldsymbol{\Psi} = \boldsymbol{L}\boldsymbol{L}'$ and $\boldsymbol{L}'\boldsymbol{Z} = 0$ (Green, 1987), and $N^* = \sum_{i=1}^{N} n_i$ is the total number of observations. Using the fact that $\boldsymbol{\theta}'\boldsymbol{\Psi}\boldsymbol{\theta} = \boldsymbol{a}'\boldsymbol{a}$, the smoothing spline estimator can be obtained equivalently by fitting the linear mixed model (Zhang et al., 1998)

$$\boldsymbol{Y} = \mathbf{1}\delta_0 + \boldsymbol{Z}\delta_1 + \boldsymbol{A}\boldsymbol{a} + \boldsymbol{\epsilon},$$

where $\boldsymbol{\epsilon} = (\boldsymbol{\epsilon}_1', \ldots, \boldsymbol{\epsilon}_N')'$, $\boldsymbol{\epsilon}_i = (\epsilon_{i1}, \ldots, \epsilon_{in_i})'$, $\boldsymbol{\epsilon} \sim N(0, \widetilde{\boldsymbol{V}})$, $\boldsymbol{a} \sim N(\boldsymbol{0}, \tau\boldsymbol{I})$, and $\tau = 1/\lambda$. Denote by $(\widehat{\boldsymbol{\delta}}, \widehat{\boldsymbol{a}})$ the best linear unbiased predictor (BLUP) estimator of the fixed effect $\boldsymbol{\delta} = (\delta_0, \delta_1)'$ and the random effects $\boldsymbol{a}$ under the linear mixed model (9.12). Then the smoothing spline estimator $\widehat{\boldsymbol{\theta}}$ in (9.11) is identical to a linear combination of the BLUP estimators $(\widehat{\boldsymbol{\delta}}, \widehat{\boldsymbol{a}})$ as

$$\widehat{\boldsymbol{\theta}} = \mathbf{1}\widehat{\delta}_0 + \boldsymbol{Z}\widehat{\delta}_1 + \boldsymbol{A}\widehat{\boldsymbol{a}}.$$

The second attractive feature of such a linear mixed-model connection with smoothing spline estimation is that one can treat the smoothing parameter τ as an extra variance component in addition to the variance components $\boldsymbol{\gamma}$ in $\boldsymbol{V}(\boldsymbol{\gamma})$, and estimate both τ and $\boldsymbol{\gamma}$ simultaneously using restricted maximum likelihood (REML) under the linear mixed model (9.12) (Zhang et al., 1998; Wang, 1998).

The third attractive feature of such a linear mixed model and spline smoothing connection is that one can test easily for a parametric model against a non-parametric smoothing spline model using variance component tests. Specifically consider a test for $H_0 : \theta(z) = \alpha_0 + \cdots + \alpha_1 z^{(r-1)}$, that is, $\theta(z)$ is an $(r-1)$th-order polynomial, versus $H_1 : \theta(z)$ is an $(2r-1)$th-order smoothing spline. By examining the spline mixed model (9.12), one can see that this test is equivalent to a variance component test for the smoothing parameter $H_0 : \tau = 0$ versus $H_1 : \tau > 0$. The null hypothesis is on the boundary of the parameter space and the model under the alternative spline mixed model is rather complicated. Score tests (Zhang and Lin, 2003) and likelihood ratio tests (Crainiceanu and Ruppert, 2004) have been proposed.

9.3.3 Theoretical properties of the smoothing spline estimator and its relationship with the SUR kernel estimator

Lin et al. (2004) studied the theoretical properties of the smoothing spline estimator (9.11) and showed that it is asymptotically equivalent to the SUR kernel estimator (9.3). Specifically, both the SUR kernel estimator and the smoothing spline estimator can be constructed iteratively using pseudo-observations. Such a construction provides insight into their asymptotic equivalence.

Denote by

$$Y_{ij}^{[l+1]} = Y_{ij} + (v^{jj})^{-1} \sum_{k \neq j} v^{jk} \left\{ Y_{ik} - \widehat{\theta}^{[l]}(Z_{ik}) \right\}$$

the pseudo-observations at the $(l+1)$th iteration, where $\widehat{\theta}^{[l]}(Z_{ik})$ is the estimator for $\theta(Z_{ik})$ at the lth iteration and v^{jk} is the (j,k)th element of $\boldsymbol{V}^{-1}$. Suppose one fits the classical non-parametric regression model for independent data,

$$Y_{ij}^{(l+1)} = \theta(Z_{ij}) + e_{ij}, \tag{9.13}$$

where the e_{ij} are assumed to be independent and follow a normal distribution with mean 0 and variance v^{jj}. If a standard average kernel estimator, such as the Nadaraya–Watson kernel estimator, is used to calculate $\theta(z)$ under (9.13) at each iteration, then at convergence $\widehat{\theta}(z)$ is the SUR kernel estimator (9.3). If a standard smoothing spline estimator is used to estimate $\theta(z)$ under (9.13) at each iteration, then at convergence $\widehat{\theta}(z)$ is the generalized smoothing spline estimator (9.11).

This iterative pseudo-observation construction provides an intuitive explanation for why the SUR kernel estimator and the generalized smoothing spline estimator are asymptotically equivalent. Specifically, under the working independence model (9.13), at each iteration, the standard average kernel estimator and the standard smoothing spline estimator are asymptotically equivalent (Silverman, 1984). At convergence they become the SUR kernel estimator $\widehat{\theta}_K^*(z)$ and the generalized smoothing spline estimator $\widehat{\theta}_S(z)$. Hence, $\widehat{\theta}_K^*(z)$ and $\widehat{\theta}_S(z)$ are asymptotically equivalent. Linton et al. (2004) proposed an estimator using different pseudo-observations that is more efficient than the kernel GEE but less efficient than the SUR kernel. Huggins (2006) gave an alternative construction of pseudo-observations that yields the SUR kernel estimator $\widehat{\theta}_K^*(z)$ at convergence.

Write the SUR kernel estimator as $\widehat{\theta}_K^*(z) = \boldsymbol{W}_K^*(z)'\boldsymbol{Y}$ and the smoothing spline estimator as $\widehat{\theta}_S(z) = \boldsymbol{W}_S(z)'\boldsymbol{Y}$. A graphical contrast of the weight functions $W_K^*(z)$ and $W_S(z)$ for cluster size $m = 3$ can be found in Figure 1 of Lin et al. (2004). This reveals that their shapes are very similar and non-local, and provides numerical evidence that they are asymptotically equivalent and that both are non-local estimators. Such a non-locality property is dramatically different from well-known local properties of non-parametric estimators for independent data.

Using the pseudo-observations in (9.13), Lin et al. (2004) studied the theoretical properties of the generalized smoothing spline estimator $\widehat{\theta}_S(z)$. Suppose $n_i = n$ and $\boldsymbol{V}_i = \boldsymbol{V}$. For the rth-order smoothing spline estimator, its asymptotic bias and variance are

$$E\{\widehat{\theta}_S(z)\} = (-1)^{r-1} h^{2r}(z) b_S(z) + o\{h^{2r}\},$$

$$\mathrm{Var}\{\widehat{\theta}_S(z)\} = (Nh)^{-1}(z) C_S(z; \boldsymbol{V}, \boldsymbol{\Sigma}) + o\{(Nh)^{-1}\},$$

where

$$h(z) = \left[\frac{\lambda}{\sum_{j=1}^n v^{jj} f_j(z)} \right]^{1/2r}$$

is the equivalent bandwidth, $f_j(z)$ is the density of Z_j, $b_S(z)$ satisfies the Fredholm integral equation of the second kind, and the constant $C_S(z; \boldsymbol{V}, \boldsymbol{\Sigma})$ is minimized when the working covariance matrix $\boldsymbol{V}$ equals the true covariance $\boldsymbol{\Sigma}$ of $\boldsymbol{Y}_i$.

These results show that the generalized smoothing spline estimator is a higher-order SUR kernel estimator; for example, the cubic smoothing spline estimator $\widehat{\theta}_S(z)$ is a fourth-order SUR kernel estimator. The generalized smoothing spline estimator is consistent, and the most efficient generalized smoothing spline estimator is obtained by accounting for the within-cluster correlation. Note that the asymptotic results for the SUR kernel estimator and the generalized smoothing spline estimator show that their asymptotic biases depend on the design density, which might sometimes be regarded as less desirable. This can be corrected by a two-step approach (Carroll et al., 2004; Lin and Carroll, 2006).

There has been limited research on the theoretical properties of the REML estimators of the smoothing parameter λ. Numerical results show that joint estimation of λ and the variance components γ using REML through the mixed-model connection (Section 9.3.2) works very well. However, the theoretical properties of such REML estimation are currently unknown. Gu and Ma (2005) studied the theoretical properties of generalized cross-validation in (9.5) and showed that it yielded optimal smoothing.

9.3.4 P-splines and regression splines for longitudinal data

A smoothing spline uses all data points as knots. Hence, for very large data sets, its computation may be expensive. A P-spline (Eilers and Marx, 1996; Ruppert, Wand, and Carroll, 2003) is an attractive alternative. As described in Section 8.2.3, P-splines do not use all the data points as knots, but assign a moderate number of knots (e.g., $L = 20$ to 30) and estimate the moderate number of regression coefficients using a penalized likelihood. The selected number of knots L is generally much smaller than the sample size. One often places knots using percentiles of the observed covariate values $\{Z_{ij}\}$, although for B-splines equally spaced knots are the rule. Denote by $\{(z - z_1)^3_+, \ldots, (z - z_L)^3_+\}$ the plus function bases constructed using the knots $z_1, \ldots, z_L$, where $a_+ = a$ if $a > 0$ and 0 otherwise. Then $\theta(z)$ can be approximated by

$$\theta(z) = \alpha_0 + \alpha_1 z + \alpha_2 z^2 + \alpha_3 z^3 + \sum_{l=1}^{L}(z - z_l)^3_+ \alpha_{3+l} = \boldsymbol{B}(z)'\boldsymbol{\alpha},$$

where $\boldsymbol{B}(z) = \{1, z, z^2, z^3, (z - z_1)^3_+, \ldots, (z - z_L)^3_+\}'$. Under model (9.5), one estimates the regression coefficients $\boldsymbol{\alpha} = (\alpha_1, \ldots, \alpha_L)'$ using the penalized log-likelihood

$$-\frac{1}{2}\sum_{i=1}^{N}\{\boldsymbol{Y}_i - \boldsymbol{B}_i\boldsymbol{\alpha}\}'V^{-1}\{\boldsymbol{Y}_i - \boldsymbol{B}_i\boldsymbol{\alpha}\} - \frac{1}{2\tau}\boldsymbol{\alpha}'_P\boldsymbol{\alpha}_P, \tag{9.14}$$

where $\boldsymbol{B}_i = \{\boldsymbol{B}(Z_{i1}), \ldots, \boldsymbol{B}(Z_{in_i})\}'$, $\boldsymbol{\alpha}_P = (\alpha_4, \ldots, \alpha_{L+3})'$, and τ is a tuning parameter. Denote by $\widehat{\boldsymbol{\alpha}}$ the maximizer of (9.14). The P-spline estimator of $\theta(z)$ is $\widehat{\theta}_P(z) = \boldsymbol{B}(z)'\widehat{\boldsymbol{\alpha}}$. A key advantage of P-splines is their computation, since estimation of $\boldsymbol{\alpha}$ involves inverting an $(L+4) \times (L+4)$ matrix. Parallel to the smoothing spline results, there is a close connection between P-splines and linear mixed models (Ruppert, Wand, and Carroll, 2003). Specifically, $\widehat{\boldsymbol{\alpha}}$ can be calculated by fitting the linear mixed model

$$Y_{ij} = \alpha_0 + \alpha_1 Z_{ij} + \alpha_2 Z^2_{ij} + \alpha_3 Z^3_{ij} + \sum_{l=1}^{L}\alpha_{l+3}(Z_{ij} - z_l)^3_+ + \epsilon_{ij},$$

where $\boldsymbol{\alpha}_P = (\alpha_4, \ldots, \alpha_{L+3})'$ is treated as random effects following $N(\boldsymbol{0}, \tau\boldsymbol{I})$ and the smoothing parameter τ is treated as an extra variance component and estimated using REML. If B-spline bases are used, the penalty uses second-order differences of the αs (Fahrmeir and Lang, 2001). The theoretical properties of P-splines are not well understood, especially for longitudinal data.

Regression splines that use a small number of knots (see Section 8.2.3) have been proposed for non-parametric regression for longitudinal data under model (9.5); see Rice and Wu (2001) and Huang, Wu, and Zhou (2002). A regression spline approximates $\theta(z)$ by $\theta(z) = \sum_{l=1}^{L} B_l(z)\alpha_l$, where L is small (e.g., $L = 3$ to 6), and $\{B_l(z)\}$ is a set of base functions (e.g., B-spline bases). Under the normal longitudinal model, one estimates α_l using weighted least squares under the model

$$Y_{ij} = \sum_{l=1}^{L} \alpha_l B_l(Z_{ij}) + \epsilon_{ij}.$$

Under the normal model (9.5), Carroll et al. (2004) compared, via simulations, the performance of the smoothing spline, the P-spline, and the regression spline estimators. In their simulations they found that smoothing splines and P-splines outperform regression splines. These authors also studied the asymptotic properties of the regression spline estimator when the $B_l(z)$ are step functions.

9.4 Extensions of single-covariate non-parametric models for longitudinal data

In this section we provide an overview of several extensions of the univariate non-parametric models discussed in Section 9.2 and Section 9.3. They include functional mixed models, varying-coefficient mixed models, generalized additive mixed models (GAMMs), and self-modeling mixed models.

9.4.1 Functional mixed models

Functional mixed models are an extension of the non-parametric mixed model (9.5) to model explicitly the within-subject correlation using random effects and random stochastic processes. They allow one to estimate a population curve and subject-specific curves (Zhang et al., 1998; Guo, 2002). In this case, the covariate z is the time t in a longitudinal study. Denote by $\theta_i(t)$ a subject-specific curve for subject i. Partition $\theta_i(t)$ as $\theta_i(t) = \theta(t) + S_i(t)$, where $\theta(t)$ is a population curve and $S_i(t)$ is a departure of the subject-specific curve $\theta_i(t)$ from the population curve $\theta(t)$. Rewriting model (9.5) by decomposing the residuals ϵ_{ij} into $S_i(t) + e_i(t)$, we have

$$Y_i(t) = \theta(t) + \boldsymbol{U}_i(t)'\boldsymbol{b}_i + S_i(t) + e_i(t), \qquad (9.15)$$

where $\boldsymbol{b}_i$ is a vector of random effects (e.g., random intercepts and slopes) which are modeled as $N\{\boldsymbol{0}, \boldsymbol{D}(\boldsymbol{\gamma}_1)\}$, $\boldsymbol{U}_i(t)$ is a covariate vector associated with $\boldsymbol{b}_i$, $S_i(t)$ is a mean-zero random Gaussian process with covariance matrix $\mathrm{Cov}\{S_i(t), S_i(s)\} = \Gamma(t, s; \boldsymbol{\gamma}_2)$ that depends on the covariance parameters $\boldsymbol{\gamma}_2$, and $e_i(t)$ is an independent measurement error following $N(0, \sigma^2)$.

Multiple choices of the covariance matrix $\Gamma(t, s)$ of $S_i(t)$ have been proposed (Taylor, Cumberland, and Sy, 1994; Shi, Weiss, and Taylor, 1996; Zeger and Diggle, 1994; Zhang et al., 1998; Guo, 2002), such as the Ornstein–Uhlenbeck (OU) process, the Wiener process, the integrated Wiener process, the integrated OU process that is suitable for growth curve analysis, and the ante-dependence process for equally spaced time points. The OU process assumes an exponential decay correlation. If the integrated Wiener process is assumed, the resulting subject-specific curves $\widehat{S}_i(t)$ are cubic smoothing splines. Zhang et al. (1998) considered spline estimation in (9.15) by using the connection between splines and mixed models given in (9.12), and approached estimation in (9.15) using an augmented mixed-effects model. BLUPs can be used to estimate the population curve $\widehat{\theta}(t)$ and subject-specific curves $\widehat{\theta}_i(t) = \widehat{\theta}(t) + \boldsymbol{U}_i(t)'\widehat{b}_i + \widehat{S}_i(t)$, where $\widehat{b}_i$ and $\widehat{S}_i(t)$ are the BLUPs of $\boldsymbol{b}_i$ and $S_i(t)$ under model (9.15). The standard errors of the smoothing spline estimator $\widehat{\theta}(t)$ can be estimated using the frequentist standard errors (SEs) by treating $\theta(t)$ as a fixed unknown smooth function or using Bayesian SEs by treating $\theta(t)$ as an unknown random function generated from the smoothing spline prior (Wahba, 1983; Zhang et al., 1998). The Bayesian SEs in fact account for the bias in $\widehat{\theta}(t)$ and reflect the MSE of $\widehat{\theta}(t)$. Several other approaches have been proposed for estimation in (9.15), such as the state-space method (Guo, 2002) and the local-polynomial kernel method (Wu and Zhang, 2002). The wavelet functional mixed model was developed by Morris and Carroll (2006).

9.4.2 Generalized additive mixed models

Generalized additive mixed models (Lin and Zhang, 1999) are an extension of generalized linear mixed models (GLMMs) (Breslow and Clayton, 1993; see also Chapter 4) to allow the parametric fixed effects to be modeled non-parametrically using additive smooth functions, in a similar spirit to Hastie and Tibshirani (1990). Specifically, suppose $\{Z_{ij1}, \ldots, Z_{ijq}\}$ are q covariates associated with Y_{ij}. GAMMs are an additive extension of the generalized non-parametric mixed model (9.8) given as

$$g(\mu_{ij}) = \beta_0 + \theta_1(Z_{ij1}) + \cdots + \theta_q(Z_{ijq}) + \boldsymbol{U}'_{ij}\boldsymbol{b}_i, \tag{9.16}$$

where $\theta_k(\cdot)$, $k = 1, \ldots, q$, is an unknown centered smooth function of the kth covariate Z_k, and $\boldsymbol{b}_i$ is a vector of random effects following $N\{\boldsymbol{0}, D(\boldsymbol{\gamma})\}$.

The smoothing spline estimators of $\{\theta_1(\cdot), \ldots, \theta_q(\cdot)\}$ maximize the penalized integrated log-likelihood

$$\sum_{i=1}^{N} \ell\{\boldsymbol{Y}_i, \boldsymbol{Z}_{i1}, \ldots, \boldsymbol{Z}_{iq}; \theta_1(\cdot), \ldots, \theta_q(\cdot)\} - \sum_{k=1}^{q} \frac{1}{2\tau_k} \int \{\theta_k^{(2)}(z)\}^2 dz, \tag{9.17}$$

where $\ell(\cdot)$ is the integrated log-likelihood that takes the same form as the right-hand side of (9.9) with $\ell(Y_{ij}|\boldsymbol{b}_i)$ defined under the GAMM (9.16). Approximating the integrated log-likelihood in the first term of (9.17) using the Laplace approximation (Breslow and Clayton, 1993) and applying the spline mixed-model connection (9.12) for each $\boldsymbol{\theta}_k = \boldsymbol{Z}_k\boldsymbol{\beta}_k + \boldsymbol{A}_k\boldsymbol{a}_k$, GAMM model (9.16) can be formulated as a GLMM,

$$g(\boldsymbol{\mu}) = \boldsymbol{1}\sum_{k=0}^{q} \boldsymbol{Z}_k\boldsymbol{\beta}_k + \sum_{k=1}^{q} \boldsymbol{A}_k\boldsymbol{a}_k + \boldsymbol{U}\boldsymbol{b},$$

where $\boldsymbol{a}_k \sim N(\boldsymbol{0}, \tau_k\boldsymbol{I})$ and $\boldsymbol{b} \sim N\{\boldsymbol{0}, \text{diag}\{D(\boldsymbol{\gamma})\}$ are random effects and τ_k is the smoothing parameter for $\theta_k(z_k)$.

Lin and Zhang (1999) calculated approximate smoothing spline estimators of the $\boldsymbol{\theta}_k$ by maximizing the double penalized quasi-likelihood (DPQL)

$$\sum_{i=1}^{N} \sum_{j=1}^{n_i} \ell(Y_{ij}|\boldsymbol{a}_1, \ldots, \boldsymbol{a}_q, \boldsymbol{b}) - \sum_{k=1}^{q} \frac{1}{2\tau_k} \boldsymbol{a}'_k\boldsymbol{a}_k - \frac{1}{2}\boldsymbol{b}'D^{-1}\boldsymbol{b},$$

with respect to $\boldsymbol{\beta}_k, \boldsymbol{a}_k$, and $\boldsymbol{b}$. The resulting DPQL estimator of $\boldsymbol{\theta}_k$ is $\widehat{\boldsymbol{\theta}}_k = \boldsymbol{Z}_k\widehat{\boldsymbol{\beta}}_k + \boldsymbol{A}_k\widehat{\boldsymbol{a}}_k$, where $\widehat{\boldsymbol{\beta}}_k$ and $\widehat{\boldsymbol{a}}_k$ are the maximum DPQL estimators. The smoothing parameters $\{\tau_k\}$ and the variance components $\boldsymbol{\gamma}$ can be estimated jointly using REML-type likelihood approximations.

Several authors have investigated alternative estimation methods in GAMMs, and have proposed several extensions of GAMMs. Zhang and Davidian (2004) proposed likelihood and conditional likelihood inference in GAMMs. Fahrmeir and Lang (2001) considered Bayesian inference in GAMMs using Markov random field priors. Wood (2006) proposed tensor estimation in GAMMs in additive thin-spline models.

9.4.3 Self-modeling non-linear mixed models

Non-parametric mixed models (9.15) and GAMMs (9.16) introduce random effects that are additive to non-parametric functions $\theta(z)$. Several authors proposed self-modeling non-linear mixed models, where unobserved random effects $\boldsymbol{b}_i$ enter inside the non-parametric function

$\theta(z)$ (Ke and Wang, 2001; Brumback and Lindstrom, 2004). Following the notation in (9.15), suppose the covariate z is time t. A popular self-modeling non-linear mixed model is

$$Y_i(t_{ij}) = \beta_0 + a_i + b_i\theta\{(t_{ij} - c_i)/d_i\} + e_i(t_{ij}), \qquad (9.18)$$

where $\{a_i, b_i, c_i, d_i\}$ are subject-specific random effects following $N(0, \boldsymbol{D})$ and $\theta(t)$ is an unknown smooth function. This model is an extension of self-modeling non-linear regression, also known as a shape-invariant model (Lawton, Sylvester, and Maggio, 1972), where $\{a_i, b_i, c_i, d_i\}$ are treated as fixed effects. If $\theta(t)$ is a parametric non-linear function, (9.18) corresponds to non-linear mixed models (Lindstrom and Bates, 1990; see also Chapter 5). Such models are useful in analyzing growth curves and speech movement data.

Ke and Wang (2001) and Brumback and Lindstrom (2004) proposed multi-stage iterative estimation procedures using several steps of approximation. Specifically, given β_0 and the random effects $\{a_i, b_i, c_i, d_i\}$, one estimates $\theta(\cdot)$ using a smoothing spline via a penalized likelihood. Given $\widehat{\theta}(\cdot)$, one treats (9.18) as a non-linear mixed model and estimates β_0, $\{a_i, b_i, c_i, d_i\}$, and variance components $\boldsymbol{D}$ using the BLUP/REML method, an approximate maximum likelihood estimation method of Lindstrom and Bates (1990). One iterates between these two steps until convergence. Lin and Zhang (2001) discussed theoretical difficulties in inference in such self-modeling non-linear mixed models.

9.4.4 Varying-coefficient models

Several authors have proposed varying-coefficient models for longitudinal data. A simple varying-coefficient model for Gaussian longitudinal outcomes can be written as (Hoover et al., 1998; Wu, Chiang, and Hoover, 1998)

$$Y_i(t_{ij}) = \boldsymbol{X}'_i(t_{ij})\boldsymbol{\theta}(t_{ij}) + e_i(t_{ij}), \qquad (9.19)$$

where $\boldsymbol{X}_i(t)$ is a $p \times 1$ vector of covariates and $\boldsymbol{\theta}(t) = \{\theta_0(t), \ldots, \theta_p(t)\}$ is a $p \times 1$ vector of unknown smooth functions. If $\boldsymbol{X}_i(t) = 1$, model (9.19) reduces to the simple non-parametric model (9.5). This model extends the varying-coefficient model of Hastie and Tibshirani (1993) to longitudinal data by allowing the errors $e_i(t_{ij})$ to be correlated within the same subject.

Several methods have been proposed to estimate the non-parametric functions $\theta_k(t)$ ($k = 0, \ldots, p$). Hoover et al. (1998), Wu, Chiang, and Hoover (1998), and Fan and Zhang (2000) considered local polynomial kernel methods by ignoring within-subject correlation. Huang, Wu, and Zhou (2002, 2004) used regression spline methods while accounting for the within-subject correlation. Eubank et al. (2004) considered smoothing spline methods. Lin and Ying (2001) used martingale methods by casting the model within a survival analysis framework. For extensions of (9.19) to mixed-effects models and non-Gaussian outcomes, see Wu and Liang (2004) and Zhang (2004).

9.5 The longitudinal progesterone data example

We illustrate the non-parametric regression methods using data from a longitudinal progesterone study (Sowers et al., 1995; Zhang et al., 1998). This study involved 34 healthy women whose urine samples were collected in a menstrual cycle. Their urinary progesterone was assayed on alternate days, with 11–28 observations over time and an average of 14.5 observations per woman. Progesterone is a reproductive hormone responsible for normal fertility and menstrual cycling. The menstrual cycle lengths of these women ranged from 23 to 56 days, with an average of 29.6 days. Following Sowers et al. (1995), each woman's menstrual cycle length was standardized uniformly to a reference 28-day cycle of interest. The standardization resulted in 98 distinct time points. A log transformation was applied

Table 9.1 Estimates of the Variance Components and Smoothing Parameter in the Progesterone Data

Parameter	Estimate	SE
τ	5.74	4.21
D_{11}	0.47	0.20
D_{12}	-0.04	0.18
D_{22}	0.003	0.32
ξ	0.90	0.34
σ^2	0.55	0.04

to the progesterone level to make a normality assumption more plausible. The scientific interest was to estimate the time course of the progesterone level in a menstrual cycle.

We fit a functional mixed-effects model (see Section 9.4.1),

$$Y_{ij} = \theta(t_{ij}) + b_{0i} + t_{ij}b_{1i} + S_i(t_{ij}) + e_i(t_{ij}), \tag{9.20}$$

to the data, where Y_{ij} is log progesterone for subject i at the jth time point t_{ij}, $\theta(\cdot)$ is an unknown smooth function, (b_{0i}, b_{1i}) is a vector distributed as $N(\mathbf{0}, \mathbf{D})$, $S_i(\cdot)$ is a random mean-zero integrated Wiener stochastic process with $\mathrm{Cov}\{S_i(t), S_i(t')\} = \xi\Gamma(t, t') = \xi t^2(3t' - t)/6$ for $t' \geq t$, and $e_i(\cdot)$ follows $N(0, \sigma^2)$. We estimate $\theta(\cdot)$ using a smoothing spline by using its mixed-model representation (9.12) and fit (9.20) using an augmented mixed model. The integrated Wiener process of $S_i(\cdot)$ allows the population curve $\theta(t)$ and subject-specific curves $\theta_i(t) = \theta(t) + b_{0i} + b_{1i}t + S_i(t)$ to be estimated as smoothing splines. The smoothing parameters and the variance components are estimated simultaneously using REML (Zhang et al., 1998).

Table 9.1 gives the estimated smoothing parameters and the variance components. The results show strong within-subject correlation and significant between-subject variability. Figure 9.1 shows the estimated population curve $\widehat{\theta}(t)$ and its pointwise 95% frequentist and Bayesian CIs. The frequentist CIs use the frequentist SEs that treat $\theta(t)$ as an unknown smooth function, while the Bayesian CIs use the Bayesian SEs that treat $\theta(t)$ as an unknown random function from a smoothing spline prior and reflect the MSE of $\widehat{\theta}(t)$.

The results show that progesterone level is low and stable in the first half of a menstrual cycle and quickly increases after ovulation, reaching a peak around the 23rd reference day and then decreasing. A plausible biological explanation of the latter bell-shape curve is

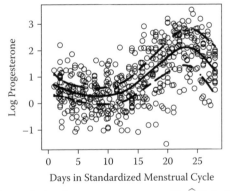

Figure 9.1 The estimated population log progesterone profile $\widehat{\theta}(t)$ (solid line) over a standardized menstrual cycle for the progesterone data and estimated pointwise Bayesian CIs (dotted lines) and frequentist CIs (dashed lines). The Bayesian CIs and frequentist CIs overlap, indicating they are similar for this data set.

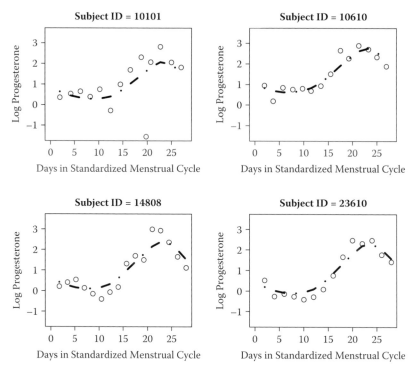

Figure 9.2 The observed log progesterone data and the estimated BLUP individual profiles for four randomly chosen subjects.

that evaluated progesterone level stimulates the secretion by endometrium, which provides necessary nutrients for the development of the ovum. If conception does not occur, the endometrium is sloughed and presented as menstrual bleeding. The frequentist and Bayesian CIs are similar. The simulation study of Zhang et al. (1998) supports these observations and further shows that the Bayesian CI has a slightly better coverage than the frequentist CI. These authors analyzed the data using an alternative non-homogeneous Ornstein–Uhlenbeck process and showed similar results.

Figure 9.2 plots the observed progesterone data for four randomly chosen subjects and overlays the BLUP estimates of their individual curves calculated using $\widehat{\theta}_i(t) = \widehat{\theta}(t) + \widehat{b}_{0i} + \widehat{b}_{1i}t + \widehat{S}_i(t)$. One can see that the predicted curves capture the observed data for each individual quite well.

9.6 Semi-parametric regression for longitudinal data

The models discussed in Section 9.2 through Section 9.4 model all the covariate effects non-parametrically. It is of practical interest in many applications to model some covariate effects parametrically (e.g., discrete covariates and covariates whose effects are well understood), while other covariate effects are modeled non-parametrically. Zeger and Diggle (1994) considered CD4 count data from the Multicenter AIDS Cohort Study (MACS), where the time course of CD4 is complex. It is desirable to model the CD4 count time trajectory non-parametrically, while other covariates (e.g., age, drug use, and number of sex partners) are more naturally modeled parametrically. For details about the CD4 data, see Section 9.7. A challenge in estimation in such semi-parametric models is that some parameters are finite dimensional while other parameters are infinite dimensional. Statistical inference in such semi-parametric models should therefore be approached with caution. In Section 9.6.1 we first consider semi-parametric regression in marginal models using profile/kernel estimating equations, and extensions of such methods to likelihood settings. We then discuss semi-parametric regression with the non-parametric function estimated using splines.

9.6.1 Semi-parametric regression in marginal models using profile/kernel estimating equations

Suppose Y_{ij} is the outcome for the jth observation ($j = 1, \ldots, n_i$) on the ith subject ($i = 1, \ldots, N$), $\boldsymbol{X}_{ij}$ is a $p \times 1$ vector of covariates whose effects are modeled parametrically, and Z_{ij} is a scalar covariate whose effect is modeled non-parametrically. Consider a semi-parametric marginal model,

$$g(\mu_{ij}) = \boldsymbol{X}'_{ij}\boldsymbol{\beta} + \theta(Z_{ij}), \tag{9.21}$$

where $g(\cdot)$ is a link function, $\boldsymbol{\beta}$ is a $p \times 1$ vector of regression coefficients, and $\theta(\cdot)$ is an unknown smooth function.

Zeger and Diggle (1994) first considered this semi-parametric model for normally distributed outcomes, where they estimated $\theta(z)$ using local polynomial kernel smoothing by ignoring the within-subject correlation and estimated $\boldsymbol{\beta}$ using weighted least squares by accounting for within-subject correlation and plugging in the working independence kernel estimator $\hat{\theta}_K(z)$. Lin and Carroll (2001) considered profile/kernel estimation in the general marginal semi-parametric model (9.21), where $\theta(z)$ is estimated using the kernel GEE method (see Section 9.2) and $\boldsymbol{\beta}$ is estimated using the profile method.

Specifically, given $\boldsymbol{\beta}$, using the notation in (9.2), $\boldsymbol{\alpha}$ is estimated by solving the local polynomial kernel GEE estimating equation

$$\sum_{i=1}^{N} \boldsymbol{Z}_i(z)'\boldsymbol{\Delta}_i(z)\boldsymbol{K}_{ih}^{1/2}(z)\boldsymbol{V}_{1i}^{-1}(z)\boldsymbol{K}_{ih}^{1/2}(z\{\boldsymbol{Y}_i - \boldsymbol{\mu}_i(z;\boldsymbol{\beta})\} = 0, \tag{9.22}$$

where the jth component of $\boldsymbol{\mu}_i(z;\boldsymbol{\beta})$ is $g^{-1}\{\boldsymbol{X}'_{ij}\boldsymbol{\beta} + \boldsymbol{Z}_{ij}(z)'\boldsymbol{\alpha}\}$, and V_{1i} is a working covariance matrix. Denote by the resulting local polynomial kernel GEE estimator of $\theta(z)$ by $\hat{\theta}_K(z;\boldsymbol{\beta}) = \hat{\alpha}_0$. The profile/kernel estimator of $\boldsymbol{\beta}$ solves the profile estimating equation

$$\sum_{i=1}^{N} \frac{\partial \boldsymbol{\mu}\{\boldsymbol{X}_i\boldsymbol{\beta} + \hat{\boldsymbol{\theta}}(\boldsymbol{Z}_i;\boldsymbol{\beta})\}'}{\partial \boldsymbol{\beta}} \boldsymbol{V}_{2i}^{-1}(\boldsymbol{X}_i, \boldsymbol{Z}_i)[\boldsymbol{Y}_i - \boldsymbol{\mu}\{\boldsymbol{X}_i\boldsymbol{\beta} + \hat{\boldsymbol{\theta}}(\boldsymbol{Z}_i;\boldsymbol{\beta})\}] = 0, \tag{9.23}$$

where $\hat{\boldsymbol{\theta}}(\boldsymbol{Z}_i;\boldsymbol{\beta}) = \{\hat{\theta}(Z_{i1};\boldsymbol{\beta}), \ldots, \hat{\theta}(Z_{in_i};\boldsymbol{\beta})\}'$ and $\boldsymbol{V}_{2i}$ is a working covariance matrix. A striking result shown in Lin and Carroll (2001) is that unless one ignores the within-cluster correlation in both the kernel GEE estimating equation (9.22) and the profile estimating equation (9.23) by assuming $\boldsymbol{V}_{1i} = \boldsymbol{V}_{2i} = \boldsymbol{I}$, or undersmooths $\theta(z)$, the profile estimator of the regression coefficient vector $\boldsymbol{\beta}$ is generally not $\sqrt{N}$-consistent. If one accounts for within-cluster correlation, the sandwich method for estimating $\text{Cov}(\hat{\boldsymbol{\beta}})$ fails. These results show that the traditional profile local polynomial kernel method fails for longitudinal data if one attempts to account for within-subject correlation. Lin and Carroll (2001) derived the semi-parametric efficient score of $\boldsymbol{\beta}$ and showed that even when one undersmooths $\theta(z)$ using the kernel GEE estimating equation (9.22), the profile estimator of $\boldsymbol{\beta}$ is still not semi-parametric efficient.

In view of the failure of the general profile local polynomial kernel method for estimation in the semi-parametric model (9.21), Wang, Carroll, and Lin (2004) proposed to estimate $\theta(z)$ using the SUR kernel method instead and estimate $\boldsymbol{\beta}$ using the profile estimating equation (9.23). Specifically, one replaces the local polynomial kernel estimating equation (9.22) by the SUR kernel estimating equation (9.3). For given $\boldsymbol{\beta}$, denote by $\hat{\theta}_K^{[l]}(z;\boldsymbol{\beta})$ the

SUR kernel estimator at the lth iteration; then the updated estimator of the SUR kernel $\theta(z)$ at the $(l+1)$th iteration is $\widehat{\theta}_K^{[l+1]}(z; \boldsymbol{\beta}) = \widehat{\alpha}_0$, where $\widehat{\boldsymbol{\alpha}} = (\alpha_0, \alpha_1, \ldots, \alpha_d)'$ solves

$$\sum_{i=1}^{N} \sum_{j=1}^{n_i} K_h(Z_{ij} - z) \boldsymbol{Z}_{ij}(z)' \boldsymbol{V}_{1i}^{-1} \{ \boldsymbol{Y}_i - \mu_{i(j)}(z; \boldsymbol{\alpha}, \boldsymbol{\beta}) \} = 0,$$

where $\boldsymbol{V}_1$ is the working covariance matrix and

$$\mu_{i(j)}(z; \boldsymbol{\alpha}, \boldsymbol{\beta}) = \left\{ \widehat{\theta}_K^{[1]}(Z_{i1}; \boldsymbol{\beta}), \ldots, \widehat{\theta}_K^{[l]}(Z_{i,j-1}; \boldsymbol{\beta}), \quad \sum_{k=0}^{d} (Z_{ij} - z)^k \alpha_k, \right.$$

$$\left. \widehat{\theta}_K^{[l]}(Z_{i,j+1}; \boldsymbol{\beta}), \ldots, \widehat{\theta}_K^{[l]}(Z_{in_i}; \boldsymbol{\beta}) \right\}'.$$

Denote the resulting SUR kernel estimator at convergence by $\widehat{\theta}_K^*(z; \boldsymbol{\beta})$. The profile estimator of $\boldsymbol{\beta}$ is obtained by replacing the polynomial kernel estimator $\widehat{\theta}_K(z; \boldsymbol{\beta})$ in the profile estimating equation (9.23) by the SUR kernel estimator $\widehat{\theta}_K^*(z; \boldsymbol{\beta})$. The sandwich method can be used to estimate the covariance of the resulting profile estimator $\widehat{\boldsymbol{\beta}}^*$. Wang et al. (2004) showed that such a SUR/profile SUR kernel estimator $\widehat{\boldsymbol{\beta}}^*$ is consistent for any working covariance matrix $\boldsymbol{V}_{1i} = \boldsymbol{V}_{2i} = \boldsymbol{V}_i$ without the need for undersmoothing of $\theta(z)$ and is most efficient when $\boldsymbol{V}_i$ equals the true covariance of $\boldsymbol{Y}_i$. When $\boldsymbol{Y}_i$ is normally distributed, Wang et al. (2004) showed that the profile SUR kernel estimator $\widehat{\boldsymbol{\beta}}^*$ is semi-parametric efficient and reaches the semi-parametric efficiency bound derived by Lin and Carroll (2001). In fact, one can further show that the SUR kernel estimator $\widehat{\boldsymbol{\beta}}^*$ is semi-parametric efficient under the semi-parametric marginal mean model (9.21).

If $\boldsymbol{Y}_i$ is normally distributed, both the profile local polynomial kernel estimator $\widehat{\boldsymbol{\beta}}$ and the profile SUR kernel estimator $\widehat{\boldsymbol{\beta}}^*$ have closed-form solutions. Specifically, consider the model

$$E(Y_{ij}) = \theta(Z_{ij}) + \boldsymbol{X}_{ij}' \boldsymbol{\beta}. \tag{9.24}$$

For simplicity, consider the local kernel estimator $\widehat{\theta}_K(z; \boldsymbol{\beta})$ and the SUR kernel estimator $\widehat{\theta}_K^*(z; \boldsymbol{\beta})$. Define

$$\widehat{\boldsymbol{\theta}}_K(\boldsymbol{\beta}) = \{ \widehat{\theta}_K(Z_{11}; \boldsymbol{\beta}), \ldots, \widehat{\theta}_K(Z_{Nn_N}; \boldsymbol{\beta}) \}',$$

and $\widehat{\boldsymbol{\theta}}_K^*(\boldsymbol{\beta})$ similarly. Using the results in Section 9.2.4, their closed-form expressions can be written as $\widehat{\boldsymbol{\beta}}_K = \boldsymbol{A}_K \boldsymbol{Y}$ and $\widehat{\boldsymbol{\beta}}_K^* = \boldsymbol{A}_K^* \boldsymbol{Y}$, where the rows of A_K are given by the coefficients of the Ys in (9.6) and the rows of A_K^* are given by the coefficients of the Ys in (9.7) for the average kernel ($d = 0$). It follows that the profile local kernel estimator $\widehat{\theta}_K(z; \boldsymbol{\beta})$ and the SUR kernel estimator $\widehat{\theta}_K^*(z; \boldsymbol{\beta})$ both have closed-form solutions as

$$\widehat{\boldsymbol{\beta}}_K = \{ \boldsymbol{X}'(\boldsymbol{I} - \boldsymbol{A}_K) \widetilde{\boldsymbol{V}}^{-1}(\boldsymbol{I} - \boldsymbol{A}_K) \boldsymbol{X} \}^{-1} \boldsymbol{X}'(\boldsymbol{I} - \boldsymbol{A}_K) \widetilde{\boldsymbol{V}}^{-1} \boldsymbol{Y}, \tag{9.25}$$

$$\widehat{\boldsymbol{\beta}}_K^* = \{ \boldsymbol{X}'(\boldsymbol{I} - \boldsymbol{A}_K^*) \widetilde{\boldsymbol{V}}^{-1}(\boldsymbol{I} - \boldsymbol{A}_K^*) \boldsymbol{X} \}^{-1} \boldsymbol{X}'(\boldsymbol{I} - \boldsymbol{A}_K^*) \widetilde{\boldsymbol{V}}^{-1} \boldsymbol{Y}, \tag{9.26}$$

where $\boldsymbol{X} = (\boldsymbol{X}_{11}', \ldots, \boldsymbol{X}_{Nn_N}')'$ and $\boldsymbol{Y}$ is defined similarly, and $\widetilde{\boldsymbol{V}} = \text{diag}(V)$.

The closed form of the working independence profile local kernel estimator $\widehat{\boldsymbol{\beta}}_K$ in (9.25) is given in Fan and Li (2004), while the closed form of the profile SUR kernel estimator $\widehat{\boldsymbol{\beta}}_K^*$ in (9.26) is given in Lin and Carroll (2006). The covariances of $\widehat{\boldsymbol{\beta}}_K$ and $\widehat{\boldsymbol{\beta}}_K^*$ can be constructed easily using the sandwich method. Specifically, write $\widehat{\boldsymbol{\beta}}_K = \boldsymbol{H}_K \boldsymbol{Y}$ and $\widehat{\boldsymbol{\beta}}_K^* = \boldsymbol{H}_K^* \boldsymbol{Y}$. Suppose the true covariance of $\boldsymbol{Y}_i$ is $\boldsymbol{\Sigma}_i$. Denote $\widetilde{\boldsymbol{\Sigma}} = \text{diag}(\boldsymbol{\Sigma}_i)$; then $\text{Cov}(\widehat{\boldsymbol{\beta}}_K) = \boldsymbol{H}_K \widetilde{\boldsymbol{\Sigma}} \boldsymbol{H}_K'$ and $\text{Cov}(\widehat{\boldsymbol{\beta}}_K^*) = \boldsymbol{H}_K^* \widetilde{\boldsymbol{\Sigma}} \boldsymbol{H}_K^{*'}$. Note that $\widehat{\boldsymbol{\beta}}_K$ is consistent when working independence $\boldsymbol{V} = \boldsymbol{I}$

is assumed with regular smoothing of $\theta(\cdot)$ (e.g., using cross-validation), or when $\theta(\cdot)$ is undersmoothed. On the other hand, $\widehat{\boldsymbol{\beta}}_K^*$ is consistent for any working covariance matrix $\boldsymbol{V}$ with regular smoothing of $\theta(\cdot)$ (e.g., using cross-validation), and is semi-parametric efficient if $\boldsymbol{V} = \boldsymbol{\Sigma}$.

9.6.2 Semi-parametric regression for likelihood-based models using the profile/SUR kernel method

The estimating equation based profile SUR kernel method can be extended to semi-parametric regression in likelihood settings. A useful likelihood extension of the marginal semi-parametric model (9.21) is the generalized semi-parametric mixed model. Suppose that, conditional on subject-specific random effects $\boldsymbol{b}_i$, Y_{ij} follows an exponential family distribution with mean μ_{ij}. A generalized semi-parametric mixed model can be written as

$$g(\mu_{ij}) = \boldsymbol{X}_{ij}'\boldsymbol{\beta} + \theta(Z_{ij}) + \boldsymbol{U}_{ij}'\boldsymbol{b}_i, \qquad (9.27)$$

where $\boldsymbol{\beta}$ is a vector of regression coefficients that models the effects of covariates $\boldsymbol{X}$ parametrically, $\theta(\cdot)$ is an unknown smooth function that models the effect of the scalar covariate Z non-parametrically, and the random effects $\boldsymbol{b}_i$ follow $N\{\boldsymbol{0}, \boldsymbol{D}(\boldsymbol{\gamma})\}$, and $\boldsymbol{\gamma}$ is a vector of variance components. The likelihood function $\ell\{\boldsymbol{Y}_i; \boldsymbol{\beta}, \boldsymbol{\gamma}, \theta(Z_{i1}), \dots, \theta(Z_{in_i})\}$ takes the same form as (9.9) except that $\ell(Y_{ij}|\boldsymbol{b}_i)$ is the log-likelihood function under the semi-parametric model (9.27) and depends on $\boldsymbol{\beta}$, $\theta(\cdot)$, and $\boldsymbol{\gamma}$.

Estimation of $\theta(\cdot)$ and $(\boldsymbol{\beta}, \boldsymbol{\gamma})$ can proceed with the profile SUR kernel method. Specifically, given $(\boldsymbol{\beta}, \boldsymbol{\gamma})$, one estimates $\theta(\cdot)$ using the SUR kernel estimating equation (9.10). Denote the resulting estimator by $\widehat{\theta}_K^*(z; \boldsymbol{\beta}, \boldsymbol{\gamma})$. Then one estimates $\boldsymbol{\beta}$ by maximizing the profile log-likelihood

$$\sum_{i=1}^{N} \ell\{\boldsymbol{Y}_i; \boldsymbol{\beta}, \boldsymbol{\gamma}, \widehat{\theta}(Z_{i1}; \boldsymbol{\beta}, \boldsymbol{\gamma}), \dots, \widehat{\theta}(Z_{in_i}; \boldsymbol{\beta}, \boldsymbol{\gamma})\}.$$

Lin and Carroll (2006) showed that such a profile SUR kernel estimator $(\widehat{\boldsymbol{\beta}}_K^*, \widehat{\boldsymbol{\gamma}}_K^*)$ is consistent and semi-parametric efficient. They also showed that the more easily computed backfitting estimator shares the same asymptotic distribution as the profile estimator, although the former needs undersmoothing.

9.6.3 Semi-parametric regression using the spline method

An alternative method for estimation in semi-parametric models is to use splines to estimate the non-parametric function $\theta(z)$. We focus here on the use of splines for estimation in the semi-parametric mixed model for Gaussian data. A key advantage of the spline approach is that, as discussed in Section 9.3.2 through Section 9.3.4, both smoothing splines and P-splines have close connections with linear mixed models. Thus, one can conveniently jointly estimate all the model parameters, including regression coefficients $\boldsymbol{\beta}$, the non-parametric function $\theta(z)$, the variance components $\boldsymbol{\gamma}$, and the smoothing parameter τ within a unified linear mixed-model framework. We focus here on joint estimation by estimating $\theta(z)$ using a smoothing spline; later, we comment briefly on the method using the P-spline.

Consider the semi-parametric mixed model

$$Y_{ij} = \boldsymbol{X}_{ij}'\boldsymbol{\beta} + \theta(Z_{ij}) + \boldsymbol{U}_{ij}'\boldsymbol{b}_i + e_{ij}, \qquad (9.28)$$

where the random effects $\boldsymbol{b}_i \sim N\{\boldsymbol{0}, \boldsymbol{D}(\boldsymbol{\gamma}_1)\}$ and $\boldsymbol{e}_i = (e_{i1}, \dots, e_{in_i})'$ follow $N\{\boldsymbol{0}, \boldsymbol{R}(\boldsymbol{\gamma}_2)\}$. It follows that $\mathrm{Cov}(\boldsymbol{Y}_i) = \boldsymbol{\Sigma}_i = \boldsymbol{U}_i\boldsymbol{D}(\boldsymbol{\gamma}_1)\boldsymbol{U}_i' + \boldsymbol{R}(\boldsymbol{\gamma}_2)$. Write $\boldsymbol{\theta}_i = \{\theta(Z_{i1}), \dots \theta(Z_{in_i})\}'$ and $\boldsymbol{\theta} = (\boldsymbol{\theta}_1', \dots, \boldsymbol{\theta}_N')'$. Define $\boldsymbol{Y}_i, \boldsymbol{Y}, \boldsymbol{X}_i, \boldsymbol{X}, \boldsymbol{U}, \boldsymbol{b}, \boldsymbol{e}$ similarly. Given the variance components

$\gamma = (\gamma_1', \gamma_2')'$, joint estimation of θ and β proceeds by maximizing the penalized log-likelihood

$$-\frac{1}{2}\sum_{i=1}^{N}(\boldsymbol{Y}_i - \boldsymbol{X}_i\boldsymbol{\beta} - \boldsymbol{\theta}_i)'\boldsymbol{\Sigma}_i^{-1}(\boldsymbol{Y}_i - \boldsymbol{X}_i\boldsymbol{\beta} - \boldsymbol{\theta}_i) - \frac{1}{2}\lambda\int\{\theta''(z)\}^2 dz,$$

$$= -\frac{1}{2}(\boldsymbol{Y} - \boldsymbol{\theta} - \boldsymbol{X}\boldsymbol{\beta})'\widetilde{\boldsymbol{\Sigma}}^{-1}(\boldsymbol{Y} - \boldsymbol{\theta} - \boldsymbol{X}\boldsymbol{\beta}) - \frac{1}{2}\lambda\boldsymbol{\theta}'\boldsymbol{\Psi}\boldsymbol{\theta},$$

where $\widetilde{\boldsymbol{\Sigma}}(\gamma) = \text{diag}\{\boldsymbol{\Sigma}_i(\gamma)\}$. The resulting joint maximum penalized log-likelihood estimators of θ and β can be written as

$$\widehat{\boldsymbol{\theta}} = \{\widetilde{\boldsymbol{\Sigma}}^{-1} + \lambda\boldsymbol{\Psi}\}^{-1}\widetilde{\boldsymbol{\Sigma}}^{-1}(\boldsymbol{Y} - \boldsymbol{X}\boldsymbol{\beta}) = \boldsymbol{S}(\boldsymbol{Y} - \boldsymbol{X}\boldsymbol{\beta}),$$

$$\widehat{\boldsymbol{\beta}} = \{\boldsymbol{X}'\widetilde{\boldsymbol{V}}^{-1}(\boldsymbol{I} - \boldsymbol{S})\boldsymbol{X}\}^{-1}\boldsymbol{X}'\widetilde{\boldsymbol{\Sigma}}^{-1}(\boldsymbol{I} - \boldsymbol{S})\boldsymbol{Y},$$

where $\boldsymbol{S} = \{\widetilde{\boldsymbol{\Sigma}}^{-1} + \lambda\boldsymbol{\Psi}\}^{-1}\widetilde{\boldsymbol{\Sigma}}^{-1}$ denotes the weighted smoothing spline hat matrix. Simple calculations further show that

$$\widehat{\boldsymbol{\theta}}_S = (\boldsymbol{W}_\theta + \lambda\boldsymbol{\Psi})^{-1}\boldsymbol{W}_\theta\boldsymbol{Y},$$

$$\widehat{\boldsymbol{\beta}}_S = (\boldsymbol{X}'\boldsymbol{W}_\beta\boldsymbol{X})^{-1}\boldsymbol{X}'\boldsymbol{W}_\beta\boldsymbol{Y},$$

where

$$\boldsymbol{W}_\theta = \widetilde{\boldsymbol{\Sigma}}^{-1} - \widetilde{\boldsymbol{\Sigma}}^{-1}(\widetilde{\boldsymbol{\Sigma}}^{-1} + \lambda\boldsymbol{\Psi})^{-1}\widetilde{\boldsymbol{\Sigma}}^{-1},$$

$$\boldsymbol{W}_\beta = \widetilde{\boldsymbol{\Sigma}}^{-1} - \widetilde{\boldsymbol{\Sigma}}^{-1}(\boldsymbol{X}'\widetilde{\boldsymbol{\Sigma}}^{-1}\boldsymbol{X})^{-1}\widetilde{\boldsymbol{\Sigma}}^{-1}.$$

Using the correspondence of a smoothing spline and the BLUP as given in (9.12), Zhang et al. (1998) showed that the above joint maximum penalized log-likelihood estimator of θ and β can be equivalently obtained by fitting the augmented linear mixed model

$$\boldsymbol{Y} = \boldsymbol{1}\delta_0 + \boldsymbol{Z}\delta_1 + \boldsymbol{X}\boldsymbol{\beta} + A\boldsymbol{a} + \boldsymbol{U}\boldsymbol{b} + e, \tag{9.29}$$

where $\boldsymbol{A}$ is defined in (9.12), $\boldsymbol{\delta} = (\delta_0, \delta_1)'$ and $\boldsymbol{\beta}$ are regression coefficients, $\boldsymbol{a}$ is the random effect resulting from the smoothing spline penalty and follows $N(\boldsymbol{0}, \tau I)$, $\tau = 1/\lambda$, the random effects $\boldsymbol{b}$ from the semi-parametric mixed model (9.28) follow $N[\boldsymbol{0}, \text{diag}\{\boldsymbol{D}(\boldsymbol{\gamma}_1)\}]$, and $e \sim N[\boldsymbol{0}, \text{diag}\{\boldsymbol{R}_i(\boldsymbol{\gamma}_2)\}]$.

A key advantage of fitting the semi-parametric mixed model (9.28) using the augmented linear mixed model (9.29) is that all the parameters can be estimated within a unified linear mixed-model framework. In particular, the regression coefficients β, the smoothing spline estimator of $\theta = \delta_0 + \boldsymbol{Z}\delta_1 + A\boldsymbol{a}$, and the random effects $\boldsymbol{b}$ can be estimated using the BLUPs. The variance components γ and the smoothing parameter τ can be estimated simultaneously using REML by treating τ as an extra variance component.

Specifically, denote by $\widehat{\boldsymbol{\delta}}, \widehat{\boldsymbol{\beta}}$ and $(\widehat{\boldsymbol{a}}, \widehat{\boldsymbol{b}})$ the BLUP estimators of the fixed effects $(\boldsymbol{\delta}, \boldsymbol{\beta})$ and the random effects $(\boldsymbol{a}, \boldsymbol{b})$ under the linear mixed model (9.29). Then the joint penalized log-likelihood estimators $(\widehat{\boldsymbol{\theta}}_S, \widehat{\boldsymbol{\beta}}_S)$ in (9.29) are identical to the BLUP estimators under (9.29) in the sense that $\widehat{\boldsymbol{\theta}}_S = \boldsymbol{1}\widehat{\delta}_0 + \boldsymbol{Z}\widehat{\delta}_1 + A\widehat{\boldsymbol{a}}$ and $\widehat{\boldsymbol{\beta}}_S = \widehat{\boldsymbol{\beta}}$. Standard errors of $\widehat{\boldsymbol{\theta}}_S$ and $\widehat{\boldsymbol{\beta}}_S$ can be estimated using either a frequentist method by treating the true $\theta(z)$ as a fixed unknown smooth function, or using a Bayesian method by treating the true $\theta(z)$ as a random function drawn from an integrated Weiner process (Wahba, 1986; Zhang et al., 1998) and be easily obtained from the mixed-model SE estimators of the BLUP estimators $\widehat{\boldsymbol{\delta}}$ and $\widehat{\boldsymbol{a}}$ (e.g., from SAS `proc mixed`). The Bayesian SE in fact measures the mean squared error of the smoothing spline estimator of $\theta(z)$ by accounting for its bias and hence is often slightly larger than its frequentist counterpart.

The REML log-likelihood for $\boldsymbol{\gamma}$ and τ under the augmented mixed model (9.29) can be written as

$$\ell_R(\boldsymbol{\gamma}, \tau) = -\frac{1}{2}\log|\boldsymbol{\Sigma}_R| - \frac{1}{2}\log|\boldsymbol{X}_R \boldsymbol{\Sigma}_R^{-1} \boldsymbol{X}_R|$$
$$-\frac{1}{2}(\boldsymbol{Y} - \boldsymbol{1}\delta_0 - \boldsymbol{Z}\delta_1 - \boldsymbol{X}\boldsymbol{\beta})' \boldsymbol{\Sigma}_R^{-1}(\boldsymbol{Y} - \boldsymbol{1}\delta_0 - \boldsymbol{Z}\delta_1 - \boldsymbol{X}\boldsymbol{\beta}),$$

where $\boldsymbol{\Sigma}_R = \tau \boldsymbol{A}\boldsymbol{A}' + \tilde{\boldsymbol{\Sigma}}$ and $\boldsymbol{X}_R = (\boldsymbol{1}, \boldsymbol{Z}, \boldsymbol{X})$.

The close connection between a smoothing spline estimator and a linear mixed model makes estimation in the semi-parametric mixed model (9.28) easy to implement in practice using existing software for linear mixed models. Numerical simulation results show that this procedure works well. However, theoretical properties of such a joint penalized log-likelihood estimator of $\theta(z)$ and $\boldsymbol{\beta}$ and the joint REML estimators of τ and $\boldsymbol{\gamma}$ have not been studied. Future research is needed.

9.7 Application of the semi-parametric model to the longitudinal CD4 data

We demonstrate the semi-parametric model (9.24) by analyzing the longitudinal CD4 count data on HIV seroconverters reported by Zeger and Diggle (1994). This study involved 369 subjects whose CD4 counts were measured repeatedly during 3–6 years of follow-up after seroconversion. The number of CD4 observations per subject varied from 1 to 12, resulting in a total of 2376 CD4 observations. The interest is in estimating the time profile of CD4 counts as well as the effects of covariates, including age, smoking (packs of cigarettes per day), drug use (yes=1, no=0), number of sex partners, and the Center for Epidemiologic Studies Depression measure.

We transformed CD4 counts using a square-root transformation to make the normality assumption more plausible. We assumed the marginal semi-parametric model $E(Y_{ij}) = \boldsymbol{X}'_{ij}\boldsymbol{\beta} + \theta(T_{ij})$, where the covariates $\boldsymbol{X}_{ij}$ are given in the previous paragraph, T_{ij} is time, and $\theta(\cdot)$ is an unknown smooth function. Following Zeger and Diggle (1994), we assumed the covariance structure resulting from a random intercept and an exponential decay serial correlation by specifying the covariance structure as $\gamma_1 I + \gamma_2 J + \gamma_3 \Omega$, where J is a matrix of 1s and $\Gamma(j, k) = \exp(-\gamma_4 |T_{ij} - T_{ik}|)$. We estimated $\theta(z)$ using the SUR kernel method and estimated $\boldsymbol{\beta}$ using the profile method. The bandwidth was estimated using a "partial" leave-one-subject-out cross-validation by dropping 50 randomly selected subjects. This gave an estimated bandwidth of 1.86. We estimated the covariance parameters using the method of moments as $\hat{\boldsymbol{\gamma}} = (11.32, 3.26, 22.15, 0.23)$.

Table 9.2 gives the regression coefficient estimates of the covariates using the working independence method and the profile SUR kernel semi-parametric efficient method that accounts for the correlation. The SEs were all calculated using the sandwich method. The

Table 9.2 Regression Coefficient Estimates in the HIV Study Using the Working Independent and SUR Profile Kernel Methods

	Working Independence		Semi-Parametric Efficient	
	Estimate	SE	Estimate	SE
Age	0.014	0.035	0.008	0.032
Smoking	0.984	0.182	0.579	0.139
Drug Use	1.049	0.526	0.584	0.335
Sex Partners	−0.054	0.059	0.078	0.039
Depression	−0.033	0.021	−0.046	0.014

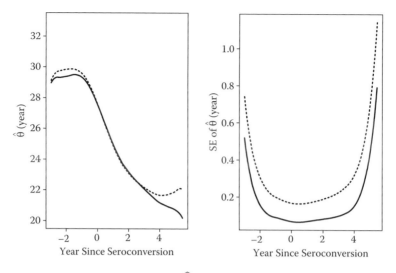

Figure 9.3 The estimated CD4 time curves $\widehat{\theta}(t)$ (left panel) and their estimated pointwise SEs (right panel). The solid and dotted curves correspond to the SUR and the working independence kernel estimates, respectively.

profile SUR kernel method yields regression coefficients with much smaller standard errors. The results show that smoking and the number of sex partners were significantly positively associated with the CD4 counts, while age, drug use, and depression had no significant effects.

Figure 9.3 shows the non-parametric estimates of the time profile using the working independence kernel GEE (dotted line) and SUR kernel (solid line) methods. Both methods show that the CD4 counts were stable before seroconversion and dropped quickly after seroconversion. The right panel of Figure 9.3 gives the estimated SEs of the curves. They show that the SUR kernel estimate, which accounts for the within-subject correlation, has a smaller SE than its working independence counterpart, which ignores the within-subject correlation.

9.8 Discussion

In this chapter we have provided an overview of non-parametric and semi-parametric regression methods for the analysis of longitudinal data. This field has enjoyed significant developments in the last 10 years. We have focused our discussions here on kernel and smoothing spline methods.

The within-subject correlation complicates the kernel and spline techniques. The classical local likelihood based kernel method for independent data is not directly applicable to longitudinal data. Researchers have hence been challenged to develop alternative kernel methods, such as SUR kernels. Although the SUR kernel estimator enjoys attractive theoretical properties, its construction is somewhat challenging and requires an iterative procedure, except for the Gaussian case where a closed form is available. There has not been systematic investigation of whether the iterative procedure converges in general settings. The optimality of the SUR kernel estimator is restricted to the particular class of kernel estimating equations. It is an open question how to define a general optimal non-parametric estimator in marginal mean models for longitudinal data, for example, whether there exists an analog minimax criterion (Fan, 1993) for longitudinal non-parametric estimators. Future research is needed.

Compared to the kernel methods, spline methods are much easier to extend from independent data to longitudinal data using the weighted penalized likelihood. A close connection between smoothing splines/P-splines and mixed-effects models provides an attractive vehicle to fit smoothing splines. However, theoretical properties of smoothing splines are only available for Gaussian data. Even for Gaussian data, theoretical properties of the REML estimator of the smoothing parameter and the joint maximum penalized log-likelihood estimators of regression coefficients and the non-parametric function in semi-parametric models are not well understood. Further research is needed.

References

Breslow, N. E. and Clayton, D. G. (1993). Approximate inference in generalized linear mixed models. *Journal of the American Statistical Association* **88**, 9–25.

Brumback, B. and Rice, J. A. (1998). Smoothing spline models for the analysis of nested and crossed samples of curves (with discussion). *Journal of the American Statistical Association* **93**, 961–1006.

Brumback, L.C. and Lindstrom, M.J. (2004). Self-modeling with flexible, random time transformations. *Biometrics* **60**, 461–470.

Carroll, R. J., Hall, P., Apanasovich, T. V., and Lin, X. (2004). Histospline method in nonparametric regression models with application to longitudinal/clustered data. *Statistical Sinica* **14**, 649–674.

Chen, K. and Jin, Z. (2005). Local polynomial regression analysis for clustered data. *Biometrika* **92**, 59–74.

Crainiceanu, C. and Ruppert, D. (2004). Restricted likelihood ratio tests in nonparametric longitudinal models. *Statistica Sinica* **14**, 713–729.

Eilers, P. H. and Marx, B. D. (1996). Flexible smoothing with B-splines and penalities (with discussion). *Statistical Science* **11**, 89–121.

Eubank, R. L., Huang, C., Maldonado, Y. M., Wang, N. Wang, S., and Buchanan, R. J. (2004) Smoothing spline estimation in varying-coefficient models. *Journal of the Royal Statistical Society, Series B* **66**, 653–667.

Fahrmeir, L. and Lang, S. (2001). Bayesian inference for generalized additive mixed models based on Markov random field priors. *Applied Statistics* **50**, 201–220.

Fan, J. (1993). Local linear regression smoothers and their minimax efficiencies. *Annals of Statistics* **21**, 196–216.

Fan, J. and Li, R. (2004). New estimation and model selection procedures for semiparametric modeling in longitudinal data analysis. *Journal of the American Statistical Association* **99**, 710–723.

Fan, J. and Zhang, J. T. (2000). Two-step estimation of functional linear models with applications to longitudinal data. *Journal of the Royal Statistical Society, Series B* **62**, 303–322.

Green, P. J. (1987). Penalized likelihood for general semi-parametric regression models. *International Statistical Review* **55**, 245–260.

Green, P. J. and Silverman, B. W. (1994). *Nonparametric Regression and Generalized Linear Models: A Roughness Penalty Approach*. London: Chapman & Hall.

Gu, C. and Ma, P. (2005). Optimal smoothing in nonparametric mixed-effect models. *Annals of Statistics* **33**, 1357–1379.

Guo, W. (2002). Functional mixed effects models. *Biometrics* **58**, 121–128.

Hastie, T. and Tibshirani, R. (1990). *Generalized Additive Models*. London: Chapman & Hall.

Hastie, T. and Tibshirani, R. (1993). Varying-coefficient models (with discussion). *Journal of the Royal Statistical Society, Series B* **55**, 757–796.

Hoover, D. R., Rice, J. A., Wu, C. O., and Yang, L.-P. (1998). Nonparametric smoothing estimates of time-varying coefficient models with longitudinal data. *Biometrika* **85**, 809–822.

Huang, J., Wu, C., and Zhou, L. (2002). Varying-coefficient models and basis function approximations for the analysis of repeated measurements. *Biometrika* **89**, 111–128.

Huang, J., Wu, C., and Zhou, L. (2004). Polynomial spline estimation and inference for varying coefficient models with longitudinal data. *Statistica Sinica* **14**, 763–788.

Huggins, R. (2006). Understanding nonparametric estimation for clustered data. *Biometrika* **93**, 486–489.

Ke, C. and Wang, Y. (2001). Semiparametric nonlinear mixed effects models and their applications (with discussions). *Journal of the American Statistical Association* **96**, 1272–1298.

Lawton, W. H. Sylvestre, E. A., and Maggio, M. S. (1972). Self-modeling nonlinear regression. *Technometrics* **14**, 513–532.

Liang, K. Y. and Zeger, S. L. (1986). Longitudinal data analysis using generalized linear models. *Biometrika* **73**, 13–22.

Lin, D. and Ying, Z. (2001). Semiparametric and nonparametric regression analysis of longitudinal data. *Journal of the American Statistical Association* **96**, 103–126.

Lin, X. and Carroll, R. J. (2000). Nonparametric function estimation for clustered data when the predictor is measured without/with error. *Journal of the American Statistical Association* **95**, 520–534.

Lin, X. and Carroll, R. J. (2001). Semiparametric regression for clustered data using generalized estimating equations, *Journal of the American Statistical Association* **96**, 1045–1056.

Lin, X. and Carroll, R. J. (2006). Semiparametric estimation in general repeated measures problems. *Journal of the Royal Statistical Society, Series B* **68**, 69–88.

Lin, X. and Zhang, D. (1999). Inference in generalized additive mixed model using smoothing splines. *Journal of the Royal Statistical Society, Series B* **61**, 381–400.

Lin, X. and Zhang, D. (2001). Discussion of "Semiparametric nonlinear mixed effects models and their applications" by C. Ke and Y. Wang. *Journal of the American Statistical Association* **96**, 1288–1291.

Lin, X., Wang, N., Welsh, A. and Carroll, R. J. (2004). Equivalent kernels of smoothing splines in nonparametric regression for clustered data. *Biometrika* **91**, 177–193.

Lindstrom, M. J. and Bates, D. M. (1990). Nonlinear mixed effects models for repeated measurement data. *Biometrics* **46**, 673–687.

Linton O. B., Mammen E., Lin, X., and Carroll, R. J. (2004). Accounting for correlation in marginal longitudinal nonparametric regression. In D. Lin and P. Heagerty (eds.), *Proceedings of the Second Seattle Symposium in Biostatistics: Analysis of Correlated Data*, pp. 23–34. New York: Springer.

McCullagh, P. and Nelder, J. (1989). *Generalized Linear Models*. London: Chapman & Hall.

Morris, J. S. and Carroll, R. J. (2006). Wavelet-based functional mixed models. *Journal of the Royal Statistical Society, Series B* **68**, 179–199.

Pepe, M. and Anderson, G. L. (1994). A cautionary note on inference for marginal regression models with longitudinal data and general correlated response data. *Communications in Statistics* **23**, 939–951.

Rice, J. A. and Wu, C. O. (2001). Nonparametric mixed effects models for unequally sampled noisy curves. *Biometrics* **57**, 253–259.

Ruppert, D., Wand, M. P., and Carroll, R. J. (2003). *Semiparametric Regression*. Cambridge: Cambridge University Press.

Shi, M., Weiss, R. E., and Taylor, J. M. G. (1996). An analysis of pediatric AIDS CD4 counts using flexible random curves. *Applied Statistics* **45**, 151–164.

Silverman, B. (1984). Spline smoothing: the equivalent variable kernel method. *Annals of Statistics* **12**, 898–916.

Sowers, M. F., Crutchfield, M., Randolph, J. F., Shapiro, B., Zhang, B., Pietra, M. L., and Schork, M. A. (1995). Urinary ovarian and gonadotrophin hormone levels in premenopausal women with low bone mass. *Journal of Bone and Mineral Research* **13**(7), 1191–1202.

Taylor, M. G., Cumberland, W. G., and Sy, J. P. (1994). A stochastic model for analysis of longitudinal AIDS data. *Journal of the American Statistical Association* **89**, 727–736.

Verbyla, A. P., Cullis, B. R., Kenward, M. G., and Welham, S. J. (1999). The analysis of designed experiments and longitudinal data using smoothing splines. *Applied Statistics* **48**, 269–311.

Wahba, G. (1983). Bayesian "confidence intervals" for the cross-validated smoothing spline. *Journal of the Royal Statistical Society, Series B* **45**, 133–150.

Wang, N., Carroll, R. J., and Lin, X. (2005). Efficient semiparametric marginal estimation for longitudinal/clustered data. *Journal of the American Statistical Association* **100**, 147–157.

Wang, N. (2003). Marginal nonparametric kernel regression accounting for within-subject correlation. *Biometrika* **90**, 43–52.

Wang, Y. (1998). Mixed effects smoothing spline analysis of variance. *Journal of the Royal Statistical Society, Series B* **60**, 159–174.

Wood, S.N. (2006). Low rank invariant tensor product smooth interactions in generalized additive mixed models. *Biometrics* **62**, 1025–1036.

Wu, C. O., Chiang, C. T., and Hoover, D. R. (1998). Asymptotic confidence regions for kernel smoothing of a varying coefficient model with longitudinal data. *Journal of the American Statistical Association* **93**, 1388–1402.

Wu, H. and Liang, H. (2004). Random varying-coefficient models with smoothing coviarates: Applications to an AIDS clinical study. *Scandinavian Journal of Statistics* **31**, 3–19.

Wu, H. and Zhang, J. (2002). Local polynomial mixed effects models for longitudinal data. *Journal of the American Statistical Association* **97**, 883–897.

Zeger, S. L. and Diggle, P. J. (1994). Semiparametric models for longitudinal data with application to CD4 cell numbers in HIV seroconverters. *Biometrics* **50**, 689–699.

Zhang, D. (2004). Generalized linear mixed models with varying coefficients for longitudinal data. *Biometrics* **60**, 8–15.

Zhang, D. and Lin, X. (2003). Hypothesis testing in semiparametric additive mixed models. *Biostatistics* **4**, 57–74.

Zhang, D., Lin, X., Raz, J., and Sowers, M. (1998). Semiparametric stochastic mixed models for longitudinal data. *Journal of the American Statistical Association* **93**, 710–719.

Zhang, D. and Davidian, M. (2004). Likelihood and conditional likelihood inference for generalized additive mixed models. *Journal of Multivariate Analysis* **91**, 90–106.

CHAPTER 10

Functional modeling of longitudinal data

Hans-Georg Müller

Contents

10.1 Introduction

Longitudinal studies are characterized by data records containing repeated measurements per subject, measured at various points on a suitable time axis. The aim is often to study change over time or time dynamics of biological phenomena such as growth, physiology, or pathogenesis. One is also interested in relating these time dynamics to certain predictors or responses. The classical analysis of longitudinal studies is based on parametric models, which often contain random effects, such as the generalized linear mixed model (GLMM) (Chapter 4), or on marginal methods such as generalized estimating equations (GEE) (Chapter 3). The relationships between the subject-specific random-effects models and the marginal, population-averaged models underlying methods such as GEE are quite complex (see Zeger, Liang, and Albert, 1988; Heagerty, 1999; Heagerty and Zeger, 2000; and the detailed discussion of this topic in Chapter 7). To a large extent, this non-compatibility of various approaches is due to the parametric assumptions that are made in these models. These include the assumption of a parametric trend (linear or quadratic in the simplest cases) over time and of parametric link functions. Specific common additional assumptions are normality of the random effects in a GLMM and a specific covariance structure ("working correlation") in a GEE. Introducing non-parametric components (non-parametric link and non-parametric covariance structure) can ameliorate the difficulties of relating various longitudinal models to each other, as it increases the inherent flexibility of the resulting longitudinal models substantially (see the estimated estimating equations approach in Chiou and Müller, 2005).

Taking the idea of modeling with non-parametric components one step further, the functional data analysis (FDA) approach to longitudinal data provides an alternative non-parametric method for the modeling of individual trajectories. The underlying idea is to view observed longitudinal trajectories as a sample of random functions, which are not parametrically specified. The observed measurements for an individual then correspond to the values of the random trajectory, corrupted by measurement error. A primary objective is to reduce the high dimension of the trajectories — considered to be elements of an infinite-dimensional function space — to finite dimension. One goal is to predict individual trajectories from the measurements made for a subject, borrowing strength from the entire sample of subjects. The necessary dimension reduction or regularization step can be implemented in various ways. For the analysis of longitudinal data, with its typically sparse and irregular measurements per subject, the method of functional principal component analysis (FPCA), which will be reviewed in Section 10.5, has recently been proposed (Yao, Müller, and Wang, 2005a, 2005b), extending previous work by James (2002). Other regularization methods that have proven useful in FDA include smoothing splines (Ke and Wang, 2001), B-splines (Rice and Wu, 2001), and P-splines (Yao and Lee, 2006); see Chapter 11 and Chapter 12.

The classical theory and applications of FDA have been developed for densely sampled or fully observed trajectories that in addition are sampled without noise. This setting is not conducive to applications in longitudinal studies, due to the common occurrence of irregular and sparse measurement times, often due to missing data. Excellent overviews on FDA for densely sampled data or fully observed trajectories can be found in the two recent books by Ramsay and Silverman (2002, 2005). Early approaches were based primarily on smoothing techniques and landmarks (e.g., Gasser et al., 1984, 1985). The connections between FDA and longitudinal data analysis have been revisited more recently, see, for example, Rice (2004), Zhao, Marron, and Wells (2004), and Müller (2005). Of interest is also a discussion that was held in 2004 at a conference dedicated to exploring these connections (Marron et al., 2004). While a number of practical procedures and also theoretical results are available, the use of FDA methodology for the analysis of longitudinal data is far from being established practice. This is an area of ongoing research.

Even the estimation of a mean trajectory is non-trivial. First, dependency of the repeated measurements coming from the same subject needs to be taken into account (Wang, 2003; Lin et al., 2004) to improve efficiency of this estimation step. Second, a problem with practical impact for the estimation of a meaningful mean trajectory is individual time variation, which often occurs in addition to amplitude variation, due to differences in time dynamics across subjects. We discuss approaches for handling this issue in Section 10.4.

We focus on an approach of applying FDA to longitudinal data that is based on FPCA and thus allows for subject-specific models that include random effects and are data-adaptive (Section 10.5). Our focus is less on marginal, population-averaged modeling, although we discuss the difficulties that arise for marginal modeling in the presence of time variation. Auxiliary quantities of interest include estimates of the underlying population-averaged mean function and of the covariance surface describing the dependency structure of the repeated measurements. These steps require smoothing methods, briefly reviewed in Section 10.3.

Once a suitable estimate of the covariance surface is available, one can obtain the eigenfunctions of the underlying stochastic process that is assumed to generate the individual random trajectories. We do not assume stationarity. Individual trajectories are represented by their first few functional principal component (FPC) scores, which play the role of random effects. Thus, functional data are reduced to a vector of scores. These scores may subsequently be entered into further statistical analysis, either serving as predictors or as responses in various statistical models, including functional regression models (Section 10.6). Section 10.2 and Section 10.3 provide background on FDA and smoothing methods. Further discussion can be found in Section 10.7.

10.2 Basics of functional data analysis

Functional data consist of a sample of random curves, which are typically viewed as independent and identically distributed realizations of an underlying stochastic process. For each subject or experimental unit, one samples one or several functions $Y(t)$, $t \in [0, T]$, for some $T > 0$. A common assumption is that trajectories are square integrable and smooth, say twice differentiable. A major difference between functional data and multivariate data is that, in the case of functional data, order and neighborhood relations are well defined and relevant, while for multivariate data, these notions do not play any role. This is illustrated by the fact that one can reorder the components of a multivariate data vector and arrive at exactly the same statistical analysis as for the data vector arranged in the original order. For functional data, the situation is entirely different, and reordering the data will disrupt the analysis.

Goals for FDA include the construction of meaningful models for basic data-descriptive measures such as a mean trajectory. If one were given a sample of entirely observed trajectories $Y_i(t)$, $i = 1, \ldots, N$, for N subjects, a mean trajectory could be simply defined as the sample average, $\overline{Y}(t) = N^{-1} \sum_{i=1}^{N} Y_i(t)$, $t \in [0, T]$. However, this relatively straightforward situation is rather the exception than the norm, as we face the following difficulties: the trajectories may be sampled at sparsely distributed times, with timings varying from subject to subject; the measurements may be corrupted by noise and are dependent within the same subject; and time variation may be present, in addition to amplitude variation, a challenge that is typical for some longitudinal data such as growth curves and is discussed further in Section 10.4. Thus, what constitutes a reasonable population mean function is much less straightforward in the FDA setting than it is in the multivariate case. Further notions of interest that require special attention include measures of variance, covariance, and correlation between curves.

Measures of correlation are of interest for studies in which several trajectories per subject are observed. An initial idea was the extension of canonical correlation from the multivariate (Hotelling, 1936) to the functional case. The resulting functional canonical correlation requires solving an "inverse problem," which necessitates some form of regularization, a feature typical of many functional techniques. Two main types of regularization have been used: regularization by an additive penalty term, usually penalizing against non-smooth curve estimates, and often used in combination with spline modeling; or truncation of a functional series expansion such as a Fourier series or wavelet expansion, at a finite number of terms, also referred to as thresholding. Both approaches depend on the choice of an appropriate regularization parameter. For functional canonical correlation, both regularization by a penalty (Leurgans, Moyeed, and Silverman, 1993) and by truncation (He, Müller, and Wang, 2004) have been proposed. One consistent finding is that functional canonical correlation is highly sensitive to the choice of the regularization parameter (size of penalty or truncation threshold). Due to the difficulties in calibrating the regularization for functional canonical correlation, alternative notions of functional correlation measures (Service, Rice, and Chavez, 1998; Heckman and Zamar, 2000; Dubin and Müller, 2005) have been proposed.

Beyond functional correlation, the problem of relating several observed curves per subject to each other or to a scalar response leads to the problem of functional regression. Functional regression models come in various flavors. For a scalar response, one may consider one or several functional predictors. There are also situations in which the response is a function, combined with scalar or multivariate predictors. Another complex case involves the simultaneous presence of functional predictors and functional responses. These models will be discussed in Section 10.6. In functional regression, one can distinguish a classic FDA approach, which requires the availability of fully observed, noise-free individual trajectories and has been well investigated in recent years (Ramsay and Dalzell, 1991; Cardot et al., 2003), and a modified approach suitable for longitudinal data that is of more recent origin. Methods extending density estimation and non-parametric regression to functional objects have also been developed in recent years (Ferraty and Vieu, 2006); such developments face theoretical challenges and are the subject of ongoing research.

The term "functional data" is used here to denote a sample of curves, rather than a single curve, as one may encounter in dose-response analysis or in non-parametric regression. However, in general, the use of the term "functional data" is not always that rigorous and often simply refers to the fact that a model contains a non-parametric curve as a component. When faced with functional data, a useful first step is simply to plot the data. In situations characterized by reasonably dense sampling of measurements for each subject, one may generate such plots by linearly interpolating the points corresponding to the repeated measurements made on the same subject (producing a so-called "spaghetti plot"). In other situations, when data are irregularly sampled or a derivative is required, as in the modeling of growth curves, a preliminary smoothing or differentiation step may be helpful (examples are provided in Section 10.5 and Section 10.6 and in the top panel of Figure 10.1). From such plots one may discern a general trend in the data, changes in sampling frequencies (e.g., caused by dropouts), and the shapes of individual trajectories and how much these shapes vary across subjects. Last, but not least, one may identify subjects with outlying trajectories; these are candidates for further study or removal before proceeding with the analysis.

10.3 Non-parametric regression

10.3.1 Kernel smoothing

Smoothing methods for non-parametric regression are an important ingredient of many functional methods. These key techniques exploit the continuity of the trajectories. We focus here on kernel-type smoothers that have proven useful due to their straightforward interpretation and the large body of accumulated knowledge about their properties, especially their asymptotic behavior. Explicit representations in terms of weighted averages in the data, which are available for this class of smoothers, greatly facilitate the investigation of asymptotic properties and also of the functional methods that utilize them. Excellent textbooks and monographs on kernel-type smoothing procedures include Silverman (1986), Wand and Jones (1995), Fan and Gijbels (1996), and Bowman and Azzalini (1997). Other smoothing methods, such as various types of splines, can often be used equally well (Eubank, 1999). See Chapter 8 for a brief review.

The goal of smoothing in the non-parametric regression setting is to estimate a smooth regression function or surface $g(x) = E(Y|X = x)$, usually assumed to be twice continuously differentiable. For the random design case, this regression function is characterized by the joint distribution of vectors (X, Y), while for fixed designs the predictor levels X_j, at which responses Y_j are recorded, are assumed to be non-random (and usually assumed to be

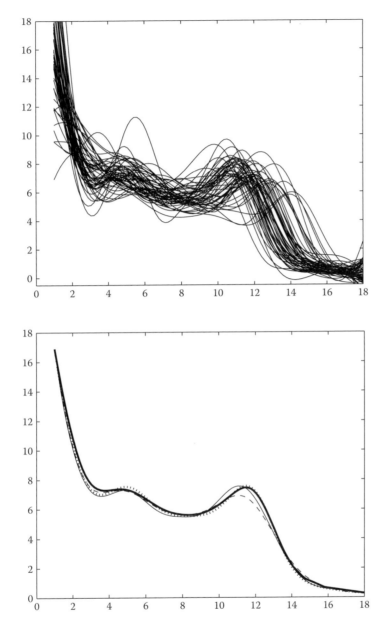

Figure 10.1 Time warping of growth curves. Top: Sample of estimated growth velocities (first derivatives) of 54 girls from the Berkeley Growth Study. Bottom: Comparison of different registration procedures, including functional convex mean using area-under-the-curve registration (solid bold line), continuous monotone registration (so-called Procrustes method, solid line), landmark registration (dotted line), and cross-sectional average (dashed line). In both panels, the x-axis is age in years, and the y-axis is velocity in centimeters per year. (Reproduced from Liu and Müller, 2004. With permission.)

generated by a "design density"). The response Y is univariate, while predictors X can be univariate or multivariate. Of interest for FDA applications are the cases $X \in \mathbb{R}$, the case of a one-dimensional regression function, and $X \in \mathbb{R}^2$, the case of a regression surface. Note that X and Y will be used in subsequent sections to denote functional objects rather than vectors or scalars.

To define a kernel smoother for a one-dimensional predictor, given n data points $\{(X_j, Y_j), j = 1, \ldots, n\}$, we need a bandwidth or window width h and a one-dimensional kernel function K_1. The bandwidth serves as a smoothing parameter and determines the trade-off between variance and bias of the resulting non-parametric regression estimators. The kernel K_1 typically is chosen as a smooth and symmetric density function; for some types of kernel estimators, such as convolution estimators, negative-valued kernels can be used to accelerate rates of convergence (Gasser, Müller, and Mammitzsch, 1985). Commonly used non-negative kernels with domain $[-1, 1]$ are rectangular (box) kernels, $K_1(x) = \frac{1}{2}$, or quadratic (Epanechnikov) kernels, $K_1(x) = \frac{3}{4}(1 - x^2)$, which enjoy some optimality properties. Popular non-negative kernels with unbounded domain are Gaussian kernels, which correspond to the standard normal density.

A classic kernel smoothing method primarily aimed at regular designs X_j is the class of convolution kernel smoothers (Priestley and Chao, 1972; Gasser and Müller, 1984). The smoothing window for estimating at predictor level x is $[x - h, x + h]$ if a kernel function K_1 with domain $[-1, 1]$ is used. Let $S_{(j)} = (X_{(j)} + X_{(j-1)})/2$, where $X_{(j)}$ is the jth order statistic of the X_j, and let $Y_{[j]}$ denote the concomitant of $X_{(j)}$. Convolution-type kernel estimators are defined as

$$\widehat{g}_C(x) = \sum_{j=1}^{n} Y_{[j]} \int_{S_{(j)}}^{S_{(j+1)}} \frac{1}{h} K_1 \left(\frac{x - s}{h} \right) ds.$$

Near the endpoints of the regression function, specially constructed boundary kernels should be used to avoid boundary bias effects (e.g., Müller, 1991; Jones and Foster, 1996).

10.3.2 Extensions and local linear fitting

We note that these smoothers can be easily extended to the case of estimating derivatives (Gasser and Müller, 1984). Convolution-type smoothers have been applied extensively to conduct non-parametric analysis of longitudinal growth studies (Gasser et al., 1984; Müller, 1988). Growth studies belong to a class of longitudinal studies for which one has relatively dense and regular measurement grids. In such situations, one can smooth each trajectory individually, independent of the other observed trajectories. This is justified by postulating asymptotically ever denser designs, where n, the number of measurements per subject increases within a fixed domain (also referred to as "in-fill asymptotics"). As $n \rightarrow \infty$, using appropriate kernels and bandwidth sequences, this approach leads to estimators of trajectories and derivatives with typical non-parametric rates of convergence of the order $n^{-(k-\nu)/(2k+1)}$. Here, ν is the order of derivative to be estimated and $k > \nu$ is the order of assumed smoothness of the trajectory (number of continuous derivatives). For an example of a sample of estimated first derivatives of growth data, see the bottom panel of Figure 10.1.

The analyses of growth data with these smoothing methods demonstrated that non-parametric regression methods are essential tools to discern features of longitudinal time courses. An example is the detection of a pre-pubertal growth spurt, which had been omitted from previously used parametric models. Once a longitudinal feature is not properly reflected in a parametric model, it can be very difficult to discover these features through a lack-of-fit analysis. A non-parametric approach should always be used concurrently with a parametric modeling approach in order to insure against omitting important features. Non-parametric methods achieve this by being highly flexible and by not reflecting preconceived notions about the shape of time courses. In the above-mentioned analysis of growth studies, first and second derivatives were estimated for each individual separately to assess the dynamics of growth. For the practically important problem of bandwidth choice, one can use cross-validation (minimization of leave-one-out prediction error), generalized cross-validation (a faster approximation), or a variety of plug-in methods aimed at minimizing mean squared error or integrated mean squared error.

Boundary adjustments are automatically included in local polynomial fitting, which is a great advantage. Local linear smoothers are particularly easy to use and have become the most popular kernel-based smoothing method. They have been around for a long time and are based on the very simple idea of localizing a linear regression fit from the entire data domain to local windows. Compared to convolution kernel estimators, this method has better conditional variance properties in random designs. A theoretical problem is that the unconditional variance of local linear estimators is unbounded and, therefore, mean squared error does not exist, in contrast to the convolution methods where it is always bounded. Practically, this is reflected by problems caused by occasional gaps in the designs, that is, for the case of a random design the probability that not enough data fall into at least one smoothing window is not negligible; see Seifert and Gasser (1996, 2000) for further discussion of these issues and improved local linear estimation.

The local linear kernel smoother (Fan and Gijbels, 1996) is obtained via the minimizers $\widehat{a}_0, \widehat{a}_1$ of

$$\sum_{j=1}^{n} K_1\left(\frac{x - X_j}{h}\right)\{Y_j - a_0 - a_1(x - X_j)\}^2.$$

Once $\widehat{a}_0 = \widehat{a}_0(x)$ has been obtained, we define the local linear kernel estimator $\widehat{g}_L$ as $\widehat{g}_L(x) = \widehat{a}_0(x)$. The older kernel methods of Nadaraya (1964) and Watson (1964) correspond to the less flexible special case of fitting constants to the data locally by weighted least squares. However, fitting local constants leads to somewhat awkward bias behavior if the designs are non-equidistant. All kernel type estimators as well as smoothing splines exhibit very similar behavior in the interior of the domain (away from the boundaries) and when the design is regular. Differences emerge for random designs and when estimating near the boundaries.

For two-dimensional smoothing, we aim at the regression function (regression surface) $g(x_1, x_2) = E(Y | X_1 = x_1, X_2 = x_2)$. Locally weighted least squares then provide a criterion for fitting local planes to the data $\{(X_{j1}, X_{j2}, Y_j), j = 1, \ldots, n\}$, leading to the surface estimate $\widehat{g}(x_1, x_2) = \widehat{a}_0$, where $(\widehat{a}_0, \widehat{a}_1, \widehat{a}_2)$ are the minimizers of the locally weighted sum of squares

$$\sum_{j=1}^{M} K_2\left(\frac{x_1 - X_{j1}}{h_1}, \frac{x_2 - X_{i2}}{h_2}\right)[Y_j - \{a_0 + a_1(X_{j1} - x_1) + a_2(X_{j2} - x_2)\}]^2. \quad (10.1)$$

Here, K_2 is a two-dimensional kernel function, usually chosen as a two-dimensional density function. A common choice for K_2 is the product of two one-dimensional densities, $K_2(u_1, u_2) = K_1(u_1)K_1(u_2)$, where K_1 is any of the one-dimensional kernels discussed above. Two bandwidths h_1, h_2 are needed; for simplicity, they are often chosen to coincide, $h_1 = h_2$, so that only one smoothing parameter needs to be selected. We note that often useful explicit formulae for the smoothing weights employed for both one- and two-dimensional smoothers can be easily obtained (e.g., Hall, Müller, and Wang, 2006, Formulae [2.5]).

10.4 Time warping and curve synchronization

10.4.1 Overview

A main motivation for considering time warping in biomedical applications is the empirical observation that individuals may progress through time at their own individual pace, referred to as biological time or "eigenzeit" (Capra and Müller, 1997). In contrast to clock time, this time is often defined by physiological processes of development, and significant events are defined by reaching certain stages of maturity. These stages are attained earlier for some individuals, and later for others. The idea is then to compare individuals at corresponding stages of their biological development, and not based on chronological age.

A typical example is human growth, where various growth spurts have been well identified and provide natural "landmarks" of development. These include the well-known pubertal growth spurt and also the so-called mid-growth spurt that has been rediscovered using non-parametric smoothing methods; see Gasser et al. (1984). As these spurts occur at a different chronological age for each individual, adequate models for corresponding longitudinal data need to reflect the presence of time variation (the variation in the timing of the spurts) in addition to amplitude variation (the variation in the size of the spurts). In the growth curve example, the presence of time variation implies that a simple cross-sectional average growth curve will often not be very meaningful. The reason is that it will not resemble any individual trajectory closely, as data obtained for non-corresponding times are averaged, which may lead to wrong impressions about the dynamics of growth. While the phenomenon of time variation is more obvious for some longitudinal data than for others, its presence is always a possibility.

Time warping has been studied primarily for densely and regularly sampled data, such as data from longitudinal growth studies, but is of potential relevance for many other longitudinal studies. Addressing the warping issue is also referred to as time synchronization, curve registration, or curve alignment (Gasser and Kneip, 1995; Ramsay and Li, 1998; Rønn, 2001; Gervini and Gasser, 2004; Liu and Müller, 2004). In warping models, reflecting the individually determined flow of time, one assumes that the time axis is individually distorted by a random time transformation function that is monotone increasing while the beginning and end point of the time domain remain unchanged, for example. We do not attempt to give comprehensive descriptions of the various curve registration methods. The reader is referred to the literature for more details; the material in this section is not needed for the remainder of this chapter.

In the presence of warping, the simultaneous random variation in amplitude and time can lead to identifiability issues, and, therefore, common warping methods often contain implicit assumptions about the nature of the time variation. When each subject follows his or her own time scale, time synchronization as a pre-processing step often improves subsequent analysis of functional data and often is a necessity for a meaningful analysis when the time dynamics of longitudinal data are of interest. Analyzing the nature of the time variation may be of interest in itself. For example, in gene time course expression analysis, gene classification can be based on a simple approach of time-shift warping (Silverman, 1995; Leng and Müller, 2006).

A specific example for growth data is shown in Figure 10.1. The top panel displays a sample of growth velocities (obtained by using local polynomial fitting to estimate derivatives) for 54 girls from the Berkeley Growth Study, while the bottom panel features a comparison of the cross-sectional mean growth curve, displaying various mean functions that were obtained by applying the traditional cross-sectional average as well as various warping (registration) methods. Among these, the landmark method, pioneered in Kneip and Gasser (1992), is known to work very well for these data and serves as a benchmark.

In landmark warping, one first identifies landmark locations, often defined as peaks or troughs in first or second derivatives of the individual trajectories. These locations are times of events, such as peak growth velocity during a growth spurt, that have a meaningful biological interpretation and betray an individual's time line, for example, accelerated or delayed development. The landmark times are considered to correspond to each other across individuals, and in the landmark approach the average curve is required to include both average location and average curve value for each of the landmark events. These characteristic points are supplemented by an interpolation step to define an average smooth curve that connects the averaged landmarks. In the growth curve example, the landmark average curve includes the point defined by (i) the time which corresponds to the average timing of the pubertal growth spurts, defined as the location of maximal growth velocity; and by (ii) the average value of all maximum growth velocities. It also goes through all points similarly defined by other landmarks.

The alternative Procrustes method (Ramsay and Li, 1998) is an iterative procedure, warping curves at each step to match the current cross-sectional mean as closely as possible. The current cross-sectional mean is then updated and serves again as a template for the next warping step, and this is repeated until convergence. A third method is area-under-the-curve registration (Liu and Müller, 2004), which synchronizes time points associated with the same relative area under the curve between the left endpoint of the domain and the respective time point. When comparing the resulting average growth curves in the bottom panel of Figure 10.1, the cross-sectional mean is found to underestimate the size of the pubertal growth spurt, and it also produces a biased location. Similar distortions are found for the mid-growth spurt, the smaller growth spurt that occurs around 5 years. While the Procrustes method is improving upon the cross-sectional mean, area-under-the-curve warping is found to mimic landmark warping most closely in this example. We note that this method is very simple to implement.

We conclude that even a simple notion, such as a mean curve, needs to be carefully reconsidered in the presence of time variation. A simulated example further demonstrating the distorting effects of warping in FDA is shown in Figure 10.2. Here, all individual trajectories have been generated as bimodal curves, but the cross-sectional mean does not reflect this shape at all and is clearly inadequate. Modifications aiming at time synchronization are needed to arrive at a representative mean when warping is present; the area-under-the-curve warping method is seen to provide these for this example and nicely recovers the original shape of individual trajectories.

Landmark identification and alignment (Gasser and Kneip, 1995) has become a gold standard for warping for those situations where landmarks are easy to identify (see Figure 10.1). Landmarks have proved useful for the analysis of longitudinal growth curves early on (Gasser et al., 1984), due to the prominence of the growth spurts. However, landmark methods do not work in situations where the individual curves are variable to the extent that they do not share common shapes. Procrustes and area-under-the-curve registration are not subject to shape requirements and are therefore more universally applicable. Alternative robust warping methods have been developed lately (Gervini and Gasser, 2005). Much work remains to be done in this area.

10.4.2 Methods for time synchronization

Simple warping transformations include time-shift warping (Silverman 1995; Leng and Müller, 2006), where one assumes in the simplest case that, for the ith trajectory, $Y_i(t) = Y_0(t - \tau_i)$, where τ_i denotes a (random) time shift for the ith subject, and Y_0 is a synchronized trajectory. Another simple variant that is often useful, especially when the sampled functions have varying domains, is scale warping. Here, one models $Y_i(t) = Y_0(t/\sigma_i)$ for scale factors $\sigma_i > 0$. Both schemes can be combined, leading to shape-invariant modeling (Lindstrom, 1995; Wang, Ke, and Brown, 2003).

A useful framework is to view warping as a time synchronization step, formalized as follows. Time for each subject is mapped from a standard or synchronized time $t \in [0, 1]$ to the individual or warped time $\widetilde{X}(t)$, where this mapping must be strictly monotone and invertible, and in most approaches is considered to be a random function. Ideally, a warping method will satisfy the boundary conditions $\widetilde{X}(0) = 0$, $\widetilde{X}(1) = T$. The sample of observed trajectories can then be viewed as being generated by a latent bivariate stochastic process in "synchronized time space" $\mathcal{S}$ (Liu and Müller, 2004), $[\{\widetilde{X}(t), \widetilde{Y}(t)\}, t \in [0, 1]] \subset L^2([0, 1]) \times L^2([0, 1])$. The observed sample then corresponds to $[\widetilde{Y}\{\widetilde{X}^{-1}(x)\}, x \in [0, T]] \subset L^2([0, T])$,

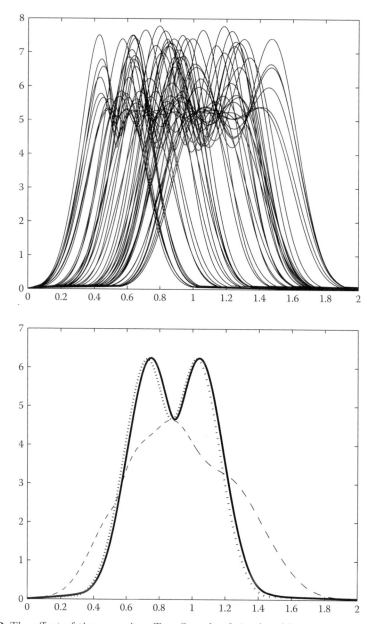

Figure 10.2 The effect of time warping. Top: Sample of simulated bimodal trajectories. Bottom: Target is the true bimodal functional warped mean (solid line). Estimated means are the naive cross-sectional mean computed from the sample curves (dashed line) and the warped functional mean computed from the sample curves, using area-under-the-curve registration (dotted line). (Reproduced from Liu and Müller, 2004. With permission.)

and the associated warping mapping is

$$\psi: \ [\{\widetilde{X}(t), \widetilde{Y}(t)\}, \ t \in [0,1]] \mapsto [\{x, Y(x)\}, \ x \in [0,T]],$$

defined by $Y(x) = \widetilde{Y}\{\widetilde{X}^{-1}(x)\}$.

The identifiability problem corresponds to the fact that this mapping does not have a unique inverse. The way this ambiguity is resolved differentiates between the various warping methods, such as the Procrustes method or landmark warping, each providing a concrete way to define an inverse mapping and thus a synchronization algorithm.

A simple and often effective warping method that is featured in Figure 10.1 and Figure 10.2 is area-under-the-curve warping. This method is designed for samples of non-negative random trajectories. One assumes that synchronized time corresponds to the relative area under each individual trajectory. The total area is normalized to 1, and, if the fraction of area under the curve is the same for two different observed times, then these are considered to correspond to the same point in individual development and therefore are mapped to the same synchronized time. Formally, to obtain the inverse warping process $\widetilde{X}^{-1}$, which corresponds to the time-synchronizing mapping, as a function of each observed trajectory process Y, one simply determines the fractions of the area under the observed curves Y and defines this to be the synchronized time,

$$\varphi(Y)(x) = \widetilde{X}^{-1}(x) = \frac{\int_0^x |Y(s)|\,ds}{\int_0^T |Y(s)|\,ds}.$$

Applying this time-synchronizing mapping is referred to as area-under-the-curve warping.

Considering the latent bivariate processes $\{(\widetilde{X}(t), \widetilde{Y}(t)), t \in [0,1]\}$, as $\widetilde{X}(\cdot)$ is constrained to be positive increasing, the space where the bivariate processes live is a convex space. This leads to a convex calculus. Given two observed processes Y_1, Y_2, and a fixed $0 \leq \pi \leq 1$, define a functional convex sum

$$\pi Y_1 \oplus (1-\pi)Y_2 = \psi\{\pi\psi^{-1}(Y_1) + (1-\pi)\psi^{-1}(Y_2)\},$$

where ψ^{-1} is the inverse mapping $\psi^{-1}(Y) = (\{\varphi^{-1}(Y)\}(t), Y[\{\varphi^{-1}(Y)\}(t)], t \in [0,1])$. The functional convex sum can be easily extended to the case of J functions, $J > 2$,

$$\bigoplus_{j=1}^{J} \pi_j Y_j = \psi\left(\sum_{j=1}^{J} \pi_j X_j, \sum_{j=1}^{J} \pi_j Y_j\right),$$

for any π_j such that $\sum \pi_j = 1$, $0 \leq \pi_j \leq 1$, $j = 1, \ldots, J$. This then leads to the warped average function (functional convex average, shown in the bottom panels of Figure 10.1 and Figure 10.2),

$$\bar{Y}_\oplus = \bigoplus_{j=1}^{n} \frac{1}{n} Y_j.$$

Similarly, a convex path connecting observed random trajectories is $\{\pi Y_1 \oplus (1-\pi)Y_2, \pi \in [0,1]\}$. Further results on this general warping framework and area-under-the-curve warping can be found in Liu and Müller (2003, 2004).

10.5 Functional principal component analysis

10.5.1 Square integrable stochastic processes

Functional principal component analysis has emerged as a major tool for dimension reduction within FDA. One goal is to summarize the infinite-dimensional random trajectories through a finite number of FPC scores. This method does not require distributional assumptions and is solely based on first- and second-order moments. It also provides eigenfunction estimates that are known as "modes of variation." These modes often have a direct biological interpretation and are of interest in their own right (Kirkpatrick and Heckman, 1989). They offer a visual tool to assess the main directions in which the functional data vary. An important application is a characterization of individual trajectories through an empirical Karhunen–Loève representation. It is always a good idea to check and adjust for warping before carrying out an FPCA.

For square integrable random trajectories $Y(t)$, we define mean and covariance functions

$$\mu(t) = E\{Y(t)\},$$
$$G(s,t) = \text{Cov}\{Y(s), Y(t)\}, \quad s, t \in [0, T],$$

and the auto-covariance operator

$$(Af)(t) = \int_0^T f(s)G(s,t)\,ds.$$

This is a linear Hilbert–Schmidt operator in the function space of square integrable functions $L^2([0,T])$ with Hilbert–Schmidt kernel G (Conway, 1985). Under minimal assumptions, this operator has orthonormal eigenfunctions ψ_k, $k = 1, 2, \ldots$, with associated ordered eigenvalues $\lambda_1 \geq \lambda_2 \geq \ldots$, that is, satisfying

$$A\psi_k = \lambda_k \psi_k, \quad k = 1, 2, \ldots.$$

The eigenfunctions of the auto-covariance operator turn out to be very useful in FDA for dimension reduction, due to the Karhunen–Loève expansion. This expansion holds under minimal assumptions (see Ash and Gardner, 1975) and converges in the L^2 sense and also pointwise. It provides an important representation of individual trajectories Y,

$$Y(t) = \mu(t) + \sum_{k=1}^{\infty} A_k \psi_k(t),$$

where the A_k are uncorrelated random variables, known as the functional principal component scores. They satisfy $E(A_k) = 0$, $\text{Var}(A_k) = \lambda_k$ and have the explicit representation

$$A_k = \int_0^T \{Y(t) - \mu(t)\}\psi_k(t)\,dt. \tag{10.2}$$

The situation is analogous to the representation of random vectors in multivariate analysis by principal components, replacing the inner product in the vector space $\mathbb{R}^d$, given by $\langle x, y \rangle = \sum_{k=1}^d x_k y_k$, by $\langle x, y \rangle = \int x(t)y(t)\,dt$, and replacing matrices by linear operators. Then a random vector can be represented in the basis defined by the eigenvectors of its covariance matrix, which is the finite-dimensional equivalent of the Karhunen–Loève expansion.

10.5.2 From Karhunen–Loève representation to functional principal components

One of the main attractions of FPCA is that knowledge of the distribution of the (uncorrelated) set of FPC scores $\{A_1, A_2, \ldots\}$ is equivalent to knowledge of the distribution of $Y - \mu$, which is a consequence of the Karhunen–Loève expansion. While this equivalence is of theoretical interest, in practice one needs to truncate the sequence of FPC scores at a suitable index (chosen data-adaptively whenever possible). This truncation then corresponds to the needed regularization step, mapping the infinite trajectories to a finite number of FPC scores. Along the way, one also needs to estimate the (smooth) mean functions and the relevant eigenfunctions. This can be done by smoothing methods as demonstrated in the following.

Alternative representations of functional data by expansions in fixed basis functions have also been considered. These include Fourier and wavelet bases (Morris and Carroll, 2006). These representations have the advantage that neither mean nor eigenfunctions need to be determined from the data. Wavelets are particularly suited for data with non-smooth trajectories, such as functions containing small jumps and sharp edges. They are less well suited to reproduce smooth trajectories. The disadvantage of fixed basis functions is that they may not be very parsimonious, and a larger number of basis functions may be needed to represent a given sample of trajectories. In addition, the estimated coefficients are not uncorrelated (which means they carry less information and are less convenient for subsequent applications such as regression).

A preliminary exploration of functional principal components for longitudinal data is due to Rao (1958) in the context of growth curves. Other key references are Castro, Lawton, and Sylvestre (1986), who introduced the notion that eigenfunctions are functional "modes of variation"; Rice and Silverman (1991), who emphasized the need for smoothing, for which they used B-splines; and Ramsay and Silverman (2005), who started with a pre-smoothing step to first generate a sample of smooth trajectories, before proceeding with FPCA.

If complete trajectories are observed, or data observed on a grid are pre-smoothed and then considered as completely observed, one typically creates an equidistant grid $\{t_1, t_2, \ldots, t_N\}$ of N design points on the domain $[0, T]$ (where N is the same as the number of sampled trajectories, which corresponds to the number of subjects), and then one treats the data as N-vectors, one for each of the N subjects. The next step is then a multivariate principal component analysis for these N-vectors, that is, one obtains mean vector, eigenvectors, and principal component scores, without any smoothing (compare Cardot, Ferraty, and Sarda, 1999). Theoretical analysis focuses on asymptotics as $N \to \infty$. However, if data are either irregularly or not densely sampled, or are contaminated with noise, this approach does not work, and smoothing is necessary. As noise-contaminated measurements are the norm rather than the exception, the case of completely observed noise-free trajectories is mainly of theoretical interest.

10.5.3 The case of longitudinal data

For sparsely sampled longitudinal data, pre-smoothing to create completely observed trajectories is a less attractive option, as it introduces bias and artificial correlations into longitudinal data. This is because scatter-plot smoothing requires relatively dense and not too irregular designs. If there are "gaps" in the predictors, bandwidths must be increased, which in turn leads to increased bias. Irregular and sparse data, as typically encountered in longitudinal studies, were first considered by James, Hastie, and Sugar (2001), who used B-splines. The B-spline approach with random coefficients, pioneered by Shi, Weiss, and Taylor (1996) and Rice and Wu (2001), can also be easily adapted to the sparse and irregular case.

The FPCA approach and Karhunen–Loève representation cannot be directly adopted in the case of longitudinal data, which from now on we assume to consist of sparse, irregular and noisy measurements of the longitudinal trajectories. According to (10.2), the FPC scores, which are the random effects in the representation, would normally be estimated by approximating the integral by a Riemann sum. This works nicely for the case of fully observed trajectories but does not work for longitudinal data, due to large discretization errors. If the data are contaminated by noise, the approximation by sums does not work consistently, even for the case of dense measurements. The case of noisy measurements in FDA was first emphasized in the work of Staniswalis and Lee (1998).

We model noisy longitudinal data as follows. Let Y_{ij} be measurements of trajectories $Y_i(\cdot)$, made at sparse and irregularly spaced time points t_{ij}, $i = 1, \ldots, N$, $j = 1, \ldots, n_i$. Then

$$Y_{ij} = Y_i(t_{ij}) + e_{ij} = \mu(t_{ij}) + \sum_{k=1}^{\infty} A_{ik} \psi_k(t_{ij}) + e_{ij}.$$

Here, the e_{ij} are independent and identically distributed measurement errors with moments $E(e_{ij}) = 0$, $E(e_{ij}^2) = \sigma^2$; and the e_{ij} are considered to be independent of the FPC scores A_{ik}, where A_{ik} denotes the score for the ith subject and kth eigenfunction.

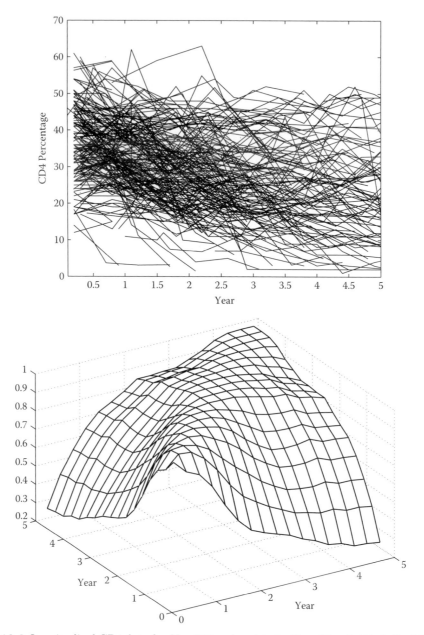

Figure 10.3 Longitudinal CD4 data for $N = 238$ male subjects. Top: "Spaghetti plot" of the data, connecting repeated measurements by straight lines. Bottom: Smooth estimate of the covariance surface, where the diagonal has been removed. (Reproduced from Yao, Müller, and Wang, 2005a. With permission.)

An example of sparse and irregular data for which this model may apply are longitudinal CD4 counts of AIDS patients (Figure 10.3, top panel); an application is presented in Section 10.5.6.

10.5.4 Principal analysis by conditional expectation

Here, we describe the principal analysis via conditional expectation (PACE) method to carry out FPCA for longitudinal (used synonymously here for sparse and irregularly sampled) data (Yao, Müller, and Wang, 2005a). The basis for this method is the principal analysis of random trajectories (PART) algorithm for obtaining the empirical Karhunen–Loève representation of smooth functional data, where measurements are contaminated with additional measurement error. This algorithm works irrespective of whether the measurements have been sampled on a dense and regular grid or on a sparse and irregular grid. Alternative algorithms that use pre-smoothing are available; see, for example, Ramsay and Silverman (2005) and the associated web site.

The PART algorithm consists of the following steps. In a first step, one pools all available measurements (t_{ij}, Y_{ij}), $i = 1, \ldots, N$, $j = 1, \ldots, n_i$, into one scatter plot and uses a one-dimensional smoother to obtain the estimate $\hat{\mu}(t)$ of the overall mean function $\mu(t)$. A technical requirement here is that the pooled locations t_{ij} over all subjects are dense on the domain or at least can be reasonably considered to become dense asymptotically. This will lead to consistency for this estimation step. Next, one forms all pairwise products

$$\{Y_{ij} - \hat{\mu}(t_{ij})\}\{Y_{il} - \hat{\mu}(t_{il})\}, \quad j \neq l,$$

which will be the responses for predictors (t_{ij}, t_{il}). These data are then entered into a two-dimensional scatter-plot smoother; for example, the local linear weighted least-squares approach (10.1). The output is the estimated covariance surface. The diagonal elements (for which $j = l$) are omitted from the input for the two-dimensional smoothing step, as they are contaminated by the measurement errors. The measurement error variance in fact can be estimated from these diagonal elements, under the assumption either that it is a fixed constant, or that it varies over time. In the latter case, one may obtain the variance function of the errors by smoothing along the direction of the diagonal.

While in our implementation we use the local linear smoothers described in Section 10.3, any alternative smoothing method can be used as well. One potential problem is that, while the estimated covariance matrix is easily seen to be symmetric (as the responses that are entered are symmetric in t_{ij}, t_{il}), it is not necessarily positive definite. This problem can be solved by projecting on positive definite surfaces, simply by truncating negative eigenvalues (for details, see Yao et al., 2003). From the estimated covariance surface, one obtains eigenfunctions and eigenvalues numerically after discretizing. The bandwidths for the smoothing steps can be obtained by cross-validation or similar procedures.

Once the mean function and eigenfunctions have been obtained, an important step in completing the empirical Karhunen–Loève representation, and thus the functional dimension reduction, is the estimation of the FPC scores. If the observations are noisy or sparse, the Riemann sums will not provide reasonable approximations to the integrals (10.2), and this is where the PACE approach to predict individual FPC scores comes in. For this approach, we make Gaussian assumptions, that is, A_{ik}, e_{ij} are assumed to be jointly normal. Define the vectors

$$\boldsymbol{Y}_i = (Y_{i1}, \ldots, Y_{in_i})',$$
$$\boldsymbol{\mu}_i = \{\mu(t_{i1}), \ldots, \mu(t_{in_i})\}',$$
$$\boldsymbol{\psi}_{ik} = \{\psi_k(t_{i1}), \ldots, \psi_k(t_{in_i})\}'.$$

The best predictors for the random effects are obtained via the conditional expectation

$$E(A_{ik}|\boldsymbol{Y}_i) = E(A_{ik}) + \mathrm{Cov}(A_{ik}, \boldsymbol{Y}_i)\mathrm{Cov}(\boldsymbol{Y}_i, \boldsymbol{Y}_i)^{-1}(\boldsymbol{Y}_i - \boldsymbol{\mu}_i)$$
$$= \lambda_k \boldsymbol{\psi}'_{ik}\Sigma_{Y_i}^{-1}(\boldsymbol{Y}_i - \boldsymbol{\mu}_i),$$

where

$$(\Sigma_{Y_i})_{j,l} = \mathrm{Cov}\{Y_i(t_{ij}), Y_i(t_{il})\} + \sigma^2 \delta_{jl},$$

with $\delta_{jl} = 1$ if $j = l$, and $\delta_{jl} = 0$ if $j \neq l$. Substituting the estimates discussed previously then leads to estimated predicted FPC scores,

$$\widehat{E}(A_{ik}|\boldsymbol{Y}_i) = \widehat{\lambda}_k \widehat{\boldsymbol{\psi}}'_{ik} \widehat{\Sigma}^{-1}_{Y_i} (\boldsymbol{Y}_i - \widehat{\boldsymbol{\mu}}_i). \tag{10.3}$$

10.5.5 Predicting individual trajectories

Once the number of included random coefficients K has been determined, we can use the predicted FPC scores (10.3) to obtain predicted individual trajectories

$$\widehat{Y}(t) = \widehat{\mu}(t) + \sum_{k=1}^{K} \widehat{E}(A_k|\boldsymbol{Y}_i)\, \widehat{\psi}_k(t). \tag{10.4}$$

An important issue is the choice of K. This corresponds to the number of FPC scores and, accordingly, the number of random effects in the model. For this choice, several options are available. A simple and fast method is the scree plot. Here, one plots the fraction of variance unexplained by the first K components against the number of included components K,

$$S(K) = 1 - \sum_{k=1}^{K} \widehat{\lambda}_k \Big/ \sum_{k=1}^{K_\infty} \widehat{\lambda}_k \,,$$

where K_∞ is a large number of components, which is clearly larger than the number to be selected. One looks for an "elbow" in this graph, that is, a value of K at which the rate of decline slows substantially, as K increases further, and the number K where this elbow occurs is the selected number.

A second promising approach is Akaike information criterion (AIC) type criteria. As no likelihood exists *a priori*, one can devise various types of pseudo-likelihood and then construct a pseudo-AIC value. For example, a pseudo-Gaussian log-likelihood is

$$\widehat{L}_1 = \sum_{i=1}^{N} \left[-\frac{n_i}{2} \log (2\pi) - \frac{1}{2} \log \{\det(\widehat{\Sigma}_{Y_i})\} - \frac{1}{2}(\boldsymbol{Y}_i - \widehat{\boldsymbol{\mu}}_i)' \widehat{\Sigma}^{-1}_{Y_i}(\boldsymbol{Y}_i - \widehat{\boldsymbol{\mu}}_i) \right],$$

while a conditional version of the likelihood, conditioning on predicted FPC scores, is given by

$$\widehat{L} = \sum_{i=1}^{N} \left\{ -\frac{n_i}{2} \log (2\pi) - \frac{n_i}{2} \log (\widehat{\sigma}^2) \right.$$
$$\left. - \frac{1}{2\widehat{\sigma}^2} \left(\boldsymbol{Y}_i - \widehat{\boldsymbol{\mu}}_i - \sum_{k=1}^{K} \widehat{A}_{ik} \widehat{\psi}_{ik} \right)' \left(\boldsymbol{Y}_i - \widehat{\boldsymbol{\mu}}_i - \sum_{k=1}^{K} \widehat{A}_{ik} \widehat{\psi}_{ik} \right) \right\}. \tag{10.5}$$

In either version, the pseudo-AIC value is then AIC $= -2\widehat{L} + 2K$, and the minimizing value for K is selected. Pseudo-BIC criteria can also be used.

A characteristic of the PACE method is that it borrows strength from the entire sample to predict individual trajectories, in contrast to the more traditional non-parametric regression analysis of longitudinal data, where each curve would be fitted separately from the others by smoothing it individually, without regard for the other curves in the sample. We note that this traditional approach has proven useful for regular designs, as encountered in growth studies, and remains recommended as an exploratory tool. However, this approach ignores the information available across the sample and is thus less efficient than the functional approaches. It is also infeasible for the commonly encountered longitudinal data where the number of observations per curve is small and the locations of the measurements are irregular. In theoretical analysis, this situation is adequately reflected by assuming that the number of repeated measurements per subject is bounded, while the number of individuals will potentially be large.

We note that, once the individual FPC scores have been obtained, they can be entered into further statistical analysis. Pairwise scatter plots of one FPC score against another, plotted for all subjects, can reveal patterns of interest, and are a very useful exploratory tool, for example to identify clusters in the data. Pairs or vectors of FPC scores have been successfully employed for more formal classification or clustering of samples of trajectories (Müller, 2005; compare also James and Sugar, 2003).

A number of asymptotic properties of functional principal components have been investigated. Most of the earlier results are due to the French school (an early paper is Dauxois, Pousse, and Romain, 1982). Assuming that more than one but at most finitely many observations are available per trajectory, and without making Gaussian assumptions, it was shown in Hall, Müller, and Wang (2006) that the eigenfunction estimators achieve the usual non-parametric rates for estimating a smooth function as sample size $N \to \infty$. For the case where entire trajectories are available without measurement error, the optimal rates are parametric (Hall and Hosseini-Nasab, 2006). The above estimators for covariance surface and mean function converge in sup-norm and so do the eigenfunctions, under longitudinal designs (Yao, Müller, and Wang, 2005a). The predicted FPC scores as obtained from the PACE method converge to the actual FPC scores as the designs get asymptotically denser (more and more measurements per subject, see Müller, 2005). Under additional Gaussian assumptions, the estimators of the predicted FPC scores converge to their targets, and pointwise/uniform confidence bands can be constructed for predicted trajectories (Yao, Müller, and Wang, 2005a).

10.5.6 Application to longitudinal CD4 data

As an illustration, the PACE method was applied to longitudinal CD4 counts obtained for 283 male AIDS patients. Potential issues with informative dropout in this study are ignored in this analysis. These data fit the description of sparse and irregular data and are shown in the top panel of Figure 10.3, where the data for each individual are connected by straight lines. The numbers of observations per subject are between 2 and 14. We aim to describe the characteristic features of the underlying longitudinal trajectories. The estimated covariance function for these data is in the bottom panel of Figure 10.3, where the diagonal has been omitted as described above. The overall mean function $\widehat{\mu}(t)$ is depicted in the bottom panels of Figure 10.4. The conditional pseudo-likelihood (10.5) based AIC criterion yielded $K = 3$, that is, three eigenfunctions are included.

Of interest is an assessment of the extremes in a sample. In the functional setting, it is not so straightforward to conceptualize what these extremes are. One possibility is to identify those subjects whose trajectories are most aligned with an eigenfunction. This device has been studied by Jones and Rice (1992), who also carried out further exploration of samples of functional data by means of the eigenfunctions. The bottom panels of Figure 10.4 display these extreme trajectories. Along with the eigenfunctions, they provide an idea about the modes of variation that are present. The first mode is a linear decline, exemplified by the subject in the left bottom panel of Figure 10.4; the second mode is a decline with a plateau in the middle, during the third year, after which a more rapid decline in CD4 counts resumes. The third and weakest mode corresponds to a leveling off toward the end, stabilizing at a low level, with a possible increase. One should not read too much into the increase at the right end of the third eigenfunction; this may simply be caused by boundary effects. Nevertheless, as this example shows, an added benefit of eigenfunctions is their interpretability as (orthogonal) modes of variation that in their entirety explain the observed

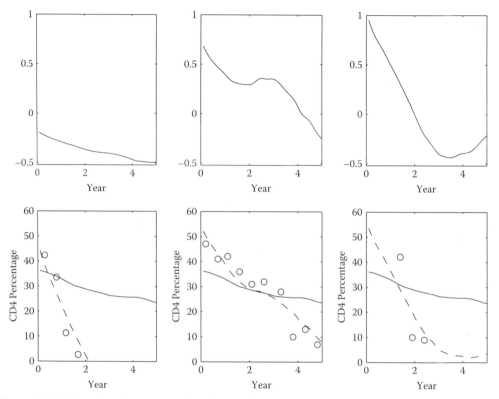

Figure 10.4 Eigenfunctions, mean function, and extreme trajectories for longitudinal CD4 data. Top (left to right): First, second, and third eigenfunction. Bottom: Mean function for all trajectories (solid line), and three individuals with data and fitted trajectories, most aligned in the directions of the three eigenfunctions. (Reproduced from Yao, Müller, and Wang, 2005a. With permission.)

variation in the functional data. The shapes of individual eigenfunctions often associate a concrete description with these modes of variation.

Finally, we are interested in modeling individual trajectories, which are obtained via the estimated FPC scores; see (10.4). The predicted trajectories for four subjects, including confidence bands, are shown in Figure 10.5, including Gaussian-based confidence bands. Open questions that will be of interest for future research include the extension of FDA methods to repeated non-Gaussian (binomial, Poisson) data, the case of varying eigenfunctions in dependency on a covariate, and incorporating informative missingness and time-to-event information. A full exploration of practical features in the context of various longitudinal studies will also be of interest.

10.6 Functional regression models

10.6.1 Overview

For longitudinal data analysis, the trajectories observed for each subject can serve as both predictors and response in a regression model. The case where they are included among the predictors has been well explored in FDA, primarily for the case where the trajectories are fully observed without noise (Cardot et al., 2003, Cai and Hall, 2006). We review here some of the available models. Linear functional models may include a random trajectory in either predictors or responses, or both. We assume here that the data are written as (X, Y), where Y stands for response and X for predictor, which could be scalar or functional. Means will

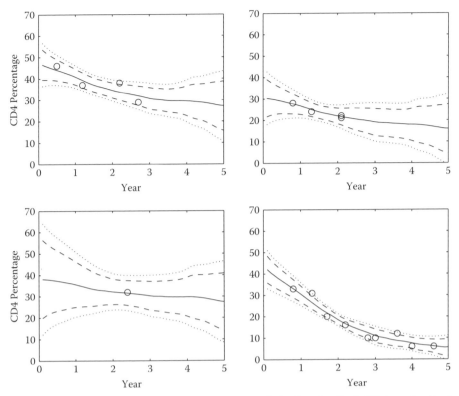

Figure 10.5 Predicted trajectories for four subjects of the longitudinal CD4 study. Each panel displays the observed data (circles), predicted trajectory (solid line), and local (dashed line) as well as uniform (dotted line) 95% confidence bands, obtained under Gaussian assumptions. (Reproduced from Yao, Müller, and Wang, 2005a. With permission.)

be denoted by μ_X, μ_Y. In the functional case, we denote eigenfunctions by ϕ_k for X and ψ_k for Y.

The linear model for a scalar response and a functional predictor is

$$E(Y|X) = \mu_Y + \int_0^S \{X(s) - \mu_X(s)\}\beta(s)\,ds,$$

where β is the regression parameter function, and $[0, S]$ is the domain of X. An extension to functional responses is the model

$$E\{Y(t)|X\} = \mu_Y(t) + \int_0^S \{X(s) - \mu_X(s)\}\beta(s, t)\,ds, \tag{10.6}$$

where now the regression parameter function has two arguments (i.e., is a surface). This model dates back to Ramsay and Dalzell (1991). It can be interpreted as an extension of the multivariate linear regression model $E(Y|X) = BX$ for a parameter matrix B to the functional case.

In such a multivariate linear regression model, a common estimation scheme proceeds via the least-squares normal equation. For $\boldsymbol{X} \in \mathbb{R}^p$, $\boldsymbol{Y} \in \mathbb{R}^q$, the normal equation is $\mathrm{Cov}(\boldsymbol{X}, \boldsymbol{Y}) = \mathrm{Cov}(\boldsymbol{X})B$, where $\mathrm{Cov}(\boldsymbol{X}, \boldsymbol{Y})$ is the $p \times q$ matrix with elements $a_{jk} = \mathrm{Cov}(X_j, Y_k)$. This equation can be solved for B if the $p \times p$ covariance matrix $\mathrm{Cov}(\boldsymbol{X})$ is invertible. The situation is much less straightforward for the functional extension. We can define an analogous "functional normal equation" (He, Müller, and Wang, 2000),

$$r_{XY} = R_{XX}\beta \quad \text{for } \beta \in L^2,$$

where $R_{XX} : L^2 \to L^2$ is the auto-covariance operator of X, defined by

$$(R_{XX}\beta)(s,t) = \int r_{XX}(s,w)\beta(w,t)dw,$$

and

$$r_{XX}(s,t) = \text{Cov}\{X(s), X(t)\}, \quad r_{XY}(s,t) = \text{Cov}\{X(s), Y(t)\}.$$

As R_{XX} is a compact operator in L^2, it is in principle not invertible. Thus, we face an inverse problem which requires regularization (compare He, Müller, and Wang, 2003).

A model that is useful in classifying longitudinal time courses is the generalized functional linear model (James, 2002; Müller and Stadtmüller, 2005; Müller, 2005). Here, the predictors are functional, and the responses are generalized variables, such as binary outcomes, which may stand for class membership or Poisson counts. With an appropriate link function h^{-1} (see Chapters 1, 3, and 4), this model can be written as

$$E(Y|X) = h^{-1}\left\{\mu + \int_0^S X(s)\beta(s)\,ds\right\},$$

coupled with a variance function $\text{Var}(Y|X) = v\{E(Y|X)\}$. This model is an extension of the common generalized linear model to the case of functional predictors. It can be implemented with both known or unknown link/variance function (see Müller and Stadtmüller, 2005).

The class of "functional response models" (Faraway, 1997; Chiou, Müller, and Wang, 2003, 2004) is of interest in functional dose-response models and similar applications. In this model, the predictor is usually a vector $\boldsymbol{Z}$, while the response is functional,

$$E\{Y(t)|\boldsymbol{Z} = \boldsymbol{z}\} = \mu(t) + \sum_{k=1}^{K} \alpha_k(\boldsymbol{\gamma}_k'\boldsymbol{z})\psi_k(t).$$

Here, the $\boldsymbol{\gamma}_k$ are single indices (i.e., vectors that project the covariates $\boldsymbol{Z}$ into one dimension), and the α_k are link functions to the random effects. Sometimes, simpler structured models such as a "multiplicative effects model" are useful, that is,

$$\mu(t, \boldsymbol{z}) = \mu_0(t)\theta(\boldsymbol{z}), \quad E\{Y(t)\} = \mu_0(t), \quad E\{\theta(\boldsymbol{Z})\} = 1,$$

for a function $\theta(\cdot)$ (see Chiou et al., 2003).

Further classes of models of interest are those with varying supports. In the regression models above, the entire predictor function is assumed to contribute to a response. In many applications, this might not be realistic. Examples of this were given in Malfait and Ramsay (2003) and Müller and Zhang (2005). In the latter paper, the response is remaining lifetime, to be predicted from a longitudinal covariate that is observed up to current time. As current time progresses, the functional regression model needs to be updated. This leads to time-varying domains and, accordingly, to time-varying coefficient functional regression models. In the extreme case, the usual varying-coefficient model

$$E\{Y(t)|X\} = \mu_Y(t) + \zeta(t)X(t)$$

(under the assumption of one predictor process) emerges as a special case; here, $\zeta(\cdot)$ is the varying-coefficient function (Fan and Zhang, 2000; Wu and Yu, 2002).

These models can be extended to the longitudinal (sparse and irregular) case, following the above PACE approach, whenever the model can be written in terms of FPC scores. In the following, we demonstrate this feature for the functional regression model (10.6).

10.6.2 Functional regression for longitudinal data

Extending the functional linear regression model (10.6) introduced previously to the case of sparse and irregular data, we assume that available measurements for predictor and response curves are given as follows, with their Karhunen–Loève representations included:

$$U_{il} = X_i(s_{il}) + e_{il} = \mu_X(s_{il}) + \sum_{m=1}^{\infty} A_{im}\phi_m(s_{il}) + \epsilon_{il}, \quad s_{il} \in [0, S],$$

$$V_{ij} = Y_i(t_{ij}) + e_{ij} = \mu_Y(t_{ij}) + \sum_{k=1}^{\infty} B_{ik}\psi_k(t_{ij}) + e_{ij}, \quad t_{ij} \in [0, T].$$

Here, the times s_{ij} are those where measurements are recorded for predictor processes X; t_{ij} are the times where measurements are recorded for the response processes Y; and these times can differ between X and Y, but are both assumed to be sparse. The random effects (FPCA scores) are denoted here by A_{im} for predictor processes and by B_{ik} for response processes. Measurements for both processes are not only sparse and irregular, but also contaminated by measurement errors ϵ_{il} and e_{ij}, respectively, which are assumed to satisfy the properties listed in Section 10.5.3.

Applying FPCA, by using the orthonormality properties of the eigenfunctions, one finds that the regression parameter function β in (10.6) can be represented by

$$\beta(s, t) = \sum_{k,m=1}^{\infty} \frac{E(A_m B_k)}{E(A_m^2)} \phi_m(s)\psi_k(t). \tag{10.7}$$

This reduces the problem of estimating β to the problem of obtaining an estimator of $E(A_m B_k)$, for which we consider

$$\widehat{E}(A_m B_k) = \int_0^T \int_0^S \widehat{\phi}_m(s)\widehat{\Gamma}_{XY}(s,t)\widehat{\psi}_k(t)\, ds dt, \tag{10.8}$$

where $\widehat{\Gamma}_{XY}(s,t)$ is a local linear smoother for the cross-covariance function $\Gamma_{XY}(s,t) = \text{Cov}\{X(s), Y(t)\}$ (Yao, Müller, and Wang, 2005b), and the integral is evaluated by numerical integration.

Once the regression parameter surface β has been obtained, one may then aim to predict individual response trajectories from the available observations of the corresponding predictor process, that is, to predict Y^* from the observations $\boldsymbol{U}^* = (U_1^*, \ldots, U_{L^*}^*)'$ available for $X^*(\cdot)$, where $\boldsymbol{U}^*$ denotes data available on the predictor process for one individual. Under Gaussian assumptions, the best predictor is given by

$$E\{Y^*(t)|X^*(\cdot)\} = \mu_Y(t) + \int_0^S \beta(s, t)\{X^*(s) - \mu_X(s)\}\, ds$$

$$= \mu_Y(t) + \sum_{k,m=1}^{\infty} \frac{E(A_m B_k)}{E(A_m^2)} A_m^* \psi_k(t).$$

An estimate for this predictor is simply obtained by plugging in estimates for the unknown quantities. Choosing K and M for the number of included components to represent processes

X and Y, respectively, we arrive at

$$\widehat{Y}_{KM}^*(t) = \widehat{\mu}_Y(t) + \sum_{m=1}^{M}\sum_{k=1}^{K} \frac{\widehat{E}(A_m B_k)}{\widehat{E}(A_m^2)}\widehat{E}(A_m^*|\boldsymbol{U}^*)\widehat{\psi}_k(t),$$

where $\widehat{E}(A_m^*|\boldsymbol{U}^*)$ is estimated by the PACE method, as described in Section 10.5.4, given observations $\boldsymbol{U}^* = (U_1^*, \ldots, U_{L^*}^*)'$ of $X^*(\cdot)$.

The theory developed by Yao, Müller, and Wang (2005b) includes consistency of the regression parameter surface estimators, as well as some basic inference, and also construction of pointwise and uniform confidence bands for predicted trajectories under Gaussian assumptions. This paper also contains extensions of the usual coefficient of determination $R^2 = \mathrm{Var}\{E(Y|X)\}/\mathrm{Var}(Y)$, which is used to measure the strength of a regression relationship, to the functional case. Applying orthonormality properties of the eigenfunctions, one such possible extension can be represented as

$$R^2 = \frac{\int_0^T \mathrm{Var}[E\{Y(t)|X\}]dt}{\int_0^T \mathrm{Var}\{Y(t)\}dt} = \frac{\sum_{k,m=1}^{\infty} E(A_m^2 B_k^2)/E(A_m^2)}{\sum_{k=1}^{\infty} E(B_k^2)}.$$

The quantities $E(A_m^2)$, $E(B_k^2)$ correspond to the eigenvalues of the X and Y processes, and $E(A_m B_k)$ can be estimated as in (10.8). Substituting the corresponding estimates then leads to the estimated functional coefficient of determination R^2.

10.6.3 Illustration with data from the Baltimore Longitudinal Study of Aging

Longitudinal measurements of body mass index (BMI) and systolic blood pressure (SBP) were obtained for 812 participants of the Baltimore Longitudinal Study on Aging (BLSA), as reported in Pearson et al. (1997). The measurements fit the description of being irregular and sparse. We provide a brief summary of the functional regression analysis conducted in Yao, Müller, and Wang (2005b). The data and mean function estimates for all subjects can be found in Figure 10.6. From this figure one can see the irregular nature of the timings as well as their sparseness. The relationship between the trajectories in the left and right panels is difficult to discern.

Running the functional regression machinery, we obtain the estimate of the regression surface function $\widehat{\beta}(\cdot, \cdot)$ for these data for the functional regression of SBP or BMI, as depicted in Figure 10.7. This function illustrates the influence of predictor functions on response trajectories. The time axis of predictor (BMI) trajectories is labeled s, running toward the right, while the time axis of response (SBP) trajectories is labeled t, running toward the left. In this functional regression model, the entire predictor trajectory influences the entire response curve. We can interpret the features of this regression parameter surface as follows. At early ages, around 60, SBP is related to an overall average of BMI. At later ages, around 80, SBP is positively correlated with what is best characterized as rate of increase in BMI. A continuous transition between these regimes occurs in between.

Finally, predicted trajectories of systolic blood pressure for four randomly selected participants are displayed in Figure 10.8. The predictors are the measurements of BMI, which are not shown in the graphs. A curious stricture occurs in the confidence bands around age 75. This is an area where apparently the variation of the response trajectories has a minimum.

The methods described above can be extended to the case of more than one predictor process, where one can use the FPC scores derived from the different predictor processes as predictors. However, the uncorrelatedness feature of the predictors will be lost in this case.

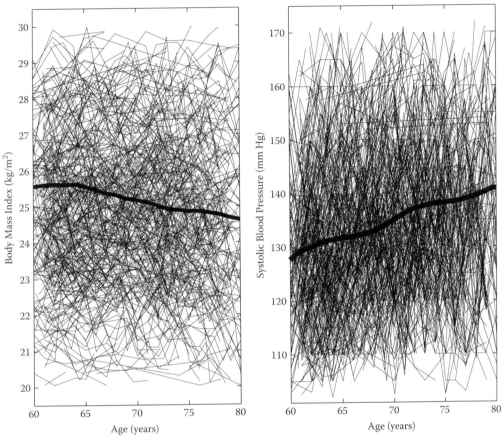

Figure 10.6 Longitudinal measurements of (left) BMI and (right) SBP for 812 participants in the BLSA. Thick solid curves are estimated mean functions. (Reproduced from Yao, Müller, and Wang, 2005b. With permission.)

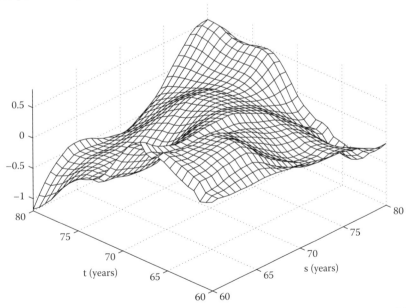

Figure 10.7 Estimated regression parameter surface β (10.7) for regressing SBP or BMI, for the BLSA data. (Reproduced from Yao, Müller, and Wang, 2005b. With permission.)

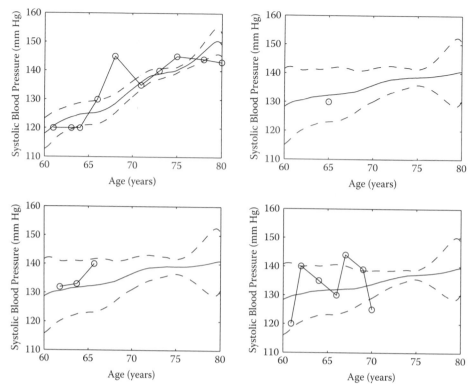

Figure 10.8 Predicted response trajectories for four BLSA participants. Shown are predicted trajectories (solid curves), data for the response (SBP) trajectories (circles, connected by straight lines), and pointwise 95% confidence intervals. Note that the data shown are not used for the prediction, which is solely based on measurements for the predictor (BMI) trajectories. (Reproduced from Yao, Müller, and Wang, 2005b. With permission.)

10.7 Concluding remarks and outlook

Functional data analysis provides an inherently non-parametric approach for the analysis of data that consist of samples of time courses or random trajectories. It is a relatively young field with a focus on modeling and data exploration under very flexible model assumptions with no or few parametric components. In this chapter, we have reviewed some of the tools of FDA, which include smoothing, functional principal components, functional linear models, and time warping. Warping or curve registration is an FDA-specific methodology that aims at adjusting for random time distortions.

While in the usual functional data analysis paradigm the sample functions are considered as continuously observed trajectories, in longitudinal data analysis one mostly deals with sparsely and irregularly observed data that also are corrupted with noise. We have described some of the adjustments of the FDA techniques that are needed to take full advantage of the FDA approach when analyzing longitudinal data. The extension of FDA toward the analysis of longitudinal data is a fairly recent undertaking that presents a promising avenue for future research.

In addition to the FDA methodology described in this chapter, several other FDA approaches have applications for longitudinal data. These include functional analysis-of-variance decompositions using smoothing spline models, proposed in Brumback and Rice (1998), and applications of P-splines (Bugli and Lambert, 2006); see Chapter 11 and Chapter 12. Another class of non- or semi-parametric models that is of interest for longitudinal studies is varying-coefficient models, where $Y(t)$ is related to a series of predictors $X_1(t), \ldots,$

$X_p(t)$. In a typical two-step procedure (Fan and Zhang, 2000), one conducts a linear regression at each time point that is an element of a grid of time points and then applies smoothing to the resulting regression coefficients to obtain smooth varying-coefficient functions. In addition, shape-invariant modeling is a promising functional method, as it combines the analysis of time variation (warping) with that of amplitude variation.

For functional inference, bootstrap methods based on resampling subjects, keeping data together that belong to the same subject, have been used, and are very promising, both for construction of confidence regions and for significance tests. Asymptotic inference is available under Gaussian assumptions, but is not yet available on a wider scale; compare Fan and Lin (1998). This chapter demonstrates that functional approaches provide a flexible alternative to common parametric models for analyzing longitudinal data. Software is currently available in the form of specialized R and Matlab programs offered by various researchers. No definitive packages have emerged as of yet.

At this time, a number of key techniques are in place, notably smoothing and differentiation of noisy data, warping and curve registration, functional principal component analysis, functional regression, and penalized regularization techniques. The unique combination of methods from functional analysis, stochastic processes, multivariate analysis, and smoothing, and the many open questions, make the interface of FDA and longitudinal data analysis a rewarding research area. Practitioners can benefit from the highly flexible FDA toolbox, which is useful for both exploratory analysis and inference and facilitates new insights into the dynamics of longitudinal data.

Acknowledgments

I am indebted to two editorial reviewers for very careful reading and numerous suggestions which led to improvements. This work was supported in part by U.S. National Science Foundation grants DMS03-54448 and DMS05-05537, and NIH Grants POI-AG08761 and POI-AG022500.

References

Ash, R. B. and Gardner, M. F. (1975). *Topics in Stochastic Processes*. New York: Academic Press.

Bowman, A. W. and Azzalini, A. (1997). *Applied Smoothing Techniques for Data Analysis: The Kernel Approach with S-Plus Illustrations*. Oxford: Oxford Science Publications.

Brumback, B. and Rice, J. (1998). Smoothing spline models for the analysis of nested and crossed samples of curves. *Journal of the American Statistical Association* **93**, 961–976.

Bugli, C. and Lambert, P. (2006). Functional ANOVA with random functional effects: An application to event-related potentials modelling for electroencephalograms analysis. *Statistics in Medicine* **25**, 3718–3739.

Cai, T. and Hall, P. (2006). Prediction in functional linear regression. *Annals of Statistics* **34**, 2159–2179.

Capra, W. B. and Müller, H. G. (1997). An accelerated-time model for response curves. *Journal of the American Statistical Association* **92**, 72–83.

Cardot, H., Ferraty, F., and Sarda, P. (1999). Functional linear model. *Statistics and Probability Letters* **45**, 11–22.

Cardot, H., Ferraty, F., Mas, A., and Sarda, P. (2003). Testing hypotheses in the functional linear model. *Scandinavian Journal of Statistics* **30**, 241–255.

Castro, P. E., Lawton, W. H., and Sylvestre, E. A. (1986). Principal modes of variation for processes with continuous sample curves. *Technometrics* **28**, 329–337.

Chiou, J. M. and Müller, H. G. (2005). Estimated estimating equations: Semiparametric inference for clustered/longitudinal data. *Journal of the Royal Statistical Society, Series B* **67**, 531–553.

Chiou, J. M., Müller, H. G., and Wang, J. L. (2003). Functional quasi-likelihood regression models with smooth random effects. *Journal of the Royal Statistical Society, Series B* **65**, 405–423.

Chiou, J. M., Müller, H. G., and Wang, J. L. (2004). Functional response models. *Statistica Sinica* **14**, 675–693.

Chiou, J. M., Müller, H. G., Wang, J. L., and Carey, J. R. (2003). A functional multiplicative effects model for longitudinal data, with application to reproductive histories of female medflies. *Statistica Sinica* **13**, 1119–1133.

Conway, J. B. (1985). *A Course in Functional Analysis*. New York: Springer-Verlag.

Dauxois, J., Pousse, A., and Romain, Y. (1982). Asymptotic theory for the principal component analysis of a vector random function: Some applications to statistical inference. *Journal of Multivariate Analysis* **12**, 136–154.

Dubin, J. and Müller, H. G. (2005). Dynamical correlation for multivariate longitudinal data. *Journal of the American Statistical Association* **100**, 872–881.

Eubank, R. (1999). *Nonparametric Regression and Spline Smoothing*. New York: Marcel Dekker.

Fan, J., and Gijbels, I. (1996). *Local Polynomial Modelling and Its Applications*. London: Chapman & Hall.

Fan, J. and Lin, S. K. (1998). Test of significance when data are curves. *Journal of the American Statistical Association* **93**, 1007–1021.

Fan, J. and Zhang, J. (2000). Two-step estimation of functional linear models with applications to longitudinal data. *Journal of the Royal Statistical Society, Series B* **62**, 303–322.

Faraway, J. J. (1997). Regression analysis for a functional response. *Technometrics* **39**, 254–262.

Ferraty, F. and Vieu, P. (2006). *Nonparametric Functional Data Analysis*. New York: Springer.

Gasser, T. and Kneip, A. (1995). Searching for structure in curve samples. *Journal of the American Statistical Association* **90**, 1179–1188.

Gasser, T. and Müller, H. G. (1984). Estimating regression functions and their derivatives by the kernel method. *Scandinavian Journal of Statistics* **11**, 171–184.

Gasser, T., Müller, H. G., and Mammitzsch, V. (1985). Kernels for nonparametric curve estimation. *Journal of the Royal Statistical Society, Series B* **47**, 238–252.

Gasser, T., Müller, H. G., Köhler, W., Molinari, L., and Prader, A. (1984). Nonparametric regression analysis of growth curves. *Annals of Statistics* **12**, 210–229.

Gasser, T., Müller, H. G., Köhler, W., Prader, A., Largo, R., and Molinari, L. (1985). An analysis of the mid-growth spurt and of the adolescent growth spurt based on acceleration. *Annals of Human Biology* **12**, 129–148.

Gervini, D. and Gasser, T. (2004). Self-modelling warping functions. *Journal of the Royal Statistical Society, Series B* **66**, 959–971.

Gervini, D. and Gasser, T. (2005). Nonparametric maximum likelihood estimation of the structural mean of a sample of curves. *Biometrika* **92**, 801–820.

Hall, P. and Hosseini-Nasab, M. (2006). On properties of functional principal components analysis. *Journal of the Royal Statistical Society, Series B* **68**, 109–126.

Hall, P., Müller, H. G., and Wang, J. L. (2006). Properties of principal component methods for functional and longitudinal data analysis. *Annals of Statistics* **34**, 1493–1517.

He, G., Müller, H. G., and Wang, J.L. (2000). Extending correlation and regression from multivariate to functional data. In M. L. Puri (ed.), *Asymptotics in Statistics and Probability: Papers in Honor of George Gregory Roussas*, pp. 301–315. Utrecht: VSP.

He, G., Müller, H. G., and Wang, J. L. (2003). Functional canonical analysis for square integrable stochastic processes. *Journal of Multivariate Analysis* **85**, 54–77.

He, G., Müller, H. G., and Wang, J. L. (2004). Methods of canonical analysis for functional data. *Journal of Statistical Planning and Inference* **122**, 141–159.

Heagerty, P. J. (1999). Marginally specified logistic-normal models for longitudinal binary data. *Biometrics* **55**, 688–698.

Heagerty, P. J. and Zeger, S. L. (2000). Marginalized multilevel models and likelihood inference. *Statistical Science* **15**, 1–26.

Heckman, N. and Zamar, R. (2000). Comparing the shapes of regression functions. *Biometrika* **87**, 135–144.

Hotelling, H. (1936). Relations between two sets of variates. *Biometrika* **28**, 321–377.

James, G. (2002). Generalized linear models with functional predictors. *Journal of the Royal Statistical Society, Series B* **64**, 411–432.

James, G., Hastie, T. G., and Sugar, C. A. (2001). Principal component models for sparse functional data. *Biometrika* **87**, 587–602.

James, G. and Sugar, C. A. (2003). Clustering for sparsely sampled functional data. *Journal of the American Statistical Association* **98**, 397–408.

Jones, M. C. and Foster, P. J. (1996). A simple nonnegative boundary correction for kernel density estimation. *Statistica Sinica* **6**, 1005–1013.

Jones, M. C. and Rice, J. (1992). Displaying the important features of large collections of similar curves. *American Statistician* **46**, 140–145.

Ke, C., and Wang, Y. (2001). Semiparametric nonlinear mixed-effects models and their applications. *Journal of the American Statistical Association* **96**, 1272–1281.

Kirkpatrick, M., and Heckman, N. (1989). A quantitative genetic model for growth, shape, reaction norms and other infinite-dimensional characters. *Journal of Mathematical Biology* **27**, 429–450.

Kneip, A. and Gasser, T. (1992). Statistical tools to analyze data representing a sample of curves. *Annals of Statistics* **16**, 82–112.

Leng, X. and Müller, H. G. (2006). Time ordering of gene co-expression. *Biostatistics* **7**, 569–584.

Leurgans, S. E., Moyeed, R. A., and Silverman, B. W. (1993). Canonical correlation analysis when the data are curves. *Journal of the Royal Statistical Society: Series B* **55**, 725–740.

Lin, X., Wang, N., Welsh, A. H., and Carroll, R. J. (2004). Equivalent kernels of smoothing splines in nonparametric regression for clustered longitudinal data. *Biometrika* **91**, 177–193.

Lindstrom, M. J. (1995). Self-modelling with random shift and scale parameters and a free knot spline shape function. *Statistics in Medicine* **14**, 2009–2021.

Liu, X. and Müller, H. G. (2003). Modes and clustering for time-warped gene expression profile data. *Bioinformatics* **19**, 1937–1944.

Liu, X. and Müller, H. G. (2004). Functional convex averaging and synchronization for time-warped random curves. *Journal of the American Statistical Association* **99**, 687–699.

Malfait, N. and Ramsay, J. O. (2003). The historical functional linear model. *Canadian Journal of Statistics* **31**, 115–128.

Marron, J. S., Müller, H. G., Rice, J., Wang, J. L., Wang, N. Y., Wang, Y. D., Davidian, M., Diggle, P., Follmann, D., Louis, T. A., Taylor, J., Zeger, S., Goetghebeur, E., Little, R., and Carroll, R. J. (Discussants) (2004). Discussion of nonparametric and semiparametric regression. *Statistica Sinica* **14**, 615–629.

Morris, J. S. and Carroll, R. J. (2006). Wavelet-based functional mixed models. *Journal of the Royal Statistical Society, Series B* **68**, 179–199.

Müller, H. G. (1988). *Nonparametric Regression Analysis of Longitudinal Data*, Lecture Notes in Statistics 46. New York: Springer.

Müller, H. G. (1991). Smooth optimum kernel estimators near endpoints. *Biometrika* **78**, 521–530.

Müller, H. G. (2005). Functional modeling and classification of longitudinal data (with discussion). *Scandinavian Journal of Statistics* **32**, 223–246.

Müller, H. G. and Stadtmüller, U. (2005). Generalized functional linear models. *Annals of Statistics* **33**, 774–805.

Müller, H. G. and Zhang, Y. (2005). Time-varying functional regression for predicting remaining lifetime distributions from longitudinal trajectories. *Biometrics* **61**, 1064–1075.

Nadaraya, E. A. (1964). On estimating regression. *Theory of Probability and Its Applications* **9**, 141–142.

Pearson, J. D., Morrell, C. H., Brant, L. J., Landis, P. K., and Fleg, J. L. (1997). Gender differences in a longitudinal study of age associated changes in blood pressure. *Journal of Gerontology: Medical Sciences* **52**, 177–183.

Priestley, M. B. and Chao, M. T. (1972). Non-parametric function fitting. *Journal of the Royal Statistical Society, Series B* **34**, 385–392.

Ramsay, J. and Dalzell, C. J. (1991). Some tools for functional data analysis. *Journal of the Royal Statistical Society, Series B* **53**, 539–572.

Ramsay, J. and Li, X. (1998). Curve registration. *Journal of the Royal Statistical Society, Series B* **60**, 351–363.

Ramsay, J. and Silverman, B. (2002). *Applied Functional Data Analysis.* New York: Springer.

Ramsay, J. and Silverman, B. (2005). *Functional Data Analysis.* New York: Springer.

Rao, C. R. (1958). Some statistical methods for the comparison of growth curves. *Biometrics* **14**, 1–17.

Rice, J. (2004). Functional and longitudinal data analysis: Perspectives on smoothing. *Statistica Sinica* **14**, 631–647.

Rice, J. and Silverman, B. (1991). Estimating the mean and covariance structure nonparametrically when the data are curves. *Journal of the Royal Statistical Society, Series B* **53**, 233–243.

Rice, J. and Wu, C. (2001). Nonparametric mixed effects models for unequally sampled noisy curves. *Biometrics* **57**, 253–259.

Rønn, B. B. (2001). Nonparametric maximum likelihood estimation for shifted curves. *Journal of the Royal Statistical Society, Series B* **63**, 243–259.

Seifert, B. and Gasser, T. (1996). Finite sample variance of local polynomials: Analysis and solutions. *Journal of the American Statistical Association* **91**, 267–275.

Seifert, B. and Gasser, T. (2000). Data adaptive ridging in local polynomial regression. *Journal of Computational and Graphical Statistics* **9**, 338–360.

Service, S. K., Rice, J. A., and Chavez, F. P. (1998). Relationship between physical and biological variables during the upwelling period in Monterey Bay, CA. *Deep-Sea Research II — Topical Studies in Oceanography* **45**, 1669–1685.

Shi, M., Weiss, R. E., and Taylor, J. M. G. (1996). An analysis of paediatric CD4 counts for acquired immune deficiency syndrome using flexible random curves. *Applied Statistics* **45**, 151–163.

Silverman, B. W. (1986). *Density Estimation for Statistics and Data Analysis.* London: Chapman & Hall.

Silverman, B. W. (1995). Incorporating parametric effects into functional principal components analysis. *Journal of the Royal Statistical Society, Series B* **57**, 673–689.

Staniswalis, J. G. and Lee, J. J. (1998). Nonparametric regression analysis of longitudinal data. *Journal of the American Statistical Association* **93**, 1403–1418.

Wand, M. P. and Jones, M. C. (1995). *Kernel Smoothing.* London: Chapman & Hall.

Wang, N. (2003). Marginal nonparametric kernel regression accounting for within-subject correlation. *Biometrika* **90**, 43–52.

Wang, Y., Ke, C., and Brown, M. B. (2003). Shape-invariant modeling of circadian rhythms with random effects and smoothing spline ANOVA decompositions. *Biometrics* **59**, 804–812.

Watson, G. S. (1964). Smooth regression analysis. *Sankhyā, Series A* **26**, 359–372.

Wu, C. O. and Yu, K. F. (2002). Nonparametric varying coefficient models for the analysis of longitudinal data. *International Statistical Review* **70**, 373–393.

Yao, F. and Lee, T. C. M. (2006). Penalized spline models for functional principal component analysis. *Journal of the Royal Statistical Society, Series B* **68**, 3–25.

Yao, F., Müller, H. G., and Wang, J. L. (2005a). Functional data analysis for sparse longitudinal data. *Journal of the American Statistical Association* **100**, 577–590.

Yao, F., Müller, H. G., and Wang, J. L. (2005b). Functional linear regression analysis for longitudinal data. *Annals of Statistics* **33**, 2873–2903.

Yao, F., Müller, H. G., Clifford, A. J., Dueker, S. R., Follett, J., Lin, Y., Buchholz, B. A., and Vogel, J. S. (2003). Shrinkage estimation for functional principal component scores with application to the population kinetics of plasma folate. *Biometrics* **59**, 676–685.

Zeger, S. L., Liang, K. Y., and Albert, P. S. (1988). Models for longitudinal data: A generalized estimating equation approach. *Biometrics*, **44**, 1049–1060.

Zhao, X., Marron, J. S., and Wells, M. T. (2004). The functional data analysis view of longitudinal data. *Statistica Sinica* **14**, 789–808.

CHAPTER 11

Smoothing spline models for longitudinal data

S. J. Welham

Contents

11.1 Introduction and overview

Smoothing splines have become an established method for fitting curves with a flexible but smooth form that is determined by the data. This is particularly useful where there is no *a priori* form for response to an underlying covariate, or where examination of the data indicates that standard linear or non-linear models are unlikely to give a good fit. In the context of longitudinal data, smoothing splines have proved useful in providing a quantitative description of treatment profiles across time. Comparison of treatment profiles then requires a good model for covariation between subjects and across measurements within subjects. The embedding of smoothing splines within the linear mixed-model framework allows an easy route to the joint modeling of mean and variance via residual or restricted maximum likelihood (REML) estimation, and this chapter discusses these methods of modeling longitudinal data. For simplicity, we assume initially that data can be considered to be normally distributed (possibly after transformation), and later discuss extensions to non-normal data.

The structure of this chapter is as follows. The cubic smoothing spline is considered first, as this is the most commonly used spline in practice. Section 11.2 defines the general cubic spline, the natural cubic spline, and gives an overview of spline methods for modeling data to distinguish the smoothing spline from regression spline and penalized spline methods. Section 11.3 defines the bases commonly used for the unconstrained and natural cubic splines and some important properties of the natural cubic spline. Section 11.4 contains an overview of linear mixed models for longitudinal data and the REML method of estimation and Section 11.5 demonstrates the representation of a smoothing spline as a best linear unbiased predictor (BLUP) from a linear mixed model. Section 11.6 describes different forms of the smoothing spline mixed model for longitudinal data; it starts with a simple smoothing spline model for an individual, and then develops a complex model that uses smoothing splines to describe common, group, and individual subject profiles, allowing for additional explanatory covariates. An example is used to demonstrate the modeling process in Section 11.7. Smoothing spline models for subject profiles may be used to generate an implicit within-subject covariance structure, and in Section 11.8 diagnostics to examine the goodness of fit of this covariance model are considered. Section 11.9 describes the wider class of L-spline mixed models, which includes the cubic smoothing spline and other polynomial smoothing splines. Section 11.10 considers cases where use of a smoothing spline is computationally difficult and describes low-rank approximations to the smoothing spline. The connection between these low-rank approximations and P-splines or penalized splines is briefly discussed. Finally, in Section 11.11, extensions to the model are discussed, including constrained splines and non-normal data. First, the development of smoothing spline methods is set into an historical context.

11.1.1 A brief history of smoothing splines

Cubic splines were introduced into statistics in the context of fitting a smooth but unknown function to data in the presence of error. Longitudinal data typically consist of a set of measurements on each of N subjects. To illustrate the basic concepts of the spline methods, first consider a single set of n_i measurements on subject i, $\boldsymbol{Y}_i = (Y_{i1}, \ldots, Y_{in_i})'$ observed at times $\boldsymbol{t}_i = (t_{i1}, \ldots, t_{in_i})'$, with $t_{i1} < t_{i2} < \cdots < t_{in_i}$, and assume that there are no other relevant explanatory covariates. Later, in Section 11.6, a model is constructed for the full set of subjects, allowing for additional covariates. A model for subject i might then be written as

$$\boldsymbol{Y}_i = g(\boldsymbol{t}_i) + \boldsymbol{e}_i, \tag{11.1}$$

where $g(t)$ is a smooth but unknown function and $\boldsymbol{e}_i = (e_{i1}, \ldots, e_{in_i})'$ is a vector of errors, assumed to have a normal distribution with zero mean and covariance matrix $V_i = V_i(\boldsymbol{\phi})$, which may be a function of unknown parameters $\boldsymbol{\phi}$ that describe covariance across time within subjects. The criterion of smoothness for function $g(t)$ is enforced by using a penalty to restrict the amount of curvature allowed in the fitted function. The estimated function $g(t)$ must then minimize the residual sum of squares subject to this roughness penalty, that is, minimize a penalized sum of squares (PSS) of the form

$$[\boldsymbol{Y}_i - g(\boldsymbol{t}_i)]'V_i^{-1}[\boldsymbol{Y}_i - g(\boldsymbol{t}_i)] + \lambda \int_{t_{\min}}^{t_{\max}} [g''(s)]^2 ds, \tag{11.2}$$

where $\lambda \geq 0$ and $t_{\min}$ and $t_{\max}$ are the times of the first and last measurements across the full set of N subjects. It can be shown that, for any given value of λ, this PSS criterion is minimized by a natural cubic spline with knots at the distinct measurement times, called the cubic smoothing spline (Green and Silverman, 1994, Chapter 2). The general introduction of smoothing splines into statistics followed the work of Wahba and others in the 1970s (e.g., Kimeldorf and Wahba, 1970a, 1970b, 1971; Boneva, Kendall, and Stefanov, 1971; Wahba 1975; see also Wahba, 1990, for an overview and full references).

The fitted smoothing spline is dependent on the value of the smoothing parameter λ. As λ increases, the influence of the roughness penalty within the PSS increases and the fitted spline becomes smoother (has less curvature) until as $\lambda \to \infty$ the fitted spline becomes linear, with zero penalty. The optimal value of the smoothing parameter for an individual data set depends on the characteristics of the data. Craven and Wahba (1979) developed generalized cross-validation (GCV) as an objective method of estimating an appropriate value of the smoothing parameter and Wahba (1985) developed the alternative technique of general maximum likelihood (GML).

Partial smoothing spline models, also known as semi-parametric or additive models, were developed to embed smoothing splines into the linear model as

$$\boldsymbol{Y}_i = X_{ri}\boldsymbol{\beta}_r + g(\boldsymbol{t}_i) + \boldsymbol{e}_i,$$

where X_{ri} is the design matrix for subject i for additional explanatory variables, and $\boldsymbol{\beta}_r$ is a vector of coefficients for these terms. Models of this type were considered by Green (1985a) and Wahba (1984) and were extended to the generalized linear model as generalized additive models by Hastie and Tibshirani (1986), also considered as generalized semi-parametric models by Green (1987). Methods based on these models have become widespread in practical data analysis; see, for example, Hastie and Tibshirani (1990) or Hastie, Tibshirani, and Friedman (2001).

Both Green (1985b) and Thompson (1985) pointed out that the simple smoothing spline model (11.1) could be written as a linear mixed model with the smoothing parameter estimated using REML (Patterson and Thompson, 1971). Speed (1991) noted that the fitted spline could be represented as a BLUP from that mixed model and also remarked that the method of GML (Wahba, 1985) for estimating the smoothing parameter was equivalent to using REML to estimate the smoothing parameter as a variance component from this mixed model. These ideas (eventually) led to partial spline models within the mixed-model framework

$$\begin{aligned} \boldsymbol{Y}_i &= X_{ri}\boldsymbol{\beta}_r + Z_{ri}\boldsymbol{b}_r + g(\boldsymbol{t}_i) + \boldsymbol{e}_i \\ &= X_{ri}\boldsymbol{\beta}_r + Z_{ri}\boldsymbol{b}_r + X_{si}\boldsymbol{\beta}_s + Z_{si}\boldsymbol{\delta} + \boldsymbol{e}_i, \end{aligned}$$

where the smoothing spline $g(t)$ is represented as the sum of the fixed intercept and linear terms $X_{si}\boldsymbol{\beta}_s$ and the random effects $Z_{si}\boldsymbol{\delta}$. The models can then be fitted by standard mixed-model software. The terms $X_{ri}\boldsymbol{\beta}_r$, $Z_{ri}\boldsymbol{b}_r$, and $\boldsymbol{e}_i$ represent other fixed and random terms in the model and the error term, respectively. General covariance structures may be assumed for the random effects and errors. Models of this type were considered by Wang (1998), Zhang et al. (1998), Brumback and Rice (1998), and Verbyla et al. (1999). The development of these models meant that spline functions could easily be incorporated into complex linear mixed models, with all sources of error variation accounted for. Because of the flexibility they introduce directly into the linear mixed-model framework, these models have become increasingly popular. Wang (1998), Brumback and Rice (1998), and Verbyla et al. (1999) all proposed that these models were particularly appropriate for longitudinal data as the splines can provide a flexible model for both the mean and covariance structure.

Before this complex model is described, it is useful to understand the background of more general spline methods.

11.2 Background: Cubic splines and spline models

A cubic spline $g(t)$ for $t_{\min} \leq t \leq t_{\max}$ with r distinct knots $\boldsymbol{\tau} = (\tau_1, \ldots, \tau_r)'$, such that $t_{\min} < \tau_1 < \cdots < \tau_r < t_{\max}$, satisfies the following conditions (Dierckx, 1993):

(1) $g(t)$ is a piecewise cubic polynomial, (i.e., specified by a polynomial of degree no more than 3), on each interval $[t_{\min}, \tau_1], [\tau_1, \tau_2], [\tau_2, \tau_3], \ldots, [\tau_{r-1}, \tau_r], [\tau_r, t_{\max}]$.

(2) $g(t)$ and its first and second derivatives are all continuous on $[t_{\min}, t_{\max}]$.

The natural cubic spline obeys an additional constraint:

(3) $g^{(j)}(t_{\min}) = g^{(j)}(t_{\max}) = 0$ for $j = 2, 3$, that is, the second and third derivatives are zero at the endpoints, so the natural cubic spline is linear outside the range of the knots.

An unconstrained cubic polynomial has four parameters. The cubic spline then has four parameters within each section between knots, with three continuity constraints at each knot point. This gives a total of $4(r+1)$ parameters with $3r$ constraints, leaving $r+4$ free parameters. A natural cubic spline has an additional four constraints, due to condition (3), leaving r free parameters.

For a cubic spline with r distinct knots $\boldsymbol{\tau}$, a set of basis functions is a set of functions $\{s_j(x); j = 1, \ldots, r+4\}$ that are themselves cubic splines with knots $\boldsymbol{\tau}$, and that span the space of cubic splines for knots $\boldsymbol{\tau}$, so that any such spline $g(t)$ can be represented as

$$g(t) = \sum_{j=1}^{r+4} c_j s_j(x),$$

for suitably chosen coefficients c_j, $j = 1, \ldots, r+4$. Similarly, a basis for a natural cubic spline can be constructed from a suitable set of r natural cubic splines.

A set of spline basis functions can be defined in many ways. Two popular bases used with penalized splines and P-splines are the truncated power function (TPF) basis and the B-spline basis, respectively. These bases are closely related to each other and also to a basis commonly used for the natural cubic spline, and will be defined in Section 11.3.

Suppose the cubic spline $g(t)$ with r knots and basis functions $s_j(x), j = 1, \ldots, q$, is to be fitted to a set of n_i data points for subject i, $\boldsymbol{Y}_i$, using model (11.1). Using an appropriate basis representation of the spline, the model can be written as

$$\boldsymbol{Y}_i = \sum_{j=1}^{q} c_{ij} s_j(\boldsymbol{t}_i) + \boldsymbol{e}_i, \tag{11.3}$$

where $\boldsymbol{c}_i = (c_{i1}, \ldots, c_{iq})'$ is a set of unknown coefficients, with $q = r+4$ for an unconstrained cubic spline and $q = r$ for a natural cubic spline.

Methods used to choose the number of knots r and estimate the coefficients $\boldsymbol{c}_i$ can be classified into three categories:

(1) A small number of knots is chosen relative to the number of distinct values of $\boldsymbol{t}_i$, (i.e., $r \ll n_i$), and the coefficients are estimated as fixed effects in a linear model. This approach is known as the regression spline method.

(2) Each distinct value in $\boldsymbol{t}_i$ is used as a knot and the coefficients $\boldsymbol{c}_i$ are estimated subject to a penalty. Smoothing spline methods fall into this category.

(3) Hybrid methods, which involve knot selection but retain a relatively large number of knots and still estimate coefficients subject to a penalty. P-splines and penalized splines fall into this category.

The main motivation for the third category of methods has been the computational burden for smoothing splines when there are large numbers of knots (distinct measurement times). These methods (in common with regression splines) have a problem of optimal knot selection, but the influence of the choice of knots is minimized by using as large a set as is practically feasible. Methods subdivide into those that use a subset of the full basis, and those that use the full basis for a reduced knot set. Luo and Wahba (1997) developed hybrid adaptive splines, which select a subset of the full basis using forward stepwise regression to minimize a GCV score. P-splines (Eilers and Marx, 1996) and penalized splines (Ruppert, Wand, and Carroll, 2003) use a full basis on a reduced knot set with a discrete approximation to the standard smoothing spline penalty.

This chapter is mainly concerned with the second category. Below, for comparison, brief details of the regression spline method are given. Cubic smoothing splines are then considered in more detail. The third category is briefly considered in Section 11.10 and in Chapter 12.

11.2.1 Regression splines

A regression spline is formed as a linear combination of a set of spline basis functions, $s_j(x)$, generated from a small set of distinct knot points $\boldsymbol{\tau} = (\tau_1, \ldots, \tau_r)'$, $r \ll n_i$, which span the range of the covariate values $\boldsymbol{t}_i = (t_{i1}, \ldots, t_{in_i})'$ with $\tau_1 > t_{\min}$ and $\tau_r < t_{\max}$. Each basis function is evaluated at the set of measurement times $(\boldsymbol{t}_i)$ to generate a covariate $s_j(\boldsymbol{t}_i)$ and hence a linear model for the data of form (11.3). Linear (mixed) models may be used to estimate the coefficients $\boldsymbol{c}_i$ and the variance parameters $\boldsymbol{\phi}$.

The potential flexibility in the regression spline, and hence the minimum amount of smoothing, is determined by the number and position of knot points chosen; that is, the amount of smoothing is effectively constrained by the user. In regions of the covariate with denser knot spacing the curve can take a more complex form, but in regions where there are few knots the curve is constrained to be smoother. There is a large literature on choice of knot position, much of it from the perspective of numerical analysis. Within the statistical literature, knots are often equally spaced or placed at quantiles of the data. Wand (2000) gave a brief survey. Criteria such as Akaike's information criterion (AIC) can be used to compare the fit of regression splines using different numbers or positions of knots (e.g., Ruppert, Wand, and Carroll, 2003).

The regression spline method has been extended in various ways for longitudinal applications. Rice and Wu (2001) extended the method to random regression splines, which uses regression splines to model both the mean and subject variation in longitudinal data. A very similar model was used for growth curves by Mackenzie, Donovan, and McArdle (2005). Pan and Goldstein (1998) used regression splines extended to include piecewise rational polynomial functions to model growth.

11.2.2 Smoothing splines

For model (11.1), a cubic smoothing spline for subject i is the natural cubic spline with knots at $\boldsymbol{t}_i$ that minimizes the PSS (11.2) for a given value of the smoothing parameter λ. In contrast with regression splines, there is then no choice of knot points to be made for smoothing splines, but a value for the smoothing parameter must be selected. The value of the smoothing parameter λ may be either chosen or estimated, usually by either GCV or REML/GML.

In the context of generalized additive models (Hastie and Tibshirani, 1990), the smoothing parameter is sometimes chosen to fix the equivalent (or effective) degrees of freedom (EDF) of the smoothing spline. The EDF is calculated as a function of the smoothing parameter using

$$\text{EDF}(\lambda) = \text{trace}[A(\lambda)],$$

where $A(\lambda)$ is the hat matrix such that the fitted spline $\tilde{g}(t)$ satisfies $\tilde{g}(t_i) = A(\lambda)Y_i$. An alternative form for this expression is given in Chapter 12. Different, but related, estimates of the EDF are sometimes used (see Hastie and Tibshirani, 1990). This method may be thought subjective, in the sense that the smoothing parameter is indirectly specified, but objective methods, such as information criteria, can be used to compare the goodness of fit for different candidate values.

The methods of cross-validation (CV) and GCV were introduced as automatic methods of smoothing parameter choice. CV aimed to minimize the mean squared error of prediction (MSEP) across a set of data points. One disadvantage of this method was that data values with high leverage could have a large influence on the choice of λ. GCV (Craven and Wahba, 1979) was developed to downweight these values in the estimation of the smoothing parameter. The GCV method obtains an estimate of the value λ that minimizes the MSEP. Efficient algorithms are available to calculate both the CV and the GCV estimates. However, both CV and GCV estimates may work poorly in the presence of correlated errors, although they are thought to be reasonably robust against misspecification of the error distribution. Gu and Ma (2005) examined the performance of GCV for non-normal longitudinal data.

As an alternative, Wahba (1985) introduced GML for estimation of the smoothing parameter, and showed that asymptotically the GML estimate tended to be larger than the GCV estimate of λ, that is, that GCV tended to deliver rougher estimates of the fitted spline than GML. Wahba (1990) dismissed use of GML largely because of its dependence on correct specification of the error distribution. However, Kohn, Ansley, and Tharm (1991) reported an extensive simulation study that compared the performance of CV, GCV, and GML in terms of MSEP in the estimation of a set of smooth functions across a large range of sample sizes and error variances, with independent errors, using both cubic and quintic smoothing splines. Their results showed that GML performed similarly to GCV, in terms of MSEP, over the full range of functions tested, even when the tails of the error distribution were either heavier or lighter than expected for the normal distribution. They did not examine robustness to skewed distributions. From the equivalence of REML and GML estimation, the simulation studies of Kohn, Ansley, and Tharm (1991) therefore showed that REML estimation of the smoothing parameter is comparable with the standard method of GCV in the case of independent errors. A major advantage of REML estimation of the smoothing parameter within the linear mixed model for longitudinal data is the easy generalization to a correlated error structure. The simple spline model can then be extended to account for other covariates and all sources of error variation.

Before returning to the smoothing spline model in Section 11.5, it is useful to review the form of different spline basis functions, the linear mixed model for longitudinal data, and REML estimation.

11.3 Basis functions and properties of natural cubic splines

In this section, brief details of different bases for the cubic spline are given. This shows the connections between the bases that determine the relationship between different models used in Section 11.10. An important property of the natural cubic spline is also described.

The TPF basis is intuitively simple, and for a cubic spline can equivalently be written as either

$$g(t) = \sum_{i=0}^{3} \beta_i t^i + \sum_{j=1}^{r} \nu_j (t - \tau_j)_+^3, \qquad (11.4)$$

where $\beta_0, \ldots, \beta_k, \nu_1, \ldots, \nu_r$ are unknown parameters with

$$(t - \tau_j)_+^3 = T_j(t) = \begin{cases} (t - \tau_j)^3 & t > \tau_j, \\ 0 & \text{otherwise,} \end{cases}$$

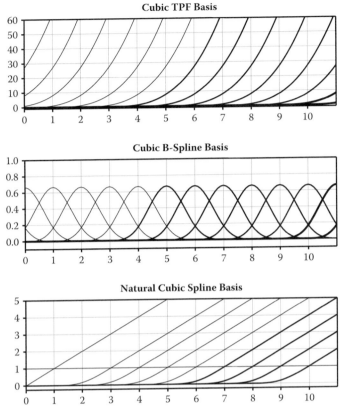

Figure 11.1 Bases for cubic splines with knots $1, \ldots, 10$ on range $[0, 11]$.

or as

$$g(t) = \sum_{j=-3}^{r} \nu_j \, (t - \tau_j)_+^3, \qquad (11.5)$$

for unknown parameters $\nu_{-3}, \ldots, \nu_r$. Knots can be added or removed from the basis by the addition or removal of the corresponding function, but the basis has bad computational properties in the sense that matrices of the form $[T_1(t_i), \ldots, T_r(t_i)]$ can be ill-conditioned and must be processed carefully to ensure stable computation. A TPF basis in the form of (11.5) is shown in Figure 11.1.

The B-spline basis is popular and widely used because of its sparse, local form and good computational properties in terms of matrix inversion. Cubic B-splines can be derived as scaled fourth divided differences of a set of cubic truncated power functions, $\{(t - \tau_j)_+^3, \ j = -3, \ldots, r + 4\}$ (de Boor, 1978; Welham et al., 2007). The cubic spline is then written as

$$g(t) = \sum_{j=-3}^{r} \nu_j \, B_{j,4}(t), \qquad (11.6)$$

where $B_{j,4}(t)$ is the jth B-spline basis function, $j = -3, \ldots r$, of order 4 (degree 3) for knots $\boldsymbol{\tau}$, defined by

$$B_{j,4}(t) = (\tau_{j+4} - \tau_j)[\tau_j, \ldots, \tau_{j+4}](t - \cdot)_+^3,$$

where $[\tau_j, \ldots, \tau_{j+4}]f$ denotes the fourth divided difference of function f (de Boor, 1978). The cubic B-splines, $\{B_{j,4}, \ j = -3, \ldots, r\}$, can also be defined recursively (de Boor, 1978,

Equations X.4 and X.5) in terms of lower-order B-splines, from

$$B_{j,1}(t) = \begin{cases} 1 & \tau_j \le t < \tau_{j+1}, \\ 0 & \text{otherwise,} \end{cases}$$

$$B_{j,k+1}(t) = \frac{t - \tau_j}{\tau_{j+k} - \tau_j} B_{j,k}(t) + \frac{\tau_{j+k+1} - t}{\tau_{j+k+1} - \tau_{j+1}} B_{j+1,k}(t), \qquad k > 0.$$

For the cubic B-spline basis, addition or removal of knots requires the amendment of the five basis functions that use the knot in their construction. A cubic B-spline basis is shown in Figure 11.1.

Neither the TPF nor B-spline basis can be used to represent a smoothing spline without further constraint, as this requires a natural cubic spline. However, understanding the close relationship between these bases makes clear the connections between penalized or P-spline models and smoothing splines described in Section 11.10.

An alternative basis is required for natural cubic splines, in which the basis functions are themselves natural cubic splines. Such basis functions can be calculated as scaled second divided differences of the truncated power functions $\{(t - \tau_j)^3_+, \ j = 1, \dots, r\}$ to give a set of $r - 2$ basis functions, $P_j(t)$, labeled by convention as $j = 2, \dots, r - 1$, with

$$6P_j(t) = \left[h_j^{-1}(t - \tau_{j+1})^3_+ - \left(h_j^{-1} + h_{j-1}^{-1} \right)(t - \tau_j)^3_+ + h_{j-1}^{-1}(t - \tau_{j-1})^3_+ \right],$$

where $h_j = \tau_{j+1} - \tau_j$. This set can be augmented by a constant and linear trend to give a complete natural basis. A natural cubic spline can then be represented as

$$g(t) = \beta_{s,0} + \beta_{s,1} t + \sum_{j=2}^{r-1} \delta_j P_j(t), \qquad (11.7)$$

for suitably chosen coefficients $\beta_{s,0}, \beta_{s,1}, \delta_2, \dots, \delta_{r-1}$. In this representation it is straightforward to verify that the unknown parameter δ_j evaluates the second derivative of the spline $g(t)$ at $t = t_j$.

The functions $P_j(t)$ can be written explicitly as

$$6P_j(t) = \begin{cases} 0 & t \le \tau_{j-1}, \\ h_{j-1}^{-1}(t - \tau_{j-1})^3 & \tau_{j-1} < t \le \tau_j, \\ h_{j-1}^{-1}(t - \tau_{j-1})^3 - (h_{j-1}^{-1} + h_j^{-1})(t - \tau_j)^3 & \tau_j < t \le \tau_{j+1}, \\ (h_{j-1} + h_j)(3t - \tau_{j-1} - \tau_j - \tau_{j+1}) & \tau_{j+1} < t, \end{cases} \qquad (11.8)$$

as given by White et al. (1998). Adding or deleting a knot changes the three functions that use the knot in their construction. The cubic TPF, B-spline, and natural basis corresponding to knots $1, \dots, 10$ on range $[0, 11]$ are shown in Figure 11.1.

The natural cubic spline has an important property which is helpful in deriving the mixed-model form of the cubic smoothing spline. It can be shown that a natural cubic spline must satisfy the following condition at the knot points (Green and Silverman, 1994, Section 2.5):

$$Q'\boldsymbol{g_\tau} = G_s^{-1}\boldsymbol{\delta}, \qquad (11.9)$$

where $\boldsymbol{g_\tau} = g(\boldsymbol{\tau}) = (g(\tau_1), \dots, g(\tau_r))'$, $\boldsymbol{\delta} = (\delta_2, \dots, \delta_{r-1})'$, Q is an $r \times (r - 2)$ matrix, and G_s^{-1} is a symmetric matrix with $r - 2$ rows. For convenience, the rows and columns of G_s^{-1} and columns of Q are labeled $2, \dots, r - 1$ and the rows of Q are labeled $1, \dots, r$. Each column of Q has three non-zero elements with the elements of Q, $q_{i,j} = [Q]_{ij}$, defined by

$$q_{j-1,j} = h_{j-1}^{-1}, \quad q_{j,j} = -h_{j-1}^{-1} - h_j^{-1}, \quad q_{j+1,j} = h_j^{-1} \quad \text{for } j = 2, \dots, r - 1,$$

$$q_{i,j} = 0 \quad \text{for } |i - j| \ge 2. \qquad (11.10)$$

The elements of matrix G_s are defined in terms of its tridiagonal inverse as

$$
\begin{aligned}
g^{i,i} &= (h_{i-1} + h_i)/3 & i &= 2, \ldots, r-1, \\
g^{i,i+1} &= g^{i+1,i} = h_i/6 & i &= 2, \ldots, r-2, \\
g^{i,j} &= 0 & |i-j| &\geq 2,
\end{aligned}
\tag{11.11}
$$

where $g^{i,j}$ is the element of G_s^{-1} in row i and column j. It follows from the piecewise cubic nature of the spline and (11.9) (Green and Silverman, 1994, Section 2.5.2) that the penalty term in the PSS (11.2) can be written for a natural cubic spline as

$$
\int_{t_{\min}}^{t_{\max}} [g''(s)]^2 ds = \boldsymbol{\delta}' G_s^{-1} \boldsymbol{\delta}.
\tag{11.12}
$$

11.4 A brief review of linear mixed models and REML estimation

A linear mixed model for longitudinal data taken from N subjects, with n_i measurements for subject i, $\boldsymbol{Y}_i = (Y_{i1}, \ldots, Y_{in_i})'$ taken at time $\boldsymbol{t}_i = (t_{i1}, \ldots, t_{in_i})'$, can be written as

$$
\boldsymbol{Y}_i = X_i \boldsymbol{\beta} + Z_i \boldsymbol{b}_i + \boldsymbol{e}_i,
\tag{11.13}
$$

where X_i is an $n_i \times p$ design matrix for subject i for a set of p covariates to be fitted as fixed effects, $\boldsymbol{\beta}$ is a vector of p unknown fixed effects, Z_i is an $n_i \times q$ design matrix for a set of q covariates to be fitted as random effects, $\boldsymbol{b}_i$ is a vector of q unknown subject-specific random effects, and $\boldsymbol{e}_i = (e_{i1}, \ldots, e_{in_i})'$ is a vector of n_i residual errors. Sets of covariates in design matrices X_i and Z_i may represent factors (i.e., qualitative covariates), and usually X_i and Z_i can be partitioned corresponding to distinct model terms, that is,

$$
X_i = [X_{i1}, X_{i2}, \ldots, X_{ib}], \quad Z_i = [Z_{i1}, Z_{i2}, \ldots, Z_{ic}],
$$

where X_{ij} is an $n_i \times p_j$ design matrix for the jth fixed term and Z_{ij} is an $n_i \times q_j$ design matrix for the jth random term, with $\sum_j p_j = p$, $\sum_j q_j = q$. The vectors $\boldsymbol{\beta}, \boldsymbol{b}_i$ are partitioned accordingly. It is assumed that the random effects are normally distributed with

$$
\boldsymbol{e}_i \sim N(\boldsymbol{0}, V_i(\boldsymbol{\phi})), \quad \mathrm{Cov}(\boldsymbol{e}_i, \boldsymbol{e}_j) = 0, \quad i \neq j,
$$
$$
\boldsymbol{b}_i \sim N(\boldsymbol{0}, G), \quad \mathrm{Cov}(\boldsymbol{b}_i, \boldsymbol{e}_j) = 0, \quad \forall \, i, j,
$$

where $i, j = 1, \ldots, N$, and $G = \mathrm{diag}\{G_1(\boldsymbol{\gamma}_1), \ldots, G_c(\boldsymbol{\gamma}_c)\} = \oplus\{G_i(\boldsymbol{\gamma}_i)\}$, that is, partitioned conformably with Z_i and $\boldsymbol{b}_i$. The vectors $\boldsymbol{\phi}$ and $\boldsymbol{\gamma} = (\boldsymbol{\gamma}_1', \ldots, \boldsymbol{\gamma}_c')'$ may contain unknown parameters.

The form (11.13) is convenient for specification of a model for an individual subject, but the assumption of only subject-specific random effects can be restrictive. Random effects shared across subjects may arise when groups of subjects have common environmental variables. For example, if a longitudinal study uses patients from a set of hospitals, which are regarded as a sample from the general population of hospitals, then hospital may be considered as a random effect, and all patients from the same hospital share that random effect. These common random effects are usually accommodated by a slight extension of the model definition to collapse common effects into additional random terms, but may also be accounted for by imposing constraints on the subject-specific form. For example, if the subset of random effects $\boldsymbol{b}_{i1}$ for $i = 1, \ldots, N$ are common across all subjects, the constraint $\boldsymbol{b}_{i1} = \boldsymbol{b}_{j1}$ for $i, j = 1, \ldots, N$ can be imposed indirectly via the covariance model

$$
\mathrm{Cov}(\boldsymbol{b}_{i1}, \boldsymbol{b}_{j1}) = G_1(\boldsymbol{\gamma}_1),
$$

for $i, j = 1, \ldots, N$. This form will be used here for notational convenience. If common effects are gathered into the first $m \leq c$ subsets of random effects, then it is assumed that

$$
\mathrm{Cov}(\boldsymbol{b}_{ik}, \boldsymbol{b}_{jk}) = G_k(\boldsymbol{\gamma}_k),
$$

for $k \leq m$ where $\boldsymbol{b}_{ik} = \boldsymbol{b}_{jk}$ (i.e., shared random effects), and $\text{Cov}(\boldsymbol{b}_{ik}, \boldsymbol{b}_{jk}) = \boldsymbol{0}$ otherwise. The covariance matrix of the data for subject i, Σ_{ii}, can be written as

$$\text{Cov}(\boldsymbol{Y}_i) = \Sigma_{ii} = Z_i G Z_i' + V_i.$$

In this model, non-zero covariances between subjects arise only from shared random effects as

$$\text{Cov}(\boldsymbol{Y}_i, \boldsymbol{Y}_j) = \Sigma_{ij} = \sum_{k=1}^{c} Z_{ik} \text{Cov}(\boldsymbol{b}_{ik}, \boldsymbol{b}_{jk}) Z_{jk}'.$$

A model for the full data set including all subjects can be written as

$$\begin{pmatrix} \boldsymbol{Y}_1 \\ \vdots \\ \boldsymbol{Y}_N \end{pmatrix} = \begin{pmatrix} X_1 \\ \vdots \\ X_N \end{pmatrix} \boldsymbol{\beta} + \begin{bmatrix} Z_1 & & \\ & \ddots & \\ & & Z_N \end{bmatrix} \begin{pmatrix} \boldsymbol{b}_1 \\ \vdots \\ \boldsymbol{b}_N \end{pmatrix} + \begin{pmatrix} \boldsymbol{e}_1 \\ \vdots \\ \boldsymbol{e}_N \end{pmatrix},$$

or more succinctly, using the obvious extensions to the notation,

$$\boldsymbol{Y} = X\boldsymbol{\beta} + Z\boldsymbol{b} + \boldsymbol{e},$$

with

$$\text{Cov} \begin{pmatrix} \boldsymbol{Y}_1 \\ \boldsymbol{Y}_2 \\ \vdots \\ \boldsymbol{Y}_N \end{pmatrix} = \begin{bmatrix} \Sigma_{11} & \Sigma_{12} & \dots & \Sigma_{1N} \\ \Sigma_{21} & \Sigma_{22} & \dots & \Sigma_{2N} \\ \vdots & \vdots & \dots & \vdots \\ \Sigma_{N1} & \Sigma_{N2} & \dots & \Sigma_{NN} \end{bmatrix},$$

or $\text{Cov}(\boldsymbol{Y}) = \Sigma$. The vector $\boldsymbol{Y} = (\boldsymbol{Y}_1', \dots, \boldsymbol{Y}_N')'$ has length N_T, where $N_T = \sum_{i=1}^{N} n_i$. For estimation and computation, the extended set of random effects $\boldsymbol{b} = (\boldsymbol{b}_1', \dots, \boldsymbol{b}_N')'$ is mapped onto a reduced set $\boldsymbol{b}^*$ which merges the shared effects. The set $\boldsymbol{b}^*$ has design matrix $Z^* = ZM$ where $\boldsymbol{b} = M\boldsymbol{b}^*$ for a rectangular matrix M of full column rank such that $\text{Cov}(\boldsymbol{b}^*) = G^*(\boldsymbol{\gamma})$ where G^* is a positive-definite matrix.

The log-likelihood function of the data is then given by

$$-2 \times L(\boldsymbol{Y}; \boldsymbol{\beta}, \boldsymbol{\phi}, \boldsymbol{\gamma}) = \log(|\Sigma|) + (\boldsymbol{Y} - X\boldsymbol{\beta})' \Sigma^{-1} (\boldsymbol{Y} - X\boldsymbol{\beta}),$$

and the maximum likelihood estimates of the fixed effects and variance parameters can be estimated by maximizing the log-likelihood function L. Maximum likelihood estimates of the variance parameters $\boldsymbol{\phi}$ and $\boldsymbol{\gamma}$ are biased downwards, as no account is taken of the degrees of freedom used in estimating the fixed effects. For this reason, Patterson and Thompson (1971) developed the REML method. The residual (restricted) log-likelihood function used to estimate the variance parameters is then the log-likelihood function of the residual contrasts, defined as the set of contrasts with zero expectation. Ignoring contrasts with non-zero expectation has the effect of removing the fixed degrees of freedom before estimation of variance parameters, so that the resulting estimates are less biased than the maximum likelihood estimates. Verbyla (1990) showed that the log-likelihood function of the residual contrasts, denoted $RL(\boldsymbol{Y})$, can be written as the log-likelihood of $L_2'\boldsymbol{Y}$ where L_2 is an $N_T \times [N_T - \text{rank}(X)]$ matrix of full column rank with $L_2'X = 0$:

$$\begin{aligned} -2 \times RL(\boldsymbol{Y}; \boldsymbol{\phi}, \boldsymbol{\gamma}) &= -2 \times L(L_2'\boldsymbol{Y}; \boldsymbol{\phi}, \boldsymbol{\gamma}) \\ &= c(X) + \log(|X'\Sigma^{-1}X|) + \log(|\Sigma|) + \boldsymbol{Y}'P\boldsymbol{Y}, \end{aligned} \quad (11.14)$$

where

$$P = \Sigma^{-1} - \Sigma^{-1} X (X'\Sigma^{-1}X)^{-1} X'\Sigma^{-1}, \quad (11.15)$$

and $c(X)$ is a term that is constant with respect to the variance parameters, but is a function of the design matrix X, that is, depends upon the structure of the fixed model. Note that the log-likelihood function RL is no longer a function of the fixed effects $\boldsymbol{\beta}$.

If the variance parameters are known, then the best linear unbiased estimate (BLUE) of the fixed effects $\boldsymbol{\beta}$ can be calculated as

$$\widehat{\boldsymbol{\beta}} = (X'\Sigma^{-1}X)^{-1}X'\Sigma^{-1}\boldsymbol{Y},$$

and the BLUP for the random effects, $\tilde{\boldsymbol{b}}^*$, is the expected value of $\boldsymbol{b}^*$ given the data, $\tilde{\boldsymbol{b}}^* = E(\boldsymbol{b}^*|L_2'\boldsymbol{Y})$,

$$\tilde{\boldsymbol{b}}^* = G^*Z^{*\prime}P\boldsymbol{Y},$$

then $\tilde{\boldsymbol{b}} = M\tilde{\boldsymbol{b}}^*$. It is straightforward to show that the fixed and random effects can be calculated from the mixed-model equations (MMEs) that take the form

$$\begin{bmatrix} X'V^{-1}X & X'V^{-1}Z^* \\ Z^{*\prime}V^{-1}X & Z^{*\prime}V^{-1}Z^* + G^{*-1} \end{bmatrix} \begin{pmatrix} \widehat{\boldsymbol{\beta}} \\ \tilde{\boldsymbol{b}}^* \end{pmatrix} = \begin{pmatrix} X'V^{-1}\boldsymbol{Y} \\ Z^{*\prime}V^{-1}\boldsymbol{Y} \end{pmatrix},$$

where $V = \operatorname{diag}(V_1 \ldots V_N)$. The matrix on the left-hand side of the MME is denoted as

$$C = \begin{bmatrix} C_{11} & C_{12} \\ C_{21} & C_{22} \end{bmatrix} = \begin{bmatrix} X'V^{-1}X & X'V^{-1}Z^* \\ Z^{*\prime}V^{-1}X & Z^{*\prime}V^{-1}Z^* + G^{*-1} \end{bmatrix}.$$

Prediction error variances for the estimated effects can be obtained by inverting the left-hand side of the MME as

$$\operatorname{Cov}\begin{pmatrix} \widehat{\boldsymbol{\beta}} - \boldsymbol{\beta} \\ \tilde{\boldsymbol{b}}^* - \boldsymbol{b}^* \end{pmatrix} = C^{-1} = \begin{bmatrix} C^{11} & C^{12} \\ C^{21} & C^{22} \end{bmatrix}. \tag{11.16}$$

In practice, the variance parameters $\boldsymbol{\phi}$, $\boldsymbol{\gamma}$ are not known and their REML estimates $(\widehat{\boldsymbol{\phi}}, \widehat{\boldsymbol{\gamma}})$ are used to construct $\widehat{V}$ and $\widehat{G}^*$, which are then used in the MMEs to get empirical BLUEs for the fixed effects, empirical BLUPs for the random effects, and estimates for the prediction error variances. Due to uncertainty in the variance parameter estimates $\widehat{\boldsymbol{\phi}}$ and $\widehat{\boldsymbol{\gamma}}$, this process may affect both the expected value and covariance matrix of the estimated effects.

Assessment of whether specific variance parameters are effectively zero (or some other fixed value) can be carried out by standard likelihood ratio tests, based on the change $-2 \times RL$ (sometimes called the deviance) under the null and alternative hypotheses. The change in deviance must be assessed while the fixed model is unchanged (Welham and Thompson, 1997). For unrestricted variance parameters, the change in deviance has an asymptotic chi-squared distribution with number of degrees of freedom equal to the number of variance parameters held fixed under the null hypothesis. Where the null hypothesis fixes parameters on the boundary of the parameter space, some adjustment to the test statistic distribution is required (see Stram and Lee, 1994). In particular, under specific conditions, where the null hypothesis fixes a single non-negative variance component at zero, the test statistic distribution is a 50:50 mixture of a χ_0^2 distribution (takes value 0 with probability 1) and a χ_1^2 distribution. However, recently Crainiceanu and Ruppert (2004a) pointed out that the conditions used by Stram and Lee (1994) to develop this result do not necessarily hold in the general mixed-model setting. The asymptotic result requires either that the data be independent and identically distributed, or that the data be partitioned into a number of independent subsets and that the number of subsets increases with the size of the data set. Crainiceanu and Ruppert (2004a) assert that the asymptotic approximation is poor if either the independence condition does not hold, or if the number of independent subsets is small. Although Crainiceanu and Ruppert (2004a) give formulae for the case of a simple model with a single variance parameter in addition to the residual variance, there is no formula for the general case, where use of the parametric bootstrap is suggested to establish the distribution of the test statistic under the null hypothesis.

Changes in RL cannot be used to investigate the evidence that effects associated with fixed model terms are different from zero. Wald tests can be used to test the null hypothesis that fixed effects are equal to zero. For a vector of estimated fixed effects $\widehat{\boldsymbol{\beta}}$ with estimated variance matrix V_β, the Wald statistic is defined as $\widehat{\boldsymbol{\beta}}'[V_\beta]^{-1}\widehat{\boldsymbol{\beta}}$. When the variance parameters are all known, then under the null hypothesis the Wald statistic has an asymptotic chi-squared distribution with degrees of freedom equal to the number of parameters estimated in $\widehat{\boldsymbol{\beta}}$. In practice, with unknown variance parameters, the distribution of the Wald statistic under the null hypothesis is closer to a scaled F-statistic. However, for unbalanced data the denominator degrees of freedom cannot easily be determined. The chi-squared approximation improves as the residual degrees of freedom in the strata where the effects are estimated increase. In small samples, the test is anti-conservative, that is, under the null hypothesis, significant results are found more often than the nominal confidence level suggests (Welham and Thompson, 1997). Adjustments to improve the standard errors to take account of uncertainty in the variance parameters, and the calculation of approximate degrees of freedom for F-tests, were presented by Kenward and Roger (1997).

Further information or results for REML estimation in the linear mixed model can be found in any standard text on the subject, such as Verbeke and Molenberghs (2000).

11.5 Representation of a cubic smoothing spline as a BLUP from a linear mixed model

It is straightforward to show that the cubic smoothing spline can be represented as a BLUP from a linear mixed model. For now, consider a cubic smoothing spline for measurements from a single individual, labeled as subject i, with model (11.1). The cubic smoothing spline for this subject is a natural cubic spline with knots at their measurement times, that is, $r = n_i$ and $\boldsymbol{\tau} = \boldsymbol{t}_i = (t_{i1}, \ldots, t_{in_i})'$. Using (11.12), the roughness penalty in the PSS (11.2) can be simplified as

$$\int_{t_{\min}}^{t_{\max}} [g''(s)]^2 ds = \boldsymbol{\delta}' G_s^{-1} \boldsymbol{\delta},$$

where now the vector $\boldsymbol{\delta}$ evaluates the second derivative of the spline at the measurement times $\boldsymbol{t}_i$, and the matrix G_s^{-1} (11.11) is evaluated in terms of $\boldsymbol{t}_i$, that is, using differences between measurement times, $h_j = t_{i(j+1)} - t_{ij}$ for $j = 1, \ldots, n_i - 1$. From (11.7), this natural cubic spline takes the form

$$g(t) = \beta_{s,0} + \beta_{s,1}t + \sum_{j=2}^{n_i-1} \delta_j P_j(t),$$

for functions $P_j(t)$ defined as in (11.8) with $\boldsymbol{\tau} = \boldsymbol{t}_i$. The PSS (11.2) for subject i can then be written as

$$(\boldsymbol{Y}_i - X_{si}\boldsymbol{\beta}_s - Z_{si}\boldsymbol{\delta})'V_i^{-1}(\boldsymbol{Y}_i - X_{si}\boldsymbol{\beta}_s - Z_{si}\boldsymbol{\delta}) + \lambda\boldsymbol{\delta}'G_s^{-1}\boldsymbol{\delta}, \tag{11.17}$$

where $X_{si} = [\mathbf{1}\ \boldsymbol{t}_i]$ is an $n_i \times 2$ matrix, $\boldsymbol{\beta}_s = (\beta_{s,0}\ \beta_{s,1})'$, and $Z_{si} = [P_2(\boldsymbol{t}_i), \ldots, P_{n_i-1}(\boldsymbol{t}_i)]$ is an $n_i \times (n_i-2)$ matrix.

Minimization of the PSS (11.17) with respect to $\boldsymbol{\beta}_s$ and $\boldsymbol{\delta}$ to get the fitted spline then requires solution of the equations:

$$\begin{bmatrix} X'_{si}V_i^{-1}X_{si} & X'_{si}V_i^{-1}Z_{si} \\ Z'_{si}V_i^{-1}X_{si} & Z'_{si}V_i^{-1}Z_{si} + \lambda G_s^{-1} \end{bmatrix} \begin{pmatrix} \widehat{\boldsymbol{\beta}}_s \\ \tilde{\boldsymbol{\delta}} \end{pmatrix} = \begin{pmatrix} X'_{si}V_i^{-1}\boldsymbol{Y}_i \\ Z'_{si}V_i^{-1}\boldsymbol{Y}_i \end{pmatrix}. \tag{11.18}$$

From the results in Section 11.4, it is clear that these are the mixed-model equations from a linear mixed model of the form

$$\boldsymbol{Y}_i = X_{si}\boldsymbol{\beta}_s + Z_{si}\boldsymbol{\delta} + \boldsymbol{e}_i, \tag{11.19}$$

with fixed effects $\boldsymbol{\beta}_s$, random effects $\boldsymbol{\delta}$, and residual $\boldsymbol{e}_i$. The mixed-model assumption $\boldsymbol{e}_i \sim N(\boldsymbol{0}, V_i)$ is retained. The default assumption for the random effects in the mixed model that produces (11.18) is $\boldsymbol{\delta} \sim N(\boldsymbol{0}, \sigma_s^2 G_s)$ with $\lambda = 1/\sigma_s^2$ and hence $\sigma_s^2 \geq 0$. For the smoothing spline model this assumption is not necessarily required, as the justification of the fitting procedure is solely in the coincidence of the solution for the PSS (11.17) and the MMEs (11.18). It is therefore often more appropriate to regard $\boldsymbol{\delta}$ as a vector of effects with inverse penalty matrix G_s and associated (unknown) scalar $\sigma_s^2 = 1/\lambda$. The fitted spline at the measurement times can then be evaluated as

$$\tilde{\boldsymbol{g}} = \tilde{g}(\boldsymbol{t}_i) = X_{si}\widehat{\boldsymbol{\beta}}_s + Z_{si}\tilde{\boldsymbol{\delta}}. \tag{11.20}$$

The notation $\tilde{\boldsymbol{g}}$ is used to indicate that the fitted spline is a BLUP. This derivation assumed that the smoothing parameter λ had been pre-specified. In practice, it can be estimated by REML, either directly or via the variance component σ_s^2.

Equation (11.19) gives one basic form of the mixed-model smoothing spline for one subject. A model that applies to the full set of N subjects is required, but first alternative forms of this simple mixed-model spline and inference are considered.

Recall that, for a natural cubic spline with knots at the measurement times, (11.9) showed that $Q'\boldsymbol{g} = G_s^{-1}\boldsymbol{\delta}$, where now Q (11.10) and G_s^{-1} (11.11) have been constructed in terms of knots $\boldsymbol{\tau} = \boldsymbol{t}_i$. From $\boldsymbol{g} = g(\boldsymbol{t}_i) = X_{si}\boldsymbol{\beta}_s + Z_{si}\boldsymbol{\delta}$ it follows that

$$Q'\boldsymbol{g} = Q'Z_{si}\boldsymbol{\delta},$$

since $Q'X_{si} = 0$. As $g(t)$ is a natural cubic spline, it must then be true that

$$Q'Z_{si} = G_s^{-1},$$

which holds if and only if

$$Z_{si} = Q(Q'Q)^{-1}G_s^{-1} + X_{si}B,$$

for some $2 \times (n_i - 2)$ matrix B. Any basis for the natural cubic spline with knots at $\boldsymbol{\tau} = \boldsymbol{t}_i$ can be represented in this form. The fitted spline is independent of the basis chosen, but the component fitted as linear trend via the fixed effects will vary between bases. For interpretation, it is often helpful to use a basis with $B = 0$ so that $X'_{si}Z_{si} = 0$, that is, using spline design matrix

$$Z_{0i} = Q(Q'Q)^{-1}G_s^{-1}. \tag{11.21}$$

This design matrix can be interpreted in terms of underlying basis functions

$$V_j(t) = P_j(t) - a_j - b_j t,$$

where

$$\begin{pmatrix} a_j \\ b_j \end{pmatrix} = (X'_{si}X_{si})^{-1}X'_{si}P_j(\boldsymbol{t}_i),$$

for $j = 2, \ldots, n_i - 1$ as shown by White et al. (1998). In this model the set of random effects associated with the spline, $\boldsymbol{\delta}$, is still the set of second derivative values at the knots $\boldsymbol{t}_i$, and the term $X_{si}\boldsymbol{\beta}_s$ now represents all of the linear trend for subject i. This form of the model gives an unambiguous interpretation of the fixed component of the spline. In addition, it can be useful to transform the basis so that the estimated random effects are independent by using spline design matrix

$$Z_{ui} = Q(Q'Q)^{-1}G_s^{-0.5}, \tag{11.22}$$

with associated random effects $\boldsymbol{u}_s = G_s^{-0.5}\boldsymbol{\delta}$, with inverse penalty matrix I and associated scalar σ_s^2. This is the basis used by Verbyla et al. (1999). With independent random effects, the mixed-model smoothing spline can be fitted by most statistical software for mixed models.

As for all random model terms based on quantitative covariates (as opposed to 0/1 qualitative covariates or factors), the diagonal elements of the matrix $Z_{si}Z'_{si}$ are not equal to 1. The scale of the covariate t_i is thus directly related to the estimated spline variance component $\widehat{\sigma}_s^2$. For ease of interpretation, it is often helpful to present

$$\widehat{\sigma}_s^2 \times \text{trace}(Z_{si}Z'_{si})/\text{nrows}(Z_{si}),$$

which represents the average contribution of the spline term to the estimated variance of a measurement. If this approach is used for all random terms, then the estimated variance parameters are more directly comparable across terms.

For the mixed-model smoothing spline, the underlying mixed model is usually considered as a numerical tool to obtain the smoothing spline that minimizes the penalized sum of squares, that is, to obtain the BLUP for the fitted spline, $\tilde{g}(x)$, and to estimate the smoothing parameter. Given an appropriate basis, the underlying fixed linear component is interpretable as linear trend, and the random component is interpretable as smooth variation about the linear trend. However, the random component is not considered as contributing to a realistic description of the variation in the data. The status of these effects as random is used purely as a tool to get the required smoothing on the fitted spline. As remarked above, it then follows that there is no requirement for the spline random effects, δ, to conform to the underlying mixed-model assumption that random effects arise from a normal distribution. However, as this assumption is built into the mixed-model estimation, spline estimates may conform better to mixed-model properties when this is the case.

In the general mixed model, inference is usually made taking into account the influence of the population of random effects on variation in the data. When variation from the wider population is not relevant to inference on the sample of random effects obtained, it is appropriate to exclude this variation from inferences and to consider the model conditional on the random effects. Figure 12.4 in Chapter 12 demonstrates that the population of spline curves generated from the mixed-model form $\delta \sim N(0, \sigma_s^2 G_s)$ can be far from the form seen in the data, and further discussion of this point can be found in Chapter 12.

The mixed-model smoothing spline is therefore considered as

$$Y_i | \beta_s, \delta \sim N(g(t_i), V_i).$$

The fitted spline at the measurement times, $\tilde{g}$, takes the form (11.20) with

$$E(\tilde{g}|\beta_s, \delta) = g - V_i P Z_{si}\delta,$$

where P is derived for the spline model from the general form (11.15). The fitted spline is thus biased under the conditional distribution due to shrinkage. In this context, shrinkage of the fitted spline provides the required smoothing. Shrinkage, and hence bias, decreases as σ_s^2 increases (λ decreases). The limit as $\sigma_s^2 \to \infty$ corresponds to fitting the spline coefficients δ as fixed effects, resulting in an unbiased spline that interpolates the data: the bias has been eliminated, but at the cost of decreasing the smoothness of the fitted curve. This is an example of the bias–variance trade-off that is common to all smoothing methods (Hastie and Tibshirani, 1990, Section 3.3).

Interpolation of the fitted spline at time t_p is calculated from

$$\tilde{g}(t_p) = [1 \; t_p]\widehat{\beta}_s + \sum_{j=2}^{n_i-1} \tilde{\delta}_j P_j(t_p),$$

or equivalently, using matrix notation,

$$\tilde{g}(t_p) = x'_p\widehat{\beta}_s + z'_p\tilde{\delta},$$

where $\boldsymbol{x}'_p = [1 \ t_p]$ and $\boldsymbol{z}'_p = [P_2(t_p), \ldots, P_{n_i-1}(t_p)]$. Uncertainty in the fitted curve is calculated from uncertainty in the conditional distribution of the estimated effects using

$$\text{Cov}\left(\left.\begin{matrix}\widehat{\boldsymbol{\beta}}_s \\ \widetilde{\boldsymbol{\delta}}\end{matrix}\right| \boldsymbol{\beta}_s, \boldsymbol{\delta}\right) = \begin{bmatrix} C^{11} & C^{12} \\ C^{21} & C^{22} \end{bmatrix} - \lambda \begin{bmatrix} C^{12} \\ C^{22} \end{bmatrix} G_s^{-1} \begin{bmatrix} C^{21} & C^{22} \end{bmatrix},$$

where

$$\begin{bmatrix} C^{11} & C^{12} \\ C^{21} & C^{22} \end{bmatrix} = \begin{bmatrix} X'_{si} V_i^{-1} X_{si} & X'_{si} V_i^{-1} Z_{si} \\ Z'_{si} V_i^{-1} X_{si} & Z'_{si} V_i^{-1} Z_{si} + \sigma_s^{-2} G_s^{-1} \end{bmatrix}^{-1},$$

from (11.16). This quantity is not readily available from standard mixed-model software. However, because of the bias in the fitted spline, Zhang et al. (1998) and (for penalized splines) Ruppert, Wand, and Carroll (2003) suggested using the marginal mean squared error

$$E\left[\begin{pmatrix} \widehat{\boldsymbol{\beta}}_s - \boldsymbol{\beta}_s \\ \widetilde{\boldsymbol{\delta}} - \boldsymbol{\delta} \end{pmatrix}^2\right] = \text{Cov}\begin{pmatrix} \widehat{\boldsymbol{\beta}}_s - \boldsymbol{\beta}_s \\ \widetilde{\boldsymbol{\delta}} - \boldsymbol{\delta} \end{pmatrix} = \begin{bmatrix} C^{11} & C^{12} \\ C^{21} & C^{22} \end{bmatrix},$$

instead of the conditional variance. This is available from mixed-model software as the usual prediction error variance of fixed and random effects. On average, the marginal mean squared error accounts for both bias and variance, but as a pointwise estimate of conditional mean squared error it may still be inaccurate. This is an area requiring further research, and is discussed further in Chapter 12.

From the fitted spline

$$\tilde{g}(t) = \widehat{\beta}_{s,0} + \widehat{\beta}_{s,1} t + \sum_{i=2}^{n_i-1} \widetilde{\delta}_j P_j(t),$$

it can be seen that derived quantities such as first or second derivatives or integrals can be calculated as linear functions of the estimated effects, for example

$$\tilde{g}'(t) = \widehat{\beta}_{s,1} + \sum_{i=2}^{n_i-1} \widetilde{\delta}_j P'_j(t).$$

Estimated standard errors of such linear functions follow from the discussion above.

In practical situations it is often of interest to assess whether there is evidence to support the non-linear component of the smoothing spline, that is, whether the underlying linear term provides an adequate summary of the pattern in the data. Verbyla et al. (1999) stated that a likelihood ratio test based on the log-likelihood function RL (11.14) could be used to assess the significance of the random spline (non-linear) component of the model. Let RL_A denote the maximum of log-likelihood function RL under the simple spline model (11.19) and RL_0 denote the maximum under the reduced model

$$\boldsymbol{Y} = X_s \boldsymbol{\beta} + \boldsymbol{e},$$

that is, with $\sigma_s^2 = 0$ and hence $\boldsymbol{\delta} = \mathbf{0}$. The likelihood ratio test statistic is calculated as

$$-2 \times (RL_0 - RL_A).$$

Verbyla et al. (1999) argued, using results of Stram and Lee (1994), that the appropriate asymptotic distribution for this test was a 50:50 mixture of a χ_0^2 and a χ_1^2 distribution. However, they reported that limited simulation results found the test conservative. Verbyla et al. (1999) also suggested a two-stage procedure to test the hypothesis $g(t) \equiv 0$ by first testing $\sigma_s^2 = 0$ using the likelihood ratio test described above, and then, if there was no evidence of a non-linear component, using Wald tests to investigate the hypothesis $\boldsymbol{\beta} = \mathbf{0}$.

Zhang and Lin (2003) developed a score test for the null hypothesis $\sigma_s^2 = 0$ versus the alternative $\sigma_s^2 \neq 0$, and showed that under the null hypothesis the test had an asymptotic distribution that was a mixture of χ^2-distributions and provided a Satterthwaite approximation for the distribution. However, calculation of both the score statistic and the weights used in the approximation was computationally demanding.

Crainiceanu and Ruppert (2004a) returned to this issue for both the general linear mixed model and the specific case of the mixed-model spline. They argued that the conditions used by Stram and Lee (1994) to develop their results did not hold in the case of the mixed-model spline, and that the true distribution of the test statistic under the null hypothesis is a different mixture of χ^2-distributions. Crainiceanu and Ruppert (2004a) gave formulae for the case of a simple model with a single variance parameter in addition to the residual variance, but no formula for the general case, where use of the parametric bootstrap was suggested by Crainiceanu et al. (2005) to establish the distribution of the test statistic under the null hypothesis. Claeskens (2004) and Crainiceanu et al. (2005) extended the results of Crainiceanu and Ruppert (2004a) using similar decompositions to uncover the underlying structure of the model, but did not achieve results that would be any simpler to implement in practice, especially for complex models. Liu and Wang (2004) assessed various methods, including the test of Crainiceanu and Ruppert (2004a), and proposed estimation of the mixture proportions as a less computationally intense approach, but found that this did not work well for small data sets. This is an active research area, so standard procedures are not yet established.

11.6 Mixed-model cubic smoothing spline for longitudinal data

All aspects of the smoothing spline mixed model used to describe the profile of a single subject have now been considered. In practice, a model is required to simultaneously describe the profiles for a set of N subjects according to treatment effects and any relevant covariates, while providing a realistic model of within- and between-subject covariances. Observations between subjects are usually assumed to be independent, except for covariation due to shared environmental effects, such as location, or genetic relationships. Observations within subjects are expected to be correlated, with correlation usually decreasing as the lag between measurements increases. The covariance between measurements within the same subject is often modeled directly by some standard variance model applied to the residual errors for each subject, V_i (e.g., Wolfinger, 1996). The same underlying correlation structure is usually assumed to apply across subjects. Alternatively, within-subject covariances may be modeled indirectly via latent variables, as in random-coefficient regression models (Longford, 1993). A series of increasingly complex models for a set of subjects will now be considered, using smoothing splines to model profiles over time.

11.6.1 Simple spline model

In this case, a homogenous set of subjects from a common environment with no additional covariates or other structure is assumed. Then a reasonable model for the data consists of a common profile across time, with individual variation about the common profile that is correlated within subject across time. The common profile can be modeled using a smoothing spline, $g(t)$. Recall that, because it minimizes the penalized sum of squares (11.2), the smoothing spline will provide a smooth curve that follows the trend in the profiles. The smoothness of the curve is controlled by the smoothing parameter, which is to be estimated from the data by REML. It is assumed that within-subject covariances can be modeled using some standard covariance model $\text{Cov}(\boldsymbol{e}_i) = V_i(\boldsymbol{\phi})$, with $\text{Cov}(\boldsymbol{e}_i, \boldsymbol{e}_j) = 0$ for $i \neq j$, $i, j = 1, \ldots, N$. This model can be written as

$$Y_{ij} = g(t_{ij}) + e_{ij},$$

for $i = 1, \ldots, N, j = 1, \ldots, n_i$. In matrix notation for subject i this becomes model (11.1), with the model for the full data set written as

$$Y = g(t) + e, \tag{11.23}$$

where $Y = (Y_1', \ldots, Y_N')'$, $t = (t_1', \ldots, t_N')'$, and $e = (e_1', \ldots, e_N')'$. The smoothing spline is now a natural cubic spline with knots at the ordered set of distinct measurement times combined across the full set of N subjects, that is, the set of r knots is defined as the maximal set

$$\tau = (\tau_1, \ldots, \tau_r)' = \{\tau_i \in t : \tau_i < \tau_{i+1}\}.$$

The model can then be written as

$$Y_{ij} = \boldsymbol{\beta}_{s,0} + \boldsymbol{\beta}_{s,1} t_{ij} + \sum_{j=2}^{r-1} \delta_j P_j(t_{ij}) + e_{ij},$$

for $i = 1, \ldots, N$, $j = 1, \ldots, n_i$. In matrix notation for subject i this becomes

$$Y_i = X_{si}\boldsymbol{\beta}_s + Z_{si}\boldsymbol{\delta} + e_i,$$

with the model for the full data set written as

$$Y = X_s\boldsymbol{\beta}_s + Z_s\boldsymbol{\delta} + e,$$

where vector $\boldsymbol{\delta}$ has inverse penalty matrix G_s with associated scalar σ_s^2. The functions $P_j(t)$, the $N_T \times 2$ matrix $X_s = (X_{s1}', \ldots, X_{sN}')'$, the $N_T \times (r-2)$ matrix $Z_s = (Z_{s1}', \ldots, Z_{sN}')'$, and the symmetric matrix G_s^{-1} with $r - 2$ rows use the same general form as in previous definitions, but are now defined in terms of the combined set of r knots τ. In terms of the general mixed model of Section 11.4, the spline random effects are held in common so that $b_i = \boldsymbol{\delta}$ for $i = 1, \ldots, N$.

When measurement times within subjects are distinct, then $\max\{n_i; i = 1, \ldots, N\} \leq r \leq N_T$, with $r = N_T$ only when every subject is observed at different times. This model reduces to the spline model for an individual subject used in Section 11.5 when the design is balanced in the sense that all subjects are measured at the same times, with $n_i = n$ and $t_i = (t_1, \ldots, t_n)'$ for $i = 1, \ldots, N$.

As the random component of the spline is not regarded as contributing to the within-subject covariance matrix, the model for within-subject variation takes the form

$$\mathrm{Cov}(Y_i \mid \boldsymbol{\beta}_s, \boldsymbol{\delta}) = V_i.$$

11.6.2 Simple spline model with lack of fit

Verbyla et al. (1999) argued strongly for the addition of a lack-of-fit term to the mixed-model smoothing spline (11.23) where replicate measurements are present. If the data Y arise from a balanced design with $n_i = n$ and common measurement times $t_i = (t_1, \ldots, t_n)'$ for $i = 1, \ldots, N$, then the pattern in subject mean profiles at measurement times can be partitioned into smooth trend, fitted by the smoothing spline, plus lack of fit. Verbyla et al. (1999) fitted this term by using a separate random effect for each time point (group of replicate measurements), and assumed that these random effects were independent with common variance. This assumption imposes no covariance across time, so that the lack-of-fit term can be interpreted as a non-smooth component of trend. Other forms for lack of fit could also be considered.

This model can be written as

$$Y_{ij} = \boldsymbol{\beta}_{s,0} + \boldsymbol{\beta}_{s,1} t_{ij} + \sum_{j=2}^{n-1} \delta_j P_j(t_{ij}) + u_{g,j} + e_{ij},$$

with $\boldsymbol{u}_g = (u_{g,1}, \ldots, u_{g,n})' \sim N(\boldsymbol{0}, \sigma_g^2 I)$. In matrix form for subject i, the model takes the form

$$\boldsymbol{Y}_i = X_{si}\boldsymbol{\beta}_s + Z_{si}\boldsymbol{\delta} + \boldsymbol{u}_g + \boldsymbol{e}_i.$$

In the notation of the general mixed model (11.13), then $Z_i = [Z_{si} \ I_n]$ and $\boldsymbol{b}_i = (\boldsymbol{\delta}' \ \boldsymbol{u}_g')'$ for $i = 1, \ldots, N$, so that all random effects are held in common across subjects.

In the general unbalanced case, there may be some aliasing between the lack-of-fit term ($\boldsymbol{u}_g$) and the residual ($\boldsymbol{e}_i$) so that the model becomes difficult to fit numerically and the lack-of-fit term is less straightforward to interpret. In particular, if $V_i = \sigma_e^2 I$, then the random effects $\boldsymbol{u}_g$ and $\boldsymbol{e}_i$ cannot be distinguished unless several subjects are measured at the same times.

If the lack-of-fit term is regarded as random non-smooth variation about the smooth trend, then it is perhaps reasonable that it should be considered as contributing to within-subject variation. In that case, the within-subject covariance model becomes

$$\text{Cov}(\boldsymbol{Y}_i \mid \boldsymbol{\beta}_s, \boldsymbol{\delta}) = \sigma_g^2 I + V_i.$$

Note also that $\text{Cov}(\boldsymbol{Y}_i, \boldsymbol{Y}_j \mid \boldsymbol{\beta}_s, \boldsymbol{\delta}) = \sigma_g^2 I$ for $i \neq j$, as the shared random effects $\boldsymbol{u}_g$ introduce covariance between subjects due to common deviation from the smooth trend.

Inclusion of the lack-of-fit term in the spline model is directly analogous to the lack-of-fit term used to measure deviations from the proposed model in regression, although in that case the term is fitted as a set of fixed effects. In terms of analysis of variance, inclusion of the lack-of-fit term means that the smooth trend can be tested against the lack-of-fit term rather than against the residual variation. The spline is then assessed as to whether it provides a dominant component of the time trend. The change in log-likelihood, RL, on removal of the lack-of-fit term from the model can be used to assess the presence of non-smooth trend over time. It can be argued that, in order to preserve the hierarchical structure introduced by the random lack-of-fit term, this term should be tested after the random component of the spline. A detailed critical assessment of the impact and performance of the lack-of-fit term in mixed-model smoothing splines has not yet been made.

11.6.3 Spline model with additional covariates

The model can now be extended to allow additional covariates which may describe attributes of the subjects or their environments that may affect the measurements. Some of these covariates, such as age or sex, may be fitted as fixed, and some, such as location or environmental effects, may be fitted as random. If the additional fixed and random covariates are amalgamated into design matrices X_{ri} and Z_{ri}, respectively, for subject i, the extended model can be written in matrix notation for each subject as

$$\boldsymbol{Y}_i = X_{ri}\boldsymbol{\beta}_r + Z_{ri}\boldsymbol{b}_r + X_{si}\boldsymbol{\beta}_s + Z_{si}\boldsymbol{\delta} + \boldsymbol{e}_i,$$

with $\text{Cov}(\boldsymbol{b}_r) = G_r$, and

$$\text{Cov}(\boldsymbol{Y}_i \mid \boldsymbol{\beta}_r, \boldsymbol{\beta}_s, \boldsymbol{\delta}) = Z_{ri} G_r Z_{ri}' + V_i.$$

11.6.4 Spline model with separate subgroup profiles

The model can now be extended to allow for different smooth profiles across time for different groups of subjects. This will often be required when treatments have been applied to subsets of individuals, or if some intrinsic character of subjects, such as sex or age group, may affect their progress over time. It is then useful to fit a common spline, to describe the component of trend common to all individuals, and to also fit separate splines for the subgroups, in order to explore differences between them. The model for measurement j on subject i in subgroup k can then be written as

$$Y_{ijk} = g(t_{ij}) + h_k(t_{ij}) + e_{ij},$$

where $h_k(t)$ is a cubic smoothing spline modeling smooth deviations from the common spline for subgroup k, sometimes called an interaction spline. The common smoothing spline $g(t)$ is defined as in Section 11.6.1, and the subgroup smoothing spline $h_k(t)$ is defined as

$$h_k(t) = \zeta_{sk,0} + \zeta_{sk,1}t + \sum_{j=2}^{r-1} v_{k,j} P_j(t),$$

where the basis functions $P_j(t)$, $j = 2, \ldots, r-1$, are still defined in terms of the knots $\tau_1, \ldots, \tau_r$ that are the ordered unique measurement times combined across all subjects. For the purposes of mixed-model estimation, the coefficients $\boldsymbol{v}_k = (v_{k,2}, \ldots, v_{k,r-1})'$ are assumed to have inverse penalty matrix G_s, with associated scalar σ_v^2, with $\sigma_v^2 \geq 0$ and independence (no penalty) between subgroups. In matrix form, the model for subject i in subgroup k can be written as

$$\boldsymbol{Y}_{ik} = X_{si}(\boldsymbol{\beta}_s + \boldsymbol{\zeta}_{sk}) + Z_{si}(\boldsymbol{\delta} + \boldsymbol{v}_k) + \boldsymbol{e}_i,$$

where $\boldsymbol{\zeta}_k = (\zeta_{sk,0} \ \zeta_{sk,1})'$. Clearly there will be some redundancy between the common intercept and linear trend and the subgroup intercept and linear trend parameters. This is resolved by using parameter constraints as usual in regression models. The within-subject covariance matrix becomes

$$\text{Cov}(\boldsymbol{Y}_{ik} \mid \boldsymbol{\beta}_s, \boldsymbol{\zeta}_k, \boldsymbol{\delta}, \boldsymbol{v}_k) = V_i.$$

For balanced data, if there are several subjects in each subgroup then a lack-of-fit term for the subgroups can be included in the model for both the common and subgroup splines. The model becomes

$$Y_{ijk} = g(t_{ij}) + h_k(t_{ij}) + u_{g,j} + u_{gk,j} + e_{ij},$$

with

$$\boldsymbol{Y}_{ik} = X_{si}(\boldsymbol{\beta}_s + \boldsymbol{\zeta}_{sk}) + Z_{si}(\boldsymbol{\delta} + \boldsymbol{v}_k) + \boldsymbol{u}_g + \boldsymbol{u}_{gk} + \boldsymbol{e}_i,$$

where $\boldsymbol{u}_g$ was defined above and $\boldsymbol{u}_{gk} = (u_{gk,1}, \ldots, u_{gk,n})'$ with $\text{Cov}(\boldsymbol{u}_{gk}) = \sigma_{g2}^2 I$ and $\text{Cov}(\boldsymbol{u}_{gk}, \boldsymbol{u}_{gl}) = 0$ for $k \neq l$. Then

$$\text{Cov}(\boldsymbol{Y}_{ik} \mid \boldsymbol{\beta}_s, \boldsymbol{\zeta}_k, \boldsymbol{\delta}, \boldsymbol{v}_k) = (\sigma_g^2 + \sigma_{g2}^2)I + V_i.$$

Both lack-of-fit terms can be fitted only when replicate measurements are made at the same times on subjects across several subgroups. Again, interpretation of this term is only straightforward for balanced data.

The interaction splines $h_k(t)$ can be used to investigate whether patterns of smooth trend differ between subgroups. Setting the variance component σ_v^2 equal to zero corresponds to dropping the random component of the interaction splines from the model. The change in log-likelihood RL on dropping the random interaction spline terms from the model can then be used as the basis for a likelihood ratio test, although the provisos on this test statistic discussed in Section 11.5 still apply. If there is no evidence for different curvature between subgroups as represented by the random component of the spline, then approximate F-tests can be used to evaluate the evidence for separate linear trend between subgroups.

Kenward and Welham (1995) suggested that separate variance components σ_{vk}^2 might be used for the subgroups. This would allow the degree of smoothing to be estimated separately for each subgroup and might be desirable if the smoothness of profiles varied between groups. This suggestion does not appear to have been critically evaluated.

If the set of unique measurement times is different between subgroups, then it is sensible to consider whether the set of unique measurement times for each subgroup should be used as knots to construct the subgroup splines. In fact, the same smoothing spline and hence the same fitted splines are obtained in each case. The use of the combined set of measurement times means that the model can be written more simply as the same basis functions are used throughout. This issue becomes more important for the subject splines described in the next section.

The extension to higher-order structures for nested effects or factorial treatment structures follows directly, as does the inclusion of other relevant fixed or random covariates into this model.

11.6.5 Spline model with separate subject profiles

Random-coefficient regression has become a popular method for jointly modeling mean and covariance in longitudinal data. In the simple case, subjects are expected to follow a common linear profile with some random variation in subject intercepts and slopes. The linear random-coefficient regression model for subject i is written as

$$Y_{ij} = \boldsymbol{\beta}_{s,0} + \eta_{si,0} + (\boldsymbol{\beta}_{s,1} + \eta_{si,1})t_{ij} + e_{ij}$$

or, in matrix form,

$$\boldsymbol{Y}_i = (\boldsymbol{\beta}_s + \boldsymbol{\eta}_{si})X_{si} + \boldsymbol{e}_i, \tag{11.24}$$

where $X_{si} = [\mathbf{1}\ \boldsymbol{t}_i]$ as previously, $\boldsymbol{\eta}_{si} = (\eta_{si,0}\ \eta_{si,1})'$ with $\boldsymbol{\eta}_{si} \sim N(\mathbf{0}, G_c)$,

$$G_c = \begin{pmatrix} \sigma_{11} & \sigma_{12} \\ \sigma_{21} & \sigma_{22} \end{pmatrix},$$

and $\mathrm{Cov}(\boldsymbol{\eta}_{si}, \boldsymbol{\eta}_{sj}) = 0$ for $i \neq j$. The covariances across time within individuals then take the form

$$\mathrm{Cov}(Y_{ij}) = \sigma_{11} + 2\sigma_{12}t_{ij} + \sigma_{22}t_{ij}^2 + V_i[j,j]$$
$$\mathrm{Cov}(Y_{ij}, Y_{ik}) = \sigma_{11} + \sigma_{12}(t_{ij} + t_{ik}) + \sigma_{22}t_{ij}t_{ik} + V_i[j,k],$$

where $V_i[j,k]$ is the (j,k)th element of matrix V_i. In matrix form, the model for subject i becomes

$$\mathrm{Cov}(\boldsymbol{Y}_i \mid \boldsymbol{\beta}_s) = X_{si}G_cX'_{si} + V_i.$$

In terms of the general form of the mixed model (11.13), the random effects are now truly subject-specific, with $\boldsymbol{b}_i = \boldsymbol{\eta}_{si}$ for $i = 1, \ldots, N$.

In this model, the subject-specific intercept and slope are assumed to account for within-subject covariance across time and so it is usually assumed that $V_i = \sigma_e^2 I$. Within-subject covariances are then assumed to follow a quadratic pattern across time. The covariance parameter σ_{12} increases the flexibility of the variance structure and is essential if the model is to be mathematically invariant to translations in t, for example from t to $t - t_{\min}$.

This model can be extended to higher-order polynomial forms to allow for more complex profiles across time, which leads to models for within-subject covariances that are also of higher-order polynomial form. A natural extension is to introduce a smoothing spline in place of the higher-order polynomial. The model then posits a common spline to predict the population mean profile across time, with individual smooth variation about the common profile for each subject. This is similar to the subgroup spline model described above, except that now a spline is fitted to the set of measurements for each subject separately and, in the spirit of the linear random-coefficient regression model, the intercept and slope parameters for each subject are assumed to be random deviations about the common intercept and slope. The model can be written as

$$Y_{ij} = g(t_{ij}) + l_i(t_{ij}) + e_{ij},$$

where $l_i(x)$ is a cubic smoothing spline modeling smooth deviations from the common spline for subject i, sometimes called a subject spline. The common smoothing spline $g(t)$ is defined as above, and the subgroup smoothing spline $l_i(t)$ is defined as

$$l_i(t) = \eta_{si,0} + \eta_{si,1}t + \sum_{j=2}^{r-1} w_{i,j}P_j(t),$$

where the basis functions $P_j(t)$, $j = 2, \ldots, r-1$, are defined in terms of the knots $\tau_1, \ldots, \tau_r$ that are the combined set of ordered unique measurement times across all subjects. The coefficients $\boldsymbol{w}_i = (w_{i,2}, \ldots, w_{i,r-1})'$ have inverse penalty matrix G_s with scalar σ_w^2 ($\sigma_w^2 \geq 0$) and independence between subjects. In matrix form, the model for subject i is written as

$$\boldsymbol{Y}_i = X_{si}(\boldsymbol{\beta}_s + \boldsymbol{\eta}_{si}) + Z_{si}(\boldsymbol{\delta} + \boldsymbol{w}_i) + \boldsymbol{e}_i, \qquad (11.25)$$

with $\boldsymbol{\eta}_{si}$ considered as random effects, as in (11.24). As for linear random-coefficient regression, within-subject covariances are generated by the subject-specific profiles. In this context, the subject spline effects are therefore usually considered as arising from a normal population, that is, $\boldsymbol{w}_i \sim N(\boldsymbol{0}, \sigma_w^2 G_s)$ with $\mathrm{Cov}(\boldsymbol{w}_i, \boldsymbol{w}_j) = 0$ for $i \neq j$. The within-subject covariances then take the form

$$\mathrm{Cov}(\boldsymbol{Y}_i \mid \boldsymbol{\beta}_s, \boldsymbol{\delta}) = X_{si}G_cX_{si}' + \sigma_w^2 Z_{si}G_sZ_{si}' + V_i,$$

and all the basis functions (linear and smooth trend) contribute to subject variation about the common population trend. In terms of the general mixed model (11.13), now $Z_i = [Z_{si} \ X_{si} \ Z_{si}]$ with $\boldsymbol{b}_i = (\boldsymbol{\delta}' \ \boldsymbol{\eta}_{si}' \ \boldsymbol{w}_i')'$, where the spline effects $\boldsymbol{\delta}$ are held in common across subjects, and $\boldsymbol{\eta}_{si}$ and $\boldsymbol{w}_i$ are subject-specific effects.

Model (11.25) was used by Verbyla et al. (1999) and Guo (2002), who both made the additional assumption $\mathrm{Cov}(\boldsymbol{\eta}_{si}, \boldsymbol{w}_i) = 0$. Wang (1998) used the same form of model, but used a different spline basis. Brumback and Rice (1998) used a similar form of model, but with the subject intercept and slope parameters considered as fixed effects.

White et al. (1998) showed that, with the assumption $\mathrm{Cov}(\boldsymbol{\eta}_{si}, \boldsymbol{w}_i) = 0$, the within-subject variance model was not invariant to a change of spline basis. This means that the goodness of fit of the variance model to the data may depend upon the spline basis used. To make the variance model invariant to the spline basis chosen, the random effects $\boldsymbol{\eta}_{si}$, $\boldsymbol{w}_i$ must have a joint covariance model

$$\mathrm{Cov}\begin{pmatrix} \boldsymbol{\eta}_{si} \\ \boldsymbol{w}_i \end{pmatrix} = \begin{bmatrix} G_c & G_{sc}' \\ G_{sc} & \sigma_s^2 G_s \end{bmatrix},$$

where G_{sc} is an $(r-2) \times r$ matrix of separate covariance parameters. As for the random-coefficient regression model (11.24), the extra parameters provide a more general covariance model as well as invariance to the choice of basis, but may prove difficult to estimate in practice.

Because the subject-specific linear trend has been fitted using random effects, the exact cubic smoothing spline is no longer obtained. In terms of the PSS (11.2), an additional penalty has been added for deviations of the subject intercept and slope from the population values. The subject splines are used to indirectly model within-subject covariances. However, although minimization of the modified penalized sum of squares seems likely to still yield a profile that follows the subject trend well, there is no part of the criterion to force the fitted function to provide an adequate model of the covariances. White et al. (1998) suggested diagnostics to check for discrepancy between the fitted variance model and covariance patterns in the data, and these are examined in Section 11.8. However, even if the variance model fits the data well, it is possible that the population of splines generated by the nominal underlying mixed model may not be realistic, as discussed in Chapter 12, where a bootstrap method is suggested as an alternative approach.

Again, in this model, the combined set of measurement times across all subjects has been used as the set of knots to construct the spline. Because the model is not invariant to change of basis, but removing knots can be considered as a change to the basis, then the use of the set of subject measurement times as knots for that subject's spline would lead to different underlying variance models for subjects with different measurement times. In order to retain a common covariance structure it is then necessary to use a common set of knots.

A lack-of-fit term cannot be included within this model, as it has been assumed that a single measurement was made at each time. Without replicate measurements, the lack-of-fit term is completely aliased with residual error.

11.6.6 Spline model with lack of fit, additional covariates, and separate subgroup and subject profiles

The components of the models described above can be combined to model features of any data set. In a complex case, a model might include the population mean, subgroup, and individual profiles to be modeled by a spline, with additional covariates in both the fixed and random models. Using previous notation, this model might be written for subject i in subgroup k as

$$\boldsymbol{Y}_{ik} = X_{si}(\boldsymbol{\beta}_s + \boldsymbol{\zeta}_{sk} + \boldsymbol{\eta}_{si}) + X_{ri}\boldsymbol{\beta}_r + Z_{ri}\boldsymbol{b}_r + Z_{si}(\boldsymbol{\delta} + \boldsymbol{v}_k + \boldsymbol{w}_i) + \boldsymbol{u}_g + \boldsymbol{u}_{gk} + \boldsymbol{e}_i, \quad (11.26)$$

with the random effects $\boldsymbol{\eta}_{si}$, $\boldsymbol{b}_r$, $\boldsymbol{\delta}$, $\boldsymbol{v}_k$, $\boldsymbol{w}_i$, $\boldsymbol{u}_g$, $\boldsymbol{u}_{gk}$, and $\boldsymbol{e}_i$ having normal distributions with zero mean and variance models as specified above. In addition, it is assumed that these sets of random effects are mutually independent. The within-subject covariance model is then

$$\mathrm{Cov}(\boldsymbol{Y}_{ik} \mid \boldsymbol{\beta}_s, \boldsymbol{\zeta}_{sk}, \boldsymbol{\beta}_r, \boldsymbol{\delta}, \boldsymbol{v}_k) = Z_{ri}G_r Z_{ri}' + X_{si}G_c X_{si}' + \sigma_w^2 Z_{si}G_s Z_{si}' + \left(\sigma_g^2 + \sigma_{g2}^2\right)I + V_i.$$

The lack-of-fit terms are regarded here as contributing to the description of variance rather than trend. This designation depends on the interpretation of the underlying model. Complex models of this type provide a very flexible framework for the joint modeling of mean profiles and within-subject variances. It is conventional to use either subject splines with homogenous independent residual error or a more complex covariance model for residual error, but not both.

11.6.7 Modeling strategy

Some care must be taken when fitting complex models of this form to data. The variance parameters are not independent, so their values may change according to the set of random terms present in the model. In addition, the values of the variance parameters are heavily dependent on whether the fixed terms present in the model fully describe systematic trend. It is therefore sensible to fit an initial model including all potential fixed explanatory covariates and the full (most complex) variance model. Likelihood ratio tests can then be used to progressively simplify the variance model, with the fixed model left unchanged. Once an adequate variance model has been established, then approximate F-tests (Kenward and Roger, 1997) can be used to simplify the fixed model. In the situation where several different types of variance models are plausible, the best model should be fitted within each of the competing types and then, assuming a common set of fixed effects, these models can be compared using an information criterion such as the AIC (Verbeke and Molenberghs, 2000, Section 6.4).

EXAMPLE 275

11.7 Example: Balanced longitudinal data

This modeling strategy is demonstrated here using the balanced longitudinal data first analyzed by Grizzle and Allen (1969). The data consist of coronary sinus potassium concentrations measured on each of 36 dogs, which were divided into four different treatment groups, containing 9, 8, 9, and 10 dogs, respectively. Seven measurements were made on each dog, every 2 minutes from 1 to 13 minutes after an event (occlusion). The aim of this analysis is to obtain a functional model to quantify the difference in profiles between treatments: a good model is therefore required for both treatment profiles and within-subject variation. Grizzle and Allen (1969) found that low-order polynomials did not fit the treatment profiles (shown in Figure 11.2) adequately, so this data set is an obvious candidate for smoothing spline models, and was analyzed in this way by Wang (1998). The spline model (11.26) with common splines, separate splines for treatment groups, and individual subject splines was fitted with lack-of-fit terms for the common and treatment group splines. In symbolic terms, extending the notation of Wilkinson and Rogers (1973), this model can be written as

$$fixed = trt + t + trt.t,$$
$$random = spl(t) + fac(t) + trt.spl(t) + trt.fac(t) + dog + dog.t + dog.spl(t) + error,$$

where trt and dog are factors indicating the allocation of measurements to treatment groups and dogs (subjects), respectively, t is a continuous variate holding the time of measurement for each unit (mean-centered for computational convenience), $spl(t)$ indicates the random component of the smoothing spline across time, and $fac(t)$ indicates a factor with separate groups for each of the seven measurement times that acts as the lack of fit. The random $spl(t)$, $fac(t)$, and $error$ terms are fitted as independent random effects (using the Z_u parameterization of [11.22] for spline terms), with the variance component constrained to remain

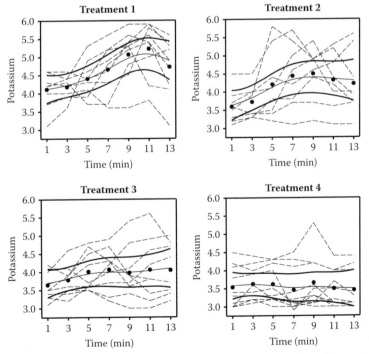

Figure 11.2 Fitted treatment profiles (solid line) with approximate 95% confidence intervals (gray-black thick line), treatment means (•), and subject profiles (gray dashed lines).

Table 11.1 Variance Parameter Estimates and Log-Likelihood (RL) from Full Model for Dog Data as Output from Software and Rescaled $(\times \mathrm{trace}(Z_i Z_i')/N_T)$

Term	Variance Parameter	Full Model Output	Full Model Rescaled
$spl(t)$	σ_s^2	0.0005	0.0028
$trt.spl(t)$	σ_v^2	0.0041	0.0248
$dog.spl(t)$	σ_w^2	0.0083	0.0498
$fac(t)$	σ_g^2	0	0
$trt.fac(t)$	σ_{g2}^2	0	0
dog	σ_{11}	0.2619	0.2619
$dog.t$	σ_{22}	0.0026	0.0419
$\mathrm{Corr}(dog, dog.t)$	$\sigma_{12}/\sqrt{(\sigma_{11}\sigma_{22})}$	0.48	0.48
$error$	σ_e^2	0.0587	0.0587
Log-Likelihood (df)		68.21 (235)	

non-negative. The dog and $dog.t$ terms (subject intercept and slope for each dog) are assumed correlated using covariance matrix G_c as above. Variance parameter estimates for this model are shown in Table 11.1, with the log-likelihood (RL) evaluated excluding term $c(X)$ in (11.14). This analysis was done in GenStat (Payne, 2003) and ASREML (Gilmour et al., 2003). The scaled variance parameters indicate that the dominant source of variation was differences in intercept between dogs, with smaller variation due to differences in slope and residual error. Both lack-of-fit terms were estimated at the lower boundary as zero. The common spline term was smaller than the treatment and dog spline terms, indicating little common trend but substantial variation between the profiles of treatment groups and individual dogs.

The lack-of-fit terms were omitted from the final model $(RL = 68.21, \mathrm{df} = 237)$. None of the other variance parameters could be omitted without a large decrease in the log-likelihood function (>4 units), except the common spline term, which was retained to keep the model for profiles more interpretable. The fitted profiles for treatment groups and individual dogs are shown with approximate 95% confidence intervals in Figure 11.2 and Figure 11.3. It can be seen that the treatment splines fit the treatment mean values well, and that the fitted subject splines follow the individual dog profiles quite closely.

To investigate the adequacy of the within-subject variance models, alternative models were fitted. Fitting a model without individual dog (subject) splines $(RL = 52.42, \mathrm{df} = 238)$ confirmed that this term gave an improved variance model compared to that from the linear random-coefficient regression. Fitting a general covariance matrix between the subject spline and intercept and slope effects (matrix G_{sc}, 10 parameters) further improved the log-likelihood $(RL = 77.55, \mathrm{df} = 227)$, although convergence of this model was difficult to achieve, and highly dependent on starting values. The change in deviance, 18.68 on 10 df, indicated that these extra parameters significantly improved the fit of the model $(p \simeq 0.04)$. Little benefit was gained by using a simplified version of this matrix with either a single covariance parameter $(RL = 69.87, \mathrm{df} = 236)$ or a common covariance parameter for slopes and intercepts separately $(RL = 70.74, \mathrm{df} = 235)$.

For comparison, a full unstructured matrix $(RL = 102.63, \mathrm{df} = 214)$ and ante-dependence models of order 1 $(RL = 92.10, \mathrm{df} = 229)$ and 2 $(RL = 94.65, \mathrm{df} = 224)$ were fitted as the within-subject residual term (V_i) without the subject spline terms. The AIC, calculated as $RL - N_v$, where N_v is the number of variance parameters fitted, was used to compare the non-nested models. On this scale, large values of AIC indicate a better fit, and these values are summarized in Table 11.2. It is clear that a much better description of the within-subject

EXAMPLE 277

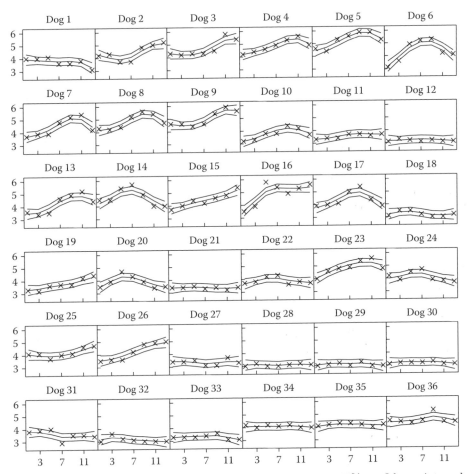

Figure 11.3 Fitted subject profiles (solid line) with approximate 95% confidence intervals and subject measurements (×).

Table 11.2 Summary of Alternative Models Fitted for Dog Data

| Components of Within-Subject Covariance Model | | | | | | # Variance | |
dog + dog.t	dog.spl(t)	G_{sc}	V_i	RL	df	Parameters	AIC
G_c	Omitted	–	$\sigma_e^2 I$	52.42	238	6	46.42
G_c	$\sigma_s^2 G_s$	$\mathbf{0}$	$\sigma_e^2 I$	68.21	237	7	61.21
G_c	$\sigma_s^2 G_s$	$\sigma_a^2 [1\ 1]$	$\sigma_e^2 I$	69.87	236	8	61.87
G_c	$\sigma_s^2 G_s$	$[\sigma_a^2\ \sigma_b^2] \otimes \mathbf{1}$	$\sigma_e^2 I$	70.74	235	9	61.74
G_c	$\sigma_s^2 G_s$	US	$\sigma_e^2 I$	77.55	227	17	60.55
Omitted	Omitted	–	AD(1)	92.10	229	15	77.10
Omitted	Omitted	–	AD(2)	94.65	224	20	74.65
Omitted	Omitted	–	US	102.63	214	30	72.63

Note: All models have the same fixed terms and include the common and treatment spline terms. US indicates an unstructured model, AD(k) indicates an ante-dependence model of order k.

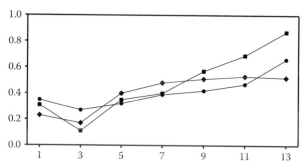

Figure 11.4 Fitted subject variances at each measurement time for subject spline model with $G_{sc} = \mathbf{0}$ (solid line with dots), with general G_{sc} (solid line with squares), and for US and AD(1) models (solid line with diamonds) shown in Table 11.2.

covariance model was achieved by the ante-dependence and unstructured models than by any of the subject spline models. In particular, the subject spline models did not provide a good description of the subject variances at each measurement time (see Figure 11.4).

In view of these results, it would appear sensible to consider alternatives to the within-subject spline variance model in balanced data where alternatives exist. For unbalanced data, more complex alternative models may be required. In these cases, the subject spline model may provide a significant improvement on alternative models based on random-coefficient regression, and the diagnostics discussed below can provide information on the appropriateness of that model.

11.8 Diagnostics for the subject spline variance model

Before moving on to consider a wider family of smoothing spline methods, it is useful to consider in more detail the form of the covariance model generated by the subject spline model, and to consider the diagnostics suggested by White et al. (1998) to assess whether the model is suitable for a given data set. White et al. (1998) identified two possible reasons for poor fit of the implicit subject spline variance model (11.25): (1) the subject intercept and slope are correlated with the spline effects, $\mathrm{Cov}(\boldsymbol{\eta}_{si}, \boldsymbol{w}_i) \neq 0$; (2) the nominal covariance for spline effects $\mathrm{Cov}(\boldsymbol{w}_i) = \sigma_w^2 G_s$ is incorrect. White et al. (1998) suggested diagnostics for these problems in the context of balanced longitudinal data, that is, where measurements were made at the same times on each subject, $n_i = n$, $\boldsymbol{t}_i = (t_1, \ldots, t_n)'$ for $i = 1, \ldots, N$. Using the orthogonal parameterization of the smoothing spline (11.21), the model for a single subject in balanced data can be written as

$$\boldsymbol{Y}_i = X_s \boldsymbol{\eta}_{si} + Z_0 \boldsymbol{w}_i + \boldsymbol{e}_i, \tag{11.27}$$

for $i = 1, \ldots, N$, where $X_s = [\mathbf{1} \; \boldsymbol{t}_i]$, $Z_0 = Q(Q'Q)^{-1}G_s^{-1}$, and Q and G_s are defined as in (11.10) and (11.11) for n knots $\boldsymbol{\tau} = (t_1, \ldots, t_n)'$, with

$$\mathrm{Cov} \begin{pmatrix} \boldsymbol{\eta}_{si} \\ \boldsymbol{w}_i \\ \boldsymbol{e}_i \end{pmatrix} = \begin{bmatrix} G_c & 0 & 0 \\ 0 & \sigma_w^2 G_s & 0 \\ 0 & 0 & \sigma_e^2 I_n \end{bmatrix}.$$

Use of an independent residual means that the subject spline variance model is intended to fully explain the within-subject covariance structure.

White et al. (1998) used a transformation to obtain two sets of orthogonal contrasts, and then examined the empirical covariances of these contrasts for correspondence with the proposed subject spline variance model. The two sets of contrasts are obtained as

$$X_s'\boldsymbol{Y}_i = X_s'X_s\boldsymbol{\eta}_{si} + X_s'\boldsymbol{e}_i,$$
$$U'Q'\boldsymbol{Y}_i = U'G_s^{-1}\boldsymbol{w}_i + U'Q'\boldsymbol{e}_i,$$

for $i = 1, \ldots, N$, since $X_s'Z_0 = 0$, where the $n \times (n-2)$ matrix U is such that

$$U'Q'QU = I_{n-2}, \quad U'G_s^{-1}U = D,$$

for some diagonal matrix $D = \text{diag}(d_1, \ldots, d_{n-2})$. Under the subject spline model (11.27), these two sets of contrasts are statistically independent, with

$$\text{Cov}(X_s'\boldsymbol{Y}_i, U'Q'\boldsymbol{Y}_i) = 0, \quad \text{Cov}(U'Q'\boldsymbol{Y}_i) = \sigma_w^2 D + \sigma_e^2 I_{n-2}, \qquad (11.28)$$

for $i = 1, \ldots, N$. Standard multivariate tests on the sample covariance matrix for the set of contrasts across subjects can be used to test the null hypotheses that (1) the linear contrasts $\{X_s'\boldsymbol{Y}_i; i = 1, \ldots, N\}$ are independent of the non-linear contrasts $\{U'Q'\boldsymbol{Y}_i; i = 1, \ldots, N\}$, that is, the subject intercepts and slopes are independent of the spline coefficients; (2) the set of non-linear contrasts are uncorrelated, that is, the form of the spline covariance matrix, G_s, is correct. Finally, the variances of the non-linear contrasts can be compared with their expected values (11.28), either graphically or more formally using a generalized linear model, as a further check on the spline covariance structure. Applying these tests to the dog data analyzed earlier showed evidence that the subject intercepts and slope coefficients were correlated with those for the spline basis functions ($p < 0.01$) and that the coefficients of the spline basis functions were not independent ($p < 0.001$). Those results match with the change in log-likelihood achieved by fitting more complex models in Table 11.2. A graph of the contrast variances was relatively uninformative with only five basis functions, but the variances did not appear incompatible with the model.

These diagnostics give some evaluation of the fit of subject spline variance models, and insight into discrepancies between the model and the data. However, further work is required to extend the diagnostics to the case of unbalanced data and alternative methods may be required. Jaffrézic, White, and Thompson (2003) used score tests to detect lack of fit in covariance models for longitudinal data. One advantage of the subject spline model is its potential to model non-stationary covariance structures. However, if the subject spline model does not provide a good description of the covariance pattern, then it may be beneficial to use alternative non-stationary models such as structured ante-dependence models (Núñez-Antón and Zimmerman, 2000) or character-process models (Pletcher and Geyer, 1999; Jaffrézic and Pletcher, 2000).

11.9 L-splines

Within the mixed-model cubic smoothing spline, the cubic spline is partitioned into a fixed linear component plus a random component, with zero expectation, representing smooth deviations about the linear trend. For a given value of the smoothing parameter, the fitted spline is determined by minimizing the residual sum of squares for the model subject to a penalty expressed in terms of the smoothing parameter and the integrated squared second derivative of the fitted curve (i.e., of deviations from linear trend). However, in many examples the underlying trend is not linear, or even polynomial, especially when some underlying periodic cycle is present in the data. The penalty may then be inappropriate. Kimeldorf and Wahba (1971) extended the penalized sum of squares to allow more general penalties corresponding to different assumptions about the underlying form of the data, and called the resulting class of functions L-splines. The amount of smoothing is again controlled by a penalty, which is constructed to penalize departures from an appropriate underlying form. Welham et al. (2006) showed that, as for the cubic smoothing spline, L-splines can be written as the BLUP from a specific linear mixed model with the smoothing parameter estimated using REML. These L-splines then provide an alternative to the cubic smoothing spline for modeling longitudinal data.

As previously, the form of the mixed-model L-spline is illustrated using data from a single subject, labeled as subject i. Suppose that n_i data values, $\boldsymbol{Y}_i = (Y_{i1}, \ldots, Y_{in_i})'$, are to be modeled as a profile across n_i distinct ordered measurement times $t_{i1} < \cdots < t_{in_i}$, which can occur in the range $[t_{\min}, t_{\max}]$. An L-spline is defined in terms of an underlying parametric form, described by a set of independent core functions $\{f_j; j = 1, \ldots, m\}$ and an associated linear differential operator (LDO) L of order m. This LDO must annihilate the core functions, that is, $Lf = 0$ if and only if f is also a linear combination of the f_j or $f \equiv 0$. For example, if a data set shows periodic cycles about zero, with no trend, then a suitable underlying form might use core functions

$$f_1(t) = \cos(\omega t), \quad f_2(t) = \sin(\omega t),$$

with associated LDO $L = \omega^2 I + D^2$ of order $m = 2$, where D is the differential operator. If the data had non-zero mean and showed trend over time, then a larger set of core functions,

$$f_1(t) = \cos(\omega t), \quad f_2(t) = \sin(\omega t), \quad f_3(t) = 1, \quad f_4(t) = t,$$

would be appropriate with associated LDO $L = \omega^2 D^2 + D^4$ of order $m = 4$. The polynomial smoothing splines (of odd degree, $k = 2m - 1$) form a subset of L-splines where $L = D^m$ for positive integers m, with $m = 2$ for cubic smoothing splines. In general, the functions f may depend on unknown parameters. Here it is assumed that all such parameters (e.g., the seasonal period parameter ω) are assumed known.

An L-spline penalizes departures from the underlying form, which is written in terms of the core functions. The simple L-spline model is again of the form (11.1). The L-spline is the function $g(t)$ that minimizes a PSS of the form

$$[\boldsymbol{Y}_i - g(\boldsymbol{t}_i)]'V_i^{-1}[\boldsymbol{Y}_i - g(\boldsymbol{t}_i)] + \lambda \int_{t_{\min}}^{t_{\max}} [Lg(s)]^2 ds, \tag{11.29}$$

for a given value of the smoothing parameter, $\lambda \geq 0$. As previously, the smoothing parameter determines the balance between fidelity to the data (measured by the first term, a residual sum of squares) and fidelity to the underlying core form (measured by the second term).

The form of the L-spline in terms of basis functions can be obtained using techniques from functional analysis. Full details can be found in Gu (2002) or Ramsay and Silverman (1997). Ramsay and Silverman (1997, Section 15.2) prove that, for any basis $\{f_j; j = 1, \ldots, m\}$ for the set of core functions, the function $g(t)$ that minimizes the PSS (11.29) has the form

$$g(t) = \sum_{j=1}^{m} \beta_{L,j} f_j(t) + \sum_{l=1}^{n_i} c_l k_2(t_{il}, t),$$

where k_2 is the reproducing kernel function for a subspace defined in terms of the operator L and a set of boundary conditions. Initial value boundary constraints were used here (see Ramsay and Silverman, 1997, Section 13.5.1). Ramsay and Silverman (1997, Section 15.3) further show that the PSS (11.29) can then be re-expressed as

$$(\boldsymbol{Y}_i - X_{Li}\boldsymbol{\beta}_L - K_i\boldsymbol{c})'V_i^{-1}(\boldsymbol{Y}_i - X_{Li}\boldsymbol{\beta}_L - K_i\boldsymbol{c}) + \lambda \boldsymbol{c}'K_i\boldsymbol{c}, \tag{11.30}$$

where the (l, j)th element of X_{Li} is $f_j(t_{il})$ for $l = 1, \ldots, n_i, j = 1, \ldots, m$, the (l, j)th element of K_i is $k_2(t_{ij}, t_{il})$ for $l, j = 1, \ldots, n_i$, $\boldsymbol{\beta}_L = (\beta_{L,1}, \ldots, \beta_{L,m})'$, and $\boldsymbol{c} = (c_1, \ldots, c_{n_i})'$. Gu (2002) showed that the symmetric matrix K is non-negative definite. Heckman and Ramsay (2000) and Dalzell and Ramsay (1993) give recipes for constructing reproducing kernel functions k_2. For $L = \omega^2 I + D^2$ and $t_{\min} = 0$ the reproducing kernel function $k_2(t_{ij}, t)$ can be written for $j = 1, \ldots, n_i$ as

$$
k_2(t_{ij}, t) = \begin{cases} \frac{1}{\omega^2}\left[t_{ij}\cos\omega(t_{ij} - t) + \frac{1}{2\omega}\sin\omega(t - t_{ij}) - \frac{1}{2\omega}\sin\omega(t + t_{ij})\right] & t < t_{ij}, \\[2mm] \frac{1}{\omega^2}\left[t\cos\omega(t - t_{ij}) + \frac{1}{2\omega}\sin\omega(t_{ij} - t) - \frac{1}{2\omega}\sin\omega(t + t_{ij})\right] & t_{ij} < t. \end{cases}
$$

This form for the basis functions is complex and obscures their underlying structure. It is straightforward (but tedious) to show that functions of this type are generated from functions defined piecewise on $[t_{i(j-1)}, t_{ij}]$ as

$$
a_{1i}\cos\omega t + a_{2i}\sin\omega t + a_{3i}t\cos\omega t + a_{4i}t\sin\omega t,
$$

for $t_{i(j-1)} \leq t \leq t_{ij}$, $j = 1, \ldots, n_i$ ($t_{i0} := t_{\min}$), with the requirement that the overall function is continuous and differentiable up to order 2 ($= 2m - 2$) at the knots $t_{i1}, \ldots, t_{in_i}$. These functions are thus piecewise (between knots) periodic with linearly varying amplitude. For $L = \omega^2 D^2 + D^4$, added piecewise cubic trend is also allowed, with the constraint that the overall function is six times differentiable at the knots. For comparison with the cubic spline, basis functions for these L-splines are shown in Figure 11.5.

11.9.1 L-spline mixed models

The L-spline is evaluated by finding the coefficient values $\widehat{\boldsymbol{\beta}}_L$ and $\tilde{\boldsymbol{c}}$ that minimize the PSS (11.30) for a given value of the smoothing parameter λ. Minimization of that expression requires solution of the equations

$$
\begin{bmatrix} X'_{Li}V_i^{-1}X_{Li} & X'_{Li}V_i^{-1}K_i \\ K_iV_i^{-1}X_{Li} & K_iV_i^{-1}K_i + \lambda K_i \end{bmatrix} \begin{pmatrix} \widehat{\boldsymbol{\beta}}_L \\ \tilde{\boldsymbol{c}} \end{pmatrix} = \begin{pmatrix} X'_{Li}V_i^{-1}\boldsymbol{Y}_i \\ K_iV_i^{-1}\boldsymbol{Y}_i \end{pmatrix}.
$$

These equations contain m implicit constraints which can be expressed as $X'_{Li}\tilde{\boldsymbol{c}} = \boldsymbol{0}$ (Wahba, 1990, Equation 1.3.17). This constraint means that the fitted L-spline takes the form of the core functions for $x < x_1$ and $x > x_n$ (Welham et al., 2006; Welham, 2005).

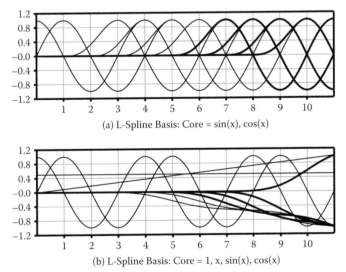

(a) L-Spline Basis: Core = sin(x), cos(x)

(b) L-Spline Basis: Core = 1, x, sin(x), cos(x)

Figure 11.5 Bases for L-splines with knots $1, \ldots, 10$ on range $[0,11]$ for (a) $L = \omega^2 I + D^2$ and (b) $L = \omega^2 D^2 + D^4$.

Following the terminology of polynomial splines, this can be called a natural L-spline. For convenience, this implicit constraint on the parameter estimates can be made explicit in the model as $X'_{Li}c = 0$. The constraint $X'_{Li}c = 0$ implies $c = C_i\delta_L$ for some vector δ_L of length $n - m$ where C_i is any $n_i \times (n_i - m)$ matrix of full column rank such that $X'_{Li}C_i = 0$. Inserting this reparameterization into the PSS (11.30) gives

$$(\boldsymbol{Y}_i - X_{Li}\boldsymbol{\beta}_L - Z_{Li}\boldsymbol{\delta}_L)'V_i^{-1}(\boldsymbol{Y}_i - X_{Li}\boldsymbol{\beta}_L - Z_{Li}\boldsymbol{\delta}_L) + \lambda\boldsymbol{\delta}'_L H_i^{-1}\boldsymbol{\delta}_L,$$

where $Z_{Li} = K_iC_i$, $H_i^{-1} = C'_iK_iC_i$. The penalty matrix H_i^{-1} can be constructed so that it is banded (Heckman and Ramsay, 2000) and positive definite, analogous to G_s^{-1} for the cubic smoothing spline.

The L-spline PSS (11.30) is thus minimized by the BLUP

$$\tilde{g}(\boldsymbol{t}_i) = X_{Li}\widehat{\boldsymbol{\beta}}_L + Z_{Li}\tilde{\boldsymbol{\delta}}_L,$$

obtained from the linear mixed model

$$\boldsymbol{Y}_i = X_{Li}\boldsymbol{\beta}_L + Z_{Li}\boldsymbol{\delta}_L + \boldsymbol{e}_i, \tag{11.31}$$

with $\mathrm{Cov}(\boldsymbol{e}_i) = V_i$, random effects $\boldsymbol{\delta}_L$ with inverse penalty matrix H_i and associated scalar σ_L^2, $\lambda = 1/\sigma_L^2$. As for the cubic smoothing spline, the smoothing parameter can be estimated by REML via the variance component σ_L^2. In addition, the L-spline mixed model (11.31) can be modified to make the fixed and random components orthogonal, in order to get better interpretation of the fixed component, and the random component can be transformed onto the scale of independent and identically distributed effects.

The transformation onto the reduced basis Z_{Li} was proposed by Welham et al. (2006) to reduce the dimension of the problem and provide a positive-definite covariance matrix H_i. This action is worth taking even when K_i is positive definite and the reduction in size is small because the matrix H_i has better numerical properties than the original matrix K_i, which is often very badly conditioned.

The model (11.31) for a single subject can be extended to obtain models for the full set of subjects exactly as for the cubic smoothing spline. For example, the simple L-spline model for a set of N subjects, with a combined set of r distinct measurement points $\boldsymbol{\tau} = (\tau_1, \ldots, \tau_r)'$, takes the form

$$\boldsymbol{Y}_i = X_{Li}\boldsymbol{\beta}_L + Z_{Li}\boldsymbol{\delta}_L + \boldsymbol{e}_i,$$

where $\mathrm{Cov}(\boldsymbol{e}_i) = V_i$, $\mathrm{Cov}(\boldsymbol{e}_i, \boldsymbol{e}_j) = 0$, random effects $\boldsymbol{\delta}_L$ have inverse penalty matrix H, and the symmetric matrix H of size $(r - m)$ is defined in terms of the knots $\boldsymbol{\tau}$. Inference proceeds in an analogous manner to the mixed-model cubic smoothing spline. The full range of models described in Section 11.6 can be formed in this way, using the appropriate L-spline in place of the cubic smoothing spline. However, similar caveats apply to the use of subject L-splines as implicit variance models. Welham et al. (2006) used a mixed-model L-spline to analyze longitudinal data from a grazing experiment across several years that showed periodic seasonal variation. They used a model with common, treatment (subgroup), and plot (subject) L-splines, and found that adding an exponential covariance model improved the description of within-plot covariance across time.

11.9.2 Which spline?

Given the wider family of mixed-model L-splines, it is necessary to choose an appropriate spline model for any data set. The log-likelihood RL was used to evaluate whether the random component of splines should be retained or omitted from the overall model, so the log-likelihood, or some related information criterion, might be used to differentiate between two different spline models with the same fixed terms. However, the logic of this proposal may depend on the interpretation of the spline model. It was stated earlier that the mixed-model representation of the spline is used merely as a numerical tool to obtain the smoothing spline, rather than as a genuine representation of a covariance model for the data, except for the special case of the subject splines. The log-likelihood, RL, measures the goodness of fit of the variance model to the data, in this case the random component of the smoothing splines, and so may not be an appropriate criterion. Welham (2005) undertook a small study to examine the performance of the log-likelihood RL in choosing between partial cubic splines with a common set of fixed effects and different penalty functions, and found that the RL criterion did not always choose the optimal model when judged against the MSEP criterion. Further work is required to assess the potential of this method. The comparison of two mixed-model splines fitted with different fixed and random effects using REML estimation is a more difficult problem. Welham and Thompson (1997) suggested a likelihood-based method for comparing different fixed models under REML estimation, but this method has not been tested in the spline context.

Welham et al. (2006) investigated whether there was any real difference in performance between cubic splines and particular L-splines for periodic data. In the context of a grassland experiment, they considered that perturbations to the underlying periodic cycle were likely to manifest as changes in the amplitude, phase, and length of the cycle due to variability in seasonal weather patterns, and slow changes in long-term trend were possible due to stock management. They used different combinations of these perturbations to generate single time series that were fitted using a cubic spline, a partial cubic spline with additional sine/cosine fixed terms, and an L-spline with $L = \omega^2 D^2 + D^4$, and examined the performance of the spline models in terms of MSEP. The cubic spline performed badly in terms of MSEP and almost always overfitted the data (undersmoothed). The partial cubic spline performed much better, and was the best model when the perturbations were in long-term trend alone. However, the L-spline consistently outperformed the partial cubic spline when perturbations to the periodic cycle were present. Given the form of the cubic spline (Figure 11.1) and L-spline basis functions (Figure 11.5), it is not surprising that the cubic spline deals well with changes in long-term trend and the L-spline deals well with changes in amplitude, as they are explicitly of an appropriate form in each case. In further work including an L-spline with $L = \omega^2 D + D^3$ (Welham, 2005), some differences in sensitivity of the different L-splines were detected for small deviations from the underlying periodic function, which indicated that as m increases, the minimum size of the deviations that can be detected reliably also increases. Further work is required to confirm this result.

In conclusion, it seems that there can be a real difference in the performance of competing mixed-model splines. Welham et al. (2007) showed similar differences between cubic splines with different penalties across a range of smooth functions. Although the mixed-model formulation of splines is used purely as a numerical tool to obtain the fitted spline as a BLUP, the results discussed above suggest that the behavior of these models may improve if the underlying mixed model is also compatible with the data. In the absence of an objective criterion, some consideration of the underlying form of the data and the structure of possible deviations may therefore indicate a suitable spline model.

11.10 Low-rank approximations: P-splines and penalized splines

For data sets with a large number of distinct measurement times, or with several spline terms in the model, the cubic spline mixed model generates a large number of random spline effects with a dense design matrix Z_s. Solution of the mixed-model equations may then require a large amount of computer workspace and processing time. Both can be reduced by using a (relatively) small number of knots, s say, defined at distinct covariate values $\tau^* = (\tau_1^*, \tau_2^*, \ldots, \tau_s^*)'$. The development here is similar to that of Brumback, Ruppert, and Wand (1999), Ruppert and Carroll (2000), and Wand (2003), who used a reduced number of knots in generating polynomial spline basis functions as a low-rank approximation to the full basis.

For a cubic spline, the set of $s - 2$ non-linear basis functions is generated as $P_j(x)$ for $j = 2, \ldots, s - 1$, using the reduced set of knots, $\boldsymbol{\tau}^*$. The spline function then takes the form

$$g(t) = \beta_{s,0} + \beta_{s,1}t + \sum_{j=2}^{s-1} \delta_j P_j(t).$$

For a simple spline model (11.23), the PSS (11.2) can be expressed for functions $g(t)$ of this form as

$$\text{PSS} = [\boldsymbol{Y} - g(t)]'V^{-1}[\boldsymbol{Y} - g(t)] + \lambda \int_{t_{\min}}^{t_{\max}} [g''(s)]^2 \, ds$$

$$= (\boldsymbol{Y} - X_s\boldsymbol{\beta}_s - Z_s^*\boldsymbol{\delta}^*)'V^{-1}(\boldsymbol{Y} - X_s\boldsymbol{\beta}_s - Z_s^*\boldsymbol{\delta}^*) + \lambda\boldsymbol{\delta}^{*'}G_s^{*-1}\boldsymbol{\delta}^*,$$

where now the symmetric matrix G_s^* with $s - 2$ rows is defined in terms of the reduced set of knots $\boldsymbol{\tau}^*$, $Z_s^* = [P_2(\boldsymbol{t}), \ldots, P_{s-1}(\boldsymbol{t})]$ evaluates the $s-2$ basis functions at the measurement times, and $\boldsymbol{\delta}^* = (\delta_2^*, \ldots, \delta_{s-1}^*)'$ is a vector of second derivative values at the knots, $\boldsymbol{\tau}^*$. Minimizing this penalized sum of squares yields mixed-model equations for the standard form of the spline mixed model

$$\boldsymbol{Y} = X_s\boldsymbol{\beta}_s + Z_s^*\boldsymbol{\delta}^* + \boldsymbol{e}, \tag{11.32}$$

where the reduced set of second derivatives, $\boldsymbol{\delta}^*$, has inverse penalty matrix G_s^* with associated scalar σ_s^2.

The fitted spline, estimated smoothing parameter, and log-likelihood function RL are not invariant to the reduced set of knots chosen, although the fitted spline is usually similar at the data points. In addition, although the fitted reduced-knot spline gives the minimum value of the PSS over the set of functions of form (11.32), it no longer has the optimality property of producing a global minimum of the PSS, which is achieved by the natural cubic spline with knots at the distinct covariate values. It may therefore be sensible to evaluate the sensitivity of the fitted spline to the set of knots chosen.

Reduced-knot splines of this type were described by Eilers and Marx (1996) and Ruppert, Wand, and Carroll (2003). In both cases, the mixed-model form of the spline follows directly from minimization of a penalized sum of squares. Eilers and Marx (1996) used the term P-splines to describe their splines, which used a B-spline basis with a discrete differencing penalty, and minimized penalized sums of squares of the form

$$(\boldsymbol{Y} - B\boldsymbol{\nu})'V^{-1}(\boldsymbol{Y} - B\boldsymbol{\nu}) + \boldsymbol{\nu}'D_d'D_d\boldsymbol{\nu},$$

where, using the notation of (11.6) for an example of a cubic spline with dth-order differencing, $\boldsymbol{\nu} = (\nu_{-3}, \ldots, \nu_s)'$, $B = [B_{-3,4}(\boldsymbol{t}), \ldots, B_{s,4}(\boldsymbol{t})]$, with D_d a dth-order differencing matrix for the extended set of knots $\tau_{-3}^*, \ldots, \tau_s^*$. The default P-spline with a cubic basis and second-order differencing penalty approximates the cubic smoothing spline model (11.32) for the same set of knots. Currie and Durban (2002) placed P-splines within the mixed-model framework. Ruppert, Wand, and Carroll (2003) described a family of penalized splines, which used a TPF basis and minimized penalized sums of squares of the form

$$(\boldsymbol{Y} - X_T\boldsymbol{\beta}_T - Z_T\boldsymbol{\nu}_T)'V^{-1}(\boldsymbol{Y} - X_T\boldsymbol{\beta}_T - Z_T\boldsymbol{\nu}_T) + \boldsymbol{\nu}_T'\boldsymbol{\nu}_T,$$

where, using the notation of (11.4) for an example of a cubic spline, $X_T = [\boldsymbol{1}, \boldsymbol{t}, \boldsymbol{t}^2, \boldsymbol{t}^3]$, $\boldsymbol{\nu}_T = (\nu_1, \ldots, \nu_s)'$, and $Z_T = [(\boldsymbol{t} - \tau_1^*)_+^3, \ldots, (\boldsymbol{t} - \tau_s^*)_+^3]$. Crainiceanu and Ruppert (2004b) showed that the penalty in this model was approximately equal to a penalty on the fourth derivatives of the spline. The use of the different bases obscures the close connections between these two models, which can be derived from the fourth difference relationship between the two bases. Welham et al. (2007) gave details of transformation between these bases, and their relationship to polynomial smoothing splines. Specifically, the cubic penalized spline is equivalent to a cubic P-spline with fourth-order differencing, and the linear penalized spline is equivalent to a linear P-spline with second-order differencing. Out of the family of penalized splines, the linear penalized spline gives the closest approximation to the cubic smoothing spline penalty. Both the cubic P-spline with second-order differencing and the linear penalized spline are used in Chapter 12.

The reduced-knot splines are particularly appealing for longitudinal data where subjects have been measured at different times, or for designed experiments where different treatments have been assessed at different covariate values. In these cases, use of a common reduced knot set across subgroups or subjects means that the same underlying spline model is applied in all cases, without requiring large numbers of knots.

Durban et al. (2005) used a penalized spline mixed model with subject-specific splines to model longitudinal data. However, their model has the same problems as the cubic smoothing spline model (11.25) in that the within-subject covariance model is specific to the basis chosen, and may not provide an adequate description of within-subject covariances. In general, the issues associated with the cubic smoothing spline mixed models described above apply similarly to P-spline and penalized spline models.

Hastie (1996) proposed pseudosplines, a class of linear smoothers constructed as the best low-rank approximation to a given smoother, then fitted using penalized regression. The pseudosplines would be expected to offer a better low-rank approximation to the PSS than a spline with an arbitrary reduced knot set, but do not easily fit within the mixed-model framework. Chapter 12 proposes the use of an effective basis, that is, a subset of the full basis that approximates the EDF of the smoothing spline.

11.11 Extensions and directions for future research

The smoothing spline mixed models allow joint modeling of the mean and covariance structure, but use the assumption of a normal distribution for errors. For non-normal data, the extension has been made to smoothing splines within the generalized linear mixed-model framework. Lin and Zhang (1999) used penalized quasi-likelihood (PQL) methods, with bias adjustment, for estimation of the smoothing parameter and showed that they performed well across a range of situations. Crainiceanu, Ruppert, and Wand (2005) and Zhao et al. (2006) used Bayesian versions of the generalized linear mixed models with a Markov chain Monte Carlo approach to estimation and inference, which avoided the approximations inherent in the PQL method. The extension to non-linear mixed models is required to generate mixed-model L-splines when there are unknown parameters in the core functions. This is required if the length of the periodic cycle is unknown, or if smooth deviations occur around a non-linear form such as an exponential function. Ke and Wang (2001) consider the case of a smoothing spline within a non-linear model as a model for longitudinal data, and again use PQL methods for estimation. These extensions provide a wider range of models for longitudinal data, while keeping the advantages of working within the mixed-model framework.

In some cases, such as periodic data, additional constraints on the fitted spline may be required. Constraints can easily be built into the smoothing spline where these constraints can be written as a linear function of the spline coefficients. Zhang, Lin, and Sowers (2000) and Wang and Brown (1996) used cubic splines constrained to be periodic. Welham et al. (2006) used an L-spline constrained to be periodic in addition to an unconstrained L-spline in order to partition deviations from an underlying regular periodic form into long-term and periodic components. Wood (2006) and Welham et al. (2007) both give general forms for the mixed-model spline with these constraints. More general constraints, such as monotonicity of the fitted spline, or the inclusion of an asymptote cannot be imposed using this method. Upsdell (1994, 1996) described a class of Bayesian splines which could build these wider constraints into the model, but it is not clear whether this type of method can be placed within the mixed-model framework.

Smoothing spline mixed models offer a flexible and data-driven method of fitting population, group, or subject profiles within the linear mixed-model framework. Although several theoretical aspects of these models remain to be resolved, it seems likely that, due to their generality and flexibility, they will continue to be widely used in practical data analysis.

Acknowledgments

This work was supported by the Australian Grains Research and Development Corporation and Rothamsted Research. Rothamsted Research receives grant-aided support from the Biotechnology and Biological Sciences Research Council.

References

Boneva, L. I., Kendall, D., and Stefanov, I. (1971). Spline transformations — three new diagnostic aids for the statistical data-analyst (with discussion). *Journal of the Royal Statistical Society, Series B* **33**, 1–70.

Brumback, B. A. and Rice, J. A. (1998). Smoothing spline models for the analysis of nested and crossed samples of curves (with discussion). *Journal of American Statistical Association* **93**, 961–994.

Brumback, B. A., Ruppert, D., and Wand, M. P. (1999). Comment on: Variable selection and function estimation in additive nonparametric regression using a data-based prior (by Shively, Kohn and Wood). *Journal of the American Statistical Association* **94**, 794–797.

Claeskens, G. (2004). Restricted likelihood ratio lack-of-fit tests using mixed model splines. *Journal of the Royal Statistical Society, Series B* **66**, 909–926.

Crainiceanu, C. M. and Ruppert, D. (2004a). Likelihood ratio tests in linear mixed models with one variance component. *Journal of the Royal Statistical Society, Series B* **66**, 165–185.

Crainiceanu, C. M. and Ruppert, D. (2004b). Restricted likelihood ratio tests in nonparametric longitudinal models. *Statistica Sinica* **12**, 713–729.

Crainiceanu, C. M., Ruppert, D., and Wand, M. P. (2005). Bayesian analysis for penalized spline regression using WinBUGS. *Journal of Statistical Software* **14**(14).

Crainiceanu, C. M., Ruppert, R., Claeskens, G., and Wand, M. P. (2005). Exact likelihood ratio tests for penalized splines. *Biometrika* **92**, 91–103.

Craven, P. and Wahba, G. (1979). Smoothing noisy data with spline functions: Estimating the correct degree of smoothing by the method of generalized cross validation. *Numerische Mathematik* **31**, 377–403.

Currie, I. D. and Durban, M. (2002). Flexible smoothing with P-splines: A unified approach. *Statistical Modelling* **4**, 333–349.

Dalzell, C. J. and Ramsay, J. O. (1993). Computing reproducing kernels with arbitrary boundary constraints. *SIAM Journal of Scientific Computing* **14**, 511–518.

de Boor, C. (1978). *A Practical Guide to Splines*. New York: Springer.

Dierckx, P. (1993). *Curve and Surface Fitting with Splines*. Oxford: Clarendon Press.

Durban, M., Harezlak, J., Wand, M. P., and Carroll, R. J. (2005). Simple fitting of subject-specific curves for longitudinal data. *Statistics in Medicine* **24**, 1153–1167.

Eilers, P. H. C. and Marx, B. D. (1996). Flexible smoothing with B-splines and penalties. *Statistical Science* **11**, 89–121.

Gilmour, A. R., Gogel, B. J., Cullis, B. R., Welham, S. J., and Thompson, R. (2002). *ASREML User Guide Release 1.0*. Hemel Hempstead: VSN International.

Green, P. J. (1985a). Linear models for field trials, smoothing and cross-validation. *Biometrika* **72**, 527–537.

Green, P. J. (1985b). Comment on: Some aspects of the spline smoothing approach to nonparametric regression curve fitting (by B. W. Silverman). *Journal of the Royal Statistical Society, Series B* **47**, 29.

Green, P. J. (1987). Penalized likelihood for general semi-parametric regression models. *International Statistical Review* **55**, 245–259.

Green, P. J. and Silverman, B. W. (1994). *Nonparametric Regression and Generalized Linear Models*. London: Chapman & Hall.

Grizzle, J. E. and Allen, D. M. (1969). Analysis of growth and dose response curves. *Biometrics* **25**, 357–381.

Gu, C. (2002). *Smoothing Spline ANOVA Models*. New York: Springer.

Gu, C. and Ma, P. (2005). Generalized nonparametric mixed-effect models: computation and smoothing parameter selction. *Journal of Computational and Graphical Statistics* **14**, 485–504.

Guo, W. (2002). Functional mixed effects models. *Biometrics* **58**, 121–128.

Hastie, T. J. (1996). Pseudosplines. *Journal of the Royal Statistical Society, Series B* **52**, 379–396.

Hastie, T. J. and Tibshirani, R. J. (1986). Generalized additive models (with discussion). *Statistical Science* **1**, 297–318.

Hastie, T. J. and Tibshirani, R. J. (1990). *Generalized Additive Models*. London: Chapman & Hall.

Hastie, T. J., Tibshirani, R. J., and Friedman, J. (2001). *The Elements of Statistical Learning*. New York: Springer.

Heckman, N. E. and Ramsay, J. O. (2000). Penalized regression with model-based penalties. *Canadian Journal of Statistics* **28**, 241–258.

Jaffrézic, F. and Pletcher, S. D. (2000) Statistical models for estimating the genetic basis of repeated measures and other function-valued traits. *Genetics* **156**, 913–922.

Jaffrézic, F., White, I. M. S., and Thompson, R. (2003). Use of the score test as a goodness of fit measure of the covariance structure in genetic analysis of longitudinal data. *Genetics Selection Evolution* **35**, 185–198.

Ke, C. and Wang, Y. (2001). Semiparametric nonlinear mixed effects models and their applications. *Journal of the American Statistical Association* **96**, 1272–1298.

Kenward, M. G. and Roger, J. H. (1997). The precision of fixed effects estimates from restricted maximum likelihood. *Biometrics* **53**, 983–997.

Kenward, M. G. and Welham, S. J. (1995). Use of splines in extending random coefficient regression models for the analysis of repeated measurements. In T. A. B. Snijders et al. (eds.), *Symposium of Statistical Software 1995*, pp. 95–112. Groningen: Interuniversitair Expertisecentrum ProGAMMA.

Kimeldorf, G. S. and Wahba, G. (1970a). A correspondence between Bayesian estimation of stochastic processes and smoothing by splines. *Annals of Mathematical Statistics* **41**, 495–502.

Kimeldorf, G. S. and Wahba, G. (1970b). Spline functions and stochastic processes. *Sankhyā, Series A* **32**, 173–180.

Kimeldorf, G. S. and Wahba, G. (1971). Some results on Tchebysheffian spline functions. *Journal of Mathematical Analysis and Applications* **33**, 82–95.

Kohn, R., Ansley, C. F., and Tharm, D. (1991). The performance of cross-validation and maximum likelihood estimators of spline smoothing parameters. *Journal of the American Statistical Association* **86**, 1042–1050.

Lin, X. and Zhang, D. (1999). Inference in generalized additive mixed models by using smoothing splines. *Journal of the Royal Statistical Society, Series B* **61**, 381–400.

Liu, A. and Wang, Y. D. (2004). Hypothesis testing in smoothing spline models. *Journal of Statistical Computation and Simulation* **74**, 581–597.

Longford, N. T. (1993). *Random Coefficient Models*. Oxford: Clarendon Press.

Luo, Z. and Wahba, G. (1997). Hybrid adaptive splines. *Journal of the American Statistical Association* **92**, 107–116.

Mackenzie, M. L., Donovan, C. R., and McArdle, B. H. (2005). Regression spline mixed models: A forestry example. *Journal of Agricultural, Biological, and Environmental Statistics* **10**, 394–410.

Núñez-Antón, V. and Zimmerman, D. L. (2000). Modeling nonstationary longitudinal data. *Biometrics* **56**, 699–705.

Pan, H. Q. and Goldstein, H. (1998). Multi-level repeated measures growth modelling using extended spline functions. *Statistics in Medicine* **17**, 2755–2770.

Patterson, H. D. and Thompson, R. (1971). Recovery of interblock information when block sizes are unequal. *Biometrika* **31**, 100–109.

Payne, R. W. (ed.) (2003). *The Guide to Genstat, Part 2: Statistics*. Hemel Hempstead: VSN International.

Pletcher, S. D. and Geyer, C. J. (1999) The genetic analysis of age-dependent traits: Modeling a character process. *Genetics* **153**, 825–833.

Ramsay, J. O. and Silverman, B. W. (1997). *Functional Data Analysis*. New York: Springer.

Rice, J. A. and Wu, C. O. (2001). Nonparametric mixed effects models for unequally sampled noisy curves. *Biometrics* **57**, 253–259.

Ruppert, D. and Carroll, R. J. (2000). Spatially adaptive penalties for spline fitting. *Australian and New Zealand Journal of Statistics* **42**, 205–223.

Ruppert, D., Wand, M. P., and Carroll, R. J. (2003). *Semiparametric Regression*. Cambridge: Cambridge University Press.

Speed, T. P. (1991). Comment on: That BLUP is a good thing: the estimation of random effects (by G. K. Robinson). *Statistical Science* **6**, 44.

Stram, D. O. and Lee, J. W. (1994). Variance components testing in the longitudinal mixed effects setting. *Biometrics* **50**, 1171–1177.

Thompson, R. (1985). Comment on: Some aspects of the spline smoothing approach to nonparametric regression curve fitting (by B. W. Silverman). *Journal of the Royal Statistical Society, Series B* **47**, 43–44.

Upsdell, M. P. (1994). Bayesian smoothers as an extension of nonlinear regression. *New Zealand Statistician* **29**, 66–81.

Upsdell, M. P. (1996). Choosing an appropriate covariance function in Bayesian smoothing. *Bayesian Statistics* **5**, 747–756.

Verbeke, G. and Molenberghs, G. (2000). *Linear Mixed Models for Longitudinal Data*. New York: Springer.

Verbyla, A. P. (1990). A conditional derivation of residual maximum likelihood. *Australian Journal of Statistics* **32**, 227–230.

Verbyla, A. P., Cullis, B. R., Kenward, M. G., and Welham, S. J. (1999). The analysis of designed experiments and longitudinal data using smoothing splines. *Applied Statistics* **48**, 269–311.

Wahba, G. (1975). Smoothing noisy data by spline functions. *Numerische Mathematik* **24**, 383–393.

Wahba, G. (1984). Partial spline models for the semi-parametric estimation of functions of several variables. In *Statistical Analysis of Time Series, Proceedings of the Japan U.S. Joint Seminar, Tokyo*.

Wahba, G. (1985). A comparison of GCV and GML for choosing the smoothing parameter in the generalized spline problem. *Annals of Statistics* **13**, 1378–1402.

Wahba, G. (1990). *Spline Models for Observational Data*. Philadelphia: SIAM.

Wand, M. P. (2000). A comparison of regression spline smoothing procedures. *Computational Statistics* **15**, 443–462.

Wand, M. P. (2003). Smoothing and mixed models. *Computational Statistics* **18**, 223–249.

Wang, Y. (1998). Mixed effects smoothing spline analysis of variance. *Journal of the Royal Statistical Society, Series B* **60**, 159–174.

Wang, Y. and Brown, M. B. (1996). A flexible model for human circadian rhythms. *Biometrics* **52**, 588–596.

Welham, S. J. (2005). Smoothing spline methods within the mixed model framework. PhD Thesis, University of London.

Welham, S. J., and Thompson, R. (1997). Likelihood ratio tests for fixed model terms using residual maximum likelihood. *Journal of the Royal Statistical Society, Series B* **59**, 701–714.

Welham, S. J., Cullis B. R., Kenward, M. G., and Thompson, R. (2006). The analysis of longitudinal data using mixed model L-splines. *Biometrics* **62**, 392–401.

Welham, S. J., Cullis B. R., Kenward, M. G., and Thompson, R. (2007). A comparison of mixed model splines. *Australian and New Zealand Journal of Statistics* **49**, 1–23.

White, I. M. S., Cullis, B. R., Gilmour, A. R., and Thompson, R. (1998). Smoothing biological data with splines. In *Proceedings of International Biometrics Conference, 1998*.

Wilkinson, G. N. and Rogers, C. E. (1973). Symbolic description of factorial models for analysis of variance. *Applied Statistics* **22**, 392–399.

Wolfinger, R. D. (1996). Heterogeneous variance-covariance structures for repeated measures. *Journal of Agricultural, Biological, and Environmental Statistics* **1**, 362–389.

Wood, S. N. (2006). Low-rank scale-invariant tensor product smooths for generalized additive mixed models. *Biometrics* **62**, 1025–1036.

Zhang, D. W. and Lin, X. H. (2003). Hypothesis testing in semiparametric additive mixed models. *Biostatistics* **4**, 57–74.

Zhang, D. W., Lin, X. H., and Sowers, M. F. (2000). Semiparametric regression for periodic longitudinal hormone data from multiple menstrual cycles. *Biometrics* **56**, 31–39.

Zhang, D., Lin, X. H., Raz, J., and Sowers, M. F. (1998). Semiparametric stochastic mixed models for longitudinal data. *Journal of the American Statistical Association* **93**, 710–719.

Zhao, Y., Staudenmayer, J., Coull, B. A., and Wand M. P. (2006). General design Bayesian generalized linear mixed models. *Statistical Science* **21**, 35–51.

CHAPTER 12

Penalized spline models for longitudinal data

Babette A. Brumback, Lyndia C. Brumback, and Mary J. Lindstrom

Contents

12.1 Introduction

In longitudinal data analysis, where the goal is to estimate population or individual-specific averages over time, it is often the case that the repeated observations from a single individual can be viewed as arising from sampling a smooth underlying individual mean curve together with non-smooth deviations. The deviations can be thought of as a combined result of measurement error and biologic variability that would fluctuate upon resampling the same individual over the same period of time. When the assumption of *smooth individual mean curves* is reasonable, it makes sense to incorporate it into the estimation process in order to capitalize on the variance reduction that typically ensues when correct assumptions are used to reduce the effective degrees of freedom used by a statistical model.

When the smooth curves can be parametrically modeled using either a polynomial basis or a regression spline basis (Section 12.1.4), or another known basis for smooth functions, then estimation and inference can proceed in a relatively straightforward fashion. However, sometimes a low-dimensional basis is difficult to select, which has led to the practice of selecting a high-dimensional basis but then penalizing the estimated coefficients, to keep the effective degrees of freedom low. When this approach is applied together with a regression spline basis, the resulting estimates are known as *penalized splines*. During the late 1990s, several papers began to appear applying this idea to problems in longitudinal data analysis; see, for example, Brumback and Rice (1998), Wang (1998), Berhane and Tibshirani (1998), Lin and Zhang (1999), Verbyla et al. (1999), and the references therein.

Many of these papers exploited the connection between penalized splines and best linear unbiased prediction (BLUP) (see Robinson, 1991) based on a corresponding linear mixed-effects (LME) model; this connection will be detailed in Section 12.3. In exploiting the connection, authors often based inferences on the LME model analog; for example, the prediction error variance (PEV) (see Chapter 11) associated with the LME model is sometimes used to estimate variability of the fitted curves. The recent book on semi-parametric regression by Ruppert, Wand, and Carroll (2003) argues in favor of using the PEV to estimate mean squared error (MSE); however, the authors focus on estimating just one curve rather than on estimating characteristics of a collection of curves. In Section 12.3, we re-examine using the PEV to estimate the MSE of a penalized spline estimator of a single curve, and we discover that it can lead to non-negligible bias. Analogously, applying this method to longitudinal data can also lead to bias. Because the PEV is identical to the posterior variance obtained with a Bayesian version of the mixed-effects model, frequentist properties of the resulting Bayesian inferences are also suspect.

Our preference is thus steered toward using simulation-based techniques, such as the bootstrap (Efron and Tibshirani, 1993) for inference, rather than employing analytic results based on LME models or a Bayesian paradigm. The bootstrap was the primary approach to inference taken by Brumback and Rice (1998), but in a limited way due to the lack of computing power. With the advances in computers we have seen over the last decade, it is now possible to go much further with the bootstrap idea. Thus, in this chapter, we will revisit the analyses of Brumback and Rice (1998) and reanalyze the hormone data using today's computers and software: we use the `lme` function in the `nlme` (Pinheiro and Bates, 2000) library of R (`http://cran.r-project.org/`), and a computationally intensive non-parametric bootstrap for nested individual curves. Brumback and Rice (1998) needed to use Matlab, derive simplifications based on eigenanalysis, and assume a common smoothing parameter for all curves to have enough computational efficiency; furthermore, the non-parametric bootstrap was simply out of computational reach.

We will also analyze longitudinal randomized clinical trial data from a study of cystic fibrosis (CF) patients in which patients were randomized to inhaled tobramycin or placebo and measures of their lung function were taken over 24 weeks. The clinical trial data are proprietary, but the hormone data set is available at StatLib (`http://lib.stat.cmu.edu/`). The hormone data set together with R code used for several of the analyses we conduct in this chapter are also available at the web site for this book given in the Preface, so that readers can use them to apply and extend the methods we describe to other problems in longitudinal data analysis. Let us now introduce the case studies.

12.1.1 Progesterone data

The hormone data consist of repeated progesterone metabolite (pregnanediol-3-glucuronide, PdG) measures from day -8 to day 15 in the menstrual cycle (day 0 is ovulation day) on a sample of 22 conceptive cycles from 22 women and 69 non-conceptive cycles from 29 women. Measures are missing for some cycles on some days; throughout, we assume that missing data are missing at random (see Little and Rubin, 1987; see also Chapter 17 of this volume). The data are described in much more detail in Brumback and Rice (1998), whose Figures 1 and 2 also display it for several individuals. We have stored the data in an R object called `pdg.dat`, which has the following structure:

```
   group id cycle day        lpdg missing
1      0  1     1  -8 -0.10651824       0
2      0  1     1  -7  0.04606832       0
```

3	0	1	1	-6	-0.02592671	0
4	0	1	1	-5	-0.09207704	0
5	0	1	1	-4	-0.05878623	0

where group is a binary indicator of conceptive cycle (group=1) or not (group=0), id equals 1 through 51 identifying women, cycle equals 1 through 91 identifying cycles, day equals -8 through 15 indicating day in cycle, lpdg contains log progesterone metabolite concentrations, and missing is a binary indicator of missing lpdg data (missing = 1) or not (missing=0). When missing equals one, lpdg is set to zero. The goals of our analysis are to compare the mean hormone profiles across conceptive and non-conceptive cycles, to decompose repeated measures from a given cycle into the sum of three smooth curves (group mean plus individual mean plus cycle mean) plus deviations, and to use the fitted curves to impute missing data.

12.1.2 Cystic fibrosis lung function data

The CF data set consists of nine repeated measures on lung function, each from 426 participants in a randomized clinical trial. Details about the trial are provided in Ramsey et al. (1999), but we will give a brief overview. CF is a genetic disorder that leads to progressive obstructive lung disease associated with bacterial infection and inflammatory response (Cantin, 1995). Participants were randomized to receive inhaled tobramycin (an antibiotic) or placebo in three on–off cycles for a total of 24 weeks. The primary goal of the trial was to determine if tobramycin improves lung function as measured by relative change from baseline to week 20 in forced expiratory volume in the first second (FEV_1), expressed as a percentage of the value predicted on the basis of age, height, and sex. Forced expiratory volume in the first second was measured at 0, 2, 4, 6, 8, 12, 16, 20, and 24 weeks after randomization. Of the 520 participants randomized, 462 participants (231 randomized to receive tobramycin and 231 randomized to receive placebo) had data available for the primary analysis. Ramsey et al. (1999) report that the mean relative change in percent-predicted FEV_1 was a 10% increase in the tobramycin group and a 2% decrease in the placebo group (p-value < 0.001 for the difference in means), and thus inhaled tobramycin improved lung function. The area under the FEV_1 curve has become another outcome of interest in studies of CF, and as such was also supplied in the Food and Drug Administration report for the tobramycin trial; however, this analysis used basic methods that did not incorporate the assumption of smooth individual mean profiles. The goal of our analysis is to incorporate this assumption in comparing mean FEV_1 profiles and mean area under the FEV_1 profiles across the two treatment groups.

12.1.3 Chapter organization

The remainder of this chapter is organized as follows. We conclude this introduction with a brief review of regression splines, which leads into a more thorough discussion of penalized spline models for data from a single individual in Section 12.2. Because penalized spline models for longitudinal data evolved from methods for smoothing data from just a single individual, it is helpful to first review this context carefully. In Section 12.3 we present the connection between penalized splines and BLUP estimates of certain linear mixed-effects models. Section 12.4 extends and adapts the methods of the previous two sections to the typical longitudinal data setting with repeated measures from multiple individuals, focusing for clarity on methods for analysis of the hormone and CF data sets. Section 12.5 concludes with a summary and discussion.

12.1.4 Regression splines

Penalized splines are penalized versions of regression splines. A regression spline is a piecewise polynomial joined at a sequence of distinct knots $\nu_1, \ldots, \nu_q$ that partition the domain of the function. The joins are smooth in the sense that the curve is continuous and often has one or more continuous derivatives. For example, regression splines derived from piecewise cubic polynomials are constrained to have two continuous derivatives at each of the knots. For modeling longitudinal data, we will represent curves as functions of time t. Whereas a basis for cubic polynomials on [0,1] is given by the functions 1, t, t^2, and t^3, the truncated power series basis for piecewise cubic regression splines on [0,1] with knots $\nu_1, \ldots, \nu_q$ is given by the $q + 4$ functions 1, t, t^2, t^3, $(t - \nu_1)_+^3$, $(t - \nu_2)_+^3, \ldots, (t - \nu_q)_+^3$, where $a_+ = a$ when $a > 0$ and 0 otherwise.

The regression spline basis functions most commonly used in practice are B-splines, which span the same linear subspace as the corresponding truncated power series basis functions, but have been re-expressed so that the functions evaluate to zero at most times. The B-spline basis for cubic regression splines on $[-8, 15]$ with knots at $-7, \ldots, 14$ is partially displayed in Figure 12.1.

A B-spline basis, $B_1(\cdot), \ldots, B_p(\cdot)$, like its corresponding truncated power series basis, is not orthogonal in the subspace of functions defined on $[a, b]$ with respect to the usual inner

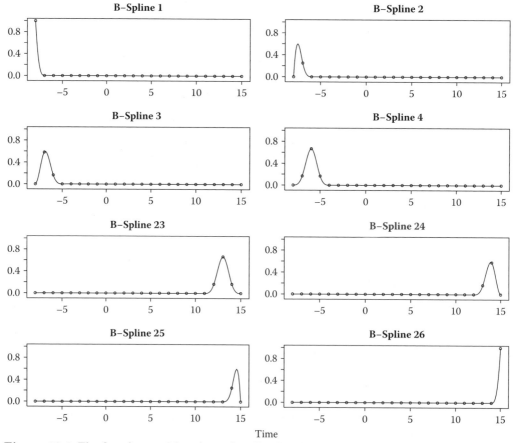

Figure 12.1 The first four and last four of 26 B-spline basis functions for the subspace of cubic regression splines on $[-8, 15]$ with knots at $-7, \ldots, 14$, sampled at times $-8, \ldots, 15$.

product

$$\langle f_1(\cdot), f_2(\cdot) \rangle = \int_a^b f_1(t) f_2(t) dt,$$

nor is its sampled counterpart $B = (B_1, \ldots, B_p)$ with entries

$$B_{jk} = B_k(t_j), \quad k = 1, \ldots, p, \quad j = 1, \ldots, n,$$

generally an orthogonal collection of vectors in the data space $\mathbb{R}^n$ with respect to the usual inner product

$$\langle a, b \rangle = \sum_{j=1}^n a_j b_j.$$

12.2 Penalized spline models for a single individual

Penalized splines are useful when observations from a single individual i can be modeled as

$$Y_{ij} = s_i(t_{ij}) + \varepsilon_{ij}, \tag{12.1}$$

where $s_i(\cdot)$ is a smooth deterministic function and the ε_{ij} are independent and identically distributed (i.i.d.) mean-zero error terms with variance σ^2. The scientific question typically will concern $s_i(\cdot)$ and relate to its value at a single time, to the area under the curve, or perhaps to the entire function over a domain of interest.

If a relatively low-dimensional basis $x_1(\cdot), \ldots, x_p(\cdot)$ can be assumed for $s_i(\cdot)$, for example a polynomial basis or a regression spline basis with just a few knots, then $s_i(t)$ can be estimated with a parametric regression of the form

$$\boldsymbol{Y}_i = X_i \boldsymbol{\beta}_i + \boldsymbol{\varepsilon}_i,$$

where X_i is the $n_i \times p$ matrix of columns $1, \ldots, p$ each representing one dimension of the basis evaluated at times $t_1, \ldots, t_{n_i}$, $\boldsymbol{\beta}_i$ is a vector of regression parameters, and $\boldsymbol{\varepsilon}_i$ is a vector of error terms. The ordinary least-squares estimator for $\boldsymbol{\beta}_i$,

$$\widehat{\boldsymbol{\beta}}_i = (X_i' X_i)^{-1} X_i' \boldsymbol{Y}_i,$$

leads to the estimator for $s_i(t)$,

$$\widehat{s}_i(t) = \sum_{k=1}^p x_k(t) \widehat{\beta}_{ik}.$$

This is the best linear unbiased estimator (BLUE), with minimum variance among all unbiased linear estimators. If the ε_{ij} are normally distributed, then $\widehat{s}_i(t)$ is the maximum likelihood estimator and hence is asymptotically efficient; under other certain distributional assumptions for the ε_{ij}, $\widehat{s}_i(t)$ is not asymptotically efficient. See Seber (1977) and the references therein for more details. If no distributional form for ε_{ij} is specified, then $\widehat{s}_i(t)$ is semi-parametric efficient among the class of regular asymptotically linear estimators (van der Vaart, 1998), and under regularity assumptions it is also, by the central limit theorem and the delta method, asymptotically normal with distribution governed by the large-sample approximation

$$\widehat{\boldsymbol{\beta}}_i \sim N(\boldsymbol{\beta}_i, \widehat{\sigma}^2 (X_i' X_i)^{-1}),$$

with

$$\widehat{\sigma}^2 = \frac{1}{n_i - p} \boldsymbol{Y}_i' (I_{n_i} - H_i) \boldsymbol{Y}_i,$$
$$H_i = X_i (X_i' X_i)^{-1} X_i',$$

where I_n is the identity matrix for $\mathbb{R}^n$.

When integrals W_k of the $x_k(\cdot)$ over a domain $[a, b]$ are known, then the area A_i under the curve $s_i(\cdot)$ over $[a, b]$ can be estimated as

$$\widehat{A}_i = \sum_{k=1}^{p} W_k \widehat{\beta}_{ik},$$

with properties analogous to those of $\widehat{s}_i(t)$.

When the estimand is one-dimensional, as is $s_i(t)$ or A_i, approximate $1 - \alpha$ confidence intervals can be derived based on the asymptotic distribution of $\widehat{\beta}_i$ and the usual formula, e.g.,

$$\widehat{s}_i(t) \pm z_{1-\alpha/2} \; \widehat{\text{s.d.}}(\widehat{s}_i(t)) \tag{12.2}$$

with $\widehat{\text{s.d.}}(\widehat{a})$ denoting the estimated standard deviation of $\widehat{a}$ and $z_{1-\alpha/2}$ the $(1-\alpha/2)$ quantile of the standard normal distribution.

For simultaneous estimation of $s_i(t)$ for all $t \in [c, d]$, confidence bands can be constructed as

$$\widehat{s}_i(t) \pm r_{1-\alpha} \; \widehat{\text{s.d.}}(\widehat{s}_i(t)), \tag{12.3}$$

with $r_{1-\alpha}$ denoting the square root of the $(1 - \alpha)$ quantile of the random variable

$$\sup_{t \in [c,d]} \frac{(\widehat{s}_i(t) - s(t))^2}{[\widehat{\text{s.d.}}(\widehat{s}_i(t))]^2},$$

which can be quickly approximated using simulation on a fine grid.

It happens that low-dimensional polynomial bases are often not flexible enough in practice, and low-dimensional regression spline bases involve a careful choice of knots that can be tedious or inaccurate. These difficulties may lead one to prefer the penalized spline approach to estimating $s_i(t)$, or perhaps another automated method of effectively reducing the dimension of the basis for the fit.

With the penalized spline approach, a higher-dimensional regression spline basis, with more knots or higher-order polynomial pieces or fewer constraints concerning the joining of the pieces, is selected to represent $s_i(t)$, and estimation is conducted using a penalty to reduce the effective dimension of the fit. Letting X_i be the same as before but with more columns (larger p) to represent the higher-dimensional spline basis, and again letting $x_k(t)$ be the kth basis function evaluated at time t, the $\widehat{\beta}_i$ minimizing the residual sum of squares plus penalty

$$\sum_{j=1}^{n_i} (Y_{ij} - X_{ij}\beta_{ij})^2 + \lambda P(\beta_i) \tag{12.4}$$

leads to the penalized spline estimate,

$$\widehat{s}_{PS,i}(t) = \sum_{k=1}^{p} x_k(t) \widehat{\beta}_{ik},$$

of $s_i(t)$. Because of their mathematical convenience, we will consider only quadratic penalties, which can be expressed as

$$P(\beta_i) = \beta_i' D \beta_i, \tag{12.5}$$

for D a symmetric positive semi-definite matrix (which thus has a square root). When the penalty function $P(\beta_i)$ is a *roughness penalty*, which downweights the "rough" or "wiggly" components of the estimated curve, the tuning parameter λ is known as a smoothing parameter, since for larger values of λ, the estimated curve will be smoother. We will focus attention on roughness penalties in this chapter, and particularly on the cubic smoothing spline penalty

$$P_{css}(\boldsymbol{\beta}_i) = \int_a^b \left(\sum_{k=1}^p x_k^{(2)}(t)\boldsymbol{\beta}_{ik} \right)^2 dt,$$

where the range of integration $[a, b]$ is any interval containing the data and $x_k^{(2)}(t)$ is the second derivative of $x_k(\cdot)$ evaluated at t. $P_{css}(\boldsymbol{\beta}_i)$ derives its name from the result that, for $x_1(\cdot), \ldots, x_{n_i+2}(\cdot)$ a cubic B-spline basis with knots at each of the interior unique times $t_2 < \ldots < t_{n_i-1}$, the penalized estimate of $s_i(t)$ is the cubic smoothing spline fit to the data, which happens to be a unique natural cubic spline with knots at the interior time points. Remarkably, the minimizer is exactly the same even when the basis $x_1(\cdot), \ldots, x_{p=\infty}(\cdot)$ is countably infinite and spans the entire Sobolev space of functions whose second derivatives are square integrable. See Hastie and Tibshirani (1990), as well as Hastie, Tibshirani, and Friedman (2001) for more details. For longitudinal data on multiple individuals observed at many times or at unidentical times, it is possible to apply $P_{css}(\boldsymbol{\beta}_i)$ in conjunction with a cubic B-spline basis with fewer knots than distinct observation times, to reduce computation. For a given spline basis, the cubic smoothing spline quadratic penalty matrix D_{css} has (k, l)th entry

$$D_{css,kl} = \int_a^b x_k^{(2)}(t)x_l^{(2)}(t)dt.$$

Suppose the basis $x_1, \ldots, x_p$ is such that $x_1(t) = 1$ and $x_2(t) = t$. Let the spectral decomposition of D_{css} be $Q_1\Psi_1Q_1' + Q_2\Psi_2Q_2'$ with $\Psi_1 = 0$ (there will be two zero eigenvalues associated with x_1 and x_2). Then redefine the basis so that $x_k^*(t) = x_k(t)$, $k = 1, 2$, and

$$x_k^*(t) = x_k(t)Q_2\Psi_2^{-1/2}, \quad k = 3, \ldots, p.$$

The penalty matrix will, in turn, be re-expressed as D_{css}^*, a diagonal matrix with

$$D_{css,kk}^* = \begin{cases} 0 & k = 1, 2, \\ 1 & k = 3, \ldots, p. \end{cases}$$

Although the cubic smoothing spline penalty is probably the most common, other penalties are also in use. Eilers and Marx (1996) take a B-spline basis and penalize higher-order finite differences of the coefficients of adjacent B-splines (see Figure 12.1, which clearly depicts the adjacency); they call their resulting estimate a P-spline. Ruppert, Wand, and Carroll (2003), following Brumback, Ruppert, and Wand (1999), make extensive use of the linear spline basis $x_1, \ldots, x_p = 1, t, (t - \nu_1)_+, \ldots, (t - \nu_q)$ for $p = q + 2$, in conjunction with the penalty

$$P_{RWC}(\boldsymbol{\beta}_i) = \sum_{k=3}^p \beta_{ik}^2,$$

which corresponds to the diagonal penalty matrix D_{RWC} with

$$D_{RWC,kk} = \begin{cases} 0 & k = 1, 2, \\ 1 & k = 3, \ldots, p. \end{cases}$$

Note that while the penalty matrix has exactly the same form as D_{css}^*, the basis is much different, and will not even be of the same dimension, leading to different estimators altogether.

The applications used to illustrate this chapter will be analyzed using either a cubic B-spline basis and penalty $P_{css}(\boldsymbol{\beta}_i)$ or a linear spline basis with $P_{RWC}(\boldsymbol{\beta}_i)$, but we hope it is evident that the choice of basis and penalty combinations one could use is abundant.

Whereas the properties of parametric regression estimators of functions of $s_i(\cdot)$ are relatively straightforward and well known, the properties of the penalized spline estimators of these quantities are more difficult to derive and are also the subject of some confusion, largely because of the algebraic correspondence between penalized spline estimators and mixed-effects model estimators, the topic of the next section.

12.3 Mixed-effects model representation

We next show that the estimator of $s_i(t)$ based on the minimizer $\widehat{\boldsymbol{\beta}}_i$ of

$$\sum_{j=1}^{n_i} (Y_{ij} - X_{ij}\boldsymbol{\beta}_{ij})^2 + \lambda \boldsymbol{\beta}'_i D \boldsymbol{\beta}_i \tag{12.6}$$

can be written as the BLUP of an LME model,

$$\boldsymbol{Y}_i = X_i^* \boldsymbol{\beta}_i^* + Z_i \boldsymbol{b}_i + \boldsymbol{\varepsilon}_i, \tag{12.7}$$

with X_i^*, $\boldsymbol{\beta}_i^*$, Z_i, and $\boldsymbol{b}_i$ to be defined and $\boldsymbol{Y}_i$ and $\boldsymbol{\varepsilon}_i$ as before but with the additional assumption of multivariate normality. To show this, we start by identifying (12.6) with the optimization criterion for BLUP solutions based on a completely random-effects model,

$$\boldsymbol{Y}_i = X_i \boldsymbol{u}_i + \boldsymbol{\varepsilon}_i, \tag{12.8}$$

where X_i is as before, $\boldsymbol{\varepsilon}_i$ is as in (12.7), and $\boldsymbol{u}_i$ is a $p \times 1$ vector of random effects, independent of $\boldsymbol{\varepsilon}_i$ and with typically improper distribution

$$p(\boldsymbol{u}_i) \propto \exp\left(-\frac{\lambda}{2\sigma^2}\boldsymbol{u}'_i D \boldsymbol{u}_i\right).$$

The BLUP equations maximize $p(\boldsymbol{u}_i|\boldsymbol{Y}_i) \propto p(\boldsymbol{Y}_i|\boldsymbol{u}_i)p(\boldsymbol{u}_i)$ and can be readily shown to minimize (12.6).

When D is non-singular, $p(\boldsymbol{u}_i)$ is a mean-zero multivariate normal density with covariance $(\sigma^2/\lambda)D^{-1}$, but when D is singular, as is usual for penalized spline models, $p(\boldsymbol{u}_i)$ is a partially improper distribution. Now, D has p columns, as does X_i, and we will denote its rank by r. Let D have the spectral decomposition $Q\Psi Q'$, where $Q = (Q_1, Q_2)$ is an orthonormal matrix with Q_1 a $p \times (p-r)$ matrix with columns the eigenvectors of D corresponding to the zero eigenvalues, and Q_2 a $p \times r$ matrix with columns containing the other eigenvectors. Ψ is the diagonal matrix of eigenvalues, with first $p - r$ diagonal entries equal to 0. Ψ thus represents the direct sum of two diagonal matrices: Ψ_1 with all entries equal to zero and Ψ_2 with the non-zero eigenvalues on its diagonal. As with all eigenvector decompositions, the columns of Q form an orthonormal basis for $\mathbb{R}^p$.

Using this decomposition for D, we can rewrite model (12.8) as

$$\begin{aligned} \boldsymbol{Y}_i &= X_i Q_1 \boldsymbol{\beta}_i^* + X_i Q_2 \Psi_2^{-1/2} \boldsymbol{b}_i + \boldsymbol{\varepsilon}_i \\ &= X_i^* \boldsymbol{\beta}_i^* + Z_i \boldsymbol{b}_i + \boldsymbol{\varepsilon}_i, \end{aligned} \tag{12.9}$$

with $\boldsymbol{\beta}_i^*$ a $(p-r)$-vector of fixed effects and $\boldsymbol{b}_i$ an r-vector of multivariate normal mean-zero random effects with covariance $\sigma^2/\lambda I_r$. Returning to (12.7), we can identify X_i^* with $X_i Q_1$ and Z_i with $X_i Q_2 \Psi_2^{-1/2}$. Models (12.8) and (12.9) are equivalent in the sense that the BLUP of $\boldsymbol{u}_i$, constructed to match the minimizer $\widehat{\boldsymbol{\beta}}_i$ of (12.6), is identical to the BLUP of $(\boldsymbol{\beta}_i^*, \boldsymbol{b}_i)$, when λ and σ^2 are known. Returning to (12.7), we can thus identify X_i^* with $X_i Q_1$ and Z_i with $X_i Q_2 \Psi_2^{-1/2}$.

The correspondence between penalized splines and mixed-effects models arises from the derivation of a penalized spline as a Bayes estimate (Kimeldorf and Wahba, 1970; Wahba, 1978). Whereas the setting of these papers was in continuous time, Silverman (1985) developed the discrete time analog, and Speed (1991) first pointed out the connection between smoothing splines on the one hand, and BLUP and restricted maximum likelihood (REML) in a mixed-effects model on the other hand.

Despite this correspondence, the high-dimensional fixed-effects model

$$Y_i = X_i\beta_i + \varepsilon_i$$

specifies that only ε_i is random, whereas the linear mixed-effects model

$$Y_i = X_i^*\beta_i^* + Z_i b_i + \varepsilon_i$$

specifies that both b_i and ε_i are random. Conditionally on b_i, the models are identical through the correspondence

$$X_i\beta_i = X_i^*\beta_i^* + Z_i b_i.$$

The dimension of β_i^* will typically be small compared to that of b_i, and the dimension of β_i will equal the sum of the dimensions of β_i^* and b_i.

Assuming that the true mean curve $s_i(\cdot)$ for participant i is smooth, and that the basis X_i is large enough to capture it, the high-dimensional fixed-effects model will correspond exactly to the sampling model for generating repeated longitudinal data from individual i over a fixed period of time. As we will show in Section 12.3.3, sampling from the LME model generates longitudinal data that look nothing like those generated by sampling from the high-dimensional fixed-effects model.

In summary, both the penalized high-dimensional fixed-effects model and the LME model lead to the same fitted $\widehat{Y}_i$ as an estimate of $\{s(t_{i1}), \ldots, s(t_{in_i})\}$ or, more generally, to the same estimator $X^a\widehat{\beta}_i$ as an estimator of $\{s(t_1^a), \ldots, s(t_n^a)\}$, where X^a is the design matrix evaluated at the alternative times $(t_1^a, \ldots, t_n^a)$. However, frequentist inference pertaining to $s_i(\cdot)$ of model (12.1) should be based on the high-dimensional fixed-effects model. In many applications, the dimension of β_i is as large as that of Y_i. In these cases, σ^2 cannot be estimated using traditional parametric techniques. In Section 12.3.4, we discuss alternative methods used to estimate it.

12.3.1 Using LME software to select λ and estimate BLUPs

Several authors have proposed selecting λ based on the LME model defined by (12.7) and (12.9), using the REML algorithm. This corresponds to the general maximum likelihood (GML) method of Wahba (1985) in the case of cubic smoothing splines. An advantage of this method is that standard LME software, such as the `lme` function available in the `nlme` library of R, can be used to select λ and estimate (β_i^*, b_i), as well as σ^2. A disadvantage is that it is somewhat illogical to select the smoothing parameter assuming that the mixed model is true but then to reject that assumption (as we will do) when calculating variability of the estimated curve. A more logical choice would be to use cross-validation. However, Wahba (1985) showed that GML and generalized cross-validation (GCV) performed somewhat similarly, and our experience has shown that a "large" change in λ is required for the resulting estimated smooth curve to appear much different.

The R code for using this method to fit a cubic smoothing spline, and also to fit a penalized spline using the linear spline basis (with knots at $-9, \ldots, 14$) and penalty of Ruppert, Wand, and Carroll (2003), to the first cycle of hormone data from the first individual is available at the web site for this book given in the Preface. Note that, for the cubic smoothing spline, the two columns of X_iQ_1 span the constant and linear functions, and we reparameterize the model to use 1 and t for the fixed-effects basis.

Figure 12.2 compares the cubic smoothing spline fit using the R function `lme` with that of using the R function `smooth.spline`; the only difference is due to use of different methods for choosing the smoothing parameter λ; `smooth.spline` uses GCV. The figure also displays the fit using `lme` with the basis and penalty of Ruppert, Wand, and Carroll (2003). The three smooths are virtually identical.

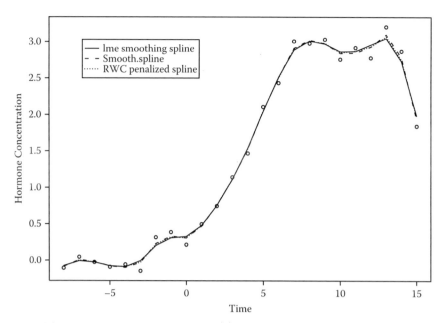

Figure 12.2 Three penalized spline fits using (a) the linear mixed-effects model in conjunction with the smoothing spline penalty (b) GCV to select the smoothing parameter (smooth.spline), and (c) the linear mixed-effects model in conjuction with the RWC penalty.

12.3.2 Eigenvector representation of the hat matrix

Once λ is selected, we can write the penalized spline estimator $\widehat{\boldsymbol{Y}}_i$ of $\{s_i(t_{i1}), \ldots, s_i(t_{in_i})\}$ as a linear transformation of $\boldsymbol{Y}_i$, that is, as

$$\widehat{\boldsymbol{Y}}_i = H_i \boldsymbol{Y}_i.$$

The linearity of the transformation follows from the BLUP equations given in Robinson (1991) and in Chapter 11. The matrix H_i is known as the "hat matrix" or the "smoother." The hat matrix, like a projection matrix, is symmetric and positive semi-definite, but it is not idempotent (i.e., $H_i^2 \neq H_i$). Because the hat matrix is symmetric, it has a spectral decomposition, which allows us to better understand its action on the data.

In general, the spectral decomposition of a symmetric $n \times n$ matrix A can be expressed as

$$A = \sum_{k=1}^{n} a_k \boldsymbol{v}_k \boldsymbol{v}'_k,$$

where the a_k are the eigenvalues and the $\boldsymbol{v}_k$ are a set of orthonormal eigenvectors of A. The matrices $\boldsymbol{v}_k \boldsymbol{v}'_k$ are each projection matrices onto the orthogonal one-dimensional subspaces spanned by each of the $\boldsymbol{v}_k$; the direct sum of the $\boldsymbol{v}_k \boldsymbol{v}'_k$ is equal to I_n. Thus, the action of the matrix A on a vector $\boldsymbol{Y}$ in $\mathbb{R}^n$ can be understood as follows: A projects $\boldsymbol{Y}$ onto its eigenvectors, multiplies each of the projections by a_k, and sums the results.

In this way, an $n \times n$ projection matrix P with a range of dimension p can be represented as

$$P = \sum_{k=1}^{p} \boldsymbol{v}_k \boldsymbol{v}'_k.$$

The collection $(\boldsymbol{v}_1, \ldots, \boldsymbol{v}_p)$ spans the range of P, and the a_k corresponding to this collection are all equal to one. For unconstrained ordinary least-squares estimation of $\boldsymbol{\beta}_i$, or for the LME model with $\boldsymbol{b}_i = 0$, the corresponding hat matrix H_i is a projection matrix. However, the hat matrix of a penalized spline estimator or of a general LME model is not a projection matrix. Its eigenvector representation will give us insight into its action on the data.

When the condition

$$X_i^{*\prime} Z_i = 0$$

is true, then the hat matrix has the eigenvector representation

$$H_i = \sum_{k=1}^{p-r} \boldsymbol{v}_k \boldsymbol{v}_k' + \sum_{k=p-r+1}^{n_i} \frac{d_k}{\lambda + d_k} \boldsymbol{v}_k \boldsymbol{v}_k', \qquad (12.10)$$

where the collection $\{\boldsymbol{v}_k\}$ are orthonormal eigenvectors of $Z_i Z_i'$, a matrix with first $p - r$ eigenvalues equal to zero and remaining $n_i - (p - r)$ eigenvalues denoted by the d_k. The eigenvalues a_k of H_i are equal to 1 for $k = 1, \ldots, p - r$ and $a_k = d_k/(\lambda + d_k) \in [0, 1)$ for $k = p - r + 1, \ldots, n_i$.

Conversely, when a hat matrix has the representation (12.10), we can write the linear mixed-effects model in *orthogonal form* as

$$\boldsymbol{Y}_i = X_i^* \boldsymbol{\beta}_i + Z_i \boldsymbol{b}_i + \boldsymbol{\varepsilon}_i, \qquad (12.11)$$

with $X_i^* = (\boldsymbol{v}_1, \ldots, \boldsymbol{v}_{p-r})$, $Z_i = (\boldsymbol{v}_{p-r+1}, \ldots, \boldsymbol{v}_{n_i}) \text{diag}(d_{p-r+q}, \ldots, d_{n_i})^{-1/2}$, $\text{diag}(a)$ denoting the diagonal matrix with diagonal entries given by the elements of a, $\boldsymbol{\beta}_i$ the vector of fixed effects, $\boldsymbol{b}_i$ the $N(0, (\sigma^2/\lambda)I)$ vector of random effects, and $\boldsymbol{\varepsilon}_i$ the $N(0, \sigma^2 I)$ vector of error terms. With this representation, $X_i^{*\prime} Z_i = 0$.

When an eigenvector decomposition of the hat matrix for fixed λ shows that all eigenvalues are between zero and one, one can use it to construct the decomposition (12.10) and mixed-effects model (12.11) for general λ. While it would be easy to verify empirically that the eigenvalues are between zero and one for a given penalized spline model and fixed λ, we next show that for all penalized spline models satisfying a certain condition, we can represent the hat matrix in form (12.10) and hence construct the mixed model with form (12.11).

Using multivariate calculus to optimize (12.4) with $P(\boldsymbol{\beta}_i)$ as in (12.5), and supposing that the matrix

$$X_i' X_i + \lambda D$$

is non-singular, as it typically will be, then the hat matrix can be written as

$$H_i = X_i (X_i' X_i + \lambda D)^{-1} X_i'. \qquad (12.12)$$

To derive the representation (12.10), we again let $Q \Psi Q'$ be the spectral decomposition of D and let $U \Phi U'$ be that of $X_i Q$. Then, because $U' Q' X_i' X_i Q U = \Phi$ is diagonal, $V = X_i Q U \Phi^{-1/2}$ has orthonormal columns. With these transformations so defined, we can rewrite (12.12) as

$$X_i Q U \Phi^{-1/2} \Phi^{1/2} (\Phi + \lambda \Psi)^{-1} \Phi^{1/2} \Phi^{-1/2} U' Q' X_i',$$

which shows that the spectral decomposition of H_i can be written $V A V'$, with V as above and A the diagonal matrix $\Phi(\Phi + \lambda \Psi)^{-1}$. It is now easy to see that, since the entries of Φ and Ψ are non-negative (those of Φ are the eigenvalues of a matrix having form $C'C$ and D has been previously defined as a positive semi-definite matrix), for $\lambda > 0$ the eigenvalues of H_i are in $[0, 1]$, and thus H_i has a representation (12.10) with $\boldsymbol{v}_k$ the columns of V and a_k the diagonal entries of A, suitably ordered.

12.3.3 Effective degrees of freedom and effective basis

Now that we have shown that the hat matrix of a penalized spline smoother can be represented via (12.10), we can use this representation to investigate the action of the penalized spline smoother on the data. Hastie and Tibshirani (1990) define the *effective degrees of freedom* (edof) of a penalized spline estimator as the trace of H_i, that is, as the sum of the eigenvalues of H_i, which by (12.10) we can write as

$$\text{edof} = p - r + \sum_{k=p-r+1}^{n_i} \frac{d_k}{\lambda + d_k}.$$

More recently, Hastie, Tibshirani, and Friedman (2001) refer to this quantity as the *effective number of parameters* of a statistical model. Each eigenvalue represents the proportion of a one-dimensional projection that is retained by H_i. For example, if the proportion were 0.70, then effectively 70% of that projection is incorporated into the statistical model; alternatively, that projection uses 70% of a parameter. Were H_i a projection matrix onto a p-dimensional subspace, p of the eigenvalues would equal one and the remainder would be zero; thus, the edof would equal the actual number of parameters of the statistical model.

Historically, the notion of effective degrees of freedom was introduced for approximating distributions of quadratic forms as chi-squared; see Giesbrecht (1982) for a review. Let R be a quadratic form with expectation A and variance B. Then, one would approximate the distribution of R as α multiplied by a chi-squared random variable with f degrees of freedom. One would set $\alpha f = A$ and $2\alpha^2 f = B$ and solve for f; the result was known as the effective degrees of freedom.

Following Hastie and Tibshirani (1990), the literature on smoothing has redefined edof to agree more closely with the notion of an effective number of parameters. Hastie and Tibshirani (1990) replaced the historical notion of effective degrees of freedom with the term *error degrees of freedom*, and though they did discuss estimating it using a two-moment correction as has been done historically, they also estimate it using a simple one-moment method. Unfortunately, due to the shrinking action of H_i, the error degrees of freedom are not equal to n_i minus the effective degrees of freedom.

The definition of edof used by Hastie and Tibshirani (1990) is now prevalent in the literature, but the companion notion of an *effective basis* is not. The basis for a parametric regression model has a dimension equal to the degrees of freedom of the model. Analogously, we consider finding an effective basis with dimension equal to the approximate edof for a penalized spline model. One such choice is given by ordering the eigenvectors of H_i so that the corresponding eigenvalues are decreasing, and then choosing the first edof of them. This choice does not use the data $\boldsymbol{Y}_i$ except for selecting λ. A choice that uses the data much more so is given by first computing the weights w_k such that $\boldsymbol{Y}_i = \sum_{k=1}^{n_i} w_k \boldsymbol{v}_k$, that is, $w_k = a_k \boldsymbol{v}_k' Y_i$, and then choosing a set of $K_{.95}$ v_k such that the sum of corresponding $|w_k|$ is a high proportion (say, at least 95%) of the the total sum of $|w_k|$. Clearly, this choice will lead to a different definition of edof; it need not be the case that $K_{.95} = \text{edof}$.

We apply this notion to model the first cycle of hormone data from the first individual using a cubic smoothing spline penalty and a B-spline basis with knots at the interior time points. Figure 12.3 presents the first 16 of 24 eigenvectors of the hat matrix, together with the eigenvalues as well as the data-based weights w_k. Using the eigenvalues to compute the edof we get 12.6, but if we use the data-based weights and a proportion of 95%, it is 8. The eigenvectors selected for the effective basis using the eigenvalues are 1 through 13, and using the w_k are 1 through 5, 8, 9, and 11. The advantage of the second method is that the selected effective basis is smaller; however, it is also more error-prone.

Using either choice of an effective basis, properties of the resulting estimator $\widehat{s}_i(\cdot)$ could possibly be well approximated using the parametric regression theory cited above, assuming that the vector of true functionals $\{s_i(t_1), \ldots, s_i(t_{n_i})\}'$ could in fact be expressed as a linear combination of the effective basis. For estimating $\widehat{s}_i(t)$ for any $t \in [a, b]$, we need to construct functions in the span of our original spline basis whose sampled versions are the $\boldsymbol{v}_k$. For example, if we have selected a cubic B-spline basis with knots at the interior time points and

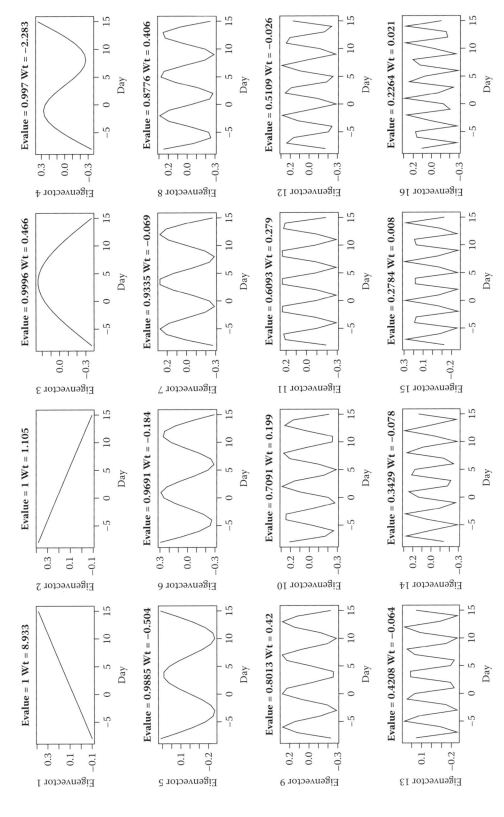

Figure 12.3 Choosing an effective basis for smoothing hormone data from the first cycle of the first individual.

if we are using the cubic smoothing spline penalty D_{css}, then because $\widehat{\boldsymbol{Y}}_i$ will be a sampled version of a natural cubic spline with knots at the interior time points, and because there is a unique natural cubic spline with those same knots which interpolates each $\boldsymbol{v}_k$ (see Green and Silverman, 1994), we can construct function-space extensions of the $\boldsymbol{v}_k$ using the natural cubic spline interpolants. It seems to us that function-space extensions of the v_k collection are not so easy to construct in general, but perhaps by using the theory presented in Demmler and Reinsch (1975) a general constructive method could be outlined.

12.3.4 Estimating variance and bias

A way to estimate the covariance of $\{\widehat{s}_i(t_1^a), \ldots, \widehat{s}_i(t_n^a)\}'$ for $(t_1^a, \ldots, t_n^a)'$, a collection of times not necessarily equal to the observation times, is as follows. We let X^a be the design matrix corresponding to the grid of time points $(t_1^a, \ldots, t_n^a)'$; then $\{\widehat{s}_i(t_1^a), \ldots, \widehat{s}_i(t_n^a)\}' = H^a Y_i$ for $H^a = X^a (X_i' X_i + \lambda D) X_i'$, and hence

$$\widehat{\mathrm{Var}}(H^a \boldsymbol{Y}_i) = \widehat{\sigma}^2 H^a H^{a\prime}$$

with

$$\widehat{\sigma}^2 = \frac{1}{\mathrm{errdof}} \boldsymbol{Y}_i'(I_{n_i} - H_i)^2 \boldsymbol{Y}_i,$$

where errdof is an estimate of the error degrees of freedom, or with σ^2 estimated using the mixed-model software. Using this estimate, confidence intervals and simultaneous bands for $s_i(\cdot)$ can be constructed as in (12.2) and (12.3). However, it has been noted, for example, by Ruppert, Wand, and Carroll (2003), that, whereas this method for estimating the covariance might be reasonable, one should account as well for the bias inherent in the penalization when constructing confidence intervals and bands. This bias is due to the shrinking that occurs when some of the a_k are less than one; if $\boldsymbol{Y}_i$ has a smooth mean with component $\alpha \boldsymbol{v}_k$, then the estimated α is shunk to zero by the factor a_k.

The solution Ruppert, Wand, and Carroll (2003) adopt for this problem is to estimate the mean squared error of $\widehat{s}_i(\cdot)$ using the prediction error variance; this is also the approach taken in Chapter 11. The PEV is equivalent to the Bayesian posterior variance assuming the Bayesian version of (12.7) with an improper prior for $\boldsymbol{\beta}_i$. Using the PEV to estimate the MSE is equivalent to estimating the squared bias of $H^a \boldsymbol{Y}_i$ via the formula

$$E_{b_i} \left[E_{e_i|b_i}[H^a \boldsymbol{Y}_i - (X_i^{*a} \boldsymbol{\beta}_i + Z_i^a \boldsymbol{b}_i)|\boldsymbol{b}_i] E_{e_i|b_i}[H^a \boldsymbol{Y}_i - (X_i^{*a} \boldsymbol{\beta}_i + Z_i^a \boldsymbol{b}_i)|\boldsymbol{b}_i]' \right], \qquad (12.13)$$

where X_i^{*a} and Z_i^a are the fixed- and random-effects design matrices evaluated at $(t_1^a, \ldots, t_n^a)'$, and $\boldsymbol{Y}_i$ is assumed to be generated according to the LME model (12.7). However, the term inside the outer expectation of (12.13) more closely approximates the true squared bias. In fact, it equals the true squared bias when $\{s_i(t_1^a), \ldots, s_i(t_n^a)\}'$ is in the span of X_i^{*a} and Z_i^a. To address the problem that $\boldsymbol{b}_i$ is unknown, those who use the PEV and hence (12.13) are averaging the true squared bias over the distribution of $\boldsymbol{b}_i$ specified by the LME model (12.7).

Intuitively, using the PEV to estimate the true MSE is better than using the naive MSE computed as

$$E_{b_i, e_i} \{(H^a \boldsymbol{Y}_i - s_i^a)(H^a \boldsymbol{Y}_i - s_i^a)'\},$$

where $s_i^a = \{s_i(t_1^a), \ldots, s_i(t_n^a)\}'$ is the non-random true mean and $\boldsymbol{Y}_i$ is assumed to be generated by the LME model (12.7).

However, the PEV will typically be a biased estimator of the true MSE because (12.13) will typically be a biased estimator of the true squared bias. The reason is that averaging the true squared bias over the distribution of $\boldsymbol{b}_i$ will typically not reproduce the true squared

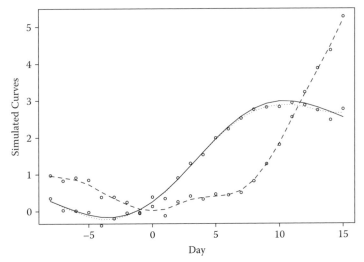

Figure 12.4 The problem with assuming that the LME model generates the data. The true mean curve $s_i(\cdot)$ corresponds to the solid line. One realization of the simulation with only ε_i random corresponds to the points surrounding the dotted line and one realization of the simulation with both $\boldsymbol{b}_i$ and ε_i random corresponds to the points surrounding the dashed line.

bias. Moreover, the joint distribution of $\boldsymbol{b}_i$ and ε_i specified by the LME model (12.7) can generate data that are radically different from the original data — even the direction of curvature can be reversed. We illustrate this problem in Figure 12.4, which displays the true mean curve $s_i(\cdot)$ (the solid line) and two simulated curves graphed as points together with smoothing spline estimates (dotted line for a simulation based on the true model in which only ε_i is random, and dashed line for the LME model simulation in which $\boldsymbol{b}_i$ and ε_i are both random).

It is informative to compare the true MSE, the PEV, and the naive MSE using a simulation study based on the same LME model that generated Figure 12.4. Specifically, we suppose that the true smooth mean curve can be described by $s_i(t) = 0.8623 + 0.1567t - 0.245Z_{23}(t)$. $Z_{23}(\cdot)$ is the 23rd column of the random-effects design matrix as created by the `single.smooth` function with $x = -8, \ldots, 15$ in the R program available at the web site for this book given in the Preface. We sampled this smooth mean curve at times $-8, \ldots, 15$, and to each sampled observation we added i.i.d. $N(0, 0.1327^2)$ deviations. We then estimated $s_i(\cdot)$ using `single.smooth` to compute a cubic smoothing spline; the `lme` function of R estimated the standard deviation of the random effects at 0.1130. Finally, we computed the MSE of $\widehat{s}_i(7)$ (with a true mean equal to $s_i(7) = 2.57$) in three ways. The first way, corresponding to the true MSE, simulates 1000 times from the model that we used to generate the data; that is, we used a deterministic $s_i(\cdot)$ and i.i.d. deviations. The second and third ways correspond to the PEV and the naive MSE, and rely on simulations from the LME model that treats the coefficients of $Z_1(t), \ldots, Z_{24}(t)$ (also created by the `single.smooth` function with $x = -8, \ldots, 15$) as i.i.d. random effects with standard deviation 0.1130, and simulates 1000 random curves $s_i(t) = 0.8263 + 0.1567t + u_1Z_1(t) + \cdots + u_{24}Z_{24}(t)$ plus i.i.d. $N(0, 0.1327^2)$ deviations. The results are presented in Table 12.1. The first way leads to a true MSE of 0.00469 and a true squared bias of 1.5×10^{-5}. The PEV leads to an estimated MSE of 0.0064, which is 1.36 times larger than the truth, and an estimated squared bias of 0.0017, which is 114 times larger than the truth. The naive MSE leads to an estimated MSE of 0.801 and an estimated squared bias of 0.382, both completely off target.

Table 12.1 Inaccuracies in Inferences Based on the Linear Mixed-Effects Model

	Truth	PEV	Naive
Mean Squared Error	0.00469	0.0064	0.801
Squared Bias	1.5×10^{-5}	0.0017	0.382

PEV = prediction error variance.

12.4 Extension to multiple individuals

Not only can inferences based on the LME model be inaccurate when the curve $s_i(\cdot)$ is considered deterministic, but even when $s_i(\cdot)$ can be rightly viewed as a random selection from a distribution of curves, as is the case with longitudinal data, there is no good reason to assume that the distribution of curves is well approximated by the LME model representation of the penalized spline. First, the linear component is not random in this model. Second, the random distribution is artificial and merely invoked for smoothing. Thus, while the LME model does represent *a* population of curves, it will typically *not* correspond to *the* population of curves from which $s_i(\cdot)$ is selected. Referring again to Figure 12.4, if the true population of curves all shared the same smooth mean and error structure, then the true population would be well approximated by simulations assuming random ε_i only. Clearly, the simulations for random $\boldsymbol{b}_i$ and ε_i would generate an artificial population of curves with radically different characteristics.

Therefore, in this section, we consider how best to extend the models and methods of penalized spline smoothing to the longitudinal data setting, wherein multiple individuals $i = 1, \ldots, N$ each contribute data that can be modeled as

$$Y_{ij} = s_i(t_{ij}) + \varepsilon_{ij},$$

where the observation times t_{ij} are possibly different for each individual. When there are data from just a single individual, it is essentially necessary to treat the curve $s_i(\cdot)$ as deterministic when conducting inference because there is no information about the population of curves to which $s_i(\cdot)$ belongs. But when multiple individuals contribute data, we can view each $s_i(\cdot)$ as a random observation from a population of curves S. Thus, in the longitudinal data setting, not only is ε_{ij} random, but $s_i(t_{ij})$ is random as well.

This randomness does not prevent us, however, from conducting our initial point estimation *conditionally* on the curves $s_i(\cdot)$. Then, for computing point estimates and confidence intervals, bands, or other measures of variability pertaining to the *population* of curves S, we will conduct inference *unconditionally* on the curves $s_i(\cdot)$. That is, we will average in a consistent and, hopefully, efficient way across the fitted $s_i(\cdot)$ and use a partially parametric or non-parametric bootstrap (Efron and Tibshirani, 1993) which is *unconditional*, that is, we assume the $s_i(\cdot)$ are random. For example, when the multiple individuals are well modeled as sampled i.i.d. from the population, and the target of inference is the population mean curve $E\{s_i(\cdot)\}$, then we would estimate this as $(1/N)\sum_{i=1}^{N}\widehat{s}_i(\cdot)$. Clearly, if we had observed the true curves $s_i(\cdot)$ for each individual in our sample, then $(1/N)\sum_{i=1}^{N}s_i(\cdot)$ would be a consistent and efficient estimate of $E\{s_i(\cdot)\}$ at each point in time. In general, our approach is to (1) pretend we have observed the true curves, (2) construct a consistent and efficient estimator of the population quantity based on the true curves, and then (3) substitute in the estimated version of the curves. The bootstraps we recommend for the i.i.d. setting are as follows. For the partially parametric bootstrap, the estimated vectors $\{\widehat{s}_i(t_{i1}), \ldots, \widehat{s}_i(t_{in_i})\}$, $i = 1, \ldots, N$, will be resampled with replacement and so will the entire collection of estimated $\widehat{\varepsilon}_{ij}$. For the non-parametric bootstrap, the collection of $\boldsymbol{Y}_i$, $i = 1, \ldots, N$, will be resampled with replacement. Within each bootstrap iteration, estimation of the collection of curves will proceed in a conditional fashion; that is, the curves will be treated as fixed curves rather than random curves. Then, variability of estimates of population quantities such as $E\{s_i(t)\}$ will be estimated using the unconditional bootstrap sample of estimated population quantities.

Intuitively, this approach should be more accurate than basing inference on corresponding LME models that are artificially constructed to induce smoothing (as done by Wang, 1998; Lin and Zhang, 1999; Verbyla et al., 1999; and, to some extent, by Brumback and Rice, 1998). Furthermore, it is clearly also better than conducting inference strictly conditionally on the sample of individuals we obtained data on; after all, the entire population is typically of most interest in longitudinal data analysis.

Also, note that our approach is to apply penalized spline smoothing to *each* individual's data; that is, to estimate each $s_i(\cdot)$ as a penalized spline. This differs from the approach taken by several others (Zhang, Lin, and Sowers, 2000; Chiang, Rice, and Wu, 2001), which estimates only $E\{s_i(\cdot)\}$ as a penalized spline and models individual deviations $s_i(\cdot) - E\{s_i(\cdot)\}$ as i.i.d. parametric stochastic processes.

Now it is possible that, by adopting an effective basis as discussed in the preceding section, we could approximate our penalized spline smoothing for longitudinal data using estimators based on a parametric model. More specifically, we could base estimation on a parametric random-effects model of the form

$$\boldsymbol{Y}_i = X_i\boldsymbol{\beta} + X_i\boldsymbol{b}_i + \boldsymbol{\varepsilon}_i,$$

with the columns of X_i representing the effective basis, $X_i\boldsymbol{\beta}$ representing a population mean curve $E\{s_i(\cdot)\}$ sampled at the observation times for individual i, and $X_i\boldsymbol{b}_i$ representing a sampled random curve, that is, the individual-specific deviation $s_i(\cdot) - E\{s_i(\cdot)\}$. We leave the development of this idea for future research.

For point estimation of curves $s_i(\cdot)$, $i = 1, \ldots, N$, one has the choice of jointly smoothing the curves or of smoothing data from one individual at a time. Joint smoothing occurs when one or more parameters are assumed shared by several individuals — perhaps just a smoothing parameter, or perhaps a smooth mean curve. Letting individuals share a smooth mean curve is a good way to smooth an individual curve with missing data while retaining a scientifically valid shape. Either way, jointly smoothing or smoothing individual curves, we will choose a common spline basis $x_k(\cdot)$, $k = 1, \ldots, p$, and penalty matrix D for each curve to be estimated. In this way, we are extending the penalized spline models from the previous section. Our extensions will be application-specific, and the R code that goes along with them will be available at the web site for this book given at the Preface; we hope that the reader will easily be able to adapt the methods presented herein for his or her own application.

12.4.1 Application 1: Penalized spline analysis of variance

The hormone data can be regarded as a nested sample of curves, where cycles are nested within women, who are in turn nested within group. Because the curves are nested, it is natural to model the observations from a given cycle i using a penalized spline ANOVA. That is, we assume

$$Y_{ij} = s_{g(i)}(t_{ij}) + s_{w(i)}(t_{ij}) + s_{c(i)}(t_{ij}) + \varepsilon_{ij},$$

where $s_{g(i)}(\cdot)$, $s_{w(i)}(\cdot)$, and $s_{c(i)}(\cdot)$ represent a smooth group mean curve, a smooth woman departure from the group mean curve, and a smooth cycle departure from the individual mean. The labels $g(i)$, $w(i)$, and $c(i)$ map cycle i to its respective group, woman, and cycle numbers. The ε_{ij} are assumed to be i.i.d. deviations with variance σ^2. Note that this decomposition is compatible with the assumption that

$$Y_{ij} = s_i(t_{ij}) + \varepsilon_{ij},$$

for $s_i(\cdot)$ a smooth curve; we have simply expressed $s_i(\cdot)$ in terms of a smooth decomposition.

Brumback and Rice (1998) prove that the estimates of the collection of group, woman, and cycle curves that minimize

$$\sum_i \sum_j [Y_{ij} - \{s_{g(i)}(t_{ij}) + s_{w(i)}(t_{ij}) + s_{c(i)}(t_{ij})\}]^2$$

$$+ \lambda_g \sum_l \int \{s_{g_l}^{(2)}(t)\}^2 dt + \lambda_w \sum_l \int \{s_{w_l}^{(2)}(t)\}^2 dt + \lambda_c \sum_l \int \{s_{c_l}^{(2)}(t)\}^2 dt, \quad (12.14)$$

where g_l indexes all groups, w_l indexes all women, c_l indexes all cycles, and all smooth curves belong to the Sobolev space of functions whose second derivatives are square integrable, are natural cubic splines with knots at the collective interior design points, and that, furthermore, these minimizers are algebraically equivalent to the BLUP solutions of the mixed-effects model

$$\boldsymbol{Y}_i = X_i^* \boldsymbol{\beta}_{g(i)} + Z_i \boldsymbol{b}_{g(i)} + X_i^* \boldsymbol{\beta}_{w(i)} + Z_i \boldsymbol{b}_{w(i)} + X_i^* \boldsymbol{\beta}_{c(i)} + Z_i \boldsymbol{b}_{c(i)} + \boldsymbol{\varepsilon}_i, \quad (12.15)$$

for $i = 1, \ldots, N$, where N is the total number of cycles; X_i is the design matrix for cycle i based on the cubic B-spline basis with knots at the collective interior time points (which happen to be $-7, \ldots, 14$ for this application); and X_i^* and Z_i are derived as in Section 12.3 using the cubic smoothing spline penalty D_{css}. The βs are all fixed effects; the $\boldsymbol{b}$s are all random effects, the $\boldsymbol{b}_{\alpha(i)}$ are independent of the $\boldsymbol{b}_{\gamma(i^a)}$ for $\alpha \neq \gamma$ or $i \neq i^a$; the collection of distinct $\boldsymbol{b}_{g(i)}$ is i.i.d. $N\{0, (\sigma^2/\lambda_g)I\}$; that of the distinct $\boldsymbol{b}_{w(i)}$ is i.i.d. $N\{0, (\sigma^2/\lambda_w)I\}$; that of the distinct $\boldsymbol{b}_{c(i)}$ is i.i.d. $N\{0, (\sigma^2/\lambda_c)I\}$; and that of the $\boldsymbol{\varepsilon}_i$ is i.i.d. $N(0, \sigma^2 I)$.

Now, contrary to what Brumback and Rice (1998) state, neither the mimimizers of (12.14) nor the BLUP solutions of (12.15) are as of yet unique because the βs are not identifiable; for identification, we constrain them to sum to zero at each level except the topmost (group). Thus, in practice, we will rewrite (12.15) as

$$\boldsymbol{Y}_i = X_i^* \boldsymbol{\beta}_i + Z_i \boldsymbol{b}_{g(i)} + Z_i \boldsymbol{b}_{w(i)} + Z_i \boldsymbol{b}_{c(i)} + \boldsymbol{\varepsilon}_i,$$

and then reconstruct the βs of the original model as defined by the constraints. A relatively simple R program that estimates the variance components, and hence smoothing parameters, as well as the fixed and random effects for this LME model, using the hormone data, is available at the web site for this book given in the Preface. It takes about 10 minutes to run on a Dell Latitude X1 laptop. The R program estimates three separate smoothing parameters, one for each level of nesting. Due to computational burden, Brumback and Rice (1998) assumed a single common smoothing parameter. However, results of the two analyses are quite similar. Figure 12.5 depicts the fitted curves for individual 11; the results are strikingly similar to Figure 4 of Brumback and Rice (1998). The solid line represents the sum of the cycle, woman, and group smooth curves; the dotted line is the sum of the woman and group smooth curves, and the dashed line is the group smooth curve. The smoothing parameters as estimated via the REML algorithm of `lme` were $\widehat{\lambda}_g = 13$, $\widehat{\lambda}_w = 1169$, and $\widehat{\lambda}_c = 95$, which appear rather dissimilar. $\widehat{\sigma}^2$ was estimated at 0.1894.

We next used a nested non-parametric bootstrap to compute 95% simultaneous confidence bands around the two estimated group mean curves, using (12.3). The R code for the nested bootstrap is also available at the web site for this book given in the Preface. For the non-conceptive group, which contains 69 cycles nested in 29 women, we first sampled 29 women with replacement, and then sampled cycles from the sampled women with replacement. The number of cycles sampled per woman was based on the number of cycles that particular

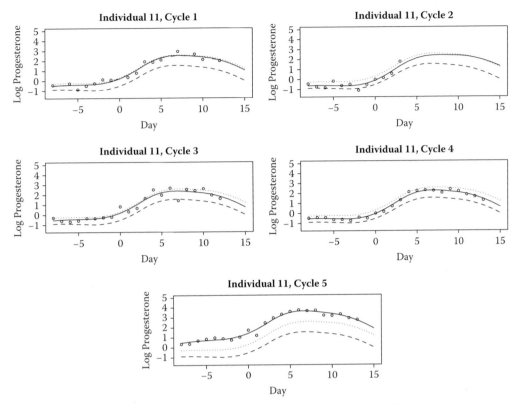

Figure 12.5 Estimated smooth curve decomposition for individual 11. The solid line represents the sum of cycle, woman, and group smooth curves; the dotted line is the sum of the woman and group smooth curves; and the dashed line is the group smooth curve.

woman contributed to the original data set. Thus, while our bootstrap sample contained data on 29 bootstrap individuals, the number of non-conceptive cycles was typically different from 69. The conceptive group contains 22 conceptive cycles from 22 women, and we simply sampled 22 of these cycles with replacement, to complete the entire bootstrap data set for one replication.

We generated 100 bootstrap samples in batches of 10, 20, 30, and 40, taking care to reset the random seed each time to a different number, and this took approximately 20 computer hours in total. For each bootstrap sample, we computed the estimated group mean curves. For the non-conceptive group the $r_{1-\alpha}$ quantile of (12.3) was estimated at 2.61, whereas for the conceptive group it was estimated at 2.70. The estimated group means together with the confidence bands are presented in Figure 12.6. We see that subsequent to day 12 post-ovulation, the confidence bands diverge completely. This makes sense because, biologically, hormones remain high following a successful conception but return to baseline when no conception occurs. But prior to day 12, the confidence bands are always overlapping.

We finish by observing that our method of estimating the smooth component curves is a joint smoothing method: smoothing parameters as well as mean curves are shared among cycles. Another approach would be to smooth each cycle separately and then recombine, but as we see with individual 11, cycle 2, some cycles contain a great deal of missing data. For these cycles, it is helpful to smooth the data jointly, so that the smooth retains a scientifically valid shape.

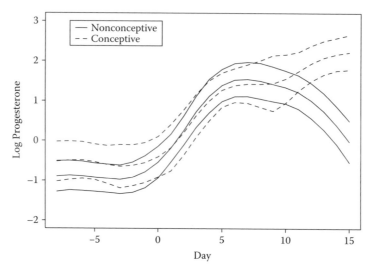

Figure 12.6 Estimated group mean curves and simultaneous 95% confidence bands. Solid lines represent the non-conceptive group and dashed lines represent the conceptive group.

12.4.2 Application 2: Modeling area under the curve

When individuals are sampled at densely spaced times and there are no missing data, the standard approach of computing area under the curve for each individual separately using Simpson's rule will be a good one. However, when individuals have missing data and the observation times are not densely spaced, it can be helpful to employ a smooth model for the individual curves.

The FEV_1 profiles from the CF trial can, like the progesterone data, be modeled as a sample of nested smooth curves plus deviations. Whereas the hormone data were represented by three levels, the FEV_1 profiles need just two to represent individuals nested in treatment group. As such, we can model the data from individual i as

$$Y_{ij} = s_i(t_{ij}) + \varepsilon_{ij}, \tag{12.16}$$

or, by using its smooth decomposition,

$$Y_{ij} = s_{g(i)}(t_{ij}) + s_{p(i)}(t_{ij}) + \varepsilon_{ij}, \tag{12.17}$$

where g denotes treatment group and p denotes individual, and the rest of the model structure is the same as for the hormone data.

This model is particularly useful when analyzing area under the curves, for two reasons: first, by assuming that individual and group mean curves are smooth, we can presumably estimate the area under these curves with more precision than we could otherwise; and second, when integrals of the basis functions for the $s_g(\cdot)$ and $s_p(\cdot)$ are easily computed, we have a ready method for computing area under a sampled curve. The standard methods involve approximating the area using step functions or Simpson's rule.

Our first approach is to model the set of observed FEV_1 values for each participant separately, and thus allow the smoothing parameters to vary across participants. Specifically, we use the LME model representation of (12.16),

$$\boldsymbol{Y}_i = X_i^* \boldsymbol{\beta}_i + Z_i \boldsymbol{b}_i + \boldsymbol{\varepsilon}_i, \tag{12.18}$$

for $i = 1, \dots 462$. Here X_i^* is a two-column matrix with first column the unit vector and second column the time points for individual i; $Z_i = X_i Q_2 \Psi_2^{-1/2}$, where X_i is the cubic B-spline basis with knots at the collective interior time points (2, 4, 6, 8, 12, 16, 20), evaluated at the time points for individual i; and Q_2 and $\Psi_2^{-1/2}$ are derived as in Section 12.3 using the cubic smoothing spline penalty D_{css}. The $\boldsymbol{\beta}_i$ are all fixed effects; the $\boldsymbol{b}_i$ are all random effects, the $\boldsymbol{b}_i$ are $N\{0, (\sigma_i^2/\lambda_i)I\}$ and independent of the $\boldsymbol{\varepsilon}_i$, which are $N\{0, \sigma_i^2 I\}$. Note that the variance of the random effects and errors, and thus the smoothing parameter, depends on participant i.

The minimum, quartiles, and maximum of the estimated smoothing parameters from the 462 participants are 8×10^{-14}, 90, 3663, 3×10^{11}, and 2×10^{14}. Figure 12.7 graphs the estimated smoothing splines (solid lines) and the repeated measures for two participants. The estimated smoothing parameters are $\widehat{\lambda} = 38$ for the top panel, and $\widehat{\lambda} = 1617$ for the bottom panel. As expected, the larger smoothing parameter yields a smoother curve.

The area under the curve is easily computed as the integral of a line plus the integral of a spline. The integral of a spline over its domain is simply a function of its coefficients and knots. In particular, the integral of a spline of degree $k - 1$ with B-spline coefficients α

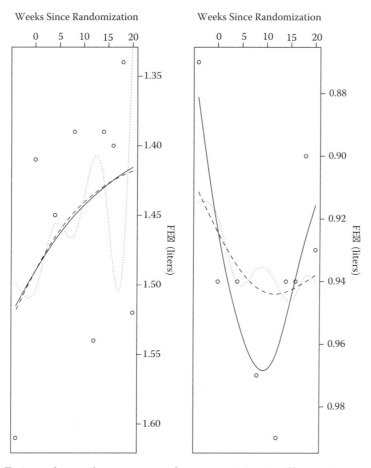

Figure 12.7 Estimated smooth mean curves for two participants. The points represent observed FEV_1 values. The solid lines represent the estimated curves from model (12.18) that allows different smoothing parameters across participants; the estimated smoothing parameters are $\widehat{\lambda} = 38$ for the left panel, and $\widehat{\lambda} = 1617$ for the right panel. The dashed lines represent estimated curves from the model that assumes a common smoothing parameter for all participants. Note that the estimated common smoothing parameter, 1096, is closer to the estimated smoothing parameter of 1617 for the participant in the right panel than that of the participant in the left panel, and thus the solid and dashed lines in the right panel are closer together. The dotted lines represent the estimated group plus individual curves based on model (12.19).

is $1/k$ multiplied by the sum of the elements of the dot product of α and γ, where the jth element of γ is $\nu_{j+k} - \nu_j$ and where ν_j is the jth element of the vector of knots with the end knots repeated k times. The derivation uses the fact that the integral of a spline of degree $k - 1$ is a spline of degree k (see de Boor, 1978). The R code for computing the area under the ith participant's curve is available at the web site for this book given in the Preface.

We estimate the mean area under the curve for each treatment as the average of individual areas; this gives 39.58 for the tobramycin group and 36.11 for the placebo group. To assess variability, we used a simple non-parametric bootstrap that resampled curves as i.i.d. within treatment groups 1000 times. The difference, 3.46, is statistically significant at the 0.05 level with a 95% confidence interval of (0.90, 6.02). Our analysis concurs with that of Ramsey et al. (1999) in concluding that tobramycin improves lung function.

One could instead estimate the group mean curves and then compute the area under these curves, which would give the same answer. Figure 12.8 presents the estimated group mean curves (solid lines) together with a graph of longitudinal averages of the observed FEV_1 values.

Our second approach is to jointly smooth the longitudinal FEV_1 values for each participant, assuming a common smoothing parameter and error variance for all participants. Specifically, we use model (12.18) and assume the $\boldsymbol{b}_i$ are $N\{0, (\sigma^2/\lambda)I\}$ and independent of the ε_i, which are $N(0, \sigma^2 I)$. The estimated common smoothing parameter is $\widehat{\lambda} = 1096$. The estimated smoothing splines for the two participants are graphed in Figure 12.7

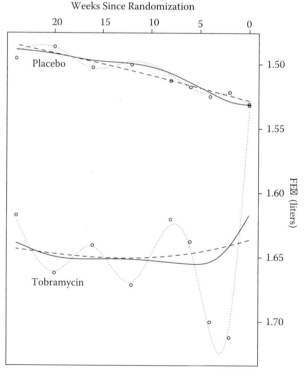

Figure 12.8 Estimated mean curves for the tobramycin and placebo groups. The solid curves represent estimates based on model (12.18), which allows smoothing parameters to vary across participant. The dashed curves correspond to the model that assumes a common smoothing parameter across participants. The dotted curves correspond to model (12.19), which decomposes the data into a smooth group curve (with common smoothing parameter across groups) and a smooth participant deviation (with common smoothing parameter across participants).

with dashed lines. Since the estimated common smoothing parameter of 1096 is relatively close to the estimated individual smoothing parameter of 1617 for the bottom panel, the estimated curves based on the common and individual smoothing parameter are similar in that panel, but not in the top one.

When we assume a common smoothing parameter and error variance for the placebo group, the estimated mean area under the curve is 39.53 for the tobramycin group and 36.14 for the placebo group, with a difference of 3.39. Interestingly, these results are very similar to the previous ones (see Table 12.2 for a summary) even though the estimated smoothing parameters vary tremendously across participants. The estimated group curves, shown as dashed lines in Figure 12.8, are also very similar.

Our third approach is to base estimation of the group mean curves on (12.17) instead of (12.16), and to then compute the area under these curves. As with the hormone example, the fixed effects are constrained to sum to zero within each treatment group. This leads to the LME model

$$\boldsymbol{Y}_i = X_i^* \boldsymbol{\beta}_i + Z_i \boldsymbol{b}_{g(i)} + Z_i \boldsymbol{b}_{p(i)} + \boldsymbol{\varepsilon}_i \tag{12.19}$$

for $i = 1, \ldots 462$, with the $\boldsymbol{b}_{\alpha(i)}$ independent of the $\boldsymbol{b}_{\gamma(i^a)}$ for $\alpha \neq \gamma$ or $i \neq i^a$; the collection of distinct $\boldsymbol{b}_{g(i)}$ i.i.d. $N\{0, (\sigma^2/\lambda_g)I\}$; and that of the distinct $\boldsymbol{b}_{p(i)}$ i.i.d. $N\{0, (\sigma^2/\lambda_p)I\}$.

The estimated group and participant smoothing parameters are $\widehat{\lambda}_g = 48$ and $\widehat{\lambda}_p = 938$. These smoothing parameters are smaller than those from the previous models; thus, the estimated curves are wigglier. See, for example, the estimated smoothing splines for the two participants' data in Figure 12.7 that are shown as dotted lines, and the estimated group curves in Figure 12.8, also shown as dotted lines. The estimated mean area under the curve for the tobramycin group and for the placebo group based on model (12.19) are 39.71 and 36.11, with a difference of 3.61. We did not compute bootstrap estimates of variability for the second and third approaches because a data set of 462 individuals is large enough that the lme function takes approximately 1 hour to jointly smooth the curves.

In this analysis of the CF trial, we have compared three penalized spline approaches. The first approach smooths curves individually, whereas the second and third approaches smooth curves jointly. Somewhat surprisingly, smoothing curves individually led to a similar comparison of group mean curves to smoothing individual curves jointly, sharing a common smoothing parameter. Also surprising was that the third analysis, which introduced smooth group mean component curves, led to undersmoothed estimates. Although introducing the group curves to the model adds one more parameter, that of the group smoothing parameter, and hence could lead to overparameterization, this should be trivial compared to the multiplicity of smoothing parameters used by the first analysis (one per participant). The real problem is more likely that the variance component for the group random effects is estimated high due to the stark difference between the two groups (refer to Figure 12.8). A larger group variance component would lead to a smaller group smoothing parameter, and hence to undersmoothing of both the group and the group plus individual curves.

Table 12.2 Estimates for Tobramycin Clinical Trial

	Mean Area under the Curve		Difference in
Model	Placebo	Tobramycin	Mean Areas
Different λ	36.11	39.58	3.46
Common λ	36.14	39.53	3.39
Smooth Decomposition	36.11	39.71	3.61

Note: 95% confidence interval for difference in mean areas estimated with model (12.18) and the bootstrap: (0.90, 6.02).

This again causes us to question the appropriateness of using the mixed-effects model and REML to select the smoothing parameters, a procedure that confounds variability of curves with smoothness of curves. We leave a thorough investigation of this phenomenon as a topic for future research.

We have also demonstrated the ease with which area under the curve can be estimated based on penalized spline models. For future research, it would be interesting to compare the efficiency of the penalized spline approach with that of standard approaches, which do not incorporate the assumption of smooth individual mean curves.

12.5 Discussion

We have used the connection between penalized splines and linear mixed-effects models to incorporate the assumption of smooth individual and population mean curves into longitudinal data analysis via estimators easily programmed with the `lme` function of R. We have argued that inferences based on the mixed-effects model can be inaccurate, whether assessing variability of $\widehat{s}_i(t)$ or $\widehat{E}\{s_i(t)\}$, and we have discovered with our analysis of the CF trial that even selecting smoothing parameters via the mixed-effects model can yield undesirable results. The source of all of these problems is easily recognized as the confounding of *variability of curves* with *smoothness of curves*, implicit in the mixed-effects model formulation. Thus, although the mixed-effects model software leads to conveniently programmed estimators, we recommend using it in conjunction with simulation (bootstrap) based methods for assessing variability.

The methodology we describe nearly fits within the SSANOVA framework of Wahba and colleagues (Wahba, 1990; Wahba et al., 1995; Wang et al., 1997; Wang, 1998) and, in fact, when the selected spline basis spans the natural cubic splines with knots at the unique interior time points and the penalty is the cubic smoothing spline penalty, the penalized spline ANOVAs we describe are special cases of SSANOVA. In general, however, the selected spline basis can be smaller than the ones implicit in SSANOVA, and thus will not lead to the same optima.

Another closely related literature pertains to functional principal component analysis (Rice and Silverman, 1991; Ramsey and Silverman, 2002, 2005; Yao, Müller, and Wang, 2005; Yao and Lee, 2006). However, functional principal component analysis selects the components of the smooth effective basis based on the covariance matrix of the observed curves, whereas the penalized spline approach we describe selects the basis components *a priori* and only adapts to the data by way of the smoothing parameter, which specifies the effective dimension.

Whereas we have focused on penalized spline models for repeated outcomes that do not require a link function (McCullagh and Nelder, 1989), generalizations that incorporate link funtions can be found in Wahba et al. (1995), Lin and Zhang (1999), Zhang (2004), and Gu and Ma (2005). However, these authors generally conduct inference assuming either that an artificial generalized linear mixed-effects model generates the data, or that a corresponding Bayesian paradigm is appropriate. Because the fitting of generalized linear mixed-effects models requires several iterations compared with linear mixed-effects models, simulation-based inference remains computationally challenging at this point in time.

References

Berhane, K. and Tibshirani, R. J. (1998). Generalized additive models for longitudinal data. *Canadian Journal of Statistics* **26**, 517–535.

Brumback, B.A. and Rice, J. A. (1998). Smoothing spline models for the analysis of nested and crossed samples of curves (with discussion). *Journal of the American Statistical Association* **93**, 961–994.

Brumback, B. A., Ruppert, D., and Wand, M. P. (1999). Comment on: Variable selection and function estimation in additive non-parametric regression using a data-based prior, by T.S. Shively, R. Kohn, and S. Wood, *Journal of the American Statistical Association* **94**, 794–797.

Cantin, A. (1995). Cystic fibrosis lung inflammation: early, sustained, and severe. *American Journal of Respiratory & Critical Care Medicine* **151**, 939–941.

Chiang, C., Rice, J. A., and Wu, C. O. (2001). Smoothing spline estimation for varying coefficient models with repeatedly measured dependent variables. *Journal of the American Statistical Association* **96**, 605–619.

de Boor, C. (1978). *A Practical Guide to Splines*. New York: Springer.

Demmler, A. and Reinsch, C. (1975). Oscillation matrices with spline smoothing. *Numerische Mathematik* **24**, 375–382.

Efron, B. and Tibshirani, R. J. (1993). *An Introduction to the Bootstrap*. London: Chapman & Hall.

Eilers, P. H. C. and Marx, B. D. (1996). Flexible smoothing with B-splines and penalties (with discussion). *Statistical Science* **11**, 89–121.

Giesbrecht, F. G. (1982). Effective degrees of freedom. In S. Kotz, N. L. Johnson, and C. B. Read (eds.), *Encyclopedia of Statistical Sciences*, Volume 2, pp. 467–468. New York: Wiley.

Green, P. J. and Silverman, B. W. (1994). *Non-parametric Regression and Generalized Linear Models*. London: Chapman & Hall.

Gu, C. and Ma, P. (2005). Generalized non-parametric mixed-effect models: Computation and smoothing parameter selection. *Journal of Computational and Graphical Statistics* **14**, 485–504.

Hastie, T. J. and Tibshirani, R. J. (1990). *Generalized Additive Models*. London: Chapman & Hall.

Hastie, T. J., Tibshirani, R. J., and Friedman, J. (2001). *The Elements of Statistical Learning: Data Mining, Inference, and Prediction*. New York: Springer.

Kimeldorf, G. S. and Wahba, G. (1970). A correspondence between Bayesian estimation on stochastic processes and smoothing by splines. *Annals of Mathematical Statistics*, **41**, 495–502.

Lin, X. and Zhang, D. (1999). Inference in generalized additive mixed models by using smoothing splines. *Journal of the Royal Statistical Society, Series B* **61**, 381–400.

Little, R. J. and Rubin, D. B. (1987). *Statistical Analysis with Missing Data*. New York: Wiley.

McCullagh, P. and Nelder, J. (1989). *Generalized Linear Models* (2nd ed.) London: Chapman & Hall.

Pinheiro, J. C. and Bates, D. M. (2000). *Mixed-Effects Models in S and S-PLUS*. New York: Springer.

Ramsey, B. W., Pepe, M. S., Quan, J. M., Otto, K. L., Montgomery, A. B., Williams-Warren, J., Vasiljev, K. M., Borowitz, D., Bowman, C. M., Marshall, B. C., Marshall, S., and Smith, A.L. (1999). Intermittent administration of inhaled tobramycin in patients with cystic fibrosis. *New England Journal of Medicine* **340**, 23–30.

Ramsay, J. O. and Silverman, B. W. (2002). *Applied Functional Data Analysis*. New York: Springer.

Ramsay, J. O. and Silverman, B. W. (2005). *Functional Data Analysis*. New York: Springer.

Rice, J. A. and Silverman, B. W. (1991). Estimating the mean and covariance structure non-parametrically when the data are curves. *Journal of the Royal Statistical Society, Series B* **53**, 233–243.

Robinson, G. K. (1991). That BLUP is a good thing: The estimation of random effects (with discussion). *Statistical Science* **6**, 15–32.

Ruppert, D., Wand, M. P., and Carroll, R. J. (2003). *Semi-parametric Regression*. Cambridge: Cambridge University Press.

Seber, G. A. F. (1977). *Linear Regression Analysis*. New York: Wiley.

Silverman, B. W. (1985). Some aspects of the spline smoothing approach to non-parametric regression curve fitting. *Journal of the Royal Statistical Society, Series B* **47**, 1–21.

Speed, T. P. (1991). Comment on: That BLUP is a good thing: the estimation of random effects, by G. K. Robinson. *Stctistical Science* **6**, 42–44.

van der Vaart, A. W. (1998). *Asymptotic Statistics*. Cambridge: Cambridge University Press.

Verbyla, A. P., Cullis, B. R., Kenward, M. G., and Welham, S. J. (1999). The analysis of designed experiments and longitudinal data by using smoothing splines (with discussion.) *Applied Statistics* **48**, 269–311.

Wahba, G. (1978). Improper priors, spline smoothing, and the problem of guarding against model errors in regression. *Journal of the Royal Statistical Society, Series B* **40**, 364–372.

Wahba, G. (1985). A comparison of GCV and GML for choosing the smoothing parameter in the generalized spline smoothing problem. *Annals of Statistics* **4**, 1378–1402.

Wahba, G. (1990). *Spline Models for Observational Data*. Philadelphia: SIAM.

Wahba, G., Wang, Y., Gu, C., Klein, R., and Klein, B. (1995). Smoothing spline ANOVA for exponential families, with application to the Wisconsin epidemiological study of diabetic retinopathy. *Annals of Statistics* **23**, 1865–1895.

Wang, Y., Wahba, G., Gu, C., Klein, R., and Klein, B. (1997). Using smoothing spline ANOVA to examine the relation of risk factors to the incidence and progression of diabetic retinopathy. *Statistics in Medicine* **16**, 1357–1376.

Wang, Y. (1998). Mixed effects smoothing spline analysis of variance. *Journal of the Royal Statistical Society, Series B* **60**, 159–174.

Yao, F., Müller, H., and Wang, J. (2005). Functional data analysis for sparse longitudinal data. *Journal of the American Statistical Association* **100**, 577–590.

Yao, F. and Lee, T. C. M. (2006). Penalized spline models for functional principal component analysis. *Journal of the Royal Statistical Society, Series B* **68**, 3–25.

Zhang, D., Lin, X., and Sowers, M. (2000). Semi-parametric regression for periodic longitudinal hormone data from multiple menstrual cycles. *Biometrics* **56**, 31–39.

Zhang, D. (2004). Generalized linear mixed models with varying coefficients for longitudinal data. *Biometrics*, **60**, 8–15.

PART IV

Joint Models for Longitudinal Data

CHAPTER 13

Joint models for longitudinal data: Introduction and overview

Geert Verbeke and Marie Davidian

Contents

13.1 Introduction

Thus far, the focus of this book has been on statistical models and methods useful for the analysis of a single outcome, measured repeatedly on each individual. In practice, however, one is often confronted with situations in which multiple outcomes, recorded simultaneously, are measured repeatedly within each subject over time. These outcomes may be of similar or disparate types, and a variety of scientific questions may be of interest, depending on the application.

In toxicological studies, interest may focus on the relationship between dose of a toxic agent and several outcomes reflecting possible deleterious effects of the agent. For example, birth weight, a continuous measure, and a binary indicator of malformation, may be recorded on each fetus in a teratogenicity study. In some studies, multiple outcomes thought to be associated with toxicity, such as continuous and discrete measures of different aspects of neurological function, are recorded longitudinally on each individual, and the goal is to characterize the nature of the relationship between dose and the simultaneous time course of these measures. A statistical model that jointly represents these relationships is an appropriate framework in which these questions may be addressed.

In other settings, longitudinal measurements on a continuous or discrete response may be recorded along with a (possibly censored) time-to-event ("survival") outcome on each subject in a clinical study. A familiar example is that of HIV studies, where measures of immunological and virological status, such as CD4 T-cell count and viral RNA copy number ("viral load"), are collected longitudinally on each participant along with the time to progression to AIDS or death, which may be administratively censored. One objective is to characterize the relationship between features of CD4 and viral load trajectories and the event time, knowledge of which has been used to evaluate the prognostic value of these longitudinal markers and, in the presence of treatment, to assess the role of these measures as potential surrogates. Similarly, the goal may be to predict renal graft failure based on longitudinally collected information on hemoglobin levels and data on time to failure of the graft from renal transplant patients. In some settings, instead of a single, possibly terminating event time (e.g., death), times to recurrent events, such as seizure episodes experienced by an epileptic subject, may be recorded along with other longitudinal outcomes, and interest focuses on the aspects of the relationship between the longitudinal and recurrent event processes. Finally, the primary objective may be to make inferences on a single longitudinal outcome, for example, to compare treatments on the basis of longitudinal rates of change, but this is complicated by potentially informative dropout, which may be viewed as a second, time-to-event outcome. In all of these examples, the substantive questions can only be fully addressed by a joint model for the longitudinal measures and the time(s) to the event.

Yet another example where joint modeling of multiple outcomes is required is provided by a study seeking to elucidate how hearing ability changes during aging based on longitudinal measurements of hearing thresholds at various frequencies, potentially measured separately for the left and right ear, respectively. Of particular interest is to evaluate whether or not the rate of loss of hearing ability is the same at different frequencies. Addressing this issue obviously requires a statistical framework in which the data from all frequencies may be represented.

A number of approaches to joint modeling of multiple outcomes, where some or all of the outcomes are ascertained longitudinally, have been proposed. In this chapter, we provide a brief introduction to the conceptual perspectives underlying many of these approaches. Subsequent chapters present detailed accounts of how these ideas are exploited in the formulation of joint models relevant to specific settings exemplified by the foregoing applications. Joint models and associated statistical methods suitable for making inferences on scientific questions involving simultaneous continuous and discrete outcomes, as in the toxicological studies discussed above, are the subject of Chapter 14. In Chapter 15, joint models for continuous longitudinal outcomes and a possibly right-censored time-to-event, as in the HIV and renal transplantation applications, are discussed in detail. Most of the emphasis in the literature has been on joint modeling and analysis of a small number of outcomes, say two or three, as in Chapter 14 and Chapter 15. However, recently, there has been a growing interest in the analysis of high-dimensional multivariate longitudinal data. These developments are discussed in Chapter 16.

13.2 Approaches to joint modeling

A broad objective of joint modeling is to provide a framework within which questions of scientific interest pertaining to systematic relationships among the multiple outcomes and between them and other factors (treatment, dose, etc.) may be formalized. To ensure valid inferences, joint models must appropriately account for the correlation among the outcomes. To this end, joint modeling approaches attempt to characterize the joint distribution of the multiple outcomes in various ways. To fix ideas, let Y_1 and Y_2 be two outcomes measured on a subject for which joint modeling is of scientific interest, where one or both may be collected longitudinally, so that, in general, Y_1 and Y_2 may be random vectors comprising longitudinal measurements or other quantities, as discussed further below. While we restrict attention to the case of two outcomes, extension to higher dimensions is straightforward. Also, in all models discussed next, we will suppress notation for any dependence on covariates.

13.2.1 Multivariate marginal models

A first approach attempts to specify directly the joint density $f(\boldsymbol{y}_1, \boldsymbol{y}_2)$ of $(\boldsymbol{Y}_1, \boldsymbol{Y}_2)$ (see Galecki, 1994; see also Molenberghs and Verbeke, 2005, Section 24.1 for an overview). Such a model will include of necessity assumptions about the marginal association among the longitudinally measured elements within each of the vectors $\boldsymbol{Y}_1$ and $\boldsymbol{Y}_2$, but also must include assumptions on the nature of the association between elements of $\boldsymbol{Y}_1$ and $\boldsymbol{Y}_2$. Especially when $\boldsymbol{Y}_1$ and $\boldsymbol{Y}_2$ are of different types (e.g., continuous and discrete, continuous and survival) and/or in the case of (highly) unbalanced data, this becomes cumbersome. Moreover, extension to dimensions higher than two involves considerable challenges, as this would require assumptions on larger covariance structures and higher-order associations. The main advantage of this class of models is that they allow for direct inferences for marginal characteristics of the outcomes, such as average evolutions. This is also reflected by the symmetric treatment of the two outcomes. This is in strong contrast to several of the other approaches, discussed next.

13.2.2 Conditional models

One way to avoid direct specification of a joint distribution for $(\boldsymbol{Y}_1, \boldsymbol{Y}_2)$ is to factorize the density as a product of a marginal and a conditional density, that is,

$$f(\boldsymbol{y}_1, \boldsymbol{y}_2) = f(\boldsymbol{y}_1 | \boldsymbol{y}_2) f(\boldsymbol{y}_2) \tag{13.1}$$

$$= f(\boldsymbol{y}_2 | \boldsymbol{y}_1) f(\boldsymbol{y}_1). \tag{13.2}$$

Such factorizations reduce the modeling tasks to the specification of models for each of the outcomes separately, with a marginally specified model for one of the outcomes and a conditionally specified model for the other. Note that, in the case where both outcomes are measured longitudinally, specification of a conditional model such as $f(\boldsymbol{y}_1 | \boldsymbol{y}_2)$ requires very careful reflection about plausible associations between $\boldsymbol{Y}_1$ and $\boldsymbol{Y}_2$, where the latter plays the role of a time-varying covariate, and different choices can lead to very different, sometimes completely opposite, results and conclusions.

Another drawback of conditional models is that they do not directly lead to marginal inferences. As an example, consider the study of hearing changes with age introduced in Section 13.1. Let $\boldsymbol{Y}_1$ and $\boldsymbol{Y}_2$ contain the longitudinally measured hearing thresholds at a high and a low frequency, respectively, and suppose that interest focuses on comparison of the average rate of decline in hearing ability between the two frequencies. Model (13.1) directly allows inferences about the marginal evolution of $\boldsymbol{Y}_2$, but additional calculations are needed in order to obtain inferences about characteristics of the marginal distribution of $\boldsymbol{Y}_1$. For example, the marginal average requires computation of

$$E(\boldsymbol{Y}_1) = E\{E(\boldsymbol{Y}_1 | \boldsymbol{Y}_2)\} = \int \left\{ \int \boldsymbol{y}_1 f(\boldsymbol{y}_1 | \boldsymbol{y}_2) \, d\boldsymbol{y}_1 \right\} f(\boldsymbol{y}_2) \, d\boldsymbol{y}_2,$$

which, depending on the actual models, may be far from straightforward. One way to circumvent this would be to fit both models (13.1) and (13.2). However, specification of both models in a compatible way often requires direct specification of the joint density $f(\boldsymbol{y}_1, \boldsymbol{y}_1)$, thus involving the problems discussed in Section 13.2.1. Further, it should be noted that the marginal mean of $\boldsymbol{Y}_1$ is not, in general, of the same form as the original conditional mean. For example, a logistic regression model for the conditional mean of $\boldsymbol{Y}_1$ given $\boldsymbol{Y}_2$ does not marginalize to a logistic regression for the marginal mean of $\boldsymbol{Y}_1$. So, with the exception of linear regression models, the marginalization, even when computationally straightforward or feasible, is not always useful or helpful.

The asymmetric treatment of the outcomes can, in some situations, be very unappealing. For example, in a clinical trial with two main outcomes, models (13.1) and (13.2) will rarely be of interest due to the conditioning on a post-randomization outcome. Moreover, when $\boldsymbol{Y}_1$ and $\boldsymbol{Y}_2$ are highly correlated and considered to be manifestations of a common underlying treatment effect, conditioning on one of the outcomes will, in general, attenuate the treatment effect on the other.

Although in the case of two outcomes $\boldsymbol{Y}_1$ and $\boldsymbol{Y}_2$ only two factorizations, (13.1) and (13.2), are possible, there are many more possible factorizations when more than two outcomes must be jointly modeled. Hence, conditional models are often not the preferred choice for the analysis of high-dimensional multivariate longitudinal data. We refer to Molenberghs and Verbeke (2005, Section 24.1) for an overview of models within this framework.

13.2.3 Shared-parameter models

It has been demonstrated in earlier chapters how random effects can be used to generate an association structure between the repeated measurements of a specific outcome. The same idea can be used to construct multivariate longitudinal models. A popular approach is to postulate a so-called *shared-parameter model*, formulated as follows.

Let $\boldsymbol{b}$ denote a vector of random effects, common to the model for $\boldsymbol{Y}_1$ and the model for $\boldsymbol{Y}_2$, and assume independence of both outcomes, conditionally on $\boldsymbol{b}$. The joint density for $(\boldsymbol{Y}_1, \boldsymbol{Y}_2)$ is then obtained from

$$f(\boldsymbol{y}_2, \boldsymbol{y}_2) = \int f(\boldsymbol{y}_1, \boldsymbol{y}_2 | \boldsymbol{b}) f(\boldsymbol{b}) \, d\boldsymbol{b} = \int f(\boldsymbol{y}_1 | \boldsymbol{b}) f(\boldsymbol{y}_2 | \boldsymbol{b}) f(\boldsymbol{b}) \, d\boldsymbol{b}, \qquad (13.3)$$

in which $f(\boldsymbol{b})$ denotes the random-effects density. Often, $\boldsymbol{b}$ is assumed normally distributed, but alternatives are possible (see Chapter 2 for a general discussion). The random effect $\boldsymbol{b}$ is a "shared parameter" that serves to induce a correlation between $\boldsymbol{Y}_1$ and $\boldsymbol{Y}_2$ through their joint dependence on $\boldsymbol{b}$. That $\boldsymbol{Y}_1$ and $\boldsymbol{Y}_2$ are conditionally independent given $\boldsymbol{b}$ may be interpreted as reflecting the belief that a common set of underlying characteristics of the individual governs both outcome processes.

One of the main advantages of this approach is that $\boldsymbol{Y}_1$ and $\boldsymbol{Y}_2$ do not have to be of the same type. For example, a linear mixed model for a continuous outcome $\boldsymbol{Y}_1$ (see Section 1.2) can easily be combined with a logistic mixed model for a binary outcome $\boldsymbol{Y}_2$ (see Chapter 4), with both models sharing the same random effects. Moreover, the parameters in the joint model have the same interpretation as in each of the "univariate" models. Finally, extending the model to more than two outcomes is straightforward and, because dimensionality of the integration in (13.3) does not increase, does not entail any additional computational burden.

A setting in which shared-parameter models have been routinely employed is in joint modeling of a longitudinal outcome and a time-to-event outcome, particularly in the context of the study of HIV infection (e.g., DeGruttola and Tu, 1994; Tsiatis, DeGruttola, and Wulfsohn, 1995; Faucett and Thomas, 1996). For definiteness, let $\boldsymbol{Y}_1$ represent a vector of longitudinal (continuous) observations on CD4 count, and let $\boldsymbol{Y}_2 = (U, \delta)'$ denote the bivariate outcome comprising $U = \min(T, C)$, where T is the potential time to the event of interest (e.g., death or progression to AIDS) and C is a potential censoring time, assumed independent; and δ is an indicator of whether or not U corresponds to the event of interest or is right-censored. A popular framework for representing the relationship between CD4 count and survival is to assume that $\boldsymbol{Y}_1$ may be represented by a linear mixed-effects model (see Section 1.3) depending on a vector of random effects whose elements may describe subject-specific features of the longitudinal trajectory, so that $f(\boldsymbol{y}_1 | \boldsymbol{b})$ in (13.3) is the corresponding multivariate normal density. For example, letting $Y_1(t)$ represent measured CD4 count at any time t, assume that the elements of $\boldsymbol{Y}_1$ are observations on $Y_1(t)$ at intermittent time points, where one might postulate a subject-specific straight-line model for $Y_1(t)$ of the form

$$
\begin{aligned}
Y_1(t) &= \beta_0 + b_0 + (\beta_1 + b_1)t + e_1(t) \\
&= \mu(t, \boldsymbol{b}) + e_1(t),
\end{aligned}
\tag{13.4}
$$

say, where the random-effects vector $\boldsymbol{b} = (b_0, b_1)'$ is independent of $e_1(t)$ and multivariate normal $N(\boldsymbol{0}, G)$, and $e_1(t)$ are independent and identically distributed $N(0, \sigma^2)$. In (13.4), then, b_1 is a subject-specific slope effect reflecting the (assumed constant) rate of change of the subject-specific mean CD4 count $\mu(t, \boldsymbol{b}) = E\{Y_1(t)|\boldsymbol{b}\}$ over time. Likewise, the event time of interest may be taken to follow a proportional hazards model conditional on the random effects, where the linear predictor depends on the longitudinal process solely through a function of random effects in (13.4). That is, one may assume a model such as

$$
\begin{aligned}
\lim_{h \to 0} \Pr\{t \leq T < t + h | T \geq t, \mu(u, \boldsymbol{b}), 0 \leq u < t\} & \\
= \lim_{h \to 0} \Pr\{t \leq T < t + h | T \geq t, \boldsymbol{b}\} & \\
= \lambda_0(t) \exp\{\alpha g(\boldsymbol{b})\}, &
\end{aligned}
\tag{13.5}
$$

where $\lambda_0(t)$ is an unspecified baseline hazard function; and $g(\boldsymbol{b})$ is a linear function of $\boldsymbol{b}$, for example $g(\boldsymbol{b}) = \mu(t, \boldsymbol{b})$, so that the hazard depends on the CD4 count history up to time t through the subject-specific mean $\mu(t, \boldsymbol{b})$ at t, or $g(\boldsymbol{b}) = b_1$ or $\beta_1 + b_1$, so that the hazard depends on the CD4 count history through the subject-specific slope. A model like (13.5) leads to an expression for the density $f(\boldsymbol{y}_2|\boldsymbol{b})$ involving α, $\boldsymbol{b}$, $\lambda_0(t)$, and δ. The question of whether there is a relationship between the feature of the subject-specific CD4 trajectories represented by $g(\boldsymbol{b})$ in (13.5) and the time-to-event may be addressed via inferences on α. This basic formulation, and extensions of and variations on it, as well as its use in addressing scientific questions like those in Section 13.1, are discussed in detail in Chapter 15; an extensive bibliography is also presented. A careful account of the philosophical considerations involved in and implicit assumptions underlying specification of models like (13.5) is given in Tsiatis and Davidian (2004).

13.2.4 Random-effects models

A key disadvantage of shared-parameter models is that they can imply very strong assumptions about the association between the outcomes modeled, a phenomenon that is particularly evident when both $\boldsymbol{Y}_1$ and $\boldsymbol{Y}_2$ are vectors of longitudinal measurements on two outcomes. As an example, suppose that the elements of $\boldsymbol{Y}_k$, $k = 1, 2$, are realizations of stochastic processes $Y_k(t)$, $k = 1, 2$, say, taken at intermittent times. To focus ideas, assume that both of these processes can be well described by random-intercept models given by

$$
\begin{aligned}
Y_1(t) &= \beta_1 + b + \beta_2 t + e_1(t), \\
Y_2(t) &= \beta_3 + \gamma b + \beta_4 t + e_2(t),
\end{aligned}
\tag{13.6}
$$

respectively, where b is a scalar, normally distributed, mean-zero random effect common to both models; $e_1(t)$ and $e_2(t)$ are normally distributed with mean zero and are independent of b and of each other for all t; and $\mathrm{Var}\{e_1(t)\} = \sigma_1^2$ and $\mathrm{Var}\{e_2(t)\} = \sigma_2^2$, constant for all t. Note that it is necessary to scale the shared random effect b (by γ) in the model for $Y_2(t)$ in (13.6). One may think of models like (13.6) as extensions to more than one outcome of the classical linear mixed model for a single outcome, discussed in Section 1.3 of this book.

Under these conditions, the joint shared parameter–random intercept model (13.6) leads to the following expressions for marginal correlations, $s \neq t$:

$$\text{Corr}\{Y_1(s), Y_1(t)\} = \frac{\text{Var}(b)}{\text{Var}(b) + \sigma_1^2},$$

$$\text{Corr}\{Y_2(s), Y_2(t)\} = \frac{\gamma^2 \text{Var}(b)}{\gamma^2 \text{Var}(b) + \sigma_2^2},$$

$$\text{Corr}\{Y_1(s), Y_2(t)\} = \frac{\gamma \text{Var}(b)}{\sqrt{\text{Var}(b) + \sigma_1^2}\sqrt{\gamma^2 \text{Var}(b) + \sigma_2^2}}$$

$$= \sqrt{\text{Corr}\{Y_1(s), Y_1(t)\}}\sqrt{\text{Corr}\{Y_2(s), Y_2(t)\}}. \tag{13.7}$$

These results show that, although (13.6) of course implies the standard assumption for single-outcome random-intercept models that correlations among all pairs of longitudinal measurements are constant (independent of s and t) and the same for all s and t, it also imposes a particularly strong additional restriction on the joint behavior. Specifically, the fact that the random intercept b is shared by both outcomes dictates that correlations between pairs of measurements from different outcomes must be equal to the product of the correlation between measurements from the first outcome and measurements from the second outcome, as shown in (13.7). Note that similar restrictions would hold in more general models that involve shared random-effect vectors beyond just a scalar random intercept.

As a consequence, shared-parameter models may be of limited value for yielding a faithful representation of dependence structure of the joint distribution of multiple longitudinal outcomes. Thus, these strong restrictions motivate consideration of joint random-effects models that imply more flexible correlation patterns, albeit at the expense of greater model complexity, as we now describe. In particular, the rigid constraint (13.7) may be relaxed by allowing the models for $\boldsymbol{Y}_1$ and $\boldsymbol{Y}_2$ to depend on separate random effects $\boldsymbol{b}_1$ and $\boldsymbol{b}_2$, which are themselves correlated. As a simple illustration, consider the models in (13.6), but modified to involve separate random intercepts for both processes, that is,

$$\begin{aligned} Y_1(t) &= \beta_1 + b_1 + \beta_2 t + e_1(t), \\ Y_2(t) &= \beta_3 + b_2 + \beta_4 t + e_2(t), \end{aligned} \tag{13.8}$$

where, as before, $e_1(t)$ and $e_2(t)$ are normally distributed with mean zero and are independent of each other for all t; and $\text{Var}\{e_1(t)\} = \sigma_1^2$ and $\text{Var}\{e_2(t)\} = \sigma_2^2$, constant for all t. Like (13.6), (13.8) is a multiple-outcome version of the linear mixed model for a single outcome. However, rather than being linked by shared dependence on a common random effect as in (13.6), the models for the two outcomes in (13.8) are joined by assuming that the random-effect vector $\boldsymbol{b} = (b_1, b_2)'$, assumed independent of $e_1(t)$ and $e_2(t)$ for all t, has a multivariate normal distribution with zero mean vector and covariance matrix G, with off-diagonal element G_{12}, say, possibly different from zero, imposing an associational structure across outcomes.

Under these assumptions, the following expressions for the relevant correlations may be obtained for $s \neq t$:

$$\text{Corr}\{Y_1(s), Y_1(t)\} = \frac{\text{Var}(b_1)}{\text{Var}(b_1) + \sigma_1^2},$$

$$\text{Corr}\{Y_2(s), Y_2(t)\} = \frac{\text{Var}(b_2)}{\text{Var}(b_2) + \sigma_2^2},$$

$$\text{Corr}\{Y_1(s), Y_2(t)\} = \frac{\text{Cov}(b_1, b_2)}{\sqrt{\text{Var}(b_1) + \sigma_1^2}\sqrt{\text{Var}(b_2) + \sigma_2^2}}$$

$$= \frac{\text{Cov}(b_1, b_2)}{\sqrt{\text{Var}(b_1)}\sqrt{\text{Var}(b_2)}} \frac{\sqrt{\text{Var}(b_1)}\sqrt{\text{Var}(b_2)}}{\sqrt{\text{Var}(b_1) + \sigma_1^2}\sqrt{\text{Var}(b_2) + \sigma_2^2}}$$

$$= \text{Corr}(b_1, b_2)\sqrt{\text{Corr}\{Y_1(s), Y_1(t)\}}\sqrt{\text{Corr}\{Y_2(s), Y_2(t)\}} \quad (13.9)$$

$$\leq \sqrt{\text{Corr}\{Y_1(s), Y_1(t)\}}\sqrt{\text{Corr}\{Y_2(s), Y_2(t)\}}, \quad (13.10)$$

where, from above, $\text{Corr}(b_1, b_2) = G_{12}/\sqrt{\text{Var}(b_1)\text{Var}(b_2)}$. Expression (13.9) demonstrates explicitly the role of the correlation between the outcome-specific random effects in dictating the between-process outcome correlation at any two time points. Moreover, (13.10) shows that the restriction imposed by the shared-parameter model (13.6) is relaxed in the sense that the model no longer assumes that the product of the within-process correlations equals the between-process correlation, thus allowing a more general dependence structure. Note that the earlier shared-parameter model (13.6) can be obtained as a special case of (13.8) by restricting the correlation between b_1 and b_2 to be equal to 1.

Random-effects models in the spirit of (13.8) are a standard framework for the joint analysis of continuous and discrete outcomes. Chapter 14 offers a detailed account of the specification and interpretation of such models and of their implementation using standard statistical software. Additional discussion and a further demonstration of implementation are offered in Molenberghs and Verbeke (2005, Chapter 24).

This general idea of joining separate mixed models by allowing their model-specific random effects to be correlated can very easily be extended to more than two outcomes. The main disadvantage is that the dimensionality of the total vector of random effects in the resulting multivariate model for all outcomes grows with the number of outcome variables. This often leads to computational problems when the number of outcomes exceeds two, and/or when some of the outcomes are best described by a generalized linear, or non-linear, mixed model (see Chapter 4 and Chapter 5). At first sight, this approach is therefore not feasible for the analysis of high-dimensional, multivariate longitudinal profiles. In Chapter 16, a solution for this numerical challenge is presented and illustrated.

13.2.5 Methods based on dimension reduction

As we have indicated, especially in cases where (many) more than two outcomes must be modeled simultaneously, some of the approaches described above will not be feasible; will involve numerical difficulties; or will be based on extremely strong, often unrealistic, assumptions about the association structure between the various outcomes in the multivariate response vector. Therefore, a number of methods have been proposed based on dimension reduction. The general idea is to use a factor-analytic, or principal-component-type, analysis to first reduce the dimensionality of the response vector. In a second stage, the principal factors are analyzed using any of the classical longitudinal models.

The main disadvantage of such approaches is that they restrict inferences to the principal factor(s), no longer allowing inferences about aspects of the original outcome variables. Also, unless extremely strong restrictions are imposed on the principal components, techniques based on dimension reduction cannot be applied in cases of highly unbalanced repeated measurements, that is, cases in which unequal numbers of measurements are available for different subjects and/or cases where observations are taken at arbitrary time points.

13.3 Concluding remarks

It is clear that there are a number of substantive research questions that are best addressed within the framework of a joint model for multiple outcomes, some or all of which may be measured repeatedly over time. In some cases, joint modeling is required because the association structure among the outcomes is of interest. In other cases, joint modeling is needed in order to be able to draw joint inferences about the different outcomes. The statistical literature on joint models is diverse and growing. Indeed, for example, reports on new approaches to joint modeling of longitudinal outcomes and event times and their application in numerous subject-matter areas continue to be forthcoming. Many of the available techniques are practically restricted to simultaneous modeling of a small number of outcomes. Heightened substantive interest in understanding relationships in high dimensions and the ability to collect high-dimensional data requires approaches that render simultaneous analysis of a large number of outcomes feasible. Although most existing methods can be generalized to accommodate higher dimensions, this is often not possible without extremely strong assumptions about the association between repeated measurements of the same outcome, and/or the association between various outcomes at specific points in time. A challenge for the future is thus continued development of joint modeling approaches focused on this problem.

Many of the ideas summarized in this chapter can be seen as outgrowths of the missing-data literature. Let Y_1 denote the vector of repeated measurements on a specific outcome of interest, and let Y_2 be the associated vector of binary responses indicating whether the corresponding measurement in Y_1 is available or missing. Unless very specific (untestable) assumptions are made about the association between Y_1 and Y_2, joint models are needed in order to obtain valid inferences, even if interest focuses on aspects of Y_1 only. Many of the approaches discussed here have missing-data counterparts, and a completely separate terminology has been developed. For example, with these definitions of Y_1 and Y_2, in the context of the conditional models in Section 13.2.2, factorization (13.1) refers to a pattern-mixture model, while factorization (13.2) is called a selection model. More details on these and other missing-data models are given in Part 5 of this book.

References

DeGruttola, V. and Tu, X. M. (1994). Modeling progression of CD-4 lymphocyte count and its relationship to survival time. *Biometrics* **50**, 1003–1014.

Faucett, C. J. and Thomas, D. C. (1996). Simultaneously modeling censored survival data and repeatedly measured covariates: A Gibbs sampling approach. *Statistics in Medicine* **15**, 1663–1685.

Galecki, A. T. (1994). General class of covariance structures for two or more repeated factors in longitudinal data analysis. *Communications in Statistics — Theory and Methods* **23**, 3105–3119.

Molenberghs, G. and Verbeke, G. (2005). *Models for Discrete Longitudinal Data*. New York: Springer.

Tsiatis, A. A. and Davidian, M. (2004). Joint modeling of longitudinal and time-to-event data: An overview. *Statistica Sinica* **14**, 809–834.

Tsiatis, A. A., DeGruttola, V., and Wulfsohn, M. S. (1995). Modeling the relationship of survival to longitudinal data measured with error: Applications to survival and CD4 counts in patients with AIDS. *Journal of the American Statistical Association* **90**, 27–37.

CHAPTER 14

Joint models for continuous and discrete longitudinal data

Christel Faes, Helena Geys, and Paul Catalano

Contents

14.1 Introduction

Measurements of both continuous and categorical outcomes appear in many statistical problems. One such example is a toxicology experiment, where a set of outcomes is recorded to study the toxicity of the substance of interest. For example, both the malformation status of a live fetus (typically recorded as a binary or ordinal outcome) and low birth weight (measured on a continuous scale) are important variables in the context of teratogenicity. Also in longitudinal studies, it is common to collect several outcomes from a different type. Perhaps the most common situation is that of a continuous, often normally distributed, and a binary or ordinal outcome.

While multivariate methods for the analysis of continuous outcomes are well known (Johnson and Wichern, 2002), methods that jointly analyze discrete and continuous outcomes and adequately account for the correlation structure in the data are not widely available. Broadly, there are two approaches. A first model approach toward a joint model of a continuous and a discrete outcome is to apply a conditioning argument that allows the joint distribution to be factorized into a marginal component and a conditional component, where the conditioning can be done either on the discrete outcome or on the continuous outcome. Conditional models have been discussed by Tate (1954), Olkin and Tate (1961), Little and Schluchter (1985), Krzanowski (1988), and Cox and Wermuth (1992, 1994). A second model approach directly formulates a joint model for both outcomes. In this context, one often starts from a bivariate continuous variable, one component of which is explicitly observed and the other one observed in a dichotomized, or generally discretized, version only. Such models presuppose the existence of an unobservable variable underlying the binary outcome. The binary event is then assumed to occur if the latent variable exceeds some threshold value. Catalano and Ryan (1992) noted that latent-variable models provide a useful and intuitive way to motivate the distribution of the discrete outcome. A common method assumes an unobservable normally distributed random variable underlying the binary outcome, resulting in a probit-type model. Alternatively, Molenberghs, Geys, and Buyse (2001) presented a model based on a Plackett–Dale approach, where a bivariate Plackett distribution is assumed.

In a longitudinal study, these bivariate endpoints are often collected repeatedly over time, introducing another hierarchy as well. Another specific hierarchy stems from clustered data, where a continuous and a categorical endpoint are measured on each member in a cluster. For example, in the context of developmental toxicity studies, interest is in the dose-response relationship for both weight and malformation, and in addition, methods should account for the correlation induced by the clustering of fetuses within litters, or the well-known "litter effect." In this context, several models have been investigated. An overview of possible models for bivariate data in the setting of clustered data can be found in Aerts et al. (2002). Catalano and Ryan (1992) and Fitzmaurice and Laird (1995) proposed a factorization model for a combined continuous and discrete outcome, both allowing for clustering of fetuses within litters. Catalano and Ryan (1992) apply the latent-variable concept to derive the joint distribution of a continuous and a discrete outcome and then extend the model, using generalized estimating equations (GEE) ideas, to incorporate clustering. They parameterize the model in a way that allows the joint distribution to be written as the product of the marginal distribution of the continuous response, and the conditional distribution of the binary response given the continuous one. The marginal distribution of the continuous response is related to covariates using a linear link function, while for the conditional distribution they use a probit link. While this method can be easily implemented, the lack of marginal interpretation of the regression parameters in the probit model and the lack of robustness toward the misspecification of the association between the binary and continuous outcomes may be considered unattractive features of this approach. Fitzmaurice and Laird (1995) factorize the joint distribution as the product of a marginal Bernoulli distribution for the discrete response, and a conditional Gaussian distribution for the continuous response given the discrete one. Under independence, their method yields maximum likelihood estimates of the marginal means that are robust to misspecification of the association between the binary and continuous response. To allow for clustering, they use the GEE methodology, thus avoiding the computational complexity of maximum likelihood in this more elaborate setting. A conceptual difficulty with this model is the interpretation of the parameters, which depends on cluster size. Catalano (1997) extended the model by Catalano and Ryan (1992) to accommodate ordinal variables. A drawback of mixed-outcome models based on factorization (as above) is that it is difficult to express the joint probability for the discrete and continuous outcome directly, and that the intra-subject correlation itself cannot be directly estimated.

Regan and Catalano (1999a) proposed a probit-type model, based on the Ochi and Prentice (1984) method, to accommodate joint continuous and binary outcomes in a clustered data context. They assume an underlying continuous variable for each binary outcome, following a normal distribution. Geys et al. (2001) used a Plackett latent variable to model bivariate endpoints to the same effect. The bivariate latent-variable models are fundamentally different in the way the association between both variables is described. The probit approach uses a correlation coefficient, while the Plackett–Dale approach makes use of an odds ratio. However, a difficulty of these joint models is the computational intractability of the full likelihood in the hierarchical setting. Regan and Catalano (1999a) proposed maximum likelihood, while Regan and Catalano (1999b) also considered GEE as an option. Geys et al. (2001) made use of pseudo-likelihood. Ordinal extensions were proposed in Regan and Catalano (2000) and in Faes et al. (2004). These methods make it easy to specify the marginal models for each outcome, to estimate the bivariate intra-subject correlation between the continuous and discrete endpoints, and to account for the association due to the hierarchy in the data.

In this chapter, we will focus on some of these models. The methods will be described for the specific setting of a continuous and a binary endpoint (Section 14.2). A probit-normal model will be described in Section 14.2.1, a bivariate generalized linear mixed model of a joint nature will be discussed in Section 14.2.2, and a Plackett–Dale model will be developed in Section 14.2.3. Extensions of these models to accommodate for longitudinal endpoints will be discussed in Section 14.3. Using data from a repeated toxicology experiment, the methods will be illustrated, and possible extensions of the models will be presented. In Section 14.4.1 we focus on the toxicity at 8 and 24 hours after the first dosage. In Section 14.4.2 we consider the two available repeated sequences of 15 components each, one discrete and one continuous.

14.2 A continuous and a binary endpoint

In this section, we present three methods for the analysis of joint continuous and discrete outcomes. The first method applies a conditioning argument that allows the joint distribution to be factorized into a marginal component and a conditional component. The two other methods use a more symmetric treatment of both outcome variables. In Section 14.3, extensions of the models to the fully hierarchical case are discussed.

Let Y_{1i} and Y_{2i} denote two outcomes measured on subject i $(i = 1, \ldots, N)$, and $\boldsymbol{Y}_i = (Y_{1i}, Y_{2i})'$. We start with a joint continuous and binary endpoint, where Y_{1i} denotes the continuous outcome and Y_{2i} the binary outcome.

14.2.1 A probit-normal model

Let $\widetilde{Y}_{2i}$ be a latent variable of which Y_{2i} is the dichotomized version, such that

$$\Pr(Y_{2i} = 1) = \Pr(\widetilde{Y}_{2i} > 0).$$

For the bivariate response, we assume that the observed continuous outcome Y_{1i} and the unobserved latent variable $\widetilde{Y}_{2i}$ for individual i together constitute a bivariate normal model of the form

$$Y_{1i} = \alpha_0 + \alpha_1 X_i + \varepsilon_{1i}, \tag{14.1}$$

$$\widetilde{Y}_{2i} = \beta_0 + \beta_1 X_i + \varepsilon_{2i}, \tag{14.2}$$

where α_0 and β_0 are fixed intercepts, α_1 and β_1 are the fixed effects of the covariate X_i on the continuous and binary endpoints, respectively, and where the variance of the residual error is assumed to be

$$\text{Var}\begin{pmatrix} \varepsilon_{1i} \\ \varepsilon_{2i} \end{pmatrix} = \begin{pmatrix} \sigma^2 & \dfrac{\rho\sigma}{\sqrt{1-\rho^2}} \\ \dfrac{\rho\sigma}{\sqrt{1-\rho^2}} & \dfrac{1}{1-\rho^2} \end{pmatrix}.$$

The special variance-structure parameterization of $\widetilde{Y}_{2i}$ is chosen for reasons that will be made clear in what follows. From this model, it is easily seen that the density of the continuous outcome Y_{1i} is univariate normal with mean $\alpha_0 + \alpha_1 X_i$ and variance σ^2, implying that the parameters α_0, α_1, and σ^2 can be determined using linear regression software with response Y_{1i} and single covariate X_i. Similarly, the conditional density of the latent variable $\widetilde{Y}_{2i}$, given the covariate X_i and the continuous endpoint Y_{1i}, is

$$\widetilde{Y}_{2i}|Y_{1i}, X_i \sim N\left\{\left(\beta_0 - \frac{\rho}{\sigma\sqrt{1-\rho^2}}\alpha_0\right) + \left(\beta_1 - \frac{\rho}{\sigma\sqrt{1-\rho^2}}\alpha_1\right)X_i + \frac{\rho}{\sigma\sqrt{1-\rho^2}}Y_{1i}, 1\right\},$$

having unit variance and thus motivating our earlier choice for the covariance matrix of Y_{1i} and $\widetilde{Y}_{2i}$. The corresponding probability equals

$$\Pr\left(Y_{2i} = 1|Y_{1i}, X_i\right) = \Phi_1\left(\lambda_0 + \lambda_X X_i + \lambda_Y Y_{1i}\right), \tag{14.3}$$

where

$$\lambda_0 = \beta_0 - \frac{\rho}{\sigma\sqrt{1-\rho^2}}\alpha_0, \tag{14.4}$$

$$\lambda_X = \beta_1 - \frac{\rho}{\sigma\sqrt{1-\rho^2}}\alpha_1, \tag{14.5}$$

$$\lambda_Y = \frac{\rho}{\sigma\sqrt{1-\rho^2}}, \tag{14.6}$$

and Φ_1 is the standard normal cumulative distribution function. The λ parameters can be found by fitting model (14.3) to Y_{2i} with covariates X_i and Y_{1i}. This can be done with standard logistic regression software if it allows the probit rather than the logit link to be specified (e.g., the `logistic` and `genmod` procedures in SAS). Given the parameters from the linear regression on Y_{1i} (α_0, α_1, and σ^2) and the probit regression on Y_{2i} (λ_0, λ_X, and λ_Y), the parameters from the linear regression on $\widetilde{Y}_{2i}$ can now be obtained from (14.4) through (14.6):

$$\beta_0 = \lambda_0 + \lambda_Y\alpha_0, \tag{14.7}$$
$$\beta_1 = \lambda_X + \lambda_Y\alpha_1, \tag{14.8}$$
$$\rho^2 = \frac{\lambda_Y^2\sigma^2}{1 + \lambda_Y^2\sigma^2}. \tag{14.9}$$

The asymptotic covariance matrix of the parameters (α_0, α_1) can be found from standard linear regression output. The estimated variance of σ^2 equals $2\widehat{\sigma}^4/N$. The asymptotic covariance of $(\lambda_0, \lambda_X, \lambda_Y)$ follows from logistic (probit) regression output. These three statements yield the covariance matrix of the six parameters upon noting that it is block-diagonal. In order to derive the asymptotic covariance of (β_0, β_1, ρ), it suffices to calculate the derivatives of (14.7) through (14.9) with respect to the six original parameters and apply the delta method. They are:

$$\frac{\partial(\beta_0, \beta_1, \rho)}{\partial(\alpha_0, \alpha_1, \sigma^2, \lambda_0, \lambda_X, \lambda_Y)} = \begin{pmatrix} \lambda_Y & 0 & 0 & 1 & 0 & \alpha_0 \\ 0 & \lambda_Y & 0 & 0 & 1 & \alpha_1 \\ 0 & 0 & h_1 & 0 & 0 & h_2 \end{pmatrix}, \tag{14.10}$$

where

$$h_1 = \frac{1}{2\rho}\frac{\lambda_Y^2}{(1 + \lambda_Y^2\sigma^2)^2},$$

$$h_2 = \frac{1}{2\rho}\frac{2\lambda_Y\sigma^2}{(1 + \lambda_Y^2\sigma^2)^2}.$$

Molenberghs, Geys, and Buyse (2001) developed a program in GAUSS that performs the joint estimation directly by maximizing the likelihood based on contributions (14.1) and (14.3). Also the SAS procedure `glimmix` allows (14.1) and (14.3) to be jointly estimated using maximum (quasi-)likelihood.

14.2.2 A generalized linear mixed model

Mixed models are probably the models most frequently used to analyze multivariate data. It is also straightforward to use a mixed model in situations where various outcomes of a different nature are observed (Molenberghs and Verbeke, 2005). For the bivariate response vector $\boldsymbol{Y}_i = (Y_{1i}, Y_{2i})'$, we can assume a general model of the form

$$\boldsymbol{Y}_i = \boldsymbol{\mu}_i + \boldsymbol{\varepsilon}_i, \tag{14.11}$$

where $\boldsymbol{\mu}_i$ is specified in terms of fixed and random effects and $\boldsymbol{\varepsilon}_i$ is the residual error term. Let

$$\boldsymbol{\mu}_i = \boldsymbol{\mu}_i(\boldsymbol{\eta}_i) = \boldsymbol{h}(X_i\boldsymbol{\beta} + Z_i\boldsymbol{b}_i),$$

in which the components of the inverse link function $\boldsymbol{h}(\cdot)$ are allowed to change with the nature of the various outcomes in $\boldsymbol{Y}_i$. For example, we can choose the identity link for the continuous component, and the logit link for the binary component. As usual, X_i and Z_i are $(2 \times p)$ and $(2 \times q)$ matrices of known covariate values, and $\boldsymbol{\beta}$ a p-vector of unknown fixed regression coefficients. Further, $\boldsymbol{b}_i \sim N(\boldsymbol{0}, G)$ are the q-dimensional random effects.

A general first-order approximate expression for the variance–covariance matrix of $\boldsymbol{Y}_i$ (Molenberghs and Verbeke, 2005) is

$$\text{Var}(\boldsymbol{Y}_i) \simeq \Delta_i Z_i G Z_i' \Delta_i' + V_i, \tag{14.12}$$

with

$$\Delta_i = \left(\frac{\partial \boldsymbol{\mu}_i}{\partial \boldsymbol{\eta}_i}\right)\Big|_{\boldsymbol{b}_i=0}$$

and $V_i \simeq \Xi_i^{1/2} A_i^{1/2} R_i(\alpha) A_i^{1/2} \Xi_i^{1/2}$, where A_i is a diagonal matrix containing the variances following from the generalized linear model specification of Y_{ij} given the random effects $\boldsymbol{b}_i = \boldsymbol{0}$, that is, with diagonal elements $v(\mu_{ij}|\boldsymbol{b}_i = \boldsymbol{0})$. Likewise, Ξ_i is a diagonal matrix with the overdispersion parameters along the diagonal. Finally, R_i is a correlation matrix.

As a result, the correlation among the two outcomes can be modeled either using the residual variance of $\boldsymbol{Y}_i$ or through a shared parameter $\boldsymbol{b}_i$. When there are no random effects in (14.11), a marginal model is obtained, which is referred to as a marginal generalized linear model (MGLM) approach. When there are no residual correlations in $R_i(\alpha)$, a conditional independence model or purely random-effects model results, which is a generalized linear mixed model (GLMM). Each of these models will be discussed in more detail for the setting of a continuous and binary endpoint.

For the special case of a continuous and binary endpoint, the model with no random effects (MGLM) is of the form

$$\begin{pmatrix} Y_{1i} \\ Y_{2i} \end{pmatrix} = \begin{pmatrix} \alpha_0 + \alpha_1 X_i \\ \frac{\exp(\beta_0 + \beta_1 X_i)}{1 + \exp(\beta_0 + \beta_1 X_i)} \end{pmatrix} + \begin{pmatrix} \varepsilon_{1i} \\ \varepsilon_{2i} \end{pmatrix},$$

in which the first component of the inverse link function $\boldsymbol{h}$ equals the identity link, and the second component corresponds with the logit link. The correlation ρ_{12} among the two endpoints is specified via the variance of $\boldsymbol{\varepsilon}_i$. The residual error variance is assumed to be

$$\text{Var}\begin{pmatrix} \varepsilon_{1i} \\ \varepsilon_{2i} \end{pmatrix} = \begin{pmatrix} \sigma^2 & \rho\sigma\sqrt{v_{i2}} \\ \rho\sigma\sqrt{v_{i2}} & v_{i2} \end{pmatrix},$$

where $v_{i2} = \pi_{i2}(\boldsymbol{b}_i = \boldsymbol{0})\{1 - \pi_{i2}(\boldsymbol{b}_i = \boldsymbol{0})\}$ and ρ denotes the correlation between ε_{1i} and ε_{2i}. Since the variance in the case of no random effects reduces to $\text{Var}(\boldsymbol{Y}_i) \simeq V_i$, ρ is exactly the correlation between the two endpoints Y_{1i} and Y_{2i}.

Another approach is to use a shared random parameter model. More specifically, for a mixed continuous/binary setting, the general linear mixed model (14.11) can be written in the form

$$\begin{pmatrix} Y_{1i} \\ Y_{2i} \end{pmatrix} = \begin{pmatrix} \alpha_0 + \alpha_1 X_i + \lambda b_i \\ \dfrac{\exp(\beta_0 + \beta_1 X_i + b_i)}{1 + \exp(\beta_0 + \beta_1 X_i + b_i)} \end{pmatrix} + \begin{pmatrix} \varepsilon_{1i} \\ \varepsilon_{2i} \end{pmatrix},$$

where b_i denotes the shared random effect of the two outcomes. Note that a scale parameter λ in the continuous component of the model is included, given the continuous and binary outcome are measured on different scales.

For this model, the variance of $\mathbf{Y}_i$ can be derived from Equation (14.12), in which

$$Z_i = \begin{pmatrix} \lambda \\ 1 \end{pmatrix}, \qquad \Delta_i = A_i = \begin{pmatrix} 1 & 0 \\ 0 & v_{i2} \end{pmatrix}, \qquad \Xi_i = \begin{pmatrix} \sigma^2 & 0 \\ 0 & 1 \end{pmatrix}.$$

This implies that the variance of $\mathbf{Y}_i$ is approximately equal to

$$\mathrm{Var}(\mathbf{Y}_i) = \begin{pmatrix} \lambda^2 & v_{i2}\lambda \\ v_{i2}\lambda & v_{i2}^2 \end{pmatrix} \tau^2 + \begin{pmatrix} \sigma^2 & \rho\sigma\sqrt{v_{i2}} \\ \rho\sigma\sqrt{v_{i2}} & v_{i2} \end{pmatrix}$$

$$= \begin{pmatrix} \lambda^2\tau^2 + \sigma^2 & v_{i2}\lambda\tau^2 + \rho\sigma\sqrt{v_{i2}} \\ v_{i2}\lambda\tau^2 + \rho\sigma\sqrt{v_{i2}} & v_{i2}^2\tau^2 + v_{i2} \end{pmatrix}.$$

As a result, the approximate correlation among the two outcomes is

$$\rho_{12} = \frac{v_{i2}\lambda\tau^2 + \rho\sigma\sqrt{v_{i2}}}{\sqrt{\lambda^2\tau^2 + \sigma^2}\sqrt{v_{i2}^2\tau^2 + v_{i2}}}. \tag{14.13}$$

This correlation depends on the fixed effects through v_{i2}. Some special cases are of interest. Under conditional independence, $\rho \equiv 0$ and (14.13) reduces to

$$\rho_{12} = \frac{v_{i2}\lambda\tau^2}{\sqrt{\lambda^2\tau^2 + \sigma^2}\sqrt{v_{i2}^2\tau^2 + v_{i2}}}, \tag{14.14}$$

which still is a function of the fixed effects. If there are no random effects, (14.13) reduces to a constant correlation, which is exactly what was obtained for the MGLM.

These calculations cannot only be easily performed for continuous and binary endpoints, but for any endpoints of arbitrary nature. Also the generalization to more than two endpoints and to general random-effects design matrices is straightforward.

14.2.3 A Plackett–Dale model

Let us denote the cumulative distribution of the continuous endpoint Y_{1i} and the binary endpoint Y_{2i} as $F_{Y_{1i}}$ and $F_{Y_{2i}}$, respectively. The dependence between the continuous and discrete outcome can be defined using a global cross-ratio,

$$\psi_i = \frac{F_{Y_{1i},Y_{2i}}\left(1 - F_{Y_{1i}} - F_{Y_{2i}} + F_{Y_{1i},Y_{2i}}\right)}{\left(F_{Y_{1i}} - F_{Y_{1i},Y_{2i}}\right)\left(F_{Y_{2i}} - F_{Y_{1i},Y_{2i}}\right)}.$$

The global cross-ratio ψ_i should be understood in terms of dichotomizing the continuous outcome Y_{1i} using a cutpoint y_{1i}. If $\psi_i = 1$ this implies no association, whereas a positive ψ_i indicates that larger values of Y_{1i} are associated with a higher probability $\mathrm{Pr}(Y_{2i} = 1)$. Using this relationship, the joint cumulative distribution $F_{Y_{1i},Y_{2i}}$ can be written as a function of the marginal distributions and the global cross-ratio (Plackett, 1965):

$$F_{Y_{1i},Y_{2i}} = \begin{cases} \dfrac{1 + \left(F_{Y_{1i}} + F_{Y_{2i}}\right)(\psi_i - 1) - S\left(F_{Y_{1i}}, F_{Y_{2i}}, \psi_i\right)}{2(\psi_i - 1)} & \psi_i \neq 1, \\ F_{Y_{1i}} F_{Y_{2i}} & \psi_i = 1, \end{cases}$$

with

$$S(F_{Y_{1i}}, F_{Y_{2i}}, \psi_i) = \sqrt{\left\{1 + (\psi_i - 1)(F_{Y_{1i}} + F_{Y_{2i}})\right\}^2 + 4\psi_i(1 - \psi_i)F_{Y_{1i}}F_{Y_{2i}}}.$$

Based on the cumulative distribution function $F_{Y_{1i}, Y_{2i}}(y_1, y_2)$, we can now derive a bivariate Plackett density function $g_i(y_1, y_2)$ for joint continuous–discrete outcomes. We can then define $g_i(y_1, y_2)$ by specifying $g_i(y_1, 0)$ and $g_i(y_1, 1)$ such that they sum to $f_{Y_{1i}}(y)$. Defining

$$g_i(y_1, 0) = \frac{\partial F_{Y_{1i}, Y_{2i}}(y_1, 0)}{\partial y_1}$$

leads to specifying the density function $g_i(y_1, y_2)$ by

$$g_i(y_1, 0) = \begin{cases} \frac{f_{Y_{1i}}(y_1)}{2} \left[1 - \frac{1 + F_{Y_{1i}}(y_1)(\psi_i - 1) - F_{Y_{2i}}(y_2)(\psi_i + 1)}{S\{F_{Y_{1i}}(y_1), F_{Y_{2i}}(y_2), \psi_i\}} \right] & \text{if } \psi \neq 1, \\[2ex] f_{Y_{1i}}(y_1)(1 - \pi_i) & \text{if } \psi = 1, \end{cases}$$

and

$$g_i(y_1, 1) = f_{Y_{1i}}(y_1) - g_i(y_1, 0).$$

One can show that the function $g_i(y_1, y_2)$ satisfies the classical density properties:

1. $g_i(y_1, y_2) \geq 0$ for all possible values of y_1 and y_2;
2. $\int \sum_{y_1} g_i(y_1, y_2) dy_2 = \int f_{Y_{1i}}(y_1) dy_1 = 1$.

Further, note that $g_i(y_1, y_2)$ factorizes as a product of the marginal density $f_{Y_{1i}}(y_1)$ and the conditional density $f_{Y_{2i}|Y_{1i}}(y_2|y_1)$. If the two endpoints are independent, the function $g_i(y_1, y_2)$ reduces to $f_{Y_{1i}}(y_1) f_{Y_{2i}}(y_2)$.

If we assume the continuous endpoint Y_{1i} to be normally distributed with mean $\mu_i = \alpha_0 + \alpha_1 X_i$ and variance σ_i^2, and denote the success probability of the binary outcome Y_{2i} as π_i, with $\text{logit}(\pi_i) = \beta_0 + \beta_1 X_i$, then the regression parameters $\boldsymbol{\nu} = (\alpha_0, \alpha_1, \sigma^2, \beta_0, \beta_1, \psi_i)$ are easily obtained by solving the estimating equations $\boldsymbol{U}(\boldsymbol{\nu}) = \boldsymbol{0}$, using a Newton–Raphson iteration scheme, where $\boldsymbol{U}(\boldsymbol{\nu})$ is given by

$$\sum_{i=1}^{n} \left(\frac{\partial \boldsymbol{\eta}_i}{\partial \boldsymbol{\nu}} \right)' \left\{ \left(\frac{\partial \boldsymbol{\eta}_i}{\partial \boldsymbol{\theta}} \right)' \right\}^{-1} \left(\frac{\partial}{\partial \boldsymbol{\theta}_i} \ln g_i(y_{1i}, y_{2i}) \right)',$$

and

$$\boldsymbol{\theta} = \begin{pmatrix} \mu_i \\ \sigma_i^2 \\ \pi_i \\ \psi_i \end{pmatrix} \quad \text{and} \quad \boldsymbol{\eta}_i = \begin{pmatrix} \mu_i \\ \ln(\sigma_i^2) \\ \text{logit}(\pi_i) \\ \ln(\psi_i) \end{pmatrix}.$$

This method provides an alternative to the frequently used multivariate normal latent variables. The main advantages are the flexibility with which the marginal densities can be chosen (normal, logistic, complementary log-log, etc.) and the familiarity of the odds ratio which is used as a measure of association, providing an alternative to correlation. Another advantage of this approach is that it allows us to directly model the bivariate association ψ, as well as the variance of the continuous endpoint σ^2, as functions of covariates of interest. Also extensions toward endpoints of a different nature are possible. For example, a Plackett–Dale model for a continuous and ordinal outcome is presented in Faes et al. (2004).

14.3 Longitudinal mixed endpoints

In the previous section we discussed three approaches toward a joint model for a continuous and a discrete outcome. Such joint outcomes can be measured repeatedly over time, or might be observed within a hierarchical context. Assume we have two sequences of n outcomes each. We denote the sequence of continuous endpoints for subject i as $\boldsymbol{Y}_{1i} = (Y_{1i1}, Y_{1i2}, \ldots, Y_{1in})$ and the one with binary endpoints as $\boldsymbol{Y}_{2i} = (Y_{2i1}, Y_{2i2}, \ldots, Y_{2in})$. Y_{1ij} and Y_{2ij} represent, respectively, the jth continuous and binary outcome for subject i. In this section we present generalizations of the previously proposed methods for joint continuous and discrete outcomes, to appropriately account for the hierarchical structure in the data.

14.3.1 A probit-normal model

One option toward a generalization of the probit-normal model to the longitudinal setting is to consider a two-step analysis, as proposed in Molenberghs, Geys, and Buyse (2001). At the first step, we can assume the model

$$\begin{pmatrix} Y_{1ij} \\ \widetilde{Y}_{2ij} \end{pmatrix} = \begin{pmatrix} \alpha_{0i} + \alpha_{1i}X_{ij} \\ \beta_{0i} + \beta_{1i}X_{ij} \end{pmatrix} + \begin{pmatrix} \varepsilon_{1ij} \\ \varepsilon_{2ij} \end{pmatrix},$$

where α_{1i} and β_{1i} are subject-specific effects of the covariate X on the endpoints for subject i, α_{0i} and β_{0i} are subject-specific intercepts, and ε_{1ij} and ε_{2ij} are correlated error terms, assumed to be mean-zero normally distributed correlated error terms for both measurements at time point j of subject i, having covariance matrix

$$V = \begin{pmatrix} \sigma^2 & \frac{\rho\sigma}{\sqrt{1-\rho^2}} \\ \frac{\rho\sigma}{\sqrt{1-\rho^2}} & \frac{1}{1-\rho^2} \end{pmatrix}.$$

In short, we use the probit formulation described in Section 14.2.1. At the second stage, we can impose a distribution on the subject-specific regression parameters and assume

$$\begin{pmatrix} \alpha_{0i} \\ \beta_{1i} \\ \alpha_{1i} \\ \beta_{2i} \end{pmatrix} = \begin{pmatrix} \alpha_0 \\ \beta_0 \\ \alpha_1 \\ \beta_1 \end{pmatrix} + \begin{pmatrix} a_{0i} \\ b_{0i} \\ a_{1i} \\ b_{1i} \end{pmatrix}, \tag{14.15}$$

where the second term on the right-hand side of (14.15) is assumed to follow a normal distribution with mean zero and dispersion matrix G. In this model, the correlation among the two endpoints per time point is described by the parameter ρ, while the correlation of the same response at different time points is described by the dispersion matrix G. Alternatively, this model can be fit as a hierarchical random-effects model.

As before, we can easily derive the marginal distribution of the observed continuous outcome Y_{2i} from this expression, as well as the conditional probability of the binary outcome Y_{1i} given the outcome Y_{2i}. The continuous outcome is normally distributed as

$$Y_{1ij} \sim N\left\{(\alpha_0 + a_{0i}) + (\alpha_1 + a_{1i})X_{ij}, \sigma^2\right\}. \tag{14.16}$$

Further, the probability of the binary outcome, given the continuous one, is described using a probit model

$$\Pr(Y_{2ij} = 1 | Y_{1ij}, X_i) = \Phi_1\left\{(\lambda_0 + u_{0i}) + (\lambda_X + u_{Xi})X_{ij} + \lambda_Y Y_{1ij}\right\}, \tag{14.17}$$

where

$$\lambda_0 = \beta_0 - \frac{\rho}{\sigma\sqrt{1-\rho^2}}\alpha_0, \, u_{0i} = b_{0i} - \frac{\rho}{\sigma\sqrt{1-\rho^2}}a_{0i},$$

$$\lambda_X = \beta_1 - \frac{\rho}{\sigma\sqrt{1-\rho^2}}\alpha_1, \, u_{Xi} = b_{1i} - \frac{\rho}{\sigma\sqrt{1-\rho^2}}a_{1i}, \tag{14.18}$$

$$\lambda_Y = \frac{\rho}{\sigma\sqrt{1-\rho^2}}.$$

These parameters can be found by maximizing the likelihoods from expressions (14.16) and (14.17). This can be done using standard mixed-model software for continuous and discrete outcomes (e.g., the mixed and genmod procedures in SAS). The SAS procedure glimmix can be used to jointly maximize the bivariate likelihood. Conversely, the parameters from the linear regression on the latent variable $\tilde{Y}_{2i}$ can be obtained from the linear regression parameters on Y_{1i} and the probit regression on Y_{2i}, using the formulae

$$\beta_0 = \lambda_0 + \lambda_Y \alpha_0, \qquad b_{0i} = u_{0i} + \lambda_Y a_{0i},$$
$$\beta_1 = \lambda_X + \lambda_Y \alpha_1, \qquad b_{1i} = u_{Xi} + \lambda_Y a_{1i},$$
$$\rho^2 = \frac{\lambda_Y^2 \sigma^2}{1 + \lambda_Y^2 \sigma^2}.$$

Using normal distribution theory, the variance of the random effects can be derived easily from these expressions. In a similar way as in Section 14.2.1, the asymptotic covariance of $(\hat{\beta}_0, \hat{\beta}_1, \hat{\rho})$ can be calculated using the delta method.

Instead of using a random-effects approach, we can include the association coming from the longitudinal structure of the model into the residual error matrix. Regan and Catalano (1999a) generalized the probit-normal model by considering the vector of multiple continuous and binary outcomes $(Y_{1i1}, Y_{1i2}, \ldots, Y_{1in}, Y_{2i1}, Y_{2i2}, \ldots, Y_{2in})$, and assuming an underlying latent variable for each of the binary endpoints. They then assumed that the resulting stochastic vector, composed of directly observed and latent outcomes, is normally distributed. This approach is natural and appealing. A difficulty, however, is the computational intractability of the full likelihood. Regan and Catalano (1999b) show how this problem can be avoided by adopting GEE methodology. They bypass fully specifying the distribution within a subject i by specifying only the marginal distribution of the bivariate outcome and applying GEE ideas to take the correlation into account.

14.3.2 A generalized linear mixed model

Model (14.11) extends straightforwardly toward inclusion of repeated measures, meta-analysis, clustered data, correlated data, etc. Both fixed- and random-effects structures can be formulated sufficiently generally so as to cover all these settings. We formulate three possible models to account for the longitudinal structure of joint continuous and binary endpoints: a fully marginal method, a conditional independence random-intercepts model, and a random-intercepts model with a correlated residual error structure. Other generalizations are possible as well.

A first method is to incorporate both the association between the two endpoints at each time point as well as the association coming from the longitudinal structure of the data in the residual error structure, by the specification of a general (e.g., unstructured) correlation matrix $R_i(\alpha)$, allowing each pair of outcomes to have its own correlation coefficient. Correlations follow in a straightforward fashion when such a purely marginal GLMM is used. While such a marginal model, with fully unstructured $2n \times 2n$ variance–covariance matrix, is very appealing because of its ease of interpretation, it can become computationally very demanding, especially when the number of measurements per subject (n) gets larger.

A second approach is to use a conditional independence random-intercepts model, with a general variance–covariance matrix G. This model can be written in the following form:

$$\begin{pmatrix} Y_{1ij} \\ Y_{2ij} \end{pmatrix} = \begin{pmatrix} \alpha_0 + \alpha_1 X_{ij} + b_{1i} \\ \dfrac{\exp(\beta_0 + \beta_1 X_{ij} + b_{2i})}{1 + \exp(\beta_0 + \beta_1 X_{ij} + b_{2i})} \end{pmatrix} + \begin{pmatrix} \varepsilon_{1ij} \\ \varepsilon_{2ij} \end{pmatrix}, \qquad (14.19)$$

where the random effects b_{1i} and b_{2i} are normally distributed as

$$\begin{pmatrix} b_{1i} \\ b_{2i} \end{pmatrix} \sim N\left(\begin{pmatrix} 0 \\ 0 \end{pmatrix}, \begin{pmatrix} \tau_1^2 & \rho \tau_1 \tau_2 \\ \rho \tau_1 \tau_2 & \tau_2^2 \end{pmatrix} \right)$$

and where ε_{1ij} and ε_{2ij} are independent. The correlation among the continuous and binary endpoints is induced by the incorporation of a correlation ρ among the two random effects. The variance of Y_{1ij} and Y_{2ij} can be derived from (14.12) in which

$$Z_i = \begin{pmatrix} 1 & 0 \\ 0 & 1 \end{pmatrix}, \qquad \Delta_i = A_i = \begin{pmatrix} 1 & 0 \\ 0 & v_{i2} \end{pmatrix},$$

$$G = \begin{pmatrix} \tau_1^2 & \rho\tau_1\tau_2 \\ \rho\tau_1\tau_2 & \tau_2^2 \end{pmatrix}, \qquad \Xi_i = \begin{pmatrix} \sigma^2 & 0 \\ 0 & 1 \end{pmatrix},$$

and where $v_{2ij} = \pi_{2ij}(\boldsymbol{b}_i = \boldsymbol{0})\{1 - \pi_{2ij}(\boldsymbol{b}_i = \boldsymbol{0})\}$. As a result, the approximate variance–covariance matrix of the two measurements for subject i at time point j is equal to:

$$\text{Var}(\boldsymbol{Y}_{ij}) = \begin{pmatrix} \tau_1^2 & \rho\tau_1\tau_2 v_{2ij} \\ \rho\tau_1\tau_2 v_{2ij} & v_{2ij}^2\tau_2^2 \end{pmatrix} + \begin{pmatrix} \sigma^2 & 0 \\ 0 & v_{2ij} \end{pmatrix}$$

$$= \begin{pmatrix} \tau_1^2 + \sigma^2 & \rho\tau_1\tau_2 v_{2ij} \\ \rho\tau_1\tau_2 v_{2ij} & v_{2ij}^2\tau_2^2 + v_{2ij} \end{pmatrix}.$$

Thus, the correlation ρ_{12} among the binary and continuous outcome follows from the fixed effects and variance components, and is approximately equal to

$$\rho_{12} = \frac{\rho\tau_1\tau_2 v_{2ij}}{\sqrt{\tau_1^2 + \sigma^2}\sqrt{v_{2ij}^2\tau_2^2 + v_{2ij}}}.$$

In the case of conditional independence ($\rho \equiv 0$), the approximate marginal correlation function ρ_{12} also equals zero. In the case of $\rho \equiv 1$, this model reduces to the shared-parameter model as presented in Section 14.2.2. The correlation between the two endpoints Y_{1ij} and Y_{2ij} is then given by (14.14) in which the scale factor λ is equal to τ_1/τ_2. This model is easily programmed in SAS proc nlmixed.

A third approach is to use a random-intercepts model for each endpoint with a correlated residual error structure to account for the association among the continuous and binary endpoints. In this model, a random effect is introduced at the level of the linear predictor after application of the link function, while the residual correlation is introduced at the level of ε_i. The model is also of the form (14.19), but where b_{1i} and b_{2i} are independent zero-mean normally distributed random effects with variance τ_1^2 and τ_2^2, respectively. The residual error is assumed to have variance–covariance matrix

$$V_{ij} = \begin{pmatrix} \sigma^2 & \rho\sigma\sqrt{v_{2ij}} \\ \rho\sigma\sqrt{v_{2ij}} & v_{2ij} \end{pmatrix}.$$

Here, ρ denotes the correlation between ε_1 and ε_2. The variance among the continuous and binary endpoints at time point j, for subject i, can be derived from the variance–covariance matrix (14.12), and is approximately equal to

$$\text{Var}(\boldsymbol{Y}_{ij}) = \begin{pmatrix} \tau_1^2 & 0 \\ 0 & \tau_2^2 v_{2ij}^2 \end{pmatrix} + \begin{pmatrix} \sigma^2 & \rho\sigma\sqrt{v_{2ij}} \\ \rho\sigma\sqrt{v_{i2}} & v_{2ij} \end{pmatrix}$$

$$= \begin{pmatrix} \tau_1^2 + \sigma^2 & \rho\sigma\sqrt{v_{2ij}} \\ \rho\sigma\sqrt{v_{2ij}} & v_{2ij}^2\tau_2^2 + v_{2ij} \end{pmatrix}.$$

As a result, the approximate correlation among the two endpoints is

$$\rho_{12} = \frac{\rho\sigma\sqrt{v_{2ij}}}{\sqrt{\tau_1^2 + \sigma^2}\sqrt{v_{2ij}^2\tau_2^2 + v_{2ij}}}.$$

In the special case of no random effects, this correlation reduces to ρ. Further, ρ_{12} equals zero in the case of conditional independence ($\rho \equiv 0$). The SAS procedure `glimmix` can be used for parameter estimation, using pseudo-likelihood methods in which pseudo-data are created based on a linearization of the mean (Wolfinger and O'Connell, 1993).

Next to the three models presented, other generalizations with shared random effects or other residual error structures are possible as well. The correlation structures for each setting can be obtained from (14.12) or specific forms derived thereof. Although we have only presented the correlation among the continuous and binary endpoints at a specific time point, the correlation structure among two continuous or among two binary endpoints measured at two different time points can be derived in a similar way.

14.3.3 A Plackett–Dale approach

In Section 14.2.3 we described the Plackett–Dale approach used by Molenberghs, Geys, and Buyse (2001) and Molenberghs and Geys (2001) to model independent bivariate endpoints in which one component is continuous and the other is binary.

In a longitudinal setting, observations at different time points are typically not independent. In this case, we avoid the computational complexity of the $2n$-dimensional full likelihood distribution of each individual i, that is, $f(y_{1i1}, \ldots, y_{1in}, y_{2i1}, \ldots, y_{2in})$, but use a pseudo-likelihood function as proposed by Arnold and Strauss (1991). The simplest method is by solely specifying the bivariate outcomes, just as before, and assembling them into a (log) pseudo-likelihood function:

$$pl = \sum_{i=1}^{N} \sum_{j=1}^{n} \ln g_{ij}(y_{1ij}, y_{2ij}).$$

As such, the association between the two endpoints measured for each individual is modeled explicitly, but the correlation structure of endpoints at different time points is left unspecified. This approach acknowledges the fact that, while the association between different endpoints on the same individual is often of scientific interest, the association due to repeated measurements is often considered a nuisance.

However, if one is interested in the correlation structure between outcomes of different measurements on the same individual, other types of pseudo-likelihood can be specified as well. For example, one can extend the previous pseudo-likelihood function by including the products of the bivariate probabilities of (i) two continuous outcomes measured at two different time points for the same individual, (ii) two discrete outcomes measured for the same individual, but at two different time points, and (iii) a continuous and discrete outcome measured at two different time points. This leads to the following log-pseudo-likelihood function:

$$pl = \sum_{i=1}^{N} \frac{1}{3} \sum_{j=1}^{n_i} \ln g_1(y_{1ij}, y_{2ij}) + \sum_{i=1}^{N} \frac{1}{3(n_i - 1)} \sum_{k<j} \ln g_2(y_{1ik}, y_{1ij})$$

$$+ \sum_{i=1}^{N} \frac{1}{3(n_i - 1)} \sum_{k<j} \ln g_3(y_{2ik}, y_{2ij}) + \sum_{i=1}^{N} \frac{1}{3(n_i - 1)} \sum_{j \neq k}^{n_i} \ln g_4(y_{1ik}, y_{2ij}),$$

with, for example, g_1, g_3, g_4 bivariate Plackett distributions, characterized by potentially different odds ratios, and g_2 a bivariate normal distribution. The factors before each contribution in the log-pseudo-likelihood function correct for the fact that each response occurs several times in the ith contribution of the pseudo-likelihood, and ensures that the pseudo-likelihood reduces to full likelihood under independence. An example of this method can be found in Faes et al. (2004).

While the pseudo-likelihood estimation method achieves important computational economies by changing the method of estimation, it provides consistent and asymptotically normal estimates (Arnold and Strauss, 1991) and it does not affect parameter interpretation. Also, the pseudo-likelihood is robust against misspecification of the correlation structure. Typically, an independence working correlation is used, leaving the longitudinal part of the correlation structure unspecified. But, of course, alternative pseudo-likelihood functions can be used as well; for example, a pseudo-likelihood model equivalent to an AR(1) correlation structure. Thus, one can change the pseudo-likelihood depending on which parameters are needed to formulate answers to the scientific questions.

14.4 Irwin's toxicity study

The data considered here come from a 3-day repeated dose-toxicity study for the evaluation of the neurofunctional effects of a psychotrophic drug. Male rats were repeatedly dosed, and several behavioral observations were recorded. In the dosed group, 15 rats received on three consecutive days 40 mg/kg-day of a chemical substance. There were five rats in the vehicle group (0 mg/kg-day). On days 0, 1, and 2, at 2, 4, 6, 8, and 24 hours after daily oral administration of the chemical substance, all rats were examined for possible neurotoxic effects. The neurofunctional integrity of the rats was assessed by specific observations based on Irwin's (1964) method. This method provides a behavioral and functional profile by observational assessment of the rat. The pupil size was measured, and it was recorded whether or not the rat responded to a toe pinch. Toe pinch is a sensorimotor reflex response, while pupil size is an autonomic feature. Figure 14.1 presents the observed data: the first part of Figure 14.1 shows the probability that the rats did not respond to a toe pinch; the second part shows the pupil size of the rats. The top and bottom figures correspond to, respectively, the control and dosed group. The response to toe pinch is smaller in the dosed group in comparison with the vehicle group. After dosage, there is always a drop in the response to toe pinch. However, the response to toe pinch increased from 2 to 24 hours post-dosage. At day 0, the pupil was widened from 2 to 8 hours post-dosage. After 24 hours, this effect was still present, though to a lesser extent in most animals. At day 1, an abnormally widened pupil was recorded in nearly all animals from 2 to 8 hours post-dosage. At day 2, a widened pupil was recorded in few animals, but this effect was no longer present in any rat after 24 hours post-dosage.

14.4.1 Toxicity at a single point in time

First, we investigate the effects of the test substance on the central nervous system at 8 and 24 hours after first administration. For each rat, it is recorded whether or not it responded to a toe pinch, and the pupil size is measured. We consider a probit-normal model as in Section 14.2.1, a Plackett–Dale model as in Section 14.2.3, and a generalized linear mixed model, with and without random effects, as in Section 14.2.2. Parameter estimates (standard errors) are displayed in Table 14.1 and Table 14.2.

Parameter estimates and standard errors for the probit-normal model are calculated based on (14.7) through (14.9) and (14.10). `proc glimmix` can be used to perform the joint estimation directly, based on (14.1) and (14.3). This procedure has the advantage that it can be extended to capture the longitudinal setting as well. The treatment seems to result in a large increase of the pupil size at 8 hours after exposure, but 24 hours post-dosage this effect seems to have weakened. There is no significant effect on the response to toe pinch at either 8 or 24 hours post-dosage. Under the probit model, the correlation between both endpoints is estimated as $\widehat{\rho} = -0.05$ at time point 8 and $\widehat{\rho} = 0.29$ at time point 24. However, none of these correlations seem to be significant.

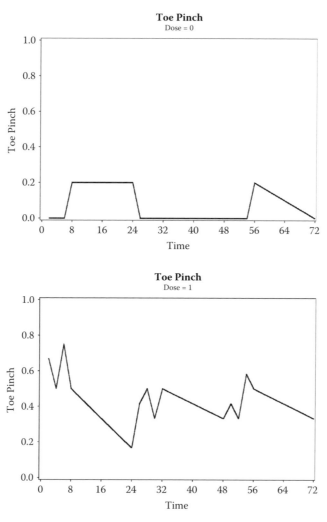

Figure 14.1 Binary variable toe pinch and continuous variable pupil size.

For the MGLM, we consider both a logit and probit link for the binary endpoint. The pseudo-likelihood estimation method as proposed by Wolfinger and O'Connell (1993) is used as an approximation method. `proc glimmix` is applied, allowing us to jointly model endpoints with different distributions and/or different link functions. Parameter estimates and standard errors for the main effect under the MGLM model with a probit link are very similar to the probit-normal model. However, the correlation estimated under the MGLM model is a bit smaller. This can be due to the use of the pseudo-likelihood estimation method, since typically there is some downward bias in the parameter estimates based on the pseudo-likelihood. That said, when comparing the estimates from a probit-normal model under maximum likelihood estimation (using `proc logistic` and `proc mixed`) and penalized quasi-likelihood estimation (using `proc glimmix`), both methods yield identical results. A more important reason for the difference is that the probit model features the correlation between the observed continuous endpoint and the latent variables, whereas the MGLM captures the correlation between the two observable outcomes.

For the Plackett–Dale model, a logit link is employed for the binary endpoint. For the binary endpoint, the parameter estimates differ somewhat, but differences are rather small.

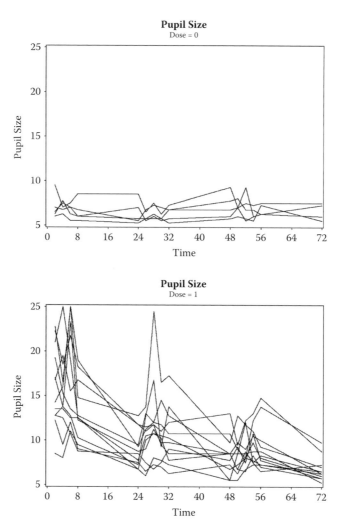

Figure 14.1 *Continued.*

The parameter estimates for the continuous endpoint agree more closely, except for the variance σ^2, which differs a bit more. The Plackett–Dale model uses the odds ratio rather than the correlation to describe the association between the continuous and binary endpoint. While at 8 hours post-dosage the odds ratio is close to 1, suggesting independence among the two outcomes, the odds ratio at 24 hours post-dosage is close to 4, indicating a positive association among the two endpoints. This is in line with previously discussed models.

Finally, we also fitted a shared random-effects model (GLMM), with either a logit or probit link for the binary endpoint. The fixed effect parameter estimates and standard errors for the continuous endpoint are almost the same as in the marginal models, while most fixed effects for the binary endpoint are larger. This follows from the fact that parameters in a marginal and in a random-effects model have completely different interpretation in the case of a non-normal distribution or in the case of a link function, which is different from the identity link. Using (14.14), the correlation among the two endpoints can be calculated. Note that this correlation differs for vehicle (ρ_0) and exposed group (ρ_1) because of its dependence on the fixed effects. However, in this case, the correlations in both treatment arms are roughly the same. The estimated correlation under the random-effects model is

Table 14.1 Toxicity at 8 Hours Post-Dosage

Effect	Par.	Probit-Normal	Plackett-Dale	MGLM Logit	MGLM Probit	GLMM Logit	GLMM Probit
				Binary Endpoint			
Intercept	μ_1	−0.84 (0.64)	−1.61 (1.10)	−1.39 (1.19)	−0.84 (0.68)	−1.39 (1.13)	−0.85 (0.78)
Treatm. Effect	α	0.84 (0.74)	1.42 (1.25)	1.29 (1.34)	0.84 (0.78)	1.39 (1.27)	0.85 (0.86)
Overdis. Par.	ϕ			1.13 (0.42)	1.13 (0.41)		
				Continuous Endpoint			
Intercept	μ_2	6.55 (1.37)	6.58 (1.09)	6.55 (1.19)	6.55 (1.37)	6.55 (1.29)	6.55 (1.29)
Treatm. Effect	β	6.47 (1.63)	6.83 (1.37)	6.47 (1.63)	6.47 (1.63)	6.47 (1.53)	6.47 (1.53)
Variance	σ^2	9.42 (3.44)	3.87 (1.35)	9.42 (3.44)	9.42 (3.44)	5.50 (93.96)	6.16 (229.67)
Inflation	λ					−12.67 (397.33)	−14.71 (1575.37)
				Association			
Odds Ratio	ψ		1.03 (1.25)				
Correlation	ρ	−0.05 (0.32)		−0.04 (0.26)	−0.04 (0.26)		
Correlation	ρ_0					−0.03 (0.20)	−0.02 (0.12)
Correlation	ρ_1					−0.04 (0.25)	−0.03 (0.15)
RI Var.	τ^2					0.02 (0.53)	0.01 (1.07)

Note: Parameter estimates (standard errors) of the bivariate analysis for pupil size (continuous endpoint) and toe pinch (binary endpoint).

Table 14.2 Toxicity at 24 Hours Post-Dosage

Effect	Par.	Probit-normal	Plackett-Dale	MGLM		GLMM	
				Logit	Probit	Logit	Probit
Binary Endpoint							
Intercept	μ_1	−0.90 (0.69)	−1.28 (1.10)	−1.39 (1.19)	−0.84 (0.68)	−1.43 (1.16)	−0.90 (0.95)
Treatm. Effect	α	−0.10 (0.81)	−0.36 (1.37)	−0.22 (1.45)	−0.13 (0.82)	−0.31 (1.43)	−0.10 (0.82)
Overdis. Par.	ϕ			1.13 (0.42)	1.13 (0.41)		
Continuous Endpoint							
Intercept	μ_2	6.40 (0.85)	6.44 (0.75)	6.40 (0.85)	6.40 (0.85)	6.40 (0.80)	6.40 (0.80)
Treatm. Effect	β	2.62 (1.01)	2.62 (0.91)	2.62 (1.01)	2.62 (1.01)	2.62 (0.95)	2.62 (0.95)
Variance	σ^2	3.61 (1.32)	1.30 (0.77)	3.61 (1.32)	3.61 (1.32)	<0.00 (<0.00)	0.20 (42.30)
Inflation	λ					3.17 (3.55)	5.82 (86.70)
Association							
Odds ratio	ψ		3.89 (4.33)				
Correlation	ρ	0.29 (0.32)		0.21 (0.25)	0.21 (0.25)		
Correlation	ρ_0					0.22 (0.24)	0.11 (0.14)
Correlation	ρ_1					0.20 (0.20)	0.10 (0.12)
RI Var.	τ^2					0.32 (0.72)	0.09 (1.38)

Note: Parameter estimates (standard errors) of the bivariate analysis for pupil size (continuous endpoint) and toe pinch (binary endpoint).

also very similar to the estimated correlation under the marginal GLM model. A disadvantage of this approach is the instability of the model due to the non-linearity induced by the inflation factor λ. This random-effects model was fit using `proc nlmixed`, by use of the general likelihood feature. The inflation factor often appears to diverge. Careful selection of initial values is important, as well as monitoring of the convergence process and fine-tuning of the optimization method.

14.4.2 Repeated dose-toxicity study

Let us now focus on the full hierarchical model. It is of interest to test whether there is a significant treatment effect on either toe pinch or pupil size. One could opt for a separate analysis at each time point. However, this is not an efficient method and might not capture all effects present in the data. We focus on a longitudinal analysis, accounting for possible association among time points.

We consider a random-intercepts probit-normal model as presented in Section 14.3.1, a pseudo-likelihood Plackett–Dale model as discussed in Section 14.3.3, and a correlated and uncorrelated random-intercepts model as given in Section 14.3.2. Both logit and probit links are used in the GLMM models.

The first column in Table 14.3 shows the results for the correlated probit-normal model. A hierarchical model approach is followed. In this model, only subject-specific Intercepts are considered. The parameters for the binary endpoint are derived from Equation (14.18). The random-intercepts variance for the latent variable (τ_2^2) is equal to $\tau_\lambda^2 + \lambda_Y^2 \tau_1^2$, where τ_λ^2 is the random-intercepts variance in the probit model on Y_{2i} and τ_1^2 the random-intercepts variance of the normal regression model on Y_{1i}. Results show a significantly decreased response on toe pinch, due to the treatment. There is also a clear treatment effect on the pupil size. The correlation among the two endpoints is very small.

The second column in Table 14.3 shows the parameter estimates from the Plackett–Dale model. This model is based on a logit model for the binary endpoint. While an independency working correlation for endpoints at different time points is assumed, the pseudo-likelihood method is used to correct for possible dependencies in the data. Results are in line with the probit model, showing a significant effect of treatment on toe pinch. Also, the response on toe pinch in time is significantly different for the two treatment groups. Extension of the pseudo-likelihood function where all pairwise associations are modeled, as in Section 14.3.3, reveals strong associations among the measurements at different time points. The correlation among two continuous outcomes for the same subject is estimated as $\hat{\rho} = 0.40$ (with standard error 0.002), while the odds ratio representing the association among two binary outcomes is estimated as $\hat{\psi} = 2.42$ (with standard error 0.001). Further, we investigated the dependency of the association on covariates of interest. For example, the odds ratio as a function of X can be investigated using $\log(\psi) = \psi_0 + \psi_1 X$. While there is no difference in correlation among the two treatment arms, there is a significant effect of time on the correlation. The correlation at 8 hours after dosage is estimated as $\hat{\psi} = 1.28$, while at 24 hours after dosage it is estimated as $\hat{\psi} = 0.52$.

Finally, GLMMs were used to analyze the data at hand. We considered a marginal model, with fully unstructured 30×30 variance–covariance matrix. However, due to the computational complexity, this model did not converge. Alternatively, uncorrelated and correlated random-intercepts models were used, as explained in Section 14.3.2. Parameter estimates can be found in Table 14.3. Results of the correlated random-intercepts model are presented in Figure 14.2. The model fit seems to closely follow the data. The fixed-effects parameter estimates and the standard errors for both the continuous endpoints are approximately the same for all models considered, while the fixed-effects estimates for the binary endpoints differ somewhat more. All logit-type models yield very similar parameter estimates

Table 14.3 Repeated Dose-Toxicity Study

Effect	Par.	Probit-Normal	Plackett–Dale	Uncorr. RI Logit	Uncorr. RI Probit	Corr. RI Logit	Corr. RI Probit
Binary Endpoint							
Intercept	β_{11}	−2.02 (0.64)	−3.33 (0.00)	−3.65 (1.36)	−2.09 (0.73)	−3.38 (1.25)	−1.97 (0.66)
Treatm. Effect	β_{12}	2.20 (0.70)	3.93 (0.62)	4.47 (1.45)	2.51 (0.78)	4.19 (1.34)	2.40 (0.71)
Time	β_{13}	0.03 (0.03)	1.06 (0.00)	0.04 (0.07)	0.02 (0.04)	0.04 (0.64)	0.02 (0.03)
Day	β_{14}	−0.46 (0.39)	−0.46 (2.38)	−1.12 (0.92)	−0.50 (0.42)	−1.10 (0.93)	−0.48 (0.39)
Time*Day	β_{15}	0.02 (0.01)	0.01 (0.04)	0.04 (0.03)	0.02 (0.02)	0.04 (0.02)	0.02 (0.01)
Treat*Time	β_{16}	−0.07 (0.03)	−1.05 (0.05)	−0.14 (0.07)	−0.08 (0.04)	−0.14 (0.07)	−0.07 (0.03)
Treat*Day	β_{17}	0.34 (0.38)	0.23 (2.14)	0.55 (0.87)	0.21 (0.40)	0.54 (0.78)	0.20 (0.37)
RI Var.	τ_1	0.65 (0.56)		2.00 (1.07)	0.68 (0.35)	2.13 (1.06)	0.74 (0.34)
Overdisp. Par.	ϕ	0.79 (0.07)				0.83 (0.08)	0.79 (0.45)
Continuous Endpoint							
Intercept	β_{21}	7.39 (1.00)	7.46 (0.00)	7.39 (0.95)	7.39 (0.95)	7.39 (1.00)	7.39 (1.00)
Treatm. Effect	β_{22}	8.66 (1.17)	8.63 (0.93)	8.67 (1.11)	8.67 (1.11)	8.66 (1.17)	8.66 (1.17)
Time	β_{23}	−0.08 (0.04)	−0.08 (0.00)	−0.08 (0.04)	−0.08 (0.04)	−0.08 (0.04)	−0.08 (0.04)
Day	β_{24}	−0.76 (0.37)	−0.70 (0.07)	−0.76 (0.36)	−0.76 (0.36)	−0.76 (0.37)	−0.76 (0.37)
Time*Day	β_{25}	0.08 (0.02)	0.08 (0.01)	0.08 (0.02)	0.08 (0.02)	0.08 (0.04)	0.08 (0.04)
Treat*Time	β_{26}	−0.18 (0.04)	−0.18 (0.03)	−0.18 (0.04)	−0.18 (0.04)	−0.18 (0.04)	−0.18 (0.04)
Treat*Day	β_{27}	−2.73 (0.37)	−2.76 (0.46)	−2.73 (0.37)	−2.73 (0.37)	−2.73 (0.37)	−2.73 (0.37)
Variance	σ^2	4.89 (0.45)	9.85 (0.04)	4.79 (0.44)	4.79 (0.44)	4.89 (0.45)	4.89 (0.45)
RI Var.	τ_2	3.62 (1.44)		3.16 (1.19)	3.16 (1.19)	3.62 (1.44)	3.62 (1.44)
Association							
Odds Ratio	ψ		1.61 (0.00)				
Correlation	ρ	0.12 (0.06)		0.39 (0.25)	0.39 (0.25)	0.005 (0.07)	0.007 (0.07)

Note: Parameter estimates (standard errors) of the bivariate analysis for pupil size (continuous endpoint) and toe pinch (binary endpoint), based on the probit-normal model, the Plackett–Dale model, the uncorrelated random-intercepts (uncorr. RI) model, and the correlated random-intercepts model (corr. RI).

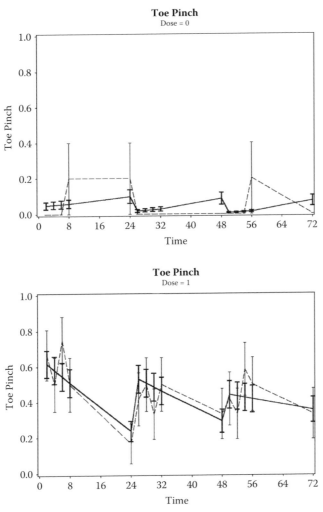

Figure 14.2 Fit of uncorrelated random-intercepts model to the binary variable toe pinch and the continuous variable pupil size.

for the binary endpoints, and this is also the case for all probit-type models. The correlation parameters ρ for the uncorrelated random-intercepts model represent the correlation among the random intercepts, while in the correlated random-intercepts model it is the correlation in the residual error structure. This estimate can be used to calculate the correlation among the continuous and discrete outcomes. This correlation depends on the binary endpoint, via the variance v_{2i}. All models are consistent in estimating the correlations among the binary and continuous endpoints as insignificant.

14.5 Conclusions

Correlated data are common in many health sciences studies, where clustered, longitudinal, hierarchical, and spatially organized data are frequently observed. When several outcomes of interest are measured on the same individual, these multivariate outcomes are also likely to be correlated. The observations under study share some common characteristics, and statistical analysis requires taking such associations into account. There exist many ways to deal with these correlation structures, ranging from the most naive one of ignoring the

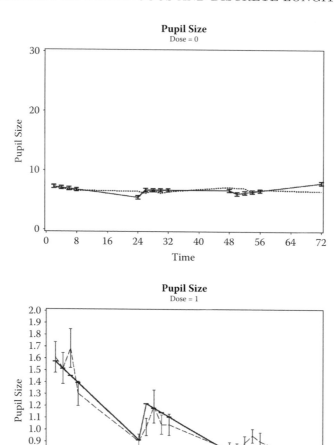

Figure 14.2 *Continued.*

associations to approaches that correct for correlations or model the correlation. Methods for longitudinal, multivariate continuous data are widely available, where the normal distribution with its elegant properties plays a prominent role. However, when the outcome variables are of a mixed type, techniques are less standard because of the lack of a discrete analog to the multivariate normal distribution.

We have considered three approaches for mixed continuous–binary outcomes from longitudinal data: the probit-normal approach, the Plackett–Dale model, and the generalized linear mixed model. The three methods are fundamentally different in the way the association between both variables is described. The question as to which model should be used is difficult to answer. Often, the answer depends on the research questions of interest. However, no formal test statistic to choose among the models is available. Of course, fitting several of these models simultaneously is an important sensitivity analysis. In the data example, it is comforting to see that the model fits are virtually identical.

Both the Plackett–Dale model and the generalized linear mixed model extend relatively straightforwardly to settings other than combined binary and continuous outcomes. The global cross-ratio, as used in the Plackett–Dale model (Section 14.2.3), is defined in terms of the cumulative distributions of both endpoints. This implies that, in general, any type of endpoint and any distribution function could be used in the Plackett–Dale model. For example, use of the Plackett–Dale model in a mixed continuous/ordinal setting is described in Faes et al. (2004). Also, the generality of the generalized linear mixed model (defined in Section 14.2.2) makes it possible to extend the model to other settings of, for example, combined discrete and continuous outcomes. Extensions to higher-dimensions are possible for both models, but are computationally challenging. For the Placket-Dale model, the joint distribution has to be specified in terms of marginal mean functions, and pairwise and higher-order global odds ratios, as described in Molenberghs and Lesaffre (1999). The generalized linear mixed-model approach can be extended by specifying a joint (higher-dimensional) distribution for the random effects. However, while (generalized linear) mixed models are very flexible in modeling multivariate data, computational issues arise when the number of endpoints gets large.

SAS code for the models considered in this chapter is available at the web site for this book given in the Preface.

References

Aerts, M., Geys, H., Molenberghs, G., and Ryan, L. (2002). *Topics in Modelling of Clustered Data*. London: Chapman & Hall.

Arnold, B. C. and Strauss, D. (1991). Pseudolikelihood estimation: Some examples. *Sankhyā, Series B* **53**, 233–243.

Catalano, P. J. (1997). Bivariate modelling of clustered continuous and ordered categorical outcomes. *Statistics in Medicine* **16**, 883–900.

Catalano, P. J. and Ryan, L. M. (1992). Bivariate latent variable models for clustered discrete and continuous outcomes. *Journal of the American Statistical Association* **87**, 651–658.

Cox, D. R. and Wermuth, N. (1992). Response models for mixed binary and quantitative variables. *Biometrika* **79**, 441–461.

Cox, D. R. and Wermuth, N. (1994). A note on the quadratic exponential binary distribution. *Biometrika* **81**, 403–408.

Faes, C., Geys, H., Aerts, M., Molenberghs, G., and Catalano, P. (2004). Modelling combined continuous and ordinal outcomes in a clustered setting. *Journal of Agricultural, Biological, and Environmental Statistics* **9**, 515–530.

Fitzmaurice, G. M. and Laird, N. M. (1995). Regression models for a bivariate discrete and continuous outcome with clustering. *Journal of the American Statistical Association* **90**, 845–852.

Geys, H., Regan, M., Catalano, P., and Molenberghs, G. (2001). Two latent variable risk assessment approaches for mixed continuous and discrete outcomes from developmental toxicity data. *Journal of Agricultural, Biological, and Environmental Statistics* **6**, 340–355.

Irwin, S. (1964). Comprehensive observational assessment. Ia. A systematic, quantitative procedure for assessing the behavioral and physiologic state of the mouse. *Physiopharmacologia* **13**, 222–257.

Johnson, R. A. and Wichern, D. W. (2002). *Applied Multivariate Statistical Analysis*, 5th ed. Upper Saddle River, NJ: Prentice Hall.

Krzanowski, W. J. (1988). *Principles of Multivariate Analysis*. Oxford: Clarendon Press.

Little, R. J. A. and Schluchter, M. D. (1985). Maximum likelihood estimation for mixed continuous and categorical data with missing values. *Biometrika* **72**, 497–512.

Molenberghs, G. and Geys, H. (2001). A review on the analysis of clustered multivariate data from developmental toxicity studies. *Statistica Neerlandica* **55**, 319–345.

Molenberghs, G., Geys, H., and Buyse, M. (2001). Evaluations of surrogate endpoints in randomized experiments with mixed discrete and continuous outcomes. *Statistics in Medicine* **20**, 3023–3038.

Molenberghs, G. and Lesaffre, E. (1999). Marginal modelling of multivariate categorical data. *Statistics in Medicine*, **18**, 2237–2255.

Molenberghs, G. and Verbeke, G. (2005). *Models for Discrete Longitudinal Data*. New York: Springer.

Ochi, Y. and Prentice, R. L. (1984). Likelihood infrence in a correlated probit regression model. *Biometrika* **71**, 531–543.

Olkin, I. and Tate, R. F. (1961). Multivariate correlation models with mixed discrete and continuous variables. *Annals of Mathematical Statistics* **32**, 448–465 (with correction in **36**, 343–344).

Plackett, R. L. (1965) A class of bivariate distributions. *Journal of the American Statistical Association* **60**, 516–522.

Regan, M. M. and Catalano, P.J . (1999a) Likelihood models for clustered binary and continuous outcomes: Application to developmental toxicology. *Biometrics* **55**, 760–768.

Regan, M. M. and Catalano, P. J. (1999b) Bivariate dose-response modeling and risk estimation in developmental toxicology. *Journal of Agricultural, Biological, and Environmental Statistics* **4**, 217–237.

Regan, M. M. and Catalano, P. J. (2000) Regression models and risk estimation for mixed discrete and continuous outcomes in developmental toxicology. *Risk Analysis* **20**, 363–376.

Tate, R. F. (1954) Correlation between a discrete and a continuous variable. *Annals of Mathematical Statistics* **25**, 603–607.

Wolfinger, R. and O'Connell, M. (1993). Generalized linear mixed models: A pseudo-likelihood approach. *Journal of Statistical Computing and Simulation* **48**, 233–243.

CHAPTER 15

Random-effects models for joint analysis of repeated-measurement and time-to-event outcomes

Peter Diggle, Robin Henderson, and Peter Philipson

Contents

15.1 Introduction

15.1.1 Motivation

We use the term *joint modeling* to refer to model-based methods for the analysis of data from a longitudinal study in which each subject produces outcome data of two kinds: a sequence of measurements at pre-specified follow-up times, and a point process of events in time.

Figure 15.1 illustrates two general scenarios that are of interest in this context. In both plots, the smooth curve represents the development over time of an unobserved measure of the "true health" of an individual. This might be something specific, such as the level of a toxin in the blood; something less well defined, such as kidney functioning after a

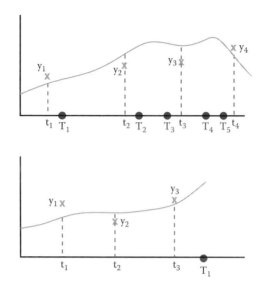

Figure 15.1 Longitudinal measurements and event times.

transplant; or something intangible, such as quality of life or mental well-being. The true state is unobserved and unobservable. Instead, snapshots $Y_1, Y_2, \ldots$ are obtained at times $t_1, t_2, \ldots$, which are determined by the experimenter. However, these snapshots do not reveal the true state because they are subject to errors. These might be physical measurement errors as, for example, would arise in measuring the level of a toxin, or may reflect an inherent error in using a surrogate such as a questionnaire instrument to measure quality of life. Together, the Y measurements provide a longitudinal profile for each subject. For the applications we have in mind, the number of observations per individual will typically be quite low, perhaps single figures or not much more. A crucial assumption is that responses from different subjects are mutually independent.

Alongside the measurement schedule, we assume that there is a stochastic process of event occurrences at $T_1, T_2, \ldots$. These are represented by the large dots in Figure 15.1. We assume that their intensity may also be related to the underlying value of the subject's "true health." The upper panel shows a sequence of recurrent events; for example, they might be rejection episodes for a kidney transplant recipient, or seizures for an epileptic patient. The lower panel shows a single event whose occurrence terminates observation of the measurement sequence. This might be because the true health process closes (e.g., if the event is death), or because the event marks a change of treatment or circumstances (e.g., rejection of a transplanted kidney may be followed by a return to dialysis), or it could simply mark removal of the subject from observation (e.g., dropout from a clinical trial). We will focus mostly on the case of a single, terminating event of clinical interest, while allowing the additional possibility of censored observation times in which the censoring time may or may not be informative.

In the remainder of this chapter, we will discuss certain *random-effects* joint models for longitudinal and event time data of the type described above. These models attempt to explain the relationship between the measurement data Y and the event time data T through their shared dependence on unobserved random effects that represent the modeled version of the "true health" curves shown in Figure 15.1; see Chapter 13 and Chapter 19 for a discussion of related such models. Before describing the models in detail, we note that the purpose of the study should always be carefully considered at the beginning of the analysis.

In the current context, the focus may be the measurement process, the time-to-event process, or the association between the two. We now give three examples to illustrate the three cases.

Example 1: A schizophrenia trial

Henderson, Diggle, and Dobson (2000) and Xu and Zeger (2001a, 2001b) analyze data from a placebo-controlled randomized trial comparing different drug treatments for chronic schizophrenia. Patients were intended to be followed up at 0, 1, 2, 4, 6, and 8 weeks after randomization. The outcome for each patient at each follow-up time was a measure of the current severity of their symptoms, the positive-and-negative symptom score (PANSS). However, of 523 patients recruited, 270 dropped out of the study at various times between 0 and 8 weeks, predominantly because of an inadequate response to therapy. In this example, the scientific focus is on the efficacy of the active treatments in alleviating symptoms; that is, on the measurement process Y corresponding to the PANSS scores. The time-to-event outcome T is dropout, which is not of direct interest but, because dropout is potentially informative, needs to be incorporated into the analysis.

Example 2: An observational study of heart surgery patients

Lim et al. (2008) report the results of an observational study of 289 cardiac patients following heart-valve implantation. Two types of heart valve, homograft and stentless, were involved. Outcome variables included survival time and three repeated measures of heart function: left ventricular mass index (LVMI), ejection fraction, and gradient. A specific objective for the analysis of these data is to describe the individual-level association between the longitudinal profile of LVMI and the hazard of death. Hence, interest is focused on the joint distribution of Y (LVMI) and T (survival time).

Example 3: A liver cirrhosis trial

Andersen et al. (1993, p. 19) discuss a trial in which 488 subjects were randomized to receive either an active steroid treatment (prednisone) or placebo and followed until death or the end of the study, some 12 years after initial recruitment to the trial. The measurement sequence for each subject was the *prothrombin index*, a measure of liver function, scheduled to be taken at 0, 3, 6, and 12 months after randomization and annually thereafter, although achieved measurement times varied. The time-to-event outcome was death. The focus of interest was the impact of treatment on survival time. Hence, survival time was the primary outcome measure, and prothrombin index could be considered as a covariate that is both time-varying and measured with error. From the perspective of joint modeling, the focus is on the conditional distribution of survival time, T, given information on the liver functionality obtainable from the longitudinal profile of the prothrombin sequence, Y.

In addition to the time-to-event outcome of interest, some of the intended measurement sequences are incomplete because of censoring. Published analyses of these data assume that censoring and time-to-death are independent. Later in the chapter we present an analysis of the data that admits the possibility that censoring may be informative.

15.1.2 Brief history

For many years, models for repeated measurements and for survival outcomes were developed separately. Book-length discussions of these two areas include Diggle et al. (2002) and Fitzmaurice, Laird, and Ware (2004) on repeated measures; and Andersen et al. (1993), Hougaard (2000), and Therneau and Grambsch (2000) on survival analysis. Within the repeated-measures context, the problem of dealing with potentially informative dropouts first received attention in the late 1980s (see, for example, Wu and Carroll, 1988). The literature on this topic grew rapidly during the 1990s; for reviews, see Little (1995), Hogan and Laird (1997), and Hogan, Roy, and Korkontzelou (2004). Diggle and Kenward (1994) introduced a class of models for informative dropout that included random and completely random dropout as special cases, thus mapping onto the hierarchy of missing-value mechanisms set out in Little and Rubin (2002, Chapter 6). Several of the discussants of Diggle and Kenward (1994) pointed out the connection between informative dropouts and informative censoring in survival analysis. At around the same time, Tsiatis, DeGruttola, and Wulfsohn (1995) considered the converse of the informative dropout problem in which a time-to-event (survival time) is the outcome of interest, and repeated measures can be considered as error-prone, time-varying covariates. Wulfsohn and Tsiatis (1997) made an important early contribution to the emerging literature on joint modeling in which the repeated measures and time-to-event outcomes were assumed to be conditionally independent given a set of unobserved, subject-level random effects. Specifically, Wulfsohn and Tsiatis assumed a random-intercept-and-slope model, as proposed by Laird and Ware (1982) in the repeated-measures setting, and developed an expectation–maximization (EM) algorithm for maximum likelihood estimation. Tsiatis and Davidian (2004) give an excellent overview of the subsequent literature on joint modeling of longitudinal and time-to-event data.

15.2 Random-effects models

15.2.1 General

To define the class of models that we consider in this chapter, we need some additional notation. Let Y_{ij} be the measurement for subject i at time t_{ij} ($i = 1, \ldots, N$, $j = 1, \ldots, n_i$). Note in particular that we do not assume a common set of measurement times for all subjects. We write $\boldsymbol{Y}_i = (Y_{i1}, \ldots, Y_{i,n_i})'$ for the associated vector of measurements on subject i. Similarly, $\boldsymbol{X_i}(t_{ij}) = (X_{ij1}, X_{ij2}, \ldots, X_{ijp})'$ denotes the vector of covariates for subject i at time t_{ij}, and X_i is the $n_i \times p$ matrix with rows $\boldsymbol{X_i}(t_{ij})'$, $j = 1, \ldots, n_i, i = 1, \ldots, N$. We write $\boldsymbol{\mu}_i = E(\boldsymbol{Y}_i \mid X_i)$ and $\Sigma_i = \mathrm{Cov}(\boldsymbol{Y}_i \mid X_i)$. We assume a linear model for the expectation, $\boldsymbol{\mu}_i = X_i\boldsymbol{\beta}$, but for the time being place no particular restriction on Σ_i.

We consider the scenario with a single, terminating event, the observed event time for subject i being $U_i = \min(T_i, C_i)$, where T_i and C_i are the underlying true event and censoring times, respectively. We use $\delta_i = 1$ or 0 to denote an actual or censored event time, respectively, and $\lambda_i(t)$ to denote the hazard for T_i at time t, conditional on any covariates or other quantities (i.e., random effects) associated with event time.

In a random-effects model, U_i and $\boldsymbol{Y}_i$ are assumed to be conditionally independent, given a q-element vector of unobserved random effects $\boldsymbol{b}_i$. Unless we place some restrictions on the $\boldsymbol{b}_i$, this is not a very useful definition. We wish to accommodate both finite-dimensional and infinite-dimensional random effects. We therefore reserve $\boldsymbol{b}_i$ to mean a finite-dimensional, time-constant vector and use $S_i(t)$ to denote an unobserved, continuous-time stochastic process associated with subject i. Then, U_i and $\boldsymbol{Y}_i$ are assumed to be conditionally independent given $\boldsymbol{b}_i$ and $\{S_i(t) : t \geq 0\}$. In practice, models of this kind are useful only when we can justify a relatively parsimonious specification for the random effects, and when at least some components of $\boldsymbol{b}_i$ and $S_i(t)$ are "shared" between the measurement and time-to-event submodels (see Chapter 13 and Chapter 19 for a discussion of related models).

In Section 15.1.1, we provided motivation for a random-effects approach. Against this, a legitimate concern about random-effects models is that parametric assumptions about unobserved variables and stochastic processes are, at best, difficult to validate from empirical data, suggesting a potential lack of robustness. We would only advocate their use when individual-level, as opposed to population-averaged, effects are of scientific interest.

15.2.2 Random-effects models and the Rubin taxonomy

Before continuing with model formulation, we digress to discuss the placing of random-effects joint models within the familiar Rubin taxonomy for missing data in which missing-data mechanisms are classified as missing completely at random (MCAR), missing at random (MAR), or not missing at random (NMAR), sometimes also called informative missingness. See, for example, Little and Rubin (2002, Chapter 6), following on from Rubin (1976), and Chapter 16 of this volume. We begin with a reminder of the standard definition of independent censoring for event history analysis (Andersen et al., 1993). This starts by distinguishing the true and complete-data world, in which all event times are seen, from the observed-data world, which is affected by a predictable censoring process that sometimes obscures observation. Censoring is defined to be independent if the chance of an event at time t, given the complete pattern of previous events, is not affected by any additional knowledge of *observation* of the previous events. Our analogy is that we have a complete-data world generated by covariates, longitudinal and event time data *and* random effects, and an observed-data world in which the random effects are not observed. In our view, it is useful to classify models in both the complete- and observed-data worlds, not just the latter.

Figure 15.2 illustrates these distinctions for a simple case in which there are just two measurement times, all subjects have the first measurement Y_1, but some subjects drop out before the second time point. Dropout is indicated by a binary variable R, while b_1 and b_2 denote unobserved random effects associated with the first and second measurement times, respectively. Plots (a) and (c) are complete-data graphs, both of which have the same marginal (observed-data world) graph (b). Graph (b) corresponds to an informative or non-random missingness (NMAR) model, because R and Y_2 are not conditionally independent given Y_1. The same is true for graph (a) in the complete-data world, where the future b_2 is allowed to influence the probability of dropout after the first measurement. In graph (c), we assume that a single random effect b_1 is associated with both Y_1 and Y_2. As the plot indicates, if b_1 could have been observed, then b_1 would have been available at the first measurement time, and R would have been conditionally independent of Y_2 given the *observed* quantity b_1. Hence, in the hypothetical complete-data world, the model is MAR, whereas in the observed-data world it is NMAR.

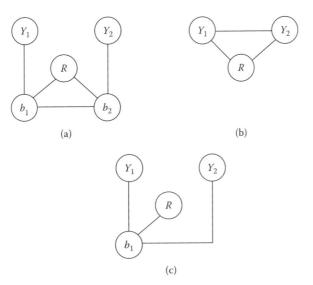

(a)

(b)

(c)

Figure 15.2 Random-effects models and the Rubin taxonomy.

In general, models that include only time-fixed random effects $\boldsymbol{b}$ will be sequentially missing at random, as defined in Hogan, Roy, and Korkontzelou (2004), in the complete-data world, but NMAR in the observed-data world. If there is additionally a time-varying random effect, $S(t)$, then the model can be NMAR in both worlds. We consider it useful to think in this way if the models are to be used for prediction, as for Example 2 above. If only time-fixed components W are included, then in principle these can be estimated accurately by making the measurement schedule more dense. If there is a stochastic component $S(t)$, then prognosis is uncertain however frequently measurements are taken. This contrast is analogous to the distinction between fixed domain and increasing domain asymptotics in spatial statistics (see, for example, Stein, 1999, Section 1.6).

15.3 Model formulation

Although a variety of models has been considered in the joint modeling literature (Tsiatis and Davidian, 2004), for the remainder of this chapter we will consider models based on what we consider to be first-choice assumptions for the two kinds of data: a Gaussian linear mixed-effects model for the longitudinal measurements and a Cox proportional hazards model for the event times conditional on the random effects.

15.3.1 Examples

We now give some specific examples, with increasing complexity, of random-effects models for the applications introduced earlier.

Linear random effects

To the best of our knowledge, the earliest use of random-effects models for what we would now call joint modeling is by Wu and Carroll (1988), whose motivation was to accommodate informative right-censoring in longitudinal studies of growth over time. In our terms, their measurement submodel was a specific case of the linear mixed-effects model

$$Y_{ij} = \boldsymbol{X}_i(t_{ij})'\boldsymbol{\beta} + \boldsymbol{Z}_i(t_{ij})'\boldsymbol{b}_i + e_{ij}, \tag{15.1}$$

with $\boldsymbol{Z}_i(t)' = (1, t)$, corresponding to the random-intercept-and-slope model. Here, and subsequently, e_{ij} is a sequence of mutually independent measurement errors. It is also assumed that $\boldsymbol{b}_i$ and e_{ij} are zero-mean Gaussian, mutually independent within each subject, and mutually independent but with a common distribution across subjects. Their hazard submodel was expressed in terms of the probability, $p_i(t)$, that subject i would experience the event of interest within the time interval $(0, t)$. They assumed that

$$p_i(t) = \mathcal{F}(\boldsymbol{\mathcal{Z}}_i(t)\boldsymbol{b}_i)$$

for some function $\mathcal{F}(\cdot)$, where $\boldsymbol{\mathcal{Z}}_i(t)$ is a vector of covariates that could be the same as $\boldsymbol{Z}_i(t)$ of (15.1), but does not not have to be. They took $\mathcal{F}(\cdot)$ to be a probit function, but commented that a Cox (1972) proportional hazards model could also have been used. Two points that we wish to emphasize are their assumptions of *shared* random effects, $\boldsymbol{b}_i$, and of conditional independence of a subject's measurement sequence and time-to-event outcome given their random effects $\boldsymbol{b}_i$.

Wulfsohn and Tsiatis (1997) proposed essentially the same model but cast it explicitly in terms of joint modeling and showed how the EM algorithm could be used to implement maximum likelihood estimation. Their measurement submodel was identical to (15.1). Their hazard submodel was

$$\lambda_i(t) = \lambda_0(t) \exp\{\alpha \boldsymbol{Z}_i(t)'\boldsymbol{b}_i\}, \tag{15.2}$$

where $\lambda_0(\cdot)$ is an arbitrary smooth function. Note the use of a single parameter, α, that is, the log relative risk for subject i is assumed to be proportional to the same subject's measurement random effect, $\boldsymbol{Z}_i(t)'\boldsymbol{b}_i$. Wulfsohn and Tsiatis (1997) also noted several possible extensions, including the addition of covariate effects to the log-linear term in the hazard submodel.

For the schizophrenia trial data (Example 1) as analyzed in the cited papers, there are no covariates other than treatment group. This means that a saturated model with a different mean for each treatment $\times$ time combination is feasible. A simple random slope and intercept model is a good starting point for the random effects, hence a candidate repeated measurement submodel is

$$Y_{ij} = \mu_{g(i)j} + b_{i0} + b_{i1}t_{ij} + e_{ij},$$

where $g(i)$ is the treatment group of person i, and $(b_{i0}, b_{i1})'$ is assumed to be bivariate Gaussian. A candidate hazard submodel is

$$\lambda_i(t) = \lambda_0(t) \exp\{\theta_{g(i)} + \alpha(b_{i0} + b_{i1}t)\}, \tag{15.3}$$

where $\theta_{g(i)}$ is a time-fixed parameter for the treatment group of subject i. Under this model, the parameter α determines both the association between the longitudinal data and the dropout rate *and* the presence of random effects (frailty) in the hazard model. In principle, a separate frailty term could be added to give a more symmetric formulation,

$$\lambda_i(t) = \lambda_0(t) \exp\{\theta_{g(i)} + \alpha(b_{i0} + b_{i1}t) + b_{i2}\},$$

with $(b_{i0}, b_{i1}, b_{i2})'$ now trivariate Gaussian. In practice, estimation of frailty is very difficult in semi-parametric models. In fact, identification is impossible without the inclusion of covariate effects (Elbers and Ridder, 1982; Heckman and Singer, 1984; Melino and Sueyoshi, 1990).

Henderson, Diggle, and Dobson (2000) found no evidence of a separate frailty term b_{i2} for the schizophrenia data, but they did find a significantly improved likelihood by using the hazard model

$$\lambda_i(t) = \lambda_0(t) \exp\{\theta_{g(i)} + \alpha_1(b_{i0} + b_{i1}t) + \alpha_2 b_{i1}\}. \tag{15.4}$$

In (15.3), the hazard is determined by the current value of the random effect. The estimated α was positive, which means that patients with higher PANSS scores (and worse mental health) were more likely to drop out. In (15.4), the hazard is determined by both the current value of the random effect and its trajectory. Estimates of both α_1 and α_2 were positive, as dropout was more likely for people who were either ill at present (high $b_{i0} + b_{i1}t$) or deteriorating from whatever baseline (high b_{i1}).

For the heart-valve patient data in Example 2, a simple random slope and intercept model is inadequate. Lim et al. (2008) note a change-point pattern to LVMI, with subject-specific decline for a period of about one year, followed by stability or more gradual change. This can be modeled by

$$Y_{ij} = \boldsymbol{X}_i(t_{ij})'\boldsymbol{\beta} + b_{i0} + b_{i1}\min(t, \tau_i) + b_{i2}\max(t - \tau_i, 0) + e_{ij},$$

where $(b_{i0}, b_{i1}, b_{i2})'$ are trivariate Gaussian, and τ_i is the subject-specific change-point. A hypothesis of clinical interest is that only the second slope, b_{i2}, is related to the hazard after the change-point. Hence, for $t > \tau_i$ we might specify

$$\lambda_i(t) = \lambda_0(t) \exp\{\boldsymbol{\mathcal{X}}_i(t)'\boldsymbol{\beta}^* + \alpha b_{i2}\},$$

where $\boldsymbol{\mathcal{X}}_i(t)$ is a vector of covariates, possibly the same as $\boldsymbol{X}_i(t)$ from the measurement submodel, but not necessarily so.

If τ_i is known, then we have a linear random-effects joint model. For unknown τ_i the problem is more difficult.

The very widely used linear mixed-effects model, with a random intercept and slope for each subject, often provides an acceptable fit to the covariance structure of short repeated measurement sequences. For longer series it tends to break down because the implicit extrapolation of the subject-specific trends becomes unrealistic. The introduction of a stationary Gaussian process (SGP), $S_i(t)$, into the random-effects model is intended to account for local variation around a long-term stable level (Diggle, 1988).

In analyzing the liver cirrhosis data (Example 3), Henderson, Diggle, and Dobson (2002) considered three different forms for the random effects within the longitudinal model:

A	Random intercept	b_{i0},
B	Random intercept and slope	$b_{i0} + b_{i1}t$,
C	Random intercept and SGP	$b_{i0} + S_i(t)$,

with

$$\text{Var}\{(b_{i0}, b_{i1})\}' = \begin{pmatrix} \sigma_0^2 & \sigma_{01} \\ \sigma_{01} & \sigma_1^2 \end{pmatrix},$$

$$\text{Cov}\{S_i(t), S_i(t+u)\} = \sigma_s^2 \exp(-|u|/\phi),$$

and measurement error variance $\text{Var}(e_{ij}) = \sigma_e^2$. The three models imply the following association structures for the longitudinal responses:

A $\quad \text{Cov}(Y_{ij}, Y_{ik}) = \sigma_0^2 + \sigma_e^2 I(j = k),$

B $\quad \text{Cov}(Y_{ij}, Y_{ik}) = \sigma_0^2 + \sigma_1^2 t_{ij} t_{ik} + \sigma_{01}(t_{ij} + t_{ik}) + \sigma_e^2 I(j = k),$

C $\quad \text{Cov}(Y_{ij}, Y_{ik}) = \sigma_0^2 + \sigma_s^2 \exp(-|t_{ij} - t_{ik}|/\phi) + \sigma_e^2 I(j = k).$

The random effect is not constant over time for models B and C. Model C, however, does induce stationarity, whereas model B does not. When limited observations per subject are available, it can be difficult, in practice, to distinguish these forms of association structure.

In each case, the total random effect was linked to the hazard through a single parameter α, as in (15.2). The log-likelihoods for the three models were $-15{,}080.053$ (A), $-14{,}992.177$ (B), and $-14{,}984.059$ (C). These represent strong evidence that B and C are preferred to A, implying that the random effects are not constant over time, and good evidence that C is preferred to B, implying local variation about a subject-specific average rather than a monotone subject-specific trend. In model C, the estimated value of ϕ was 1.915, corresponding to correlations 0.59, 0.35, and 0.21 in the SGP at lags 1, 2, and 3 years, respectively.

An obvious disadvantage of the smooth variation model involving an SGP is the increase in computational difficulty in fitting. As we shall describe in the next section, an EM algorithm can be used for estimation, treating the random effects as missing data. If there are only a small number of random effects, then the expectations required at the E-step can be found using quadrature for approximate integration. With the SGP model, the missing data strictly are the infinite-dimensional processes $S_i(t)$ for each subject and at all times t. However, one consequence of assuming a semi-parametric model for the hazard is that $S_i(t)$ appears in the likelihood only at event times. The missing data can therefore be treated as finite-dimensional, and, in principle, we could use quadrature for the E-step. In practice, the dimension can be quite high, typically of the same order as the number of subjects, and quadrature becomes computationally prohibitive. Henderson, Diggle, and Dobson (2000) proposed a nested procedure combining Monte Carlo integration with a numerical search, which is feasible but still computationally intensive.

An alternative might be to approximate the SGP using a low-rank model, such as a spline with random coefficients. For example, a cubic spline model with k knots, $s_1, \ldots, s_k$, would take the form

$$S_i(t) = \sum_{p=0}^{3} b_{ip}t^p + \sum_{j=1}^{k} b_{i,p+j}t_j^+,$$

where $t_j^+ = \max(0, t - s_j)$; and the b_{ij}, $j = 1, \ldots, p + k$, are Gaussian random effects associated with subject i. For a general discussion of low-rank models and their use in semi-parametric regression modeling, see Ruppert, Wand, and Carroll (2003).

More generally, by replacing the infinite-dimensional Gaussian process $S_i(t)$ by $B(t)'\boldsymbol{b}_i$, where $B(t)$ is a prescribed vector of basis functions, and $\boldsymbol{b}_i = (b_{i1}, \ldots, b_{im})'$ is a multivariate Gaussian vector, we may achieve a useful compromise between the requirements for both flexibility and computational feasibility. To our knowledge, this approach has not yet been explored in the context of joint modeling.

15.3.2 The general model

The models that we have considered explicitly thus far can be written, in general form, as follows.

The *measurement submodel* for subject i is

$$Y_{ij} = \boldsymbol{X}_i(t_{ij})'\boldsymbol{\beta} + \boldsymbol{Z}_i(t_{ij})'\boldsymbol{b}_i + S_i(t_{ij}) + e_{ij}, \quad j = 1, \ldots, n_i, \tag{15.5}$$

where $\boldsymbol{Z}_i(t_{ij})$ is a $q \times 1$ vector of covariates for the random-effects component $\boldsymbol{b}_i$ at time t_{ij}, with $\boldsymbol{b}_i, S_i(\cdot)$, and e_{ij} as defined earlier. We assume independence within subjects, and a common distribution, with mutual independence, across subjects for $\boldsymbol{b}_i, S_i(\cdot)$, and e_{ij}.

The *hazard submodel* for subject i is

$$\lambda_i(t) = \lambda_0(t)\exp\{\boldsymbol{\mathcal{X}_i}(t)'\boldsymbol{\beta}^* + \boldsymbol{\mathcal{Z}_i}(t)'\boldsymbol{b}_i^* + S_i^*(t)\}, \tag{15.6}$$

where $\lambda_0(\cdot)$ is an arbitrary smooth function, $\boldsymbol{\mathcal{X}}_i(t)$ and $\boldsymbol{\mathcal{Z}}_i(t)$ are vectors of covariates, and $\boldsymbol{b}_i^*$ and $S_i^*(\cdot)$ are the counterparts of $\boldsymbol{b}_i$ and $S_i(\cdot)$ in (15.5), again assumed to be Gaussian.

Finally, an *association submodel* is used to specify the joint distributions of $(\boldsymbol{b}_i, \boldsymbol{b}_i^*)$ and of $\{S_i(\cdot), S_i^*(\cdot)\}$. The association model will usually have a rather simple structure, for example, $S_i(\cdot)$ and $S_i^*(\cdot)$ may be proportional, or $\boldsymbol{b}_i^*$ may be equal to $\boldsymbol{b}_i$ (see Example 1) or may be a subvector of $\boldsymbol{b}_i$ (as was the case in Example 2).

It is possible, in principle and to some extent in practice, to cast the complete model in more general terms; for example, Berzuini and Larizza (1996) gave a very general formulation. However, the above is sufficiently general for our purposes. Also, generalization does not come without costs, both computational and inferential. We return to this point in Section 15.7.

15.4 Maximum likelihood estimation using the EM algorithm

The objective is to estimate all parameters in the model using maximum likelihood. Because (15.5) and (15.6) are standard, there would be no difficulty with estimation if the random effects $\boldsymbol{b}$ were observed. This makes it natural to consider using the EM algorithm. For simplicity, we refer to just one subject and drop the subscript i, and use $\boldsymbol{Y}$, U, and $\boldsymbol{b}$ for repeated measurements, possibly censored event times, and random effects, respectively. We combine all parameters into one vector, $\boldsymbol{\theta}$, and use $f(\cdot)$ as generic notation for distributions, densities, or likelihood contributions, as appropriate.

1. Obtain initial parameter estimates through separate analyses of the longitudinal and event time data. At this stage, we include the required random effects in the longitudinal

model but ignore them for the event time analysis. We also assume that the association parameters (the αs in Section 15.3.1) are zero.

2. Write down the combined conditional log-likelihood $l(\boldsymbol{\theta}; \boldsymbol{Y}, U, \boldsymbol{b})$ of the observed data given random effects $\boldsymbol{b}$. Any censoring of event times is assumed to be non-informative.

3. Obtain the conditional expectation of each function of $\boldsymbol{b}$ appearing in $l(\boldsymbol{\theta}; \boldsymbol{Y}, U, \boldsymbol{b})$, given $(\boldsymbol{Y}, U)$ and using the current estimates of $\boldsymbol{\theta}$.

4. Replace each function of $\boldsymbol{b}$ appearing in $l(\boldsymbol{\theta}; \boldsymbol{Y}, U, \boldsymbol{b})$ by its conditional expectation. Maximize to update the estimate of $\boldsymbol{\theta}$.

5. Iterate steps 3 and 4 until convergence.

The M-step (4) is straightforward, even under the semi-parametric model that profiles out the baseline hazard function (Wulfsohn and Tsiatis, 1997). The difficulty with the procedure is in the E-step (3), where we require functions of the form

$$E\{h(\boldsymbol{b})|\boldsymbol{Y}, U\} = \int h(\boldsymbol{b}) f(\boldsymbol{b}|\boldsymbol{Y}, U) d\boldsymbol{b} \tag{15.7}$$

for various $h(\cdot)$. In discussing joint modeling in the mid-1990s, the first two authors of this chapter considered EM estimation to this point, but were unable to make further progress because, for their models, the required integrals (15.7) are intractable. Like many of the best ideas, the solution provided by Wulfsohn and Tsiatis (1997) is simple and elegant; note that $f(U|\boldsymbol{Y}, \boldsymbol{b}) = f(U|\boldsymbol{b})$, and use two integrals instead of one:

$$
\begin{aligned}
E\{h(\boldsymbol{b})|\boldsymbol{Y}, U\} &= \int h(\boldsymbol{b}) f(\boldsymbol{b}|\boldsymbol{Y}, U) d\boldsymbol{b} \\
&= \int h(\boldsymbol{b}) \frac{f(U|\boldsymbol{Y}, \boldsymbol{b}) f(\boldsymbol{b}|\boldsymbol{Y})}{f(U|\boldsymbol{Y})} d\boldsymbol{b} \\
&= \frac{\int h(\boldsymbol{b}) f(U|\boldsymbol{b}) f(\boldsymbol{b}|\boldsymbol{Y}) d\boldsymbol{b}}{f(U|\boldsymbol{Y})} \\
&= \frac{\int h(\boldsymbol{b}) f(U|\boldsymbol{b}) f(\boldsymbol{b}|\boldsymbol{Y}) d\boldsymbol{b}}{\int f(U|\boldsymbol{b}) f(\boldsymbol{b}|\boldsymbol{Y}) d\boldsymbol{b}}.
\end{aligned}
$$

The integrands are now standard, as $\boldsymbol{b}|\boldsymbol{Y}$ is Gaussian and $U|\boldsymbol{b}$ follows proportional hazards. Hence quadrature can be used to approximate the expectations.

Note that the likelihood contribution for each subject can be obtained in a similar way, writing

$$
f(\boldsymbol{Y}, U) = \int f(\boldsymbol{Y}, U|\boldsymbol{b}) f(\boldsymbol{b}) d\boldsymbol{b} = \int f(U|\boldsymbol{b}) f(\boldsymbol{Y}|\boldsymbol{b}) f(\boldsymbol{b}) d\boldsymbol{b}
$$

$$
= f(\boldsymbol{Y}) \int f(U|\boldsymbol{b}) f(\boldsymbol{b}|\boldsymbol{Y}) d\boldsymbol{b}.
$$

Wulfsohn and Tsiatis (1997) claim that simple two-point Gauss–Hermite quadrature is sufficient for parameter estimation. In our experience, more accurate quadrature is needed for likelihood evaluation; we used 20-point quadrature for the likelihoods given in the analysis of the liver cirrhosis trial data (Example 3). Of course, the likelihood only needs to be evaluated once, after convergence of the EM algorithm. The greater accuracy is therefore obtained at little cost.

15.5 Multivariate processes

One of the advantages of random-effects models is that they allow easy extension to multivariate repeated measurements, multiple time-to-event outcomes, or both. In practice, most studies are of this kind, yet are often analyzed by extracting a single measurement or time-to-event variable as the outcome of primary interest. In Example 2, repeated measurements are made of three outcomes related to heart function, but the analysis in Lim et al. (2006) focuses on LVMI as the outcome of primary interest. In Example 1, the original data listed eight different reasons for dropout (Diggle, 1998, Table 9.1), but the cited analyses combine these into just two categories: informative (inadequate response to treatment) and non-informative (all others).

15.5.1 Multivariate repeated measurements

Let $\boldsymbol{Y} = (\boldsymbol{Y}_1, \boldsymbol{Y}_2, \ldots, \boldsymbol{Y}_r)$ generically denote r different repeated-measurement response vectors and, as previously, let U and $\boldsymbol{b}$ denote event times and random effects. An obvious approach to joint modeling is to assume conditional independence of the *different* components of $\boldsymbol{Y}$ and U given the random effects; hence,

$$f(\boldsymbol{Y}_1, \boldsymbol{Y}_2, \ldots, \boldsymbol{Y}_r, U|\boldsymbol{b}) = f(\boldsymbol{Y}_1|\boldsymbol{b})f(\boldsymbol{Y}_2|\boldsymbol{b})\ldots f(\boldsymbol{Y}_r|\boldsymbol{b})f(U|\boldsymbol{b}). \qquad (15.8)$$

A mechanistic interpretation of (15.8) is that the different components of $\boldsymbol{Y}$ are simply different measuring instruments for the same underlying "true health." A very simple example would be a model with a scalar random effect b_i for each subject and independent measurement errors,

$$Y_{rij} = \mu_r(t_{ij}) + b_i + e_{rij}, \qquad (15.9)$$

where the e_{rij} are mutually independent, zero-mean Gaussian. A more interesting example would be where the components of $\boldsymbol{Y}$ are conditionally independent of each other given $\boldsymbol{b}$, but autocorrelated in time, in which case the model for each component of $\boldsymbol{Y}$ might include a second level of component-specific random effects.

The ideas discussed earlier apply directly to models of this kind. Note in particular that each conditional distribution may depend on different components of the combined random effect vector $\boldsymbol{b}$, as, for example, would be the case in the multilevel version of (15.9). Note also that there is no requirement for all the responses to be Gaussian, although usually that would be the assumption for the random effects $\boldsymbol{b}$, as the latent-correlation structure is then easily described. If some of the $\boldsymbol{Y}$ variables are not Gaussian, then the EM algorithm summarized above becomes less straightforward because $f(\boldsymbol{b}|\boldsymbol{Y})$ is then unlikely to be available in closed form. Markov chain Monte Carlo methods provide a possible alternative fitting strategy; see Huang et al. (2001) and Xu and Zeger (2001a) for further discussion.

15.5.2 Multiple time-to-event processes

The same shared random-effect idea can be used when there are multiple time-to-event processes, $T_1, T_2, \ldots$. For example, Borgan et al. (2007) investigated the incidence and prevalence of diarrhea in 926 Brazilian infants, followed over a period of 455 days. Four processes were of interest: T_1, a recurrent event process giving the timing of diarrhea episodes; T_2, a recurrent event process giving the timing of episodes of fever; T_3, a single-event time process for dropout from the study, which could be informative; and Y_1, a repeated-measurement series measuring the daily frequency of vomiting. Borgan et al. (2007) used a dynamic covariate analysis to investigate the interrelationship between the four processes and the effects of various social, economic, and demographic risk factors, but a random-effects joint model would provide an alternative approach. Note that the processes T_1 and T_2 are observed in parallel but that the occurrence of T_3 censors further observation of the child in question. This type of competing risk or dependent censoring problem is considered in more detail in the following section.

15.6 Investigating possibly dependent censoring: a re-analysis of the liver cirrhosis trial data

The standard assumption in survival analysis, and in joint modeling, is that censoring times are independent of both the repeated-measurement process and the time-to-event process of interest. This may not be realistic unless censoring is purely administrative, such as at the end of a study follow-up period. In this section, we use a random-effects joint model to investigate the possibility of dependent censoring in the liver cirrhosis trial data of Example 3.

Before undertaking modeling, it is worthwhile to explore the data to assess whether there is any evidence of dependent censoring. The following simple method is an adaptation of the informal exploratory plots suggested by Dobson and Henderson (2003). The terminology assumes that the event of interest is death, which may be censored.

1. To remove the main covariate effects from the repeated-measurement data, fit a standard linear model to these data with no special allowance for dropout (due to death *or* censoring); if likelihood-based methods of estimation are used, this amounts to assuming that the dropout process is MAR. For the remainder of the exploratory analysis, use the residuals from this linear model rather than the original repeated measurements.

2. Choose two time points, τ_1 and $\tau_2 > \tau_1$.

3. Consider only those patients who survive at least to τ_1, and the residuals obtained at step 1 only at times before τ_1. Fit, for each subject separately, a simple linear regression model to these data.

4. Classify subjects according to what happens *between* times τ_1 and τ_2: "died" if the subject died in the interval, "censored" if the subject was censored in the interval, and "survived" if the subject was known to be alive at τ_2.

5. Determine whether there are differences among these three groups in the simple linear regression coefficients obtained at step 3.

6. Explore various choices of τ_1 and τ_2, keeping τ_1 large enough to allow a reasonable number of longitudinal measurements to be available by that time, and choosing τ_2 to make sure none of the three groups is very small.

If there is no relationship between longitudinal response and time to death, then the "died" and "survived" groups should be comparable. Similarly, if censoring is independent of the repeated-measurement response, then the "censored" and "survived" groups should be comparable.

Figure 15.3 illustrates this exploratory procedure for the liver cirrhosis data, with $\tau_1 = 3$ years and $\tau_2 = 6$ years, giving "survived," "died," and "censored" group sizes of 143, 62, and 28, respectively. Following Henderson, Diggle, and Dobson (2002), we used the following covariates in the linear model for prothrombin index Y:

- time, the follow-up time since diagnosis;
- P, an indicator for prednisone treatment;
- $P \times$ time, interaction between treatment and time;
- B, an indicator for baseline time $t = 0$, which provides a simple way of capturing sudden change in the response Y in the very early part of the follow-up period;
- $P \times B$, interaction between prednisone treatment and the baseline time indicator.

In Figure 15.3, there is a suggestion that both slope and intercept may be smaller for those who died in the interval than for those who survived beyond τ_2. This is consistent with low (worse) prothrombin being associated with higher hazards. The relationship between the censored data group and the others is less clear. To investigate further, we calculated

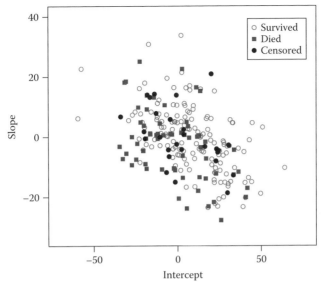

Figure 15.3 Prognostic plot for liver cirrhosis data. The plot shows coefficients from individual simple linear regressions on prothrombin data available to 3 years, after adjusting for covariate effects. The coefficients are grouped according to whether the individual survived the next three years or died or was censored within that period.

the Mahalanobis distances between the group centroids, with scaling determined by the combined variance matrix. We assessed significance by comparing the resulting Mahalanobis distances with the same quantities obtained from 1000 permutations of the group labels (Table 15.1).

Table 15.1 confirms that the group who died between 3 and 6 years already have different average longitudinal profiles between 0 and 3 years than the group who survive at least 6 years. Bearing in mind the small sample sizes, there is a suggestion that the censored group may also have a different mean profile from the survivor group.

Plots like Figure 15.3 are very easy to obtain, and provide useful information, but can be sensitive to the choice of τ_1 and τ_2. As a second way of investigating the possibility of dependent censoring, we fitted a linear random-effects joint model with the censoring indicator δ_i reversed. Because fitting the SGP model C described in Section 15.3.1 is extremely time-consuming, instead we used the random intercept and slope model B; that is, a random effect $b_{i0} + b_{i1}t$ in the repeated-measurement submodel and a proportional random effect $\alpha(b_{i0} + b_{i1}t)$ for the hazard model for censoring. We found $\widehat{\alpha} = 0.006$ and $\widehat{\alpha}/\mathrm{SE}(\widehat{\alpha}) = 1.77$. There is thus some evidence that censoring is associated with prothrombin index and, as $\widehat{\alpha}$ is positive, the suggestion is that there could be a higher probability of censoring for the healthier patients.

However, separately fitting models to the data with the original and reversed censoring indicators can lead to inconsistent conclusions. In each case, there is an implicit assumption

Table 15.1 Mahalanobis Distances between Group Centroids in Exploratory Analysis of Possibly Informative Censoring (See Text for Details)

	Mahalanobis Distance	p-Value
Died vs. Survived	0.346	0.003
Censored vs. Survived	0.319	0.076
Died vs. Censored	0.021	0.867

that times to the two types of event, censoring or death, are independent. Yet both can be found to be associated with the longitudinal data through the random effects. Consequently, a combined analysis, which explicitly allows the possibility of dependence, is preferred.

For the liver cirrhosis trial data, we assumed the following three-component model. For the repeated-measurement submodel, we took

$$Y_{ij} = \boldsymbol{X}_i(t_{ij})\boldsymbol{\beta} + b_{i0} + b_{i1}t_{ij} + e_{ij};$$

for the hazard-of-death submodel,

$$\lambda_{D_i}(t) = \lambda_{D0}(t)\exp\{\mathcal{X}_i\beta_D + \alpha_D(b_{i0} + b_{i1}t)\};$$

and for the hazard-of-censoring submodel,

$$\lambda_{C_i}(t) = \lambda_{C0}(t)\exp\{\mathcal{X}_i\beta_C + \alpha_C(b_{i0} + b_{i1}t)\}.$$

Covariates in $\boldsymbol{X}_i(\cdot)$ are as described above. In $\mathcal{X}_i$, we used only treatment group, P. Note that α_D and α_C can be of the same or opposite signs, so the model can accommodate both positive and negative association between the competing risks of death and censoring. Many competing risks models allow only positive association.

The results are presented in Table 15.2. Those for the repeated-measurement and hazard-of-death submodels are very similar to those obtained by Henderson, Diggle, and Dobson (2002). For the hazard-to-censoring submodel, we find no compelling evidence of association with the random effects, and hence with the other components of the model. Thus, we are reassured that the seldom questioned, independent censoring assumption made in Henderson, Diggle, and Dobson (2002), and in many other published analyses of these data, is tenable and that the results of these analyses are valid. It is difficult to see how

Table 15.2 Results for the Three-Component Random-Effects Analysis of the Liver Cirrhosis Trial Data

	Estimate	SE	Est/SE
Prothrombin Index			
Constant	74.601	1.611	46.294
Time	10.019	2.232	4.489
P	0.430	0.535	0.805
$P \times$ Time	-1.502	0.571	-2.631
B	-1.826	1.342	-1.361
$P \times B$	-11.536	1.997	-5.776
Var(b_0)	371.609	30.724	12.095
Var(b_1)	20.871	5.406	3.861
Corr(b_0, b_1)	0.074	0.106	0.695
Var(e)	281.772	11.642	24.204
Death Hazard			
P	-0.001	0.137	-0.010
α_D	-0.037	0.003	-11.660
Censoring Hazard			
P	0.163	0.139	1.175
α_C	0.006	0.004	1.273

this could have been established formally without a three-component joint model of some kind.

15.7 Discussion

We have discussed several versions of the model (15.5) and (15.6), and of course many others are available. There are also many other approaches to the joint analysis of longitudinal and event time data, including casting the model into a Bayesian framework, that have not been mentioned here. Papers in this area published in recent years include de Bruijne et al. (2001), Rochon and Gillespie (2001), Billingham and Abrams (2002), Do (2002), Dupuy and Mesbah (2002), Lin et al. (2002), Song, Davidian, and Tsiatis (2002a, 2002b), Troxel (2002), Brown and Ibrahim (2003a, 2003b), Ha, Park, and Lee (2003), Hashemi, Jacqmin-Gadda, and Commenges (2003), Dufoil, Brayne, and Clayton (2004), Guo and Carlin (2004), Ibrahim, Chen, and Sinha (2004), Law, Taylor, and Sandler (2004), Ratcliffe, Guo, and Ten Have (2004), Yu et al. (2004), Vonesh, Greene, and Schluchter (2006), Chi and Ibrahim (2006), Hsieh, Tseng, and Wang (2006), and Zeng and Lin (2007).

A common feature of almost all approaches is the reliance for inference upon modeling assumptions that cannot be tested on the observable data. Accepted wisdom is that sensitivity analyses should be performed to assess robustness of conclusions to at least some forms of uncertainty. We add to this a reminder that some modeling assumptions *can* be tested on the observable data and, although no model can be ruled *in*, some models can be ruled *out* through careful diagnostics. This is an area that has perhaps not had sufficient attention in the literature and is worth further development. We also feel that sensitivity to model choice is at least as important as sensitivity within a model to the value of one or a small number of parameters. Again, more attention could be paid to this. A final comment on sensitivity to assumptions is that models should, so far as possible, be informed by the scientific context as well as by empirical analysis of the data. As noted earlier, this argument carries more force if individual-level effects, rather than population-averaged effects, are of direct scientific interest.

Another consideration in choosing a strategy for joint modeling problems is the balance among computational load, inferential efficiency, and model complexity. A complex model may be theoretically more attractive, and is obviously more general than a simpler special case thereof. Against this an unnecessarily complex model inevitably results in less efficient estimation of the important parameters (Altham, 1984). A slightly different point is that model-fitting can itself form part of an exploratory analysis. When many possible models need to be fitted in the course of analyzing a set of data, the ability to fit models quickly and routinely may be important, in which case computationally simple, if potentially statistically inefficient, models and methods of estimation may be preferred to more sophisticated models and methods that require careful tuning to each application.

Finally, we note that although there has been increasing attention in the statistical literature in recent years to joint modeling of longitudinal and event time data, there is still rather limited evidence of the methods finding their way into common use in applications. Part of the reason for this may be a natural time-lag between new methodological developments and their adoption into routine practice. For example, the Cox proportional hazards model only started to have significant numbers of applied citations some 10 or more years after the original paper (Cox, 1972) appeared. Another reason is that the more recent methods have not found their way into easy-to-use standard software. There have been attempts to make software available (e.g., Guo and Carlin, 2004; Vonesh, Greene, and Schluchter, 2006) but, in general, implementation of the kinds of models discussed in this chapter is still far from routine.

Acknowledgments

This work was supported by the U.K. Engineering and Physical Sciences Research Council through the award of a Senior Fellowship to the first author (GR/S48059/01), and by the U.K. Medical Research Council through research grant no. G0400615.

References

Altham, P. M. E. (1984). Improving the precision of estimation by fitting a model. *Journal of the Royal Statistical Society, Series B* **46**, 118–119.

Andersen, P. K., Borgan, Ø., Gill, R. D., and Keiding, N. (1993). *Statistical Models Based on Counting Processes.* New York: Springer.

Berzuini, C. and Larizza, C. (1996). A unified approach for modelling longitudinal and failure time data, with application in medical monitoring. *IEEE Transactions on Pattern Analysis and Machine Intelligence* **18**, 109–123.

Billingham L. J. and Abrams K. R. (2002). Simultaneous analysis of quality of life and survival data. *Statistical Methods in Medical Research* **11**, 25–48.

Borgan, Ø., Fiaccone, R. L., Henderson, R., and Barreto, M. L. (2007). Dynamic analysis of recurrent event data with missing observations, with application to infant diarrhoea in Brazil. *Scandinavian Journal of Statistics* **34**, 53–69.

Brown, E. R. and Ibrahim, J. G. (2003a). Bayesian approaches to joint cure-rate and longitudinal models with applications to cancer vaccine trials. *Biometrics* **59**, 686–693.

Brown, E. R. and Ibrahim, J. G. (2003b). A Bayesian semiparametric joint hierarchical model for longitudinal and survival data. *Biometrics* **59**, 221–228.

Chi, Y. Y. and Ibrahim, J. G. (2006). Joint models for multivariate longitudinal and multivariate survival data. *Biometrics* **62**, 432–445.

Cox, D. R. (1972). Regression models and life tables. *Journal of the Royal Statistical Society, Series B,* **34**, 187–220.

de Bruijne, M. H. J., le Cessie, S., Kluin-Nelemans, H. C., and van Houwelingen, H. C. (2001). On the use of Cox regression in the presence of an irregularly observed time-dependent covariate. *Statistics in Medicine* **20**, 3817–3829.

Diggle, P. J. (1988). An approach to the analysis of repeated measurements. *Biometrics* **44**, 959–971.

Diggle, P. J. and Kenward, M. G. (1994). Informative drop-out in longitudinal data analysis (with discussion). *Applied Statistics* **43**, 49–94.

Diggle, P. J. (1998). Dealing with missing values in longitudinal studies. In B. S. Everitt and G. Dunn (eds.), *Recent Advances in the Statistical Analysis of Medical Data*, pp. 203–228. London: Arnold.

Diggle, P. J., Heagerty, P., Liang, K. Y., and Zeger, S. L. (2002). *Analysis of Longitudinal Data*, 2nd ed. Oxford: Oxford University Press.

Do, K.-A. (2002). Biostatistical approaches for modeling longitudinal and event time data. *Clinical Cancer Research* **8**, 2473–2474.

Dobson, A. and Henderson, R. (2003). Diagnostics for joint longitudinal and dropout time modeling. *Biometrics* **59**, 741–751.

Dufoil, C., Brayne, D., and Clayton, D. (2004). Analysis of longitudinal studies with death and drop-out: a case study. *Statistics in Medicine* **23**, 2215–2226.

Dupuy, J. F. and Mesbah, M. (2002). Joint modelling of event time and nonignorable missing longitudinal data. *Lifetime Data Analysis* **8**, 99–115.

Elbers, C. and Ridder, G. (1982). True and spurious duration dependence: The identifiability of the proportional hazard model. *Review of Economic Studies* **49**, 403–409.

Fitzmaurice, G. M., Laird, N. M., and Ware, J. H. (2004). *Applied Longitudinal Analysis.* Hoboken, NJ: Wiley.

Guo, X. and Carlin, B. P. (2004). Separate and joint modeling of longitudinal and event time data using standard computer packages. *American Statistician* **58**, 16–24.

Ha, I. D., Park, T. S., and Lee, Y. J. (2003). Joint modelling of repeated measures and survival time data. *Biometrical Journal* **45**, 647–658.

Hashemi, R., Jacqmin-Gadda, H., and Commenges, D. (2003). A latent process model for joint modeling of events and marker. *Lifetime Data Analysis* **9**, 331–343.

Heckman, J. and Singer, B. (1984). The identifiability of the proportional hazard model. *Review of Economic Studies* **51**, 231–243.

Henderson, R., Diggle, P., and Dobson, A. (2000). Joint modelling of measurements and event time data. *Biostatistics* **1**, 465–480.

Henderson, R., Diggle, P., and Dobson, A. (2002). Identification and efficacy of longitudinal markers for survival. *Biostatistics* **3**, 33–50.

Hogan, J. W. and Laird, N. M. (1997). Model-based approaches to analysing incomplete longitudinal and failure time data. *Statistics in Medicine* **16**, 259–272.

Hogan, J. W., Roy, J., and Korkontzelou, C. (2004). Tutorial in biostatistics — handling drop-out in longitudinal studies. *Statistics in Medicine* **23**, 1455–1497.

Hougaard, P. (2000). *Analysis of Multivariate Survival Data*. New York: Springer.

Hsieh, F., Tseng, Y. K., and Wang, J. L. (2006). Joint modelling of survival and longitudinal data: Likelihood approach revisited. *Biometrics* **62**, 1037–1043.

Huang, W. Z., Zeger, S. L., Anthony J. C., and Garrett, E. (2001). Latent variable model for joint analysis of multiple repeated measures and bivariate event times, *Journal of the American Statistical Association* **96**, 906–914.

Ibrahim, J. G., Chen, M. H., and Sinha, D. (2004) Bayesian methods for joint modeling of longitudinal and survival data with applications to cancer vaccine trials. *Statistica Sinica* **14**, 863–883.

Laird, N. M. and Ware, J. H. (1982). Random-effects models for longitudinal data. *Biometrics* **38**, 963–974.

Law, N. J., Taylor, J. M. G., and Sandler, H. (2002). The joint modeling of a longitudinal disease progression marker and the failure time process in the presence of cure. *Biostatistics* **3**, 547–563.

Lim, E., Ali, A., Theodorou, P., Sousa, I., Ashrafian, H., Chamageorgakis, A. D., Henein, M., Diggle, P., and Pepper, J. (2008). A longitudinal study of the profile and predictors of left ventricular mass regression after stentless aortic valve replacement. *Annals of Thoracic Surgery*. To appear.

Lin, H., Turnbull, B. W., McCulloch, C. E., and Slate, E. H. (2002). Latent class models for joint analysis of longitudinal biomarker and event process data. *Journal of the American Statistical Association* **97**, 53–65.

Little, R. J. A. (1995). Modeling the drop-out mechanism in repeated-measures studies. *Journal of the American Statistical Association* **90**, 1112–1121.

Little, R. and Rubin, D. (2002). *Statistical Analysis with Missing Data*, 2nd ed. New York: Wiley.

Melino, A. and Sueyoshi, G. (1990). A simple approach to the identifiability of the proportional hazards model. *Economics Letters* **33**, 63–68.

Ratcliffe, S. J., Guo, W., and Ten Have, T. R. (2004). Joint modelling of longitudinal and survival data via a common frailty. *Biometrics* **60**, 892–899.

Rochon, J. and Gillespie, B. W. (2001). A methodology for analysing a repeated measures and survival outcome simultaneously. *Statistics in Medicine*, **20**, 1173–1184.

Rubin, D. B. (1976). Inference and missing data. *Biometrika* **63**, 581–592.

Ruppert, D., Wand, M. P., and Carroll, R. J. (2003). *Semiparametric Regression*. Cambridge: Cambridge University Press.

Song, X. A., Davidian, M., and Tsiatis, A. A. (2002a). An estimator for the proportional hazards model with multiple longitudinal covariates measured with error. *Biostatistics* **3**, 511–528.

Song, X. A., Davidian, M., and Tsiatis, A. A. (2002b). A semiparametric likelihood approach to joint modeling of longitudinal and time-to-event data. *Biometrics* **58**, 742–753.

Stein, M. L. (1999). *Statistical Interpolation of Spatial Data: Some Theory for Kriging*. New York: Springer.

Therneau, T. M. and Grambsch, P. M. (2000). *Modeling Survival Data: Extending the Cox Model*. New York: Springer.

Troxel, A. B. (2002). Techniques for incorporating longitudinal measurements into analyses of survival data from clinical trials. *Statistical Methods in Medical Research* **11**, 237–245.

Tsiatis, A. A. and Davidian, M. (2004). Joint modelling of longitudinal and time-to-event data: an overview. *Statistica Sinica* **14**, 809–834.

Tsiatis, A. A., DeGruttola, V., and Wulfsohn, M. S. (1995). Modelling the relationship of survival to longitudinal data measured with error. Applications to survival and CD4 counts in patients with AIDS. *Journal of the American Statistical Association* **90**, 27–37.

Vonesh, E. F., Greene, T., and Schluchter, M. D. (2006). Shared parameter models for the joint analysis of longitudinal data and event times. *Statistics in Medicine* **25**, 143–163.

Wu, M. C. and Carroll, R. J. (1988). Estimation and comparison of changes in the presence of informative right censoring by modelling the censoring process. *Biometrics*, **44**, 175–88.

Wulfsohn, M. S. and Tsiatis, A. A. (1997). A joint model for survival and longitudinal data measured with error. *Biometrics* **53**, 330–339.

Xu, J. and Zeger, S. L. (2001a). The evaluation of multiple surrogate endpoints. *Biometrics* **57**, 81–87.

Xu, J. and Zeger, S. L. (2001b). Joint analysis of longitudinal data comprising repeated measures and time to events. *Applied Statistics* **50**, 375–388.

Yu, M. G., Law, N. J, Taylor J. M. G., and Sandler, H. M. (2004). Joint longitudinal-survival-cure models and their application to prostate cancer. *Statistica Sinica* **14**, 835–862.

Zeng, D. and Lin, D. Y. (2007). Maximum likelihood estimation in semiparametric regression models with censored data. *Journal of the Royal Statistical Society, Series B*, **69**, 1–30.

CHAPTER 16

Joint models for high-dimensional longitudinal data

Steffen Fieuws and Geert Verbeke

Contents

16.1 Introduction

In many areas of the applied sciences, longitudinal studies are conducted in which some characteristics are measured repeatedly over time for a set of study participants. When a single characteristic is measured longitudinally, we refer to *univariate* longitudinal data. Throughout this chapter, the words *outcome*, *response*, and *marker* will be used interchangeably to refer to such a characteristic. Often, multiple characteristics are measured at each moment in time, instead of a single one. Such a collection of longitudinal measurements will be called *multivariate* longitudinal data. Indeed, at each moment in time, multiple outcomes are measured, and this occurs repeatedly over time. To give a flavor of the type of data, consider first the following examples of multivariate longitudinal profiles.

16.1.1 Hearing data

In a hearing test, hearing threshold sound pressure levels (in decibels) are determined at different frequencies to evaluate the hearing performance of a subject. A hearing threshold is the lowest signal intensity a subject can detect at a specific frequency. In this study, hearing thresholds measured at 11 different frequencies (125 Hz, 250 Hz, 500 Hz, 750 Hz, 1000 Hz, 1500 Hz, 2000 Hz, 3000 Hz, 4000 Hz, 6000 Hz, and 8000 Hz), obtained on 603 male participants from the Baltimore Longitudinal Study of Aging (BLSA) (Shock et al., 1984), are considered. Hearing thresholds are measured at the left as well as at the right ear, leading

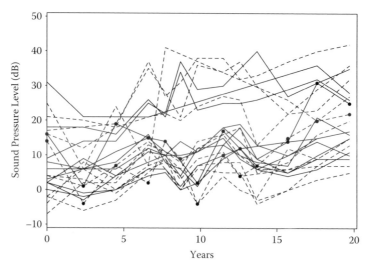

Figure 16.1 Ear- and frequency-specific profiles of hearing thresholds of one randomly selected subject. Dotted and solid lines represent the frequency-specific profiles at the left ear and right ear, respectively. Two profiles are marked with filled circles. These represent the hearing thresholds at both ears for frequency 250 Hz.

to 22 outcomes measured repeatedly over time. The number of visits per subject varies from 1 to 15 (a median follow-up time of 6.9 years). Visits are unequally spaced. Analyses of the hearing data collected in the BLSA study can be found in Brant and Fozard (1990), Morrell and Brant (1991), and Pearson et al. (1995). Ear- and frequency-specific profiles (for all 22 outcomes) for one randomly chosen subject are shown in Figure 16.1.

16.1.2 Family functioning data

An example of multivariate longitudinal binary data has been discussed by O'Brien and Fitzmaurice (2004). The data arise from an interventional trial in psychiatry, designed to study the impact of an intervention program on the transmission of affective disorders to other family members. From each family member (father, mother, child) a binary measure of family functioning was obtained at five equally spaced time points. As such, this is an example of a trivariate longitudinal setting, with the result of father, mother, and child considered as a repeated measurement of family functioning at each point in time. Observe that in this example, the unit for which multivariate longitudinal data are collected is the family and not the subject.

16.1.3 Renal graft data

A third example considers multivariate longitudinal data of mixed types, originating from patients who, between January 21, 1983 and August 16, 2000, underwent a primary renal transplantation with a graft from a deceased or living donor at the University Hospital Gasthuisberg of the Catholic University of Leuven (Belgium). These patients were intensively monitored during the years after transplant. The time intervals between subsequent clinic visits are different within and between the patients. Here, we consider data from 341 patients who survived with a functional transplant for a period of 10 years after the transplantation. Each clinic visit yields a set of biochemical and physiological markers containing valuable information to monitor the health status of the patient. The considered markers are serum creatinine, urine proteinuria, mean of systolic and diastolic blood pressure, and blood hematocrit level. All markers are continuous, except for urine proteinuria which is binary. More details can be found in Fieuws et al. (2007), who discuss analyses on an extended version of this data set, also considering patients with a renal failure within the period of 10 years after transplantation.

The mere availability of multivariate longitudinal data does not imply that a joint model needs to be constructed. It might happen that univariate longitudinal models for each outcome separately answer all research questions. In other examples, a joint modeling strategy is inevitable to answer these questions. Moreover, joint modeling sometimes has additional advantages over the separate analyses of the different outcomes. First, the association structure can be of importance. Consider as an example the renal graft data. A possible question might be how the association between the markers evolves over time ("evolution of the association"). It is known that there exists a relation between the level of hematocrit and the glomerular filtration rate. But does this relation remain constant over time? Another question is how marker-specific evolutions are related to each other ("association of the evolutions"). Do patients with a decrease in hematocrit level show a decrease in filtration rate? Analyses focusing on such questions have been described by Fieuws and Verbeke (2004, 2006) in a bivariate and multivariate setting using the hearing data. Next, a joint model is also needed when interest is in testing fixed effects referring to a set of outcomes simultaneously. As an example, consider the following questions for the family functioning data, where the effect of the intervention can be captured by comparing the evolution of family functioning between the control and the intervention group: "Is there an effect of intervention?" "Is the intervention effect different when functioning is evaluated by the father, the mother, or the child?" The first question could be answered by using separate models for each of the three outcomes. However, a joint model allows a global test for the effect of intervention, considering simultaneously the scores of father, mother, and child. Obviously, the second question cannot be answered without a joint model. Further, a joint model can have an added value for prediction purposes. Consider the situation where various longitudinally measured markers contribute information to discriminate between groups. An example of this can be found in Fieuws et al. (2007), who showed that using multiple longitudinally measured markers and taking into account the correlations between the various markers clearly improved the discrimination between patients with and without a renal graft failure. When parameters are shared by the outcomes, a joint model also often leads to a gain in efficiency. Even when all parameters are outcome-specific, there is a potential gain in efficiency when missing data are present. Finally, construction of a joint model might be motivated by the need to meet assumptions about the missingness process. For example, when the missing at random assumption (see Chapter 17) is valid for an analysis on the multivariate longitudinal data, this might not be true for analyses on each outcome separately, potentially leading to bias.

The core of this chapter is an overview of existing models designed or suitable for multivariate longitudinal data. To structure this overview, a typology of models will be introduced in the next section. This typology roughly distinguishes four model types, based on the use of latent variables to structure the information in the outcome and in the time dimension. Specific sections (16.3 through 16.6) are devoted to discussion of each of the four model types. In Section 16.7 an application on multivariate longitudinal profiles will be discussed, using the previously introduced hearing data. Section 16.8 contains a discussion.

16.2 Typology of joint models

Different approaches exist to model multivariate longitudinal data of which examples have been given in the previous section. We will give an overview of existing models that originate from various modeling traditions; their construction can be motivated by different arguments and they also might differ with respect to a number of formal characteristics, such as the structure of the data, the way the association between and across outcomes is modeled, the scale of the observed outcomes (continuous, ordinal, binary), and the use of latent variables to structure the amount of information at the subject level. Since the latter feature is strongly related to the research questions motivating the data analysis, this feature will be used to distinguish between some broad categories of models.

When modeling multivariate longitudinal data, one option is to use one or more latent variables for the outcome dimension. It is then assumed that the observed outcomes are measuring one or more underlying concepts. Consider as an example the concept of "subjective physical well-being." This concept can be measured using a set of items. Each item score can be considered as an observed outcome, measuring the underlying construct. The other option is to use a latent variable for the time dimension, that is, assuming that the observed measurements are reflecting a latent evolution for each of the outcomes. Whether or not latent variables are assumed for the time dimension and/or the outcome dimension results in four broad types of models. These four types are presented in Figure 16.2 for the hypothetical situation where physical well-being has been measured by three items

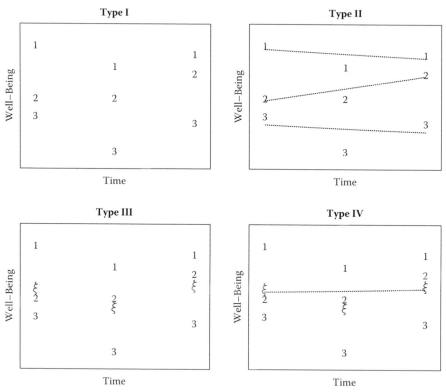

Figure 16.2 Each panel visualizes the application of one of the four model types on a set of nine measurements. The nine measurements originate from three outcomes which have been measured at three points in time. The observed measurements for the three outcomes are denoted with labels 1, 2, and 3. The four model types result from whether or not a latent variable has been used for the outcome and/or the time dimension. If a latent variable is used for the outcome dimension (Type III and IV), the level of the latent construct at a specific point in time is labeled as ξ. Dotted lines indicate that a latent evolution is used to model the observed outcomes (Type II) or the levels of the latent constructs (Type IV).

(outcomes) at three different points in time. We will refer to these models as models for the evolution of observed variables (Type I), models for the latent evolution of observed variables (Type II), models for the evolution of latent variables (Type III), and models for the latent evolution of latent variables (Type IV).

Obviously, the choice of using latent variables is strongly related to the aims of the analysis, that is, the set of research questions. For example, when interest lies in detecting associations between patterns of change of various outcomes, it is obvious that assuming one or more latent variables for the outcome dimension is not appropriate. Note that in this overview we will not consider methods that first summarize the observed information before applying a model to lower-dimensional data. An example of such a strategy based on summary statistics is to take time-specific averages over all outcomes and subsequently use a model for univariate longitudinal profiles. Another strategy would be to use a summary statistic for the longitudinal information (e.g., area under the curve) and to compare this summary statistic between the different outcomes. As pointed out by Heitjan and Sharma (1997), such methods result in loss of information and induce problems when there are missing data. Moreover, such a strategy is not possible when the outcomes are measured on different scales.

As will become clear in the overview of the various models, there exist links between the model type and the type of data structure the model can handle. For all henceforth models discussed, let $Y_{ijk} = Y_{ik}(t_{ijk})$ denote the measurement of subject i for outcome k at the jth time point t_{ijk}, with time intervals between successive t_{ijk} possibly different. Allowing for very general data structures, let $i = 1, \ldots, N$, $k = 1, \ldots, m$, and $j = 1, \ldots, n_{ik}$. We will denote this data structure as fully unbalanced. An important characteristic of a joint model is the extent to which it can deal with such data structures. Many models will assume restricted versions. In a fully balanced structure, $j = 1, \ldots, n, \forall i, k$ and $t_{ijk} = t_j, \forall i, k$. Depending on the restrictions, various grades of unbalancedness can be considered. For example, the time points t at which the measurements of a specific subject are taken can be equal for all m outcomes, but might differ between the N subjects. Following Raudenbush (2001), we can make a rough distinction between three situations: (i) The observed data are balanced, meaning that the number of measurements and the time points are the same for each subject. Moreover, the structure is the same for all outcomes. (ii) The complete data are balanced, meaning that the intended data structure is balanced as in (i), but missingness might occur. (iii) The complete data are unbalanced.

Each model for multivariate longitudinal profiles implies specific assumptions about (i) the association between the longitudinal measurements within an outcome; (ii) the association between the outcomes at a specific time point; and (iii) the association between outcomes at different points in time. These assumptions will follow partially from the strategy chosen to reduce the information (Figure 16.2). The choice of a specific association structure can be driven by the data, by the research question, or by the chosen estimation procedure. Indeed, when interest is not in the association structure itself, the parameters for the association are sometimes considered a nuisance.

Some modeling frameworks only consider one specific type of outcome (e.g., continuous, ordinal, binary). Other frameworks can handle various types of outcomes, or even contain models able to join different types of outcomes.

Although some features that are typical of time series models are used by some of the models presented in this chapter, we do not discuss the literature on multivariate time series. The models proposed in such frameworks typically consider very long series of measurements on one single unit. Some relevant references include Molenaar (1985) and Jørgensen et al. (1996).

16.3 Evolution in observed variables (Type I)

In this section, models not relying on latent variables are discussed. To a large extent, this type of model (Type I) coincides with so-called marginal models. In marginal models (see Chapter 3) parameters characterize the marginal expectation of outcome k at time point j, without conditioning on the value of that outcome at other time points, on other outcomes, or on latent variables. Another class of models discussed in this section are conditionally specified models — more specifically, extensions of transition models for univariate longitudinal data — that incorporate association parameters linking the different outcomes.

16.3.1 Marginal models

Assume the regression model

$$h(E[Y_{ijk}|\boldsymbol{x}_{ijk}]) = \boldsymbol{x}'_{ijk}\boldsymbol{\beta}_{jk}, \tag{16.1}$$

where $\boldsymbol{x}_{ijk}$ is the design vector, which might contain covariates apart from the columns referring to time, $\boldsymbol{\beta}_{jk}$ is the vector of fixed coefficients, and $h(\cdot)$ is an appropriate link function. A functional form, common to all outcomes, might be assumed for the evolution over time. Often, all covariate values in $\boldsymbol{x}_{ijk}$ will be the same for all outcomes.

When the data are Gaussian and $h(\cdot)$ is the identity link, (16.1) reduces to the well-known multivariate linear regression model. When the complete data have a fully balanced structure, the covariance matrix V_i of the vector $\boldsymbol{Y}_i$ of all measurements of subject i can have an unstructured form. Such a matrix would contain $mn(mn + 1)/2$ covariance parameters. Unless the covariance matrix depends on covariates, V_i typically only depends on subject i through the dimension of $\boldsymbol{Y}_i$. If the observed data are complete, $V_i=V$. In many practical situations, the sample size will not be large enough to estimate all covariance parameters. Moreover, an unstructured covariance matrix does not take into account the specific longitudinal character of the data. Various strategies are possible to obtain a more parsimonious structure or a structure more suitable for longitudinally gathered data. By modeling V_i as a Kronecker product of the covariance matrix for, respectively, the m outcomes and the n time points (Galecki, 1994), the number of covariance parameters can be substantially reduced. An application of this idea can be found in O'Brien and Fitzmaurice (2005), who considered measurements taken at five time points for three different outcomes. Replacing the unstructured covariance matrix by a Kronecker product version, the number of covariance parameters was reduced from 120 to 21. Since cross-correlations, that is, correlations between distinct outcomes at various points in time, are products of the marginal correlations specified for the m outcomes and the n time points, respectively, the authors indicate that this approach implies an association structure which might be too restrictive. For example, the correlation between $y_{ij'k}$ and $y_{ijk'}$, conditional on y_{ijk}, will equal zero; that is, to predict a future observation for outcome k at time point t_j, information on another outcome at the current time point has no value if one is already using the current value of outcome k. Still, in the spirit of marginal modeling, that is, with the focus on structuring V_i, alternatives have been proposed by these authors to somewhat relax these restrictions. To this end, they introduced random effects that model the evolution at the subject level and at the outcome level, the latter nested within subject. A seminal version of this idea was discussed for multivariate longitudinal data by Reinsel (1982), who assumed a compound-symmetric structure for V by using an outcome-specific random effect. Since latent variables have been used for putting a more flexible structure on V_i, these models can also be considered as special cases of Type II or Type III models (see Section 16.4 and Section 16.5).

Another marginal modeling approach for multivariate linear regression models was proposed by Carey and Rosner (2001). Designed for irregularly timed observations, they model the marginal covariance matrix V_i by assuming that the intra-outcome and inter-outcome correlations over time follow a damped autoregressive correlation structure. The intra-outcome and inter-outcome correlation functions are, respectively, given by

$$\text{Corr}(Y_{ik}(t), Y_{ik}(t + s)) = \rho_k(s) = \alpha^{|s|^{\theta_k}}$$

and

$$\text{Corr}(Y_{ik}(t), Y_{ik'}(t + s)) = \rho_{kk'}(s) = \alpha^{(|s|+1)^{\theta_{kk'}}},$$

where θ is a time-invariant damping factor and s denotes the distance in time. For example, a first-order autoregressive correlation is obtained when $\theta = 1$. It is noteworthy that the inter-outcome correlation function can be asymmetric, that is, $\rho_{kk'}(s) \neq \rho_{k'k}(s)$. Their approach implies the strong assumption that the variances as well as the correlations between the outcomes remain constant over time. The model is motivated by a specific interest in the magnitude and durability of the correlation between the outcomes.

When the data are discrete such that $h(\cdot)$ in (16.1) is not the identity link function, likelihood-based marginal models (see, for example, Molenberghs and Verbeke, 2005) are very difficult to implement when designed for multivariate longitudinal data, unless nm is relatively small. Indeed, where specification of V_i suffices for the multivariate normal linear regression models, likelihood-based marginal models for multivariate longitudinal discrete data require specification of all higher-order moments. Assuming a balanced structure, $2^{nm} - 1$ multinomial probabilities need to be modeled in the case of binary data. Examples of likelihood-based marginal models designed for a longitudinal context do exist, but only for small nm. For example, Daskalakis, Laird, and Murphy (2002) proposed full maximum likelihood estimation for a marginal model for balanced complete data. The application of their so-called multivariate logistic regression model concerned two binary outcomes, each measured twice in time. The association structure is modeled using odds ratios, its estimation following directly from the likelihood specification.

Specification of the joint distribution for discrete data can be avoided by using a generalized estimating equations (GEE) approach (see Chapter 3). Even when the association within a subject is misspecified in the so-called working correlation matrix, this approach still yields consistent estimates of the regression parameters relating the mean to the set of covariates (Liang and Zeger, 1986; Prentice, 1988). Valid standard errors for these parameters are obtained using the so-called "sandwich" variance estimator. To capture the association, correlations or odds ratios can be used. Carey, Zeger, and Diggle (1993) used odds ratios to model the association between multivariate repeated binary data. The proposed model is a combination of a marginal logistic model for the mean structure and a conditional logistic model to obtain the association parameters. The resulting so-called alternating logistic regression avoids the computational burden associated with second-order GEE (Zhao and Prentice, 1990) when odds ratios are used. The approach can be applied in longitudinal settings, as far as restrictions reasonable within a longitudinal context can be put on all possible pairwise odds ratios. O'Brien and Fitzmaurice (2004) proposed some specific examples indicating how to obtain a parsimonious set of marginal odds ratios within a multivariate longitudinal context. Without considering any restriction, there are $\binom{nm}{2}$ pairwise odds ratios, that is, $m\binom{n}{2}$ intra-outcome odds ratios, $n\binom{m}{2}$ inter-outcome odds ratios, and $\binom{m}{2}(n^2 - n)$ cross odds ratios, the latter referring to the association between two outcomes at different time points. Two proposals have been made to reduce the number of parameters. The first concerns the use of a regression model for the odds ratios, hence applying the idea of alternating logistic regression. The second concerns the Kronecker product approach, leading to the construction of an $n \times n$ matrix and an $m \times m$ matrix for the odds ratios for the time points and outcomes, respectively. The odds ratios are transformed on a scale between -1 and 1 to avoid the fact that the cross-associations would by definition be stronger than both marginal associations.

Ten Have and Morabia (1999) combined a likelihood-based marginal model for bivariate longitudinal binary data with ideas from random-effects models (see also Section 16.4). They used a Dale (1986) model as marginal model, consisting of two univariate logit components and a log odds ratio component, the latter directly modeling the association between the two outcomes at a specific point in time. The longitudinal correlation is modeled by incorporating a random intercept into the model for the log odds ratio component and a random intercept modeling each of the univariate logit components. By using random effects for modeling the longitudinal association, the approach can easily handle unbalanced complete data.

A model for bivariate longitudinal measurements of mixed type, more specifically for the combination of a binary with a continuous outcome, was proposed by Rochon (1996). He specified a GEE model for each of the outcomes and combined the pair of GEE models into an overall GEE framework. The main motivation for the joint model is the construction of a global test involving both outcomes simultaneously. The proposed working correlation structure contains autoregressive features to model the intra- and inter-outcome dependence over time. Building on earlier work for strictly continuous outcomes (Gray and Brookmeyer, 1998), Gray and Brookmeyer (2000) proposed within a marginal framework a time-accelerated model combining longitudinal outcomes of different types. The model deals with the situation where interest lies in estimating a treatment effect. The key assumption is that the treatment and the control group follow the same time trajectory, but at a different rate. This alteration of the time scale may not only concern an acceleration or deceleration, but also a change in the direction of the time effect. The advantage of the approach is that the effect of treatment can easily be compared across the various outcomes, irrespective of the metric of the outcome, and that a common treatment effect can be estimated. They considered combinations of continuous, discrete, and time-to-event outcomes. The association is treated as nuisance and modeled using association indices appropriate for the situation (correlation, odds ratios, etc.). A first example considered the simultaneous analysis of 32 longitudinally measured ordinal quality of life (QOL) items. Global odds ratios were used to model the within-outcome association. Associations between outcomes across time points were assumed to be zero. A second example considered the influence of baseline depression on the rate of Alzheimers disease progression. The latter was measured by two continuous and one time-to-event outcome. Non-linear models and a log-linear model were used to describe the mean evolution of the two continuous outcomes and the time-to-event outcome, respectively. Again, simple working association models were used.

16.3.2 Conditional models

Within a conditional modeling framework, one could model a subject's measurement on outcome k at time point j, conditional on all other $mn - 1$ measurements. Examples of this approach in a non-longitudinal context with binary data can be found, among others, in Geys, Molenberghs, and Ryan (1999). However, this approach clearly does not preserve the longitudinal nature of the data. Within a longitudinal context it is more natural not to condition on the future. Such an approach can be found in transition models which condition only on the (previous) measurements of the outcome itself. In a transition model for univariate discrete longitudinal data (Diggle et al., 2002), the time course is considered as a sequence of states. Transition probabilities can be modeled using a Markov process: the probability of being in a specific state at a specific time point will depend on the state at the previous time point(s) as well as on a set of covariates. Zeng and Cook (2004) present a generalization of this transitional approach for multivariate longitudinal binary data. A transition model is then specified for each outcome separately. The association between the outcomes is captured by modeling the association between the transition behavior of the various outcomes. More specifically, consider as an example the association between the transition from state 0 to state 1 at time point t_j for one outcome k and the transition from state 1 to state 0 at the same time point for another outcome k'. Such associations are modeled using odds ratios, which might also depend on covariates. Again, alternating logistic regression is used to solve the estimating equations involved. In their approach, the focus is on the association between the change in two or more processes, possibly depending on covariates. Although missing data can be handled, the set of possible time points is assumed common to all subjects (balanced complete data). The approach can also easily be extended to discrete data with more than two categories. Another example of a transition model can be found in Liang and Zeger (1989), who proposed a logistic regression model incorporating the effect of baseline covariates, the $m - 1$ binary outcomes observed at the same time point, and all binary outcomes of previous time points, the latter depending on the order of the assumed Markov chain.

16.4 Latent evolution in observed variables (Type II)

The key feature of the models discussed in this section is that the evolution over time of each of the m observed outcomes is modeled assuming they emanate from a latent subject-specific trajectory. More specifically, it will be assumed that $\boldsymbol{Y}_{ik}$, the n_{ik}-vector of longitudinal measurements for subject i on outcome k, satisfies

$$\boldsymbol{Y}_{ik} \mid \boldsymbol{b}_{ik} \sim F_{ik}(\boldsymbol{\theta}_k, \boldsymbol{b}_{ik}), \tag{16.2}$$

that is, conditional on subject-specific effects, $\boldsymbol{b}_{ik}$, $\boldsymbol{Y}_{ik}$ follows a pre-specified n_{ik}-dimensional distribution F_{ik}, possibly depending on covariates, and parameterized through $\boldsymbol{\theta}_k$, a vector of parameters common to all subjects, and the subject-specific vector $\boldsymbol{b}_{ik}$ characterizing the latent trajectory. Within this class of models, it is often assumed that conditional on $\boldsymbol{b}_{ik}$, all measurements in $\boldsymbol{Y}_{ik}$ are independent, such that the distribution function F_{ik} in (16.2) becomes a product over the n_{ik} elements in $\boldsymbol{Y}_{ik}$. The likelihood contribution of $\boldsymbol{Y}_{ik}$ is $L(\boldsymbol{Y}_{ik} \mid \boldsymbol{\theta}_k)$, given by

$$L(\boldsymbol{Y}_{ik} \mid \boldsymbol{\theta}_k) = \int f(\boldsymbol{Y}_{ik} \mid \boldsymbol{\theta}_k, \boldsymbol{b}_{ik}) g(\boldsymbol{b}_{ik} \mid \boldsymbol{\theta}_k) d\boldsymbol{b}_{ik}. \tag{16.3}$$

In the models discussed here, it is assumed that $g(\cdot)$ is a zero-mean centered normal distribution with covariance matrix D_k. Let $\boldsymbol{b}_i = (\boldsymbol{b}'_{i1}, \boldsymbol{b}'_{i2}, \dots, \boldsymbol{b}'_{im})'$ be the vector stacking all outcome-specific random effects $\boldsymbol{b}_{ik}$; then a joint model for the m outcomes is constructed by specifying a joint distribution for $\boldsymbol{b}_i$, which will be multivariate normal with covariance matrix D. This matrix consists of $m \times m$ blocks, with the D_k on the main diagonal. The association between the m outcomes is thus modeled through the parameters in the off-diagonal blocks of D (and possibly through additional parameters if conditional independence does not hold). A clear advantage of this joint modeling approach is that appropriate outcome-specific models can be constructed which are then tied together. First, models will be discussed where the functional form of the latent evolution is assumed to be known in advance. These models will be referred to as confirmatory. Second, models (partially) exploring the form of the latent change will be introduced.

In the statistical literature, linear mixed, generalized linear mixed, and non-linear mixed models have been introduced as flexible tools to model unbalanced longitudinal measurements (Laird and Ware, 1982; Breslow and Clayton, 1993; Davidian and Giltinan, 1995; Verbeke and Molenberghs, 2000; Molenberghs and Verbeke, 2005). This class of models is also known under a number of other names, including multilevel models (Goldstein, 1995), hierarchical linear models (Bryk and Raudenbush, 1992), and variance components models (Longford, 1987). The term *mixed* refers to the presence of random as well as fixed effects in the model. Random effects can be considered as subject-specific regression coefficients, modeling the variability between subjects. Besides possibly other model parameters (e.g., parameters describing serial correlation) needed when conditional independence does not hold, the presence of random effects accounts for the correlation between the measurements within a subject. The choice of the type of mixed model depends on the type of outcome (continuous, ordinal, categorical) and on the functional form of the relation between the outcome and the covariates in the model. Due to the flexibility of the models and the widespread availability of commercial software to fit them, mixed models have developed into a popular tool for analyzing longitudinal data within the biostatistical literature.

Recently, many generalizations of mixed models for multivariate longitudinal data can be found in the context of linear mixed models. The basic idea is that the evolution of the m normally distributed outcomes for each subject can be represented by a linear regression model. The model might differ between the outcomes, but is assumed to be the same for all subjects. As an example, assuming a simple model of linear change for each of the m outcomes, we have for the ith subject

$$
\begin{cases}
Y_{i1}(t) = \beta_{i01} + \beta_{i11}t + \varepsilon_{i1}(t) \\
Y_{i2}(t) = \beta_{i02} + \beta_{i12}t + \varepsilon_{i2}(t) \\
\quad \vdots \\
Y_{ik}(t) = \beta_{i0k} + \beta_{i1k}t + \varepsilon_{ik}(t) \\
\quad \vdots \\
Y_{im}(t) = \beta_{i0m} + \beta_{i1m}t + \varepsilon_{im}(t)
\end{cases}
\tag{16.4}
$$

where, for the kth outcome, $\beta_{i0k} = \beta_{0k} + b_{i0k}$ and $\beta_{1ki} = \beta_{1k} + b_{i1k}$ are the subject-specific intercept and slope, with (b_{i0k}, b_{i1k}) the subject-specific deviations from the average profile described by the fixed intercept and slope, β_{0k} and β_{1k}, respectively. The ε_{ik} are zero-mean normally distributed error components with covariance matrix Σ_{ik}, independent of the random effects. If it is also assumed that conditional on the random effects, all measurements are independent, then Σ_{ik} is diagonal. Using matrix notation, model (16.4) assumes for outcome k that

$$
\boldsymbol{Y}_{ik} \mid \boldsymbol{b}_{ik} \sim N(X_{ik}\boldsymbol{\beta}_k + Z_{ik}\boldsymbol{b}_{ik}, \Sigma_{ik}),
\tag{16.5}
$$

where $N(\cdot, \cdot)$ denotes an n_{ik}-dimensional normal distribution, X_{ik} and Z_{ik} are design matrices for the fixed and subject-specific effects, respectively, and $\boldsymbol{\beta}_k$ is a vector of unknown regression coefficients. Note that X_{ik} can include other (baseline) covariates than the terms referring to the evolution over time. The outcome-specific linear mixed models specified by (16.5) are combined by specifying a joint distribution for $\boldsymbol{b}_i = (b_{i01}, b_{i11}, b_{i02}, b_{i12}, \ldots, b_{i0m}, b_{i1m})$. It is important to realize that the resulting model is still a linear mixed model, but of a much higher dimension. We will denote this joint model for $\boldsymbol{Y}_i$, stacking all outcome-specific $\boldsymbol{Y}_{ik}$, as a multivariate linear mixed model (MLMM). Reinsel (1984) introduced the MLMM model formulation, but estimation was restricted to balanced observed data. In an observational study on elderly people, Beckett, Tancredi, and Wilson (2004) modeled the joint linear evolution of four aspects of cognitive functioning, with the association between the deterioration of the aspects as the focus of interest. The implied $\boldsymbol{b}_i$ vector was eight-dimensional. An example in a bivariate highly unbalanced setting can be found in Chakraborty et al. (2003). They modeled the joint linear evolution of blood and semen HIV-1 RNA, with age and CD4 count as baseline covariates. Other examples of MLMMs can be found in MacCallum et al. (1997) and Matsuyama and Ohashi (1997). Shah, Laird, and Schoenfeld (1997) discussed a more general MLMM. On top of using the random-effects distribution to join the m univariate mixed models, they allowed the error components of the m outcomes to be correlated at each occasion. Although each Σ_{ik} is still diagonal, the covariance matrix Σ_i of the multivariate normal distribution for $\boldsymbol{Y}_i \mid \boldsymbol{b}_i$ will not be diagonal anymore, but will contain non-zero off-diagonal elements referring to covariances between outcomes at the same time point. They applied the approach in a bivariate longitudinal setting. Provided m is sufficiently small, estimation of all these MLMMs can be done using, for example, the SAS procedure `proc mixed`. An illustration can be found in Thiébaut et al. (2002). An extension of the MLMM is presented by Sy, Taylor, and Cumberland (1997). They added a stochastic process for each outcome to the model formulation in expression (16.4). As such, Σ_{ik} is not diagonal anymore. In their model, the different outcomes are joined not only by the random-effects distribution, but also by the multivariate version of the stochastic process. Application of this extension was restricted to a bivariate longitudinal setting, characterized by the requirement of long series of measurements per outcome and subject. Similar in spirit, Heitjan and Sharma (1997) used an additive combination of a random subject effect and a vector autoregressive process to model the error terms.

When MLMMs are fitted within a likelihood framework, observe that they can also be considered as instances of Type I models. Indeed, although the mixed model is built from a hierarchical perspective, the fitted model is a marginal one; that is, the covariance parameters for the random effects and the error components are only variance components used to put a structure on the marginal covariance matrix.

The above examples all consider models that join a set of linear mixed models. The same rationale, however, can be applied in the context of non-linear mixed models, generalized linear mixed models, and combinations thereof. However, from a computational point of view, this generalization is less straightforward because numerical approximation methods are now needed in the calculation of the likelihood in (16.3). This quickly becomes cumbersome or even impossible as the number of random effects increases. As a consequence, examples in the literature of multivariate non-linear mixed models (MNLMMs) or multivariate generalized linear mixed models (MGLMMs) are sparse. Ribaudo and Thompson (2006) compared two treatments for lung cancer using longitudinal measurements of QOL. They modeled the longitudinal evolution for each of the six binary QOL items using a random-intercept logistic regression. The six resulting GLMMs were joined by the six-dimensional distribution for the random effects. They had to recourse to first-order penalized quasi-likelihood estimation to facilitate the estimation process. Agresti (1997) developed a multivariate extension of the Rasch model for a non-longitudinal context. As discussed in Section 16.3, Ten Have and Morabia (1999) proposed for bivariate longitudinal binary data a model integrating a marginal model and random-effects models for the cross-sectional and longitudinal association, respectively. Marshall et al. (2006) discussed a bivariate non-linear mixed model. In general, the m conditional distributions (16.2) themselves might be of a different type. In a non-longitudinal setting, Gueorguieva (2001) presented a general framework to jointly model repeated outcomes in the exponential family. For example, she considered a combination of a gamma and a binomial conditional distribution to jointly model the duration and the number of contractions observed in a myoelectric activity study in ponies. However, we are not aware of studies using such combinations of conditional distributions in a longitudinal setting.

Fieuws and Verbeke (2006) recently proposed a method to circumvent the computational complexity induced by multivariate mixed models of high dimension. The approach boils down to fitting all possible bivariate mixed models. Parameters for the joint model are obtained by simply averaging over the results from the $\binom{m}{2}$ pairwise models. For inferential purposes, pseudo-likelihood theory is used. The method can be applied irrespective of the number of outcomes involved in the joint model, and the type of univariate mixed models (linear, generalized linear, non-linear) that are combined into the joint model. It has been indicated that in most practical situations the efficiency loss will be minor (Fieuws, 2006). The method will be illustrated in Section 16.7 in a 22-variate longitudinal setting. Analyses using the pairwise approach for a joint model combining several generalized linear mixed models can be found in Fieuws et al. (2007). Also, the method has been applied for a joint model combining mixed models of different types (Fieuws et al., 2007).

So far, all models discussed in this section have assumed that $g(\cdot)$ in (16.3) is a multivariate normal distribution. A departure from this normality assumption was presented by Thum (1997) who proposed a multivariate t-distribution for $g(\cdot)$ in a non-longitudinal context. Nagin and Land (1993) and Nagin (1999) assumed a multinomial distribution for $g(\cdot)$ to identify different classes of subjects with respect to their evolution over time (the so-called trajectory groups). The likelihood contribution in (16.3) then involves a sum over the unknown classes, and estimation of the random effects $\boldsymbol{b}_{ik}$ then reduces to estimation of probabilities of class membership. The approach handles normal, censored normal, Poisson, and dichotomous data for $f(\cdot)$ in (16.3), and the evolution is modeled using a polynomial function of time. Depending on the measurement scale, the approach is also known as latent-class growth analysis (Muthén, 2002) or latent-profile analysis. This latent-class approach was extended to a bivariate setting by Nagin and Tremblay (2001), who were especially interested in studying the association pattern between two longitudinally measured outcomes. For this purpose, their approach yields probabilities of membership in the trajectory group of one outcome given membership in a trajectory group of the other outcome, and joint probabilities of membership in trajectory groups across outcomes. This approach can combine outcomes measured on different scales and puts no restrictions on the timing of the measurements.

In the latent-class growth approach discussed so far, the observed outcomes are assumed to be independent given the latent class. Allowing a random effect to have variability within a class relaxes this assumption of homogeneity within a class. However, applications of such a growth mixture modeling approach (Muthén and Shedden, 1999; Li et al., 2001) have not, to our knowledge, been generalized beyond the univariate longitudinal setting.

Typically, a specific functional form for the latent process underlying the longitudinally observed outcomes is assumed within the mixed-models tradition. However, this assumption does not have to be made *a priori*. Emerging from a factor-analysis or latent-variable framework, Meredith and Tisak (1990) have presented latent-curve models for studying inter-individual differences in intra-individual change. In a latent-curve model, the longitudinal measurements for the ith subject on the kth outcome are represented in matrix notation as

$$\boldsymbol{Y}_{ik} = \boldsymbol{\Gamma}_k \boldsymbol{\beta}_{ik} + \boldsymbol{\varepsilon}_{ik}, \tag{16.6}$$

where the columns of $\boldsymbol{\Gamma}_k$ contain basis functions representing aspects of change which are not fixed *a priori*. Hence, $\boldsymbol{\theta}_k$ in (16.2) will also contain the parameters in $\boldsymbol{\Gamma}_k$. Different approaches exist to specify the elements, also denoted as factor loadings, in $\boldsymbol{\Gamma}_k$. At one extreme, a fully exploratory approach is taken, yielding an exploratory factor analysis model. At the other extreme, the basis functions are specified in advance. Suppose a latent-curve model is constructed for linear change such that one basis function represents the intercept and the other the slope. In that case, the formulation of the latent-curve model is equivalent to a linear mixed model or a confirmatory factor analysis model. Recently, it has indeed been recognized that many models studying change have equivalent representations in the latent-curve and the mixed-models framework (e.g., MacCallum et al., 1997; Rovine and Molenaar, 2001). An intermediate approach in the framework for non-linear latent-curve models is described by Browne and Du Toit (1991) and Browne (1993). They define the basis functions in $\boldsymbol{\Gamma}_k$ as the partial first-order derivatives of a target function with respect to the parameters of that function. Each basis function then represents a particular aspect of non-linear change. The elements in $\boldsymbol{\Gamma}_k$ are not specified in advance to have fixed numerical values, but are constrained to lie on a smooth curve. This approach has been illustrated for various non-linear curves such as the exponential, logistic, and Gompertz.

The analysis of change within a latent-curve framework has the benefit of allowing a less restrictive view on the shape of the subject-specific evolutions over time. However, the other side of the coin is a severe loss of flexibility with respect to data structures the method can handle. Indeed, estimation of latent-curve models often proceeds within a structural equation modeling (SEM) framework, using software for fitting covariance structure models. Since the likelihood function is then defined in terms of the moments (means and covariances) of the observed data, usually the possibly unequally spaced time points are required to be common to all subjects. Hence, every Y_i from a given subpopulation will have a common covariance matrix, a restriction that clearly does not apply for the MLMMs handling unbalanced data structures. Indeed, in that situation, the marginal covariance matrix for Y_i, which equals $Z_i D Z_i' + \Sigma_i$, depends on i. As a consequence, applications encountered within the covariance structure analysis framework will typically have a much lower number of time points. Another important limitation of the SEM framework is that covariance structure analysis only applies when $L(Y_{ik} \mid \theta_k)$ is a multivariate normal density. This only holds if both $f(\cdot)$ and $g(\cdot)$ are normal, and if the model is linear in both the fixed- and random-effects parameters.

The extension to the multivariate longitudinal setting is straightforward and proceeds in an analogous manner as in the context of MLMMs. First, latent-curve models are constructed for each of the m outcomes separately, which are then joined by allowing the factor scores $(\beta_{i1}', \beta_{i2}', \ldots, \beta_{ik}', \ldots, \beta_{im}')'$ on the basis functions to be correlated. As with MLMMs, the basis functions might differ between the various outcomes. Applications of the latent-curve framework for multivariate longitudinal data can be found, among others, in Stoolmiller (1994), MacCallum et al. (1997), and Willett and Keiley (2000). As in the mixed-model framework, extensions have been considered. Curran and Bollen (2001) extended, in a bivariate longitudinal setting, the latent-curve approach with an autogressive structure (bivariate autoregressive latent-curve model) and cross-lagged effects between the two outcomes. A similar model was proposed by Ferrer and McArdle (2003). These authors also discuss a so-called latent-difference scores model where cross-lagged effects on the latent level are included.

Another framework, where no prior assumptions are made about the functional form of the subject-specific evolution of a continuous outcome over time, can be found within self-modeling regression (SEMOR) models or spline smoothing approaches. Coull and Staudenmayer (2004) incorporated the latent-variable idea into self-modeling regression. Using spline regression, they assumed for each subject a non-parametric latent curve underlying the evolutions of each of the m outcomes, thereby accounting for the correlation between the outcomes. Analogous to the idea of a factor loading, each outcome-specific curve is considered as a function of the latent one. The correlation within an outcome is modeled using serial correlation for the residuals. Another example of the use of the SEMOR approach for multivariate longitudinal data can be found in Wang, Guo, and Brown (2000) who assumed for each of the m outcomes a latent curve common to all subjects. A subject-specific curve for the jth outcome is considered a function of the common latent curve for outcome j. These authors modeled the correlation among the outcomes by specifying a general covariance structure for the residuals. As such, their approach rather fits within the set of Type I models. The type of research questions underpinning the approaches of Coull and Staudenmayer (2004) on the one hand and Wang, Guo, and Brown (2000) on the other hand are clearly different. For example, Coull and Staudenmayer are *a priori* interested in outcomes that show a similar shape of evolution over time within a subject.

16.5 Evolution in latent variables (Type III)

The defining feature of the Type III models is that they reduce the outcome dimension by assuming one or more underlying latent variables for the outcomes. Building on seminal work by Meredith and Tisak (1990) in the factor-analytic tradition, Oort (2001) discussed a model which can be considered as a typical example of the Type III modeling approach. Let r denote the number of latent variables measured by the m outcomes; then the model for the m observed outcomes of subject i at time point j is given by

$$Y_{ij} = \tau_j + \Lambda_j \xi_{ij} + \Delta_j \varepsilon_{ij},$$
$$\xi_{ij} \sim N(\kappa_j, \Phi_j),$$
$$\varepsilon_{ij} \sim N(0, \Sigma_j),$$

where τ_j is an m-vector of outcome-specific intercepts, Λ_j is an $m \times r$ matrix of (common) factor loadings, and Δ_j is an $m \times m$ matrix of (residual) factor loadings. Further, ξ_{ij} is a random r-vector of scores on the latent variables (the *common factors*), normally distributed with mean κ_j and unstructured covariance matrix Φ_j. ε_{ij} is an m-vector of zero-mean centered uncorrelated scores on the residual factors, the classical error components in regression models, which are also not correlated with ξ_{ij}. Combining the information over all time points yields

$$Y_i = \tau + \Lambda \xi_i + \Delta \varepsilon_i, \tag{16.7}$$

where Y_i, τ, and ε_i are mn-dimensional stacked vectors, ξ_i is an nr-dimensional stacked vector, and Λ and Δ are block-diagonal matrices, composed of the n Λ_js and Δ_js. Model (16.7) implies a well-known general mean and covariance structure, $E(Y_i) = \tau + \Lambda \kappa$ and $\text{Var}(Y_i) = \Lambda \Phi \Lambda' + \Delta \Sigma \Delta'$, where κ stacks the κ_j vectors and Φ and Σ are symmetric matrices consisting of, respectively, all Φ_j and Σ_j matrices. Since Σ is symmetric and each Σ_j is diagonal, the residual factors (the error components) of the same observed variable are allowed to be correlated between time points but not at the same time point. Model (16.7) reflects a general stochastic three-mode model (Oort, 1999) where the three modes are subjects, time points, and outcomes. Such models can be fitted using software for mean and covariance structure analysis, after choosing a set of restrictions for model identification. However, within a longitudinal context, the meaning of the latent variable underlying the m repeatedly measured outcomes should remain the same. Therefore, the specific restrictions $\Lambda_1 = \Lambda_2 = \ldots = \Lambda_j = \ldots = \Lambda_n$ and $\tau_1 = \tau_2 = \ldots = \tau_j = \ldots = \tau_n$ apply, which leads to the so-called longitudinal three-mode model (L3MM; Oort, 1999). Hence, the mean evolution over time of the observed outcomes is reflected by the changes in κ_j. Specific versions of the L3MM assume that the factor scores ξ_{ij} evolve over time along a predefined structure which is common to all subjects, but which may vary between outcomes. Examples can be found in the autoregressive variants of the model where the scores for a specific factor at time point j can be modeled using the scores on another factor at the same time point, and the antecedent scores on the same factor and/or another factor (crossed-lagged effects). Since the models proposed by Oort (2001) are cast into the covariance structure analysis framework, the approach suffers from the limitations already mentioned with respect to the data structure which can be handled. Oort discusses an application considering eight outcomes measuring three underlying constructs at four time points. In addition, Sivo (2001) presented time series models for latent-factor scores, in which autoregressive, moving average, and autoregressive moving average models for the ξ_{ij} have been considered.

The L3MM proposed by Oort (2001) can be considered as a specific application of factor analysis techniques for three-mode data. An obvious alternative is to extend the other well-known reduction technique, principal component analysis (PCA), into the time dimension. Analogous to the factor analysis framework, multivariate longitudinal data can be considered as examples of three-mode data. However, in a PCA framework, the three modes are considered fixed, whereas in the factor analysis approach distributional assumptions are made for one mode. This assumption typically pertains to the subjects; such that the information is aggregated over the subjects; that is, no component matrix is constructed for the subjects. As such, compared to a factor-analytic approach, a three-way component analysis is more exploratory. As an example of this approach, consider various simultaneous component analysis (SCA) models proposed specifically for multivariate longitudinal data, possibly unbalanced and measured at interval scale. The term *simultaneous* refers to the fact that the multivariate time series of more than one subject are modeled simultaneously. Let Y_i denote the $n_i \times m$ matrix of n_i measurements of the ith subject on m variables. Then the multivariate longitudinal profiles of a subject are decomposed as

$$Y_i = F_i B' + E_i, \tag{16.8}$$

where F_i is an $n_i \times q$ matrix containing the scores of the subject on q components, B is a time- and subject-invariant $m \times q$ loading matrix, and E_i is an $n_i \times m$ matrix of residuals. Typically, the measurements in Y_i are preprocessed, for example, centered across occasions for each subject–variable combination separately and possibly normalized subsequently to eliminate scale differences between the variables. Note that this preprocessing only removes the constants (for each subject–variable combination), not the longitudinal trend. Model (16.8) is known as a model for SCA with invariant pattern (SCA-P), originally introduced by Kiers and ten Berge (1994), where the term *invariant* refers to the loading matrix. Depending on the degree of interindividual differences, more or fewer restrictions can be put on the matrix with component scores, leading to different versions of model (16.8) (Timmerman and Kiers, 2003). If no restrictions are used, the variances of the component scores and the covariances between the component scores within subjects may vary across subjects. Restricted versions of model (16.8) have been discussed by Kiers, ten Berge, and Bro (1999) and by Timmerman and Kiers (2003). The intra- and inter-subject variability can be studied by plotting the component scores versus time. Interpretation of the component(s) is based on the loading matrix. Note that the constraints imposed only concern the covariance structure of the component scores and not the scores themselves. Smoothness constraints or even functional forms can be imposed on the evolutions of the component scores over time (Timmerman and Kiers, 2002).

The models discussed developed within the factor analysis and PCA framework only handle continuous outcomes. Ilk and Daniels (2003) considered one latent variable for m binary outcomes. However, they modeled the evolution over time using a Markov structure for the observed outcomes, contrary to the ideas of Oort (2001) and Sivo (2003) who modeled the within-subject dependence over time on the latent level.

Liu and Hedeker (2006) introduced a model for multivariate longitudinal ordinal data. Their model has been proposed within a psychometric context, considering two-parameter item response theory (IRT) models. Let y_{ijk} be the observed ordinal value of subject i at time point j on item (outcome) k. Let u_{ijk} be the corresponding underlying continuous latent variable determining, based on a set of unknown thresholds, the value of the observed y_{ijk}. Note that this notion of latent variable should not be confused with the latent variables used elsewhere in this chapter. The latent variable u_{ijk} is only introduced due to the measurement scale of the observed y_{ijk}. The authors present a probit and logit formulation for the observed y_{ijk}, linking them with u_{ijk}. The model for u_{ijk} is given by

$$u_{ijk} = \boldsymbol{x}'_{ijk}\boldsymbol{\beta} + \boldsymbol{z}'_{ijk}\boldsymbol{b}_i + \lambda_k \xi_{ij} + \varepsilon_{ijk}, \tag{16.9}$$

where $\boldsymbol{x}_{ijk}$ and $\boldsymbol{z}_{ijk}$ are the design vectors for $\boldsymbol{\beta}$, the fixed coefficients, and $\boldsymbol{b}_i$, the zero-mean normally distributed random coefficients, respectively. ξ_{ij}, the latent variable underlying the m items, refers to the first parameter of the two-parameter IRT model, that is, the subjects' ability or proficiency. Hence, only one underlying common factor is assumed. As before, λ_k is the loading of the kth item on the common factor. This is equivalent to the variance of the latent variable ξ_{ij} or the second parameter of the two-parameter IRT model, that is, the item discrimination parameter. The ε_{ijk} are the residuals which follow a normal or logistic zero-mean distribution, depending on the type of regression used (probit or logistic formulation, respectively). By assuming a latent variable measured to a different extent by a set of m outcomes, and the meaning of the latent variable remaining constant over time, the analogy with the L3MM is obvious. However, the correlation between the observed outcomes over time has been modeled by correlated residuals in the L3MM, whereas in model (16.9) random effects $\boldsymbol{b}_i$ (e.g., a random intercept and slope) are used for this purpose. As such, model (16.9) is on the verge of a Type II and a Type III model. Moreover, model (16.9) easily incorporates additional covariates, as opposed to the L3MM. As for the L3MM, model (16.9) assumes a common set of time points for all subjects, although under appropriate missingness assumptions, time points or items might be missing for a specific subject (thus, balanced complete data are required).

A multinomial latent variable ξ_{ij} can also be assumed instead of a continuous one, yielding a set of so-called latent classes. In such an approach a latent class is based on the m outcomes at a specific time point. Note that in Section 16.4 a latent-class approach has been discussed where the classes differed with respect to evolution over time (Nagin and Tremblay, 2001). Reboussin and Reboussin (1998) defined latent classes at a specific time point based on a set of binary outcomes. They modeled the evolution of ξ_{ij}, the transition between the health-risk states, with a first-order stationary transition model. Hence, the current state only depends on the previous one, and this dependence remains constant over time. Analogous to constraining the λ_k in the L3MM, the probability of class membership, conditional on the outcome, is time-invariant to ensure that the meaning of the latent variable does not change.

16.6 Latent evolution in latent variables (Type IV)

Type IV models combine features of Type II and Type III models. On the one hand, one or more latent variables are assumed underlying the m observed outcomes, as has been done in Type III models. On the other hand, a latent trajectory is assumed, not for each of the m observed outcomes, but for the latent variable(s) underlying the m outcomes. Within a latent-curve modeling framework, Type IV models are often referred to as *second-order* latent-growth models, because they model change in a latent variable, as opposed to the *first-order* latent-growth models, modeling change in observed variables. Examples of the latter are discussed in Section 16.4. In a second-order latent-growth model, the latent variables underlying the m outcomes are the first-order factors, whereas the latent variables used to model the evolution of these first-order factors are called the second-order factors. Second-order latent-growth models were introduced by McArdle (1988) and Duncan and Duncan (1996). These models have recently been attracting more attention since they have been formulated as models for mean and covariance structure analysis, which facilitates parameter estimation (e.g., Sayer and Cumsille, 2001; Hancock, Kuo, and Lawrence 2001; Muthén, 2002).

Building further on the L3MM described in Section 16.5 and Browne's (1993) work on first-order latent-growth models (see Section 16.4), Oort (2001) proposes a second-order latent-growth model, which he refers to as a latent-curve three-mode model (LC3MM) and where the scores on a specific factor r are modeled as

$$\boldsymbol{x_i}_{ir} = \Gamma_r \boldsymbol{\eta}_{ir} + \boldsymbol{v}_{ir}, \tag{16.10}$$

where, analogous to expression (16.6), the columns of $\boldsymbol{\Gamma}_r$ contain the basis functions depending on the used target function, now describing the evolution of the latent variable. The $\boldsymbol{\eta}_{ir}$

are the normally distributed random coefficients and $\boldsymbol{v}_{ik}$ is a random vector of residuals, representing deviations of the latent curve, which are uncorrelated with $\boldsymbol{\eta}_{ir}$. Suppose the latent variable is modeled using a linear change model; then $\boldsymbol{\eta}_{ir}$ contains a random intercept and a slope. Again, due to the covariance structure analysis framework, a vast range of possibilities exist for the implied association structure. As indicated before, the price to pay for this flexibility is a restriction with respect to the data structure the model can handle.

An approach similar in spirit to the second-order latent-growth models, but more flexible with respect to the data structure, has been proposed by Roy and Lin (2000). They proposed the so-called latent-variable linear mixed model (LVLMM) for multivariate longitudinal continuous outcomes, assuming that the m outcomes are measuring one underlying latent concept. The motivation for the development of the LVLMM is the inability of the MLMM (see Section 16.4) to estimate a common effect on different outcomes. Although in an MLMM it is possible to assume a common covariate effect on all outcomes, this assumption will be violated rapidly when the outcomes are measured on different scales and units. The similarity and differences with the second-order latent-growth models will become clear if we also assume for the latter only one common factor measured by the m outcomes. The two parts of Oort's L3MM are given by

$$Y_{ijk} = \tau_k + \xi_{ij}\lambda_k + \varepsilon_{ijk},$$
$$\xi_{ij} = \boldsymbol{\gamma}_j'\boldsymbol{\eta}_i + v_{ij},$$

and the two parts of the LVLMM of Roy and Lin are given by

$$Y_{ijk} = \tau_k + \xi_{ij}\lambda_k + b_{ik} + \varepsilon_{ijk},$$
$$\xi_{ij} = X_{ij}'\alpha + \boldsymbol{\gamma}_j'\boldsymbol{\eta}_i + v_{ij}.$$

In the first part of each model Y_{ijk} is the observed measurement at time point j on outcome k for subject i, τ_k is an outcome-specific intercept, ξ_{ij} is the time-specific score on the latent factor, and λ_k is the factor loading of outcome k which remains constant over time. In the LVLMM outcome-specific zero-mean normally distributed random intercepts b_{ik}, assumed to be independent between outcomes, are used to model possibly correlated measurement errors of Y_{ijk} for ξ_{ij} over time.

The second part specifies a model for the latent variable ξ_{ij}, where $\boldsymbol{\gamma}_j'$ refers to a time-specific row in Γ, the latter containing the basis functions (as in [16.6]), and $\boldsymbol{\eta}_i$ is the vector of corresponding random coefficients. In the L3MM, the elements in Γ do not have to be fixed at specific values, such that more exploratory approaches can be taken or patterns of non-linear growth can be handled. If a linear change is assumed for the latent variable, then $\boldsymbol{\gamma}_j' = (1, t_{ij})'$ where t_{ij} expresses the time at which the jth measurement is taken for subject i. These t_{ij}s are assumed to be common to all subjects in the L3MM. Due to this balanced structure, ε_{ijk} and v_{ij}, the zero-mean normally distributed error components of, respectively, the first and the second parts of the model, being both uncorrelated with ξ_{ij} and $\boldsymbol{\eta}_i$, are allowed to be correlated between the different time points in the L3MM, as opposed to the LVLMM. Since in the LVLMM the b_{ik} are assumed independent between outcomes, the cross-sectional correlation between the observed outcomes at a specific point in time is entirely due to the latent variable ξ_{ij}, an assumption which also holds in the L3MM. An important difference stems from the specification of $X_{ij}'\alpha$ in the second part of the LVLMM. Inclusion of additional covariates in the model for the latent variable proceeds in a straightforward way and allows assessment of a common covariate effect (e.g., effect of treatment) on all m outcomes. The latter would only be possible in the L3MM and other conventional second-order latent-growth models by so-called multi-sample analyses (e.g., Hancock, Kuo, and Lawrence, 2001), which are less straightforward to implement, especially for an increasing number of covariates and covariate levels.

In summary, although emerging from completely different traditions, both models share to a large extent the same modeling flavor. Whereas the LVLMM can handle unbalanced data sets and can incorporate easily other covariates in the model for the latent variable, the L3MM offers more flexibility with respect to the number of latent variables for the m outcomes, the underlying evolution for the latent variable, and the assumed association structure (given the data structure is balanced). Note also that the previously mentioned models, with the L3MM and the LVLMM as instances, only deal with continuous outcomes, with the LVLMM even restricted to a linear setting.

16.7 Application

As an illustration of an analysis on multivariate longitudinal data, we consider the hearing data introduced in Section 16.1. Two research questions drive the analysis. The first pertains to whether the deterioration in hearing ability is frequency-specific. The second concerns the association between the evolution in hearing ability at various frequencies. More specifically, is the change in hearing ability for a specific frequency related to the changes for the other frequencies?

The use of latent variables to model the subject-specific evolution over time seems the most natural approach in this setting. Indeed, we are explicitly interested in the association between the evolutions. Also, the complete data are highly unbalanced. Further, a Type III or Type IV model would be too restrictive, since these models would assume perfect correlation between the evolutions of the various frequencies. In a latent-curve framework, often the basis functions representing the aspects of change are estimated (see [16.6]), which would be cumbersome in the current situation because the data are unbalanced. Moreover, previous analyses on the hearing data (Verbeke and Molenberghs, 2000) indicate that the hearing ability for each frequency evolves linearly over time. Therefore, we will use an MLMM as joint model for the hearing data. Verbeke and Molenberghs (2000) proposed the following linear mixed model for a single frequency. Let $Y_i(t)$ denote the hearing threshold at some frequency for a subject i taken at time t: the model is specified as

$$Y_i(t) = (\beta_1 + \beta_2 \text{Age}_i + \beta_3 \text{Age}_i^2 + a_i) + (\beta_4 + \beta_5 \text{Age}_i + b_i)t + \beta_6 F(t) + \varepsilon_i(t), \quad (16.11)$$

in which t is time expressed in years from entry in the study and Age_i equals the age of subject i at the time of entry. The binary time-varying covariate F represents a learning effect from the first to the subsequent visits. Finally, a_i and b_i are, respectively, the random intercept and slope, with $(a_i, b_i)' \sim N(\mathbf{0}, D)$, independent of the error components $\varepsilon_i(t)$, which are independently and identically distributed as $N(0, \sigma^2)$.

Considering all outcomes, let $Y_{1i}(t), Y_{2i}(t), \dots, Y_{22i}(t)$ denote the hearing thresholds of the 11 frequencies at the left as well as at the right ear for a subject i taken at time t. Each of these 22 responses is described using the linear mixed-effects model (16.11). More specifically,

$$\begin{aligned} Y_{1i}(t) &= \mu_1(t) + a_{1i} + b_{1i}t + \varepsilon_{1i}(t), \\ Y_{2i}(t) &= \mu_2(t) + a_{2i} + b_{2i}t + \varepsilon_{2i}(t), \\ &\vdots \\ Y_{22i}(t) &= \mu_{22}(t) + a_{22i} + b_{22i}t + \varepsilon_{22i}(t), \end{aligned} \qquad (16.12)$$

where $\mu_1(t), \mu_2(t), \dots, \mu_{22}(t)$ refer to the average evolutions. The 44 random effects $a_{1i}, a_{2i}, \dots, a_{22i}, b_{1i}, b_{2i}, \dots, b_{22i}$ jointly follow a zero-mean normal distribution with covariance matrix denoted by D. Further, the error components follow a 22-dimensional zero-mean normal distribution with covariance matrix R. The total number of parameters in D and R is $990 + 253 = 1243$. We allow all fixed-effects parameters to be outcome-specific. Model (16.12) is an MLMM of high dimension. Using SAS `proc mixed`, we were only able

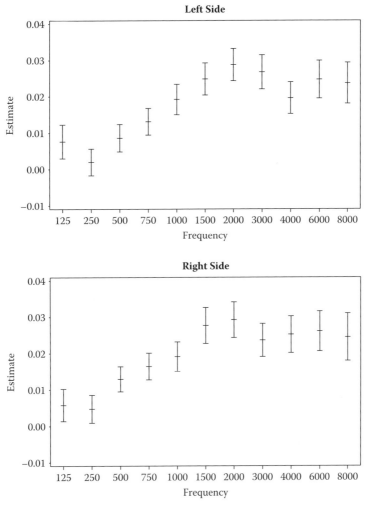

Figure 16.3 Outcome-specific estimates and their 95% confidence intervals for the interaction between the linear evolution and age at entry.

to fit an MLMM on the hearing data involving four frequencies at one ear. To obtain estimates for the parameters in model (16.12), all 231 bivariate mixed models implied by the 22-variate LMM have been fitted. This pairwise fitting approach results in 21 estimates for each fixed effect, each variance of the random effects, each covariance between random effects of the same outcome, and each variance of the error components. A unique estimate for the parameter in the MLMM is then obtained by simply averaging the 21 estimates. For each covariance parameter between the error components and each covariance parameter between random effects of different outcomes, there is only one single estimate available. For inferential purposes, pseudo-likelihood arguments were used. More details on this new computational strategy can be found in Fieuws and Verbeke (2006).

The first research question pertains to the deterioration of the hearing ability, which is represented by β_5, the interaction between age at entry and the linear evolution, in expression (16.11). We are interested in knowing if this deterioration is related to the frequency. Figure 16.3 shows the outcome-specific estimates of this interaction and their 95% confidence intervals. Except for 205 Hz at the left side, all estimates are significantly higher than zero, confirming that age accelerates hearing loss. More important is that at both sides the

Table 16.1 Correlation Matrix of the Random Slopes at the Left Ear

	(1)	(2)	(3)	(4)	(5)	(6)	(7)	(8)	(9)	(10)	(11)
(1)	1.00										
(2)	0.99	1.00									
(3)	0.34	0.43	1.00								
(4)	0.42	0.26	0.88	1.00							
(5)	−0.70	−0.19	0.61	0.77	1.00						
(6)	−0.71	0.05	0.39	0.51	0.80	1.00					
(7)	−0.61	0.09	0.30	0.42	0.50	0.85	1.00				
(8)	0.17	−0.06	0.10	0.03	0.05	0.11	0.34	1.00			
(9)	−0.22	0.10	0.17	0.02	0.02	0.04	0.11	0.74	1.00		
(10)	0.29	0.15	0.08	−0.13	−0.05	0.03	0.06	0.52	0.75	1.00	
(11)	0.45	0.02	−0.16	−0.12	−0.18	−0.10	−0.04	0.19	0.40	0.85	1.00

Note: (1)=125 Hz, (2)=250 Hz, (3)=500 Hz, (4)=750 Hz, (5)=1000 Hz, (6)=1500 Hz, (7)=2000 Hz, (8)=3000 Hz, (9)=4000 Hz, (10)=6000 Hz, (11)=8000 Hz.

increasing trend suggests that this age-related hearing loss is more severe for higher frequencies. Wald-type tests indeed confirm that the estimates for β_5 are significantly different between the frequencies, at the left side ($\chi^2_{10} = 90.4$, $p < 0.0001$) as well as at the right side ($\chi^2_{10} = 110.9$, $p < 0.0001$).

The result for the association of the subject-specific evolutions can be found in the correlation matrix of the random effects, more specifically the correlation between the random slopes. Table 16.1 presents this correlation matrix for the random slopes at the left side. To summarize the results for the association between the evolutions, Figure 16.4 shows the results of a PCA on this correlation matrix of the random effects. This figure plots the component loadings of the slopes on the first and second principal component, such that the cosine of the angle between two arrows reflects the strength of the association in this reduced representation. The result indicates that the closer the frequencies, the more strongly

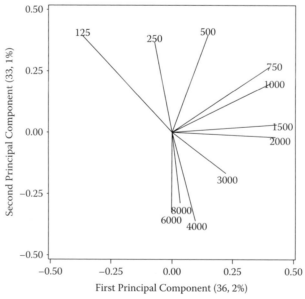

Figure 16.4 Component loadings for the first and second principal components of a PCA on the 11×11 correlation matrix of the random slopes at the left side.

their evolutions are correlated. Very similar results were obtained for the right-hand side (results not shown). The frequency-specific correlation between the slope of the left ear and the slope of the right ear ranged from 0.53 to 0.93. This indicates that a simplified model, assuming the subject-specific slopes to be common for the left and right ear, is not plausible.

16.8 Discussion

In this chapter, we have discussed a variety of models for multivariate longitudinal data. A typology has been presented to roughly distinguish between four model types. The typology is based on the absence or presence of latent variables to model the outcome and/or time dimension at the subjects level.

Building a joint model for multivariate longitudinal data often involves a trade-off between an increased computational complexity on the one hand and a gain in information on the other hand. As a result, many of the joint models discussed in this chapter are limited to situations of relatively low dimension. Still, counterexamples can be found in the various model types. A typical setting where joint models are considered is when interest lies in the estimation of a treatment effect, captured by a set of outcomes. A possible gain in efficiency or a need to report one single conclusion via a global test might motivate the construction of a joint model in such situations. Within a marginal modeling framework (Type I models) Gray and Brookmeyer (1998, 2000) proposed an attractive approach to build a joint model for these purposes: not only can a large set of outcomes be modeled jointly, but the outcomes can be of different types as well. Obviously, a setting where a large set of outcomes can be handled is when the outcomes are assumed to measure a single latent variable. This approach can be found in the Type III as well as in the Type IV models. As an example, consider the LVLMM of Roy and Lin (2000). The implied assumption is that the latent variables measured by each of the observed outcomes are perfectly correlated with each other. Instead of accepting this assumption a priori, a multivariate mixed model (Type II model), allowing outcome-specific random effects, can be used to verify the plausibility of this assumption. If it becomes computationally not feasible to fit this model, the pairwise approach (Fieuws and Verbeke, 2006) can be used. This approach was illustrated on the hearing data in Section 16.7.

When the association structure is not of interest, the association parameters can be considered as nuisance, and valid inferences for the fixed effects of interest are often still possible. However, when the focus of the analysis pertains explicitly to the association structure, the construction of the joint model becomes more complex. The association structure for multivariate longitudinal data consists of three parts: the within-outcome, the between-outcome, and the cross-outcome association. Each joint model either contains parameters governing these associations, or makes assumptions about them. It is of importance to recognize the latter when a joint modeling strategy is chosen. Consider as an example the hearing data, which were analyzed using a mixed-model approach. Fieuws and Verbeke (2004) analyzed in a bivariate setting two aspects of the association structure. First, they verified the strength of the relation between the evolution of one frequency and the evolution of another frequency ("association of the evolutions"). Second, they studied how the association between two frequencies evolved over time ("evolution of the association"). They indicated that assumptions about the association between the error components have a strong impact on the drawn conclusions.

The choice of model type will often be guided by formal characteristics of the specific problem, for example, the structure of the data or the measurement scale of the outcomes considered. Also, the researcher's expertise with a specific model class will influence this choice. However, in the overview presented it became clear that various models can differ heavily in the assumptions they make and the research questions they answer. It seems natural that the choice of a specific class of models should predominantly be guided by a specific research question. However, the fact that the models presented stem from various research traditions will turn this into a challenge for the individual researcher.

References

Agresti, A. (1997). A model for repeated measurements of a multivariate binary response. *Journal of the American Statistical Association* **92**, 315–321.

Beckett, L. A., Tancredi, D. J., and Wilson, R. S. (2004). Multivariate longitudinal models for complex change processes. *Statistics in Medicine* **23**, 213–239.

Brant, L. J. and Fozard, J. L. (1990). Age changes in pure-tone hearing thresholds in a longitudinal study of normal human aging. *Journal of the Acoustical Society of America* **88**, 813–820.

Breslow, N. E. and Clayton, D. G. (1993). Approximate inference in generalized linear mixed models. *Journal of the American Statistical Association* **88**, 9–25.

Browne, M. W. (1993). Structured latent curve models. In C. M. Caudras and C. R. Rao (eds.), *Multivariate Analysis: Future Directions 2*, pp. 171–198. Amsterdam: Elsevier.

Browne, M. W. and Du Toit, S. H. C. (1991). Models for learning data. In L. M. Collins and J. L. Horn (eds.), *Best Methods for the Analysis of Change: Recent Advances, Unanswered Questions, Future Directions*, Chapter 4. Washington, DC: American Psychological Association.

Bryk, A. and Raudenbush, S. (1992). *Hierarchical Linear Models in Social and Behavioral Research: Applications and Data Analysis Methods*. Newbury Park, CA: Sage.

Carey, V. and Rosner, B. A. (2001). Analysis of longitudinally observed irregularly timed multivariate outcomes: Regression with focus on cross-component correlation. *Statistics in Medicine* **20**, 21–30.

Carey, V., Zeger, S., and Diggle, P. (1993). Modelling multivariate binary data with alternating logistic regressions. *Biometrika* **80**, 517–526.

Chakraborty, H., Helms, R. W., Sen, P. K., and Cohen, M. S. (2003). Estimating correlation by using a general linear mixed model: Evaluation of the relationship between the concentration of HIV-1 RNA in blood and semen. *Statistics in Medicine* **22**, 1457–1464.

Coull, B. A. and Staudenmayer, J. (2004). Self-modeling regression for multivariate curve data. *Statistica Sinica* **14**, 695–711.

Curran, P. J. and Bollen, K. A. (2001). Combining autoregressive and latent curve models. In L. M. Collins and A. G. Sayer (eds.), *New Methods for the Analysis of Change*, Chapter 4. Washington, DC: American Psychological Association.

Dale, J. R. (1986). Global cross-ratio models for bivariate, discrete, ordered responses. *Biometrics* **42**, 721–727.

Daskalakis, C., Laird, N. M., and Murphy, J. M. (2002). Regression analysis of multiple-source longitudinal outcomes: A Stirling County depression study. *American Journal of Epidemiology* **155**, 88–94.

Davidian, M. and Giltinan, D. M. (1995). *Nonlinear Models for Repeated Measurement Data*, London: Chapman & Hall.

Diggle, P. J., Heagerty, P., Liang, K. Y., and Zeger, S. L. (2002). *Analysis of Longitudinal Data*, 2nd ed. Oxford: Oxford University Press.

Duncan, S. C. and Duncan, T. E. (1996). Multivariate latent growth curve analysis of adolescent substance abuse. *Structural Equation Modeling* **3**, 323–347.

Ferrer, E. and McArdle, J. J. (2003). Alternative structural models for multivariate longitudinal data. *Structural Equation Modeling* **10**, 493–524.

Fieuws, S. and Verbeke, G. (2004). Joint modelling of multivariate longitudinal profiles: Pitfalls of the random-effects approach. *Statistics in Medicine* **23**, 3093–3104.

Fieuws, S. and Verbeke, G. (2006). Pairwise fitting of mixed models for the joint modelling of multivariate longitudinal profiles. *Biometrics* **62**, 424–431.

Fieuws, S. (2006). Mixed models for multivariate longitudinal data, (Doctoral Thesis). Leuven: Katholieke Universiteit Leuven (unpublished).

Fieuws, S., Verbeke, G., Boen, F., and Delecluse, C. (2006). High-dimensional multivariate mixed models for binary questionnaire data. *Applied Statistics* **55**, 449–460.

Fieuws, S., Verbeke G., Maes, B., and Vanrenterghem, Y. (2007). Predicting renal graft failure using multivariate longitudinal profiles. *Biostatistics* (doi:10.1093/biostatistics/kxm041).

Galecki, A. T. (1994). General class of covariance structures for two or more repeated factors in longitudinal data analysis. *Communications in Statistics — Theory and Methods* **23**, 3105–3119.

Geys, H., Molenberghs, G., and Ryan, L. (1999). Pseudolikelihood modeling of multivariate outcomes in developmental studies. *Journal of the American Statistical Association* **94**, 734–745.

Goldstein, H. (1995). *Multilevel Statistical Models*. New York: Wiley.

Gray, S. M. and Brookmeyer, R. (1998). Estimating a treatment effect from multidimensional longitudinal data. *Biometrics* **54**, 976–988.

Gray, S. M. and Brookmeyer, R. (2000). Multidimensional longitudinal data: Estimating a treatment effect from continuous, discrete or time-to-event response variables. *Journal of the American Statistical Association* **95**, 396–406.

Gueorguieva, R. (2001). A multivariate generalized linear mixed model for joint modelling of clustered outcomes in the exponential family. *Statistical Modelling* **1**, 177–193.

Hancock, G. R., Kuo, W., and Lawrence, F. R. (2001). An illustration of second-order latent growth models. *Structural Equation Modeling* **8**, 470–489.

Heitjan, D. F. and Sharma, D. (1997). Modelling repeated-series longitudinal data. *Statistics in Medicine* **16**, 347–355.

Ilk, O. and Daniels, M. (2003). Marginalised transition random effects models for multivariate longitudinal binary data. In *Proceedings of the ASA, Section on Biometrics*, pp. 1927–1931. Alexandria, VA: American Statistical Association.

Jørgensen, B., Lundbye-Christensen, S., Song, P. X. K., and Sun, L. (1996). State-space models for multivariate longitudinal data of mixed types. *Canadian Journal of Statistics* **24**, 385–402 (correction **25**, 425).

Kiers, H. A. L. and ten Berge, J. M. F. (1994). Hierarchical relations between methods for simultaneous component analysis and a technique for rotation to a simple simultaneous structure. *British Journal of Mathematical and Statistical Psychology* **47**, 109–126.

Kiers, H. A. L., ten Berge, J. M. F., and Bro, R. (1999). PARAFAC2-Part 1: A direct fitting algorithm for the PARAFAC2 model. *Journal of Chemometrics* **13**, 275–294.

Laird, N. M. and Ware, J. H. (1982). Random-effects models for longitudinal data. *Biometrics* **38**, 963–974.

Li, F., Duncan, T. E., Duncan, S. C., and Acock, A. (2001). Latent growth modeling of longitudinal data: A finite growth mixture modeling approach. *Structural Equation Modeling* **8**, 493–530.

Liang, K. Y. and Zeger, S. L. (1986). Longitudinal data analysis using generalized linear models. *Biometrika* **73**, 13–22.

Liang, K. Y. and Zeger, S. L. (1989). A class of logistic regression models for multivariate binary time series. *Journal of the American Statistical Association* **84**, 447–451.

Liu, L. C. and Hedeker, D. (2006). A mixed-effects regression model for longitudinal multivariate ordinal data. *Biometrics* **62**, 261–268.

Longford, N. T. (1987). A fast scoring algorithm for maximum likelihood estimation in unbalanced mixed models with nested random effects. *Biometrika* **74**, 817–827.

MacCallum, R., Kim, C., Malarkey, W., and Kiecolt-Glaser, J. (1997). Studying multivariate change using multilevel models and latent curve models. *Multivariate Behavioral Research* **32**, 215–253.

Matsuyama, Y. and Ohashi, Y. (1997). Mixed models for bivariate response repeated measures data using Gibbs sampling. *Statistics in Medicine* **16**, 1587–1601.

Marshall, G., De la Cruz-Mésia, R., Baron, A., Rutledge, J. H., and Zerbe, G. O. (2006). Nonlinear random effects model for multivariate responses with missing data. *Statistics in Medicine* **25**(16), 2817–2830.

McArdle, J. J. (1988). Dynamic but structural equation modeling of repeated measures data. In J. R. Nesselroade and R. B. Catell (eds.), *Handbook of Multivariate Experimental Psychology*, pp. 561–614, Washington, DC: American Psychological Association.

Meredith, W. and Tisak, J. (1990). Latent curve analysis. *Psychometrika* **55**, 107–122.

Molenaar, P. C. M. (1985), A dynamic factor model for the analysis of multivariate time series. *Psychometrika* **50**, 181–202.

Molenberghs, G. and Verbeke, G. (2005). *Models for Discrete Longitudinal Data*. New York: Springer.

Morrell, C. H. and Brant, L. J. (1991). Modelling hearing thresholds in the elderly. *Statistics in Medicine* **10**, 1453–1464.

Muthén, B. (2002). Beyond SEM: General latent variable modeling. *Behaviormetrika* **29**, 81–117.

Muthén, B. and Shedden, K. (1999). Finite mixture modeling with mixture outcomes using the EM-algorithm. *Biometrics* **55**, 463–469.

Nagin, D. S. (1999). Analysing developmental trajectories: Semi-parametric, group-based approach. *Psychological Methods* **4**, 139–177.

Nagin, D. S. and Land, K. C. (1993). Age, criminal careers, and population heterogeneity: Specification and estimation of a nonparametric, mixed Poisson model. *Criminology* **31**, 327–362.

Nagin, D. S. and Tremblay, R. E. (2001). Analyzing developmental trajectories of distinct but related behaviors: A group-based method. *Psychological Methods* **6**, 18–34.

O'Brien, L. M. and Fitzmaurice, G. M. (2004). Analysis of longitudinal multiple-source binary data using generalized estimating equations. *Applied Statistics* **53**, 177–193.

O'Brien, L. M. and Fitzmaurice, G. M. (2005). Regression models for the analysis of longitudinal Gaussian data from multiple sources. *Statistics in Medicine* **24**, 1725–1744.

Oort, F. J. (1999). Stochastic three-mode models for mean and covariance structures. *British Journal of Mathematical and Statistical Psychology* **52**, 243–272.

Oort, F. J. (2001). Three-mode models for multivariate longitudinal data. *British Journal of Mathematical and Statistical Psychology* **54**, 49–78.

Pearson, J. D., Morrell, C. H., Gordon-Salant, S., Brant, L. J., Metter, E. J., Klein, L. L., and Fozard, J. L. (1995). Gender differences in a longitudinal study of age-associated hearing loss. *Journal of the Acoustical Society of America* **97**, 1196–1205.

Prentice, R. L. (1988). Correlated binary regression with covariates specific to each binary observation. *Biometrics* **44**, 1033–1048.

Raudenbush, S. W. (2001). Toward a coherent framework for comparing trajectories of individual change. In L. M. Collins and A. G. Sayer (eds.), *New Methods for the Analysis of Change*, Chapter 2. Washington, DC: American Psychological Association.

Reboussin, B. A. and Reboussin, D. M. (1998). Latent transition modeling of progression of health-risk behavior. *Multivariate Behavioral Research* **33**, 457–478.

Reinsel, G. (1982). Multivariate repeated-measurement or growth curve models with multivariate random-effects covariance structure. *Journal of the American Statistical Association* **77**, 190–195.

Reinsel, G. (1984). Estimation and prediction in a multivariate random effects generalized linear model. *Journal of the American Statistical Association* **79**, 406–414.

Ribaudo, H. J. and Thompson, S. G. (2002). The analysis of repeated multivariate binary quality of life data: A hierarchical model approach. *Statistical Methods in Medical Research* **11**, 69–83.

Rochon, J. (1996). Analyzing bivariate repeated measures for discrete and continuous outcome variables. *Biometrics* **52**, 740–750.

Rovine, M. J. and Molenaar, P. C. M. (2001). A structural equations modeling approach to the general linear mixed model. In L. M. Collins and A. G. Sayer (eds.), *New Methods for the Analysis of Change*, Chapter 3. Washington, DC: American Psychological Association.

Roy, J. and Lin, X. (2000). Latent variable models for longitudinal data with multiple continuous outcomes. *Biometrics* **56**, 1047–1054.

Sayer, A. G. and Cumsille, P. E. (2001). Second-order latent growth models. In L. M. Collins and A. G. Sayer (eds.), *New Methods for the Analysis of Change*, Chapter 6. Washington, DC: American Psychological Association.

Shah, A., Laird, N., and Schoenfeld, D. (1997). A random-effects model for multiple characteristics with possibly missing data. *Journal of the American Statistical Association* **92**, 775–779.

Shock, N. W., Greullich, R. C., Andres, R., Arenberg, D., Costa, P. T., Lakatta, E. G., and Tobin, J. D. (1984). Normal human aging: The Baltimore Longitudinal Study of Aging. National Institutes of Health publication 84-2450.

Sivo, S. A. (2001). Multiple indicator stationary time series models. *Structural Equation Modeling* **8**, 599–612.

Stoolmiller, M. (1994). Antisocial behavior, delinquent peer association and unsupervised wandering for boys: Growth and change from childhood to early adolescence. *Multivariate Behavioral Research* **29**, 263–288.

Sy, J. P., Taylor, J. M. G., and Cumberland, W. G. (1997). A stochastic model for the analysis of bivariate longitudinal AIDS data. *Biometrics* **53**, 542–555.

Ten Have, T. R. and Morabia, A. (1999). Mixed effects models with bivariate and univariate association parameters for longitudinal bivariate binary response data. *Biometrics* **55**, 85–93.

Thiébaut, R., Jacqmin-Gadda, H., Chêne, G., Leport, C., and Commenges, D. (2002). Bivariate linear mixed models using SAS PROC MIXED. *Computer Methods and Programs in Biomedicine*, **69**, 249–256.

Thum, Y.M. (1997). Hierarchical linear models for multivariate outcomes. *Journal of Educational and Behavioral Statistics* **22**, 77–108.

Timmerman, M. E. and Kiers, H. A. L. (2002). Three-way component analysis with smoothness constraints. *Computational Statistics and Data Analysis* **40**, 447–470.

Timmerman, M. E. and Kiers, H. A. L. (2003). Four simultaneous component models for the analysis of multivariate time series from more than one subject to model intraindividual and interindividual differences. *Psychometrika* **1**, 105–121.

Verbeke, G. and Molenberghs, G. (2000). *Linear Mixed Models for Longitudinal Data*. New York: Springer.

Wang, Y., Guo, W., and Brown, M. B. (2000). Spline smoothing for bivariate data with applications to association between hormones. *Statistica Sinica* **10**, 377–397.

Willett, J. B. and Keiley, M. G. (2000). Using covariance structure analysis to model change over time. In H. E. A. Tinsley and S. D. Brown (eds.), *Handbook of Applied Multivariate Statistics and Mathematical Modelling*, Chapter 23, pp. 665–669, San Diego, CA: Academic Press.

Zeng, L. and Cook, R. J. (2004). Transition models for multivariate longitudinal binary data. Working Paper 2004–03, University of Waterloo Working Paper Series. http://www.stats.uwaterloo.ca/stats_navigation/techreports/04WorkingPapers/2004-03.pdf.

Zhao, L. P. and Prentice, R. L. (1990). Correlated binary regression using a quadratic exponential model. *Biometrika* **77**, 642–648.

PART V

Incomplete Data

CHAPTER 17

Incomplete data: Introduction and overview

Geert Molenberghs and Garrett Fitzmaurice

Contents

17.1 Introduction

Although most longitudinal studies are designed to collect data on every individual in the sample at each time of follow-up, many studies have some missing observations. With longitudinal studies problems of missing data are far more acute than in cross-sectional studies because non-response can occur at any occasion. The term "non-response," as used in this context, denotes that *intended* observations are missing. An individual's response can be missing at one follow-up time and then be measured at a later follow-up time, resulting in a large number of distinct missingness patterns. At the same time, longitudinal studies often suffer from the problem of attrition or "dropout," that is, some individuals "drop out" or withdraw from the study before its intended completion. In either case, the term "missing data" is used to indicate that intended measurements could not be obtained. Although missing data are ubiquitous in longitudinal studies when the study participants are human subjects, and much of the statistical literature has focused on these settings, similar problems with missing data also arise in longitudinal studies in the biological sciences, agriculture, and veterinary medicine.

When longitudinal data are incomplete, there are important implications for their analysis. Clearly, with incomplete data, there is necessarily a loss of information and a reduction in the precision with which longitudinal change can be estimated. This reduction in precision is directly related to the amount of missing data and is influenced, to a certain extent, by the method of analysis. But of somewhat greater concern is the potential for bias. In certain circumstances, missing data can introduce bias and thereby lead to misleading inferences about changes in the response over time. It is this last feature, the potential for serious bias, that complicates the analysis of incomplete longitudinal data. When there are missing data, the validity of any method of analysis will require that certain assumptions about the reasons for missingness, often referred to as the *missing-data mechanism*, be tenable.

In general, the key issue is whether the reasons for missingness are related to the outcome of interest. When missingness is unrelated to the outcome, the impact of missing data is relatively benign and does not unduly complicate the analysis. On the other hand, when it is related to the outcome, somewhat greater care is required because there is potential for bias when individuals with missing data differ in important ways from those with complete data. Consequently, when longitudinal data are incomplete, the reasons for any missingness must be carefully considered.

In this chapter, we review a general taxonomy of missing-data mechanisms, originally introduced by Rubin (1976). These missing-data mechanisms differ in terms of assumptions about whether missingness is related to observed and unobserved responses. We also discuss the implications of these missing-data mechanisms for analysis of longitudinal data. Finally, we briefly review some alternative methods for handling missing data in longitudinal studies. These, and many other topics, are discussed in much greater detail in Chapter 18 through Chapter 23.

17.2 Hierarchy of missing-data mechanisms

To obtain valid inferences from partially missing longitudinal data, we must consider the nature of the "missing-data mechanism." Ordinarily, the missing-data mechanism is not under the control of the investigators; consequently, it is often not well understood. Instead, assumptions are made about the missing-data mechanism, and the validity of the analysis depends on whether these assumptions hold for the data at hand.

Before describing a general classification of missing-data mechanisms, we introduce some notation and terminology. We assume that for each of N individuals, we intend to take n repeated measures of the response variable on the same individual. In the following, we focus exclusively on missingness in the response; although missingness can also arise in covariates, and raises similar considerations, it is not considered here. A subject with a *complete* set of responses has an $n \times 1$ response vector denoted by $\boldsymbol{Y}_i = (Y_{i1}, Y_{i2}, \ldots, Y_{in})'$. In addition, associated with $\boldsymbol{Y}_i$ is an $n \times p$ matrix of covariates, X_i. Because we do not consider missingness in the covariates, we assume that any time-varying covariates are fixed by the study design. Thus, the so-called "complete data" are the $n \times 1$ vector of *intended* responses (and the covariates). Because of missingness, some of the components of $\boldsymbol{Y}_i$ are not observed for at least some individuals. We let $\boldsymbol{R}_i$ be an $n \times 1$ vector of "response indicators," $\boldsymbol{R}_i = (R_{i1}, R_{i2}, \ldots, R_{in})'$, with $R_{ij} = 1$ if Y_{ij} is observed and $R_{ij} = 0$ if Y_{ij} is missing.

Given $\boldsymbol{R}_i$, the "complete data" (i.e., the set of intended responses, $\boldsymbol{Y}_i = (Y_{i1}, \ldots, Y_{in})'$) can be partitioned into two components $\boldsymbol{Y}_i^o$ and $\boldsymbol{Y}_i^m$, corresponding to those responses that are observed and missing, respectively. Here, $\boldsymbol{Y}_i^o$ denotes the vector of *observed* responses on the ith subject, and $\boldsymbol{Y}_i^m$ denotes the complementary set of responses that are missing. These two components are often referred to as the "observed data" and "missing data," respectively. Note that the random vector $\boldsymbol{R}_i$ is recorded for all individuals; in the statistical literature on missing data, the "complete data" together with the missingness indicators are referred to as the "full data." Except when all elements of $\boldsymbol{R}_i$ equal 1, the full-data components are never jointly observed.

The missing-data mechanism describes the probability that a response is observed or missing at any occasion. Specifically, it specifies a probability model for the distribution of the response indicators, $\boldsymbol{R}_i$, conditional on $\boldsymbol{Y}_i^o$, $\boldsymbol{Y}_i^m$, and X_i. A hierarchy of three different types of missing-data mechanisms can be distinguished by considering how $\boldsymbol{R}_i$ is related to

Y_i (and X_i): (i) *missing completely at random* (MCAR); (ii) *missing at random* (MAR); and (iii) *not missing at random* (NMAR). Rubin (1976) introduced this hierarchy of missing-data mechanisms, and it is useful because the type of missing-data mechanism determines the appropriateness of different methods of longitudinal analysis.

17.2.1 Missing completely at random

Data are said to be missing completely at random when the probability that responses are missing is unrelated to either the specific values that, in principle, should have been obtained or the set of observed responses. That is, longitudinal data are MCAR when R_i is independent of both Y_i^o and Y_i^m, the observed and unobserved components of Y_i, respectively. As such, missingness in Y_i is simply the result of a chance mechanism that does not depend on observed or unobserved components of Y_i. An example where partially missing longitudinal data can validly be assumed to be MCAR is the "rotating panel" study design. In such a study design, commonly used in longitudinal surveys to reduce response burden, individuals rotate in and out of the study after providing a predetermined number of repeated measures. However, the number and timing of the measurements are determined by design and are not related to the vector of responses, Y_i.

In the statistical literature, there does not appear to be universal agreement on whether the definition of MCAR also assumes no dependence of missingness on the covariates, X_i. To streamline the discussion of certain points we wish to make, and following Little (1995), we restrict the use of the term MCAR to the case where

$$\Pr(R_i | Y_i^o, Y_i^m, X_i) = \Pr(R_i).$$

When missingness depends on X_i, but is conditionally independent of Y_i,

$$\Pr(R_i | Y_i^o, Y_i^m, X_i) = \Pr(R_i | X_i),$$

a subtle, but important, issue arises. The conditional independence of Y_i and R_i, given X_i, may not hold when conditioning on only a subset of the covariates. Consequently, when an analysis is based on a subset of X_i that excludes a covariate predictive of R_i, Y_i is no longer unrelated to R_i. To avoid any potential ambiguity, Little (1995) suggests that MCAR be reserved for the case where there is no dependence of R_i on Y_i or X_i; when there is dependence on X_i alone, he suggests that the missing-data mechanism be referred to as "covariate-dependent" missingness. However, we caution the reader that this subtle distinction is often not made in the statistical literature on missing data in longitudinal studies. It is therefore wise to always carefully check what an author means when referring to the concept of MCAR.

The essential feature of MCAR is that the observed data can be thought of as a random sample of the complete data. Consequently, all moments, and even the joint distribution, of the observed data do not differ from the corresponding moments or joint distribution of the complete data. This has the following three implications. First, the so-called "completers" (i.e., those subjects with no missing data) can be regarded as a random sample from the target population; moreover, any method of analysis that yields valid inferences in the absence of missing data will also yield valid, albeit inefficient, inferences when the analysis is restricted to the "completers." The latter is often referred to as a "complete-case" analysis. We caution the reader not to confuse the terms *completers* and *complete-case* analysis with the so-called *complete data*. Recall that the latter term refers to the data we would have obtained in the absence of any missingness, while "complete-case" analysis refers to the subset of individuals with no missing data ("completers"). Second, a similar result holds for subjects with any non-response pattern. The conditional distribution of Y_i^o for these subjects coincides with the distribution of the same components of Y_i in the target population.

Third, the distribution of $\boldsymbol{Y}_i^m$ for subjects with any non-response pattern coincides with the distribution of the same components of $\boldsymbol{Y}_i$ for the "completers." Consequently, all available data can be used to obtain valid estimates of moments such as means, variances, and covariances.

An MCAR mechanism has important consequences for the analysis of longitudinal data. In general, all methods for longitudinal analysis that yield valid inferences in the absence of missing data will also yield valid inferences when the analysis is based on all available data, or even when it is restricted to the "completers." Finally, we note that the validity of the assumption of MCAR (or "covariate-dependent" missingness) can be checked empirically from the data at hand, against the alternative of the union of MAR and NMAR. Specifically, we can compare and formally test equality of the distribution of the observed responses across patterns of missingness. In the case of "covariate-dependent" missingness, equality of the distribution of the observed responses across patterns of missingness can be tested after suitable adjustment for the covariates.

17.2.2 Missing at random

In contrast to MCAR, data are said to be missing at random when the probability that responses are missing depends on the set of observed responses, but is further unrelated to the specific missing values that, in principle, should have been obtained. Put another way, if subjects are stratified on the basis of similar values for the responses that have been observed, missingness is simply the result of a chance mechanism that does not depend on the values of the unobserved responses. In particular, longitudinal data are MAR when $\boldsymbol{R}_i$ is conditionally independent of $\boldsymbol{Y}_i^m$, given $\boldsymbol{Y}_i^o$,

$$\Pr(\boldsymbol{R}_i|\boldsymbol{Y}_i^o,\boldsymbol{Y}_i^m,X_i) = \Pr(\boldsymbol{R}_i|\boldsymbol{Y}_i^o,X_i).$$

An example where longitudinal data are MAR arises when a study protocol requires that a subject be removed from the study once the value of an outcome variable falls outside of a certain range of values. In that case, missingness in $\boldsymbol{Y}_i$ is under the control of the investigator and is related to observed components of $\boldsymbol{Y}_i$ only.

Because the missing-data mechanism depends upon $\boldsymbol{Y}_i^o$, the distribution of $\boldsymbol{Y}_i$ in each of the distinct strata defined by the patterns of missingness is not the same as the distribution of $\boldsymbol{Y}_i$ in the target population. The MAR assumption has the following implications. First, the "completers" are a biased sample from the target population; consequently, an analysis restricted to the "completers" is not valid. Furthermore, the conditional distribution of $\boldsymbol{Y}_i^o$ for subjects with any non-response pattern does not coincide with the distribution of the same components of $\boldsymbol{Y}_i$ in the target population. Therefore, the sample means, variances, and covariances based on either the "completers" or the available data are biased estimates of the corresponding moments in the target population. With MAR, the observed data cannot be viewed as a random sample of the complete data.

However, the MAR assumption has the following important implication for the distribution of the missing data. The conditional distribution of an individual's missing values, $\boldsymbol{Y}_i^m$, given the observed values, $\boldsymbol{Y}_i^o$, is the same as the conditional distribution of the corresponding observations for the "completers," conditional on the "completers" having the same values as $\boldsymbol{Y}_i^o$. Put another way, upon stratification on values of $\boldsymbol{Y}_i^o$, the distribution of $\boldsymbol{Y}_i^m$ is the same as the distribution of the corresponding observations in the "completers" and also in the target population. When expressed this way, it should be apparent that the validity of the MAR assumption cannot be checked empirically from the data at hand, against NMAR, unless a very specific alternative model is assumed.

When data are MAR, certain methods of analysis no longer provide valid estimates of longitudinal change without correctly specifying the joint distribution of the responses or correctly modeling the missing-data mechanism. For example, likelihood-based methods can yield valid inferences because, in a certain sense, maximum likelihood estimation implicitly allows the missing values to be validly "predicted" or "imputed" using the observed data and a correct model for the joint distribution of the responses. Alternatively, methods that do not require specification of the joint distribution of the responses can be adjusted to account for the non-response; the necessary adjustments, however, require that a model for $\Pr(\boldsymbol{R}_i|\boldsymbol{Y}_i^o, X_i)$ must be explicitly specified and estimated. Next, we consider each of these two approaches in more detail.

For likelihood-based inferences, the assumption that data are MAR implies that the likelihood contribution for the ith subject can be factored as

$$f(\boldsymbol{Y}_i^o, \boldsymbol{R}_i|X_i, \boldsymbol{\gamma}, \boldsymbol{\psi}) = f(\boldsymbol{R}_i|\boldsymbol{Y}_i^o, X_i, \boldsymbol{\psi}) \times \int f(\boldsymbol{Y}_i^o, \boldsymbol{Y}_i^m|X_i, \boldsymbol{\gamma})d\boldsymbol{Y}_i^m$$
$$= f(\boldsymbol{R}_i|\boldsymbol{Y}_i^o, X_i, \boldsymbol{\psi})f(\boldsymbol{Y}_i^o|X_i, \boldsymbol{\gamma}), \tag{17.1}$$

because $f(\boldsymbol{R}_i|\boldsymbol{Y}_i^o, X_i, \boldsymbol{\psi})$ then does not depend on $\boldsymbol{Y}_i^m$ and consequently is independent of the integrator. Here, $\boldsymbol{\gamma} = (\boldsymbol{\beta}', \boldsymbol{\alpha}')'$ and $\boldsymbol{\psi}$ denotes parameters of the missing-data mechanism. Thus, if $\boldsymbol{\gamma}$ and $\boldsymbol{\psi}$ are variation-independent, the likelihood for $\boldsymbol{\gamma}$ is proportional to the likelihood obtained by ignoring the missing-data mechanism. When $\boldsymbol{\gamma}$ and $\boldsymbol{\psi}$ are variation independent and the data are MAR, the missing-data mechanism is referred to as "ignorable" (Rubin, 1976). Specifically, Rubin (1976) showed that likelihood-based inferences can be based on the likelihood ignoring the missing-data mechanism, obtained by integrating the missing responses from the joint distribution, $f(\boldsymbol{Y}_i|X_i, \boldsymbol{\gamma})$:

$$L(\boldsymbol{\gamma}; \boldsymbol{Y}_i^o, X_i) = \text{constant} \times \prod_{i=1}^{N} \int f(\boldsymbol{Y}_i^o, \boldsymbol{Y}_i^m|X_i, \boldsymbol{\gamma})d\boldsymbol{Y}_i^m.$$

Thus, when data are MAR, the missing values can be "predicted" using the observed data and a model for the joint distribution of $\boldsymbol{Y}_i$. However, the validity of the predictions of the missing values rests upon correct specification of the entire joint distribution. With MAR, one does not require a model for $\Pr(\boldsymbol{R}_i|\boldsymbol{Y}_i^o, X_i)$, only a model for $\boldsymbol{Y}_i$ given X_i. Since MCAR is a special case of MAR, the same is also true of MCAR; for this reason, MCAR and MAR are often referred to as *ignorable* mechanisms; the ignorability refers to the fact that once we establish that $\Pr(\boldsymbol{R}_i|\boldsymbol{Y}_i, X_i)$ does not depend on missing observations, we can ignore $\Pr(\boldsymbol{R}_i|\boldsymbol{Y}_i, X_i)$ and obtain a valid likelihood-based analysis provided we have correctly specified the model for $f(\boldsymbol{Y}_i|X_i)$. While these considerations hold for likelihood-based inferences, and also for Bayesian inferences, they do not for frequentist methods (Rubin, 1976).

Alternatively, methods for analyzing longitudinal data that only require a model for the mean response (e.g., methods based on generalized estimating equations [GEE]), but do not specify the joint distribution of the response vector, can be adapted to provide a valid analysis by explicitly modeling $\Pr(\boldsymbol{R}_i|\boldsymbol{Y}_i^o, X_i)$ The standard GEE approach (Liang and Zeger, 1986) requires that we have a model for the expected value of the responses given the covariates; see Chapter 3 for a comprehensive description of GEE methods. When the data are MAR, this model for the mean response will generally not hold for the observed data. Consequently, the validity of GEE methods applied to the available data is compromised. Recently, methods have been devised for making adjustments to the analysis by using weights. The weights have to be estimated using a model for $\Pr(\boldsymbol{R}_i|\boldsymbol{Y}_i^o, X_i)$, hence the non-response model must be explicitly specified and estimated, although the joint distribution of the responses need not be. These weighting methods are discussed in greater detail in Chapter 20 and Chapter 23.

Finally, we note that the definition of MAR is sometimes confused with MCAR, and the choice of terminology has not helped. However, the subtle distinction between MAR and

MCAR should not be overlooked because it has very important implications for the validity of different methods of analysis of longitudinal data. The MAR assumption is far less restrictive on the missing-data mechanism and may be considered a more plausible assumption about missing data in many applications. Arguably, the MAR assumption should be the default assumption for the analysis of partially missing longitudinal data unless there is a strong and compelling rationale to support the MCAR assumption.

17.2.3 Not missing at random

Missing data are said to be not missing at random when the probability that responses are missing is related to the specific values that should have been obtained, in addition to the ones actually obtained. That is, the conditional distribution of $\boldsymbol{R}_i$, given $\boldsymbol{Y}_i^o$, is related to $\boldsymbol{Y}_i^m$ and $\Pr(\boldsymbol{R}_i|\boldsymbol{Y}_i^o, \boldsymbol{Y}_i^m, X_i)$ depends on at least some components of $\boldsymbol{Y}_i^m$. An NMAR mechanism is often referred to as *non-ignorable* missingness because the missing-data mechanism cannot be ignored when the goal is to make inferences about the distribution of the complete longitudinal responses. For example, in likelihood-based methods we can no longer factorize the individual likelihood contributions as in (17.1). In general, any valid inferential method under NMAR requires specification of a model for the missing-data mechanism.

The term *informative* is sometimes used to describe data that are NMAR, especially for the monotone pattern of missingness owing to dropout. In this context, missingness is said to be "informative" in the sense that when a component of $\boldsymbol{R}_i$ is equal to 0 this provides us with some information about the distribution of the missing observations. Specifically, the distribution of $\boldsymbol{Y}_i^m$, conditional on $\boldsymbol{Y}_i^o$ (and X_i), is not the same as that in the "completers" or in the target population. Rather, the distribution of $\boldsymbol{Y}_i^m$ depends upon $\boldsymbol{Y}_i^o$ and on $\Pr(\boldsymbol{R}_i|\boldsymbol{Y}_i, X_i)$. Thus, the model assumed for $\Pr(\boldsymbol{R}_i|\boldsymbol{Y}_i, X_i)$ is critical and it must be included in the analysis. Moreover, the specific model chosen for $\Pr(\boldsymbol{R}_i|\boldsymbol{Y}_i, X_i)$ can drive the results of the analysis. Note that, in principle, the term "informative" is not necessary, given the concept of (non-)ignorability. However, because it is quite frequently encountered in the literature, it is appropriate to make mention of it. We must also stress that, short of tracking down the missing data, any assumptions made about the missingness process are wholly unverifiable from the data at hand. That is, the observed data provide no information that can either support or refute one NMAR mechanism over another. Recognizing that, without additional information, identification is driven by unverifiable assumptions, many authors have discussed the importance of conducting sensitivity analyses (e.g., Rosenbaum and Rubin, 1983, 1985; Nordheim, 1984; Little and Rubin, 1987; Laird, 1988; Vach and Blettner, 1995; Scharfstein, Rotnitzky, and Robins, 1999; Copas and Eguchi, 2001; Molenberghs, Kenward, and Goetghebeur, 2001; Verbeke et al., 2001). Therefore, when missingness is thought to be NMAR, it is important to assess carefully the sensitivity of inferences to a variety of plausible assumptions concerning the missingness process. Sensitivity analysis under different assumptions about missingness is the topic of Chapter 22.

Next, we consider the implications of NMAR for longitudinal analysis. When data are NMAR, almost all standard methods of longitudinal analysis are invalid. For example, standard likelihood-based methods that ignore the missing-data mechanism yield biased estimates of mean response trends. To obtain valid estimators, joint models for the response vector and the missing-data mechanism are required. In the following section we briefly review the literature on joint models and distinguish two model-based approaches: *selection* and *pattern-mixture* models. The main focus is on likelihood-based methods. However, we note that alternative methods for handling NMAR have been developed that do not require specification of the joint distribution of the longitudinal responses; as stated before, these weighting methods are discussed in detail in Chapter 20 and Chapter 23.

17.3 Joint models for non-ignorable missingness

Joint models are often used in longitudinal analyses to correct for non-ignorable non-response. Little and Rubin (1987) and Little (1993) identified two broad classes of joint models for the longitudinal data and response indicators: *selection* models and *pattern-mixture* models (see Chapter 18 for a discussion of these two factorizations). In selection models, one uses a complete-data model for the longitudinal outcomes, and then the probability of non-response is modeled conditional on the possibly unobserved outcomes. For identifiability, the set of outcomes is usually restricted in some way. That is, with selection models identification comes from unverifiable models for the dependence of the non-response probabilities on the unobserved outcomes. In contrast, with pattern-mixture models, one uses a model for the conditional distribution of the outcomes given non-response patterns and then a model for non-response. It is immediately clear that the distribution of outcomes given patterns of non-response is not completely identifiable because for all but the "completers" pattern, certain response variables are not observed. Hence, restrictions must be built into the model to ensure that there are links among the distributions of the outcomes conditional on the patterns of non-response (Little 1993, 1994; Little and Wang, 1996). In the following sections, we briefly review the early literature on these two types of models and highlight some of the potential advantages and disadvantages of each of the approaches.

17.3.1 Selection models

Some of the earlier work on methods for handling non-ignorable non-response in longitudinal studies was conducted by Wu and colleagues. Focusing on the problem of non-ignorable dropouts, Wu and Carroll (1988) proposed a selection modeling approach used by many subsequent researchers. It assumes that the continuous responses follow a simple linear random-effects model (Laird and Ware, 1982; see also Chapter 1 and Chapter 2) and that the dropout process depends upon an individual's random intercept and slope. Models where the dropout probabilities depend indirectly upon the unobserved responses, via the random effects, are often referred to as *shared-parameter* models (see Chapter 13 and Chapter 19); shared-parameter models are discussed in detail in Chapter 19. Some authors treat them as a separate category, next to selection models and pattern-mixture models. This is a matter of taxonomy, and which classification is preferred may depend on the context. Without taking a strong position, we merely wish to point this out to the reader. An alternative selection modeling approach, based on earlier work in the univariate setting by econometricians (e.g., Heckman, 1976), was proposed by Diggle and Kenward (1994) who allowed the probability of non-response to depend directly on the unobserved outcome rather than on underlying random effects. Selection models, where the non-response probabilities depend directly or indirectly (via dependence on random effects) on unobserved responses, have also been extended to discrete longitudinal data (e.g., Baker, 1995; Fitzmaurice, Molenberghs, and Lipsitz, 1995; Fitzmaurice, Laird, and Zahner, 1996; Molenberghs, Kenward, and Lesaffre, 1997; Ten Have et al., 1998, 2000).

Finally, there is closely related work on selection models where the target of inference is the time-to-event (e.g., dropout time) distribution, rather than the distribution of the repeated measures. Schluchter (1992) and DeGruttola and Tu (1994) independently developed an alternative selection model which assumes the (transformed) event time and the repeated responses to have a joint multivariate normal distribution, with an underlying random-effects structure, and proposed full likelihood approaches, utilizing the EM algorithm (Dempster, Laird, and Rubin, 1977). Tsiatis, DeGruttola, and Wulfsohn (1995) extended this approach to permit a non-parametric time-to-event distribution with censoring. Note that the primary objective of this latter research is inferences about the time-to-event distribution and the ability of the repeated measures to capture treatment effects on, for example, survival; a more detailed discussion of these and other models is presented in Chapter 13 and Chapter 15. In contrast, the more usual focus of selection models for longitudinal data is on estimation of the mean time trend of the repeated measures, and its relation to covariates, regarding the dropout time or non-response patterns as a "nuisance" characteristic of the data.

17.3.2 Pattern-mixture models

Because of the complexity of model fitting in the selection modeling approach proposed by Wu and Carroll (1988), Wu and Bailey (1988, 1989) suggested approximate methods for inference about the time course of the continuous response. Their pattern-mixture modeling approach was based on method-of-moments type fitting of a linear model to the least-squares slopes, conditional on dropout time, then averaging over the distribution of dropout time. Other than assuming dropouts occur at discrete times, this work made no distributional assumption on the event times. Hogan and Laird (1997) extended this pattern-mixture model by permitting censored dropout times, as might arise when there are late entrants to a trial and interim analyses are performed. Follmann and Wu (1995) generalized Wu and Bailey's (1988) conditional linear model to permit generalized linear models without any parametric assumption on the random effects. Other related work on pattern-mixture models is described in Rubin (1977), Glynn, Laird, and Rubin (1986), Mori, Woodworth, and Woolson (1992), Hedeker and Gibbons (1997), Ekholm and Skinner (1998), Molenberghs et al. (1999), Park and Lee (1999), Michiels, Molenberghs, and Lipsitz (1999), and Fitzmaurice and Laird (2000).

There is an additional avenue of research on pattern-mixture models that can be distinguished. Little (1993, 1994) considered pattern-mixture models that stratify the incomplete data by the pattern of missing values and formulate *distinct* models within each stratum. In these models, additional assumptions about the missing-data mechanism have to be made to yield supplemental restrictions that identify the models. Interestingly, Little (1994) relied on a *selection* model for missingness to motivate a set of identifying restrictions for the pattern-mixture model. These identifying restrictions are unverifiable from the observed data; thus Little (1994) recommended conducting a sensitivity analysis for a range of plausible values. Thijs et al. (2002) provided a taxonomy for the various ways to conceive of and fit pattern-mixture models.

17.3.3 Contrasting selection and pattern-mixture models

Pattern-mixture and selection models each have their own distinct advantages and disadvantages. One key advantage of selection models is that they directly model the marginal distribution of the complete data, the usual target of inference in a longitudinal study. In addition, selection models often seem more intuitive to statisticians, as it is straightforward to formulate hypotheses about the non-response process. While assumptions about the non-response process are transparent in selection models, what is less clear is how these translate into assumptions about the distributions of the unobserved outcomes. Furthermore, with selection models, identification comes from postulating unverifiable models for the dependence of the non-response process on the unobserved outcomes. However, except in very simple cases, it can be very difficult to determine the identifying restrictions that must be placed on the model (Glonek, 1998; Bonetti, Cole, and Gelber, 1999). Finally, selection models for longitudinal data can be computationally demanding.

In contrast, pattern-mixture models are often as easy to fit as standard models that assume non-response is ignorable. With pattern-mixture models, it is immediately clear that the distribution of the outcomes given patterns of non-response is not completely identifiable because for all but the "completers" pattern, certain response variables are not observed. Identification comes from postulating unverifiable links among the distributions of the outcomes conditional on the patterns of non-response. But, in contrast to selection models, it is relatively straightforward to determine the identifying restrictions that must be imposed. However, pattern-mixture models do have one very important drawback that has, so far, limited their usefulness in longitudinal data analysis. The main drawback of pattern-mixture models is that the natural parameters of interest are not immediately available; they require marginalization of the distribution of outcomes over non-response patterns. That is, because pattern-mixture models typically parameterize the mean of the longitudinal responses conditional upon covariates and the indicators of non-response patterns, the interpretation of the model parameters is unappealing due to the conditioning or stratification on non-response patterns. In many longitudinal studies the target of inference is the marginal distribution of the repeated measures (i.e., *not* conditional on non-response patterns). Although this marginal distribution can be obtained by averaging over the distribution of the non-response patterns (Wu and Bailey, 1989; Hogan and Laird, 1997; Fitzmaurice and Laird, 2000), the assumed form for the model for the conditional means no longer holds for the marginal means when a non-linear link function has been adopted.

There has been much recent research on selection and pattern-mixture models for handling missing data; this topic is discussed in greater detail in Chapter 18. However, it is worth emphasizing that all models for handling non-ignorable non-response, whether a selection or pattern-mixture modeling approach is adopted, are fundamentally non-identifiable unless some arbitrary modeling assumptions are imposed. That is, inference is possible only once some unverifiable assumptions about the non-response process or the distributions of the missing responses have been made. As a result, considerable caution is required in drawing any conclusions from the analysis. In general, it is important to conduct sensitivity analyses, with results reported for analyses conducted under a range of plausible assumptions about non-ignorable non-response. Sensitivity analysis under different assumptions about missingness is the topic of Chapter 22.

17.4 Methods for handling missing data in longitudinal data analysis

In this final section, we very briefly review three commonly used methods for handling missing data in longitudinal analysis: (i) imputation methods; (ii) likelihood-based methods; and (iii) weighting methods. In a very general way, these approaches are interrelated in the sense that they "impute" certain values for the missing data; the difference is that in (i) the imputation is explicit, whereas in (ii) and (iii) it is implicit. We also discuss the assumptions about missingness required for each of the methods to yield valid inferences in the longitudinal data setting.

17.4.1 Imputation methods

One approach for handling missing data that is widely used in practice is some form of imputation. The basic idea behind imputation is very simple: substitute or fill in the values that were not recorded with imputed values. One of the chief attractions of imputation methods is that, once a filled-in data set has been constructed, standard methods for complete data can be applied. However, methods that rely on just a single imputation, creating only a single filled-in data set, fail to acknowledge the uncertainty inherent in the imputation of the unobserved responses. Multiple imputation circumvents this difficulty. In multiple imputation, the missing values are replaced by a set of M plausible values, thereby acknowledging the uncertainty about what values to impute for the missing responses. The M filled-in data sets produce M different sets of parameter estimates and their standard errors. These are then appropriately combined to provide a single estimate of the parameters of interest, together with standard errors that reflect the uncertainty inherent in the imputation of the unobserved responses. Typically, a small number of imputations (e.g., $5 \leq M \leq 10$) is sufficient to obtain realistic estimates of the sampling variability. Imputation methods are discussed in greater detail in Chapter 21.

The main idea behind multiple imputation is very simple; what is less clear-cut is how to produce the imputed values for the missing responses. Many imputation methods aim to draw values of $\boldsymbol{Y}_i^m$ from the conditional distribution of the missing responses given the observed responses, $f(\boldsymbol{Y}_i^m|\boldsymbol{Y}_i^o, X_i)$. (Note that such distributions are evidently conditional on a vector of parameters, which is suppressed from notation for ease of display.) For example, with the monotone missing-data patterns produced by dropouts, it is relatively straightforward to impute missing values by drawing values of $\boldsymbol{Y}_i^m$ from $f(\boldsymbol{Y}_i^m|\boldsymbol{Y}_i^o, X_i)$ in a sequential manner. A variety of imputation methods can be used to draw values from $f(\boldsymbol{Y}_i^m|\boldsymbol{Y}_i^o, X_i)$; two commonly used methods are propensity score and predictive mean matching (see Chapter 21 for a more detailed discussion of different imputation methods.) When missing values are imputed from $f(\boldsymbol{Y}_i^m|\boldsymbol{Y}_i^o, X_i)$, regardless of the particular imputation method adopted, subsequent analyses of the observed and imputed data are valid when missingness is MAR (or MCAR). Furthermore, multiple imputation ensures that the uncertainty is properly accounted for.

17.4.2 Likelihood-based methods

Likelihood-based methods, regardless of whether the missing-data mechanism is ignored or modeled, can also be thought of as imputation methods. For example, when missingness is assumed to be ignorable, likelihood-based methods are effectively imputing the missing values by modeling and estimating parameters for the joint distribution of $\boldsymbol{Y}_i$, $f(\boldsymbol{Y}_i|X_i, \boldsymbol{\gamma})$. As we discuss in Section 17.2.2, when missingness is ignorable, likelihood-based methods can be used based solely on the marginal distribution of the observed data. That is, maximum likelihood (ML) estimates can be obtained by maximizing $f(\boldsymbol{Y}_i^o|X_i, \boldsymbol{\gamma})$, the ordinary marginal distribution of the particular subset of $\boldsymbol{Y}_i$ determined by $\boldsymbol{Y}_i^o$. In a certain sense, the missing values are validly predicted by the observed data via the model for the conditional mean, $E(\boldsymbol{Y}_i^m|\boldsymbol{Y}_i^o, X_i, \boldsymbol{\gamma})$. This form of "imputation" becomes more transparent when a particular implementation of ML, known as the *EM algorithm* (Dempster, Laird, and Rubin, 1977) is adopted. In the EM algorithm, a two-step iterative algorithm alternates between filling in missing values with their conditional means, given the observed responses and parameter estimates from the previous iteration (the expectation or E-step), and maximizing the likelihood for the resulting "complete data" (the maximization or M-step).

For example, if the responses are assumed to have a multivariate normal distribution, then "predictions" of the missing values in the E-step of the EM algorithm are based on the conditional mean of $\boldsymbol{Y}_i^m$, given $\boldsymbol{Y}_i^o$,

$$E(\boldsymbol{Y}_i^m|\boldsymbol{Y}_i^o, \boldsymbol{\gamma}) = \boldsymbol{\mu}_i^m + \Sigma_i^{mo}\Sigma_i^{o-1}(\boldsymbol{Y}_i^o - \boldsymbol{\mu}_i^o),$$

where $\boldsymbol{\mu}_i^m$ and $\boldsymbol{\mu}_i^o$ denote those components of the mean response vector corresponding to $\boldsymbol{Y}_i^m$ and $\boldsymbol{Y}_i^o$, and Σ^o and Σ_i^{mo} denote those components of the covariance matrix corresponding to the covariance among the elements of $\boldsymbol{Y}_i^o$ and the covariance between $\boldsymbol{Y}_i^m$ and $\boldsymbol{Y}_i^o$.

Thus, when missingness is ignorable, likelihood-based inference does not require specification of the missing-data mechanism, but does require full distributional assumptions about $\boldsymbol{Y}_i$. Furthermore, the model for $f(\boldsymbol{Y}_i|X_i, \boldsymbol{\gamma})$ must be correctly specified. In the example above, any misspecification of the model for the covariance will, in general, yield biased estimates of the mean response trend. When missingness is non-ignorable, a joint model (e.g., selection or pattern-mixture model) is required and inferences are sensitive to model assumptions.

17.4.3 Weighting methods

An alternative approach for handling missing data is to weight the observed data in some appropriate way. In weighting methods, the underrepresentation of certain response profiles in the observed data is taken into account and corrected. A variety of different weighting methods that adjust for missing data have been proposed. These approaches are often called propensity weighted or inverse probability weighted methods and are the main topic of Chapter 20.

In weighting methods, the underlying idea is to base estimation on the observed responses but weight them to account for the probability of non-response. Under MAR, the propensities for non-response can be estimated as a function of the observed responses and also as a function of the covariates and any additional variables that are thought likely to predict non-response. Moreover, dependence on unobserved responses can also be incorporated in these methods (Rotnitzky and Robins, 1997; see also Chapter 20).

Weighting methods are especially simple to apply in the case of monotone missingness due to dropout. When there is dropout, we can replace the vector of response indicators, R_i, with a simple dropout indicator variable, D_i, with $D_i = k$ if the ith individuals drops out between the $(k-1)$th and kth occasions. In that case, the required weights can be obtained sequentially as the product of the propensities for dropout,

$$w_{ij} = (1 - \pi_{i1}) \times (1 - \pi_{i2}) \times \cdots \times (1 - \pi_{ij}),$$

where w_{ij} denotes the probability that the ith subject is still in the study at the jth occasion and $\pi_{ik} = \Pr(D_i = k | D_i \geq k)$ can be estimated from those remaining at the $(k-1)$th occasion, given the recorded history of all available data up to the $(k-1)$th occasion. Then, given estimated weights, say $\widehat{w}_{ij}$ (based on the $\widehat{\pi}_{ij}$s), a weighted analysis can be performed where the available data at the jth occasion are weighted by $\widehat{w}_{ij}^{-1}$. For example, the GEE approach can be adapted to handle data that are MAR by making adjustments to the analysis for the propensities for dropout, or for patterns of missingness more generally (Robins, Rotnitzky, and Zhao, 1995).

The intuition behind the weighting methods is that each subject's contribution to the weighted analysis is replicated w_{ij}^{-1} times, to count once for herself and $w_{ij}^{-1} - 1$ times for those subjects with the same history of prior responses and covariates, but who dropped out. Thus, weighting methods can also be thought of as methods for imputation where the observed responses are counted more than once (w_{ij}^{-1} times).

Inverse probability weighted methods have a long and extensive history of use in statistics. They were first proposed in the sample survey literature (Horvitz and Thompson, 1952), where the weights are known and based on the survey design. In contrast to sample surveys, here the weights are not ordinarily known, but must be estimated from the observed data (e.g., using a repeated sequence of logistic regressions for the π_{ij}s). Therefore, the variance of inverse probability weighted estimators must also account for estimation of the weights. In general, weighting methods are valid provided the model that produces the estimated w_{ij}s is correctly specified.

17.5 Concluding remarks

In longitudinal studies, missing data are the rule, not the exception, and pose a major challenge for data analysis. Since the seminal paper on missing data by Rubin (1976), this has been a fertile area for methodological research. There now exists an extensive body of literature on both likelihood-based and multiple-imputation methods for handling missing data in longitudinal studies; this literature is synthesized in Chapters 18, 19, and 21. More recently, there have been important advances in methods for handling missing data in semi-parametric models for longitudinal data. The theory for estimation of semi-parametric models with missing data is comprehensively described in Chapter 20.

Finally, we note that causal inference from longitudinal studies can be construed as a "missing-data problem." For example, in the so-called *counterfactual* formulation of causation, causal inference under models for *potential outcomes* can be regarded as a problem of making inference from missing data. Chapter 23 provides an extensive and comprehensive review of methods for the estimation of the causal effect of a time-varying exposure or treatment intervention in longitudinal studies. In particular, Chapter 23 describes recent methods for dynamic, time-varying treatment/exposure regimes and also tackles the thorny problem of deducing the optimal regime.

Acknowledgments

This work was supported by grants GM 29745 and MH 54693 from the U.S. National Institutes of Health, and by the Belgian Science Policy IAP Research Network #P6/03.

References

Baker, S. G. (1995). Marginal regression for repeated binary data with outcome subject to non-ignorable non-response. *Biometrics* **51**, 1042–1052.

Bonetti, M., Cole, B. F., and Gelber, R. D. (1999). A method-of-moments estimation procedure for categorical quality-of-life data with non-ignorable missingness. *Journal of the American Statistical Association* **94**, 1025–1034.

Copas, J. and Eguchi, S. (2001). Local sensitivity approximations for selectivity bias. *Journal of the Royal Statistical Society, Series B* **63**, 871–895.

DeGruttola, V. and Tu, X. M. (1994). Modeling progression of CD4 lymphocyte count and its relationship to survival time. *Biometrics* **50**, 1003–1014.

Dempster, A. P., Laird, N. M., and Rubin, D. B. (1977). Maximum likelihood with incomplete data via the EM algorithm. *Journal of the Royal Statistical Society, Series B* **39**, 1–38.

Diggle, P. J. and Kenward, M. G. (1994). Informative dropout in longitudinal data analysis (with discussion). *Applied Statistics* **43**, 49–93.

Ekholm, A. and Skinner, C. (1998). The Muscatine children's obesity data reanalysed using pattern-mixture models. *Applied Statistics* **47**, 251–263.

Fitzmaurice, G. M. and Laird, N. M. (2000). Generalized linear mixture models for handling non-ignorable dropouts in longitudinal studies. *Biostatistics* **1**, 141–156.

Fitzmaurice, G. M., Laird, N. M., and Zahner, G. E. P. (1996). Multivariate logistic models for incomplete binary responses. *Journal of the American Statistical Association* **91**, 99–108.

Fitzmaurice, G. M., Molenberghs, G., and Lipsitz, S. R. (1995). Regression models for longitudinal binary responses with informative drop-outs. *Journal of the Royal Statistical Society, Series B* **57**, 691–704.

Follmann, D. and Wu, M. (1995). An approximate generalized linear model with random effects for informative missing data. *Biometrics*, **51**, 151–168.

Glonek, G. F. V. (1998). On identifiability in models for incomplete binary data. *Statistics and Probability Letters* **41**, 191–197.

Glynn, R. J., Laird, N. M., and Rubin, D. B. (1986). Selection modelling versus mixture modelling with non-ignorable non-response. In H. Wainer (ed.), *Drawing Inferences from Self-Selected Samples*. New York: Springer.

Heckman, J. (1976). The common structure of statistical models of truncation, sample selection and limited dependent variables, and a simple estimator for such models. *Annals of Economic Social Measurements* **5**, 475–492.

Hedeker, D. and Gibbons, R. D. (1997). Application of random-effects pattern-mixture models for missing data in longitudinal studies. *Psychological Methods* **2**, 64–78.

Hogan, J. W. and Laird, N. M. (1997). Mixture models for the joint distribution of repeated measures and event times. *Statistics in Medicine* **16**, 239–258.

Horvitz, D. G. and Thompson, D. J. (1952). A generalization of sampling without replacement from a finite universe. *Journal of the American Statistical Association* **47**, 663–685.

Laird, N. M. (1988). Missing data in longitudinal studies. *Statistics in Medicine* **7**, 305–315.

Laird, N. M. and Ware, J. H. (1982). Random effects models for longitudinal data. *Biometrics* **38**, 963–974.

Liang, K.-Y. and Zeger, S. L. (1986). Longitudinal data analysis using generalized linear models. *Biometrika* **73**, 13–22.

Little, R. J. A. (1993). Pattern-mixture models for multivariate incomplete data. *Journal of the American Statistical Association* **88**, 125–134.

Little, R. J. A. (1994). A class of pattern-mixture models for normal incomplete data. *Biometrika* **81**, 471–483.

Little, R. J. A. (1995). Modeling the drop-out mechanism in repeated-measures studies. *Journal of the American Statistical Association* **90**, 1112–1121.

Little, R. J. A. and Rubin, D. B. (1987). *Statistical Analysis with Missing Data.* New York: Wiley.

Little, R. J. A. and Wang, Y. (1996). Pattern-mixture models for multivariate incomplete data with covariates. *Biometrics* **52**, 98–111.

Michiels, B., Molenberghs, G., and Lipsitz, S. R. (1999). A pattern-mixture odds ratio model for incomplete categorical data. *Communications in Statistics — Theory and Methods* **28**, 2843–2869.

Molenberghs, G., Kenward, M. G., and Goetghebeur, E. (2001). Sensitivity analysis for incomplete contingency tables: The Slovenian plebiscite case. *Applied Statistics* **50**, 15–29.

Molenberghs, G., Kenward, M. G., and Lesaffre, E. (1997). The analysis of longitudinal ordinal data with nonrandom drop-out. *Biometrika* **84**, 33–44.

Molenberghs, G., Michiels, B., Kenward, M. G., and Diggle, P. J. (1999). Missing data mechanisms and pattern-mixture models. *Statistica Neerlandica* **52**, 153–161.

Mori, M., Woodworth, G. G., and Woolson, R. F. (1992). Application of empirical Bayes inference to estimation of rate of change in the presence of informative right censoring. *Statistics in Medicine* **11**, 621–631.

Nordheim, E. V. (1984). Inference from nonrandomly missing categorical data: An example from a genetic study in Turner's syndrome. *Journal of the American Statistical Association* **79**, 772–780.

Park, T. and Lee, S. L. (1999). Simple pattern-mixture models for longitudinal data with missing observations: Analysis of urinary incontinence data. *Statistics in Medicine* **18**, 2933–2941.

Robins, J. M., Rotnitzky, A., and Zhao, L. P. (1995). Analysis of semiparametric regression models for repeated outcomes in the presence of missing data. *Journal of the American Statistical Association* **90**, 106–121.

Rosenbaum, P. R. and Rubin, D. B. (1983). Assessing sensitivity to an unobserved binary covariate in an observational study with binary outcome. *Journal of the Royal Statistical Society, Series B* **45**, 212–218.

Rosenbaum, P. R. and Rubin, D. B. (1985). The bias due to incomplete matching. *Biometrics* **41**, 103–116.

Rotnitzky, A. and Robins, J. M. (1997). Analysis of semi-parametric regression models with non-ignorable non-response. *Statistics in Medicine* **16**, 81–102.

Rubin, D. B. (1976). Inference and missing data. *Biometrika* **63**, 581–592.

Rubin, D. B. (1977). Formalizing subjective notions about the effect of nonrespondents in sample surveys. *Journal of the American Statistical Association* **72**, 538–543.

Scharfstein, D., Rotnitzky, A., and Robins, J. M. (1999). Adjusting for non-ignorable drop-out using semiparametric non-response models (with discussion). *Journal of the American Statistical Association* **94**, 1096–1146.

Schluchter, M. D. (1992). Methods for the analysis of informatively censored longitudinal data. *Statistics in Medicine* **11**, 1861–1870.

Ten Have, T. R., Kunselman, A. R., Pulkstenis, E. P., and Landis, J. R. (1998). Mixed effects logistic regression models for longitudinal binary response data with informative drop-out. *Biometrics* **54**, 367–383.

Ten Have, T. R., Miller, M. E., Reboussin, B. A., and James, M. K. (2000). Mixed effects logistic regression models for longitudinal ordinal functional response data with multiple-cause drop-out from the longitudinal study of aging. *Biometrics* **56**, 279–287.

Thijs, H., Molenberghs, G., Michiels, B., Verbeke, G., and Curran, D. (2002). Strategies to fit pattern-mixture models. *Biostatistics* **3**, 245–265.

Tsiatis, A. A., DeGruttola, V., and Wulfsohn, M. S. (1995). Modeling the relationship of survival to longitudinal data measured with error: applications to survival and CD4 counts in patients with AIDS. *Journal of the American Statistical Association* **90**, 27–37.

Vach, W. and Blettner, M. (1995). Logistic regression with incompletely observed categorical covariates: Investigating the sensitivity against violations of the missing at random assumption. *Statistics in Medicine* **12**, 1315–1330.

Verbeke, G., Molenberghs, G., Thijs, H., Lesaffre, E., and Kenward, M. G. (2001). Sensitivity analysis for non-random dropout: a local influence approach. *Biometrics* **57**, 7–14.

Wu, M. C. and Bailey, K. R. (1988). Analyzing changes in the presence of informative right censoring caused by death and withdrawal. *Statistics in Medicine* **7**, 337–346.

Wu, M. C. and Bailey, K. R. (1989). Estimation and comparison of changes in the presence of informative right censoring: Conditional linear model. *Biometrics* **45**, 939–955.

Wu, M. C. and Carroll, R. J. (1988). Estimation and comparison of changes in the presence of informative right censoring by modeling the censoring process. *Biometrics* **44**, 175–188.

CHAPTER 18

Selection and pattern-mixture models

Roderick Little

Contents

18.1 Introduction: Theoretical framework

Missing data are a common problem in longitudinal data sets, as the overview in Chapter 17 discussed. This chapter considers likelihood-based methods for handling this problem, based on parametric models for the data and missing-data mechanism. These models can also form the basis for multiple imputation approaches discussed in Chapter 21. Approaches based on estimating equations other than the likelihood, including inverse probability weighting methods, are discussed in Chapter 20. A useful tutorial that discusses both likelihood-based and estimating equations approaches is Hogan, Roy, and Korkontzelou (2004).

Unless missing data are a deliberate feature of the study design, it is important to try to limit them during data collection, since any method for compensating for missing data requires unverifiable assumptions that may or may not be justified. Since data are still likely to be missing despite these efforts, it is important to try to collect covariates that are predictive of the missing values, so that an adequate adjustment can be made. In addition, the process that leads to missing values should be determined during the collection of data if possible, since this information helps to model the missing-data mechanism when the incomplete data are analyzed.

We first briefly review parametric likelihood methods in the absence of missing data, as discussed in earlier chapters in this book. We suppose there are N individuals, and $\boldsymbol{Y}_i = (Y_{i1}, \dots, Y_{in_i})$ is a vector of repeated measurements planned for individual i, and write $Y = \{\boldsymbol{Y}_1, \dots, \boldsymbol{Y}_N\}$. Also associated with individual i at time j is a $(p \times 1)$ vector of covariates $\boldsymbol{X}_{ij} = (X_{ij1}, \dots, X_{ijp})'$, with $X_i = (\boldsymbol{X}_{i1}, \dots, \boldsymbol{X}_{in_i})$ the resulting $(n_i \times p)$ matrix of covariates, and $X = \{X_1, \dots, X_N\}$. Likelihood-based methods assume a model for the distribution $f(Y|X, \boldsymbol{\gamma})$ of Y given X with unknown parameters $\boldsymbol{\gamma}$. Assuming the individuals i are independent, this distribution factors into a product of distributions over the individuals i:

$$f(Y|X,\gamma) = \prod_{i=1}^{N} f(\boldsymbol{Y}_i|X_i,\boldsymbol{\gamma}),$$

where $f(\boldsymbol{Y}_i|X_i,\boldsymbol{\gamma})$ is the distribution of $\boldsymbol{Y}_i$ given X_i (density function for continuous $\boldsymbol{Y}_i$). The likelihood of $\boldsymbol{\gamma}$ given data $\{(\boldsymbol{Y}_i, X_i) : i = 1, \ldots, N\}$ is

$$L(\gamma|Y,X) = c \prod_{i=1}^{N} f(\boldsymbol{Y}_i|X_i,\boldsymbol{\gamma}),$$

considered as a function of the parameters $\boldsymbol{\gamma}$, where c is an arbitrary factor that does not depend on $\boldsymbol{\gamma}$. The maximum likelihood (ML) estimate $\widehat{\gamma}$ of γ is the value that maximizes $L(\boldsymbol{\gamma}|Y,X)$. Large-sample ML inferences under the model are based on the normal approximation

$$(\boldsymbol{\gamma} - \widehat{\gamma}) \sim N(\mathbf{0}, C), \tag{18.1}$$

where $N(\mathbf{0}, C)$ denotes the multivariate normal distribution with mean 0 and covariance matrix C, and C is one of several estimates, for example the sample covariance matrix from the bootstrap distribution of the estimates, or the inverse observed information matrix $\{-\partial^2 \log L(\boldsymbol{\gamma}|Y,X)/\partial\gamma\partial\gamma'\}^{-1}$. Following Little and Rubin (2002), Equation (18.1) is written to have both a frequentist interpretation, where γ is fixed, $\widehat{\gamma}$ is random, and the equation represents the asymptotic sampling distribution of $\widehat{\gamma}$, or a Bayesian interpretation, where $\widehat{\gamma}$ is fixed, γ is random, and the equation represents a large-sample approximation to the posterior distribution of γ. Bayesian inference adds a prior distribution $p(\gamma)$ for the parameters, and bases inference on the posterior distribution $p(\gamma|Y,X) = c\,p(\gamma)L(\gamma|Y,X)$, where c is a normalizing constant.

Now suppose there are gaps in the data Y on the repeated measures. I consider an unobserved value to be *missing* if there is a true underlying value that is meaningful for analysis. This may seem obvious, but is not always the case. For example, in a study of a behavioral intervention for people with heart disease, it is not meaningful to consider a quality of life measure to be missing for subjects who die prematurely during the course of the study. Rather, it is preferable to restrict the analysis to the quality of life measures of individuals while they are alive. This issue — whether values are truly considered missing or not — has implications for the choice of missing-data model, as discussed further below.

Let R_{ij} be the missing-data indicator for Y_{ij}, with value 1 if Y_{ij} is observed and 0 if Y_{ij} is missing, and $\boldsymbol{R}_i = (R_{i1}, \ldots, R_{in_i})$. The vector $\boldsymbol{Y}_i^o$ denotes the set of observed values for individual i, and $\boldsymbol{Y}_i^m$ the set of missing values. Unless stated otherwise, we assume that X_i is observed for all i, so the covariates do not contain missing values. This is not an innocuous assumption; for example, if covariates are measured repeatedly over time, then they are typically also missing after an individual drops out. Some comments on the case where covariates are also missing are provided in Section 18.7. As in Chapter 17, the problem is then to make inferences about γ based on the set of incomplete data $(R, Y^o, X) = \{(\boldsymbol{R}_i, \boldsymbol{Y}_i^o, X_i) : i = 1, \ldots, N\}$, rather than the complete data (Y, X).

With incomplete data, in general we need a model for the joint distribution of Y and R, with density $f(R, Y|X, \boldsymbol{\theta})$ indexed by parameters $\boldsymbol{\theta} = (\gamma, \phi)$, where γ characterizes the model for the data Y and ϕ the model for the missing-data indicators R. Again assuming independence over individuals, this density can be written as $f(R, Y|X, \boldsymbol{\theta}) = \prod_{i=1}^{N} f(\boldsymbol{R}_i, \boldsymbol{Y}_i|X_i, \boldsymbol{\theta})$. With no missing values, likelihood inferences would be based on the *complete-data likelihood*

$$L(\boldsymbol{\theta}|R, Y, X) = c \prod_{i=1}^{N} f(\boldsymbol{R}_i, \boldsymbol{Y}_i|X_i, \boldsymbol{\theta}),$$

where, as before, c is an arbitrary constant independent of $\boldsymbol{\theta}$. With missing data, likelihood inferences are based on the *observed-data likelihood*, or simply the likelihood, given data $\{(\boldsymbol{R}_i, \boldsymbol{Y}_i^o, X_i) : i = 1, \dots, N\}$, which is obtained formally by integrating the missing data $\boldsymbol{Y}_i^m$ out of the density of $(\boldsymbol{R}_i, \boldsymbol{Y}_i)$:

$$L(\boldsymbol{\theta}|R, Y^o, X) = c \prod_{i=1}^{N} \int f(\boldsymbol{R}_i, \boldsymbol{Y}_i | X_i, \boldsymbol{\theta}) d\boldsymbol{Y}_i^m. \tag{18.2}$$

In principle, inferences for $\boldsymbol{\theta}$ can then proceed in the same way as for inferences about $\boldsymbol{\gamma}$ in the case of complete data. That is, the ML estimate $\widehat{\boldsymbol{\theta}}$ of $\boldsymbol{\theta}$ is the value that maximizes $L(\boldsymbol{\theta}|R, Y^o, X)$. Large-sample ML inferences under the model are based on the normal approximation

$$(\boldsymbol{\theta} - \widehat{\boldsymbol{\theta}}) \sim N(\boldsymbol{0}, C_\theta),$$

where C_θ is an estimate of the large-sample covariance matrix, for example the inverse observed information matrix $\{-\partial^2 \log L(\boldsymbol{\theta}|R, Y^o, X)/\partial\boldsymbol{\theta}\partial\boldsymbol{\theta}'\}^{-1}$. Bayesian inference is based on the posterior distribution of $\boldsymbol{\theta}$, obtained by multiplying the likelihood $L(\boldsymbol{\theta}|R, Y^o, X)$ by a prior distribution $p(\boldsymbol{\theta})$ for the parameters. ML estimates under a correctly specified model are fully efficient, and in particular make use of information in the incomplete cases that is lost when the incomplete cases are dropped from the analysis. The Bayesian approach shares the optimal large-sample properties of ML, and can yield better small-sample inferences. See, for example, Little and Rubin (2002, Chapter 6).

Missing data complicate likelihood-based inferences in a number of ways:

(i) As described above, in general a model for the joint distribution of R and Y is needed, rather than simply a model for the distribution of Y. Specifying a model for R requires knowledge about the process leading to missing values, about which little is often known. Results for the parameters of interest $\boldsymbol{\gamma}$ tend to be sensitive to the assumptions contained in this model, so a bad specification of this model can lead to poor inferences, even if the model for Y is correctly specified. When the missing-data mechanism is *ignorable* for likelihood inference, inferences can be based on the ignorable likelihood

$$L_{\text{ign}}(\boldsymbol{\gamma}|Y^o, X) = c \prod_{i=1}^{N} \int f(\boldsymbol{Y}_i|X_i, \boldsymbol{\gamma}) d\boldsymbol{Y}_i^m = c \prod_{i=1}^{N} f(\boldsymbol{Y}_i^o|X_i, \boldsymbol{\gamma}). \tag{18.3}$$

This function is generally much easier to deal with than the full likelihood (18.2). The integral in the latter complicates the computation; the ignorable likelihood (18.3) does not require a model for R, which can be difficult to specify; furthermore, the parameters in the full likelihood tend to be at best weakly identified, making inference problematic.

For these reasons, most likelihood analyses for incomplete longitudinal data with dropouts or intermittent missingness are currently based on (18.3) rather than the full likelihood (18.2). As discussed in Chapter 17 and Chapter 20, the key condition for this simplification to occur is that the missing data are *missing at random* (MAR), in that missingness only depends on the data through observed values (Y^o, X):

$$f(\boldsymbol{R}_i|\boldsymbol{Y}_i, X_i, \boldsymbol{\phi}) = f(\boldsymbol{R}_i|\boldsymbol{Y}_i^o, X_i, \boldsymbol{\phi}) \text{ for all } \boldsymbol{Y}_i^m \tag{18.4}$$

(Rubin, 1976; Little and Rubin, 2002, Chapter 6). There is also a more technical separability condition, which states that the parameters $\boldsymbol{\gamma}$ and $\boldsymbol{\phi}$ have distinct parameter spaces, but that condition is less important and often reasonable in applications. The next section considers ML methods based on (18.3) under the assumption that the missing-data mechanism is ignorable. I then consider models that incorporate mechanisms that are not missing at random (NMAR).

(ii) The key role of (18.4) in justifying an inference based on (18.3) makes MAR very desirable, and it is worth collecting covariate information W_i that makes the MAR assumption plausible, and incorporating this information in the analysis. When it is appropriate to condition on W_i in the final analysis, the likelihood-based analysis is straightforward, because W_i is just incorporated in the covariate matrix X_i. When it is not appropriate to condition on W_i in the final analysis, the likelihood-based methods considered here are trickier, because the W_i then need to be assigned a distribution and integrated out for final inferences. While this is quite possible in principle, the multiple imputation methods in Chapter 21 provide a more convenient solution, because the imputations can be based on an imputation model that conditions on W_i, but the final analysis simply omits these variables.

(iii) The missing data may render some parameters in the likelihood function unidentified, in the sense that unique ML estimates are not available, or weakly identified. As a trivial example, the mean of Y_{ij} is not identified if all individuals drop out before measure j. As a more complex example, Example 1 below concerns a longitudinal study measuring the growth of children between ages 3 and 12, which initially recruits children of various ages and follows them for at most 8 years, or until they reach age 12. The correlation between growth measures at ages 3 and 12 is not identified, since these measures are never recorded for the same child. A general covariance matrix for the repeated measures is thus not identified from this data structure.

Lack of identifiability yields inferential problems, such as estimates with a high degree of uncertainty, and computational problems, such as iterative ML algorithms failing to converge or converging painfully slowly. Such problems are particularly prevalent when the missing-data mechanism is non-ignorable. When parameters are poorly identified, it may be better to conduct a sensitivity analysis, where answers are computed for a range of plausible values of these parameters, rather than trying to estimate them from the data. We provide some examples of analyses of this kind below.

(iv) The missing data may increase the sensitivity of inferences to misspecification of the model for the data Y. For example, incorrectly assuming a linear relationship between an outcome and a covariate may lead to more serious bias when missingness depends on the value of the covariate than when it does not. Interestingly, the cases where inferences are sensitive to the model tend to be the cases where including the incomplete cases in the analysis has the greatest payoff, in terms of reduced bias and increased precision.

(v) The observed-data likelihoods are typically more complicated than likelihoods based on the complete data, and have greater potential to be multimodal and non-normal in shape. Thus, larger samples are needed for asymptotic methods like ML and information matrix based standard errors to be satisfactory, than is the case with complete data.

(vi) Missing data complicate the estimation of standard errors of parameters. In particular, the popular EM algorithm for finding ML estimates does not yield standard errors of parameters as part of its output. The information matrix of the parameters is more complicated than with complete data, and is harder to calculate and invert. For example, with many repeated-measures models, the complete-data information matrix is block-diagonal between parameters characterizing the mean and covariance structures. This means that if standard errors are only needed for parameters characterizing the mean structure, these can be obtained by inverting the submatrix of the information matrix corresponding to those parameters. With missing data, the block diagonal structure is lost under MAR, so the full matrix needs to be inverted. Alternative approaches to computing standard errors are to base standard errors on draws from the Bayesian posterior distribution of estimates, or to compute bootstrap or jack-knife standard errors (Little and Rubin 2002, Chapter 5 and Chapter 9). Bootstrap samples can be obtained as simple random samples with replacement of size N from the sampled individuals, ignoring pattern. Stratifying on pattern is also a possibility, although gains from doing this seem unclear.

(vii) Computation of ML estimates and associated standard errors, and posterior distributions for Bayesian inference, is typically more challenging, requiring iterative methods such as the EM algorithm and Markov chain Monte Carlo simulation. I focus here on features of the models themselves rather than on computational aspects. For discussions of computation, see Tanner (1991), Schafer (1997), and Little and Rubin (2002). For many users not interested in developing their own programs, choices are often limited to missing-data methods that are available in widely available statistical software packages. These choices are increasing, but gaps still remain, as indicated below.

18.2 Ignorable maximum likelihood methods

ML inference based on the ignorable likelihood (18.3) is formally similar to ML with complete data: a model is not required for the missing-data mechanism, and in large samples, hypothesis tests and confidence intervals can be based on the ML estimates of parameters and asymptotic standard errors, as for complete data. We discuss the ignorable likelihood approach for a simple bivariate normal model with dropouts, a more general mixed model suitable for many repeated-measures problems, and extensions to non-normal models available in current statistical software.

Model 1. A normal model for two repeated measures with MAR dropout. Suppose there are just two repeated measures ($n_i = 2$ for all i), and (Y_{i1}, Y_{i2}) are observed and $R_i = 1$ for $i = 1, \ldots, r$, and Y_{i1} is observed, Y_{i2} is missing, and $R_i = 0$ for $i = r + 1, \ldots, N$. Let $\boldsymbol{\mu} = (\mu_1, \mu_2)$, $\mu_j = E(Y_{ij})$, and $\Sigma = (\sigma_{jk})$ be the covariance matrix of (Y_{i1}, Y_{i2}); often the covariance matrix for repeated-measures data is assigned a special structure, such as compound symmetry, but here I assume this matrix is unrestricted. Suppose interest concerns the difference in means between the two time points, $\mu_{\text{dif}} = \mu_2 - \mu_1$. Naive estimates of μ_{dif} include (a) the *complete-case* (CC) estimate $\bar{y}_2 - \bar{y}_1$, where $\bar{y}_j = \sum_{i=1}^r Y_{ij}/r$ is the sample mean of Y_j from the cases with both variables observed, and (b) the *available-case* (AC) estimate $\bar{y}_2 - \hat{\mu}_1$, where $\hat{\mu}_1 = \sum_{i=1}^N Y_{i1}/N$ is the sample mean of Y_1 from all the cases; the latter is obtained when the missing values of Y_{i2} are imputed by the CC mean $\bar{y}_2$. We consider ML estimates for the following normal model:

$$(Y_{i1}, Y_{i2}) \sim_{ind} N(\boldsymbol{\mu}, \Sigma).$$

The MAR assumption (18.4) here implies that missingness of Y_{i2} can depend on Y_{i1}, but conditional on Y_{i1} it does not depend on Y_{i2}, since that variable is missing for $i = r + 1, \ldots, N$. The likelihood assuming ignorable non-response is

$$L_{\text{ign}}(\boldsymbol{\mu}, \Sigma | Y^o) = \prod_{i=1}^r |\Sigma|^{-1/2} \exp\left\{-0.5(\boldsymbol{Y}_i - \boldsymbol{\mu})' \Sigma^{-1} (\boldsymbol{Y}_i - \boldsymbol{\mu})\right\}$$

$$\times \prod_{i=r+1}^N \sigma_{11}^{-1/2} \exp\{-0.5(Y_{i1} - \mu_1)^2/\sigma_{11}\}. \tag{18.5}$$

Anderson (1957) showed that the ML estimates are easily derived by an elegant trick: instead of attempting to maximize (18.5) directly, the likelihood is factored according to the marginal distribution of Y_{i1} and the conditional distribution of Y_{i2} given Y_{i1} (see also Little and Rubin, 2002, Chapter 7). The ML estimates of the marginal mean and variance of Y_1 are sample means and variances from all N cases:

$$\widehat{\mu}_1 = \sum_{i=1}^{N} y_{i1}/N, \quad \widehat{\sigma}_{11} = \sum_{i=1}^{N}(y_{i1} - \widehat{\mu}_1)^2/N,$$

and the ML estimates of the slope, intercept, and residual variance of the regression of Y_2 on Y_1 are their least-squares estimates based on the r complete cases. The corresponding ML estimates of $(\mu_2, \sigma_{12}, \sigma_{22})$ are:

$$\widehat{\mu}_2 = \bar{y}_2 + \widehat{\beta}_{21\cdot 1}(\widehat{\mu}_1 - \bar{y}_1), \qquad\qquad (18.6)$$
$$\widehat{\sigma}_{12} = \widehat{\beta}_{21\cdot 1}\widehat{\sigma}_{11},$$
$$\widehat{\sigma}_{22} = s_{22\cdot 1} + \widehat{\beta}_{21\cdot 1}^2(\widehat{\sigma}_{11} - s_{11}),$$

where $s_{jk} = \sum_{i=1}^{r}(y_{ij} - \bar{y}_j)(y_{ik} - \bar{y}_k)/r$ are sample variances $(j = k)$ and covariances $(j \neq k)$ from the r complete cases, and $\widehat{\beta}_{21\cdot 1} = s_{12}/s_{11}$ is the regression coefficient of Y_{i1} from the regression of Y_{i2} on Y_{i1}, based on the complete cases. The ML estimate (18.6) of μ_2 is called the *regression estimate* of the mean, and is also the average of observed and imputed values when the missing values of y_{i2} are imputed with predictions from the regression of Y_{i2} on Y_{i1} computed using the complete cases. The ML estimate of $\mu_{\text{dif}} = \mu_2 - \mu_1$ is

$$\widehat{\mu}_{\text{dif}} = \widehat{\mu}_2 - \widehat{\mu}_1.$$

Large-sample inference requires estimates of standard errors of these parameters. Large-sample standard errors can be based on the observed information matrix, or can be computed by bootstrapping the observed data.

Another approach is to add a prior distribution and simulate draws from the posterior distribution of the parameters. With the non-informative prior

$$f(\mu_1, \sigma_{11}, \beta_{20\cdot 1}, \beta_{21\cdot 1}, \sigma_{22\cdot 1}) \propto \sigma_{11}^{-1}\sigma_{22\cdot 1}^{-1},$$

draws $(\mu_1^{(d)}, \sigma_{11}^{(d)}, \beta_{20\cdot 1}^{(d)}, \beta_{21\cdot 1}^{(d)}, \sigma_{22\cdot 1}^{(d)})$ from the posterior distribution of $(\mu_1, \sigma_{11}, \beta_{20\cdot 1}, \beta_{21\cdot 1}, \sigma_{22\cdot 1})$ are easily obtained as follows:

1. Draw independently x_{1d}^2 and x_{2d}^2 from chi-squared distributions with $N-1$ and $r-2$ degrees of freedom, respectively. Also, draw three standard normal deviates z_{1d}, z_{2d}, and z_{3d}.

2. Compute

$$\sigma_{11}^{(d)} = N\widehat{\sigma}_{11}/x_{1d}^2,$$
$$\sigma_{22\cdot 1}^{(d)} = rs_{22\cdot 1}/x_{2d}^2,$$
$$\mu_1^{(d)} = \widehat{\mu}_1 + z_{1d}\left(\sigma_{11}^{(d)}/N\right)^{1/2},$$
$$\beta_{21\cdot 1}^{(d)} = \widehat{\beta}_{21\cdot 1} + z_{2d}\left(\sigma_{22\cdot 1}^{(d)}/rs_{11}\right)^{1/2},$$
$$\beta_{20\cdot 1}^{(d)} = \bar{y}_2 - \widehat{\beta}_{21\cdot 1}\bar{y}_1 + z_{3d}\left(\sigma_{22\cdot 1}^{(d)}/r\right)^{1/2}.$$

Draws $(\mu_2^{(d)}, \sigma_{12}^{(d)}, \sigma_{22}^{(d)})$ from the posterior distribution of $(\mu_2, \sigma_{12}, \sigma_{22})$ are then obtained by replacing the ML estimates $(\widehat{\mu}_1, \widehat{\sigma}_{11}, \widehat{\beta}_{20\cdot 1}, \widehat{\beta}_{21\cdot 1}, \widehat{\sigma}_{22\cdot 1})$ in (18.6) by the draws $(\mu_1^{(d)}, \sigma_{11}^{(d)}, \beta_{20\cdot 1}^{(d)}, \beta_{21\cdot 1}^{(d)}, \sigma_{22\cdot 1}^{(d)})$. Also $\mu_{\text{dif}}^{(d)} = \mu_2^{(d)} - \mu_1^{(d)}$ is a draw from the posterior distribution of μ_{dif}. For more details, see Little and Rubin (2002, Chapter 7).

This factored likelihood approach is readily extended to n repeated measures $(Y_{i1}, \ldots, Y_{in})$ with an n-variate $N(\boldsymbol{\mu}, \boldsymbol{\Sigma})$ distribution, and a general monotone pattern with Y_{ij} observed if Y_{ik} is observed for all $j < k$. For an unrestricted mean and covariance matrix the factored likelihood idea leads to explicit expressions for ML estimates and draws from the posterior

distribution. Non-monotone patterns, or monotone patterns with more restricted parame-
terizations as in the next model, require iterative algorithms.

**Model 2. A normal mixed model for repeated measures with MAR missing
data.** A model that is more tuned to repeated-measures data is the linear mixed model

$$\boldsymbol{Y}_i|\boldsymbol{b}_i \sim N(X_i\boldsymbol{\beta} + Z_i\boldsymbol{b}_i, V_i) \tag{18.7}$$
$$\boldsymbol{b}_i \sim N(\boldsymbol{0}, G),$$

where $\boldsymbol{b}_i$ are unobserved random effects for individual i, and X_i and Z_i are known fixed
matrices that characterize how the repeated measures depend on fixed and random factors
in the model (Hartley and Rao, 1967; Laird and Ware, 1982; Jennrich and Schluchter, 1986;
Schluchter, 1988). The matrices V_i and G characterize the covariance matrix of the repeated
measures. Programs like `proc mixed` in SAS (Littell et al., 1996; SAS, 2004) and S-Plus
(Pinheiro and Bates, 2000; Huet et al., 2004) include flexible choices of X_i, Z_i, V_i, and G,
allowing a wide range of repeated-measures models to be fitted. Asymptotic inferences for
this model are similar to the complete-data case.

Extensions to non-normal errors include models for multivariate t errors that downweight
outliers (Lange, Little, and Taylor, 1989), and ML for generalized linear mixed models
(McCulloch and Searle, 2001), which are implemented in `proc nlmixed` in SAS (2004).
For non-linear mixed models, see Vonesh and Chinchilli (1997). The Bayesian approach is
attractive in small samples where asymptotic assumptions are not appropriate. An early
discussion of Bayesian methods for normal models is Gelfand et al. (1990). These methods
can be implemented using the Bayesian modeling software in the BUGS project; see the
BUGS Web site (BUGS, 2006) for details.

Example 1. Longitudinal study of lung function. Lavange and Helms (1983) analyzed data from a longitudinal study of lung function conducted on 72 children aged 3 to
12 years. A measure of maximum expiratory flow rate was measured annually, and differences in the resulting curve were related to between-subject covariates such as race and
gender. The indexing variable of interest here is age rather than time. The number of actual
measurements recorded on each child ranged from 1 to 8, with an average of 4.2 per child. A
primary reason for the missing data was that children entered the study at different ages —
some values were missing because the child was over age 3 at the start of the study, or less
than age 12 at the end of the study. There was also some attrition from the study.

When, as here, the missing data are caused by features of the study design, rather than the
behavior of the study subjects, the MAR assumption may be quite plausible. The missing-
data mechanism depends on cohort, and if cohort is included as a covariate in the model,
it is a special form of MAR, which we call covariate-dependent missingness. On the other
hand, subjects who drop out prematurely may do so for reasons related to the outcome
measures. For example, they may move out of the area because of respiratory problems.
Such a mechanism is only MAR if the recorded values $\boldsymbol{Y}_i^o$ characterize the respiratory
problems that prompted the move.

Example 2. Dropouts in a hypertension trial. Murray and Findlay (1988) described
data from a large multicenter trial of metoprolol and ketanserin, two anti-hypertensive
agents for patients with mild to moderate hypertension, with diastolic blood pressure an
outcome measure of interest. The double-blind treatment phase lasted 12 weeks, with clinic
visits scheduled for weeks 0, 2, 4, 8, and 12. The protocol stated that patients with diastolic
blood pressure exceeding 110 mmHg at either the 4- or 8-week visit should "jump" to an
open follow-up phase — a form of planned dropout. In total, 39 of the 218 metaprolol
patients and 55 of the 211 ketanserin patients jumped to open follow-up. A further 17
metoprolol patients and 20 ketanserin patients had missing data for other reasons, of which

3 and 7, respectively, dropped out with side effects. Analyses of the observed data indicate clearly that the dropouts differed systematically from cases remaining in the study, as can be predicted by the protocol for jumping to the open phase.

18.3 Non-ignorable models for the joint distribution of Y and R

We now relax the assumption of ignorable non-response, and consider models for the joint distribution of $\boldsymbol{R}_i$ and $\boldsymbol{Y}_i$. These can be specified in a variety of ways, depending how the joint distribution is factorized. We first consider fixed-effect models that do not include random effects for subjects, and then consider mixed-effect models that use random-effect terms to model the longitudinal structure.

Selection models specify the joint distribution of $\boldsymbol{R}_i$ and $\boldsymbol{Y}_i$ through models for the marginal distribution of $\boldsymbol{Y}_i$ and the conditional distribution of $\boldsymbol{R}_i$ given $\boldsymbol{Y}_i$:

$$f(\boldsymbol{R}_i, \boldsymbol{Y}_i | X_i, \boldsymbol{\gamma}, \boldsymbol{\phi}) = f_Y(\boldsymbol{Y}_i | X_i, \boldsymbol{\gamma}) f_{R|Y}(\boldsymbol{R}_i | X_i, \boldsymbol{Y}_i, \boldsymbol{\phi}), \qquad (18.8)$$

where $\boldsymbol{\theta} = (\boldsymbol{\gamma}, \boldsymbol{\phi})$. *Pattern-mixture* models (Glynn, Laird, and Rubin 1986; Little 1993) specify the marginal distribution of $\boldsymbol{R}_i$ and the conditional distribution of $\boldsymbol{Y}_i$ given $\boldsymbol{R}_i$:

$$f(\boldsymbol{R}_i, \boldsymbol{Y}_i | X_i, \boldsymbol{\nu}, \boldsymbol{\delta}) = f_R(\boldsymbol{R}_i | X_i, \boldsymbol{\delta}) f_{Y|R}(\boldsymbol{Y}_i | X_i, \boldsymbol{R}_i, \boldsymbol{\nu}), \qquad (18.9)$$

where $\boldsymbol{\theta} = (\boldsymbol{\nu}, \boldsymbol{\delta})$. Applications of models of the form (18.9) for categorical outcomes include Ekholm and Skinner (1998) and Birmingham and Fitzmaurice (2002). Pattern-set mixture models (Little 1993), which are mixtures of these two types, can also be formulated.

For comparisons of selection and pattern-mixture models, see Glynn, Laird, and Rubin (1986), Little (1995), Kenward and Molenberghs (1999), and Michiels, Molenberghs, and Lipsitz (1999). Both of these modeling approaches have useful features. Attractive features of selection models include the following:

1. Selection models (18.8) are a natural way of factoring the model, with f_Y the model for the data in the absence of missing values, and $f_{R|Y}$ the model for the missing-data mechanism that determines what parts of Y are observed. Substantively it seems more natural to consider relationships between Y and X in the full target population of interest, rather than in subpopulations defined by missing-data pattern. In particular, the term $\boldsymbol{\gamma}$ in the distribution f_Y usually contains the parameters of substantive interest, and inferences for these parameters are available directly from the selection model analysis.

2. If the MAR assumption is plausible, the selection model formulation leads directly to the ignorable likelihood — the distribution $f_{R|Y}$ for the missing-data mechanism is not needed for likelihood inferences, which can be based solely on the model for f_Y. Thus, if MAR is viewed as reasonable, NMAR models are not contemplated, and inferences are required for the population aggregated over the missing-data patterns, then the selection modeling approach seems compelling, and I see little reason for considering a pattern-mixture formulation. For a discussion of MAR from a pattern-mixture model perspective, see Molenberghs et al. (1998).

   Pattern-mixture models have some desirable features when NMAR situations are contemplated:

3. For situations where it is not substantively meaningful to consider non-response as missing data, it may make better sense to restrict the inference to the subpopulation of cases with values observed. For example, if Y_{ij} is a measure of quality of life at age j, and $R_{ij} = 1$ for survivors at age j and $R_{ij} = 0$ for individuals who die before age j, then it appears more meaningful to consider the distribution of Y_{ij} given $R_{ij} = 1$ rather than the

marginal distribution of Y_{ij}, which effectively implies imputed quality-of-life measures for non-survivors. The pattern-mixture model formulation targets the distribution of substantive interest in this situation, and indeed a selection model that fails to condition on response is not sensible.

4. From an imputation perspective (see Section 17.4.1), missing values $\boldsymbol{Y}_i^m$ should be imputed from their predictive distribution given the observed data including $\boldsymbol{R}_i$, that is, $f(\boldsymbol{Y}_i^m|\boldsymbol{Y}_i^o, \boldsymbol{R}_i, X_i)$. Under MAR this equals $f_Y(\boldsymbol{Y}_i^m|\boldsymbol{Y}_i^o, X_i)$, which is a conditional distribution derived from the selection model distribution of Y given X. However, if data are not MAR, the predictive distribution of $\boldsymbol{Y}_i^m$ given $\boldsymbol{Y}_i^o$ and $\boldsymbol{R}_i$ is modeled directly in the pattern-mixture formulation (18.9), but it is related to the components of the selection model by the complex expression

$$f(\boldsymbol{Y}_i^m|\boldsymbol{Y}_i^o, \boldsymbol{R}_i, X_i) = \frac{f_Y(\boldsymbol{Y}_i^m|\boldsymbol{Y}_i^o, X_i) f_{R|Y}(\boldsymbol{R}_i|X_i, \boldsymbol{Y}_i)}{\int f_{R|Y}(\boldsymbol{R}_i|X_i, \boldsymbol{Y}_i) f_Y(\boldsymbol{Y}_i^m|\boldsymbol{Y}_i^o, X_i) d\boldsymbol{Y}_i^m}.$$

The more direct relationship between the pattern-mixture formulation and the predictive distribution for imputations yields gains in transparency and computational simplicity in some situations.

5. The selection model factorization does not require full specification of the model for the missing-data mechanism when the data are MAR, but it does if the data are NMAR. Some pattern-mixture models, such as Model 4 below, avoid specification of the model for the missing-data mechanism in NMAR situations, by using assumptions about the mechanism to yield restrictions on the model parameters.

18.4 Bivariate data with dropouts

We illustrate these general points with some examples of models, starting simple and then adding complexities.

Model 3. A normal selection model for two repeated measures with non-MAR dropouts. Suppose, as for model 1, there are just two repeated measures ($n_i = 2$ for all i), and (Y_{i1}, Y_{i2}) are observed and $R_i = 1$ for $i = 1, \ldots, r$, and Y_{i1} is observed, Y_{i2} is missing, and $R_i = 0$ for $i = r+1, \ldots, N$. Let $\boldsymbol{\mu} = (\mu_1, \mu_2)$, $\mu_j = E(Y_{ij})$ and $\Sigma = (\sigma_{jk})$ be the covariance matrix of (Y_{i1}, Y_{i2}), and suppose that interest lies in the difference in means $\mu_{\text{dif}} = \mu_2 - \mu_1$. We consider ML estimates for the following normal selection model:

$$(Y_{i1}, Y_{i2}) \sim N(\boldsymbol{\mu}, \Sigma),$$

$$(R_i|Y_{i1}, Y_{i2}) \sim \text{Ber}(P(\phi(Y_{i1}, Y_{i2}))), \tag{18.10}$$

$$\text{logit}\{P(\phi(Y_{i1}, Y_{i2}))\} = \phi_0 + \phi_1 Y_{i1} + \phi_2 Y_{i2},$$

where $\text{Ber}(\pi)$ represents the Bernoulli distribution with probability π. Replacing the logit specification for the response mechanism by a probit specification yields a simple case of the Heckman (1976) selection model. The likelihood for this model is

$$L(\boldsymbol{\mu}, \Sigma, \boldsymbol{\phi}|R, Y^o) = \prod_{i=1}^{r} |\Sigma|^{-1/2} \exp\left\{-0.5(\boldsymbol{Y}_i - \boldsymbol{\mu})'\Sigma^{-1}(\boldsymbol{Y}_i - \boldsymbol{\mu})\right\} P[\phi(Y_{i1}, Y_{i2})]$$

$$\times \prod_{i=r+1}^{N} \sigma_{11}^{-1/2} \int \exp\left\{-0.5(\boldsymbol{Y}_i - \boldsymbol{\mu})'\Sigma^{-1}(\boldsymbol{Y}_i - \boldsymbol{\mu})\right\} \{1 - P[\phi(Y_{i1}, Y_{i2})]\} dY_{i2}. \tag{18.11}$$

Joint ML estimation of $\boldsymbol{\gamma} = (\boldsymbol{\mu}, \Sigma)$ and $\boldsymbol{\phi} = (\phi_0, \phi_1, \phi_2)$ requires an iterative method like the EM algorithm (Little and Rubin, 2002, Example 15.7). However, the model is weakly identified, and identification is strongly dependent on the normality assumptions. Thus, preferred approaches are to make additional assumptions about the form of the mechanism, such as $\phi_1 = 0$ or $\phi_2 = 0$, or to do a sensitivity analysis by fitting the model for a variety of plausible choices of ϕ.

If $\phi_2 = 0$ then the missing data are MAR, since missingness of Y_{i2} depends on Y_{i1}, which is observed for all i. The full likelihood (18.11) then simplifies to

$$L(\mu, \Sigma, \phi | R, Y^o) = L_{\text{ign}}(\mu, \Sigma | Y^o)\{P[\phi(Y_{i1}, Y_{i2})]\}^r \{1 - P[\phi(Y_{i1}, Y_{i2})]\}^{N-r},$$

where $L_{\text{ign}}(\mu, \Sigma | Y^o)$ is given by (18.5), and ML inference for (μ, Σ) can be based on the ignorable likelihood (18.5), as discussed in Model 1.

Model 4. A normal pattern-mixture model for two repeated measures with non-MAR dropouts. An unrestricted normal pattern-mixture model for the data just described is

$$(Y_{i1}, Y_{i2} | R_i = k) \sim N(\mu^{(k)}, \Sigma^{(k)})$$
$$(R_i) \sim \text{Ber}(\delta). \tag{18.12}$$

This model implies that the marginal mean of (Y_{i1}, Y_{i2}) averaged over patterns is $\mu = (1 - \delta)\mu^{(0)} + \delta\mu^{(1)}$, and the parameter of interest is

$$\mu_{\text{dif}} = (1 - \delta)(\mu_2^{(0)} - \mu_1^{(0)}) + \delta(\mu_2^{(1)} - \mu_1^{(1)}), \tag{18.13}$$

the weighted average of the differences in means in the two patterns. Equation (18.13) is an example where the parameter of interest is not a parameter of the pattern-mixture model, but is easily expressed as a function of the model parameters; ML estimates or draws from the Bayesian posterior distribution are obtained by substituting ML estimates or Bayesian draws of the pattern-mixture model parameters in this expression.

The model (18.12) is clearly underidentified: there are 11 parameters, namely two means, two variances, and one covariance for the patterns of complete and incomplete cases, and the probability δ that $R_i = 1$. On the other hand, only eight parameters $(\delta, \mu^{(1)}, \Sigma^{(1)}, \mu_1^{(0)}, \sigma_{11}^{(0)})$ can be estimated from the data. The ML estimates of these parameters are easily shown to be $\widehat{\delta} = r/N$, $\widehat{\mu}^{(1)} = \bar{y}, \widehat{\Sigma}^{(1)} = S$, $\widehat{\mu}_1^{(0)} = \bar{y}_1^{(0)}$, and $\widehat{\sigma}_{11}^{(0)} = s_{11}^{(0)}$, where $\bar{y} = (\bar{y}_1, \bar{y}_2)$ and $S = (s_{jk})$ are the sample mean and covariance matrix of the complete cases, and $\bar{y}_1^{(0)}$ and $s_{11}^{(0)}$ are the sample mean and variance of Y_1 for the incomplete cases. The identification issue is more immediately evident with the model (18.12) than with the selection model (18.10), but it is a key issue regardless of how the joint distribution of Y and R is factorized.

Two possible resolutions of the identification issue are (a) to place restrictions on the model parameters, based on assumptions about the nature of the missing-data mechanism or the model for Y, or (b) to relate the unidentified parameters to identified parameters via Bayesian prior distributions. A simple illustration of (a) is that if the missing-data mechanism is assumed missing completely at random, then $\mu^{(1)} = \mu^{(0)} = \mu, \Sigma^{(1)} = \Sigma^{(0)} = \Sigma$, and model (18.12) is identical to the selection model (18.10) with $\phi_1 = \phi_2 = 0$. Likelihood inference for the parameters (μ, Σ) of interest is then the same as for the MAR model in Example 1 — it is not affected by the additional constraint that $\phi_1 = 0$.

Another restriction is to set $\mu^{(0)} = \mu^{(1)} + \alpha, \Sigma^{(0)} = C\Sigma^{(1)}C'$ for pre-chosen values of α and C. That is, offsets are introduced to characterize differences in the mean and covariance matrix between complete and incomplete patterns (Daniels and Hogan, 2000). One possible approach is to assess sensitivity to pre-chosen values of α and C. A severe disadvantage of this strategy is that even for this simple missing-data pattern α and C contain five distinct quantities, and prespecifying values in a five-dimensional space seems impractical, and unnecessary since only three parameter restrictions are needed to identify the model. Daniels and Hogan note that some of these parameters can be estimated from the data, providing assumptions are made about the missing-data mechanism. They describe sensitivity analyses based on this approach on longitudinal clinical trial data involving growth hormone treatments.

Little (1994) analyzes model (18.12) under the assumption that

$$\Pr(R_i = 0|Y_{i1}, Y_{i2}) = g(Y_{i1}^*), \quad Y_{i1}^* = Y_{i1} + \lambda Y_{i2}, \qquad (18.14)$$

where λ is prespecified and g is an arbitrary function. Under that assumption, the conditional distribution of Y_{i2} given Y_{i1}^* is independent of R_i, and is normal with, say, mean $\beta_{20\cdot1}^* + \beta_{20\cdot1}^* Y_{i1}^*$ and variance $\sigma_{22\cdot1}^*$. The fact that the intercept, slope, and residual variance of this distribution are the same in the two patterns yields three constraints on $(\mu^{(0)}, \Sigma^{(0)})$ that are just sufficient to identify the model. Little (1994) shows that the resulting ML estimates of (μ_1, μ_2) are $\widehat{\mu}_1 = \sum_{i=1}^N y_{i1}/N$, and

$$\widehat{\mu}_2^{(\lambda)} = \bar{y}_2 + \widehat{\beta}_{21\cdot1}^{(\lambda)}(\widehat{\mu}_1 - \bar{y}_1), \quad \widehat{\beta}_{21\cdot1}^{(\lambda)} = \frac{s_{12} + \lambda s_{22}}{s_{11} + \lambda s_{12}}.$$

The corresponding estimate of the difference in means is

$$\widehat{\mu}_{\text{dif}}^{(\lambda)} = \widehat{\mu}_2^{(\lambda)} - \widehat{\mu}_1 \qquad (18.15)$$
$$= \bar{y}_2 - [(\widehat{\beta}_{21\cdot1}^{(\lambda)} + (1 - \widehat{\beta}_{21\cdot1}^{(\lambda)})(r/N)]\bar{y}_1 + (1 - r/N)(1 - \widehat{\beta}_{21\cdot1}^{(\lambda)})\bar{y}_1^{(0)},$$

where $\bar{y}_1^{(0)}$ is the mean of Y_1 for the cases missing Y_2. Various estimates $\widehat{\mu}_{\text{dif}}^{(\lambda)}$ are obtained for different choices of λ, including the CC and AC estimates previously mentioned in Model 1. The following comments assume Y_1 and Y_2 are positively correlated for the complete cases:

(a) When $\lambda = 0$ the data are MAR, and $\widehat{\mu}_{\text{dif}}^{(\lambda=0)} = \widehat{\mu}_{\text{dif}}$, the ML estimate for the ignorable selection model.

(b) As λ increases from 0 to $(s_{11} - s_{12})/(s_{22} - s_{12})$, the ML estimate puts less and less weight on $\bar{y}_1^{(0)}$, the mean of Y_1 for the cases that drop out. When $\lambda = (s_{11} - s_{12})/(s_{22} - s_{12})$, $\widehat{\beta}_{21\cdot1}^{(\lambda)} = 1$ and $\widehat{\mu}_{\text{dif}}^{(\lambda)} = \bar{y}_2 - \bar{y}_1$, the CC estimate. This value of λ reduces to $\lambda = 1$ when $s_{11} = s_{22}$. The implication is that when missingness depends on the average value of the Y at the two time points, the CC estimate of the change is optimal.

(c) As λ increases from $(s_{11} - s_{12})/(s_{22} - s_{12})$ toward ∞, $\widehat{\beta}_{21\cdot1}^{(\lambda)} \to s_{22}/s_{12}$, and the slope of the regression predictions of missing values of Y_2 is the inverse of the slope of the regression of Y_1 on Y_2. This calibration-like estimate reverses the regressions, and is increasingly unstable as the correlation between Y_1 and Y_2 tends to zero.

(d) As λ decreases from 0 to $-s_{12}/s_{22}$, the ML estimate puts increasing weight on $\bar{y}_1^{(0)}$. When $\lambda = -s_{12}/s_{22}$, $\widehat{\mu}_{\text{dif}}^{(\lambda)} = \bar{y}_2 - \widehat{\mu}_1$, the AC estimate.

(e) When $\lambda = -s_{11}/s_{12}$, the estimate $\widehat{\mu}_{\text{dif}}^{(\lambda)}$ is indeterminate since $\widehat{\beta}_{21\cdot1}^{(\lambda)}$ has a denominator equal to zero.

(f) Suppose $s_{11} = s_{22}$ and the sample correlation for the complete cases is r. Then $\lambda = -r$ leads to the AC estimate $\bar{y}_2 - \widehat{\mu}_1$, and $\lambda = -1/r$ leads to an indeterminate estimate. These two possibilities bracket $\lambda = -1$, where missingness depends on the change $Y_2 - Y_1$.

This relatively simple example illustrates the role of the missing-data mechanism in the properties of estimates. It is tempting to attempt to estimate λ from the data, but unfortunately the various choices of λ noted above all give the same fits to the observed data. Hence, we are reduced to a sensitivity analysis where λ is varied over a plausible range. This approach is illustrated for a more complex model in Example 3 to follow.

Model 5. Selection model for n repeated measures with non-MAR dropouts and covariates. The previous two models are easily extended to data on n repeated measures $\boldsymbol{Y}_i = (Y_{i1}, \ldots, Y_{in})$ with the last measure Y_{in} subject to dropout, and $R_i = 1$ if Y_{in} is observed, and $R_i = 0$ if Y_{in} is missing. We also include a set of fully observed covariates $\boldsymbol{X}_i$, and for simplicity assume linear relationships between the repeated measures and these covariates. An NMAR selection model that extends model 3 is

$$(\boldsymbol{Y}_i | \boldsymbol{X}_i) \sim N(\boldsymbol{X}_i \boldsymbol{\beta}, \Sigma),$$

$$(R_i | \boldsymbol{Y}_i, \boldsymbol{X}_i) \sim \mathrm{Ber}(P(\phi(\boldsymbol{Y}_i, \boldsymbol{X}_i)),$$

$$\mathrm{logit}\{P(\phi(\boldsymbol{Y}_i, X_i))\} = \phi_0 + \phi_1 Y_{i1} + \ldots + \phi_{n-1} Y_{in-1} + \phi_n Y_{in} + \phi'_{n+1} \boldsymbol{X}_i.$$

This model allows the response mechanism to depend on the values of $Y_{i1}, \ldots, Y_{in}$ as well as the covariates. The MAR model corresponds to $\phi_n = 0$. The response propensity model could also be extended to allow interactions between the covariates and observed components of $\boldsymbol{Y}_i$.

Model 6. Pattern-mixture model for n repeated measures with non-MAR dropouts and covariates. A pattern-mixture analog of Model 5 is

$$(\boldsymbol{Y}_i | \boldsymbol{X}_i, R_i = k) \sim N(\boldsymbol{X}_i \boldsymbol{\beta}^{(k)}, \Sigma^{(k)}), \tag{18.16}$$

$$(R_i | \boldsymbol{X}_i) \sim \mathrm{Ber}(\delta(\boldsymbol{X}_i)\}.$$

where the parameters $(\boldsymbol{\beta}^{(k)}, \Sigma^{(k)})$ of the multivariate normal distribution are different for each pattern. The following example applies this model.

Example 3. A dose-comparison study for schizophrenia treatments. Little and Wang (1996) used a model of the form (18.16) to analyze data from a clinical trial to compare three alternative dose regimens of haloperidol for schizophrenia patients. Sixty-five patients with DSM-III diagnosis of schizophrenia were assigned to receive 5, 10, or 20 mg/day of haloperidol for 4 weeks. The outcome variable Y was the Brief Psychiatric Rating Scale Schizophrenia (BPRSS) factor, measured at $j = 1$ (baseline), $j = 2$ (week 1), and $j = 3$ (week 4). The main parameters of interest were the average change in BPRSS between baseline and week 4 for each dose group. Twenty-nine patients dropped out of the study at $j = 3$, with dropout rates varying across dose groups. Accordingly, $R_i = 1$ if Y_{i3} is observed and $R_i = 0$ if Y_{i3} is missing. A poor BPRSS outcome may cause patients to leave the study, particularly if combined with unpleasant side effects associated with the drug, particularly at high doses. Thus, models were fit where missingness of BPRSS at week 4 depended not only on the dosage, but also on the BPRSS values at week 4 and at previous times. Little and Wang fitted the following pattern-mixture model:

$$(\boldsymbol{Y}_i | \boldsymbol{X}_i, R_i = k) \sim N_3(\boldsymbol{X}_i \boldsymbol{\beta}^{(k)}, \Sigma^{(k)}),$$

$$(R_i | X_i) \sim \mathrm{MNOM}(\pi(\boldsymbol{X}_i, \boldsymbol{\delta})), \tag{18.17}$$

$$\mathrm{logit}(\pi(\boldsymbol{X}_i, \delta)) = \boldsymbol{\delta}' \boldsymbol{X},$$

where $\boldsymbol{X}_i$ represents three treatment dummies, and MNOM denotes the multinomial distribution. Thus, for pattern k, $\boldsymbol{Y}_i$ has a trivariate normal linear regression on $\boldsymbol{X}_i$ with (3×3) coefficient matrix $B^{(k)}$ and covariance matrix $\Sigma^{(k)}$. The parameters of (18.17) are the (3×1) vector $\boldsymbol{\delta}$, estimated by the vector of observed non-response rates at week 4 for each dose group, and $(\boldsymbol{\beta}^{(k)}, \Sigma^{(k)})$ for $k = 0, 1$.

This model is underidentified, in that there are no data to estimate directly the six parameters of the distribution of Y_{i3} given Y_{i1}, Y_{i2}, and $\boldsymbol{X}_i$ for the dropout pattern $R_i = 0$. These parameters are identified by assumptions about the missing-data mechanism. Specifically, suppose it is assumed that

$$\Pr(R_i = 1) = g(c_{i1}Y_{i1} + c_{i2}Y_{i2} + c_{i3}Y_{i3}, \boldsymbol{X}_i), \qquad (18.18)$$

where g is an arbitrary unspecified function, and $\boldsymbol{c}_i = (c_{i1}, c_{i2}, c_{i3})$ are prespecified coefficients. When $c_{i3} = 0$ in (18.18), the conditional distribution of Y_{i3} given $(Y_{i1}, Y_{i2}, \boldsymbol{X}_i)$ is the same for the complete and incomplete cases, and the data are MAR. The effect of non-ignorable non-response was assessed by computing ML and Bayes estimates for various other choices of $\boldsymbol{c}_i$.

Specifically, Table 18.1 shows estimates of the difference in mean BPRSS between baseline and week 4 for the three treatment groups, for the following methods:

(1) CC analysis, where incomplete cases are dropped from the analysis.

(2) Ignorable ML, where missingness is assumed to depend on the BPRSS scores at baseline and week 1. These results are ML under the ignorable pattern-mixture model or the ignorable selection model. Standard errors are the standard deviation of estimates from 1000 bootstrap samples.

(3) ML under the pattern-mixture model (18.17) and (18.18), with the following alternative choices of $\boldsymbol{c}_i$: A. $\boldsymbol{c}_i = (0.4, 0.4, 0.2)$; B. $\boldsymbol{c}_i = (0.3, 0.4, 0.4)$; C. $\boldsymbol{c}_i = (0.1, 0.1, 0.8)$; and D. $\boldsymbol{c}_i = (0, 0, 1)$. These represent progressively more extreme departures from MAR, with A being closest to the ignorable assumption $c_{i3} = 0$ corresponding to method (2). ML estimates were computed using an EM algorithm, and asymptotic standard errors were computed using the SEM algorithm (Meng and Rubin, 1991), which provides a numerical approximation to the inverse of the observed covariance matrix.

(4) For each of the models in (3), the mean and variance of the posterior distribution of the parameters, based on a non-informative prior. The posterior distributions were simulated by Gibbs sampling.

It can be seen from Table 18.1 that (a) the CC estimates deviate noticeably from estimates from the other methods, a common finding when the amount of missing data is substantial; (b) the ML/SEM and Bayes estimates for the pattern-mixture models are broadly similar, and the asymptotic standard errors are somewhat smaller than the posterior standard errors, particularly for pattern D; the posterior standard errors are preferred because they do not assume large samples; (c) the size of treatment effects is only moderately sensitive to the choice of pattern-mixture model (ignorable, A–D); the effect of choice of model is more pronounced in the high-dose group than in the other groups, reflecting the higher dropout

Table 18.1 Example 3: Estimates (Standard Errors) of Differences of Means of BPRSS between Baseline and Week 4, under Various Models

Methods	Treatment Group		
	Dose 5	Dose 10	Dose 20
(1) Complete Cases	3.70 (1.03)	4.35 (0.73)	5.67 (1.33)
(2) Ignorable ML	3.29 (0.90)	4.09 (0.62)	6.46 (1.04)
(3) Pattern-Mixture Models: ML			
Mechanism A	3.28 (0.90)	4.14 (0.62)	6.53 (1.05)
Mechanism B	3.25 (0.91)	4.18 (0.63)	6.61 (1.07)
Mechanism C	3.18 (0.95)	4.25 (0.66)	6.81 (1.16)
Mechanism D	3.14 (0.97)	4.27 (0.68)	6.91 (1.21)
(4) Pattern-Mixture Models: Bayes			
Mechanism A	3.23 (0.99)	4.07 (0.71)	6.46 (1.19)
Mechanism B	3.21 (1.02)	4.13 (0.72)	6.56 (1.22)
Mechanism C	3.13 (1.12)	4.23 (0.77)	6.81 (1.39)
Mechanism D	3.08 (1.19)	4.26 (0.82)	6.96 (1.53)

rate for that group; (d) as missingness becomes increasingly dependent on the missing week 4 BPRSS value, the mean treatment effects decrease slightly for the low- and moderate-dose groups, and increase somewhat more for the high-dose group. The net effect of this change in assumed mechanism is to slightly increase the differentials in treatment effects by size of dose; (e) the standard errors of the pattern-mixture model estimates increase from models A though D, reflecting a loss of information with increasing degree of non-ignorable non-response.

18.5 Mixed models with dropouts

The scope of models for MAR repeated-measures data was expanded in Model 2 by including unobserved within-subject random effects $\boldsymbol{b}_i$ in the model. With NMAR data, the selection and pattern-mixture formulations can be expanded to allow the possibility that the missing-data mechanism depends on these random effects (Little, 1995). This leads to a rich class of models based on various factorizations of the joint distribution of $\boldsymbol{R}_i$, $\boldsymbol{Y}_i$, and $\boldsymbol{b}_i$. There are six ways this joint distribution can be factored, and three of them condition the distribution of $\boldsymbol{b}_i$ on $\boldsymbol{Y}_i$, which is not sensible here. The remaining three factorizations yield *mixed-effect selection models* of the form

$$f(\boldsymbol{R}_i, \boldsymbol{Y}_i, \boldsymbol{b}_i | X_i, \boldsymbol{\gamma}) = f_B(\boldsymbol{b}_i | X_i, \boldsymbol{\gamma}_1) f_{Y|B}(\boldsymbol{Y}_i | X_i, \boldsymbol{b}_i, \boldsymbol{\gamma}_2) f_{R|Y,B}(\boldsymbol{R}_i | X_i, \boldsymbol{Y}_i, \boldsymbol{b}_i, \boldsymbol{\gamma}); \quad (18.19)$$

mixed-effect pattern-mixture models of the form

$$f(\boldsymbol{R}_i, \boldsymbol{Y}_i, \boldsymbol{b}_i | X_i, \boldsymbol{\delta}, \boldsymbol{\nu}) = f_R(\boldsymbol{R}_i | X_i, \boldsymbol{\delta}) f_{B|R}(\boldsymbol{b}_i | X_i, R_i, \boldsymbol{\nu}_1) f_{Y|B,R}(\boldsymbol{Y}_i | X_i, \boldsymbol{b}_i, R_i, \boldsymbol{\nu}_2); \quad (18.20)$$

and *mixed-effect hybrid models* of the form

$$f(\boldsymbol{R}_i, \boldsymbol{Y}_i, \boldsymbol{b}_i | X_i, \boldsymbol{\gamma}, \boldsymbol{\delta}) = f_B(\boldsymbol{b}_i | X_i, \boldsymbol{\gamma}_1) f_{R|B}(\boldsymbol{R}_i | X_i, \boldsymbol{b}_i, \boldsymbol{\delta}) f_{Y|B,R}(\boldsymbol{Y}_i | X_i, \boldsymbol{b}_i, R_i, \boldsymbol{\nu}_2). \quad (18.21)$$

Models based on (18.21) have not, to my knowledge, been considered in the literature. I consider some examples of (18.19) and (18.20), focusing on various assumptions about the missing-data mechanism.

18.5.1 Covariate-dependent dropout

The two factorizations (18.19) and (18.20) become equivalent under the strong assumption that the dropout mechanism does not depend on outcome values $\boldsymbol{Y}_i$ or the random effects $\boldsymbol{b}_i$, but depends only on the values of fixed covariates X_i, that is, for (18.19),

$$f_{R|Y,B}(\boldsymbol{R}_i | X_i, \boldsymbol{Y}_i, \boldsymbol{b}_i, \boldsymbol{\phi}) = f_R(\boldsymbol{R}_i | X_i, \boldsymbol{\phi}) \quad (18.22)$$

This model is a strong form of MAR that allows dependence of dropout on both between-subject and within-subject covariates that can be treated as fixed in the model. In particular, dropout can depend on treatment-group indicators or other baseline covariates that are included in the model. Diggle and Kenward (1994) called assumption (18.22) "completely random dropout," and viewed it as a special case of Rubin's (1976) missing completely at random assumption. Little (1995) reserves the term "missing completely at random" for the case when missingness does not depend on X_i as well as $\boldsymbol{Y}_i$ and $\boldsymbol{b}_i$. Assumption (18.22) is capable of some empirical verification, by comparing empirical distributions of observed outcomes $\boldsymbol{Y}_i^o$ across patterns after adjusting for the covariates. For example, if there are two outcomes, and Y_{i1} is fully observed and Y_{i2} has missing values, then one can compare the adjusted mean of Y_{i1} given X_i for the complete and incomplete cases. One way of implementing this is to regress Y_{i1} on X_i and the indicator R_{i2} for whether Y_{i2} is missing, and test whether the coefficient of R_{i2} is different from zero.

Under covariate-dependent missingness, analysis of the complete cases is not biased, although it is subject to a loss of efficiency — indeed in Example 1 above it is not feasible, since the design of the study results in no complete cases. Methods that use all the data, such as ML or GEE, are generally more efficient and yield consistent estimators under the usual assumptions of these methods.

18.5.2 MAR dropout

The MAR assumption in the context of (18.19) assumes that dropout depends on $\boldsymbol{Y}_i$ and $\boldsymbol{b}_i$ only through the observed data $\boldsymbol{Y}_i^o$, that is,

$$f_{R|Y,B}(\boldsymbol{R}_i|X_i, \boldsymbol{Y}_i, \boldsymbol{b}_i, \boldsymbol{\phi}) = f_{R|Y}(\boldsymbol{R}_i|X_i, \boldsymbol{Y}_i^o, \boldsymbol{\phi}).$$

The clinical trial of two anti-hypertensive agents in Example 2 provides an illustration of a case where MAR is plausible, since dropout, namely moving to the open phase of the protocol, depends on a blood pressure value that is recorded. Although MAR, the ML analysis is vulnerable to misspecification of the relationship between outcomes and blood pressure because this can only be estimated for the recorded blood pressures, which are lower than the blood pressures for cases after dropout. Dropping-out because of side effects would also be MAR if the side effects were recorded and included in the analysis via likelihood methods. As noted in comment (ii) in the Introduction, multiple imputation based on a model that includes the side-effect data is perhaps the most convenient approach for achieving this. This was not done in Murray and Findlay (1988), but the number of such cases is small. The nature of the mechanisms for the 37 patients who dropped out for "other reasons" is not discussed, but at least the predominant dropout mechanism here is plausibly MAR.

Under MAR and the distinctness condition noted above, the missing-data mechanism is ignorable, and ML or Bayes inference can be based on models such as Model 2. In contrast, other methods such as CC analysis or GEE generally require the stronger assumption (18.22) to yield consistent estimators (Fitzmaurice, Laird, and Rotnitzky 1993; Kenward, Lesaffre, and Molenberghs 1994). This is an important advantage of likelihood-based inference, although the methods require adequate specification of the model. The GEE approach can be modified to yield consistent estimators under the MAR assumption, by multiplying GEE weights by the inverse of estimated selection probabilities. (Robins, Rotnitsky, and Zhao, 1995). For a related use of estimated selection probabilities to adjust for random dropout in a simpler situation, see Heyting, Tolboom, and Essers (1992).

18.5.3 Non-ignorable outcome-dependent dropout

In other settings, dropout may depend on missing components $\boldsymbol{Y}_i^m$ of $\boldsymbol{Y}_i$, such as the (unrecorded) value of the outcome at the time when the subject drops out. Little (1995) calls the resulting assumption "outcome-dependent dropout":

$$f_{R|Y,B}(\boldsymbol{R}_i|X_i, \boldsymbol{Y}_i, \boldsymbol{b}_i, \boldsymbol{\phi}) = f_{R|Y}(\boldsymbol{R}_i|X_i, \boldsymbol{Y}_i^o, \boldsymbol{Y}_i^m, \boldsymbol{\phi}). \tag{18.23}$$

Diggle and Kenward (1994) used a model of the form (18.23) to analyze data from a longitudinal milk protein trial. Cows were randomly allocated to one of three diets (barley, mixed barley and lupins, and lupins) and assayed the protein content of milk samples taken weekly for a period of 20 weeks. Dropout corresponded to cows that stopped producing milk before the end of the experiment. The complete-data model $f_{Y|B}f_B$ specified a quadratic model for the mean protein content over time, with an intercept that depended on diet (thus modeling an additive effect of treatment). The covariance structure was assumed to be a combination of an autoregressive structure with an added independent measurement error. The dropout distribution $f_{R|Y}$ process was modeled as depending on the current and previous value of protein content, specifically:

$$\text{logit}\{\Pr(R_{it} = 1|R_{it-1} = 1, \boldsymbol{Y}_i, X_i, \boldsymbol{\phi})\} = \phi_{0t} + \phi_1 Y_{it-1} + \phi_2 Y_{it}. \tag{18.24}$$

The resulting ML estimates $\widehat{\phi}_1 = 12.0, \widehat{\phi}_2 = -20.4$ of the coefficients ϕ_1 and ϕ_2 suggested that the probability of dropout increases when the prevailing level of protein is low or the increment between the last and current protein content is high. As noted in the discussion of the article, underidentifiability is a serious problem with this model. A controversial issue concerns whether the parameters in (18.24) can be simultaneously estimated with the parameters of the distributions of f_B and $f_{Y|B}$. A sensitivity analysis might consider ML estimates of the parameters for a variety of plausible alternative choices of ϕ_1 and ϕ_2.

Models of the form (18.23) have been considered for non-normal data. In particular for repeated-measures ordinal data, Molenberghs, Kenward, and Lesaffre (1997) combined a multivariate Dale model for the outcomes with a model for the dropout mechanism analogous to that of Diggle and Kenward (1994). Problems of identification arise for these models too.

18.5.4 Non-ignorable random-coefficient dependent dropout

Another form of non-ignorable dropout model assumes dropout at time t depends on the value of $\boldsymbol{b}_i$, that is,

$$f_{R|Y,B}(\boldsymbol{R}_i|X_i, \boldsymbol{Y}_i, \boldsymbol{b}_i, \boldsymbol{\phi}) = f_{R|B}(\boldsymbol{R}_i|X_i, \boldsymbol{b}_i, \boldsymbol{\phi}). \qquad (18.25)$$

Examples of models of dropout of the form (18.25) include Wu and Carroll (1988), Shih, Quan, and Chang (1994), Mori, Woolson, and Woodworth (1994), Schluchter (1992), and the following:

Example 4. Longitudinal AIDS data. DeGruttola and Tu (1994) modeled the relationship between the progression of CD4 lymphocyte count and survival for patients enrolled in a clinical trial of two alternative doses for zidovudine. Here, a vector of log CD4 counts for subject i is modeled via a mixed model of the form (18.7). The main cause of dropout is death, which is measured as survival time, and modeled as a continuous, normally distributed random variable with a mean that is a linear function of covariates. ML estimation is accomplished using an EM algorithm (Dempster, Laird, and Rubin 1977), with standard errors computed using the method of Louis (1982). A drawback with this approach is that the selection-model factorization effectively treats the CD4 counts after death as missing values, which is not in accord with the definition of missing data provided above. In my view, a better analysis would condition the analysis of CD4 counts at any time on individuals who have survived up to that time.

18.5.5 Shared-parameter models

A number of models have been formulated that assume both the outcome process and the dropout process depend on shared latent variables. These are called *shared-parameter models*, and examples include Ten Have et al. (1998, 2002), Albert et al. (2002), and Roy (2003). They are special cases of (18.19) and (18.21) where $\boldsymbol{Y}_i$ and $\boldsymbol{R}_i$ are assumed independent given $\boldsymbol{b}_i$:

$$f(\boldsymbol{R}_i, \boldsymbol{Y}_i, \boldsymbol{b}_i|X_i, \boldsymbol{\gamma}, \boldsymbol{\phi}) = f_B(\boldsymbol{b}_i|X_i, \boldsymbol{\gamma}_1)f_{Y|B}(\boldsymbol{Y}_i|X_i, \boldsymbol{b}_i, \boldsymbol{\gamma}_2)f_{R|B}(\boldsymbol{R}_i|X_i, \boldsymbol{b}_i, \boldsymbol{\phi}).$$

Example 5. A shared-parameter model for heroin addiction treatment data with missing data. Albert et al. (2002) analyzed data from a clinical trial of treatments of heroin addiction that randomized patients into one of two treatment groups: buprenorphine ($n = 53$) and methadone ($n = 55$). Patients were scheduled for urine tests three times a week on Monday, Wednesday, and Friday for 17 weeks post-randomization (51 scheduled responses). Urine tests were scored as positive or negative for the presence of opiates at each follow-up visit. A primary scientific objective of the study was to compare the marginal proportion of positive urine tests over follow-up between the two treatment arms. Plots suggested that the frequency of positive urine tests was relatively constant over follow-up, with the buprenorphine arm having a lower proportion of positive urine tests than the methadone arm. Thus, the analyses focused on comparing the marginal proportions, assumed constant over time, across the two treatment arms.

The analysis was complicated by the unequally spaced observations and the large amount of missing data, which took the form of dropouts and intermittent missing data. A number of subjects withdrew from the study due to poor compliance or because they were offered places in treatment programs that gave unmasked treatment and long-term care. Intermittent missingness may be more closely associated with the response process, as patients may not show up when they are taking opiates. The missing-data mechanism appeared different in the two treatment arms. The proportion of patients dropping out by the end of the 17-week period was 80% in the methadone group and 59% in the buprenorphine group. In addition, patients had a sizable amount of intermittent missing data, with the proportion of intermittent missing data being higher in the buprenorphine arm than the methadone arm. The Spearman rank correlation between the proportion of positive tests and the time to dropout was -0.44 in the buprenorphine arm and -0.10 in the methadone arm. The correlations between the proportion of positive tests and the proportion of intermittent missing visits before dropout in the buprenorphine and methadone arms were 0.40 and 0.29, respectively.

These calculations suggest that addicts who are more likely to use drugs are both more likely to dropout and to have a higher frequency of intermittent missing data before dropout than addicts who use opiates less frequently. These associations are consistent with an NMAR missing-data mechanism. The differences in the magnitude of these correlations between treatment arms suggest that the informative missing data may be greater in the buprenorphine arm than in the methadone arm.

For the ith patient, let $y_{it_1}, y_{it_2}, \ldots, y_{it_n}$ be the sequence of n intended binary measurements, where t_j is the time of the jth follow-up, and let $\boldsymbol{Y}_i^o$ be the vector of observed binary responses. Denote $\boldsymbol{R}_i = (R_{it_1}, R_{it_2}, \ldots, R_{it_n})$ as indicators of intermittent missingness or dropout at each follow-up time, where $R_{it_j} = 0$ if y_{it_j} is observed, $R_{it_j} = 1$ if y_{it_j} is intermittently missing, and $R_{it_j} = 2$ if y_{it_j} is a value after dropout. Both $\boldsymbol{Y}_i$ and $\boldsymbol{R}_i$ are modeled conditional on a latent process, $\{b_{it}\}$, and a time-dependent covariate vector, $\{\boldsymbol{X}_{it}\}$. Specifically, a shared-parameter model of the following form is assumed (notation differs slightly from previous examples):

$$\text{logit}\{\Pr(y_{it_j} = 1|b_{it_j})\} = \boldsymbol{X}_{it_j}^T \boldsymbol{\beta} + b_{it_j}, \tag{18.26}$$

$$\Pr(R_{it_j} = \ell | b_{it_j}, R_{it_{j-1}} \neq 2) = \begin{cases} \dfrac{1}{1 + \sum_{\ell=1}^{2} \exp(\boldsymbol{\nu}_{\ell it_j} \boldsymbol{\eta}_i + \psi_i b_{it_j})}, & \ell = 0, \\[4mm] \dfrac{\exp(\boldsymbol{\nu}_{\ell it_j} \boldsymbol{\eta}_i + \psi_i b_{it_j})}{1 + \sum_{\ell=1}^{2} \exp(\boldsymbol{\nu}_{\ell it_j} \boldsymbol{\eta}_i + \psi_i b_{it_j})}, & \ell = 1, 2, \end{cases}$$

where $\boldsymbol{\nu}_{\ell it_j}$ are vectors of covariates and η_i their corresponding regression coefficients, and ψ_i are parameters that relate the missingness (intermittent missing and dropout) with the outcome data. Since $R_{it} = 2$ denotes dropout, which is an absorbing state, $\Pr(R_{it_{j+1}} = 2|R_{it_j} = 2) = 1$. The shared random parameters $\{b_{it}\}$ are modeled as a Gaussian Ornstein–Uhlenbeck process (Feller, 1971), with mean zero and covariance

$$\text{Cov}(b_{it}, b_{it'}) = \sigma^2 \exp(-\theta|t - t'|), \quad \text{where } \theta > 0.$$

Table 18.2 Parameter Estimates and Standard Errors (SE) for Shared-Parameter Model Fitted to Heroin Addiction Treatment Data of Example 5

Parameter	Methadone		Buprenorphine	
	Estimate	SE	Estimate	SE
β_G	1.44	0.43	−0.10	0.37
ν_{1G}	−1.89	0.22	−1.71	0.18
ν_{2G}	−3.42	0.21	−4.11	0.31
ψ_{1G}	0.29	0.13	0.43	0.09
ψ_{2G}	0.22	0.15	0.48	0.10
σ_G	2.84	0.55	2.77	0.50
θ_G	0.014	0.008	0.012	0.004
$P_G(y_{it} = 1)$	0.67	0.040	0.49	0.039
$P_G(R_{it} = 1)$	0.13	0.023	0.15	0.023
$P_G(R_{it} = 2)$	0.027	0.005	0.014	0.004

Source: Albert et al. (2002).

In the opiate clinical trial application, a version of (18.26) with no time effects was fitted separately in the two treatment groups, namely:

$$\text{logit}\{\Pr(y_{it_j} = 1|b_{it_j})\} = \beta_G + b_{it_j},$$

$$\Pr(R_{it_j} = \ell|b_{it_j}, R_{it_{j-1}} \neq 2) = \begin{cases} \frac{1}{1+\sum_{\ell=1}^{2}\exp(\eta_{\ell G}+\psi_{\ell G}b_{it_j})}, & \ell = 0, \\ \frac{\exp(\eta_{iG}+\gamma_{iG}b_{it_j})}{1+\sum_{\ell=1}^{2}\exp(\eta_{\ell G}+\psi_{\ell G}b_{it_j})}, & \ell = 1, 2, \end{cases}$$

$$\text{Cov}(b_{it}, b_{it'}) = \sigma_G^2 \exp(-\theta_G|t - t'|), \quad \text{where } \theta > 0,$$

where $G = 0$ and $G = 1$ index parameters in the methadone and buprenorphine groups, respectively. ML estimation was accomplished using a Monte Carlo EM algorithm, and standard errors estimated by the bootstrap with 250 bootstrap samples. Table 18.2 shows the resulting parameter estimates and standard errors. The parameter estimates for the buprenorphine group show a significant positive relationship between response, dropout, and intermittent missingness (estimates of ψ_{11} and ψ_{21} were highly significant), suggesting that the missing-data mechanism is non-ignorable for this group. The corresponding estimates for the methadone group were smaller in magnitude and not statistically significant at the 0.05 level, although they were positive.

18.6 Mixed-effect pattern-mixture models

Mixed-effect pattern-mixture models are based on the factorization (18.20). This approach stratifies the sample by pattern of missing data (e.g., by the time of dropout) and then models differences in the distribution of $\boldsymbol{Y}_i$ over these patterns. As with the selection models, these models can be formulated for the case where missingness depends on $\boldsymbol{Y}_i$,

$$f(\boldsymbol{R}_i, \boldsymbol{Y}_i, \boldsymbol{b}_i|X_i, \boldsymbol{\delta}, \boldsymbol{\nu}) = f_R(\boldsymbol{R}_i|X_i, \boldsymbol{\delta})f_B(\boldsymbol{b}_i|X_i, \boldsymbol{\nu}_1)f_{Y|B,R}(\boldsymbol{Y}_i|X_i, \boldsymbol{R}_i, \boldsymbol{b}_i, \boldsymbol{\nu}_2), \quad (18.27)$$

and for the case where missingness depends on $\boldsymbol{b}_i$,

$$f(\boldsymbol{R}_i, \boldsymbol{Y}_i, \boldsymbol{b}_i|X_i, \boldsymbol{\delta}, \boldsymbol{\nu}) = f_R(\boldsymbol{R}_i|X_i, \boldsymbol{\delta})f_{B|R}(\boldsymbol{b}_i|X_i, \boldsymbol{R}_i, \boldsymbol{\nu}_1)f_{Y|B,R}(\boldsymbol{Y}_i|X_i, \boldsymbol{b}_i, \boldsymbol{\nu}_2).$$

The latter models have the computational advantage that parameters of the distribution $f_{Y|B,R}(\boldsymbol{Y}_i|X_i, \boldsymbol{R}_i, \boldsymbol{b}_i, \boldsymbol{\nu}_2)$ can be estimated using existing mixed-model software, such as SAS `proc mixed`, by including the dropout indicators $\boldsymbol{R}_i$ as covariates in the model.

Fitzmaurice, Laird, and Schneyer (2001) discussed forms of (18.27) that are parameterized to enhance interpretability of the parameters, and apply it to data from an asthma trial with normal repeated measures. Hogan, Lin, and Herman (2004) apply a model of this form to AIDS clinical trial data, where the fixed-effects parameters in a mixed model for $\boldsymbol{Y}_i$ given X_i and $\boldsymbol{b}_i$ are allowed to depend non-parametrically on the dropout time, which may be categorical or continuous. Two examples of models of this type are now presented:

Model 7. A straight-line pattern-mixture model. Suppose that X_i is scalar (e.g., time or age), and

$$
\begin{aligned}
(Y_{ij}|X_i, \boldsymbol{b}_i, R_i = k, \nu_2) &\sim N(b_{0i} + b_{1i}X_i, \sigma^2),\\
(b_{0i}, b_{1i}|\nu_1)^T &\sim N((b_0^{(k)}, b_1^{(k)})^T, \Gamma),\\
\Pr(R_i = k) &= \pi_k,
\end{aligned}
\tag{18.28}
$$

which models $\{Y_{ij}\}$ with a linear regression on X_i with random slope b_{1i} and intercept b_{0i}, which are in turn distributed about a line with the same intercept $b_0^{(k)}$ and slope $b_1^{(k)}$ for each pattern k. This can be modeled via a standard mixed model by including as covariates dummy variables for each pattern. At least two repeated measures are needed to allow estimation of the slope and intercept for each pattern. If the quantities of interest are the expected intercept and slope, averaged over missing-data pattern, that is, $b_0 = \sum_{k=1}^{K} \pi_k b_0^{(k)}$ and $b_1 = \sum_{k=1}^{K} \pi_k b_1^{(k)}$, ML estimates of these parameters are obtained as a weighted sum of the ML estimates of the expected intercept and slope for pattern k, with weights given as the proportion of cases with pattern k. This contrasts with an MAR model, where estimates for each pattern are effectively weighted by their precision. This model can yield estimates with poor precision (Wang-Clow et al., 1995), and, to address this, additional structure might be specified for the relationship between the slope and intercepts and pattern. For example, one might assume the expected intercept is independent of pattern, and the expected slope is linearly related to the dropout time t_k for pattern:

$$
b_0^{(k)} = b_0, \quad b_1^{(k)} = \nu_0 + \nu_1 t_k.
$$

This model is easily extended to include other covariates, such as indicators of treatment group, yielding formalizations of the conditional linear model of Wu and Bailey (1989).

Model 8. An LOCF pattern-mixture model. A common method for handling dropouts in longitudinal data, sometimes called "last observation carried forward" (LOCF) imputation, is to impute the missing values with last observation prior to dropout. This imputation method implements the idea that an individual's outcome is unchanged after dropout, an assumption that needs to be checked for plausibility in real settings. Aside from the realism of the implied model for dropout, the LOCF method has the problem that if the outcome has some within-subject variation due to random fluctuations or measurement error, then imputing exactly the same value as that recorded just before dropout is not realistic. As a consequence, analysis of the data imputed by LOCF does not propagate imputation uncertainty, and hence does not yield valid inferences, even if the underlying model of no change after dropout is reasonable.

This problem can be addressed by formalizing the idea of LOCF as a pattern-mixture model, where individuals are stratified by pattern of dropout, and the individual mean outcome is assumed constant after dropout, but values after dropout can fluctuate around that mean. The key feature of the model is that each individual i has an underlying profile of expected values μ_{ij}, $j = 1, \ldots, n$, that would be observable in the absence of measurement error. If the individual drops out at some time τ_i, then $\mu_{ij} = \mu_{i\tau_i}$ for all $j > \tau_i$; that is, the underlying mean remains unchanged after dropout. As a simple example of a model of this kind, consider a homogeneous sample of N individuals with at most n repeated measures $\{y_{ij}, j = 1, \ldots, n\}$. Let d_i be the number of measures for individual i, and assume that

$$(y_{ij}|\mu_{ij}, \sigma^2, d_i = d) \sim N(\mu_{ij}, \sigma^2)$$

$$\mu_{ij} = \begin{cases} \beta_{0i} + \beta_{1i}j, & \text{if } j < d, \\ \beta_{0i} + \beta_{1i}d, & \text{if } j \geqslant d, \end{cases}$$

$$(\beta_{0i}, \beta_{1i}|\nu_1)' \sim N((\beta_0, \beta_1)', \Gamma).$$

This is an LOCF model with a linear profile up to the time of dropout. Again, extensions to include baseline covariates like treatment are readily formulated. This model could be used to multiply impute values of Y after dropout, yielding inferences that propagate imputation uncertainty.

18.7 Conclusion

I have reviewed likelihood-based methods for the analysis of models for longitudinal data with missing values. An important distinction is between models that ignore the missing-data mechanism, and hence assume the data are MAR, and models that relax the MAR assumption and incorporate assumptions about the missing-data mechanism. In many respects, ML and Bayes inferences for ignorable models are similar to corresponding inferences with complete data. The difference is that the likelihood is often more complicated, making computation more of a challenge, results are potentially more sensitive to model misspecification, and asymptotic results may be less valid because the log-likelihood function is not quadratic. Consequently, Bayesian inference based on the posterior distribution and relatively non-informative priors is attractive because it is less dependent on large sample sizes and deals in an appropriate way with nuisance parameters.

Non-ignorable models are more challenging because problems with lack of identifiability of the parameters are often severe, and assumptions about the mechanism leading to missing values need to be incorporated in the analysis. In selection models this requires an explicit parametric model for R given Y; in certain pattern-mixture models the form of the model does not have to be explicit because assumptions about the mechanism are incorporated implicitly through restrictions on the parameters across patterns. Successful modeling requires realistic assumptions about the mechanism, which implies that information about the reasons why values are missing should be determined when possible and included in the analysis. For example, if some cases that drop out are plausibly MAR but others are not, it is better to build a model that reflects these different mechanisms than to assume the same MAR or NMAR mechanism for all dropouts. In general I think non-MAR situations are often best handled by relatively simple sensitivity analyses, where the assumptions are transparent. For example, if a subset of the dropouts are thought to have an NMAR mechanism, the model might assume the mean of the predictive distribution of those values deviates from the distribution assumed under MAR by some specified amount, say 0.2 or 0.5 times the residual standard deviation given known variables for that case. The results from "tilting" the MAR model in this way can then be assessed. Others (e.g., Horowitz and Manski, 2000) have advocated a sensitivity analysis over the full range of possible values of the missing values. This conservative approach is only feasible for missing variables that have a restricted range, such as binary or ordinal data, and the results are arguably too dispersed to be very useful unless there is a small number of missing values. A Bayesian analysis based on a subjective prior distribution relating distributions for non-respondents from distributions for respondents is in my view conceptually more satisfying (e.g., Rubin, 1977), although the challenge remains of incorporating reasonable departures from MAR in this prior specification.

The models considered here have assumed that missing data are confined to the repeated measures Y, and covariate information X is fully observed. Current ML software does not allow for missing values in X, so if values of covariates are missing then some additional work is needed to address that problem. One option is simply to drop values with missing covariates, which has advantages when the missingness mechanism depends on the values of the missing covariates themselves, but is wasteful of information and can result in bias if the mechanism is MAR. Another approach is to impute values of the missing covariates based on a joint model for the missing covariates given the observed covariates and Y. Multiple imputation is recommended if this option is contemplated, so that imputation error is propagated. With relatively small numbers of missing covariates, a relatively simple model might suffice; as the fraction of missing values increases, more attention to specifying this model correctly is needed. The sequential multiple imputation (MI) methods discussed in Chapter 21 provide a useful tool for this multiple imputation step. Once the covariates are filled in, longitudinal models as discussed above can be fitted to the filled-in data sets, and results combined using MI combining rules discussed in Chapter 21.

Multiple imputation of missing values of Y under an explicit parametric model, as discussed in Chapter 21, is closely related to Bayesian inference based on the posterior distribution for that model. An advantage of MI is that the model for generating the multiple imputes may be different from the model used in the analysis; for example, in a clinical trial setting, the MI model may condition on information about side effects that are not part of the substantive models of interest, which focus on primary clinical trial outcomes.

References

Albert, P. S., Follmann, D. A., Wang, S. A., and Suh, E. B. (2002). A latent autoregressive model for longitudinal binary data subject to informative missingness. *Biometrics* **58**, 631–642.

Anderson, T. W. (1957). Maximum likelihood estimation for the multivariate normal distribution when some observations are missing. *Journal of the American Statistical Association* **52**, 200–203.

Birmingham, J. and Fitzmaurice, G. M. (2002). A pattern-mixture model for longitudinal binary responses with nonignorable nonresponse. *Biometrics* **58**, 989–996.

BUGS (2006). *The BUGS Project.* http://www.mrc-bsu.cam.ac.uk/bugs/.

Daniels, M. J. and Hogan, J. W. (2000). Reparameterizing the pattern-mixture model for sensitivity analysis under informative dropout. *Biometrics* **56**, 1241–1248.

DeGruttola, V. and Tu, X. M. (1994). Modelling progression of CD4-lymphocyte count and its relationship to survival time. *Biometrics* **50**, 1003–1014.

Dempster, A. P., Laird, N. M., and Rubin, D. B. (1977). Maximum likelihood from incomplete data via the EM algorithm. *Journal of the Royal Statistical Society, Series B* **39**, 1–38.

Diggle, P. and Kenward, M. G. (1994). Informative drop-out in longitudinal data analysis. *Applied Statistics* **43**, 49–73.

Ekholm, A. and Skinner, C. (1998). The Muscatine children's obesity data reanalysed using pattern mixture models. *Applied Statistics* **47**, 251–263.

Feller, W. (1971). *An Introduction to Probability Theory and Its Applications* (Vol. 2). New York: Wiley.

Fitzmaurice, G. M., Laird, N. M., and Rotnitzky, A. G. (1993). Regression models for discrete longitudinal responses. *Statistical Science* **8**, 284–309.

Fitzmaurice, G. M., Laird, N. M., and Schneyer, L. (2001). An alternative parameterization of the general linear mixture model for longitudinal data with non-ignorable drop-outs. *Statistics in Medicine* **20**, 1009–1021.

Gelfand, A. E., Hills, S. E., Racine-Poon, A., and Smith, A. F. M. (1990). Illustration of Bayesian inference in normal data models using Gibbs sampling. *Journal of the American Statistical Association* **85**, 972–985.

Glynn, R. J., Laird, N. M., and Rubin, D. B. (1986). Selection modelling versus mixture modelling with nonignorable nonresponse. In H. Wainer (ed.), *Drawing Inferences from Self-Selected Samples*, pp. 115–142. New York: Springer.

Hartley, H. O. and Rao, J. N. K. (1967). Maximum likelihood estimation for the mixed analysis of variance model. *Biometrika* **54**, 93–108.

Heckman, J. (1976). The common structure of statistical models of truncation, sample selection and limited dependent variables and a simple estimator for such models. *Annals of Economic and Social Measurement* **5**, 475–492.

Heyting, A., Tolboom, J. T. B. M., and Essers, J. G. A. (1992). Statistical handling of drop-outs in longitudinal clinical trials. *Statistics in Medicine* **11**, 2043–2063.

Hogan, J. W., Lin, X., and Herman, B. (2004). Mixtures of varying coefficient models for longitudinal data with discrete or continuous nonignorable dropout. *Biometrics* **60**, 854–864.

Hogan, J. W., Roy, J., and Korkontzelou, C. (2004). Biostatistics tutorial: Handling dropout in longitudinal data. *Statistics in Medicine* **23**, 1455–1497.

Horowitz, J. L. and Manski, C. F. (2000). Nonparametric analysis of randomized experiments with missing covariate and outcome data (with discussion). *Journal of the American Statistical Association* **95**, 77–88.

Huet, S., Bouvier, A., Poursat, M.-A., and Jolivet, E. (2004). *Statistical Tools for Nonlinear Regression. A Practical Guide with S-PLUS and R Examples*, 2nd ed. New York: Springer.

Jennrich, R. I. and Schluchter, M. D. (1986), Unbalanced repeated-measures models with structured covariance matrices. *Biometrics* **42**, 805–820.

Kenward, M. G. and Molenberghs, G. (1999). Parametric models for incomplete continuous and categorical longitudinal data. *Statistical Methods in Medical Research* **8**, 51–83.

Kenward, M. G., Lesaffre, E., and Molenberghs, G. (1994). An application of maximum likelihood and estimating equations to the analysis of ordinal data from a longitudinal study with cases missing at random. *Biometrics* **50**, 945–953.

Laird, N. M. and Ware, J. H. (1982). Random-effects models for longitudinal data. *Biometrics* **38**, 963–974.

Lange, K. L., Little, R. J. A., and Taylor, J. M. G. (1989). Robust statistical modeling using the *t* distribution. *Journal of the American Statistical Association* **84**, 881–896.

Lavange, L. M. and Helms, R. W. (1983). The analysis of incomplete longitudinal data with modeled covariance structures. Mimeo 1449, Institute of Statistics, University of North Carolina.

Littell, R. C., Milliken, G. A., Stroup, W. W., and Wolfinger, R. D. (1996). *SAS System for Mixed Models*. Cary, NC: SAS Institute Inc.

Little, R. J. A. (1993). Pattern-mixture models for multivariate incomplete data. *Journal of the American Statistical Association* **88**, 125–134.

Little, R. J. A. (1994). A class of pattern mixture models for normal missing data. *Biometrika* **81**, 471–483.

Little, R. J. A. (1995). Modeling the drop-out mechanism in repeated-measures studies. *Journal of the American Statistical Association* **90**, 1112–1121.

Little, R. J. A. and Rubin, D. B. (2002). *Statistical Analysis with Missing Data*, 2nd ed. New York: Wiley.

Little, R. J. A. and Wang, Y. (1996). Pattern-mixture models for multivariate incomplete data with covariates. *Biometrics* **52**, 98–111.

Louis, T. A. (1982). Finding the observed information matrix using the EM algorithm. *Journal of the Royal Statistical Society, Series B* **44**, 226–233.

McCulloch, C. E. and Searle, S. R. (2001). *Generalized, Linear, and Mixed Models*. New York: Wiley.

Meng, X. L. and Rubin, D. B. (1991). Using EM to obtain asymptotic variance-covariance matrices: the SEM algorithm. *Journal of the American Statistical Association* **86**, 899–909.

Michiels, B., Molenberghs, G., and Lipsitz, S. R. (1999). Selection models and pattern-mixture models for incomplete data with covariates. *Biometrics* **55**, 978–983.

Molenberghs, G., Kenward, M. G., and Lesaffre, E. (1997). The analysis of longitudinal ordinal data with nonrandom drop-out. *Biometrika* **84**, 33–44.

Molenberghs, G., Michiels, B., Kenward, M. G., and Diggle, P. J. (1998). Monotone missing data and pattern-mixture models. *Statistica Neerlandica* **52**, 153–161.

Mori, M., Woolson, R. F., and Woodworth, G. G. (1994). Slope estimation in the presence of informative censoring: Modeling the number of observations as a geometric random variable. *Biometrics* **50**, 39–50.

Murray, G. D. and Findlay, J. G. (1988). Correcting for the bias caused by drop-outs in hypertension trials. *Statistics in Medicine* **7**, 941–946.

Pinheiro, J. C. and Bates, D. M. (2000). *Mixed-Effects Models in S and S-PLUS*. New York: Springer.

Robins, J., Rotnitsky, A., and Zhao, L. P. (1995). Analysis of semiparametric regression models for repeated outcomes in the presence of missing data. *Journal of the American Statistical Association* **90**, 106–121.

Roy, J. (2003). Modeling longitudinal data with nonignorable dropouts using a latent dropout class model. *Biometrics* **59**, 829–836.

Rubin, D. B. (1976). Inference and missing data. *Biometrika* **63**, 581–592.

Rubin, D. B. (1977). Formalizing subjective notions about the effect of nonrespondents in sample surveys. *Journal of the American Statistical Association* **72**, 538–543.

SAS (2004). *SAS OnlineDoc® 9.1.3*. Cary, NC: SAS Institute Inc.

Schafer, J. L. (1997). *Analysis of Incomplete Multivariate Data*. London: Chapman & Hall.

Schluchter, M. D. (1988). Analysis of incomplete multivariate data using linear models with structured covariance matrices. *Statistics in Medicine* **7**, 317–324.

Schluchter, M. D. (1992). Methods for the analysis of informatively censored longitudinal data. *Statistics in Medicine* **11**, 1861–1870.

Shih, W. J., Quan, H., and Chang, M. N. (1994). Estimation of the mean when data contain non-ignorable missing values from a random effects model. *Statistics and Probability Letters* **19**, 249–257.

Tanner, M. A. (1991). *Tools for Statistical Inference: Observed Data and Data Augmentation Methods*. New York: Springer.

Ten Have, T. R., Pulkstenis, E., Kunselman, A., and Landis, J. R. (1998). Mixed effects logistic regression models for longitudinal binary response data with informative dropout. *Biometrics* **54**, 367–383.

Ten Have, T. R., Reboussin, B. A., Miller, M. E., and Kunselman, A. (2002). Mixed effects logistic regression models for multiple longitudinal binary functional limitation responses with informative drop-out and confounding by baseline outcomes. *Biometrics* **58**, 137–144.

Vonesh, E. F. and Chinchilli, V. M. (1997). *Linear and Nonlinear Models for the Analysis of Repeated Measurements*. New York: Marcel Dekker.

Wang-Clow, F., Lange, M., Laird, N. M., and Ware, J. H. (1995). Simulation study of estimators for rate of change in longitudinal studies with attrition. *Statistics in Medicine* **14**, 283–297.

Wu, M. C. and Bailey, K. R. (1989). Estimation and comparison of changes in the presence of informative right censoring: Conditional linear model. *Biometrics* **45**, 939–955.

Wu, M. C. and Carroll, R. J. (1988). Estimation and comparison of changes in the presence of informative right censoring by modeling the censoring process. *Biometrics* **44**, 175–188.

CHAPTER 19

Shared-parameter models

Paul S. Albert and Dean A. Follmann

Contents

19.1 Introduction

Modeling longitudinal data subject to missingness has been an active area for biostatistical research. The missing-data mechanism is said to be missing completely at random if the probability of missing is independent of both observed and unobserved data, missing at random (MAR) if the probability of missing is independent of only the observed data, and not missing at random (NMAR) if the probability of missingness depends on the unobserved data (Rubin, 1976; Little and Rubin, 1987). It is well known that naive methods may result in biased inferences under an NMAR mechanism. The use of shared (random) parameter models has been one approach to accounting for non-random missingness. In this formulation, a model for the longitudinal response measurements is linked with a model for the missing-data mechanism through a set of random effects that are shared between the two processes. For example, in a study of repeated measurements of lung function, Wu and Carroll (1988) proposed a model whereby the response process, which was modeled with a linear mixed model with a random intercept and slope, was linked with the censoring process by including an individual's random slope as a covariate in a probit model for the censoring process. When the probit regression coefficient for the random slope is not zero, there is dependence between the response and missing-data processes. Failure to account for this dependence can lead to biased estimation. Shared-parameter models (Follmann and Wu, 1995) induce a type of non-randomly missing-data mechanism that has been called "informative missingness" (Wu and Caroll, 1988). For review and comparison with other methods, see Little (1995), Hogan and Laird (1997b), and Vonesh, Greene, and Schlucher (2006).

A simple probability mechanism can be constructed that represents the above stochastic behavior. Denote Y_{ij} as the jth longitudinal observation at time t_j for the ith subject. Further, denote R_{ij} as an indicator of whether the jth measurement for the ith individual is observed. For a dropout process where an individual leaves the study between the $(d-1)$th and the dth time point, $R_{i1} = R_{i2} = \cdots = R_{id-1} = 1$ and $R_{id} = 0$. Suppose that Y_{ij} follows

Yearly Drop in Y

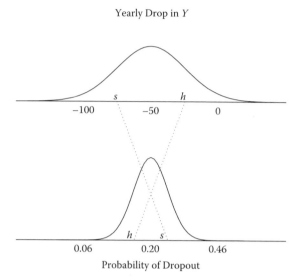

Figure 19.1 Illustration of how a shared random effect governs both the probability of dropping out and the decline in outcome. A sick patient, s, draws a single random effect that is 1 standard deviation worse than the mean of the distribution of slopes and inverse probit probability of dropping out. Her probability of dropping out on any visit is about $\Phi(\alpha_0 + \alpha_1 1) = 0.38$, while her slope is $\beta_0 + \beta_1 \times 1 = -75$. A healthy patient has a better random effect, h. With covariates, the mean of these distribution changes so the random effect reflects sickness (or health) relative to the covariate adjusted mean.

a simple linear mixed model, $Y_{ij} = \beta_0 + \beta_1 t_{ij} + b_{0i} + b_{1i} t_{ij} + \epsilon_{ij}$, where $(b_{0i}, b_{1i}) \sim N(\mathbf{0}, G)$ and $\epsilon_{ij} \sim N(0, \sigma_\epsilon^2)$ independent of (b_{0i}, b_{1i}). The random effects b_{0i} and b_{1i} reflect individual departures from the average intercept and slope, respectively. The dropout mechanism can be modeled with a geometric distribution in which, conditional on the random slope b_{1i}, $R_{ij}|(R_{ij-1} = 1, b_{1i})$ is Bernoulli with probability $P(R_{ij} = 0) = \Phi(\alpha_j + \theta b_{1i})$. This model is portrayed conceptually in Figure 19.1.

Each person draws a random slope from a Gaussian distribution. This slope governs both their expected rate of decline in the response Y and the probability of dropping out at each visit. Two individuals are illustrated here. Person h is one standard deviation better in terms of Y and also in terms of the inverse probit probability of dropping out. Data generated from such a model is given in Figure 19.2. The expected response slope for each person is given by a dotted line with data points given by s and h. The data from s stops after the seventh visit as the person dies at that point.

A simple analysis of such data can lead to biased estimates. This is illustrated in Figure 19.3 for a study with two follow-up visits. The distribution of slopes for the population follows a Gaussian distribution, but not all patients provide two measurements, and those with negative slopes tend to have fewer visits. A naive analysis will tend to weight the complete cases more heavily than they should, leading to a biased estimate of the overall slope.

Shared-parameter models make assumptions which can only in part be examined based on the data. For example, we can examine whether, for the observed data, change in Y follows a linear function of time or a quadratic function of the log of time, using standard diagnostic techniques. However, changes in Y for the unobserved data, which are imposed by the shared-parameter model, are inherently not verifiable. The assumption that the random effect is the same number of standard deviations away from the mean on both the outcome and missingness model can be weakened by introducing additional random effects. For example, additional random effects can be added to either the response or missing-data

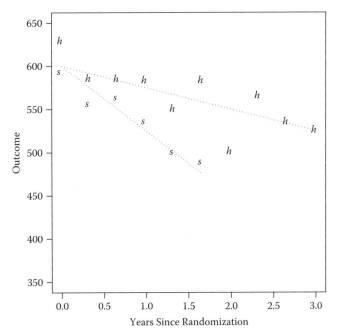

Figure 19.2 Data are simulated from the two patients. Their mean curves are given by the dashed lines, h and s denote simulated outcomes. The sicker patient declines more rapidly and dies before follow-up is complete.

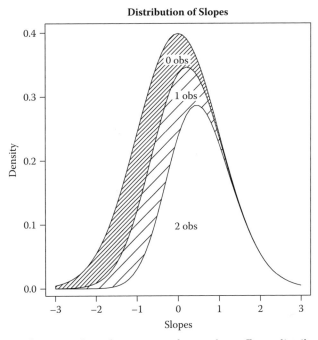

Figure 19.3 Impact of non-random dropout on the random-effects distribution in a two-visit study. Each person draws a random slope b_i from a normal distribution. People with 0, 1, and 2 follow-up visits are probabilistically culled from the distribution of b_i. The shape of the "2 obs" region is proportional to the density of $b_i | d_i = 2$ and its area is the expected proportion of complete cases. At each slope value, the heights between the curves are proportional to the probability a patient with that slope provides 2, 1, or 0 observations. Note the heavier shading at -1 compared to 1, indicating that dropout is non-random.

mechanisms that are unique to that mechanism. However, if there are few observations, and some have no follow-up data, the overall distribution of slopes relies, as shown in Figure 19.3, on what we assume for the heavily shaded region. We get no information about this. It could be that the observations that are missed would all be 0. Implicitly, we assume that what we do see tells us something about what we do not see.

In Section 19.2 we set up the general class of shared-parameter models for longitudinal data, and discuss assumptions and interpretation of the models. Approaches to parameter estimation are the subject of Section 19.3. We discuss simple methods and show how a conditional model can be constructed that approximates the shared-parameter model. It is easy to implement with standard software packages. In Section 19.4 we discuss the application to a trial in patients with respiratory illness where change in lung function over time was the primary outcome and a substantial number dropped out of the study. In Section 19.5 we discuss an application to an epilepsy clinical trial, where patients had difficulty adhering to the protocol and tended to drop out of the trial. The primary endpoint for this study was daily seizure counts over a two-week follow-up period. In Section 19.6 we present an application to an opiates clinical trial where repeated binary urine test results were the outcome. Substantial numbers of patients missed visits and dropped out of the study. We discuss various models for the underlying response process, including a shared-parameter model for binary responses. Section 19.7 follows with a discussion where we compare the shared random-effects modeling approach with other models that account for non-random missing data.

19.2 Model formulation

Let $\mathbf{Y}_i = (Y_{i1}, Y_{i2}, \ldots, Y_{iJ})'$ be a vector of longitudinal outcomes for the ith subject ($i = 1, 2, \ldots, I$) observed on J occasions $t_1, t_2, \ldots, t_J$, and let $\mathbf{R}_i = (R_{i1}, R_{i2}, \ldots, R_{iJ})'$ be a vector of random variables reflecting the missing-data status (e.g., $R_{ij} = 0$ denoting a missed visit). Further, let $\mathbf{b}_i = (b_{i1}, b_{i2}, \ldots, b_{iL})'$ be an L-vector of random effects for the ith subject which can be shared between the response and missing-data mechanism. We assume that $\mathbf{b}_i$ is multivariate normal with mean $\mathbf{0}$ and variance G. We also allow for covariates $\mathbf{X}_{ij}$ to be measured, which can influence both Y_{ij} and R_{ij}.

The joint distribution of $\mathbf{Y}_i, \mathbf{R}_i, \mathbf{b}_i$ can be written as

$$f(\mathbf{y}_i, \mathbf{r}_i, \mathbf{b}_i) = g(\mathbf{y}_i | \mathbf{b}_i, \mathbf{r}_i) m(\mathbf{r}_i | \mathbf{b}_i) h(\mathbf{b}_i).$$

Note that potential conditioning on covariates has been suppressed in the notation of all above densities. Further, we make the assumption that, conditional on the random effects, the responses do not depend on the missing-data status, thus $g(\mathbf{y}_i | \mathbf{b}_i, \mathbf{r}_i) = g(\mathbf{y}_i | \mathbf{b}_i)$. Furthermore, the elements of $\mathbf{Y}_i$ are conditionally independent given $\mathbf{b}_i$. By conditional independence, the density for the response vector $\mathbf{Y}_i$ conditional on $\mathbf{b}_i$, $g(\mathbf{y}_i | \mathbf{b}_i)$, can be decomposed into the product of the densities for the observed and unobserved values of $\mathbf{Y}_i$, $g(\mathbf{y}_i | \mathbf{b}_i) = g(\mathbf{y}_i^o | \mathbf{b}_i) g(\mathbf{y}_i^m | \mathbf{b}_i)$, where $\mathbf{y}_i^o$ and $\mathbf{y}_i^m$ are vectors of observed and missing-data responses, respectively, for the ith subject. Under conditional independence, the density for the observed random variables is

$$
\begin{aligned}
f(\mathbf{y}_i^o, \mathbf{r}_i) &= \int_{\mathbf{y}^m} \int_{\mathbf{b}} f(\mathbf{y}_i^o, \mathbf{y}_i^m, \mathbf{r}_i) d\mathbf{b} d\mathbf{y}_i^m \\
&= \int_{\mathbf{y}^m} \int_{\mathbf{b}} g(\mathbf{y}_i^o | \mathbf{b}_i) g(\mathbf{y}_i^m | \mathbf{b}_i) m(\mathbf{r}_i | \mathbf{b}_i) h(\mathbf{b}_i) d\mathbf{b} d\mathbf{y}_i^m \\
&= \int_{\mathbf{b}} g(\mathbf{y}_i^o | \mathbf{b}_i) m(\mathbf{r}_i | \mathbf{b}_i) h(\mathbf{b}_i) \left\{ \int_{\mathbf{y}_i^m} g(\mathbf{y}_i^m | \mathbf{b}) d\mathbf{y}_i^m \right\} d\mathbf{b}_i \\
&= \int_{\mathbf{b}} g(\mathbf{y}_i^o | \mathbf{b}_i) m(\mathbf{r}_i | \mathbf{b}_i) h(\mathbf{b}_i) d\mathbf{b}_i.
\end{aligned}
\tag{19.1}
$$

Although the conditional independence of $\boldsymbol{Y}_i|\boldsymbol{b}_i$ is relatively easy to verify when there are no missing data (e.g., examining for residual correlation), it is difficult to verify for shared-parameter models. As we will discuss with an example in Section 19.6, this assumption can be weakened as correlation between measurements can be incorporated by replacing the shared random effects $\boldsymbol{b}_i$ by a shared serially correlated random process $\boldsymbol{b}_i = (b_{i1}, b_{i2}, \ldots, b_{iJ})'$.

Tsiatis and Davidian (2004) provide a precise discussion of how the joint density is obtained for the case where missingness is monotone (i.e., patients only drop out of the study) and measured in continuous time.

The choice of a density function g depends on the type of longitudinal response data being analyzed. For Gaussian longitudinal data, g can be specified as a Gaussian distribution, and the model formulation can be specified as a linear mixed model (Laird and Ware, 1982). A simple linear mixed model that can be used as an illustration is

$$Y_{ij}|X_i, b_i = \beta_0 + \beta_1 X_i + b_i + \epsilon_{ij}, \tag{19.2}$$

where X_i is a subject-specific covariate such as treatment group, b_i is a random effect which is often assumed normally distributed, and ϵ_{ij} is an error term which is assumed normally distributed. Alternatively, for discrete or dichotomous longitudinal responses, g can be formulated as a generalized linear mixed model (Follmann and Wu, 1995; Ten Have et al., 1998; Aitkin and Alfo, 1998; Albert and Follmann, 2000; Alfo and Aitkin, 2000). We discuss these and some other longitudinal responses in subsequent sections of this chapter.

The choice of the density for the missing-data indicators, m, depends on the type of missing data being incorporated. When missing data are a discrete time to dropout, a "monotone" missing-data mechanism, then a geometric distribution can be used for m (Mori, Woolson, and Woodworth, 1994). For example, the probability of dropping out is

$$\Phi^{-1}\{\Pr(R_{ij} = 0|R_{ij} = 1)\} = \alpha_0 + \alpha X_i + \theta b_i. \tag{19.3}$$

Various authors have proposed shared-parameter models for the case where dropout is a continuous event time (Schluchter, 1992; Schluchter et al., 2001; DeGruttola and Tu, 1994; Tsiatis, DeGruttola, and Wulfsohn, 1995; Wulfsohn and Tsiatis, 1997; Tsiatis and Davidian, 2001; Vonesh et al., 2002). When missing data include only intermittently missed observations without dropout, then the product of Bernoulli densities across each of the potential observations may be a suitable density function for g. Alternatively, when multiple types of missing data such as both intermittent missingness and dropout need to be incorporated, a multinomial density function for g can be incorporated (Albert et al., 2002). We will illustrate various types of missing-data mechanisms with examples later in this chapter.

In a randomized clinical trial with no missing data, choice of baseline covariates to use in (19.2) is more a matter of taste as randomization assures the validity of the test of treatment effect. With missing data, randomized trials become more like observational studies. The model must be correctly specified for correct inference. In particular, choice of covariates in (19.2) and (19.3) must be considered carefully. Note that the shared-parameter model enforces $E(b_i) = 0$. Centering the random-effects distribution allows one to interpret b_i as an individual's departure from the typical response process given a set of fixed-effect covariates. It is this value which is linked between the response and dropout processes. Thus, it is the deviations from the covariate adjusted mean that are assumed to be shared (b_i), rather than the mean itself $(\beta_0 + \beta_1 b_i)$. Figure 19.4 illustrates this point graphically. The solid top curve is the overall distribution of mean responses while the dashed curves provide the distributions for $X = 0, 1$, respectively. A patient with mean response M is one standard deviation larger than the overall mean, but typical for a patient with $X = 1$. Since deviations from the mean govern the probability of missing, we see that this patient has a large (typical) probability of missing when we do not adjust (do adjust). Importantly, X

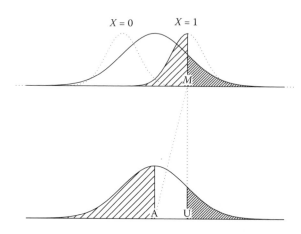

Inverse Probit of the Missingness Probabilities

Figure 19.4 With a shared-parameter model, choice of covariates in the response model matters. The solid density on the top represents the overall distribution of random effects for the treatment group. The dashed densities are random-effects distributions for the two levels of an important binary baseline covariate X. Consider the mean response M for some person with $X = 1$. This is one standard deviation larger than the unadjusted mean but average if we adjust. Whether we adjust has consequences for the probability of missing a visit.

might be a strong predictor so that if we unthinkingly choose to include X solely based on, say, the improvement in the log-likelihood we might decide to include X without recognizing its additional effect on missingness probabilities. Of course, one might also consider using $\beta_0 + \beta_1 b_i$ rather than b_i in (19.3).

The shared-parameter model incorporates an NMAR data mechanism, which can be seen with the following argument. Suppressing the index i for notational simplicity, suppose that the random effect b is a scalar with R_j indicating whether Y_j is observed. MAR implies that the conditional density of R_j given the complete data $\boldsymbol{Y}$ does not depend on $\boldsymbol{Y}^m$, while NMAR implies that this conditional density depends on $\boldsymbol{Y}^m$. The conditional density of R_j given $\boldsymbol{Y} = (\boldsymbol{Y}^o, \boldsymbol{Y}^m)$ is

$$f(r_j | \boldsymbol{y}^m, \boldsymbol{y}^o) = \frac{\int g(r_j | b) g(\boldsymbol{y}^m, \boldsymbol{y}^o | b) h(b) db}{\int g(\boldsymbol{y}^m, \boldsymbol{y}^o | b) h(b) db}$$

$$\int g(r_j | b) h(b | \boldsymbol{y}^o, \boldsymbol{y}^m) db.$$

The shared-parameter model incorporates an NMAR data mechanism since the conditional density depends on $\boldsymbol{y}^m$ since $h(b | \boldsymbol{y}^o, \boldsymbol{y}^m)$ depends on $\boldsymbol{y}^m$. Note that $h(b | \boldsymbol{y}^o, \boldsymbol{y}^m) db$ can be viewed as an empirical Bayes posterior distribution. Since posteriors depend on the entire likelihood (i.e., depend on the contributions due to both $\boldsymbol{y}^o$ and $\boldsymbol{y}^m$), it is easy to see this dependence.

19.3 Estimation approaches

There are various approaches for parameter estimation. First, maximization of the likelihood $L = \prod_{i=1}^{I} f(\boldsymbol{y}_i^o, \boldsymbol{r}_i)$, where f is given by (19.1), can be used to obtain the maximum likelihood estimates. Maximizing the likelihood may be computationally intensive since it involves integrating over the random-effects distribution. For a high-dimensional random-effects distribution, this involves the numerically difficult evaluation of a high-dimensional integral. Approaches such as Monte Carlo EM (McCulloch, 1997) or Laplace approximations of the likelihood (Gao, 2004) may be good alternatives to direct evaluation of the integral. Fortunately, many applications involve only one or two shared random effects where the integral can be evaluated more straightforwardly with Gaussian quadrature, adaptive Gaussian quadrature, or other numerical integration techniques. Maximization of the likelihood typically involves writing special software for a particular application. A generalized estimating equations approach is discussed by Vonesh et al. (2002).

A second, and much simpler, approach is to conduct an unweighted analysis. In this case, regression models are fit separately for each subject, with the resulting parameter estimates averaged across subjects. For example, for the linear mixed-effects model, $Y_{ij} = \beta_0 + \beta_1 t_j + \beta_2 X_i + \beta_3 X_i t_j + b_{0i} + b_{1i} t_j + \epsilon_{ij}$, where each patient provides us with at least two observations, we can fit a least-squares regression of Y on t for each individual. The within-group average of the slopes provides an unbiased estimate of the mean slope β_1 $(\beta_1 + \beta_3)$ for group $X = 0$ $(X = 1)$, under the NMAR data mechanism induced by a shared-parameter model. Thus, to compare two groups, one can perform a t-test using the individually estimated slopes as data. Although the unweighted estimation approach is unbiased, particularly for highly unbalanced data, it may be inefficient (Wu and Carroll, 1988; Wang-Clow et al., 1995). The unweighted approach averages individual estimates with equal weight across individuals. This is inefficient relative to a shared-parameter model, which effectively weights individual estimates according to their precision (e.g., individuals with more repeated measurements are weighted more heavily than those with fewer measurements).

An alternative approach for parameter estimation, which conditions on R_i, has been proposed (Wu and Bailey, 1989; Follmann and Wu, 1995; Hogan and Laird, 1997a; Albert and Follmann, 2000). This is relatively simple and can be more efficient than the unweighted approach. In developing this approach, first note that the joint distribution of (Y_i^o, R_i, b_i) can be rewritten as

$$f(y_i^o, r_i, b_i) = f(y_i^o, b_i | r_i) m(r_i)$$
$$= f(y_i^o | b_i, r_i) h(b_i | r_i) m(r_i)$$
$$= g(y_i^o | b_i) h(b_i | r_i) m(r_i).$$

Thus, the conditional likelihood for $y_i^o | r_i$ is given by $L = \prod_{i=1}^{I} \int g(y_i^o | b_i) h(b_i | r_i) db_i$. Note that this conditional model can be directly viewed as a pattern-mixture model as

$$f(y^o, r) = \int g(y^o | b) h(b | r) db \, m(r)$$
$$= p(y | r) m(r)$$

(Little, 1993). To implement the conditional approach, we approximate the distribution of $b_i | r_i$ as multivariate normal with the mean approximated by additive functions of the missing-data indicators. For example, $\mu_{b_i | r_i} = \omega_0 + \omega_1 d_i$ where $d_i = \sum_j r_{ij}$ is the time to dropout and $\Sigma_{b | r_i}$ approximated by τ. Thus, this conditional approach has been referred to as the approximate conditional approach.

The conditional approach was introduced by Heckman (1979) in an econometric setting with $J = 1$. Heckman did not identify b_i as a shared random effect, but his model can be written as

$$Y_i = x_i' \beta + b_i + e_{i1},$$
$$Y_i^* = w_i' \alpha + b_i + e_{i2},$$

where b_i, e_{i1}, e_{i2} are independent normals each with mean 0 and variances $\tau^2, \sigma_1^2, \sigma_2^2$, and $\boldsymbol{x}_i, \boldsymbol{w}_i$ are covariate vectors for the ith participant. The response is Y_i, while Y_i^* is the "utility" for responding, and we see Y_i if $Y_i^* > 0$. If $\tau^2 > 0$, then individuals who tend to have large b_is are more likely to respond and to have large Y_is. This model implies that

$$\Pr(R_i = 1) = \Pr(Y_i^* > 0) = \Phi(\boldsymbol{w}_i'\boldsymbol{\alpha}/\sqrt{\tau^2 + \sigma_2^2}),$$

and it can be shown that $E[b_i|R_i = 1] \propto \lambda(-\boldsymbol{w}_i'\boldsymbol{\alpha}/\sqrt{\tau^2 + \sigma_2^2})$, where $\lambda(x) = \phi(x)/\Phi(-x)$ and ϕ, Φ are the standard normal density and cdf, respectively. Heckman proposes an initial probit regression of R_i on $\boldsymbol{w}_i$ to obtain an estimate of $\boldsymbol{\alpha}^* = \boldsymbol{\alpha}/\sqrt{\tau^2 + \sigma_2^2}$. He then fits the equation

$$Y_i = \boldsymbol{x}_i'\boldsymbol{\beta} + \omega\lambda(-\boldsymbol{w}_i'\widehat{\boldsymbol{\alpha}^*}) + \epsilon$$

using least squares. This contrasts with the approximate conditional shared-parameter models used in the biostatistical setting where $J > 1$ and the conditional expectation of $b_i|\boldsymbol{R}_i$ is approximated and not formally derived.

We illustrate the conditional approach with a simple non-random dropout model given by (19.2) and (19.3). In this case, $b_i|d_i$, can be approximated by $N(\omega_0 + \omega_1 d_i, \tau^2)$, and the conditional model can be fit by fitting a linear mixed model of the form

$$Y_{ij}|d_i, X_i = \beta_0^* + \beta_1 X_i + \omega_1 d_i + b_i + \epsilon_{ij}, \tag{19.4}$$

where $\beta_0^* = \beta_0 + \omega_0$, using standard software.

A subtle yet crucial point is that the model parameters of (19.4) are conditional on the dropout time d, while interest focuses on the marginal distribution of Y. Naively interpreting β_1 as the effect of X_i leads to incorrect conclusions. As an analogy, suppose that d and Y were systolic and diastolic blood pressure measurements with correlation induced by the shared random parameter. If we were interested in the effect of treatment (X_i) on diastolic blood pressure (Y) it would be misleading to use β_1 from (19.4) as here β_1 is the effect of treatment *controlling* for systolic blood pressure. Thus, if treatment affects both systolic and diastolic, this conditional β_1 is likely to substantially underestimate the marginal effect of treatment on diastolic blood pressure.

Fortunately, marginal parameters can be obtained by averaging over the distribution of $d|X$. For example,

$$E(Y|x) = E\{E(Y|d, x)\} = \beta_0^* + \beta_1 x + \omega_1 E(d|x)$$

where $E(d|x) = \sum_d d\Pr(d|x)$, and $\Pr(d|x)$ is the conditional distribution of d given x. If X is a treatment indicator as in a two-arm clinical trial, one can use the empirical distribution of the d_is in each arm to estimate $E(d|x)$. This produces the within-arm sample averages $\widehat{E}(d|0) = \bar{d}_0$ and $\widehat{E}(d|1) = \bar{d}_1$. The estimated treatment effect is

$$E(Y|x = 1) - E(Y|x = 0) = \beta_1 + \omega_1(\bar{d}_1 - \bar{d}_0),$$

so if we assume the effect of dropout on the mean response is the same in the two groups as in (19.4) and the average dropout time is the same in the two groups, the marginal effect of treatment is the same as the conditional effect of treatment. For clinical trials it seems prudent to fit the model separately in the two groups, or at least allow for an interaction between X and d.

For continuous x the linear conditional model may imply a non-linear effect of x on Y in the marginal mean as

$$E(Y|x) = \beta_0 + \beta_1 x + \omega E(d|x),$$

which is only linear in x if $E(d|x)$ is linear in x. In general, to obtain parameters with a marginal interpretation requires specification and estimation of the distribution of $d|x$.

All three methods require estimated variances of the parameter estimates for inference. For the maximum likelihood approaches, approximate asymptotic variances can be obtained by the matrix of observed Fisher information. Alternatively, the bootstrap (Efron and Tibshirani, 1993) can be used for all three methods: the shared-parameter, unweighted, and conditional approaches.

In the next section, we illustrate the use of a shared-parameter model with continuous longitudinal data with missingness due to dropout.

19.4 Gaussian longitudinal data with non-random dropout: An analysis of the IPPB clinical trial data

The Intermittent Positive Pressure Breathing (IPPB) Trial (Intermittent Positive Pressure Breathing Trial Group, 1983) was a controlled clinical trial evaluating the effect of an intervention (IPPB) as compared with the standard compressor nebulizer therapy on changes in lung function in patients with obstructive pulmonary disease. The primary objective of this trial was to compare the rate of change in forced expiratory volume in one second (FEV_1) across treatment groups. FEV_1 measurements were taken at baseline and at 3-month intervals over a 3-year follow-up. There were 984 patients in this trial, with a substantial amount of dropout due to death and loss to follow-up. Specifically, 39% of patients dropped out of the study prior to 3 years of follow-up. For illustrative purposes, we will not distinguish between the different types of dropout in this analysis. More elaborate models can be developed to account for different types of dropout (see Ten Have et al., 2000, for example).

Figure 19.5 shows the relationship between average intercept and dropout time, and between average slope and dropout time, for IPPB patients. It shows a possibly weak relationship between the response process and dropout time, whereby there is a tendency for patients who dropped out early to have lower FEV_1 at baseline and a more rapid decline over time than patients who dropped out later or completed the trial.

We examined the effect of treatment in linear change on log-transformed FEV_1 with a linear mixed model (which does not account for non-random dropout) as well as the shared-parameter, conditional, and unweighted approaches discussed in Section 19.3. For each group, the model for the responses is

$$Y_{ij} = \beta_0 + \beta_1 t_j + b_{0i} + b_{1i} t_j + \epsilon_{ij}, \tag{19.5}$$

where $\epsilon_{ij} \sim N(\mathbf{0}, \sigma_\epsilon^2)$, $\mathbf{b}_i = (b_{i0}, b_{i1})$ are independent and identically distributed as $N(0, G)$, and the diagonal elements of G are $\sigma_{b_0}^2$ and $\sigma_{b_1}^2$.

Table 19.1(a) shows parameter estimates by treatment groups for the random-effects model. The mean slope was significantly different from zero for both treatment groups. However, there was no evidence for a difference between mean slope by treatment group (diff $= -0.00017$, with standard error [SE] 0.00049). Table 19.1(b) shows the parameter estimates by treatment group for the shared-parameter model, in which we link the random effects corresponding to the intercept and slope in (19.5) with a probit model for dropout, $\Pr(R_{ij} = 0) = \Phi(\alpha + \theta_0 b_{0i} + \theta_1 b_{1i})$. For both treatment groups, the negative estimates of θ_0 suggest that healthier patients with larger FEV_1 at baseline are less likely to drop out of the study early. As reflected in the estimates of θ_1, there is no significant relationship between linear change and dropout time in either group. The difference between average slope for the two treatment groups is -0.00031 (SE $= 0.00053$), which is similar to the estimate we obtained using the random-effects model.

Table 19.1(c) shows the parameter estimates obtained with the conditional model, $Y_{ij} = \beta_0 + \beta_1 t_j + \beta_2 d_i + \beta_3 t_j d_i + b_{i0} + b_{i1} t_j + \epsilon_{ij}$, where $d_i = \sum_j r_{ij}$. The results suggest that, in the IPPB group, those with later dropout times have larger (less negative) slopes. This relationship was not found in the control group. For each group, the average slope can be

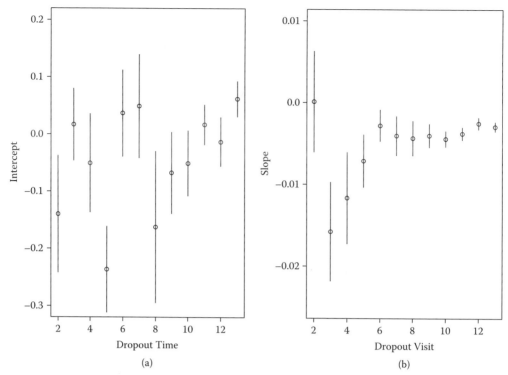

Figure 19.5 For patients in the IPPB group, the linear mixed model (19.5) was fit for patients dropping out at the same visit (2 to 12) as well as for patients completing the trial (13). Panel (a) shows the estimate ($\pm$ standard error) for the intercept (β_0) versus dropout visit. Panel (b) shows the estimate for the slope (β_1) versus dropout visit.

estimated by $\widehat{\beta}_1 + \widehat{\beta}_3 \bar{d}$, where $\bar{d} = \frac{1}{I} \sum_i d_i$. For the conditional model, the difference in the average slopes between the two treatment groups is estimated as -0.00023 (SE $= 0.00044$).

The last approach considered is the unweighted approach, whereby we compare the average individually estimated slope across the two groups. For the unweighted approach, the difference in average slope between the two groups is -0.000061 (SE $= 0.0010$). The large standard errors relative to all other approaches highlight the inefficiency of this approach.

Although there was some evidence for a relationship between the time of dropout and slope in the IPPB group, which was detected with the conditional model and not with the shared-parameter model, none of the approaches demonstrated any difference in the average slope between treatment groups.

In the next section, we consider an example with repeated counts subject to non-random dropout.

19.5 Repeated-count data with non-random dropout: An analysis of an epilepsy clinical trial

The interest in this epilepsy clinical trial (Theodore et al., 1995) was to determine if felbamate, an experimental treatment for epilepsy, reduced the frequency of seizures relative to a placebo. Patients were titrated off prior medications and randomized to either placebo ($n = 19$) or treatment ($n = 21$) and were followed over a 17-day period to assess seizure frequency. Patients had difficulty remaining in this study, perhaps due to the rapid titration off their prior medications. Approximately half of the patients dropped out before completing

Table 19.1 Parameter Estimates for the Random-Effects, Shared-Parameter, and Conditional Models for the IPPB Study

(a) Random-Effects Model

Parameter	Standard		IPPB	
	Estimate	SE	Estimate	SE
β_0	−0.005	0.018	−0.027	0.017
β_1	−0.0033	0.0004	−0.0035	0.0003

(b) Shared-Parameter Model

Parameter	Standard		IPPB	
	Estimate	SE	Estimate	SE
β_0	−0.050	0.010	−0.033	0.013
β_1	−0.0033	0.0004	−0.0036	0.0004
α	−1.35	0.027	−1.41	0.037
θ_0	−0.26	0.073	−0.20	0.077
θ_1	−0.0069	0.11	−5.20	6.0
σ_{b_1}	0.0055	0.0003	0.0054	0.0003

(c) Conditional Model

Parameter	Standard		IPPB	
	Estimate	SE	Estimate	SE
β_0	−0.27	0.042	−0.16	0.042
β_1	−0.0027	0.002	−0.0075	0.002
β_2	0.025	0.0043	0.015	0.0042
β_3	−0.00006	0.0002	0.0004	0.0002

Note: Model parameters are presented separately by treatment group (standard or IPPB groups). In most cases, only the mean structure parameters estimates are shown. Note that for (c), the β parameters have a conditional interpretation and cannot be directly compared to the marginal parameters of (a) and (b).

the follow-up (11/19 in the placebo group and 8/21 in the treatment group). Figure 19.6 shows average daily seizure frequency versus the number of days in the study for the felbamate and placebo groups. It appears that patients with the largest seizure frequency tend to drop out of the study earlier than patients with lower seizure frequency.

Let Y_{ij} denote the observed count for the ith subject at the jth day of follow-up and, as in the IPPB analysis, let d_i be the dropout time for the ith subject. Denote λ_{ij} as the conditional mean of Y_{ij} given the random effects and p_{ij} as the conditional probability of dropout during the jth interval given that the individual remained in the study given the $(j-1)$th interval and the random effects.

We will compare three different approaches for parameter estimation, including a random-effects model which does not model the dropout process, a shared-parameter model, and the conditional approach. Plots of the mean average seizure frequency over time suggested that seizure frequency was relatively constant over time for both the treatment and placebo arms (not shown). Thus, no time trend term was included in the mean model.

As in Albert and Follmann (2000), we fit a random-effects model with the mean structure,

$$\lambda_{ij} = \exp\{\beta_0 + \beta_1 G_i + b_{i1}G_i + b_{i2}(1 - G_i)\}, \tag{19.6}$$

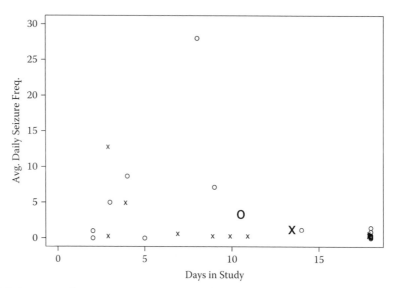

Figure 19.6 Average seizure counts versus time to dropout in the two treatment groups. Felbamate patients are denoted by crosses and placebo patients by circles. Also presented are the overall mean seizure rate at the mean number of days in the study (a large cross and a large circle for felbamate and placebo groups, respectively).

where $b_{li} \sim N(0, \sigma_l^2)$, $l = 1$ or 2, G_i is an indicator of whether the ith subject was randomized to felbamate ($G_i = 1$) or placebo ($G_i = 0$), and β_1 assesses the difference between felbamate and placebo on log seizure frequency. Further, we assume that Y_{ij} is Poisson conditional on the random effects. For the shared-parameter model, we combine (19.6) with a model for the dropout process. We model the dropout process for the felbamate clinical trial by specifying p_{it} as

$$ p_{it} = \Pr(R_{ij} = 0 | R_{i,j-1} = 1, \boldsymbol{b}_i) = \text{logit}^{-1}\{\gamma_0 + \gamma_1 G_i + \theta_1 b_{1i} G_i + \theta_2 b_{2i}(1 - G_i)\}. \quad (19.7) $$

The model specified by (19.6) and (19.7) is equivalent to fitting the shared-parameter models separately in the two groups. Note that the shared parameter $b_{\ell i}$ induces between-subject variation in the two groups as well as providing a link between the response and dropout mechanisms. Large positive values of θ_1 and θ_2 correspond to the situation where patients in the treatment and placebo group with a large seizure frequency have a tendency to drop out of the trial earlier than those patients with a smaller seizure frequency. Estimation proceeds by maximum likelihood using (19.1), (19.6), and (19.7). This maximization required an approach similar to adaptive Gaussian quadrature, which is discussed in detail in Albert and Follmann (2000).

The conditional model is formulated as

$$ \log E[Y_{it} | \boldsymbol{b}_i, d_i] = \beta_0 + \beta_1 G_i + \omega_1 G_i \log(d_i) + \omega_2(1 - G_i) \log(d_i) + G_i b_{1i} + (1 - G_i)b_{2i} \quad (19.8) $$

and can also be fit by maximum likelihood.

Table 19.2 shows parameter estimates for the random-effects (19.6), shared-parameter (19.6) and (19.7), and conditional models (19.8). Note that the estimate of β_0 (corresponding to the average seizure frequency in the placebo group) is slightly lower for the naive random effects as compared with the shared-parameter model (-0.58 versus -0.50). This is expected, since not accounting for a positive relationship between seizure frequency and time to dropout would result in an underestimate of seizure frequency. The estimate of β_1 is nearly identical under the random-effects and shared-parameter models. In part, this is due to the fact that $\widehat{\theta}_1$ and $\widehat{\theta}_2$ are close in value. The conditional model shows a statistically

Table 19.2 Parameter Estimation for the Random-Effects, Shared-Parameter, and Conditional Models for the Felbamate Study

Parameter	Random Effects		Shared Parameter		Conditional	
	Estimate	SE	Estimate	SE	Estimate	SE
β_0	−0.58	0.52	−0.50	0.56	0.57	1.27
β_1	−2.15	0.80	−2.13	0.85	2.60	1.85
σ_1	2.18	0.49	2.31	0.46	1.23	0.36
σ_2	1.98	0.42	2.15	0.42	1.87	0.41
ω_1					−2.81	0.52
ω_2					−0.51	0.55
γ_0			−2.77	0.33		
γ_1			−0.68	0.51		
θ_1			0.27	0.26		
θ_2			0.21	0.19		

Note: Estimated model parameters are presented separately by treatment group.

significant relationship between dropout time and seizure frequency in the felbamate group (ω_1 is significantly different from zero), but not in the placebo arm. Interestingly, this dependence was not detected with the shared-parameter model, where θ_1 was not significantly different from zero.

As mentioned in Section 19.2, the parameters of the conditional model cannot be directly interpreted for assessing treatment effect. One needs to marginalize over the dropout times for appropriate inference. One way of doing this is to compare the average log-transformed seizure frequency across groups. This can be estimated from the conditional model as

$$\widehat{\Delta} = \sum_{i \in \mathcal{G}_1} \left\{ \log \widehat{E}(Y_{i1}|\boldsymbol{b}_i = 0, d_i) \right\} / 21 - \sum_{i \in \mathcal{G}_0} \left\{ \log \widehat{E}(Y_{i1}|\boldsymbol{b}_i = 0, d_i) \right\} / 19,$$

where $\mathcal{G}_g$ is the set of indices for patients in the gth group. Estimates of Δ for the conditional model were −1.579 (SE = 0.878), which was not significantly different from zero using a Z-test. A direct comparison of $\widehat{\Delta}$ and $\widehat{\beta}_1$ obtained from the random-effects and shared random-effects models is appropriate. Estimates of β_1 under both models were larger in magnitude as compared with $\widehat{\Delta}$. Further, both estimates of β_1 were statistically significant, while the estimate of Δ was not. The smaller estimated treatment effect under the conditional model may be partly due to the non-random dropout mechanism that was detected in the felbamate arm (ω_1). This contrasts with the lack of a significant link in the shared-parameter model. This difference in the importance of dropout on the response resulted in a larger estimate of the group-specific mean seizure frequency in the felbamate group, and therefore a smaller estimate of treatment effect. Albert and Follmann (2000) conducted simulations to provide insight into this apparent discrepancy. They found that the conditional model was more robust and had better small-sample properties than the shared-parameter model. The simulations lead to favoring the use of the conditional model for making inference in the felbamate trial.

In the next section, we discuss a trial with binary longitudinal data with both intermittent missingness and dropout.

19.6 Longitudinal binary data with non-monotonic missing data: An analysis of an opiate clinical trial

Johnson, Jaffe, and Fudala (1992) discussed a randomized clinical trial that compared buprenorphine to methadone for the treatment of opiate dependence. The trial randomized

162 patients into one of three treatment arms (i.e., buprenorphine and methadone at two dose levels). Of particular interest was the comparison of the buprenorphine group ($n = 53$) with the low-dose methadone group ($n = 55$).

Patients were scheduled for urine tests three times a week on Monday, Wednesday, and Friday for 17 weeks after randomization for a total of 51 scheduled responses. Urine tests were scored as positive or negative for the presence of opiates at each follow-up visit. An important objective was to compare the proportion of positive responses over time across treatment groups.

There was a large amount of missing data in this trial. Of 8262 planned responses, only 3966 (or 48%) were actually observed. Missing data were classified as missing either due to a patient formally withdrawing from the trial (dropout) or due to failing to show up for a scheduled visit (intermittent missing). The reasons for intermittent missingness and dropout were different. A number of subjects withdrew from the study due to poor compliance or because they were offered places in treatment programs that gave unmasked treatment and long-term care. Intermittent missingness may have been more closely related to the response process since patients may not have shown up for a particular visit when they were taking opiates, knowing they would test positive. The Spearman rank correlation between the proportion of positive tests and the time to dropout was -0.44 in the buprenorphine group and -0.10 in the methadone group, suggesting that this relationship differed by treatment group. Further, the correlation between the proportion of positive tests and the proportion of intermittent missing visits before dropout in the buprenorphine and methadone arms was 0.40 and 0.29, respectively. These estimates are consistent with the notion that, especially in the buprenorphine group, addicts who are more likely to use drugs are both more likely to drop out and to have a higher frequency of intermittent missingness before dropping out as compared with addicts who use opiates less frequently.

The empirical proportions of a positive urine test appeared relatively constant across follow-up (see Follmann and Wu, 1995), so we fit models for the response process that did not contain time effects. We postulate a simple random-effects model separately in the two treatment arms,

$$\text{logit}\{\Pr(Y_{it} = 1|b_i)\} = \beta_0 + b_i. \tag{19.9}$$

To distinguish between dropout and intermittently missed observations before dropout, we denote $R_{ij} = 1, 2,$ and 3 when an individual is observed, is intermittently missed, or has dropped out at time t_j, respectively. We model the dropout process with a multinomial regression model,

$$\Pr(R_{ij} = l|b_i, R_{i,j-1} \neq 3) = \begin{cases} 1 / \left\{ 1 + \sum_{l=2}^{3} \exp(\eta_l + \theta_l b_i) \right\} & l = 1, \\ \exp(\eta_l + \theta_l b_i) / \left\{ 1 + \sum_{l=2}^{3} \exp(\eta_l + \theta_l b_i) \right\} & l = 2, 3, \end{cases} \tag{19.10}$$

where b_i is normal with mean zero and variance σ. Values of η_2 and η_3 dictate the probability of an intermittent missed visit and dropout for the "average" individual. Values of θ_2 or θ_3 different from zero reflect informative missingness. Positive values of these parameters correspond to the situation in which individuals with large proportions of opiate use are those individuals who have a large fraction of intermittent missingness and who drop out early.

In this trial, interest is in comparing the marginal probability of a positive urine test across treatment groups. The marginal proportions for each treatment group can be expressed as

$$\Pr(Y_{ij} = 1) = E\{E(Y_{ij} = 1|b_i)\}$$
$$= E[\exp(\beta_0 + b_i) / \{(1 + \exp(\beta_0 + b_i))\}], \tag{19.11}$$

Table 19.3 Parameter Estimation for a Shared-Parameter Model for the Opiates Study

Parameter	Methadone		Buprenorphine	
	Estimate	SE	Estimate	SE
β_O	1.66	0.33	0.09	0.26
η_2	−1.63	0.19	−1.44	0.21
η_3	−3.25	0.18	−3.80	0.28
θ_2	0.37	0.19	0.44	0.19
θ_3	0.30	0.17	0.58	0.19
σ	2.15	0.36	1.93	0.30
Marginal Mean	0.73	0.030	0.52	0.040

Note: Models are fit separately for the methadone and buprenorphine groups.

where $b_i \sim N(0, \sigma^2)$, and where the integral required for the expectation can be evaluated using Gaussian quadrature (Abramowitz and Stegun, 1972). The estimated marginal proportion is the average of $\widehat{\Pr}(Y_{ij} = 1)$ over j.

Table 19.3 shows the parameter estimates (along with marginal mean responses) for both the methadone and the buprenorphine groups. The estimates of θ_2 and θ_3 are substantially larger for the buprenorphine group than for the methadone group, suggesting that the informative missing-data mechanism may be more pronounced in the buprenorphine group than in the methadone group. In fact, Z-tests on the parameters θ_2 and θ_3 are significant for the buprenorphine group but not for the methadone group. The marginal proportions in the buprenorphine and methadone groups are estimated as 0.52 (SE = 0.040) and 0.73 (SE = 0.030). A Z-test for comparing these marginal means is $Z = -3.80$, which provides strong evidence that the marginal mean is significantly smaller in the buprenorphine group than the methadone group. We also fit an ignorable model to the opiate clinical trial data. Specifically, we fit (19.9) only using the observed data. The marginal proportions were estimated as 0.43 (SE = 0.062) and 0.71 (SE = 0.048) for the buprenorphine and methadone groups, respectively. In both groups, for the ignorable model, marginal means were smaller than those presented for the shared-parameter model. This is consistent with the observation that those who have the highest proportion of positive urine tests are the ones who have the most intermittent missing data and who drop out earliest. Failing to properly account for the non-ignorable missing-data mechanism will result in those individuals who have large opiate-use proportions being weighted too little (since they are more likely to have missed visits and drop out early) in the analysis. The attenuated estimated proportion was most pronounced in the buprenorphine group as compared to the methadone group. This is consistent with the results presented in Table 19.3, which shows a stronger informative missing-data mechanism for the buprenorphine group.

The shared-parameter model (19.9) and (19.10) imposes strong assumptions on the correlation structure among repeated responses on the same individual; conditional on the random effects, observations on the same person are uncorrelated. This may not be a good assumption for responses such as in the opiate dependence trial, where an individual who fails a urine test on Monday may be more likely to fail the same test on Wednesday of the same week than when he or she takes the test many weeks later. The shared-parameter model also imposes similar strong assumptions on the correlation for the missing-data indicators. Namely, conditional on the random effects, missing an observation at one time is not associated with missing at a closely related time. Figure 19.7 shows the variance structure in the opiate trial response data. The correlation structure is shown with an empirical variogram in which we plot the lagged distance between two observations versus the mean

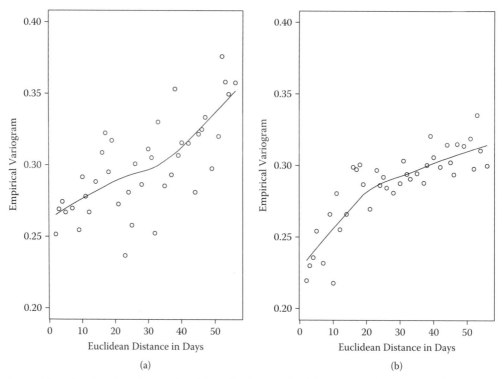

Figure 19.7 Empirical variogram for (a) methadone and (b) buprenorphine groups. The empirical variogram estimates $E[(Y_t - Y_{t+U})^2]$ for $U \geq 0$. Circles denote empirical variogram, while a solid line denotes the LOWESS smoothed curve through the empirical variogram.

squared difference between the two binary measurements. An empirical variogram which increases with distance between observations suggests serial correlation between observations. The empirical variogram presented in Figure 19.7 suggests that, in both treatment groups, there is strong positive serial correlation whereby observations close together on the same individual are more highly correlated than observations further apart.

Albert et al. (2002) proposed a shared random process which links the response and the missing-data mechanism with a shared correlated process as compared with random effects. The proposed random process allows for serial correlation in the response and missing-data processes. For this model, denote $\boldsymbol{b}_i = (b_{i1}, b_{i2}, \ldots, b_{iL})$, where L is the number of follow-up measurements ($L = 51$) for the trial, where the jth visit is observed at t_j days from randomization. In this approach, the response process in each group can be modeled as

$$\text{logit}\{\Pr(Y_{ij} = 1|b_{ij})\} = \beta_0 + b_{ij}. \tag{19.12}$$

Similar to the random-effects model, informative missing data are modeled by allowing the random process b_{ij} to be shared between the response and missing-data mechanism. The models for the missing-data mechanism have the same formulation as those presented for the random-effects model (19.10) except that the random process b_{ij} replaces the random effect b_i. Thus, an underlying propensity b_{ij} governs the chances of a positive test, missing intermittently, and dropout. The random process b_{ij} induces serial correlation in both the response and missing-data mechanism. Specifically, it is assumed that b_{ij} is a Gaussian process with mean zero and an exponential correlation structure given by

$$\text{Cov}(b_{ij}, b_{ik}) = \sigma^2 \exp(-\omega|t_j - t_k|). \tag{19.13}$$

Small positive values of ω correspond to strong long-term serial correlation, while larger values are reflective of serial correlation which "dies" off quickly with the separation of time between measurements. If time is continuous, (19.13) has been called the Gaussian Ornstein–Uhlenbeck process (Karlin and Taylor, 1981) and allows for observations to be unequally spaced such as the Monday, Wednesday, and Friday spacing of the opiate trial. As for the random-effects models, interest is in comparing marginal means across groups. As before, the average (over time) marginal (over people) means can be estimated by using (19.11) with estimates obtained by fitting the shared process model. For a model with no time effects, averaging over time is unnecessary.

In general, parameter estimation is substantially more difficult for the latent-process model as compared with the random-effects model. The former approach involves evaluating a 51-dimensional integral using a Monte-Carlo EM algorithm (Albert et al., 2002), while the latter approach involves a one-dimensional integral.

We fit model (19.10) and (19.12), replacing b_i with b_{ij}. As for the random-effects models, we fit the data separately in the buprenorphine and methadone groups. Table 19.4 shows the parameter estimates and marginal means in the two groups. The estimates of ω in the two groups suggest that the correlation in the autoregressive process decreases slowly with an increased number of days between visits. For example, for the buprenorphine group, the correlation in the random process b_{it} is 0.94, 0.74, and 0.54 for observations 5, 25, and 51 days apart on the same individual. For both groups, the parameter estimates are similar for the random-effects and random-process models. Within each group, marginal means were slightly lower for the random-process models as compared with the random-effects models. For the random-process models, the marginal means in the buprenorphine and methadone groups were 0.49 (SE = 0.039) and 0.67 (SE = 0.040), respectively. A test of the difference between the marginal means was highly significant using the random process model, $Z = -3.22$.

In this analysis, assuming a shared-parameter model (i.e., ignoring the apparent serial correlation) did not affect the inferences on treatment effect. Albert et al. (2002) showed that this will not always be the case. Assuming a shared-parameter model when the true mechanism is a shared latent process can result in severe bias when the informative missingness and serial correlation are sizable. This is in contrast to generalized linear mixed models without informative missingness in which inferences on marginal means are insensitive to misspecification in the serial dependence in a random process (Heagerty and Kurland, 2001). As we saw from Figure 19.4, specification of the correct shared-parameter model is necessary for valid inference.

Table 19.4 Parameter Estimation for a Shared Latent-Process Model for the Opiates Study

Parameter	Methadone		Buprenorphine	
	Estimate	SE	Estimate	SE
β_0	1.44	0.43	−0.10	0.37
η_2	−1.89	0.22	−1.71	0.18
η_3	−3.42	0.21	−4.11	0.31
θ_2	0.29	0.13	0.43	0.09
θ_3	0.22	0.15	0.48	0.10
σ	2.84	0.55	2.77	0.50
ω	0.014	0.008	0.012	0.004
Marginal Mean	0.67	0.040	0.49	0.039

Note: Models are fit separately for the methadone and buprenorphine groups.

19.7 Discussion

This chapter shows the usefulness of shared-parameter models for analyzing longitudinal data subject to non-random missing data. It should be recognized that, although these models account for NMAR data, it is impossible to distinguish between MAR and NMAR data within this class of models. We show how maximum likelihood estimation can be implemented along with simpler approximations to it. The conditional model, originally proposed by Wu and Bailey for Gaussian longitudinal data, and extended to binary and count data by Follmann and Wu (1995) and Albert and Follmann (2000), provides a useful approximation to the shared-parameter model which is easy to implement with standard software packages. We also show that, in some cases (e.g., the IPPB and felbamate trials), the conditional model may be more sensitive to certain types of missing-data mechanisms than maximum likelihood estimation under a shared-parameter modeling approach. Although the conditional model can be viewed as an approximation to the shared parameter, it can also be viewed as a pattern-mixture modeling framework in its own right.

Our focus in this chapter was on making marginal inference (i.e., marginal over the random effects) subject to non-random missing data. In some situations transition models may be more appropriate. For example, Albert and Follmann (2003) propose a shared-parameter Markov model for comparing the mean number of visits to a positive urine test after four weeks of treatment (another objective of the opiate clinical trial). Gallop et al. (2006) propose a similar shared random-effects model for a continuous time Markov process which is appropriate when measurements are highly irregularly spaced.

Shared-parameter models can be compared with other models that incorporate non-random missing-data mechanisms. For example, selection models explicitly model the probability of an observation being missed given the response (observed or unobserved). The shared-parameter and selection models are different in how they relate the probability of a missing observation and the response process. Shared random-effects models link the two by relating an individual's propensity to response with his or her propensity to miss a visit, while selection models directly model the probability of missing a visit as a function of the response. The choice of a modeling framework may depend on the scientific problem. Longitudinal data in which missingness is believed to be related to an individual's disease process and not to a particular realization of this process may be more appropriately modeled by a shared-parameter model than a selection model. Shared random effects are particularly appropriate for modeling longitudinal data in which the response is highly variable over time. The IPPB and epilepsy clinical trial data are good examples of this. Longitudinal data in which missingness is believed to be related to the actual observation (such as the opiate trial in which addicts may miss visits because they took drugs and know that there is a high likelihood that they would test positive) may be more appropriately modeled with a selection model than a shared-parameter model. A selection model fit to the opiate clinical trial data (Albert, 2000) gave similar results to the approaches discussed in this chapter. Although one approach may be more suitable than another, it is impossible to empirically verify which modeling framework is better. Fitting these two classes of models may provide a type of sensitivity analysis.

References

Abramowitz, M. and Stegun, I. A. (1972). *Handbook of Mathematical Functions*. New York: Dover.

Aitkin, M. and Alfo, M. (1998). Regression models for binary longitudinal responses. *Statistics and Computing*, **8**, 289–307.

Albert, P. S. (2000). A transitional model for longitudinal binary data subject to nonignorable missing data. *Biometrics* **56**, 602–608.

Albert, P. S. and Follmann, D. A. (2000). Modeling repeated count data subject to informative dropout. *Biometrics* **56**, 667–677.

Albert, P. S. and Follmann, D. A. (2003). A random effects transition model for longitudinal binary data with informative missingness. *Statistica Neerlandica* **57**, 100–111.

Albert, P. S., Follmann, D. A., Wang, S. A., and Suh, E. B. (2002). A latent autoregressive model for longitudinal binary data subject to informative missingness. *Biometrics* **58**, 631–642.

Alfo, M. and Aitkin, M. (2000) Random coefficient models for binary longitudinal responses with attrition. *Statistics and Computing* **10**, 279–287.

DeGruttola, V. and Tu, X. M. (1994). Modeling progression of CD4-lymphocyte count and its relationship to survival time. *Biometrics* **50**, 1003–1014.

Efron, B. and Tibshirani, R. J. (1993). *An Introduction to the Bootstrap.* London: Chapman & Hall.

Follmann, D. and Wu, M. (1995). An approximate generalized linear model with random effects for informative missing data. *Biometrics* **51**, 151–168.

Gallop, R. J., Ten Have, T. R., and Crits-Christoph, P. (2006) A mixed effects Markov model for repeated binary outcomes with non-ignorable dropout. *Statistics in Medicine* **25**, 2398–2426.

Gao, S. (2004). A shared random effect parameter approach for longitudinal dementia data with non-ignorable missing data. *Statistics in Medicine* **23**, 211–219.

Heagerty, P. J. and Kurland, B. F. (2001). Misspecified maximum likelihood estimates and generalized linear mixed models. *Biometrika* **88**, 973–985.

Heckman, J. J. (1979). Sample selection bias as a specification error. *Econometrica* **47**, 153–161.

Hogan, J. W. and Laird, N. M. (1997a). Mixture models for the joint distribution of repeated measures and event times. *Statistics in Medicine* **16**, 239–257.

Hogan, J. W. and Laird, N. M. (1997b). Model-based approaches to analysing incomplete longitudinal and failure time data. *Statistics in Medicine* **16**, 259–272.

Intermittent Positive Pressure Breathing Trial Group (1983). Intermittent positive pressure breathing therapy of chronic obstructive pulmonary disease. *Annals of Internal Medicine* **99**, 612–630.

Johnson, R. E., Jaffe, J. H., and Fudala, P. J. (1992). A controlled trial of buprenorphine treatment for opiate dependence. *Journal of the American Medical Association* **267**, 2750–2755.

Karlin, S. and Taylor, H.M. (1981). *A Second Course in Stochastic Processes.* San Diego, CA: Academic Press.

Laird, N. M. and Ware, J. H. (1982). Random-effects models for longitudinal data. *Biometrics* **38**, 963–974.

Little, R. J. A. (1993) Pattern-mixture models for multivariate incomplete data. *Journal of the American Statistical Association* **88**, 125–134.

Little R. J. A. (1995). Modeling the drop-out mechanism in repeated-measures studies. *Journal of the American Statistical Association* **90**, 1112–1121.

Little, R. J. A. and Rubin, D. B. (1987) *Statistical Analysis with Missing Data.* New York: Wiley.

McCulloch, C. E. (1997). Maximum-likelihood algorithms for generalized linear mixed models. *Journal of the American Statistical Association* **92**, 162–170.

Mori, M., Woolson, R. F., and Woodworth, G. G. (1994). Slope estimation in the presence of informative censoring: Modeling the number of observations as a geometric random variable. *Biometrics* **50**, 39–50.

Rubin, D. B. (1976). Inference and missing data. *Biometrika* **63**, 581–592.

Schluchter, M. D. (1992). Methods for the analysis of informative censored longitudinal data. *Statistics in Medicine* **11**, 1861–1870.

Schluchter, M. D., Greene, T., and Beck, G. J. (2001). Analysis of change in the presence of informative censoring: Application to a longitudinal clinical trial of progressive renal disease. *Statistics in Medicine* **20**, 989–1007.

Ten Have, T. R., Kunselman, A., Pulkstenis, E., and Landis, J. R. (1998). Mixed effects logistic regression models for longitudinal binary response data with informative dropout. *Biometrics* **54**, 367–383.

Ten Have, T. R., Miller, M. E., Reboussin, B. A., and James, M. K. (2000). Mixed effects logistic regression models for longitudinal ordinal functional response data with multiple cause drop-out from the Longitudinal Study of Aging. *Biometrics* **56**, 279–287.

Theodore, W. H., Albert, P. S., Stertz, B., Malow, B., Ko, D., White, S., Flamini, R., and Ketter, T. (1995). Felbamate monotherapy: Implications for antiepileptic drug development. *Epilepsia* **36**, 1105–1110.

Tsiatis, A. A. and Davidian, M. (2001). A semiparametric estimator for the proportional hazards model with longitudinal covariates measured with error. *Biometrika* **88**, 447–459.

Tsiaits, A. A. and Davidian, M. (2004) Joint modeling of longitudinal and time-to-event data: An overview. *Statistica Sinica* **14**, 809–834.

Tsiatis, A. A., DeGruttola, V., and Wulfsohn, M. S. (1995). Modeling the relationship of survival to longitudinal data measured with error. Applications to survival and CD4 counts in patients with AIDS. *Journal of the American Statistical Association* **90**, 27–37.

Vonesh, E. F., Greene, T., and Schluchter, M. D. (2006). Shared parameter models for the joint analysis of longitudinal data with event times. *Statistics in Medicine* **25**, 143–163.

Vonesh, E. F., Wang, H., Nie, L., and Majumdar, D. (2002). Conditional second-order generalized estimating equations for generalized linear and nonlinear mixed-effects models. *Journal of the American Statistical Association* **97**, 271–283.

Wang-Clow, F., Lange, M., Laird, N. M., and Ware, J. H. (1995). A simulation study of estimators for rates of change in longitudinal studies with attrition. *Statistics in Medicine* **14**, 283–297.

Wu, M. C. and Bailey, K. R. (1989). Estimation and comparison of changes in the presence of informative right censoring: Conditional linear model. *Biometrics* **45**, 939–955.

Wu, M. C. and Carroll, R. J. (1988). Estimation and comparison of changes in the presence of informative right censoring by modeling the censoring process. *Biometrics* **44**, 175–188.

Wulfson, M. S. and Tsiatis, A. A. (1997). A joint model for survival and longitudinal data measured with error. *Biometrics* **53**, 330–339.

CHAPTER 20

Inverse probability weighted methods

Andrea Rotnitzky

Contents

20.1 Introduction

In most epidemiological and clinical longitudinal studies the intended (full) data — the data that the study investigators wish to collect — are incompletely observed. In modern studies, the intended data are typically high-dimensional, usually comprised of many baseline and time-varying variables. Scientific interest, however, often focuses on a low-dimensional parameter of the distribution of the full data.

Specification of realistic parametric models for the mechanism generating high-dimensional data is most often very challenging, if not impossible. Non- and semi-parametric models, in which the data generating process is characterized by parameters ranging over a large, non-Euclidean, space and possibly also a few meaningful real-valued parameters, meet the analytic challenge posed by these high-dimensional data because they do not make assumptions about the components of the full-data distribution that are of little scientific interest. Inference about the parameters of interest under semi-parametric models is therefore protected from biases induced by misspecification of models for these secondary components. Inverse probability weighted (IPW) and inverse probability weighted augmented (IPWA) estimating functions provide a general methodology for constructing estimators of smooth (i.e., estimable at rate $\sqrt{N}$) parameters under non- or semi-parametric models for the full data and a semi-parametric or parametric model for the missingness mechanism.

The IPW and IPWA methodology has its roots in the survey sampling literature. IPW estimators are closely connected to the Horvitz and Thompson (1952) estimators of finite population totals and means. On the other hand, IPWA estimators are close relatives of the so-called difference and regression estimators of these parameters. These estimators often vastly improve over the efficiency of the Horvitz–Thompson estimator.

IPW and IPWA estimators for parameters of longitudinal data models were originally introduced by Robins and Rotnitzky (1992) as part of a general estimating function methodology in coarsened (i.e., incompletely observed) data models under non- or semi-parametric models for arbitrary full-data configurations when the data are subject to coarsening at random (CAR) (Heitjan and Rubin, 1991; Jacobsen and Keiding, 1995; Gill, van der Laan, and Robins, 1997) and the coarsening (i.e., censoring or missingness) mechanism is either known or correctly modeled. Robins and Rotnitzky (1992) derived their methodology drawing from the modern theory of semi-parametric efficiency due to van der Vaart (1988, 1991), Bickel et al. (1993), and Newey (1990), among others. Rotnitzky and Robins (1997) and Robins, Rotnitzky, and Scharfstein (2000) later extended this general methodology to non-CAR models and described IPWA estimation in this setting.

IPWA estimation has appealing properties. When missingness is by design, IPWA estimators are guaranteed to be unbiased. In contrast, fully parametric likelihood-based (FPLB) estimators are inconsistent if the parametric model is misspecified. In addition, clever choices of the augmentation term yield IPWA estimators which can be substantially more efficient than non-augmented Horvitz–Thompson-type IPW estimators. When missingness is by happenstance, certain IPWA estimators are substantially less sensitive to model misspecification than FPLB estimators. Specifically, if missingness is non-CAR then consistency of any IPWA estimator requires a correctly specified model for the non-response probabilities. In contrast, an FPLB estimator additionally requires a correctly specified model for the full-data distribution. When missingness is CAR, special types of IPWA estimators, namely the doubly robust IPWA estimators, have an appealing robustness property: they are consistent if either a model for the non-response probabilities or a model for the full-data distribution can be correctly specified (but not necessarily both). Therefore, doubly robust IPWA estimators give the data analyst two chances of getting correct inference. In contrast, the consistency of FPLB estimators requires that the model for the full data be correctly specified.

The IPWA methodology has been applied in the last 15 years to derive estimators in models for a broad class of incomplete longitudinal data structures, with and without the CAR assumption. IPWA estimators have been derived in models for endpoints of (i) repeated outcomes; (ii) continuous time-to-event; (iii) discrete time-to-event; and (iv) quality-of-life-adjusted time-to-event types, and for missingness patterns determined by (i) dropout recorded at discrete times; (ii) dropout recorded in continuous time; and (iii) intermittent non-response. Another important application of the IPWA methodology is for estimation of parameters of certain causal inference models for observational or randomized follow-up studies under the assumption of no unmeasured confounders, specifically the so-called *marginal structural models* (MSMs) (Robins, 1999). A key element in these models is the notion of potential or counterfactual outcomes (Rubin 1978; Robins, 1986) — the outcomes that the subject would have had if, possibly contrary to fact, he had followed a certain treatment plan. MSMs stipulate assumptions on the dependence of some aspect of the marginal distribution of each treatment-specific potential outcome vector on the treatment plan and, possibly, on baseline covariates. In an observational or randomized study, all treatment-specific potential outcomes of any subject are necessarily missing, with the exception of the potential outcome associated with the treatment history that the subject actually followed.

The IPWA methodology is applicable because one can regard inference in an MSM as a missing-data problem (see Chapter 23).

In this chapter we provide an overview of the basic elements of the IPWA methods as they apply to the analysis of incomplete, repeated-measures data. With the goal of giving the IPW methods an historical perspective and providing a heuristic motivation, in Section 20.2 we give a brief review of IPW and IPWA estimation in survey sampling. In Section 20.3 we introduce a general data structure, the coarsened data structure, which is adequate to describe incomplete data in longitudinal studies. In Section 20.4 we define the CAR assumption. In Section 20.5 we motivate and discuss doubly robust methods in CAR and non-CAR models. In Section 20.6 we define the IPWA estimating functions and estimators. In Section 20.7 we describe algorithms leading to reasonably efficient IPWA estimators. In Section 20.8 we revisit doubly robust estimation and provide a fundamental theoretical result. In Section 20.9 we provide a survey of the literature and in Section 20.10 we discuss open problems.

20.2 Inverse probability weighted methods in survey sampling

IPW estimators rely on an old principle whose origin dates back to the work of Horvitz and Thompson (1952) in the context of inference about finite-population parameters under probability sampling. This framework assumes a sampling plan in which a sample is to be selected according to a known (i.e., prespecified) selection probability. This sampling plan in turn determines for each unit in the population a known probability π that the unit is sampled. The probability π often depends on auxiliary variables w known before sampling. The Horvitz–Thompson estimator of the population total $T = \sum_{i=1}^{N} y_i$ is

$$\widehat{T}_{HT} = \sum_{i=1}^{N} \Delta_i \pi_i^{-1} y_i,$$

where N is the number of units in the population and Δ_i is the random binary selection indicator, $\Delta_i = 1$ if the ith unit is selected in the sample and $\Delta_i = 0$ otherwise. The key idea in the computation of $\widehat{T}_{HT}$ is that each sampled unit is weighted by π^{-1} so as to represent the remaining $\pi^{-1} - 1$ non-sampled population units which have the same value of w as the sampled unit. If $\pi > 0$ for all units in the population, then no value of w can prevent the unit from being sampled. This ensures that $\widehat{T}_{HT}$ is unbiased for T. Hansen and Hurwitz (1943) also exploited a similar idea in the context of sampling with replacement.

IPWA estimators also have their origins in the survey sampling literature. These estimators are based on a proxy $\mu(w)$ for y, where $\mu(\cdot)$ is some function of the known variable w that is either specified by the data analyst or estimated from the sample. Given $\mu(\cdot)$, an estimator of the population total T is computed as

$$\widehat{T}_{\mathrm{aug}} = T_{\mathrm{pred}} + \widehat{M},$$

where $T_{\mathrm{pred}} = \sum_{i=1}^{N} \mu(w_i)$ and $\widehat{M} = \sum_{i=1}^{N} \Delta_i \pi_i^{-1} \{y_i - \mu(w_i)\}$ (Cassel, Sarndal, and Wretman, 1976; Sarndal, 1980, 1982; Isaki and Fuller, 1982; Wright, 1983; Sarndal, Swensson, and Wretman, 1992). The idea behind the estimator $\widehat{T}_{\mathrm{aug}}$ is simple. The unknown population total T is equal to $T_{\mathrm{pred}} + M$; T_{pred} is known, but the difference $M = \sum_{i=1}^{N} \{y_i - \mu(w_i)\}$ is unknown. The estimator $\widehat{T}_{\mathrm{aug}}$ is constructed by replacing M with its Horvitz–Thompson estimator. Survey sampling statisticians call $\widehat{T}_{\mathrm{aug}}$ the *difference estimator* of T when $\mu(\cdot)$ is a prespecified function and the *regression estimator* of T when $\mu(\cdot)$ is a function estimated from the sample. The difference estimator is exactly unbiased for T, and the regression estimator is approximately unbiased when the sample size is large. Both can be substantially more efficient than $\widehat{T}_{HT}$ if $\mu(w)$ is a good proxy for y. An extreme situation illustrates this point: if $\mu(w)$ is a perfect proxy for y, then $\widehat{T}_{\mathrm{aug}}$ is identically equal to T and hence infinitely more efficient than $\widehat{T}_{HT}$.

The recognition of the potential efficiency benefits of $\widehat{T}_{\mathrm{aug}}$ when $\mu(w)$ approximates y well has led survey statisticians to compute regression estimators in which $\mu(w)$ is the fitted value from a *working* regression model for the dependence of the mean of y on w. In the sampling literature, such estimators are referred to as *model-assisted* estimators (Sarndal, Swensson, and Wretman, 1992). The name is a reminder that the basic properties of the estimators (the unbiasedness, the validity of the variance formulae, etc.) are not dependent on the validity of the working regression model. This model is only used to assist the data analyst in the construction of estimators that hopefully outperform the Horvitz–Thompson estimator. The locally efficient IPWA estimators described in Section 20.7.1 in essence use the same idea as the model-assisted regression estimators.

Straightforward algebra yields an IPWA representation for $\widehat{T}_{\mathrm{aug}}$, namely

$$\widehat{T}_{\mathrm{aug}} = \widehat{T}_{HT} + \sum_{i=1}^{N}(1 - \Delta_i \pi_i^{-1})\mu(w_i).$$

This alternative representation shows that $\widehat{T}_{\mathrm{aug}}$ is equal to the Horvitz–Thompson estimator of T plus an *augmentation term* $\sum_{i=1}^{N}(1 - \Delta_i \pi_i^{-1})\mu(w_i)$.

20.3 The coarsened data structure and the goal of inference

In its most general formulation, a coarsened data structure is defined as follows. A random sample of size N of a vector $\boldsymbol{Z}$ taking values in a set $\mathbb{H}$ represents the intended (i.e., full) data. The full data are incompletely observed. The observed data are a sample of a random subset $\mathfrak{Z} \subseteq \mathbb{H}$ in which $\boldsymbol{Z}$ is known to lie. The random set $\mathfrak{Z}$ is said to be a coarsening of $\boldsymbol{Z}$. The distribution $F_{\boldsymbol{Z}}$ of $\boldsymbol{Z}$ is usually referred to as the full-data distribution. The conditional distribution of $\mathfrak{Z}$ given $\boldsymbol{Z}$, denoted throughout with G, is usually referred to as the *coarsening mechanism*. Often, $\mathfrak{Z}$ can be described in a one-to-one way as a function of some random vector $\boldsymbol{O}$ and in such a case G also stands, in an abuse of notation, for the distribution of $\boldsymbol{O}$ given $\boldsymbol{Z}$. Suppose that $\boldsymbol{\beta}(F_{\boldsymbol{Z}})$ is a finite-dimensional, say $p \times 1$, parameter of the full-data distribution $F_{\boldsymbol{Z}}$ which admits consistent and asymptotically normal estimators in the absence of missing data under a non- or semi-parametric model $\mathcal{M}_{\mathrm{full}}$ for $F_{\boldsymbol{Z}}$. IPW methods provide estimators of $\boldsymbol{\beta}(F_{\boldsymbol{Z}})$ based on a random sample of the incomplete data $\boldsymbol{O}$.

In our discussion of IPW methods we will focus on the following special coarsened data configuration, adequate for describing incomplete data in longitudinal studies with repeated-outcome endpoints, which throughout we refer to as the multivariate coarsened data (MCD) structure. The intended (full) data that the investigators wish to collect are a random sample, that is, N independent and identically distributed (i.i.d.) copies, of a vector $\boldsymbol{Z} = (\boldsymbol{Z}_0, \boldsymbol{Z}_1, \ldots, \boldsymbol{Z}_n)$, where for $j \geq 1, \boldsymbol{Z}_j = (\boldsymbol{W}_j, Y_j)$ with Y_j an outcome of interest and $\boldsymbol{W}_j$ a vector recording auxiliary variables and, possibly, time-dependent covariates. The components of $\boldsymbol{Z}_0$ may or may not be of the same type as those of $\boldsymbol{Z}_j$ for $j > 0$. The vector $\boldsymbol{Z}_0$ is partitioned into a component $\boldsymbol{Z}_{0,0}$ which is always observed and a component $\boldsymbol{Z}_{0,1}$ which is missing in some study participants. In addition, for each sample unit, each $\boldsymbol{Z}_j, j > 0$, is either fully observed or fully missing. Define the binary variable R_j, which takes the value 1 if and only if $\boldsymbol{Z}_j$ is fully observed, $j \geq 0$. The observed data are N i.i.d. copies of a vector $\boldsymbol{O} = (\boldsymbol{O}_0, \ldots, \boldsymbol{O}_n)$ where $\boldsymbol{O}_0 = R_0 \boldsymbol{Z}_0$ if $R_0 = 1$, $\boldsymbol{O}_0 = R_0 \boldsymbol{Z}_{0,0}$ if $R_0 = 0$ and, for $j > 0, \boldsymbol{O}_j = R_j$ if $R_j = 0$ and $\boldsymbol{O}_j = (R_j, \boldsymbol{Z}_j)$ if $R_j = 1$.

MCD structures arise in longitudinal studies whose designs stipulate that measurements of $\boldsymbol{Z}_j$ are to be obtained at predetermined time points $t_j, j = 0, \ldots, n$, on a random sample of size N and in which the intended data are not fully recorded on a subset of the study participants at some study cycles. In these studies, the components of $\boldsymbol{Z}_{0,0}$ often encode known baseline treatment and subject characteristics and/or always observed, non-subject specific, time-dependent covariates such as air pollution at each study cycle. The components of $Z_{0,1}$ often include exposure variables difficult or costly to measure. The following are two examples of this situation.

Example 1(a). Estimation of marginal mean models for repeated outcomes in the presence of dropout: Suppose that Z denotes the intended data to be recorded on a participant of a two-arm follow-up AIDS clinical trial, the primary goal of which is to compare the evolution over time of the CD4 count mean in the two arms. In this setting Y_j is CD4 count at t_j and the vector W_j denotes other laboratory and clinical measurements at t_j where t_j denotes the scheduled time of the jth study cycle. The baseline vector $Z_0 = (X, W_0)$ is always observed and is comprised of a treatment arm indicator X and baseline subject variables W_0. Suppose that the investigator's model $\mathcal{M}_{\text{full}}$ stipulates that the arm-specific mean of Y_j at each occasion t_j satisfies

$$E_{F_Z}(Y_j|X) = g_j\{X; \beta(F_Z)\}, \tag{20.1}$$

where $g_j(\cdot; \cdot)$ is a known smooth function and $\beta(F_Z)$ is an unknown $p \times 1$ parameter. For instance, the choice $g_j(X; \beta) = \beta_1 + \beta_2 X + \beta_3 t_j + \beta_4 X t_j$ corresponds to assuming that mean CD4 count depends linearly on time with arm-specific slope and intercept. In the absence of missing data, $\beta(F_Z)$ is estimable at rate $\sqrt{N}$. For example, the working independence generalized estimating equations (GEE) estimator (Liang and Zeger, 1986) is consistent and asymptotically normal for $\beta(F_Z)$. Now, suppose that the longitudinal study suffers from loss to follow-up, so that for some study subjects Y_j is missing from some random time t_j on. We conceptualize the target population as comprised of two types of subjects: the "study-completers," that is, individuals who would complete the study if they were to participate in the trial, and the "study dropouts," the remaining individuals. An analysis adhering to the intention-to-treat philosophy aims to compare the evolution of the arm-specific CD4 count means in the entire study population, not just on the study completers. Thus, $\beta(F_Z)$ remains the parameter of interest even in the presence of dropout and the methodological challenge is to estimate it from just the observed data.

Example 1(b). Estimation of marginal mean models for repeated outcomes with missing covariates: Consider a prospective 5-year follow-up study on determinants of cardiovascular disease whose design stipulates that at each year j a sample of high-risk subjects will fill out a questionnaire providing information on: (i) the occurrence Y_j over the previous year of a peak of high blood pressure, dichotomized as yes/no; and (ii) dietary habits W_j. The study design also stipulates that at start of follow-up, blood serum samples are to be obtained on all study participants. In addition, demographic information $Z_{0,0}$, including age, gender, and ethnicity, are obtained from existing records on all study participants. Suppose that the study investigators wish to investigate the effect of baseline antioxidant serum vitamin A, $Z_{0,1}$, available from the baseline blood sample, on subsequent risk of high blood pressure peaks. To do so, the investigators wish to estimate the parameter $\beta = (\beta_{1,1}, \ldots, \beta_{1,5}, \beta'_{2,1}, \ldots, \beta'_{2,5})'$ of the marginal logistic regression model (20.1) where $X = (Z_{0,0}, Z_{0,1})$ with, say, $g_j(X; \beta) = \{1 + \exp(-\beta_{1,j} - \beta'_{2,j} X)\}^{-1}$, $j = 1, \ldots, 5$. Suppose that Y_j is always observed because all subjects return the yearly questionnaires but $Z_{0,1}$ is missing in some study participants because they have not agreed to have a blood sample taken. In this case, $R_j = 1$ with probability 1 for $j > 0$ but $R_0 = 0$ is an event with positive probability.

20.4 The coarsening at random assumption

Any pair (F_Z, G) gives rise to a distribution F_O for the observed data O. The map $(F_Z, G) \rightarrow F_O$ is many-to-one when its domain is the set of all pairs (F_Z, G). Given F_O, it is possible to find two pairs (F_Z, G) and (F_Z^*, G^*) that give rise to F_O and such that F_Z and F_Z^*

differ in the marginal distribution of the components of $\boldsymbol{Z}$ that, under F_O, have positive probability of not being observed. An important consequence of this result is that $\boldsymbol{\beta}(F_{\boldsymbol{Z}})$ is generally not identified, and hence not consistently estimable, from a random sample of the observed data $\boldsymbol{O}$ under a model that places no assumptions on either $F_{\boldsymbol{Z}}$ or G. (The only exception is when $\boldsymbol{\beta}(F_{\boldsymbol{Z}})$ depends only on the distribution of always observed components of $\boldsymbol{Z}$.)

Coarsening at random is an assumption about the coarsening mechanism G which suffices for identification of $\boldsymbol{\beta}(F_{\boldsymbol{Z}})$. Specifically, Gill, van der Laan, and Robins (1997) proved that $\boldsymbol{\beta}(F_{\boldsymbol{Z}})$ is identified by the observed data $\boldsymbol{O}$ under a model that assumes CAR for G and places no assumptions on $F_{\boldsymbol{Z}}$. These authors also showed that a model for $(F_{\boldsymbol{Z}}, G)$ that assumes only CAR is, for estimation purposes, essentially a non-parametric model for the distribution F_O, i.e., a model that imposes no restrictions on the distribution F_O. A practical consequence of this theoretical result is that CAR is not subject to empirical verification because the empirical distribution of any finite sample of coarsened data can be exactly fit to a model that assumes just CAR.

The precise definition of CAR for general coarsened data structures requires careful measure-theoretic considerations that take care of conditioning on null sets. In essence, CAR stipulates that the conditional distribution of $\boldsymbol{O}$ given $\boldsymbol{Z}$ does not depend on the values z taken by $\boldsymbol{Z}$ except for the restriction implied by $\boldsymbol{O}$ being a coarsening of $\boldsymbol{Z}$. When $\boldsymbol{Z}$ takes values in a finite set, so does $\boldsymbol{O}$ and CAR is precisely defined by the condition

$$\Pr(\boldsymbol{O} = o | \boldsymbol{Z} = z) = \Pr(\boldsymbol{O} = o) \text{ for all } z \text{ that can give rise to the observation } o.$$

With MCD structures, CAR is easily formulated in terms of the missingness probabilities. Specifically, CAR is equivalent to the assumption that for every vector $\boldsymbol{r} \in \{0,1\}^{n+1}$,

$$\Pr(\boldsymbol{R} = \boldsymbol{r} | \boldsymbol{Z}) = \Pr(\boldsymbol{R} = \boldsymbol{r} | \boldsymbol{Z}_{(\boldsymbol{r})}) \quad \text{with probability 1 (w.p. 1)},$$

where $\boldsymbol{Z}_{(\boldsymbol{r})}$ denotes the vector whose elements are the components of $\boldsymbol{Z}$ that are observed when $\boldsymbol{R} = \boldsymbol{r}$.

CAR has an appealing equivalent formulation under a special type of coarsening known as monotone coarsening. By definition, a coarsening $\mathfrak{Z}$ is monotone if and only if, for any two possible realizations $\mathfrak{Z}_1$ and $\mathfrak{Z}_2$, either $\mathfrak{Z}_1 \subset \mathfrak{Z}_2$, $\mathfrak{Z}_2 \subset \mathfrak{Z}_1$, or $\mathfrak{Z}_1$ and $\mathfrak{Z}_2$ are disjoint sets. With MCD structures monotonicity is equivalent to the condition that $R_j = 0$ implies $R_{j+1} = 0$, possibly after a reordering of the components $\boldsymbol{Z}_j$ of $\boldsymbol{Z}$. In this setting, information about the observed data pattern is encoded in the single scalar "censoring" variable $D = \sum_{j=0}^{n} R_j$ denoting the index of the first missing component of $\boldsymbol{Z}$ ($D = n+1$ if $\boldsymbol{Z}$ is fully observed). For ease of reference, we shall use the acronym MMCD to stand for monotone MCD. Examples of longitudinal studies with MMCD structures are: studies with always observed baseline covariates $\boldsymbol{Z}_0$ when the only reason for non-response is study dropout (in this setting D records the dropout occasion); and studies with missing baseline covariates $\boldsymbol{Z}_{0,1}$ but with always observed outcomes and auxiliaries. With MMCD structures, CAR is equivalent to the condition

$$\Pr(D = j | D > j - 1, \boldsymbol{Z}) = \Pr(D = j | D > j - 1, \overline{\boldsymbol{Z}}_{j-1}), \quad \text{w.p. 1, } j = 0, \dots, n, \quad (20.2)$$

where, by convention, $\boldsymbol{Z}_{-1} = \boldsymbol{Z}_{0,0}$ and for any vector $\boldsymbol{A} = (\boldsymbol{A}_{-1}, \boldsymbol{A}_0, \dots, \boldsymbol{A}_n)$, $\overline{\boldsymbol{A}}_j$ stands for the subvector $(\boldsymbol{A}_{-1}, \dots, \boldsymbol{A}_j)$. In longitudinal studies with dropout and always observed baseline covariates, condition (20.2) stipulates that the discrete hazard of dropout at time t_j is conditionally independent of the current and future data given the recorded data up to time t_{j-1}. Thus, CAR would hold when the recorded past data are all the simultaneous predictors of non-response at time t_j and of the future outcomes.

20.4.1 The likelihood principle and IPWA estimation under CAR

In CAR models, the likelihood $\mathcal{L}_N(F_{\mathbf{Z}}, G)$, based on N i.i.d. copies of $\mathbf{O}$, factorizes as

$$\mathcal{L}_N(F_{\mathbf{Z}}, G) = \mathcal{L}_{1,N}(F_{\mathbf{Z}})\mathcal{L}_{2,N}(G) \tag{20.3}$$

(Gill, van der Laan, and Robins, 1997). Thus, under any model in which G and $F_{\mathbf{Z}}$ are variation-independent, any method that obeys the likelihood principle must result in the same inference about $\boldsymbol{\beta}$ regardless of whether G is known, completely unknown, or known to follow some model. However, Robins and Ritov (1997) have shown that with high-dimensional CAR data any method of inference under large (i.e. non-stringent) models for $F_{\mathbf{Z}}$ that obeys the likelihood principle and thus ignores G must perform poorly in realistic sample sizes due to the curse of dimensionality. The following example illustrates this point.

Example 2. Estimation of the mean of an outcome in a cross-sectional study under CAR: Consider a cross-sectional observational study with full data $\mathbf{Z} = (\mathbf{Z}_0, Z_1)$ where $\mathbf{Z}_0 = \mathbf{W}$ is an always observed vector of baseline variables and $Z_1 = Y$ is a scalar outcome missing by happenstance on some subjects. Investigators conducting the study and wishing to make the CAR assumption would need to collect in $\mathbf{W}$ data on a vast number of variables so as to try to make the CAR condition (20.2) at least approximately true. It would not be unusual in realistic epidemiological studies that the sample size N be between 500 and 2000 and yet for $\mathbf{W}$ to be 10- to 20-dimensional.

Under CAR, (20.3) holds with $\mathcal{L}_{1,N}(F_{\mathbf{Z}}) = \prod_{i=1}^N f_{Y|\mathbf{W}}(Y_i|\mathbf{W}_i)^{\Delta_i} f_{\mathbf{W}}(\mathbf{W}_i)$, where $\Delta_i = 1$ if Y_i is observed and 0 otherwise. Suppose that we are interested in estimating $\beta = E(Y)$ under a model that assumes CAR but places no restrictions on the full-data distribution $F_{\mathbf{Z}}$ and hence it is non-parametric for the observed-data law $F_{\mathbf{O}}$. Because $\beta = \beta(F_{\mathbf{Z}}) = E_{F_{\mathbf{W}}}\{E_{F_{Y|\mathbf{W}}}(Y|\mathbf{W})\}$ depends only on $F_{\mathbf{Z}}$, its non-parametric maximum likelihood estimator (NPMLE), whenever it exists, must be equal to $\beta(\widehat{F}_{\mathbf{Z}})$ where $\widehat{F}_{\mathbf{Z}}$ is the NPMLE of $F_{\mathbf{Z}}$. The NPMLE of $F_{\mathbf{W}}$ is the empirical c.d.f. of $\mathbf{W}$. Furthermore, under CAR, $E(Y|\mathbf{W}, \Delta) = E(Y|\mathbf{W}, \Delta = 1)$. Thus, when $\mathbf{W}$ is discrete, the NPMLE of β is equal to $N^{-1}\sum_{i=1}^N \widehat{E}(Y|\mathbf{W} = \mathbf{W}_i, \Delta = 1)$ where $\widehat{E}(Y|\mathbf{W} = \mathbf{w}, \Delta = 1)$ is the empirical mean of Y in the subsample $\{i : \mathbf{W}_i = \mathbf{w}, \Delta_i = 1\}$. When $\mathbf{W}$ is continuous, each subsample consists of at most one observation, so $\widehat{E}(Y|\mathbf{W} = \mathbf{W}_i, \Delta = 1)$ is undefined for the values of $\mathbf{W}_i$ corresponding to missing outcomes Y_i. Thus, the NPMLE of β is also undefined. One could impose smoothness conditions on the conditional mean of Y given $\mathbf{W}$ and estimate it using smoothing techniques. However, when $\mathbf{W}$ is high-dimensional, this conditional expectation would not be well estimated with the moderate sample sizes found in practice, essentially because no two units would have values of $\mathbf{W}$ close enough to each other to allow the borrowing of information needed for smoothing. Indeed, unrealistically large sample sizes would be required for any estimator of β to have a well-behaved sampling distribution, that is, an approximately centered normal sampling distribution with variance small enough to be of substantive use.

IPWA estimators of $\beta(F_{\mathbf{Z}})$ under CAR rely on a model for G, and thus violate the likelihood principle. However, they are well behaved with moderate sample sizes. The continuation of the preceding example illustrates this point.

Example 2 (continued). Suppose that CAR holds and $\Pr(\Delta = 1|\mathbf{W}) > 0$ w.p. 1. Then we can represent β in terms of the observed-data distribution as

$$\beta = E\{\Delta Y / \Pr(\Delta = 1|\mathbf{W})\}.$$

This representation motivates the IPW estimator $\widetilde{\beta}$ which is computed as follows. We specify a parametric model $\pi(W; \boldsymbol{\alpha})$, for $\pi(\mathbf{W}) \equiv \Pr(\Delta = 1|\mathbf{W})$, for example the linear logistic regression model $\pi(\boldsymbol{W}; \boldsymbol{\alpha}) = \{1 + \exp(-\boldsymbol{\alpha}'\boldsymbol{W})\}^{-1}$, and calculate the maximum likelihood estimator $\widehat{\boldsymbol{\alpha}}$ of $\boldsymbol{\alpha}$ under the model. The estimator $\widetilde{\beta}$ is defined as

$$\widetilde{\beta} = \left\{ \sum_{i=1}^{N} \Delta_i Y_i / \pi(\boldsymbol{W}_i; \widehat{\boldsymbol{\alpha}}) \right\} \bigg/ \left\{ \sum_{i=1}^{N} \Delta_i / \pi(\boldsymbol{W}_i; \widehat{\boldsymbol{\alpha}}) \right\}.$$

An IPWA estimator of β is of the form

$$\widehat{\beta} = \widetilde{\beta} + \left[\sum_{i=1}^{N} \{1 - \Delta_i / \pi(\boldsymbol{W}_i; \widehat{\boldsymbol{\alpha}})\} m(\boldsymbol{W}_i) \right] \bigg/ \left\{ \sum_{i=1}^{N} \Delta_i / \pi(\boldsymbol{W}_i; \widehat{\boldsymbol{\alpha}}) \right\}$$

for some function $m(\cdot)$ that is generally estimated from the observed-data subsample. Note the close connections between $\widetilde{\beta}$ and $\widehat{T}_{HT}$, and between $\widehat{\beta}$ and $\widehat{T}_{\mathrm{aug}}$, in Section 20.2. The estimators $\widetilde{\beta}$ and $\widehat{\beta}$ are consistent and asymptotically normal provided the model $\pi(\boldsymbol{W}; \boldsymbol{\alpha})$ is correctly specified and $\Pr(\Delta = 1|\boldsymbol{W}) > \sigma$ for some $\sigma > 0$. Furthermore, they have a well-behaved, roughly Gaussian shaped, sampling distribution with moderate samples if the variance of the estimated weights $1/\pi(\boldsymbol{W}; \widehat{\boldsymbol{\alpha}})$ is not too large. Because the estimators $\widetilde{\beta}$ and $\widehat{\beta}$ depend on the model for $\Pr(\Delta = 1|\boldsymbol{W})$, they violate the likelihood principle.

20.5 Double robustness

Example 2 illustrates an important general point. It may occur that even if $\beta(F_{\boldsymbol{Z}})$ is identified from the coarsened data under some weak assumptions about G, estimators of it that are well-behaved with the realistic sample sizes found in practice may not exist. In Example 2 we exhibited well behaved IPWA estimators under models that additionally impose a dimension reducing parametric model on the coarsening mechanism. An alternative dimension reducing modeling strategy is also generally possible. Specifically, well-behaved estimators of $\beta(F_{\boldsymbol{Z}})$ also exist under models that place sufficiently stringent assumptions on the full-data distribution $F_{\boldsymbol{Z}}$. The simple data structure in Example 2 illustrates this point.

Example 2 (continued). Under CAR, β admits a second representation,

$$\beta = E\{E(Y|\boldsymbol{W}, \Delta = 1)\}.$$

This representation motivates the so-called model-based estimator $\widehat{\beta}_{\mathrm{mb}}$ which is computed as follows. We postulate a parametric regression model $\mu(\boldsymbol{W}; \tau)$ for $\mu(\boldsymbol{W}) \equiv E(Y|\boldsymbol{W}, \Delta = 1)$ where $\mu(\cdot; \cdot)$ is a known function with unknown parameter τ. Using the subsample with observed Y, we fit the regression model and compute a (possibly weighted) least-squares estimator $\widehat{\tau}$. Finally, the estimator $\widehat{\beta}_{\mathrm{mb}}$ is defined as $N^{-1} \sum_{i=1}^{N} \mu(\boldsymbol{W}_i; \widehat{\tau})$. The estimator $\widehat{\beta}_{\mathrm{mb}}$ is consistent, asymptotically normal and has a well-behaved sampling distribution with moderate samples provided the model $\mu(\boldsymbol{W}; \tau)$ is correctly specified.

Inference in large non-CAR models also suffers from the curse of dimensionality. As with CAR models, often two alternative dimension-reducing strategies are possible. The following example illustrates this point.

Example 3. Estimation of the mean of an outcome in a cross-sectional study under a non-CAR model: With the data structure of Example 2, suppose

$$\Psi\{\Pr(\Delta = 1|\boldsymbol{W}, Y)\} = \eta(\boldsymbol{W}) + r(Y, \boldsymbol{W}), \tag{20.4}$$

where $\Psi(\cdot) : [0, 1] \to \mathbb{R}$ is a known one-to-one link function, such as $\Psi(x) = \log\{x/(1-x)\}$. In (20.4) the so-called *selection bias function* $r(Y, \boldsymbol{W})$ governs the residual dependence between the missing-data indicator and the outcome Y after adjusting for the observed covariates $\boldsymbol{W}$. Assuming CAR is tantamount to specifying *a priori* that r is known and equal to the zero function. Consequently, in particular, $\beta = E(Y)$ is identified under a model that only stipulates that $r = 0$ and this model is non-parametric for the observed-data distribution. Scharfstein, Rotnitzky, and Robins (1999) extended this result and proved that models defined solely by any non-zero *a priori* specification of r are non-CAR models that identify β and are non-parametric for the observed-data distribution. Under such models and when $\Psi(x) = \log\{x/(1-x)\}$, β can be represented in terms of the observed-data distribution as

$$\beta = E\{\Delta Y + (1 - \Delta)\mu(\boldsymbol{W})\}, \tag{20.5}$$

where $\mu(\boldsymbol{W}) = E[Y \exp\{-r(Y, \boldsymbol{W})\}|\Delta = 1, \boldsymbol{W}]/E[\exp\{-r(Y, \boldsymbol{W})\}|\Delta = 1, \boldsymbol{W}]$. Furthermore, $\mu(\boldsymbol{W})$ is equal to $E(Y|\Delta = 0, \boldsymbol{W})$. Models defined solely by the specification of the function r in (20.4) are ideal for conducting a sensitivity analysis that appropriately adjusts for informative missingness due to measured factors $\boldsymbol{W}$ while simultaneously exploring the impact of different non-identifiable assumptions about the selection bias function r, and hence about the degree of residual dependence between Y and Δ due to unmeasured factors.

As in the CAR situation, calculation of the NPMLE of β requires that one replace the conditional means in $\mu(\boldsymbol{W})$ with their empirical counterparts. When $\boldsymbol{W}$ is continuous, the NPMLE of β is undefined because the empirical conditional means are undefined for units i with missing Y_i. Furthermore, as with CAR, estimation of the conditional means by smoothing techniques fails in practice when $\boldsymbol{W}$ is high-dimensional. The representation (20.5) suggests that when $\Psi(x) = \log\{x/(1-x)\}$ we can estimate β with a model-based type estimator $\widehat{\beta}_{\mathrm{mb}}$ computed as follows. We specify a parametric model $\mu(\boldsymbol{W}; \tau)$ for the ratio $\mu(\boldsymbol{W})$ and estimate τ with the consistent and asymptotically normal estimator $\widehat{\tau}$ which solves the estimating equations $\sum_{i=1}^{N} \Delta_i \{\partial\mu(\boldsymbol{W}_i; \tau)/\partial\tau\} \exp\{-r(Y_i, \boldsymbol{W}_i)\}\{Y_i - \mu(\boldsymbol{W}_i; \tau)\} = 0$. Finally, we compute $\widehat{\beta}_{\mathrm{mb}}$ as $N^{-1} \sum_{i=1}^{N} \{\Delta_i Y_i + (1 - \Delta_i)\mu(\boldsymbol{W}_i; \widehat{\tau})\}$.

The alternative representation

$$\beta = E\{\Delta Y/\Pr(\Delta = 1|\boldsymbol{W}, Y)\}$$

suggests a second estimator, an IPW estimator $\widetilde{\beta}$, which is computed as follows. We specify a parametric model $\pi(Y, \boldsymbol{W}; \alpha) \equiv \Psi^{-1}\{\eta(\boldsymbol{W}; \alpha) + r(Y, \boldsymbol{W})\}$ for $\Pr(\Delta = 1|\boldsymbol{W}, Y)$, for example $\eta(\boldsymbol{W}; \alpha) = \alpha'\boldsymbol{W}$, and calculate the estimator $\widehat{\alpha}$ of α solving the estimating equations $\sum_{i=1}^{N} b(\boldsymbol{W}_i)\{1 - \Delta_i/\pi(Y_i, \boldsymbol{W}_i; \alpha)\} = 0$, where $b(\boldsymbol{W})$ is a vector function of the same dimension as α. (When $\pi(Y, \boldsymbol{W}; \alpha)$ is correctly specified, these estimating equations are unbiased because $E[b(\boldsymbol{W}_i)\{1 - \Delta_i/\pi(Y_i, \boldsymbol{W}_i; \alpha)\}|Y_i, \boldsymbol{W}_i] = 0$ at the true α.) Finally, we compute the IPW estimator

$$\widetilde{\beta} = \left\{\sum_{i=1}^{N} \Delta_i Y_i/\pi(Y_i, \boldsymbol{W}_i; \widehat{\alpha})\right\} \bigg/ \left\{\sum_{i=1}^{N} \Delta_i/\pi(Y_i, \boldsymbol{W}_i; \widehat{\alpha})\right\}.$$

Augmented IPW estimators of β also exist in this problem. They are computed as

$$\widehat{\beta} = \widetilde{\beta} + \left[\sum_{i=1}^{N} \{1 - \Delta_i/\pi(Y_i, \boldsymbol{W}_i; \widehat{\alpha})\}m(\boldsymbol{W}_i)\right] \bigg/ \left\{\sum_{i=1}^{N} \Delta_i/\pi(Y_i, \boldsymbol{W}_i; \widehat{\alpha})\right\}$$

for some function $m(\cdot)$ which is either pre-specified or estimated from the sample.

In Example 2 and Example 3, neither the model-based estimator $\widehat{\beta}_{\mathrm{mb}}$ nor the IPW estimator $\widetilde{\beta}$ are entirely satisfactory because $\widehat{\beta}_{\mathrm{mb}}$ is inconsistent if the predictive model $\mu(\boldsymbol{W}; \tau)$ is misspecified and $\widetilde{\beta}$ is inconsistent if the non-response model is misspecified. There has been considerable controversy in the missing-data literature as to which estimator is to be preferred. This controversy could be resolved if there existed an estimator of β which was consistent, asymptotically normal, and well behaved so long as any one of the two models, the predictive or the non-response model, were correctly specified. Remarkably, in Example 3 with $\Psi(x) = \log\{x/(1-x)\}$ and in Example 2 with any model for $\pi(\boldsymbol{W})$, the IPWA estimator $\widehat{\beta}$ that uses the specific function $m(\boldsymbol{W}_i) = \mu(\boldsymbol{W}_i; \widehat{\tau}) - \widehat{\beta_{mb}}$ in the augmentation term has this property; the key to the validity of this property being that

$$E[\Delta(Y - \beta)/\pi(Y, \boldsymbol{W}; \boldsymbol{\alpha}) + \{1 - \Delta/\pi(Y, \boldsymbol{W}; \boldsymbol{\alpha})\}\{\mu(\boldsymbol{W}; \tau) - \beta^*\}] = 0$$

if $\pi(Y, \boldsymbol{W}; \boldsymbol{\alpha}) = \Pr(\Delta = 1 | \boldsymbol{W}, Y)$ or if $\mu(\boldsymbol{W}; \tau) = \mu(\boldsymbol{W})\beta = \beta^*$.

An estimator which, like the one just described, is consistent when one of two working dimension reducing models is correct (but not necessarily both) is often referred to as doubly robust. Just as in Example 2, in CAR models for longitudinal monotone missing-data patterns certain IPWA estimators are doubly robust under two specific working dimension reducing models. These estimators are described in Section 20.8. Doubly robust estimators in CAR models for non-monotone missing-data patterns can be constructed in principle, but in practice, they are difficult to implement because they require recursive algorithms to approximate solutions of integral equations. It is an open question whether doubly robust estimators can be constructed in non-CAR models for repeated measures. Section 20.9 surveys the literature on doubly robust estimation.

20.6 IPWA estimation

IPWA estimation is a general methodology for estimation in coarsened data models in which the probability of observing full data is bounded away from 0, that is, when $\Pr(\Delta = 1 | \boldsymbol{Z}) > \sigma$ for some $\sigma > 0$, where throughout Δ is defined as the indicator that $\boldsymbol{Z}$ is fully observed. In this chapter we describe IPWA estimation as it applies to MCD structures. In Section 20.9 we provide a survey of the literature on both the theory and application of IPWA estimation in general models for coarsened data.

20.6.1 IPWA estimating functions

The key element of IPWA estimation is the IPWA estimating function. An IPWA estimating function is computed by adding two terms, the so-called IPW and augmentation terms, each requiring an input from the data analyst. The input to the IPW term is a full-data unbiased estimating function for the parameter $\boldsymbol{\beta}$. The input to the augmentation term is a (usually vector-valued) function h of the observed data chosen by the data analyst. By a full-data unbiased estimating function for $\boldsymbol{\beta}$ we mean a function $d(\boldsymbol{Z}; \boldsymbol{\beta})$ depending on $\boldsymbol{\beta}$ satisfying $E_{F_{\boldsymbol{Z}}}[d\{\boldsymbol{Z}; \boldsymbol{\beta}(F_{\boldsymbol{Z}})\}] = 0$ for all $F_{\boldsymbol{Z}}$ in the full-data model $\mathcal{M}_{\text{full}}$. Though these functions do not exist in some full-data models and/or for some parameters $\boldsymbol{\beta}$, to facilitate the presentation and focus on the essential aspects of IPWA estimation we will assume in this chapter that the model and parameter of interest are such that unbiased estimating functions $d(\boldsymbol{Z}; \boldsymbol{\beta})$ for $\boldsymbol{\beta}$ exist. The theory of IPWA estimation does indeed apply also to settings in which the full-data estimating function depends on (possibly infinite-dimensional) nuisance parameters $\boldsymbol{\psi}$. The book by van der Laan and Robins (2003) provides a detailed treatment of this theory.

For MCD structures, each subject's contribution to an IPWA estimating function is of the form

$$s(\boldsymbol{O}, \boldsymbol{\beta}, G; d, h) \equiv u\{d(\boldsymbol{Z}; \boldsymbol{\beta}); G\} + a(\boldsymbol{O}, h; G),$$

for some full-data unbiased estimating function $d(\boldsymbol{Z}; \boldsymbol{\beta})$ where the IPW term has the form

$$u\{d(\boldsymbol{Z}; \boldsymbol{\beta}); G\} = \frac{\Delta d(\boldsymbol{Z}; \boldsymbol{\beta})}{\Pr_G(\Delta = 1 | \boldsymbol{Z})},$$

and the augmentation term has the form

$$a(\boldsymbol{O}, h; G) \equiv \sum_{\substack{\boldsymbol{r} \in \{0,1\}^{n+1} \\ \boldsymbol{r} \neq (1,\dots,1)}} \left\{ I_{\boldsymbol{r}}(\boldsymbol{R}) - \frac{\Delta \Pr_G(\boldsymbol{R} = \boldsymbol{r} | \boldsymbol{Z})}{\Pr_G(\Delta = 1 | \boldsymbol{Z})} \right\} h(\boldsymbol{r}, \boldsymbol{Z}_{(\boldsymbol{r})}). \tag{20.6}$$

If $\Pr_G(\Delta = 1 | \boldsymbol{Z}) > 0$, then the IPW term is an observed-data unbiased estimating function of $\boldsymbol{\beta}$ because $E_G[u\{d(\boldsymbol{Z}; \boldsymbol{\beta}); G\} | \boldsymbol{Z}] = d(\boldsymbol{Z}; \boldsymbol{\beta})$ for all $\boldsymbol{\beta}$. The augmentation term satisfies that $E_G\{a(\boldsymbol{O}, h; G) | \boldsymbol{Z}\} = 0$ for any h. Indeed, an easy calculation shows that any function of the observed data $\boldsymbol{O}$ that has conditional mean zero given the full data $\boldsymbol{Z}$ is of the form (20.6) for some h. Thus, simply put, augmentation terms are functions of the observed data that have conditional mean zero given the full data. Augmentation terms are added to the IPW terms with the hope of increasing the efficiency with which $\boldsymbol{\beta}$ is estimated.

Only subjects with complete data contribute a non-null value to the IPW term since Δ is non-null only for them. In contrast, all subjects contribute to the augmentation term. Those with incomplete data, that is, with $\boldsymbol{R} = \boldsymbol{r}$ for $\boldsymbol{r} \neq (1, \dots, 1)$, contribute to this term with $h(\boldsymbol{r}, \boldsymbol{Z}_{(\boldsymbol{r})})$, while those with complete data contribute with the sum of $-\Pr_G(\boldsymbol{R} = \boldsymbol{r} | \boldsymbol{Z}) h(\boldsymbol{r}, \boldsymbol{Z}_{(\boldsymbol{r})}) / \Pr_G(\Delta = 1 | \boldsymbol{Z})$.

When the missing-data patterns are monotone, the augmentation term simplifies to

$$a(\boldsymbol{O}, h; G) = \sum_{j=0}^{n} I(D > j - 1) \left\{ I(D = j) - \frac{\Delta \lambda_G(j | \boldsymbol{Z})}{\Pr_G(D = n + 1 | D > j - 1, \boldsymbol{Z})} \right\} h_j(\overline{\boldsymbol{Z}}_{j-1}) \tag{20.7}$$

where, throughout, $\lambda_G(j | \cdot) = \Pr_G(D = j | D > j - 1, \cdot)$ is the conditional discrete hazard of "censoring" at j given $\cdot$ and where, in a slight abuse of notation, $h_j(\overline{\boldsymbol{Z}}_{j-1})$ stands for the value of $h(\boldsymbol{r}, \boldsymbol{Z}_{(\boldsymbol{r})})$ where $\boldsymbol{r}$ is the binary vector corresponding to the value $D = j$.

In the special case in which the MCD structure is monotone and CAR, $\lambda_G(j | \boldsymbol{Z})$ depends only on $\overline{\boldsymbol{Z}}_{j-1}$, an observable vector for those censored at j. Indeed, in this case, functions of the observed data with mean zero given the full data can also be written as

$$a^*(\boldsymbol{O}, b; G) = \sum_{j=0}^{n} I(D > j - 1) \{ I(D = j) - \lambda_G(j | \overline{\boldsymbol{Z}}_{j-1}) \} b_j(\overline{\boldsymbol{Z}}_{j-1}). \tag{20.8}$$

Thus, in this special case, augmentation terms admit the two expressions (20.7) or (20.8). The functions b_j are related to the functions h_j via the recursive relationship for $j = 0, \dots, n$,

$$\sum_{l=0}^{j-1} -\lambda_G(l | \overline{\boldsymbol{Z}}_{l-1}) b_l(\overline{\boldsymbol{Z}}_{l-1}) + \{ 1 - \lambda_G(j | \overline{\boldsymbol{Z}}_{j-1}) \} b_j(\overline{\boldsymbol{Z}}_{j-1}) = h_j(\overline{\boldsymbol{Z}}_{j-1}).$$

Example 1(a) (continued). A full-data unbiased estimating function for the parameter $\boldsymbol{\beta}$ is of the form $d(\boldsymbol{Z}; \boldsymbol{\beta}) = m(X; \boldsymbol{\beta})'(\boldsymbol{Y} - g(X; \boldsymbol{\beta}))$, where g denotes the mean vector $(g_1, \dots, g_n)'$ and $m(X; \boldsymbol{\beta})$ is any user-specified $n \times p$ vector function. For example, the choice $m(X; \boldsymbol{\beta}) = \{ \partial g(X; \boldsymbol{\beta}) / \partial \boldsymbol{\beta}' \}$ yields a full-data working independence GEE estimating function. The general form of the IPW term is therefore

$$\frac{\Delta}{\Pr_G(\Delta = 1 | \boldsymbol{Z})} m(X; \boldsymbol{\beta})' \{ \boldsymbol{Y} - g(X; \boldsymbol{\beta}) \}. \tag{20.9}$$

The augmentation term has the form (20.6) for arbitrary missing-data patterns. This expression further simplifies to (20.7) because in this example we have assumed that missingness is caused only by dropout.

Example 1(b) (continued). In this example, the only possibly missing vector is $\boldsymbol{Z}_{0,1}$ and therefore Δ is the indicator that $\boldsymbol{Z}_{0,1}$ is observed. The IPW term remains equal to (20.9) but the augmentation term simplifies to

$$a(\boldsymbol{O}, h; G) = \left\{ 1 - \Delta - \frac{\Delta \Pr_G(\Delta = 0|\boldsymbol{Z})}{\Pr_G(\Delta = 1|\boldsymbol{Z})} \right\} h(\boldsymbol{Z}_{\text{obs}}),$$

where $\boldsymbol{Z}_{\text{obs}} = (\boldsymbol{Z}_{0,0}, \boldsymbol{Z}_1, \ldots, \boldsymbol{Z}_n)$.

20.6.2 IPWA estimators

IPWA estimation relies on a model for the coarsening mechanism which can be non-parametric, semi-parametric, or fully parametric. With longitudinal data, non- and semi-parametric models for the coarsening mechanism are used primarily to model continuous time censoring processes. Section 20.8 provides references on this topic. Here we describe the IPWA estimation methodology as it applies to MCD structures under a parametric model $\mathcal{G} = \{G_\alpha : \boldsymbol{\alpha} \in \Omega \subseteq \mathbb{R}^q\}$ for the coarsening mechanism, or equivalently under a parametric model indexed by $\boldsymbol{\alpha}$ for the missingness probabilities $\Pr(\boldsymbol{R} = \boldsymbol{r}|\boldsymbol{Z})$, with $\boldsymbol{\alpha}$ assumed to be variation independent of the parameters indexing the full-data model $\mathcal{M}_{\text{full}}$.

The IPWA estimating functions can be used to derive not only estimators of the full-data parameter $\boldsymbol{\beta}$ but also of the coarsening parameter $\boldsymbol{\alpha}$. Specifically, because $E_G\{a(\boldsymbol{O}, h; G_\alpha)\} = 0$ when $G = G_\alpha$, then $a(\boldsymbol{O}, h; G_\alpha)$, viewed as a function of the data and $\boldsymbol{\alpha}$, is an unbiased estimating function for the coarsening parameter $\boldsymbol{\alpha}$. This suggests estimators $(\widehat{\boldsymbol{\alpha}}(d, h), \widehat{\boldsymbol{\beta}}(d, h))$ of $(\boldsymbol{\alpha}, \boldsymbol{\beta})$ simultaneously solving the estimating equations

$$\sum_{i=1}^{N} s(\boldsymbol{O}_i, \boldsymbol{\beta}, G_\alpha; d, h) = 0,$$

where d and h are vector-valued functions of the same dimension as $(\boldsymbol{\alpha}', \boldsymbol{\beta}')'$, with d a user-specified unbiased estimating function of $\boldsymbol{\beta}$ and h an arbitrary, user-specified, function of $\boldsymbol{O}$. The estimators $(\widehat{\boldsymbol{\alpha}}(d, h), \widehat{\boldsymbol{\beta}}(d, h))$ are referred to as IPWA estimators.

If the full data and the coarsening models are correctly specified, then under regularity conditions and provided $\Pr(\Delta = 1|\boldsymbol{Z}) > \sigma$ w.p. 1 for some $\sigma > 0$,

$$\sqrt{N}\{(\widehat{\boldsymbol{\alpha}}(d, h), \widehat{\boldsymbol{\beta}}(d, h))' - (\boldsymbol{\alpha}^*, \boldsymbol{\beta}^*)'\} \to N(0, \Sigma(\boldsymbol{\alpha}^*, \boldsymbol{\beta}^*; d, h)),$$

where $(\boldsymbol{\alpha}^*, \boldsymbol{\beta}^*)$ are the true parameter values and

$$\Sigma(\boldsymbol{\beta}, \boldsymbol{\alpha}; d, h) = \{I(\boldsymbol{\beta}, \boldsymbol{\alpha}; d, h)\}^{-1} \Gamma(\boldsymbol{\beta}, \boldsymbol{\alpha}; d, h) \{I(\boldsymbol{\beta}, \boldsymbol{\alpha}; d, h)\}^{-1'},$$

with

$$I(\boldsymbol{\beta}, \boldsymbol{\alpha}; d, h) = \partial E\{s(\boldsymbol{O}, \boldsymbol{\beta}, G_\alpha; d, h)\}/\partial(\boldsymbol{\alpha}, \boldsymbol{\beta})', \quad \Gamma(\boldsymbol{\beta}, \boldsymbol{\alpha}; d, h) \equiv E\{s(\boldsymbol{O}, \boldsymbol{\beta}, G_\alpha; d, h)^{\otimes 2}\},$$

and, by convention, $A^{\otimes 2} = AA'$ for any matrix A. A sandwich variance estimator is obtained by replacing population means in the factors defining $\Sigma(\boldsymbol{\beta}, \boldsymbol{\alpha}; d, h)$ with their empirical means and evaluating the expressions at $(\boldsymbol{\alpha}, \boldsymbol{\beta}) = (\widehat{\boldsymbol{\alpha}}(d, h), \widehat{\boldsymbol{\beta}}(d, h))$. The regularity conditions include the non-trivial requirement that $E_G\{a(O, h_1; G_\alpha)\}/\partial\boldsymbol{\alpha}'|_{\boldsymbol{\alpha}=\boldsymbol{\alpha}^*}$ be positive definite for at least one subvector h_1 of h of the same dimension as $\boldsymbol{\alpha}$. In some non-CAR models it is impossible to meet this requirement because even though $\boldsymbol{\alpha}$ is identified, it is not estimable at rate $\sqrt{N}$. Rotnitzky and Robins (1997) provide examples of such models.

One salient feature of the IPWA methodology is the fact that the class of all IPWA estimators is comprised, up to asymptotic equivalence, of all consistent and asymptotically normal estimators of $(\boldsymbol{\alpha}, \boldsymbol{\beta})$. In particular, for MCD structures, it follows from the theory of Robins and Rotnitzky (1992) for CAR models and of Rotnitzky and Robins (1997) for non-CAR models that any consistent and asymptotically normal (more precisely, regular asymptotically linear) estimator $(\widehat{\boldsymbol{\beta}}, \widehat{\boldsymbol{\alpha}})$ of $(\boldsymbol{\beta}, \boldsymbol{\alpha})$ under model $(\mathcal{M}_{\text{full}}, \mathcal{G})$ computed from a random sample of $\boldsymbol{O}$ has asymptotic variance equal to $N(0, \Sigma(\boldsymbol{\alpha}^*, \boldsymbol{\beta}^*; d, h))$ for some d and h and, consequently, has the same asymptotic distribution as $(\widehat{\boldsymbol{\alpha}}(d, h), \widehat{\boldsymbol{\beta}}(d, h))$. An important practical implication of this result is that one's search for efficient estimators that maximally exploit the information in the observed data about the parameter β may be restricted to IPWA estimators, since as one varies h and d one will certainly find an IPWA estimator with the same asymptotic variance as that of any other asymptotically normal estimator of $\boldsymbol{\beta}$.

20.7 Efficiency considerations

The choice of d and h impacts the efficiency with which $(\boldsymbol{\alpha}, \beta)$ is estimated. In general, the optimal choices of d and h in the sense of minimizing the variance of the limiting distribution of $(\widehat{\boldsymbol{\alpha}}(d, h), \widehat{\boldsymbol{\beta}}(d, h))$ are not available for data analysis because they depend on unknown, usually infinite-dimensional, parameters like conditional means or variances. One possible strategy to overcome this difficulty would be to estimate the unknown infinite-dimensional parameters under dimension-reducing working submodels for them and then to compute the solutions to the IPWA estimating equations that use the estimated optimal choices of d and h. The resulting estimators of $(\boldsymbol{\alpha}, \boldsymbol{\beta})$ would (i) be consistent and asymptotically normal regardless of whether or not the working submodels are correct and (ii) have the smallest asymptotic variance among all IPWA estimators if the working models are correct. Estimators satisfying properties (i) and (ii) are referred to as locally efficient IPWA (LEIPWA) estimators at the working submodels in the class of all IPWA estimators. In essence, the philosophy behind local IPWA estimation is the same as that of the model-assisted estimators described in Section 20.2. The working models *assist* the data analyst in coming up with clever choices for the augmentation term and the full-data estimating function but they do not jeopardize the validity of inferences. The ease with which LEIPWA estimators can be implemented in practice depends on whether or not closed-form expressions for the optimal choices of d and h exist. In this section we discuss the existence of closed-form expressions for these functions separately for CAR and non-CAR models and provide an overview of LEIPWA estimation and alternative strategies to achieve estimators with reasonable efficiency properties when LEIPWA estimation cannot be easily implemented.

20.7.1 Efficient IPWA estimation in CAR models

It follows from (20.3) that in CAR models $\boldsymbol{\alpha}$ enters the likelihood based on the observed data only through the factor $\mathcal{L}_{2,N}(G_{\alpha})$. The ith contribution to this factor is $f_{\boldsymbol{O}|\boldsymbol{Z}}(\boldsymbol{O}_i|\boldsymbol{Z}_i; \boldsymbol{\alpha})$ (Gill, van der Laan, and Robins, 1997) which, under CAR, is by definition a function of $\boldsymbol{O}_i$ only. Noting that $d \log f_{\boldsymbol{O}|\boldsymbol{Z}}(\boldsymbol{O}_i|\boldsymbol{Z}_i; \boldsymbol{\alpha})/d\boldsymbol{\alpha}$ is the score for $\boldsymbol{\alpha}$ based on the observation $(\boldsymbol{Z}_i, \boldsymbol{O}_i)$ and recalling that conditional scores (evaluated at the true parameters) have conditional mean zero, we conclude that $d \log f_{\boldsymbol{O}|\boldsymbol{Z}}(\boldsymbol{O}_i|\boldsymbol{Z}_i; \boldsymbol{\alpha})/d\boldsymbol{\alpha}$ is a function of the observed data that has mean zero given the full data $\boldsymbol{Z}_i$ when the true coarsening mechanism is G_{α}. We conclude that $d \log f_{\boldsymbol{O}|\boldsymbol{Z}}(\boldsymbol{O}|\boldsymbol{Z}; \boldsymbol{\alpha})/d\boldsymbol{\alpha}$ must be an augmentation term $a(\boldsymbol{O}, h_{\alpha,\text{opt}}; G_{\alpha})$ for some function $h_{\alpha,\text{opt}}$. Thus, the maximum likelihood estimator $\widehat{\boldsymbol{\alpha}}_{\text{eff}}$ of $\boldsymbol{\alpha}$ is indeed an IPWA estimator solving $\sum_{i=1}^{N} a(\boldsymbol{O}, h_{\alpha,\text{opt}}; G_{\alpha}) = 0$. This in turn implies that the IPWA estimator of β with smallest asymptotic variance must solve

$$\sum_{i=1}^{N} u\{d(\boldsymbol{Z}_i; \boldsymbol{\beta}); G_{\widehat{\alpha}_{\text{eff}}}\} + a(\boldsymbol{O}_i, h; G_{\widehat{\alpha}_{\text{eff}}}) = 0 \qquad (20.10)$$

for specific vector-valued functions $d = d_{\beta,\text{opt}}$ and $h = h_{\beta,\text{opt}}$ of the same dimension as $\boldsymbol{\beta}$. To derive these functions one first derives for each fixed d, the optimal choice h^d of h that results in the solution $\widehat{\widehat{\boldsymbol{\beta}}}(d,h)$ of (20.10) with the smallest asymptotic variance. Then, one finds $d_{\beta,\text{opt}}$ as the d that yields $\widehat{\widehat{\boldsymbol{\beta}}}(d,h^d)$ with the smallest variance. The optimal choices for the functions d and h are $d_{\beta,\text{opt}}$ and $h^{d_{\beta,\text{opt}}}$.

An application of the Cauchy–Schwarz inequality shows that the optimal choice h^d is uniquely determined by the condition

$$E_{F_Z,G}\{s(\boldsymbol{O},\boldsymbol{\beta},G;d,h^d)a(\boldsymbol{O},h;G)\} = 0, \quad \text{for all } h, \tag{20.11}$$

where $\boldsymbol{\beta}$ is evaluated at $\boldsymbol{\beta}(F_{\boldsymbol{Z}})$ and h^d and h are vector-valued functions of the same dimension as $\boldsymbol{\beta}$. Closed-form expressions for h^d do not generally exist for non-monotone missing-data patterns. However, the choice h^d satisfying the preceding condition exists in closed form with MMCD structures; this function is easily written in terms of the optimal functions b^d corresponding to the alternative expression $a^*(\boldsymbol{O},b^d;G)$ of the augmentation term $a(\boldsymbol{O},h^d;G)$ as defined in (20.8). Specifically, Robins and Rotnitzky (1992) showed that $b_j^d(\overline{\boldsymbol{Z}}_{j-1}) = b_{\boldsymbol{\beta}(F_{\boldsymbol{Z}}),j}^d(\overline{\boldsymbol{Z}}_{j-1})$ where, for each $\boldsymbol{\beta}$,

$$b_{\boldsymbol{\beta},j}^d(\overline{\boldsymbol{Z}}_{j-1}) = \frac{\mu_j(\overline{\boldsymbol{Z}}_{j-1};\boldsymbol{\beta})}{\pi_{j,G}(\overline{\boldsymbol{Z}}_{j-1})}, \quad j = 0,\dots,n,$$

with $\pi_{j,G}(\overline{\boldsymbol{Z}}_{j-1}) = \prod_{k=0}^{j}\{1 - \lambda_G(k|\overline{\boldsymbol{Z}}_{k-1})\}$ and $\mu_j(\overline{\boldsymbol{Z}}_{j-1};\boldsymbol{\beta}) = E_{F_Z}\{d(\boldsymbol{Z};\boldsymbol{\beta})|\overline{\boldsymbol{Z}}_{j-1}\}$. The functions $\mu_j(\overline{\boldsymbol{Z}}_{j-1};\boldsymbol{\beta})$ satisfy, for $j = n,\dots,0$, the identities

$$\mu_j(\overline{\boldsymbol{Z}}_{j-1};\boldsymbol{\beta}) = E_{F_Z}\{\mu_{j+1}(\overline{\boldsymbol{Z}}_j;\boldsymbol{\beta})|\overline{\boldsymbol{Z}}_{j-1}\} \tag{20.12}$$

$$= E\{\mu_{j+1}(\overline{\boldsymbol{Z}}_j;\boldsymbol{\beta})|D > j,\overline{\boldsymbol{Z}}_{j-1}\}, \tag{20.13}$$

where $\mu_{n+1}(\overline{\boldsymbol{Z}}_n;\boldsymbol{\beta})$ is set to $d(\boldsymbol{Z};\boldsymbol{\beta})$. Equality (20.12) is true by definition of μ_j and double conditional expectations, and (20.13) holds under CAR. As illustrated in Example 1(a) to follow, (20.13) is exploited for calculation of locally efficient IPWA estimators of $\boldsymbol{\beta}$.

Results in Newey and McFadden (1994) imply that solving the IPWA estimating equations for $\boldsymbol{\beta}$ using b^d (which depends on the unknown $\boldsymbol{\beta}(F_{\boldsymbol{Z}})$) or $b_{\boldsymbol{\beta}}^d$ yields solutions with the same asymptotic distribution. That is, consider the equation

$$\sum_{i=1}^{N} u\{d(\boldsymbol{Z}_i;\boldsymbol{\beta});G_{\widehat{\alpha}_{\text{eff}}}\} + a^*(\boldsymbol{O}_i,b_{\boldsymbol{\beta}}^d;G_{\widehat{\alpha}_{\text{eff}}}) = 0 \tag{20.14}$$

in which the unknown $\boldsymbol{\beta}$ appears not only in the IPW term but also in the augmentation term through the function $b_{\boldsymbol{\beta}}^d$. Then under regularity conditions, Equation (20.14) has a solution $\widetilde{\boldsymbol{\beta}}(d,b_{\boldsymbol{\beta}}^d)$ with the same limiting distribution as $\widehat{\widehat{\boldsymbol{\beta}}}(d,h^d)$.

The estimating equations (20.14) are not available for data analysis because the functions $b_{\boldsymbol{\beta}}^d$ depend on the unknown full-data distribution $F_{\boldsymbol{Z}}$ through the unknown conditional expectations (20.12). Nevertheless, the following two-stage procedure produces estimators $\widetilde{\widehat{\boldsymbol{\beta}}}(d,\widehat{b}_{\boldsymbol{\beta}}^d)$ with the same asymptotic distribution as $\widetilde{\boldsymbol{\beta}}(d,b_{\boldsymbol{\beta}}^d)$ under working models for the unknown functions $\mu_j(\overline{\boldsymbol{Z}}_{j-1};\boldsymbol{\beta})$.

Stage 1. Postulate working models

$$E_{F_Z}\{d(\boldsymbol{Z};\boldsymbol{\beta})|\overline{\boldsymbol{Z}}_{j-1}\} = \mu_j(\overline{\boldsymbol{Z}}_{j-1};\boldsymbol{\beta},\boldsymbol{\tau}_j), \quad j = 0,\dots,n. \tag{20.15}$$

For $j = n,\dots,0$ and each $\boldsymbol{\beta}$, recursively compute the (possibly non-linear) least-squares estimator $\widehat{\tau}_j(\boldsymbol{\beta})$ of $\boldsymbol{\tau}_j$ in the regression model $\mu_j(\overline{\boldsymbol{Z}}_{j-1};\boldsymbol{\beta},\boldsymbol{\tau}_j)$ with outcome $\mu_{j+1}(\overline{\boldsymbol{Z}}_j;\boldsymbol{\beta},\widehat{\tau}_{j+1}(\boldsymbol{\beta}))$

and covariate $\overline{\mathbf{Z}}_{j-1}$ based on units with $D > j$. Compute

$$\widehat{b}_{\boldsymbol{\beta},j}^d(\overline{\mathbf{Z}}_{j-1}) = \mu_j(\overline{\mathbf{Z}}_{j-1}; \boldsymbol{\beta}, \widehat{\boldsymbol{\tau}}_{\boldsymbol{j}}(\boldsymbol{\beta}))/\pi_{j,G_{\widehat{\alpha}_{\text{eff}}}}(\overline{\mathbf{Z}}_{j-1}).$$

Stage 2. Compute $\widetilde{\widetilde{\boldsymbol{\beta}}}(d, \widehat{b}_{\boldsymbol{\beta}}^d)$, solving (20.14) with $\widehat{b}_{\boldsymbol{\beta}}^d$ instead of $b_{\boldsymbol{\beta}}^d$.

It follows from the identity (20.13) that the parameter estimators $\widehat{\tau}_j(\boldsymbol{\beta})$ consistently estimate τ_j when the models (20.15) are correctly specified. Consequently, under regularity conditions, $\widetilde{\widetilde{\boldsymbol{\beta}}}(d, \widehat{b}_{\boldsymbol{\beta}}^d)$ has the same asymptotic distribution as $\widetilde{\widetilde{\boldsymbol{\beta}}}(d, b_{\boldsymbol{\beta}}^d)$ (and hence as $\widehat{\widehat{\boldsymbol{\beta}}}(d, h^d)$) when the working models (20.15) are correctly specified, and remain consistent and asymptotically normal even if the working models are misspecified.

Examples 1(a) and 1(b) (continued). Suppose that CAR holds. Recall that in these examples, $d(\mathbf{Z}; \boldsymbol{\beta}) = m(X; \boldsymbol{\beta})'\{\mathbf{Y} - g(X; \boldsymbol{\beta})\}$.

In Example 1(a), X is always observed and is indeed a component of $\mathbf{Z}_{-1}$ and, consequently, of $\overline{\mathbf{Z}}_{j-1}$ for all j. Therefore, $\mu_j(\overline{\mathbf{Z}}_{j-1}; \boldsymbol{\beta}) = m(X; \boldsymbol{\beta})'\{\mu_j^*(\overline{\mathbf{Z}}_{j-1}) - g(X; \boldsymbol{\beta})\}$, where for each j, $\mu_j^*(\overline{\mathbf{Z}}_{j-1})$ satisfies the recursive relationship (20.13) but with μ^* instead of μ everywhere and with $\mu_n^*(\overline{\mathbf{Z}}_{n-1}) = E(\mathbf{Y}|D > n, \overline{\mathbf{Z}}_{n-1})$. We thus obtain that $u\{d(\mathbf{Z}; \boldsymbol{\beta}); G\} + a^*(\mathbf{O}, b_{\boldsymbol{\beta}}^d; G) = m(X; \boldsymbol{\beta})'\varepsilon(\boldsymbol{\beta})$ where

$$\varepsilon(\boldsymbol{\beta}) = \Delta(\mathbf{Y} - g(X; \boldsymbol{\beta}))/\Pr_G(\Delta = 1|\mathbf{Z})$$
$$+ \sum_{j=0}^n I(D > j - 1)\{I(D = j) - \lambda_G(j|\overline{\mathbf{Z}}_{j-1})\}\{\mu_j^*(\overline{\mathbf{Z}}_{j-1}) - g(X; \boldsymbol{\beta})\}. \quad (20.16)$$

Now, to compute the estimators $\widetilde{\widetilde{\boldsymbol{\beta}}}(d, \widehat{b}_{\boldsymbol{\beta}}^d)$ we postulate working models

$$E_{F_Z}\{\mu_{j+1}^*(\overline{\mathbf{Z}}_j)|\overline{\mathbf{Z}}_{j-1}\} = \mu_j^*(\overline{\mathbf{Z}}_{j-1}; \boldsymbol{\tau_j}), \quad j = 0, \dots, n, \quad (20.17)$$

where $\mu_{n+1}^*(\overline{\mathbf{Z}}_n) = \mathbf{Y}$. For $j = n, \dots, 0$, we recursively compute estimators $\widehat{\boldsymbol{\tau}}_{\boldsymbol{j}}$ by fitting, to the units with $\overline{\mathbf{Z}}_j$ observed, the regression models $\mu_j^*(\overline{\mathbf{Z}}_{j-1}; \boldsymbol{\tau_j})$ with outcomes $\mu_{j+1}^*(\overline{\mathbf{Z}}_j; \widehat{\boldsymbol{\tau}}_{\boldsymbol{j+1}})$ and covariates $\overline{\mathbf{Z}}_{j-1}$. Next we define $\widehat{\varepsilon}(\boldsymbol{\beta})$ like $\varepsilon(\boldsymbol{\beta})$ but with $\Pr_G(\Delta = 1|\mathbf{Z})$, $\lambda_G(j|\overline{\mathbf{Z}}_{j-1})$, and $\mu_j^*(\overline{\mathbf{Z}}_{j-1})$ replaced with

$$\Pr_{G_{\widehat{\alpha}_{\text{eff}}}}(\Delta_i = 1|\mathbf{Z}_i), \quad \lambda_{G_{\widehat{\alpha}_{\text{eff}}}}(j|\overline{\mathbf{Z}}_{j-1,i}), \quad \mu_j^*(\overline{\mathbf{Z}}_{j-1,i}; \widehat{\boldsymbol{\tau}}_{\boldsymbol{j}}).$$

Finally, we solve $\sum_{i=1}^N m(X_i; \boldsymbol{\beta})'\widehat{\varepsilon}_i(\boldsymbol{\beta}) = 0$ and obtain $\widetilde{\widetilde{\boldsymbol{\beta}}}(d, \widehat{b}_{\boldsymbol{\beta}}^d)$.

In Example 1(b), to define the censoring variable D, we must reorder the components of $\mathbf{Z}$ so that its last component corresponds to baseline vitamin A, the only variable missing in some subjects. We can do so, for example, by letting $\mathbf{Z}_{-1}$ be the always observed baseline covariates, $\mathbf{Z}_{j-1}$ be the variables $(Y_j, \mathbf{W}_j)$ recorded at time $t_j, j = 1, \dots, n$, and $\mathbf{Z}_n$ be baseline vitamin A. Since only $\mathbf{Z}_n$ can be missing, then D takes only the values n (when vitamin A is missing) or $n + 1$ (when vitamin A is recorded). Thus, the form of the augmentation terms simplifies to $a^*(\mathbf{O}, b; G) = \{I(D = n) - \lambda_G(n|\overline{\mathbf{Z}}_{n-1})\}b_n(\overline{\mathbf{Z}}_{n-1})$ and the optimal $b_{\boldsymbol{\beta},n}^d(\overline{\mathbf{Z}}_{n-1})$ is equal to $E\{m(X; \boldsymbol{\beta})'(\mathbf{Y} - g(X; \boldsymbol{\beta}))|D = n + 1, \overline{\mathbf{Z}}_{n-1}\}$.

The optimal full-data unbiased estimating function $d_{\boldsymbol{\beta},\text{opt}}$ solves an integral equation that depends on the semi-parametric efficient score for $\boldsymbol{\beta}$ in the full-data model $\mathcal{M}_{\text{full}}$ (Robins and Rotnitzky, 1992; Robins, Rotnitzky, and Zhao, 1994). This integral equation may or may not have a closed-form solution depending on the full-data model, and on the coarsened data configuration. For example, in the following we provide closed-form expressions for $d_{\boldsymbol{\beta},\text{opt}}$ in Example 1(a). These expressions exist because the full-data model is a marginal mean model, the coarsened data patterns are monotone and the covariates of the marginal mean model are always observed. In contrast, closed-form expressions for $d_{\beta,\mathrm{opt}}$ do not exist in Example 1(b) because the missing variables are covariates of the marginal mean model. Robins, Rotnitzky, and Zhao (1994) provide the integral equation for $d_{\beta,\mathrm{opt}}$ in this problem.

Example 1(a) (continued). Assuming as before that CAR and monotonicity hold, Robins and Rotnitzky (1995) showed that $d_{\beta,\mathrm{opt}}(Z;\beta) = m_{\beta,\mathrm{opt}}(X;\beta)'\{Y - g(X;\beta)\}$, where

$$m_{\beta,\mathrm{opt}}(X;\beta) = E\left\{\frac{\partial}{\partial\beta'}\varepsilon(\beta)\,\middle|\,X\right\}\mathrm{Var}\{\varepsilon(\beta)|X\}^{-1}$$

$$= -\left\{\frac{\partial g(X;\beta)}{\partial\beta'}\right\}\mathrm{Var}\{\varepsilon(\beta)|X\}^{-1},$$

with $\varepsilon(\beta)$ defined in (20.16).

In situations in which $d_{\beta,\mathrm{opt}}$ and $h_{\beta,\mathrm{opt}}$ exist in closed form, one can compute simple (i.e., non-recursive) LEIPWA estimators of β. When $d_{\beta,\mathrm{opt}}$ and/or $h_{\beta,\mathrm{opt}}$ do not exist in closed form, it is still possible to derive algorithms for computing LEIPWA estimators, but they are difficult to implement because they require complicated and computationally demanding recursive algorithms to numerically approximate the solutions of integral equations.

Example 1(a) (continued). We postulate, in addition to model (20.17) above, the working model

$$\mathrm{Var}\{\varepsilon(\beta)|X\} = v(X;\phi), \tag{20.18}$$

where $v(\cdot;\cdot)$ is a known function and ϕ is an unknown parameter vector. We compute $\widehat{\phi}$ by fitting the regression model $v(X;\phi)$ with outcomes $\widehat{\varepsilon}(\beta)^2$ and covariates X. Finally, we compute $\widehat{\beta}_{\mathrm{loc.eff}}$, the solution of

$$\sum_{i=1}^{n}\left\{\frac{\partial g(X_i;\beta)}{\partial\beta'}\right\}v(X_i;\widehat{\phi})\widehat{\varepsilon}_i(\beta) = 0.$$

The estimator $\widehat{\beta}_{\mathrm{loc.eff}}$ is a locally efficient IPWA estimator at the working submodels (20.17) and (20.18).

20.7.2 Efficient IPWA estimation in non-CAR models

Implementation of algorithms leading to locally efficient estimators in models for longitudinal MCD structures with non-CAR data is difficult. One source of difficulty, though not the only one, is the fact that their computation requires finding the solution h^d to Equation (20.11) for d and h vector-valued and of the same dimension as (α,β) (Rotnitzky and Robins, 1997). In non-CAR models for longitudinal MMCD data in which the probability that the component Z_j is missing depends on Z only through $\overline{Z}_{j+1}$, closed-form expressions for h^d exist but they are messy to compute. In other non-CAR models for longitudinal MCD structures, h^d does not generally exist in closed form. In non-CAR models for cross-sectional data, h^d has a simple closed-form expression. Rotnitzky and Robins (1997) have used this expression to derive LEIPWA estimators of regression parameters with missing outcomes.

A computationally simple algorithm can be implemented to obtain estimators of α and β with good efficiency properties in a special family of non-CAR models for monotone data that is specially suitable for conducting sensitivity analysis (Robins, Rotnitzky, and Scharfstein, 2000). Each model in the class is an extension of the model described in Example 3 for cross-sectional data and assumes, possibly in addition to a model $\mathcal{M}_{\mathrm{full}}$ for the full data

$F_{\mathbf{Z}}$, that the missing-data mechanism is restricted by the condition

$$\log\left\{\frac{\Pr(D > j | D > j - 1, \overline{\mathbf{Z}}_n)}{\Pr(D = j | D > j - 1, \overline{\mathbf{Z}}_n)}\right\} = \eta_j(\overline{\mathbf{Z}}_{j-1}; \boldsymbol{\alpha}) + r_j(\overline{\mathbf{Z}}_{j-1}, \underline{\mathbf{Z}}_j)$$

where the function $r_j(\overline{\mathbf{Z}}_{j-1}, \underline{\mathbf{Z}}_j)$ is a known function of the data $\overline{\mathbf{Z}}_{j-1}$ observed when $D = j$ and the unobserved data $\underline{\mathbf{Z}}_j = (\mathbf{Z}_j, \dots, \mathbf{Z}_n)$, the function $\eta_j(\overline{\mathbf{Z}}_{j-1}; \cdot)$ is known and the finite-dimensional parameter $\boldsymbol{\alpha}$ is unknown. Thus, each model in the family is determined by a specific parametric model for the dropout probabilities indexed by $\boldsymbol{\alpha}$ and a non- or semi-parametric model for the full data. In practice these models are used to repeatedly conduct inference about $\boldsymbol{\beta}$, varying the functions r_j, $j = 1, \dots, n$, in a range that includes the nil functions so as to determine the sensitivity of inferences to departures from the CAR assumption.

In these models one possible IPWA estimation strategy is to choose the augmentation term and the IPW terms so as to yield a LEIPWA estimator of $(\boldsymbol{\alpha}, \boldsymbol{\beta})$ when the functions r_j are all equal to 0. This strategy should yield reasonably efficient estimators if the functions r_j encode non-response processes that are not too far from CAR. Under this strategy, $\boldsymbol{\alpha}$ is estimated with the solution $\widehat{\boldsymbol{\alpha}}$ of $\sum_{i=1}^{N} a(\mathbf{O}_i, h_{\alpha}^*; G_{\alpha}) = 0$ where $h_{\alpha,j}^*(\overline{\mathbf{Z}}_{j-1})$ are the functions that result in the maximum likelihood estimator of $\boldsymbol{\alpha}$ when r_j are the nil functions. Specifically,

$$h_{\boldsymbol{\alpha},j}^*(\overline{\mathbf{Z}}_{j-1}) = \sum_{l=0}^{j-1} \text{expit}\left\{-\eta_l(\overline{\mathbf{Z}}_{l-1}; \boldsymbol{\alpha})\right\} b_{\alpha,l}^*(\overline{\mathbf{Z}}_{l-1}) + \left[1 - \text{expit}\left\{-\eta_j(\overline{\mathbf{Z}}_{j-1}; \boldsymbol{\alpha})\right\}\right] b_{\alpha,j}^*(\overline{\mathbf{Z}}_{j-1})$$

where

$$b_{\boldsymbol{\alpha},j}^*(\overline{\mathbf{Z}}_{j-1}) = \frac{\partial}{\partial \boldsymbol{\alpha}'} \log\left\{\frac{\eta_j(\overline{\mathbf{Z}}_{j-1}; \boldsymbol{\alpha})}{1 - \eta_j(\overline{\mathbf{Z}}_{j-1}; \boldsymbol{\alpha})}\right\}$$

and $\text{expit}(x) = \{1 + \exp(-x)\}^{-1}$.

The parameter $\boldsymbol{\beta}$ is estimated as the solution of $\sum_{i=1}^{N} s(\mathbf{O}, \boldsymbol{\beta}, G_{\widehat{\boldsymbol{\alpha}}}; \widehat{d}, \widehat{h}_{\boldsymbol{\beta}}^{\widehat{d}}) = 0$ where $\widehat{d}$ is an estimator of some user-specified full-data estimating function, preferably the optimal function $d_{\boldsymbol{\beta},\text{opt}}$ under CAR if this exists in closed form. The function $\widehat{h}_{\boldsymbol{\beta}}^{\widehat{d}}$ is computed for each j as

$$\widehat{h}_{\boldsymbol{\beta},j}^{\widehat{d}}(\overline{\mathbf{Z}}_{j-1}) = \sum_{l=0}^{j-1} \text{expit}\left\{-\eta_l(\overline{\mathbf{Z}}_{l-1}; \widehat{\boldsymbol{\alpha}})\right\} \widehat{b}_{\boldsymbol{\beta},l}^{\widehat{d}}(\overline{\mathbf{Z}}_{l-1}) + \left[1 - \text{expit}\left\{-\eta_l(\overline{\mathbf{Z}}_{j-1}; \widehat{\boldsymbol{\alpha}})\right\}\right] \widehat{b}_{\boldsymbol{\beta},j}^{\widehat{d}}(\overline{\mathbf{Z}}_{j-1})$$

where $\widehat{b}_{\boldsymbol{\beta},j}^{\widehat{d}}(\overline{\mathbf{Z}}_{j-1})$ are the estimators, computed as in the two-stage procedure of the previous subsection, of the optimal functions $b_{\boldsymbol{\beta},j}^{\widehat{d}}$ associated with $\widehat{d}$.

20.7.3 Approximating the efficient IPWA function by optimal combination of several IPWA functions

An idea that dates back to Hansen (1982) can be applied quite generally when locally efficient estimators are computationally difficult. Given a finite number, say J, of different $(q + p) \times 1$ IPWA estimating functions, $s(\mathbf{O}, \boldsymbol{\beta}, G_{\boldsymbol{\alpha}}; d_j, h_j)$, $j = 1, \dots, J$ (where p is the dimension of $\boldsymbol{\beta}$ and q is the dimension of $\boldsymbol{\alpha}$), one can consider solving the estimating equation

$$\sum_{i=1}^{N} \sum_{j=1}^{J} \omega_j s(\mathbf{O}_i, \boldsymbol{\beta}', G_{\boldsymbol{\alpha}'}; d_j, h_j) = 0 \qquad (20.19)$$

where ω_j is an arbitrary $(q+p) \times (q+p)$ matrix. Each choice of matrices $\{\omega_j : j = 1, \ldots, J\}$ yields an IPWA estimator $(\widehat{\boldsymbol{\alpha}}^\omega, \widehat{\boldsymbol{\beta}}^\omega)$. Hansen showed that among all $(\widehat{\boldsymbol{\alpha}}^\omega, \widehat{\boldsymbol{\beta}}^\omega)$, the one with the smallest asymptotic variance has $\omega_{\mathrm{opt}} = (\omega_1, \ldots, \omega_J)$ equal to

$$\omega_{\mathrm{opt}} = \frac{\partial}{\partial(\boldsymbol{\alpha}', \boldsymbol{\beta}')'} E\{S(\boldsymbol{\alpha}, \boldsymbol{\beta})\}\Big|_{=(\boldsymbol{\alpha}, \boldsymbol{\beta})} \times \mathrm{Var}[S(\boldsymbol{\alpha}, \boldsymbol{\beta})]^{-1}, \qquad (20.20)$$

where $S(\boldsymbol{\alpha}, \boldsymbol{\beta}) \equiv (S_1(\boldsymbol{\alpha}, \boldsymbol{\beta})', \ldots, S_J(\boldsymbol{\alpha}, \boldsymbol{\beta})')'$ and $S_j(\boldsymbol{\alpha}, \boldsymbol{\beta}) \equiv s(\boldsymbol{O}, \boldsymbol{\beta}, G_{\boldsymbol{\alpha}}; d_j, h_j)$. Consistent estimators $\widehat{\omega}_{\mathrm{opt}}$ of ω_{opt} can be obtained by replacing the mean and variance in (20.20) by their empirical versions evaluated at a preliminary (possibly inefficient) $\sqrt{N}$-consistent estimator of $(\boldsymbol{\alpha}, \boldsymbol{\beta})$. Results in Newey and McFadden (1994) imply that because the estimating function (20.19) is unbiased for any choice of matrices ω_j, then $(\widehat{\boldsymbol{\alpha}}^{\widehat{\omega}_{\mathrm{opt}}}, \widehat{\boldsymbol{\beta}}^{\widehat{\omega}_{\mathrm{opt}}})$ has the same asymptotic distribution as $(\widehat{\boldsymbol{\alpha}}^{\omega_{\mathrm{opt}}}, \widehat{\boldsymbol{\beta}}^{\omega_{\mathrm{opt}}})$. If the functions d_j and h_j are polynomials of increasing order, then with an appropriately chosen number J of estimating functions, the asymptotic variance of $(\widehat{\boldsymbol{\alpha}}^{\omega_{\mathrm{opt}}}, \widehat{\boldsymbol{\beta}}^{\omega_{\mathrm{opt}}})$ can approximate that of the most efficient IPWA estimator of $(\boldsymbol{\alpha}, \boldsymbol{\beta})$ to any desired degree. In practice, however, to avoid poor finite-sample performance one can only use small values of J, say $J = 2$ or 3. Newey (1993) discusses choosing J by cross-validation.

20.8 Doubly robust IPWA estimation in CAR models

In general, IPWA estimators of $\boldsymbol{\beta}$ fail to be consistent if the missing-data model $\mathcal{G}$ is incorrectly specified. However, the two-stage IPWA estimators $\widetilde{\boldsymbol{\beta}}(d, \widehat{b}_{\boldsymbol{\beta}}^d)$ of Section 20.7.1 have the following remarkable property: they are consistent and asymptotically normal if either the missing-data model $\mathcal{G}$ is correctly specified or if the working models (20.17) are correctly specified, but not necessarily both. This fact is a consequence of the following general result.

Let $h_{\boldsymbol{\beta}, F_{\boldsymbol{Z}}, G}^d$ denote the solution of (20.11) for given choices of $\boldsymbol{\beta}$, $F_{\boldsymbol{Z}}$, and G. Suppose that we specify a low-dimensional submodel $\mathcal{M}_{\mathrm{full,low}} = \{F_{\boldsymbol{Z};\tau} : \tau \in \Upsilon\}$ of $\mathcal{M}_{\mathrm{full}}$, and a low-dimensional submodel $\mathcal{G} = \{G_{\boldsymbol{\alpha}} : \boldsymbol{\alpha} \in \Omega\}$ of the model that just assumes CAR. Suppose that $\widehat{\tau}$ converges in probability to the true value of τ when model $\mathcal{M}_{\mathrm{full,low}}$ is correctly specified and $\widehat{\boldsymbol{\alpha}}$ converges in probability to the true value of $\boldsymbol{\alpha}$ when model $\mathcal{G}$ is correctly specified. Under regularity conditions, the IPWA estimator $\widehat{\boldsymbol{\beta}}_{DR} \equiv \widehat{\boldsymbol{\beta}}(d, h_{\boldsymbol{\beta}, F_{\boldsymbol{Z},\widehat{\tau}}, G_{\widehat{\boldsymbol{\alpha}}}}^d)$ solving

$$\sum_{i=1}^{N}\left[u\{d(\boldsymbol{Z}_i; \boldsymbol{\beta}); G_{\widehat{\boldsymbol{\alpha}}}\} + a\left(\boldsymbol{O}, h_{\boldsymbol{\beta}, F_{\boldsymbol{Z},\widehat{\tau}}, G_{\widehat{\boldsymbol{\alpha}}}}^d; G_{\widehat{\boldsymbol{\alpha}}}\right)\right] = 0 \qquad (20.21)$$

is consistent and asymptotically normal in the union model $\mathcal{M}_{\mathrm{full,low}} \cup \mathcal{G}$. That is to say, $\sqrt{N}(\widehat{\boldsymbol{\beta}}_{DR} - \boldsymbol{\beta})$ converges to a normal distribution if either model $\mathcal{M}_{\mathrm{full,low}}$ or model $\mathcal{G}$ is correctly specified, but not necessarily both. The estimator $\widehat{\boldsymbol{\beta}}_{DR}$ is referred to as a doubly robust estimator in the union model $\mathcal{M}_{\mathrm{full,low}} \cup \mathcal{G}$.

In the computation of the estimators $\widetilde{\boldsymbol{\beta}}(d, \widehat{b}_{\boldsymbol{\beta}}^d)$ of Section 20.7.1, the working models (20.17) determine the submodel $\mathcal{M}_{\mathrm{full,low}}$ and the assumed model for the missingness probabilities determines $\mathcal{G}$. The estimators $\widetilde{\boldsymbol{\beta}}(d, \widehat{b}_{\boldsymbol{\beta}}^d)$ solve precisely equations of the form (20.21) and are therefore doubly robust.

The consistency of $\widehat{\boldsymbol{\beta}}_{DR}$ under either model $\mathcal{M}_{\mathrm{full,low}}$ or $\mathcal{G}$ is essentially a consequence of the following remarkable property proved in Robins, Rotnitzky, and van der Laan (2000). In CAR models,

$$E_{F_Z, G}\left[u\{d(\boldsymbol{Z}; \boldsymbol{\beta}(F_{\boldsymbol{Z}})); G\} + a\left(\boldsymbol{O}, h_{\boldsymbol{\beta}(F_{\boldsymbol{Z}}), F_{\boldsymbol{Z}}^*, G}^d; G\right)\right] = 0, \quad \text{for all } F_{\boldsymbol{Z}}^*,$$

and

$$E_{F_Z,G}\left[u\{d(\boldsymbol{Z};\boldsymbol{\beta}(F_{\boldsymbol{Z}}));G^*\} + a\left(\boldsymbol{O},h^d_{\boldsymbol{\beta}(F_{\boldsymbol{Z}}),F_{\boldsymbol{Z}},G^*};G^*\right)\right] = 0, \quad \text{for all } G^*.$$

That is, if in a CAR model we calculate an IPWA function satisfying (20.11) and evaluate it under an incorrect coarsening mechanism G^* but under the correct F_Z, we obtain an unbiased estimating function for $\boldsymbol{\beta}$. Conversely, if we evaluate it under an incorrect full-data law F_Z but under the correct G, we also obtain an unbiased estimating function for $\boldsymbol{\beta}$.

20.9 A survey of the literature

The IPWA methodology has been applied to estimation in a variety of settings — not only those specific to the analysis of incomplete longitudinal data, but also notably estimation in causal inference models and in complex randomized and epidemiological designs, such as two-stage studies. The books by van der Laan and Robins (2003) and Tsiatis (2006) contain a comprehensive treatment of the IPWA methodology in CAR models. Here, we will give a review of a subset of the IPWA literature restricting attention to papers discussing IPWA methods for the analysis of incomplete repeated measures and right-censored failure time data in non-counterfactual longitudinal data models.

Robins and Rotnitzky (1992) and Robins, Rotnitzky, and Zhao (1994) constructed LEIPWA estimators of regression models with missing covariates that are missing at random. Robins, Rotnitzky, and Zhao (1995) and Robins and Rotnitzky (1995) derived LEIPWA estimators of parameters of marginal models for the means of incomplete repeated outcomes in CAR models.

Robins and Rotnitzky (1992), Robins (1993), and Robins and Finkelstein (2000) constructed LEIPWA estimators of the survival distribution of a right-censored failure time and of regression parameters of Cox proportional hazards models and accelerated failure time models for the conditional distribution of the failure time given baseline covariates. Robins (1996) constructed IPWA estimators of median regression models for right-censored failure time data and Robins, Rotnitzky, and Zhao (1994) and Nan, Emond, and Wellner (2004) described IPWA estimation of Cox proportional hazards regression parameters with missing covariates. Hu and Tsiatis (1996) used the IPWA methodology to construct estimators of a survival function from right-censored data subject to reporting delays. Zhao and Tsiatis (1997, 1999, 2000) and Van der Laan and Hubbard (1999) constructed IPWA estimators of the quality-of-life-adjusted survival time distribution from right-censored failure time data. Bang and Tsiatis (2000, 2002) and Strawderman (2000) derived IPWA estimators of a median regression model for medical costs from right-censored data and of the mean of an increasing stochastic process, respectively. Van der Laan, Hubbard, and Robins (2002) and Quale, van der Laan, and Robins (2006) constructed LEIPWA estimators of a multivariate survival function when failure times are subject to common censoring and to failure-time-specific censoring, respectively. In the same setting, Keles, van der Laan, and Robins (2004) derived IPWA estimators that are easier to compute and almost as efficient as the Quale, van der Laan, and Robins (2006) estimators. All the preceding papers assumed that censoring is informative but explainable by the past recorded data and hence CAR.

In a series of papers, Rotnitzky, Robins, and Scharfstein (1998), Robins, Rotnitzky, and Scharfstein (2000), Scharfstein, Rotnitzky, and Robins (1999), Scharfstein and Irrizarry (2003), and Birmingham, Rotnitzky, and Fitzmaurice (2003) applied the theory in Rotnitzky and Robins (1997) to develop IPWA estimators in non-CAR models for the analysis of measured outcomes (repeated or measured at the end of the longitudinal study) with monotone missing-data patterns. Lin, Scharfstein, and Rosenheck (2003) and Vansteelandt, Rotnitzky, and Robins (2007) described IPWA methods for the analysis of longitudinal studies with intermittent non-response. Scharfstein et al. (2001) and Scharfstein and Robins (2002) extended the IPWA methodology to allow estimation of the marginal survivor function of a discrete and continuous right-censored failure time T respectively assuming censoring is informative and non-explainable by measured covariates.

A fairly complete theory exists about doubly robust estimation in CAR models. The book by van der Laan and Robins (2003) and the discussion article by Robins, Rotnitzky, and van der Laan (2000) present this theory. The doubly robust estimators of Example 2 and Example 3 are discussed in Scharfstein, Rotnitzky, and Robins (1999). This paper also demonstrates that a certain LEIPWA doubly robust estimator has an alternative "regression representation." This idea is further developed in Robins (2000) and Bang and Robins (2005), who describe an algorithm for constructing doubly robust estimators in longitudinal monotone missing data and causal inference models (more specifically, marginal mean models). This algorithm ultimately gives an estimator algebraically identical to a LEIPWA estimator. However, because it relies on a sequential regression representation of the estimator, it is easy to implement with off-the-shelf regression software. Further important papers describing doubly robust estimators in CAR models are Lipsitz, Ibrahim, and Zhao (1999), Lunceford and Davidian (2004), and Neugebauer and van der Laan (2005). Much less is known about doubly robust estimation in non-CAR models. Robins and Rotnitzky (2001) give the most comprehensive theory available up to date on doubly robust estimation in arbitrary non- and semi-parametric models, including non-CAR models. These authors show, for instance, that a doubly robust estimator in the non-CAR model of Example 3 exists only when $\Psi(x) = \log(x/(1-x))$.

20.10 Discussion: A look into the future

IPWA methods, and primarily, doubly robust LEIPWA methods, provide an appealing tool for analyzing incomplete high-dimensional longitudinal data. Although a great deal of progress has been made in the last 15 years in the theory and applications of the IPWA methodology, a number of issues remain unsolved. First, in CAR models for monotone data patterns, different IPWA-type procedures exist that provide doubly robust estimators with the same asymptotic properties (see Bang and Robins, 2005). The relative merits of these procedures in finite samples are unknown. Second, in many non-CAR models doubly robust IPWA estimators do not exist. It is still an open question to characterize the entire class of models in which doubly robust estimation is feasible. Third, in some models, such as CAR models with non-monotone data, doubly robust estimators could in principle be constructed, but their implementation is not clear. Fourth, and most importantly, doubly robust inference, even if feasible, does not entirely solve the dilemma posed by the curse of dimensionality. Doubly robust estimators fail to be consistent if, as is inevitable with high-dimensional data, both (i) the model for the coarsening mechanism and (ii) the model for the distribution of the full data are misspecified. To address this problem, Robins and colleagues (Robins, 2004, Section 9; Li et al., 2006; Tchetgen et al., 2006; Robins et al., 2007) have made important first steps in the development of a new theory of estimation based on higher-order influence functions. This new theory extends the theory of $\sqrt{N}$-consistent IPWA estimation and yields (a) consistent doubly robust estimators and (b) valid confidence intervals shrinking to zero with sample size, under weaker assumptions on the models (i) and (ii) than those required by $\sqrt{N}-$ consistent doubly robust estimators, at the price of convergence of β being at a rate slower than the usual $1/\sqrt{N}$ rate. This promising theory is just emerging and new developments and applications will no doubt emerge in the future.

Not much is available for the analysis of semi-parametric models of longitudinal studies with intermittent non-response. One key difficulty is that realistic models for the missingness mechanism are not obvious. As argued in Robins and Gill (1997) and Vaansteelandt,

Rotnitzky, and Robins (2007), the CAR assumption with non-monotone data patterns is hard to interpret and rarely realistic. A recent model discussed by Lin et al. (2003) and van der Laan and Robins (2003, Chapter 6) relies on a more plausible assumption about the missingness process which nonetheless assumes no selection on unobservables for the marginal distribution of the responses. This model can be viewed as a special case of the model presented in Vaansteelandt, Rotnitzky, and Robins (2007), which allows for selection on unobservables. More investigation into realistic, easy-to-interpret models for intermittent non-response is certainly needed. Aside from the difficulties in the formulation of realistic models for non-response, implementation of efficient IPWA methods is complicated with intermittent non-response by the fact that the functions h^d solving (20.11) do not exist in closed form. More investigation is needed in this area. Rather than being a nuisance, intermittent non-response in observational studies may offer the possibility of investigating questions of interest in clinical practice. Specifically, in clinical settings, it is often the case that at each clinic visit the physician makes decisions not only on the treatment but also on the timing of the next visit. Usually, the doctor does not stipulate a precise returning date but instead indicates that no more than a certain period can elapse before the patient returns to the clinic. Naturally, the patient is free to return earlier if he or she needs to do so. It is therefore of practical interest to develop methods for estimating the effects of regimes that specify the timing of the next clinic visit, allowing for the possibility of unplanned interim clinic visits. Observational studies are well suited for the investigation of these regimes because of the variability in clinic visit timing found in them. This variability exists because different physicians make different decisions in the face of similar patient histories. IPWA methodology may prove useful for developing adequate estimators of parameters of models for the effect of the timing of visits.

Acknowledgment

This work was supported by U.S. NIH grant R01-GM48704.

References

Bang, H. and Robins, J. (2005). Double-robust estimation in missing data and causal inference models. *Biometrics* **61**, 692–972.

Bang, H. and Tsiatis, A. A. (2000). Estimating medical cost with censored data. *Biometrika* **87**, 329–343.

Bang, H. and Tsiatis A. A. (2002). Median regression with censored medical cost data. *Biometrics* **58**, 643–650.

Bickel, P. J., Klaassen, C. A. J., Ritov, Y., and Wellner, J. A. (1993). *Efficient and Adaptive Inference in Semiparametric Models*. Baltimore, MD: Johns Hopkins University Press.

Birmingham, J., Rotnitzky, A., and Fitzmaurice, G. M. (2003). Pattern-mixture and selection models for analyzing monotone missing data. *Journal of the Royal Statistical Society, Series B* **65**, 275–297.

Cassel, C. M., Sarndal, C. E., and Wretman, J. H. (1976). Some results on generalized difference estimation and generalized regression estimation for finite populations. *Biometrika* **63**, 615–620.

Gill, R. D., van der Laan, M. J., and Robins, J. M. (1997). Coarsening at random: Characterizations, conjectures and counterexamples. In D. Y. Lin and T. R. Fleming (eds.), *Proceedings of the First Seattle Symposium in Biostatistics: Survival Analysis*, pp. 255–294. New York: Springer.

Hansen, L. P. (1982). Large sample properties of generalized method of moments estimators. *Econometrica* **50**, 1029–1054.

Hansen, M. H. and Hurwitz, W. N. (1943). On the theory of sampling from finite populations. *Annals of Mathematical Statistics* **14**, 333–362.

Heitjan, D. F. and Rubin, D. B. (1991). Ignorability and coarse data. *Annals of Statistics* **19**, 2244–2253.

Horvitz, D. G. and Thompson, D. J. (1952). A generalization of sampling without replacement from a finite universe. *Journal of the American Statistical Association* **47**, 663–685.

Hu, P. H. and Tsiatis, A. A. (1996). Estimating the survival function when ascertainment of vital status is subject to delay. *Biometrika* **83**, 371–380.

Isaki, C. T. and Fuller, W. A. (1982). Survey design under the regression superpopulation model. *Journal of the American Statistical Association* **77**, 89–96.

Jacobsen, M. and Keiding, N. (1995). Coarsening at random in general sample spaces and random censoring in continuous time. *Annals of Statistics* **23**, 774–786.

Keles, S., van der Laan, M. J., and Robins, J. (2004). Estimation of the bivariate survival function with right censored data structures. In N. Balakrishnan and C. R. Rao (eds.), *Advances in Survival Analysis: Handbook of Statistics, Volume 23*. Amsterdam: Elsevier North-Holland.

Li, L., Tchetgen, E., van der Vaart, A. W., and Robins, J. (2006). Robust inference with higher order influence functions. Part II. *Proceedings of the Joint Statistical Meetings. American Statistical Association.*

Liang, K.-Y. and Zeger, S. L. (1986). Longitudinal data analysis using generalized linear models. *Biometrika* **73**, 13–22.

Lin, H., Scharfstein, D. O., and Rosenheck, R. A. (2003). Analysis of longitudinal data with irregular, informative follow-up. *Journal of the Royal Statistical Society, Series B* **66**, 791–813.

Lipsitz, S. R., Ibrahim, J. G., and Zhao, L.P. (1999). Weighted estimating equation for missing covariate data with properties similar to maximum likelihood. *Journal of the American Statistical Association* **94**, 1147–1160.

Lunceford, J. K. and Davidian, M. (2004). Stratification and weighting via the propensity score in estimation of causal treatment effects: A comparative study. *Statistics in Medicine* **23**, 2937–2960.

Nan, B., Emond, M., and Wellner, J. (2004). Information bounds for Cox regression models with missing data. *Annals of Statistics* **32**, 723–753.

Neugebauer, R. and van der Laan, M. J. (2005). Why prefer double-robust estimates? *Journal of Statistical Planning and Inference* **129**, 405–426.

Newey, W. K. (1990). Semiparametric efficiency bounds. *Journal of Applied Econometrics* **5**, 99–135.

Newey, W. K. (1993). Efficient estimation of models with conditional moment restrictions. In G. S. Madala, C. R. Rao, and H. D. Vinod (eds.), *Econometrics: Handbook of Statistics, Volume 11.* Amsterdam: North Holland.

Newey, W. K. and McFadden, D. (1994). Large sample estimation and hypothesis testing. In D. McFadden and R. Engle (eds.), *Handbook of Econometrics, Volume 4.* Amsterdam: North-Holland.

Quale, C. M., van der Laan, M. J., and Robins, J. M. (2006). Locally efficient estimation with bivariate right censored data. *Journal of the American Statistical Association* **101**, 1076–1084.

Robins, J. M. (1986). A new approach to causal inference in mortality studies with sustained exposure periods — Application to control of the healthy worker survivor effect. *Mathematical Modelling* **7**, 1393–1512.

Robins, J. M. (1993). Information recovery and bias adjustment in proportional hazards regression analysis of randomized trials using surrogate markers. *Proceedings of the Biopharmaceutical Section*, pp. 24–33. Alexandria, VA: American Statistical Association.

Robins, J. M. (1996). Locally efficient median regression with random censoring and surrogate markers. In N. P. Jewell, A. C. Kimber, M. L. T. Lee, and G. A. Whitmore (eds.), *Lifetime Data: Models in Reliability and Survival Analysis*. Dordrecht: Kluwer Academic Publishers.

Robins, J. M. (1999). Marginal structural models versus structural nested models as tools for causal inference. In *Statistical Models in Epidemiology: The Environment and Clinical Trials*. M. E. Halloran and D. Berry, (eds.), IMA Volume 116. New York: Springer-Verlag, pp. 95–134.

Robins, J. M. (2000). Robust estimation in sequentially ignorable missing data and causal inference models. *Proceedings of the 1999 Joint Statistical Meetings.*

Robins, J. M. (2004). Optimal structural nested models for optimal sequential decisions. In *Proceedings of the Second Seattle Symposium in Biostatistics.* New York: Springer.

Robins, J. M. and Finkelstein, D. (2000). Correcting for non-compliance and dependent censoring in an AIDS clinical trial with inverse probability of censoring weighted (IPCW) log-rank tests. *Biometrics* **56**, 779–788.

Robins, J. M. and Gill, R. D. (1997). Non-response models for the analysis of non-monotone ignorable missing data. *Statistics in Medicine* **16**, 39–56.

Robins, J. M. and Ritov, Y. (1997). A curse of dimensionality appropriate (CODA) asymptotic theory for semiparametric models. *Statistics in Medicine* **16**, 285–319.

Robins, J. M. and Rotnitzky, A. (1992). Recovery of information and adjustment for dependent censoring using surrogate markers. In N. P. Jewell, K. Dietz, and V. Farewell (eds.), *AIDS Epidemiology — Methodological Issues.* Boston: Birkhäuser.

Robins, J. M. and Rotnitzky, A. (1995). Semiparametric efficiency in multivariate regression models with missing data. *Journal of the American Statistical Association* **90**, 122–129.

Robins, J. M. and Rotnitzky, A. (2001). Discussion of "Inference for semiparametric models: Some questions and an answer" by P. J. Bickel and J. Kwon. *Statistica Sinica,* **11**, 863–960.

Robins, J. M., Rotnitzky, A., and Scharfstein, D. O. (2000). Sensitivity analysis for selection bias and unmeasured confounding in missing data and causal inference models. In E. Halloran and D. Berry (eds.), *Statistical Models for Epidemiology, the Environment, and Clinical Trials.* New York: Springer.

Robins, J. M., Rotnitzky, A., and van der Laan, M. J. (2000). Discussion of "On profile likelihood" by S. A. Murphy and A. W. van der Vaart. *Journal of the American Statistical Association* **95**, 477–482.

Robins, J., Rotnitzky, A., and Zhao, L. P. (1994). Estimation of regression coefficients when some of the regressors are not always observed. *Journal of the American Statistical Association* **89**, 846–866.

Robins, J. M., Rotnitzky, A., and Zhao, L. P. (1995). Analysis of semiparametric regression models for repeated outcomes in the presence of missing data. *Journal of the American Statistical Association* **90**, 106–121.

Robins, J. M., Li, L., Tchetgen, E., and van der Vaart, A. W. (2007). *Asymptotic Normality of Degenerate U-statistics,* IMS Lecture Notes — Monograph Series. Beachwood, OH: Institute of Mathematical Statistics. To appear.

Rotnitzky, A. and Robins, J. M. (1997). Analysis of semiparametric regression models with non-ignorable non-response. *Statistics in Medicine* **16**, 81–102.

Rotnitzky, A., Robins, J. M., and Scharfstein, D. (1998). Semiparametric regression for repeated outcomes with nonignorable nonresponse. *Journal of the American Statistical Association* **93**, 1321–1339.

Rubin, D. B. (1978). Bayesian inference for causal effects: The role of randomization. *Annals of Statistics* **6**, 34–58.

Sarndal, C. E. (1980). On π inverse weighting versus best linear unbiased weighting in probability sampling. *Biometrika* **67**, 639–650.

Sarndal, C. E. (1982). Implications of survey design for generalized regression estimation of linear functions. *Journal of the American Statistical Association* **7**, 155–170.

Sarndal, C. E., Swensson, B., and Wretman, J. (1992). *Model Assisted Survey Sampling.* New York: Springer.

Scharfstein, D. O. and Irrizarry, R. (2003). Generalized additive selection models for the analysis of studies with potentially nonignorable missing outcome data. *Biometrics* **59**, 601–613.

Scharfstein, D. O. and Robins, J. M. (2002). Estimation of the failure time distribution in the presence of informative censoring. *Biometrika* **89**, 617–634.

Scharfstein, D. O., Rotnitzky, A., and Robins, J. M. (1999). Adjusting for nonignorable drop-out using semiparametric nonresponse models (with discussion). *Journal of the American Statistical Association* **94**, 1096–1120.

Scharfstein, D. O., Robins, J. M., Eddings, W., and Rotnitzky, A. (2001). Inference in randomized studies with informative censoring and discrete time-to-event endpoints. *Biometrics* **57**, 404–413.

Strawderman, R. L. (2000). Estimating the mean of an increasing stochastic process at a censored stopping time. *Journal of the American Statistical Association* **95**, 1192–1208.

Tchetgen, E., Li, L., van der Vaart, A. W., and Robins, J. M. (2006). Robust inference with higher order influence functions. Part I. *Proceedings of the Joint Statistical Meetings*. American Statistical Association.

Tsiatis, A. A. (2006). *Semiparametric Theory and Missing Data*. New York: Springer.

van der Laan, M. J. and Hubbard, A. (1999). Locally efficient estimation of the quality adjusted lifetime distribution with right-censored data and covariates. *Biometrics* **55**, 530–536.

van der Laan, M. J., Hubbard, A., and Robins, J. (2002). Locally efficient estimation of a multivariate survival function in longitudinal studies. *Journal of the American Statistical Association* **97**, 494–507.

van der Laan, M.J. and Robins J. M. (2003). *Unified Methods for Censored and Longitudinal Data and Causality*. New York: Springer.

van der Vaart, A. W. (1988). *Statistical Estimation in Large Parameter Spaces*, CWI Tract. Amsterdam: Centre for Mathematics and Computer Science.

van der Vaart, A. W. (1991). On differentiable functionals. *Annals of Statistics* **19**, 178–204.

Wright, R. L. (1983). Finite population sampling with multivariate auxiliary information. *Journal of the American Statistical Association* **78**, 879–884.

Vansteelandt, S., Rotnitzky, A., and Robins, J. M. (2007). Estimation of regression models for the mean of repeated outcomes under non-ignorable non-monotone non-response. *Biometrika* **94**, 841–860.

Zhao, H. and Tsiatis A. A. (1997). A consistent estimator for the distribution of quality adjusted survival time. *Biometrika* **84**, 339–348.

Zhao, H. and Tsiatis A. A. (1999). Efficient estimation of the distribution of quality adjusted survival time. *Biometrics* **55**, 1101–1107.

Zhao, H. and Tsiatis, A. A. (2000). Estimating mean quality adjusted lifetime with censored data. *Sankhyā, Series B* **62**, 175–188.

CHAPTER 21

Multiple imputation

Michael G. Kenward and James R. Carpenter

Contents

21.1 Introduction

Following its introduction nearly 30 years ago (Rubin, 1978), multiple imputation (MI) has become an important and influential approach for dealing with the statistical analysis of incomplete data. It now has a very large bibliography, including several reviews and texts (e.g., Rubin, 1987, 1996; Rubin and Schenker, 1991; Schafer, 1997a, 1999; Horton and Lipsitz, 2001; Molenberghs and Kenward, 2007; Kenward and Carpenter, 2007; Horton and Kleinman, 2007). Online bibliographies are provided by van Buuren (2007) and Carpenter (2007).

During this period, the range of application of MI has spread from sample surveys to include many diverse areas such as the analysis of observational data from public health research and clinical trials. In parallel with these developments, tools for MI have been incorporated into several mainstream statistical packages. Although we are mainly concerned here with the analysis of incomplete longitudinal data, we provide first, in Section 21.2, an introduction to the method in its generic form, and this is followed in Section 21.3 by an outline justification. In Section 21.4 and Section 21.5 we discuss how the problem of making *proper* imputations can be approached. Section 21.6 through Section 21.8 describe some specific applications in a longitudinal setting, in particular illustrating the role of MI in sensitivity analysis. Finally, in Section 21.9 we provide some possible directions for future developments.

21.2 The basic procedure

Suppose that we are faced with a conventional estimation problem for a statistical model with a $(p \times 1)$-dimensional parameter vector $\boldsymbol{\beta}$. We call this the *substantive* model, and the aim is to make valid inferences about some or all of the elements of $\boldsymbol{\beta}$. We assume that, if no data were missing (the *complete data*), a consistent estimator of $\boldsymbol{\beta}$ is obtained as the solution to the estimating equation

$$U(\widehat{\boldsymbol{\beta}}; \boldsymbol{Y}) = \boldsymbol{0}, \tag{21.1}$$

where the data represented by $\boldsymbol{Y}$ include both outcome variables and covariates. In the usual way, we can also calculate, from the complete data, a consistent estimator of the covariance matrix of $\widehat{\boldsymbol{\beta}}$.

Suppose now that some data are missing, and define $\boldsymbol{R}$ to be the vector of binary random variables for which an element takes the value 0 if the corresponding element of $\boldsymbol{Y}$ is missing and 1 otherwise. Denote by $\boldsymbol{Y}^m$ the vector of elements of $\boldsymbol{Y}$ that are missing (i.e., for which the corresponding element of $\boldsymbol{R}$ is zero), and by $\boldsymbol{Y}^o$ the complement of $\boldsymbol{Y}^m$ in $\boldsymbol{Y}$, the observed data. A consistent estimator of $\boldsymbol{\beta}$ from the incomplete data can then be obtained from the estimating equation:

$$E_{\boldsymbol{Y}^m | \boldsymbol{Y}^o, \boldsymbol{R}} \{ U(\widehat{\boldsymbol{\beta}}; \boldsymbol{Y}^o, \boldsymbol{Y}^m) \} = \boldsymbol{0}. \tag{21.2}$$

For the special case where $U(\cdot)$ in (21.1) is the likelihood score function with the complete data, then (21.2) similarly is the score function with the incomplete data. We call the conditional distribution over which the expectation is taken in (21.2) the *conditional predictive distribution of* $\boldsymbol{Y}^m$. A consistent estimator of the covariance matrix of $\boldsymbol{\beta}$ from (21.2) can then be obtained using Louis's formula (Louis, 1982) which expresses the information matrix for the observed data as follows in terms of expectations over complete-data quantities:

$$I_O(\boldsymbol{\beta}) = E\{ I_C(\boldsymbol{\beta}) \} - E\{ U(\widehat{\boldsymbol{\beta}}; \boldsymbol{Y}) U(\widehat{\boldsymbol{\beta}}; \boldsymbol{Y})' \} + E\{ U(\widehat{\boldsymbol{\beta}}; \boldsymbol{Y}) \} E\{ U(\widehat{\boldsymbol{\beta}}; \boldsymbol{Y}) \}',$$

for $I_C(\cdot)$ the information matrix based on the complete data with the expectations taken over the conditional predictive distribution $f(\boldsymbol{Y}^m \mid \boldsymbol{Y}^o, \boldsymbol{R})$. Although (21.2) and the accompanying estimate of precision provide a general scheme for dealing with conventional regression problems when data are missing, its practical value varies greatly from problem to problem. In some very particular settings, such as with the multivariate normal linear model under the missing at random (MAR) assumption (see Section 17.2 for a formal definition of MAR), the relevant expectations can be solved analytically and a practical implementation results. In many other settings, there is no such simple route. The expectations require the calculation of so-called *incomplete-data* quantities for which there do not exist sufficiently straightforward and general methods of calculation. In such settings, one-off solutions, often quite complex, are required.

MI provides an alternative, indirect route, to solving this problem which, most importantly, uses in the analysis phase only *complete-data* quantities; that is, those that arise in the solution to (21.1) and the accompanying estimator of precision. Crudely stated, it reverses the order of expectation and solution in (21.2). The key idea is to replace each missing value with a set of M plausible values. Each value is a *Bayesian* draw from the conditional predictive distribution of the missing observation, made in such a way that the set of imputations properly represents the information about the missing value that is contained in the observed data for the chosen model. The imputations produce M "completed" data sets, each of which is analyzed using the method that would have been appropriate had the data been complete. The model used to produce the imputations, that is, to represent the conditional predictive distribution, is called the *imputation* model.

One great strength of the MI procedure is that these two models, substantive and imputation, can be considered, to some extent, separately, although certain relationships between them do need to be observed. MI is most straightforward to use under the MAR
assumption and most software implementations make this assumption. However, it is quite
possible to apply it in not missing at random (NMAR) settings (again, see Section 17.2
for a formal definition of NMAR) using, for example, reweighting of the imputations (e.g.,
Carpenter, Kenward, and White, 2007) and using certain classes of pattern-mixture model
for the imputation model (e.g., Little and Yau, 1996; Thijs et al., 2002; Carpenter and
Kenward, 2007).

MI involves three distinct phases or, using Rubin's (1987) terminology, tasks. It is assumed
for this that the complete-data estimator has a distribution which is asymptotically normal.

1. The missing values are filled in M times to generate M complete data sets.

2. The M complete data sets are analyzed by using standard, complete data, procedures.

3. The results from the M analyses are combined to produce a single MI estimator and to
   draw inferences.

More formally, the missing data are replaced by their corresponding imputation samples,
producing M completed data sets. Denoting by $\widetilde{\beta}_k$ and V_k, respectively, the estimate of
β and its covariance matrix from the kth completed data set, $(k = 1, \ldots, M)$, the MI
estimate of β is the simple average of the estimates:

$$\widehat{\beta}_{MI} = \frac{1}{M} \sum_{k=1}^{M} \widetilde{\beta}_k. \tag{21.3}$$

We also need a measure of precision for $\widehat{\beta}_{MI}$ that properly reflects the uncertainty in the
imputations. Rubin (1987) provides the following simple expression for the covariance matrix
of $\widehat{\beta}_{MI}$ that can be applied very generally and uses only complete-data quantities. Define

$$W = \frac{1}{M} \sum_{k=1}^{M} V_k \tag{21.4}$$

to be the average within-imputation covariance matrix, and

$$B = \frac{1}{M-1} \sum_{k=1}^{M} (\widetilde{\beta}_k - \widehat{\beta}_{MI})(\widetilde{\beta}_k - \widehat{\beta}_{MI})' \tag{21.5}$$

to be the between-imputation covariance matrix of $\widetilde{\beta}_k$. Then, an estimate of the covariance
matrix of $\widehat{\beta}$ is given by

$$\widehat{V}_{MI} = W + \left(\frac{M+1}{M}\right) B. \tag{21.6}$$

Apart from an adjustment to accommodate the finite number of imputations used, this is
a very straightforward combination of between- and within-imputation variability.

Although the foregoing is based on a Bayesian argument (see Section 21.3), as presented
here the final step of MI is a frequentist one. Tests and confidence intervals are based on
the approximate pivot

$$P = (\widehat{\beta}_{MI} - \beta)' \widehat{V}_{MI}^{-1} (\widehat{\beta}_{MI} - \beta).$$

In a conventional, regular problem this would have an asymptotic χ_p^2 distribution. However,
in the MI setting, the asymptotic results and hence the appropriateness of the χ^2 reference distribution do not solely depend on the sample size N, but also on the number of
imputations M. Therefore, Li, Raghunathan, and Rubin (1991) propose the use of an $F_{p,w}$

reference distribution for the scaled statistic

$$F = \frac{P}{p(1+r)},$$

with

$$w = 4 + (\tau - 4)\left[1 + \frac{(1 - 2\tau^{-1})}{r}\right]^2,$$

$$r = \frac{1}{p}\left(1 + \frac{1}{M}\right)\operatorname{tr}(BW^{-1}),$$

$$\tau = p(M - 1).$$

Here, r is the average relative increase in variance due to missingness across the components of $\boldsymbol{\beta}$. The limiting distribution of F, as $M \to \infty$, is the χ_p^2 distribution. This procedure is applicable for any vector of parameters from the substantive model, or any linear combination of these. For inference about a scalar, this reduces in an obvious way to a t approximation for the ratio

$$\frac{\widehat{\beta}_{MI}}{\sqrt{\widehat{V}_{MI}}}.$$

MI is attractive because it can be highly efficient, even for small values of M. In some applications, merely 3 to 5 imputations are sufficient to obtain excellent results. Rubin (1987, p. 114) shows that the efficiency of an estimate based on M imputations is approximately

$$\left(1 + \frac{\gamma}{M}\right)^{-1},$$

where γ is the fraction of missing information for the quantity being estimated. In terms of the average relative increase in missingness (r) and approximate degrees of freedom (w) defined above, γ can be written as

$$\gamma = \frac{r + 2/(w + 3)}{r + 1}.$$

This provides a quantification of how much more precise the estimate might have been if no data had been missing. The efficiencies achieved for various values of M and rates of missing information are shown in Table 21.1. This table shows that gains rapidly diminish after the first few imputations. In many situations, there simply is little advantage to producing and analyzing more than a few imputed data sets. However, if MI analyses of identical form are to be sufficiently reproduceable to satisfy the audience for the results, for example regulatory authorities or scientific referees, then Carpenter and Kenward (2007) show that substantially more imputations may be needed to sufficiently stabilize the results. In addition, increasing the number of imputations also improves the precision of the between-imputation covariance

Table 21.1 Relative Efficiency (Percentage) of MI Estimation by Number of Imputations M and Fraction of Missing Information γ

m			γ		
	0.1	0.3	0.5	0.7	0.9
2	95	87	80	74	69
3	97	91	86	81	77
5	98	94	91	88	85
10	99	97	95	93	92
20	100	99	98	97	96

matrix B and, when M is not small, may be necessary to provide a non-singular estimate of this.

21.3 An outline justification

At the heart of the MI method is a Bayesian argument. Suppose we have a problem with two parameters, θ_1, θ_2, and data y. In a Bayesian analysis these have a joint posterior distribution

$$f(\theta_1, \theta_2 \mid y).$$

Now suppose that our focus is on θ_2, with θ_1 being regarded as a nuisance. Because the posterior can be partitioned as

$$f(\theta_1, \theta_2 \mid y) = f(\theta_1 \mid y) f(\theta_2 \mid \theta_1, y),$$

it can be seen that the marginal posterior for θ_2 can be expressed as

$$f(\theta_2 \mid y) = E_{\theta_1}\{f(\theta_2 \mid \theta_1, y)\}.$$

In particular, the posterior mean and variance for θ_2 can be written

$$E(\theta_2 \mid y) = E_{\theta_1}\{E_{\theta_2}(\theta_2 \mid \theta_1, y)\},$$

$$\mathrm{Var}(\theta_2 \mid y) = E_{\theta_1}\{\mathrm{Var}_{\theta_2}(\theta_2 \mid \theta_1, y)\} + \mathrm{Var}_{\theta_1}\{E_{\theta_2}(\theta_2 \mid \theta_1, y)\}.$$

These can be approximated using empirical moments. Let θ_1^k, $k = 1, \ldots, M$, be draws from the marginal posterior distribution of θ_1. Then, approximately,

$$E(\theta_2 \mid y) \simeq \frac{1}{M} \sum_{k=1}^{M} \{E_{\theta_2}(\theta_2 \mid \theta_1^k, y)\} = \widetilde{\theta}_2,$$

say, and

$$\mathrm{Var}(\theta_2 \mid y) = \frac{1}{M} \sum_{k=1}^{M} \mathrm{Var}_{\theta_2}(\theta_2 \mid \theta_1^M, y) + \frac{1}{M-1} \sum_{k=1}^{k} \{E_{\theta_2}(\theta_2 \mid \theta_1^k, y) - \widetilde{\theta}_2\}^2.$$

These formulae can be generalized in an obvious way for vector-valued parameters.

The final link between these expressions and the MI procedure is then to use θ_2 to represent the parameters of the *substantive model* (β) and θ_1 to represent the *missing data* $\mathbf{Y}^m$.

It is assumed that in sufficiently large samples the conditional posterior moments for θ_2 can be approximated by maximum likelihood, or equivalent, efficient estimators from the completed data sets. This makes it clear why we need to use imputation (conditional predictive) draws from a proper Bayesian posterior. In this way, we can see that the MI estimates of the parameters of interest and their covariance matrix approximate the first two moments of the posterior distribution in a fully Bayesian analysis. The large-sample approximation of the posterior also suggests that MI should be applied to a parameter on the scale for which the posterior is better approximated by the normal distribution, for example using log odds ratios when the parameter of interest is an odds ratio.

A rigorous justification for the final *frequentist* step is surprisingly subtle and difficult to provide in practice with any generality. Rubin (1987, p. 119) defines conditions for so-called "proper" imputation in terms of the complete-data statistics which are, for a scalar setting, respectively, $\widehat{\beta}_C$ and $\widehat{V}_C$, the estimate of the parameter of the substantive model and associated variance calculated using the complete data. Rubin gives three conditions that need to be satisfied to justify the frequentist properties of the MI procedure. These are expressed using the MI quantities defined in (21.3) through (21.6). The first two conditions are laid out in terms of repeated sampling of the missing-value process $\boldsymbol{R}$, given the complete data Z as fixed: first,

$$\frac{\widehat{\beta}_{MI} - \widehat{\beta}_C}{\sqrt{B}} \overset{M \to \infty}{\sim} N(0,1);$$

and second, $\widehat{\text{Var}}_{MI}$ is consistent for $\widehat{\text{Var}}_C$ as $M \to \infty$. The third condition is that, as $M \to \infty$, $\text{Var}(\widehat{\beta}_{MI})$ is of lower order than $\text{Var}(\widehat{\beta}_C)$ where the expectations are now taken over repeated sampling of the complete data $\boldsymbol{Y}$. For further details, we refer to Rubin (1987, Section 4.2) and Schafer (1987, Section 4.5.5). Wang and Robins (1998) and Robins and Wang (2000) approach the same problem in terms of the properties of regular asymptotic linear estimators and compare the properties of MI estimators under both proper Bayesian and improper imputation schemes. The latter correspond to repeated imputation under *fixed* consistent estimates of the parameters of the imputation model, that is, when new draws of these parameters are *not* made for each set of imputations. They show that, although for finite M the "improper" estimators are the more efficient ones, Rubin's variance estimator (21.6) is an overestimate of the variability of these. Unfortunately, it appears that no such simple variance expression exists for the "improper" estimators. Robins and Wang (2000) also consider the properties of MI estimators when the imputation model is not compatible with the substantive model, a point that we return to in Section 21.8. A more accessible account of these developments is provided by Tsiatis (2006, Chapter 14).

21.4 Drawing proper imputations

Unsurprisingly, many of the practical issues with MI concern the choice of, and Bayesian draws from, the imputation model. We outlined above the formal requirements for an MI to be valid. For the simplest settings, it is possible to check formally that these conditions hold. For more realistic problems such a justification is difficult to construct, and broader guidelines are needed. Rubin (1987, pp. 125–226) provides the following:

> If imputations are drawn to approximate repetitions from a Bayesian posterior distribution of Y_{mis} under the posited response mechanism and appropriate model for the data, then in large samples the imputation method is proper [...] There is little doubt that if this conclusion were formalized in a particular way, exceptions to it could be found. Its usefulness is not as a general mathematical result, but rather as a guide to practice. Nevertheless, in order to understand why it might be expected to hold relatively generally, it is important to provide a general heuristic argument for it.

In general, the imputation model should contain both variables known to be predictive of missingness and related to the missing variables, and should accommodate structure, for example interactions and powers of variables, contained in the substantive model. In particular, for imputing covariates, the outcome variable must be included as an explanatory variable in the model. Failure to accommodate the structure appropriately can cause bias in the resulting analysis (Fay, 1992). It has been suggested that inferences are fairly robust to the choice of imputation distribution itself. Clearly, this depends very much on the setting, in particular on the substantive model and the nature, pattern, and quantity of the missing data.

Rubin (1996) also makes the point that the rules governing the choice of imputation model do not necessarily follow those familiar from prediction problems, stating

> Our actual objective is valid statistical inference not optimal point prediction under some loss function

and

> Judging the quality of missing data procedures by their ability to recreate the individual missing values (according to hit-rate, mean squared error, etc.) does not lead to choosing procedures that result in valid inference, which is our objective.

One consequence of this is that it can be argued that the introduction of redundant variables into the imputation model need not have serious consequences. For example, according to Rubin (1996),

> The possible lost precision when including unimportant predictors is usually viewed as a relatively small price to pay for the general validity of analyses of the resultant multiply-imputed data base.

Although a simulation study by Collins, Schafer, and Kam (2001) confirms this for the multivariate normal linear model setting, it seems likely that the truth of this more broadly will depend very much on the number of units in the data set and the number and type of incomplete variables. For example, with many incomplete categorical variables, large data sets will be needed to support imputation models with corresponding high-order interactions.

In formulating an imputation model and considering the implications of this for the final results of the analysis, it can be seen that one is balancing the level of missing information in the sample and the sensitivity of the results to the choice of imputation model. As the proportion of missing information rises, one would expect this sensitivity to increase. Thus, for example, an approximation that is perfectly adequate with 5% missing information may lead to non-trivial departures from the nominal properties of frequentist procedures with 30% missing.

21.5 Practical methods for drawing proper imputations

21.5.1 The generic problem

Suppose in a given problem that a unit provides (potentially) p variables $\boldsymbol{y} = (y_1, \ldots, y_p)'$. Divide these into those observed and those (potentially) missing: $\boldsymbol{y}^o$ and $\boldsymbol{y}^m$. Each of these may be an outcome or an explanatory variable in the substantive model, or may be in the imputation model only (as an explanatory variable). In its most general form, the imputation model provides a (typically multivariate) regression of $\boldsymbol{y}^m$ on $\boldsymbol{y}^o$ with conditional distribution

$$f(\boldsymbol{y}^m \mid \boldsymbol{y}^o, \boldsymbol{r}),$$

where $\boldsymbol{r}$, a vector of indicator variables, is the observed pattern of missingness for this particular unit. If the missing-data mechanism is assumed to be random, in the sense of MAR, then $\boldsymbol{R}$ can be dropped from this to give

$$f(\boldsymbol{y}^m \mid \boldsymbol{y}^o),$$

which provides considerable simplification. In particular, for simpler modeling structures the required imputation model can be estimated from the *completers* only, that is, those units without missing data.

Care needs to be taken in conceptualizing MAR when the missingness pattern is *non-monotone* (Robins and Gill, 1997). We first define a monotone missing-value pattern. For this we must assume that the potentially missing variables are ordered. The missing-data pattern is said to be monotone if a missing observation in one variable (from one subject) implies that all variables of higher order will also be missing. This is most common in longitudinal data with attrition or dropout, for which the absence of an earlier measurement implies that all later values are also missing. A missing-data pattern is non-monotone if this does not hold. In this setting, it is difficult to conceive of plausible missing-value mechanisms that maintain the MAR assumption. One route is to assume that different mechanisms apply to different groups of subjects, but this can appear rather contrived in some settings. In practice, to keep the problem manageable, this issue is usually put to one side at this stage, and r is simply omitted from the imputation model. This does not, of course, mean that the nature of the missing-value mechanism can be ignored altogether when the overall analysis is considered.

The main problem is then to construct and fit an appropriate imputation model $f(\boldsymbol{y}^m \mid \boldsymbol{y}^o)$, and make posterior Bayesian draws from it. This means, among other things, that uncertainty about the parameters of this model must be properly incorporated. Such draws can be made in many different ways, often involving some form of approximation. In most settings, but especially longitudinal studies with missing outcomes and/or covariates, the missing data will potentially be *multivariate*. The imputation model will need to reflect this, so all methods of drawing imputations that are practically relevant need to handle multivariate data. In some contexts the missing variables will be of different types, for example measured, discrete, binary, or ordinal. Joint modeling of such disparate variables in a flexible and convenient way is itself an important statistical problem. Broadly, two approaches to this have emerged in the MI setting. We consider each in turn.

A third approach, which is less relevant in the current setting, uses non-parametric methods through Bayesian versions of hot-deck imputation. Herzog and Rubin (1983) and Heitjan and Little (1991) are comparatively early examples. Such methods tend to have a more appropriate role in the analysis of large surveys, which takes us away from the present focus.

21.5.2 Using a multivariate distribution

First, for all but nominal variables, a joint multivariate normal distribution is assumed. This requires considerable approximation for very non-normal variables, such as binary, but some authors have justified the use of such approximations. See, for example, Schafer (1997a, Chapters 4 and 5). Some success has even been reported using such an approach for the extreme case of missing binary data (e.g., Shafer, 1997a, Section 5.1), although predictably issues arise when probabilities are extreme (Schafer and Schenker, 2000). Various methods can then be used to fit and make Bayesian draws from the joint distribution.

In many settings, simpler approximate imputation draws can be used. A range of such methods for generating proper imputations is given in Little and Rubin (2002, Section 10.2.3). In sufficiently large samples, we can often do acceptably well by approximating the posterior predictive distribution using a multivariate normal distribution with mean and covariance matrix taken from the maximum likelihood estimates. In some settings we can obtain these from the completers only, which may be acceptably precise when the proportion of missing data is not great.

We illustrate this using a simple normal regression model. Suppose that one incomplete variable, y_1, is to be imputed using the values of a second one, y_2. Again, we make no distinction here between outcome variables and covariates. The imputation model can then be written, for the ith subject,

$$y_{i1} \mid y_{i2} \sim N(\xi_0 + \xi_1 z_{i2}, \tau^2). \tag{21.7}$$

Under MAR, we can use maximum likelihood to get consistent estimators, $\widehat{\boldsymbol{\xi}}$ and $\widehat{\tau}^2$ say, of $\boldsymbol{\xi} = (\xi_0, \xi_1)'$ and τ^2, from the complete pairs (i.e., subjects with both observations). In this particular setting this reduces to simple least squares.

Approximate draws from a Bayesian posterior for these parameters can then be made as follows:

$$\widetilde{\tau}^2 = (m-2)\widehat{\tau}^2/X \quad \text{for } X \sim \chi^2_{m-2}, \tag{21.8}$$

$$\widetilde{\boldsymbol{\xi}} \sim N[\widehat{\boldsymbol{\xi}}; \widetilde{\tau}^2(F'F)^{-1}], \tag{21.9}$$

where m is the number of complete pairs , and F is the $m \times 2$ matrix with rows consisting of $(1, y_{i2})$ from the completers. Finally, the missing y_{i1}s are imputed from

$$y_{i1} \sim N[(1, z_{i2})\widetilde{\boldsymbol{\xi}}; \widetilde{\tau}^2].$$ (21.10)

More generally, such approaches can be applied with maximum likelihood estimators from other classes of regression models where it is assumed that the large-sample posterior can be approximated by the maximum likelihood estimator and its large-sample covariance matrix. These need not be confined to the use of the data from the completers only. Such methods can lead to problems, however, when the posterior is not well approximated by the multivariate normal distribution with mean and covariance matrix given by the maximum likelihood estimators.

This approach, sometimes called the *regression* method, is especially simple to extend for longitudinal data with dropout (or withdrawal or attrition), thus generating a *monotone* pattern of missingness. A sequential imputation procedure can be used, starting with the first measurement (assumed complete) in which, at each time point, previously imputed values (possibly a subset of these) are used in the imputation model as predictors for future values. As an illustration of this, suppose that it is planned to collect n longitudinal outcomes from each of N subjects at n common times. Denote these *complete* data for the ith subject by $\boldsymbol{y}_i = (y_{i1}, \ldots, y_{in})'$. For simplicity, suppose further that there is a single baseline covariate x_i, for example treatment group, associated with the ith subject. For the missingness indicator, $\boldsymbol{R}_i$, introduced earlier, define the time of dropout for the ith subject to be

$$D_i = 1 + \sum_{j=1}^{n} r_{ij}.$$

For those subjects who do drop out (i.e., for whom $D_i < n+1$), the vector of responses can again be partitioned into observed and missing components and written as

$$\boldsymbol{y}_i = (\boldsymbol{y}_i^{o'}, \boldsymbol{y}_i^{m'})'.$$ (21.11)

The sequential imputation method works as follows. Assuming that the data are complete at the first time point, the simple imputation method summarized in (21.7) through (21.10) can be used to impute the data missing at the second time of measurement. The imputation model takes the form

$$y_{i2} \mid y_{i1} \sim N(\xi_{2,0} + \xi_{2,1} y_{1i} + \zeta_2 x_i, \tau_2^2),$$ (21.12)

for all subjects missing the second observation. The imputed observations $\widetilde{\boldsymbol{y}}_{i2}$ are then used to "complete" the second set of outcomes:

$$y_{i2}^* = \begin{cases} y_{i2} & \text{if observed,} \\ \widetilde{y}_{i2} & \text{if missing.} \end{cases}$$ (21.13)

This process is repeated at the third time point, using the imputation model,

$$y_{i3} \mid y_{i1}, y_{i2}^* \sim N(\xi_{3,0} + \xi_{3,1} y_{1i} + \xi_{3,2} y_{i2}^* + \zeta_3 x_i, \tau_3^2),$$ (21.14)

and sequentially up to the final time point,

$$y_{in} \mid y_{i1}, y_{i2}^* \cdots y_{i,n-1}^* \sim N(\xi_{n,0} + \xi_{n,1} y_{1i} + \xi_{n,2} y_{i2}^* + \cdots + y_{i,n-1}^* \xi_{n,n-1} + \zeta_n x_i, \tau_n^2).$$ (21.15)

For each time of measurement, previously imputed values are introduced as explanatory variables into the regression model. One pass through the sequence generates a single set of imputations, and the whole process is repeated M times to obtain M completed data sets. In principle, the normal-based regression model can be replaced by appropriate models for other types of outcome, for example logistic regressions for binary data, or proportional odds models for ordinal data. Other covariates, baseline or time-dependent, can be introduced into the imputation models. Note that the the variances, τ_j^2, are here held constant at each time; these should be allowed to differ with respect to covariates, however, if such dependence occurs in the substantive model.

This method has the great advantage of simplicity, and the set of normal regression models upon which it is based corresponds to an underlying joint multivariate normal distribution for the outcomes. Unfortunately, the method breaks down in principle when the missing-data pattern is non-monotone. If there are relatively few intermediate missing values in the longitudinal setting, then the method can still be used as acceptable approximation by imputing the intermediate missing values using the same sequential method, but replacing the imputed values by the observed ones once these again become available. However, a more widely applicable treatment of the imputation problem then requires methods that are fully appropriate for non-monotone missingness. We consider such methods now.

For general missing-data patterns, a full Gibbs sampler can be constructed comparatively simply for the multivariate normal distribution with saturated means and unstructured co-variance matrix (Schafer, 1997a, Section 5.4). This method is implemented in the SAS procedure `proc mi` and we illustrate its use in Section 21.7. Multivariate nominal discrete data can be handled using a log-linear model (Schafer, 1997a, Section 8). Extensions of these methods to structured multivariate outcomes, such as appropriate for unbalanced longitu-dinal data (i.e., lacking common times of measurements among subjects) and hierarchical data, are less well developed. For such a setting Liu, Taylor, and Belin (2000) describe a Gibbs (MCMC) sampling approach for MI in the general linear mixed model when both response variables and covariates may be missing, and similar facilities are available in Schafer's S-Plus package `PAN` (Schafer, 1997b). An alternative, but similar tool has been developed by Carpenter and Goldstein (2004), based on the MCMC sampler in the package MLwiN, and this allows imputations to be drawn from a potentially complex-structured multivariate normal distribution. Although this is currently being extended to incorporate other types of variable (binary, ordinal, categorical), in general the use of explicit joint im-putation models for mixtures of variable types is awkward, and the second approach, which we describe now, provides a practically attractive alternative to this.

21.5.3 Using a set of univariate conditional distributions

In this approach, instead of starting with a fully defined joint distribution for the imputation model, a univariate conditional model is constructed for each potentially missing variable that is appropriate to the type. Examples might be a logistic regression for a binary vari-able, or a log-linear model for a nominal categorical variable. The other potentially missing variables are then used as explanatory variables in each of these univariate conditional im-putation models. Using the previous notation, this model, for the kth missing variable (of p), would be for the conditional distribution

$$f(y_k \mid y_1, \ldots, y_{k-1}, y_{k+1}, \ldots, y_q, \boldsymbol{y}^o), \quad k = 1, \ldots, p.$$

A Gibbs *type* sampling scheme (Gilks, Richardson, and Spiegelhalter, 1996) is then used with these univariate models, but, in contrast to a conventional Gibbs sampler, each model is fitted to units only with an *observed* value for the outcome variable y_k. Cycling through all p models in turn, univariate posterior draws are then made, given current values of the other variables. Starting values are needed for the first cycle. The whole cycle is then repeated, usually a small number of times, typically 10 to 20. One set of MIs is taken from the final cycle. The whole process is then repeated M times. Most implementations use approximate large-sample draws from the univariate models as described in (21.7) through (21.10). The great advantage of this approach is that each type of variable is modeled appropriately, and it is relatively simple to include interactions, non-linear terms, and other dependencies among variables. The disadvantage is that, in general, there is no guarantee that the distribution of the draws will converge to a valid joint posterior distribution, although this is less of an issue for missing-data patterns that are close to monotone. Some simulation studies do seem to suggest that the method can perform quite well in certain settings (e.g., van Buuren, 2007). The overall approach was developed independently by Taylor et al. (2002) who called it the *sequential regression imputation method* and van Buuren, Boshuizen, and Knook (1999) who used the term *multiple chained equations*. A comprehensive review of such methods is given by van Buuren (2007), who identifies at least nine earlier (some non-MI) applications of the basic idea, with a variety of different names. This approach is potentially of great practical value for complex sets of incomplete cross-sectional data or, more generally, settings where the imputations can be reduced to a cross-sectional form, such as with the sequential method for balanced longitudinal data with attrition. It has not been shown as yet, however, how the method can be extended for realistically complex *structured* imputation models as might be used with hierarchical or other forms of longitudinal data.

21.6 Sensitivity analysis using non-random pattern-mixture models

Suppose that, assuming a multivariate normal model for continuous repeated responses, instead of using a rather conventional direct-likelihood analysis, we apply MI for the missing responses using a multivariate normal imputation model that is wholly consistent with the chosen substantive model. How would we expect the results to differ from the likelihood analysis? In practice, we would expect very similar results; indeed, as the number of imputations (M) increases the results from the two analyses should converge (apart from small differences due to the asymptotic approximations involved). With finite M, the maximum likelihood estimators are actually more efficient, although the difference may be very small in practice. An illustration of this is given by Collins, Schafer, and Kam (2001).

This raises the question: "When does MI offer advantages over direct-likelihood analyses?" When covariates are missing, it offers an intuitively attractive and very manageable method for dealing with potentially very complex problems, for which likelihood analyses may be impracticable or, at least, very awkward. When responses only are missing, and likelihood analyses are relatively simple to do, as with the linear mixed model for example, it offers little. However, there are situations in which MI can be used to carry out analyses that go beyond such conventional MAR-based likelihood procedures. In this section, and the following two, we describe three examples of such MI analyses that are particularly relevant to the longitudinal setting. As with most analyses involving MI, there do exist alternative routes for achieving the same goals, but often these do not have the same flexibility or convenience.

We begin in this section by looking at sensitivity analysis for MAR-based likelihood analyses that use non-random pattern mixture models to create the imputations. The setting we are considering is that of a balanced longitudinal trial or study; that is, one in which the times of measurement are assumed to be common for all subjects.

As an illustration, we use the following example from a multicenter trial to compare the effects of an active treatment and placebo on depression in stroke patients. For simplicity, we are ignoring the centers and other baseline covariates. The primary response is the Montgomery–Åsberg Depression Rating Scale (MADRS), which we treat as continuous. A lower MADRS score indicates a better patient condition. The score was nominally collected at baseline and at weeks 1, 2, 4, 6, and 8, following randomization to treatment group. The comparison of interest for the purposes of this illustration is the active versus placebo treatment difference in MADRS score at week 8, adjusted for baseline MADRS score. There were 114 and 111 subjects, respectively, in the placebo and active arms at the start of the trial and of these approximately 20% dropped out during the trial. We have no information

Table 21.2 Patterns of Missing Values from the Stroke Patient Trial

Pattern					Number of Subjects	
1	2	4	6	8	Placebo	Active
1	1	1	1	1	91	92
1	1	1	1	0	5	3
1	1	1	0	0	6	3
1	1	0	0	0	3	4
1	0	0	0	0	4	6
0	0	0	0	0	3	1
0	1	1	1	1	1	0
1	0	1	1	1	1	0
1	0	1	0	0	1	1
1	1	1	0	1	2	3

Note: 1 denotes observed, 0 denotes missing.

on the reasons for these dropouts. The exact pattern of missing data is shown in Table 21.2. Note that there are nine subjects with non-monotone missingness patterns, but only one of these is missing the observation in week 8.

The individual observed profiles from the completers and dropouts are shown in Figure 21.1 and Figure 21.2.

A primary analysis for such data might be based on a likelihood analysis using a multivariate normal linear model. To be precise, we assume that for Y_{ij} the observation from subject i at visit j, with y_{i0} the baseline observation,

$$E(y_{ij}) = \mu + \pi_j + \theta_j y_{i0} + \tau_j D_i, \quad j = 1, \ldots, 5, \quad i = 1, \ldots, 125, \qquad (21.16)$$

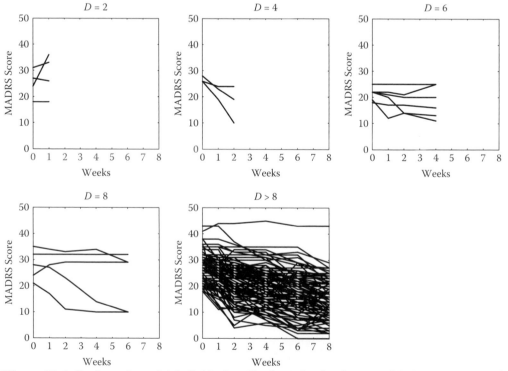

Figure 21.1 Stroke patient trial: individual profiles from the placebo group (D: dropout indicator).

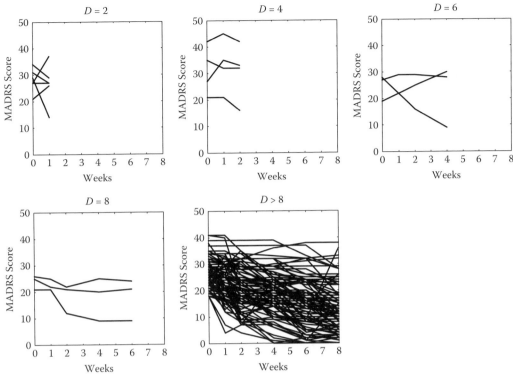

Figure 21.2 Stroke patient trial: individual profiles from the active treatment group (D: dropout indicator).

for D_i an indicator variable taking the value 1 if subject i is in the active group and 0 otherwise. The parameters τ_j and θ_j are, respectively, the treatment effect (active versus placebo) and coefficient of the baseline value at visit j, π_j is the effect associated with visit j, and μ is the intercept. The parameter that corresponds to the comparison of interest is τ_5. Further, it is assumed that the sequence of five observations from each subject follows a multivariate normal distribution with unconstrained covariance matrix, which is allowed to differ between the two treatment groups. This model is chosen to avoid making unnecessary constraints within the multivariate normal regression framework and it can be fitted in a straightforward way using standard software such as SAS `proc mixed`. If the MAR assumption holds, then, provided measures of precision are derived from the *observed* information matrix, likelihood analysis under this model will lead to consistent estimators and valid inferences. Such an analysis produces an estimated adjusted treatment difference of 2.401 (with standard error [SE] 1.040).

An analogous MI analysis can be constructed simply using the sequential regression method (21.11) through (21.15) based on the imputation procedure set out in (21.7) through (21.10). Here, the jth imputation model can be written, for treatment group $g = 1, 2$,

$$y_{ij} \mid y_{i0}, \ldots, y_{i,j-1} \sim N \left(\mu_j^g + \sum_{k=0}^{j-1} \xi_{j,k}^g y_{ik};\ \sigma_{g,j}^2 \right), \quad j = 1, \ldots, 5, \qquad (21.17)$$

for μ_j^g and (ξ_{jk}^g) the adjusted intercept and regression coefficients and $\sigma_{g,j}^2$ the corresponding residual variance from the linear regression of Y_{ij} on $y_{i0}, \ldots, y_{i,j-1}$. This is actually a reparameterization of the original model introduced in (21.16), in which each treatment group has distinct parameters. A small approximation is introduced when this is applied as well to the subjects with intermediate missing values, as future observations are ignored when the imputations are made, but with so small a proportion of such individuals the impact of this is negligible. An alternative approach that makes no approximation would be to use a full multivariate sampler such as that implemented in SAS `proc mi`. Applying both the sequential regression method and the full multivariate sampler separately to each treatment group in the current setting leads, respectively, to estimated treatment differences of 2.448 (SE 1.076) and 2.405 (SE 1.040). As expected, both results are very similar to each other and to that from the restricted maximum likelihood (REML) analysis above.

In what follows, we focus on the issues surrounding sensitivity to the assumptions concerning dropout, and assume that the intermediate missing data are missing completely at random (MCAR). The very small proportion of the latter implies that assumptions about these will have negligible impact on the conclusions.

The results obtained so far are based on an underlying MAR assumption for dropout, which cannot be assessed from the data under analysis. This has led many to argue that any such primary analysis should be augmented by an appropriate sensitivity analysis; see, for example, Scharfstein, Rotnizky, and Robins (1999), Molenberghs and Kenward (2007, Part V), Carpenter and Kenward (2007, Part III), and Chapter 22 of this volume. There are many ways of approaching such sensitivity analyses and the route taken will depend greatly on how departures from the MAR assumption are formulated. We consider one such approach here that has the advantage of being relatively transparent. It follows from the pattern-mixture formulation of the MAR assumption (Molenberghs et al., 1998). Under MAR, future statistical behavior is the same whether or not a subject drops out in the future, conditional on past response measurements and covariates. The conditional representation of the response in (21.17) provides a convenient and transparent way of viewing this assumption. It is clear that for those subjects who drop out at time j the conditional distribution of $y_{ij} \mid y_{i0}, \ldots, y_{i,j-1}$ is unobserved. We can assign any distribution to this, compatible with the support of the outcomes, and it will not affect the fit of model to the observed data. Under MAR, it is assumed to be the same for all subjects, whether or not they drop out. In a non-random model, in contrast, it would be different for dropouts. These lead to particular examples of *pattern-mixture* models (Little, 1993; Verbeke and Molenberghs, 2000; Chapter 18 and Chapter 20 of this volume). We can therefore assess sensitivity to the MAR assumption by changing this distribution for those who drop out, that is, by using a suitable NMAR pattern-mixture model. There are, of course, many ways that such a distribution might be obtained, and one route involves modification of an MAR model. An important class of such models, the members of which are particularly easy to use in practice, consists of those that are constructed directly from components of the MAR model. For a full MI analysis for such NMAR models, it is sufficient to fit the MAR model to the observed data, draw Bayesian imputations for the model parameters using conventional methods, and then construct the NMAR model for the dropouts from these components. MI draws for the missing data are then made from this constructed model in the usual way.

To make this concrete, a simple illustration is now given using a particular form of intention to treat (ITT) analysis. In an ITT analysis, subjects are assigned to their randomization group, irrespective of the treatment actually taken. For a discussion of such analyses see, for example, Little and Yau (1996). The ITT treatment comparison represents in some sense the consequences of assignment to treatment rather than the comparative effect of taking the treatment. If a dropout in such a trial also withdraws from treatment, then it is important that an ITT analysis accommodates this because the ITT analysis needs to reflect the actual compliance of subjects, irrespective of randomization. The MAR analysis above rests on the assumption that dropouts and completers with the same shared history follow the same model in the future, and by implication the same treatment. If dropout is associated with treatment termination, then this MAR analysis is *not* an ITT analysis. If we know that dropouts withdraw from active treatment, or wish to perform an analysis under that assumption, then an NMAR model is needed that reflects this. A very simple model for this uses the conditional placebo-group model for the future behavior of dropouts from the active treatment group. It is one of the interesting aspects of the missing-data problem that for an ITT analysis, which ignores compliance in the substantive model, we have to acknowledge compliance in the imputation model. For the simple setting considered here, the procedure works as follows. At each stage of the sequential regression imputation procedure (21.11) through (21.15), missing data from both the placebo and active groups are imputed using the placebo model:

$$y_{ij} \mid y_{i0}, \ldots, y_{i,j-1} \sim N\left(\widetilde{\mu}_j^1 + \sum_{k=0}^{j-1} \widetilde{\xi}_{j,k}^1 y_{ik};\ \widetilde{\sigma}_{1,j}^2\right), \quad j = 1, \ldots, 5.$$

The resulting MI analysis corresponds to the postulated NMAR model, even though we have not at any point had to explicitly fit an NMAR model to the data. This just reflects the convenience of using these forms of NMAR pattern-mixture model that are constructed from components of the original MAR model. In the current example, this leads to an estimated treatment effect of 2.037 (SE 1.113) compared with 2.448 (SE 1.076) from the MAR MI analysis. As might be expected, the ITT analysis leads to the smaller treatment difference because dropouts in the treatment group reflect placebo behavior after withdrawal. This is illustrated in Figure 21.3, where the fitted group profiles from the two MI analyses (MAR and ITT) are plotted. As expected, the two placebo profiles are virtually identical; the only difference here is due to the additional random variation introduced through the MI procedure. Under the MAR likelihood analysis, these two profiles would be identical. It can also be seen how the ITT profile from the active group is shifted toward the profile from the placebo group, with the shift increasing as time passes, that is, as more subjects drop out.

We have applied this approach here in a very simple way, but many other alternatives are possible within this framework. Little and Yau (1996) develop the approach for more general ITT settings, and Kenward, Molenberghs, and Thijs (2003) apply a similar approach using NMAR pattern-mixture models that are subject to certain constraints, in particular that the corresponding *selection model* should not allow dropout to depend on future outcomes. Other possibilities exist as potential sensitivity analyses, for example the MAR profiles could

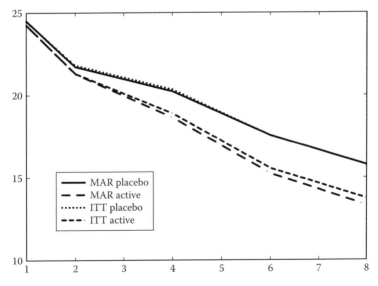

Figure 21.3 Stroke patient trial: fitted mean profiles from the MAR and ITT MI analyses.

be modified directly to explore the consequences of postulated alternative future behavior of dropouts. The advantage of the approach is that it is relatively simple computationally because it does not require the fitting of an NMAR selection model to the data, and it is usually simple to communicate the alternative assumptions being made in the NMAR pattern-mixture models that underpin the sensitivity analyses.

21.7 Introduction of additional predictors

Continuing with the setting of the previous section, that is, a longitudinal clinical trial with common times of measurement, we may in some circumstances have variables that are predictive of missingness for which we do *not* wish to adjust the treatment comparison. In other words, these variables are not to be included in the substantive model. These are sometimes called *auxiliary* variables, and a good example are post-randomization predictors of dropout in an ITT analysis. This goal can be achieved by *jointly* modeling the outcome and auxiliary variables, but this can lead to considerable additional complications and removes the intuitive appeal of keeping the substantive model that would have been used had the data been complete.

A straightforward alternative that does keep the original substantive model is to use MI in which these extra variables are included in the imputation model. Collins, Schafer, and Kam (2000) explore the use of such variables in missing settings, employing both maximum likelihood and MI. Other examples of this use of MI are given by Carpenter and Kenward (2007, Chapter 4) and we illustrate the approach with an example from this source: the Isolde study (Burge et al., 2000) in which 751 patients with chronic obstructive pulmonary disease (COPD) were randomized to receive either $50\,\mathrm{mg/day}$ of fluticasone propionate (FP) or an identical placebo. Patients were followed up for 3 years, and their FEV_1 (liters) was recorded every 3 months, although here we only use the 6-month measures. Patients with COPD are also liable to suffer acute exacerbations, and the number occurring between follow-up visits was also recorded. The primary comparison of interest is the treatment difference at year 3 (final follow-up time), adjusted for baseline.

There were, however, a large number of withdrawals. As Table 21.3 shows, only 45% of FP patients completed the trial, compared with 38% of placebo patients. Of these, many had interim missing values.

To identify key predictors of withdrawal, a logistic regression of the probability of completion was carried out on the available baseline variables together with the post-randomization exacerbation rate and rate of change in FEV_1. Table 21.4 shows the results after excluding variables with p-values greater than 0.06. The effects of age, sex, and BMI are all in line with expectations from other studies. After adjusting for these, it is clear that the response to treatment is a key predictor of patient withdrawal. In particular, discussions with the

Table 21.3 Isolde Study: Number of Patients Attending Follow-Up Visits, by Treatment Group

Visit	Number of Patients Attending Visit In	
	Placebo Arm	FP Arm
Baseline	376	374
6 month	298	288
12 month	269	241
18 month	246	222
24 month	235	194
30 month	216	174
36 month	168	141

Table 21.4 Isolde Study: Adjusted Odds Ratios for Withdrawal

Variable	Odds Ratio	(95% CI)	p-Value
Exacerbation Rate (no/year)	1.51	$(1.35, 1.69)$	< 0.001
BMI (kg/m^2)	0.95	$(0.92, 0.99)$	0.025
FEV_1 slope (ml/year)	0.98	$(0.96, 0.99)$	0.003
Age (years)	1.03	$(1.01, 1.06)$	0.011
Sex (male versus female)	1.51	$(0.99, 2.32)$	0.057

investigators suggested that high exacerbation rates were likely to act as a direct trigger for withdrawal.

It follows that the unseen data could have a potentially large impact on the treatment comparison at year 3 and a complete cases analysis is not appropriate. So, the primary analysis is based on the assumptions that the unobserved data from patients who withdraw are MAR and withdrawal is not associated with treatment discontinuation. In this way, we can produce an estimate of the original "intended" effect of treatment, adjusted only for baseline. However, to be valid, our model has to take into account the information about withdrawal in the additional variables. As discussed in the previous section, it may well be advisable to relax these assumptions, especially those surrounding treatment discontinuation, in follow-up sensitivity analyses; here we focus on the primary per-protocol analysis.

In the following, we illustrate how this can be done using baseline BMI and post-randomization exacerbation rate (square root) as auxiliary variables. MI does not represent the only way of achieving this, but the setting provides a comparatively simple platform for illustrating the technique in a longitudinal setting and also allows us to make a comparison with a full likelihood analysis.

We use the multivariate normal MCMC-based sampler in SAS `proc mi` for the actual calculations. This implies that a multivariate normal linear model with saturated means and unstructured covariance matrix is being used for the joint imputation model. This nine-dimensional model contains as variables: baseline, the six follow-up data, and the two auxiliary variables — BMI and (square root) mean number of exacerbations. We follow the same route as for the the MAR analyses in the previous section; in particular, the imputations are done completely separately for the two treatment groups, implying that the imputation model has distinct means and unstructured covariance matrices in the two groups. Because of the comparisons we wish to make here, we would like the Monte Carlo error to be less than the precision at which the estimates of precision are represented. For this reason, the number of imputations has been set at 50, a larger number than would usually be required.

In addition to this analysis, we also present results from the completers-only analysis and the full likelihood (REML) analog of the MI analysis. In this particular setting, we note that this can be done through the use of the two auxiliary variables as additional response variables. The adjusted final time point comparisons, and associated SEs, are presented in Table 21.5.

In this example, we actually see very little substantive difference in the analyses with respect to the treatment comparison. The small difference in treatment effect that is observed

Table 21.5 Isolde Study: Final Time Point Comparisons

Analysis	Treatment Effect	SE
Completers Only	-0.0889	0.0231
Multiple Imputation	-0.0892	0.0219
Likelihood (REML)	-0.0895	0.0203

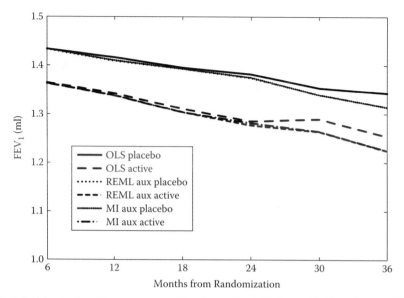

Figure 21.4 Isolde study: fitted mean profiles from the REML and MI analyses with auxiliary variables (BMI and square root exacerbation rate) and from the completers (ordinary least-squares) analysis.

is consistent with expectation, however. There is a greater withdrawal in the placebo arm and, from the results of the logistic regression displayed in Table 21.4, we see that this is associated with a greater decrease in FEV_1. Hence, the treatment effect among the completers would be expected to be smaller than that in the two analyses that make some correction for this. The estimated precision is a little higher in the two analyses that include the data from the withdrawals and the auxiliary variables (MI, REML) compared with that from the completers only, and, as expected, the likelihood estimate is slightly more precise than that obtained using MI. A greater difference between the completers and other two analyses is seen when comparing the mean estimated FEV_1 profiles as shown in Figure 21.4. In this plot, the completers profiles are estimated, for each time, using the baseline adjusted group means from those subjects who provided measurements at that time. The difference between the completers profiles (which are too high) and the adjusted profiles is particularly marked at the final two times of measurement. The MI- and REML-based analyses provide, as expected, almost coincident profiles.

In this setting, there is little to be gained from using MI compared with likelihood, because a joint multivariate normal distribution can be used with all the variables. However, with other types of auxiliary variable, such as binary and ordinal, likelihood-based analyses become more awkward, in which case the tools developed for handling such mixtures of variables (Section 21.5) using MI provides one convenient method. Further, as seen earlier, MI also provides an accessible and practical route for developing sensitivity analysis in such settings (see also Chapter 22).

21.8 Combining estimating equations and multiple imputation

When data are MAR, likelihood-based analyses have the advantage of providing valid inferences (Rubin, 1976). In the context of missing data, therefore, we need to be particularly careful when moving away from the likelihood framework. One setting in which non-likelihood methods have a particular advantage, and are widely used, is when fitting marginal (or population-averaged) models to dependent discrete data. The longitudinal setting is one very common manifestation of this, for which the use of generalized estimating equations (GEEs) (Zeger, Liang, and Albert, 1988) provides an important alternative to likelihood. These methods are valid, however, only under MCAR. One way in which validity under MAR can be realized is to use appropriate inverse probability weighting of the estimating equations; see, for example, Robins and Rotnizky (1995) and Chapter 20 and Chapter 23 of this volume. The most efficient form of this uses weights associated with each observation, so-called *observation-based* weights, but these can be awkward to calculate for non-monotone missingness. An alternative route is to use MI in which the imputation model is consistent with the MAR assumption, but not necessarily congenial with the chosen marginal substantive model. Indeed, if an imputation model were chosen that *were* fully consistent with the substantive model, we would be led back to the original problem of formulating and fitting a fully specified marginal model for the longitudinal data, the avoidance of which was the original motivation for the use of GEE. The combination of MI and GEE in the longitudinal setting is described in Molenberghs and Kenward (2007, Chapter 11) and Carpenter and Kenward (2007, Chapter 4).

There exists a range of possibilities for the imputation model in such settings. We explore these here for the common example of a binary outcome. Similar models can be formulated for other outcomes, such as ordinal or nominal. A distinctive feature of these imputation models is that they are *uncongenial*. This term was introduced and formally defined by Meng (1994). In a broad sense we can say that the imputation model is *congenial* with respect to the substantive model if there exists a proper joint model from which the two models, substantive and imputation, can be derived. In the absence of such a model, the models are said to be uncongenial.

Consider first the situation in which there is only dropout. Then a logistic version of the sequential regression method as expressed in (21.11) through (21.15) can be used. Each linear regression model is replaced by an analogous logistic model so, for example, the analog of (21.12) would be, in the absence of covariates,

$$\text{logit}\,\{P(y_{i2}=1 \mid y_{i1})\} = \xi_{2,0} + \xi_{2,1}y_{1i}.$$

The sequence of models can be fitted to the observed data using maximum likelihood and approximate Bayesian draws of the model parameters made assuming a multivariate normal posterior with mean and covariance matrix equal, respectively, to the maximum likelihood estimates of the model parameters and information-based covariance matrix of these. In the absence of covariates, the resulting model saturates the joint distribution of the longitudinal outcomes and so cannot be inconsistent with the marginal substantive model. Covariates can be introduced into these sequential models and, in a trial setting, it might be advisable to have separate imputation models in each arm. The resulting joint imputation model will no longer be saturated and so in general will not be congenial with the substantive model, but a sufficiently rich model should lead to negligible bias. Numerical difficulties can arise if *external aliasing* (in the sense of McCullagh and Nelder, 1989, Section 3.5.3) occurs in one or more of the sequential model fits, and this is most likely to be a problem if the number of units in some of the dropout groups is relatively small. It may be necessary, then, to set some of the earlier dependence parameters $\xi_{k,k'}$ to zero, effectively imposing a conditional independence structure on the data. An extreme version of this is a first-order Markov, or transition, model, in which all these parameters are zero except for neighboring observations, that is, $\xi_{k,k'} = 0$ for $k > k' + 1$ and $k' > 0$.

This sequential approach is less well suited to non-monotone missing-data patterns. For this, a log-linear imputation model can be used. This can be written, for $n = 3$ times, and again in the absence of covariates:

$$\ln\{P(y_1, y_2, y_3)\} = \lambda_0 + \lambda_1 y_1 + \lambda_2 y_2 + \lambda_3 y_3 + \lambda_{12}y_1 y_2 + \lambda_{13}y_1 y_3 + \lambda_{23}y_2 y_3 + \lambda_{123}y_1 y_2 y_3.$$

Covariates can be introduced into the model if required, as is likely in most practical settings. Schafer (1997a, Section 8.5) develops a procedure for fitting and making Bayesian draws from such models. The model has a relatively simple logistic representation in terms of conditional probabilities, for example, for $y_1 \mid y_2, y_3$:

$$\text{logit}\left\{\text{P}(y_1 \mid y_2, y_3)\right\} = \lambda_1 + \lambda_{12}y_2 + \lambda_{13}y_3 + \lambda_{123}y_2y_3.$$

This shows how simple it is to make the required draws for a single intermediate missing value from such a model. A little more work is required for multiple missing values, but nevertheless this shows that, while such log-linear models are not especially suitable as substantive models for longitudinal data, chiefly because of the conditional interpretation of the parameters, the same property makes them well suited as imputation models in the selfsame setting. As with the sequential logistic regression approach above, in the simple case without covariates, this model saturates the joint probabilities, which avoids any potential bias in the substantive model estimators following MI, but as with the sequential logistic approach, such a fully parameterized model can lead to numerical difficulties and in practice it may be advisable to restrict the log-linear model to low-order interaction terms (e.g., just two-factor terms). In the three-time-point setting, this would imply

$$\text{logit}\left\{\text{P}(y_1 \mid y_2, y_3)\right\} = \lambda_1 + \lambda_{12}y_2 + \lambda_{13}y_3,$$

and similarly for $y_2 \mid y_1, y_3$ and $y_3 \mid y_1, y_2$. It might be expected that this would, in many settings, provide a reasonable compromise between parsimony and flexibility.

Another potential imputation model would be to use a generalized linear mixed model (Molenberghs and Verbeke, 2005, Chapter 14) for which a full joint distribution of the repeated measurements is simply expressed, although such models have a less convenient conditional structure for making the imputations. However, models of this class are conveniently formulated and fitted in a fully Bayesian framework using a package such as WinBUGS, and potentially this provides a practical route for producing proper Bayesian MI draws in this setting.

21.9 Discussion

It is probably true that in most examples suitable analyses with missing data can be constructed that do not require the use of MI. Where such analyses are convenient to conduct, such as with likelihood-based analyses for longitudinal studies with attrition, MI offers little advantage under MAR. For smaller problems, fully Bayesian analyses are often quite feasible using a tool such as WinBUGS. However, MI has (at least) three distinct and important advantages. First, it can be applied very generally to very large data sets with complex patterns of missingness among covariates, and uses only complete-data quantities with very simple rules of combination. This makes it especially attractive for observational studies. Second, even when the substantive analysis under MAR is relatively straightforward, MI provides a relatively flexible and convenient route for investigating sensitivity to postulated NMAR mechanisms. Third, the imputation model may include variables not in the substantive model, which can lead to additional efficiency, or, most importantly in the clinical trial setting, allow post-randomization covariates in the imputation model if they are predictive of dropout.

Many of the issues and difficulties with MI surround the construction and use of the imputation model, in particular for mixtures of types of variable. There have been many important developments in this since the original introduction of MI, and two important routes have become delineated, joint modeling through the multivariate normal, with approximation for non-normal variables, and the use of conditional univariate models, without rigorous formal justification. Many variations on these basic themes exist.

With implementations of MI in several mainstream statistical packages, and a growing and realistic appreciation of its potential value, its use in longitudinal settings is very likely to increase. However, there remains a wide range of issues that are open to further development. These include the following.

1. Does the semi-automatic use of MI require new forms of model selection and diagnostic procedures (e.g., to detect influential points, impossible imputations, and non-linearities in the imputation models)? If so, what form should these take?

2. A more rigorous theoretical basis is needed for the chained equation approach.

3. Reliable and flexible imputation methods are needed for mixtures of types of variables from structured (i.e., hierarchical and/or longitudinal) data. This could be done using formal joint models. Using latent multivariate normal structures to tackle this is particularly appealing and would parallel developments in other areas of Bayesian modeling. Alternatively, it may also be possible to tackle this using the chained equation approach.

Acknowledgments

This work was partly supported by the ESRC research methods program, award H333 25 0047. We are grateful to GlaxoSmithKline for permission to use the data from the stroke patient and Isolde trials. We would also like to thank the editors of this volume for constructive comments on earlier versions of this chapter.

References

Burge, P. S., Calverley, P. M. A., Jones, P. W., Spencer, S., Anderson, J. A., and Maslen, T. K. (2000). Randomised, double blind placebo controlled study of fluticasone propionate in patients with moderate to severe chronic obstructive pulmonary disease. *British Medical Journal* **320**, 1297–1303.

Carpenter, J. R. (2007) Annotated bibliography on missing data. http://www.lshtm.ac.uk/msu/missingdata/biblio.html (accessed March 20, 2007).

Carpenter, J. R. and Goldstein, H. (2004). Multiple imputation in MLwiN. *Multilevel Modelling Newsletter* **16**, 9–18.

Carpenter, J. R., and Kenward, M. G. (2007). *Missing Data in Clinical Trials: A Practical Guide.* UK National Health Service, National Centre for Research on Methodology.

Carpenter, J. R., Kenward, M. G., and White, I. R. (2007). Sensitivity analysis after multiple imputation under missing at random: A weighting approach. *Statistical Methods in Medical Research* **16**, 259–276.

Collins, L. M., Schafer, J. L., and Kam, C.-M. (2001). A comparison of inclusive and restrictive strategies in modern missing data procedures. *Psychological Methods* **6**, 330–351.

Fay, R. E. (1992) When are inferences from multiple imputation valid? In *Proceedings of the Survey Research Methods Section of the American Statistical Association*, pp. 227–232.

Gilks, W. R., Richardson, S., and Spiegelhalter, D. J. (1996). Introducing Markov chain Monte Carlo. In W. R. Gilks, S. Richardson, and D. J. Spiegelhalter (eds.), *Markov Chain Monte Carlo in Practice.* London: Chapman & Hall.

Heitjan, D. and Little, R. J. A. (1991). Multiple imputation for the fatal accident reporting system. *Applied Statistics* **40**, 13–29.

Herzog, T. and Rubin, D. (1983). Using multiple imputations to handle nonresponse in sample surveys. In M. G. Madow, H. Nisselson, and I. Olkin (eds.), *Incomplete Data in Sample Surveys, Volume 2: Theory and Bibliographies*, pp. 115–142. New York: Academic Press,

Horton, N. J. and Kleinman, K. P. (2007). Much ado about nothing: A comparison of missing data methods and software to fit incomplete data regression models. *American Statistician* **61**, 79–90.

Horton, N. J. and Lipsitz, S. R. (2001). Multiple imputation in practice: Comparison of software packages for regression models with missing variables. *American Statistician* **55**, 244–254.

Kenward, M. G. and Carpenter, J. R. (2007). Multiple imputation: Current perspectives. *Statistical Methods in Medical Research* **16**, 199–218.

Kenward, M. G., Molenberghs, G., and Thijs, H. (2003). Pattern-mixture models with proper time dependence. *Biometrika* **90**, 53–71.

Li, K. H., Raghunathan, T. E., and Rubin, D. B. (1991). Large-sample significance levels from multiple imputed data using moment-based statistics and an F reference distribution. *Journal of the American Statistical Association* **86**, 1065–1073.

Little, R. J. A. (1993). Pattern-mixture models for multivariate incomplete data. *Journal of the American Statistical Association* **88**, 125–134.

Little, R. J. A. and Rubin, D. B. (2002). *Statistical Analysis with Missing Data*, 2nd ed. Chichester: Wiley.

Little, R. J. A. and Yau, L. (1996). Intent-to-treat analysis for longitudinal studies with drop-outs. *Biometrics* **52**, 1324–1333.

Liu, M., Taylor, J. M. G., and Belin, T. R. (2000). Multiple imputation and posterior simulation for multivariate missing data in longitudinal studies. *Biometrics* **56**, 1157–1163.

Louis, T. A. (1982). Finding the observed information matrix when using the EM algorithm. *Journal of the Royal Statistical Society, Series B* **44**, 226–233.

McCullagh, P. and Nelder, J. A. (1989). *Generalized Linear Models*, 2nd ed. London: Chapman & Hall.

Meng, X.-L. (1994). Multiple-imputation inferences with uncongenial sources of input. *Statistical Science* **9**, 538–558.

Molenberghs, G. and Kenward, M. G. (2007). *Missing Data in Clinical Studies*. Chichester: Wiley.

Molenberghs, G. and Verbeke, G. (2005). *Models for Discrete Longitudinal Data*. New York: Springer.

Molenbeghs, G., Michiels, B., Kenward, M. G., and Diggle, P. J. (1998). Missing value mechanisms and pattren-mixture models. *Statistica Neerlandica* **52**, 153–161.

Robins, J. M. and Gill, R. (1997). Non-response models for the analysis of non-monotone ignorable missing data. *Statistics in Medicine* **16**, 39–56.

Robins, J. M. and Rotnitzky, A. (1995). Semiparametric efficiency in multivariate regression models with missing data. *Journal of the American Statistical Association* **90**, 122–129.

Robins J. M. and Wang N. (2000). Inference for imputation estimators. *Biometrika* **85**, 113–124.

Rubin, D. B. (1976) Inference and missing data. *Biometrika* **63**, 581–592.

Rubin, D. B. (1978). Multiple imputations in sample surveys: A phenomenological Bayesian approach to nonresponse. *Proceedings of the Survey Research Methods Section of the American Statistical Association*, 20–34.

Rubin, D. B. (1987). *Multiple Imputation for Nonresponse in Surveys*. Chichester: Wiley.

Rubin, D. B. (1996). Multiple imputation after 18+ years. *Journal of the American Statistical Association* **91**, 473–490.

Rubin, D. B. and Schenker, N. (1991). Multiple imputation in health-care databases: An overview and some applications. *Statistics in Medicine* **10**, 585–598.

Schafer, J. L. (1997a). *Analysis of Incomplete Multivariate Data*. London: Chapman & Hall.

Schafer, J. L. (1997b). Imputation of missing covariates under a general linear mixed model. Technical Report, Department of Statistics, Penn State University.

Schafer, J. L. (1999). Multiple imputation: A primer. *Statistical Methods in Medical Research* **8**, 3–15.

Schafer, J. L. and Schenker, N. (2000). Inference with imputed conditional means. *Journal of the American Statistical Association* **95**, 144–154.

Scharfstein, D. O., Rotnizky, A., and Robins, J. M. (1999). Adjusting for nonignorable drop-out using semiparametric nonresponse models (with discussion). *Journal of the Americal Statistical Association* **94**, 1096–1146.

Taylor, M. G., Kooper, K. L., Wei, J. T., Sarma, A. V., Raghunathan, T. E., and Heeringa, S. G. (2002). Use of multiple imputation to correct for nonresponse bias in a survey of urologic symptoms among African-American men. *American Journal of Epidemiology* **156**, 774–782.

Thijs, H., Molenberghs, G., Michiels, B., Verbeke, G., and Curran, D. (2002). Strategies to fit pattern-mixture models. *Biostatistics* **3**, 245–265.

Tsiatis, A. (2006). *Semiparametric Theory and Missing Data*. New York: Springer.

van Buuren, S. (2007). Multiple imputation of discrete and continuous data by fully conditional specification. *Statistical Methods in Medical Research* **16**, 219–242.

van Buuren, S. (2007). Multiple imputation online. http://www.multiple-imputation.com (accessed March 20, 2007).

van Buuren, S., Boshuizen, H., and Knook, D. (1999). Multiple imputation of missing blood pressure covariates in survival analysis. *Statistics in Medicine* **18**, 681–694.

Verbeke, G. and Molenberghs, G. (2000). *Linear Mixed Models for Longitudinal Data*. New York: Springer.

Wang N. and Robins J. M. (1998). Large-sample theory for parametric multiple imputation procedures. *Biometrika* **85**, 935–948.

Zeger, S. L., Liang, K.-Y., and Albert, P. S. (1988). Models for longitudinal data: A generalized estimating approach. *Biometrics* **44**, 1049–1060.

CHAPTER 22

Sensitivity analysis for incomplete data

Geert Molenberghs, Geert Verbeke, and Michael G. Kenward

Contents

22.1 Introduction

Throughout empirical research, not all measurements planned are taken in actual practice. It is important to reflect on the nature and implications of such incompleteness, or missingness, and to accommodate it properly in the modeling process. When referring to the missing-value process, we will use terminology of Little and Rubin (2002, Chapter 6; see also Chapter 17 of this volume). A non-response process is said to be *missing completely at random* (MCAR) if the missingness is independent of both unobserved and observed data and *missing at random* (MAR) if, conditional on the observed data, the missingness is independent of the unobserved measurements. A process that is neither MCAR nor MAR is termed *non-random* (NMAR).

Given MAR, a valid analysis that ignores the missing-value mechanism can be obtained, within a likelihood or Bayesian framework, given mild regularity conditions. This situation is termed "ignorable" by Rubin (1976) and Little and Rubin (2002) and leads to considerable simplification in the analysis (Verbeke and Molenberghs, 2000; Molenberghs and Verbeke, 2005; Molenberghs and Kenward, 2007). There is a strong trend, nowadays, to prefer this kind of analysis, in the likelihood context also termed *direct-likelihood* analysis, over *ad hoc* methods such as *last observation carried forward* (LOCF), *complete-case* (CC) analysis, or simple forms of imputation (Molenberghs et al., 2004; Jansen et al., 2006a). In practice, this means that conventional tools for longitudinal and multivariate data, such as the linear and generalized linear mixed-effects models (Verbeke and Molenberghs, 2000; Molenberghs and Verbeke, 2005; see also Chapter 1 and Chapter 4) can be used in exactly the same way as with complete data. Standard software tools facilitate this paradigm shift. Another viable route is via so-called weighted generalized estimating equations (WGEE) and their extensions, which follow from the theory of Robins, Rotnitzky, and Zhao (1994). See also Tsiatis (2006) for an excellent, insightful treatment of the topic.

One should be aware, however, that, in spite of the flexibility and elegance the direct-likelihood and WGEE methods bring, there are fundamental issues when selecting a model and assessing its fit to the observed data, which do not occur with complete data. Such issues occur in the MAR case, but they are compounded further under NMAR.

Indeed, one can never fully rule out NMAR, in which case the missingness mechanism needs to be modeled alongside the mechanism generating the responses. In light of this, one approach could be to estimate from the available data the parameters of a model representing an NMAR mechanism. It is typically difficult to justify the particular choice of missingness model, and the data do not necessarily contain information on the parameters of the particular model chosen (Jansen et al., 2006b). For example, different NMAR models may fit the observed data equally well, but have quite different implications for the unobserved measurements, and hence for the conclusions to be drawn from the respective analyses. Without additional information, one can only distinguish between such models using their fit to the observed data, and so goodness-of-fit tools alone do not provide a relevant means of choosing between such models. These points will be taken up in Section 22.4, after the introduction of motivating case studies in Section 22.2 and of the notational framework in Section 22.3.

A consequence of the previous result is that one cannot definitively distinguish between MAR and NMAR models. In particular, in Section 22.5 it is demonstrated that for every NMAR model, there is exactly one MAR counterpart with exactly the same fit to the observed data but, of course, with a different prediction for the unobserved data, given the observed ones.

All of these considerations point to the necessity of sensitivity analysis. In a broad sense, we can define a sensitivity analysis as one in which several statistical models are considered simultaneously and/or where a statistical model is further scrutinized using specialized tools, such as diagnostic measures. This rather loose and very general definition encompasses a wide variety of useful approaches. The simplest procedure is to fit a selected number of (NMAR) models, all of which are deemed plausible; alternatively, a preferred (primary) analysis can be supplemented with a number of modifications. The degree to which conclusions (inferences) are stable across such ranges provides an indication of the confidence that can be placed in them. Modifications to a basic model can be constructed in different ways. One obvious strategy is to consider various dependencies of the missing-data process on the outcomes and/or on covariates. One can choose to supplement an analysis within the selection modeling framework (Chapter 18), say, with one or several in the pattern-mixture modeling framework (Chapter 18), which explicitly models the missing responses at any occasion given the previously observed responses. Alternatively, the distributional assumptions of the models can be altered.

It is simply impossible to give a definitive overview of all missing-data tools. Research in this area is vast and disparate. This is not a negative point; rather, it reflects the broad awareness of the need for sensitivity analysis. This is a relatively recent phenomenon, earlier work having been focused instead on the formulation of ever more complex models. The paradigm shift to sensitivity analysis is, therefore, welcome. We have already referred to the use of pattern-mixture models. Alternatively, consideration of shared-parameter models (Wu and Carroll, 1988; Molenberghs and Kenward, 2007; Chapter 13 and Chapter 19 of this volume) is a possibility. Not only the modeling framework, but also the distinction between parametric and non-parametric approaches (Scharfstein, Rotnitzky, and Robins, 1999) enables classification of missing-data methods. Whereas in the parametric context one is often interested in quantifying the impact of model assumptions, semi-parametric and non-parametric modelers aim to formulate models that have a high level of robustness against the impact of the missing-data mechanism. Furthermore, a number of authors have aimed to quantify the impact of one or a few observations on the substantive and missing-data mechanism related conclusions (Copas and Li, 1997; Troxel, Harrington, and Lipsitz, 1998; Verbeke et al., 2001).

Early references pointing to the aforementioned sensitivities and responses to them include Nordheim (1984), Laird (1994), Little (1994b), Rubin (1994), Fitzmaurice, Molenberghs, and Lipsitz (1995), Vach and Blettner (1995), Kenward (1998), Kenward and Molenberghs (1999), and Molenberghs et al. (1999). Many of these are to be considered potentially useful but *ad hoc* approaches. Whereas such informal sensitivity analyses are an indispensable step in the analysis of incomplete longitudinal data, it is desirable to have more formal frameworks within which to develop such analyses. Such frameworks can be found in Scharfstein, Rotnitzky, and Robins (1999), Thijs, Molenberghs, and Verbeke (2000), Kenward, Goetghebeur, and Molenberghs (2001), Molenberghs et al. (2001), Molenberghs, Kenward, and Goetghebeur (2001), Van Steen et al. (2001), Verbeke et al. (2001), and Jansen et al. (2006b). It is simply impossible to do full justice to all of these methods. Therefore, two routes to sensitivity analysis will be discussed in some detail: the so-called *interval of ignorance* method in Section 22.6 and the use of influential observations analysis in Section 22.7. A brief perspective on a few selected other approaches is offered in Section 22.8.

22.2 Motivating examples

We will introduce three examples that will be used throughout the remainder of the chapter. For the first and the second, the orthodontic growth data and the Slovenian Public Opinion Survey, initial analyses are provided here as well.

22.2.1 The orthodontic growth data

These data, introduced by Pothoff and Roy (1964), contain growth measurements for 11 girls and 16 boys. For each subject, the distance from the center of the pituitary to the pterygomaxillary fissure was recorded at ages 8, 10, 12, and 14. The data were used by Jennrich and Schluchter (1986) to illustrate estimation methods for unbalanced data, where imbalance is now to be interpreted in the sense of an unequal number of boys and girls. Individual profiles and sex group by age means are plotted in Figure 22.1.

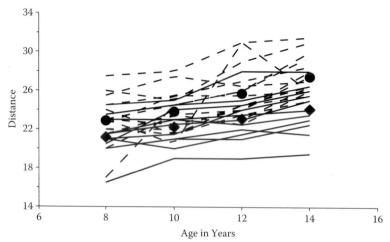

Figure 22.1 The orthodontic growth data. Observed profiles and group by age means. Solid lines and diamonds are for girls, dashed lines and bullets are for boys.

Little and Rubin (2002) deleted nine of the $(11 + 16) \times 4$ observations, thereby producing nine subjects with incomplete data, in particular, with a missing measurement at age 10. Their missingness generating mechanism was such that subjects with a low value at age 8 were more likely to have a missing value at age 10. We focus first on the analysis of the original, complete data set.

Jennrich and Schluchter (1986), Verbeke and Molenberghs (2000), and Little and Rubin (2002) each fitted the same eight models, which can be expressed within the general linear mixed-models family (Verbeke and Molenberghs, 2000) as

$$\mathbf{Y}_i = X_i\boldsymbol{\beta} + Z_i\boldsymbol{b}_i + \boldsymbol{e}_i,$$

where $\boldsymbol{b}_i \sim N(\mathbf{0}, G)$, $\boldsymbol{e}_i \sim N(\mathbf{0}, V_i)$, and $\boldsymbol{b}_i$ and $\boldsymbol{e}_i$ are statistically independent. Here, $\mathbf{Y}_i$ is the (4×1) response vector; X_i is a $(4 \times p)$ design matrix for the fixed effects; $\boldsymbol{\beta}$ is a vector of unknown fixed regression coefficients; Z_i is a $(4 \times q)$ design matrix for the random effects; $\boldsymbol{b}_i$ is a $(q \times 1)$ vector of normally distributed random effects, with covariance matrix G; and $\boldsymbol{e}_i$ is a normally distributed (4×1) random error vector, with covariance matrix V_i. Estimation and inference are traditionally obtained from likelihood principles based on the marginal distribution $\mathbf{Y}_i \sim N(X_i\boldsymbol{\beta}, Z_iGZ_i' + V_i)$.

In our example, every subject contributes exactly four measurements at exactly the same time points. It is therefore possible to drop the subscript i from the error covariance matrix V_i unless, for example, sex is thought to influence the residual covariance structure. The random error $\boldsymbol{e}_i$ encompasses both within-subject variability and serial correlation. The mean $X_i\boldsymbol{\beta}$ will be a function of age, sex, and/or the interaction between both.

Table 22.1 summarizes model fitting and comparison for the eight models originally considered by Jennrich and Schluchter (1986). The initial Model 1 assumes an unstructured group by time model, producing eight mean parameters. In addition, the variance–covariance matrix is left unstructured, yielding an additional 10 parameters. First, the mean structure is simplified, followed by the covariance structure. Models 2 and 3 consider the mean profiles to be non-parallel and parallel straight lines, respectively. While the second model fits adequately, the third one does not, based on conventional likelihood ratio tests. Thus, the crossing lines will be retained. Models 4 and 5 assume the variance–covariance structure to be of a banded (Toeplitz) and first-order autoregressive (AR(1)) type, respectively.

Table 22.1 The Orthodontic Growth Data: Original and Trimmed Data Set

Model	Mean	Covar.	# Par.	Ref.	Original Data -2ℓ	Original Data p-Value	Trimmed Data -2ℓ	Trimmed Data p-Value
1	unstr.	unstr.	18		416.5		386.96	
2	$\neq$ slopes	unstr.	14	1	419.5	0.563	393.29	0.176
3	= slopes	unstr.	13	2	426.2	0.010	397.40	0.043
4	$\neq$ slopes	Toepl.	8	2	424.6	0.523	398.03	0.577
5	$\neq$ slopes	AR(1)	6	2	440.7	0.007	409.52	0.034
6	$\neq$ slopes	RI + RS	8	2	427.8	0.215	400.45	0.306
7	$\neq$ slopes	CS (RI)	6	6	428.6	0.510	401.31	0.502
8	$\neq$ slopes	simple	5	7	478.2	<0.001	441.58	<0.001

Note: Model fit summary (# Par., number of model parameters; -2ℓ, minus twice log-likelihood; Ref., reference model for likelihood ratio test).

Model 6 assumes the covariance structure to arise from correlated random intercepts (RI) and random slopes (RS). In Model 7, a compound-symmetry (CS) structure is assumed, which can be seen as the marginalization of a random-intercepts model. Finally, Model 8 assumes uncorrelated measurements. Of these, Models 4, 6, and 7 are well-fitting. Model 7, being the most parsimonious one, will be retained.

Fits of the same eight models to the trimmed, incomplete version of the data set, as presented by Little and Rubin (2002), using direct-likelihood methods, are also summarized in Table 22.1. This implies that the same models are fitted with the same software tools, but now to a reduced set of data. Note that the same Model 7 is selected. Verbeke and Molenberghs (1997) discuss some issues arising when the covariance matrix differs between boys and girls.

22.2.2 The Slovenian Public Opinion Survey

In 1991, Slovenians voted for independence from former Yugoslavia in a plebiscite. To prepare for this result, the Slovenian government collected data in the so-called Slovenian Public Opinion (SPO) Survey, a month prior to the plebiscite. Rubin, Stern, and Vehovar (1995) studied the three fundamental questions added to the SPO and, in comparing it to the plebiscite's outcome, drew conclusions about the missing-data process.

The three questions added were: (1) Are you in favor of Slovenian independence? (2) Are you in favor of Slovenia's secession from Yugoslavia? (3) Will you attend the plebiscite? In spite of their apparent equivalence, questions (1) and (2) are different because independence would have been possible in confederal form as well. Question (3) is highly relevant because the political decision was taken that not attending was treated as an effective no to question (1). Thus, the primary estimand is the proportion θ of people who will be considered as voting yes, which is the fraction of people answering yes to both the attendance and independence question. The raw data are presented in Table 22.2.

The data were used by Molenberghs, Kenward, and Goetghebeur (2001) to illustrate their sensitivity analysis tool, the interval of ignorance. Molenberghs et al. (2007) used the data to exemplify results about the relationship between MAR and NMAR models. An overview of various analyses can be found in Molenberghs and Kenward (2007). These authors used the model proposed by Baker, Rosenberger, and DerSimonian (1992) for the setting of two-way contingency tables subject to non-monotone missingness. Such data take the form of counts $Z_{r_1,r_2,jk}$, where $j, k = 0, 1$ reference the two categories and $r_1, r_2 = 0, 1$ are the missingness indicators for each. The corresponding probabilities are $\nu_{r_1,r_2,jk}$, describing a

Table 22.2 Results of the Slovenian Public Opinion
Survey

Secession	Attendance	Independence		
		Yes	No	*
Yes	Yes	1191	8	21
	No	8	0	4
	*	107	3	9
No	Yes	158	68	29
	No	7	14	3
	*	18	43	31
*	Yes	90	2	109
	No	1	2	25
	*	19	8	96
Collapsed	Yes	1439	78	159
	No	16	16	32
	*	144	54	136

Note: The "don't know" category is indicated by *.

four-way classification, and modeled as

$$\nu_{10,jk} = \nu_{11,jk}\beta_{jk},$$
$$\nu_{01,jk} = \nu_{11,jk}\alpha_{jk},$$
$$\nu_{00,jk} = \nu_{11,jk}\alpha_{jk}\beta_{jk}\gamma.$$

The α (β) parameters describe missingness in the independence (attendance) question, and γ captures the interaction between both. The subscripts are missing from γ because Baker, Rosenberger, and DerSimonian (1992) have shown that this quantity is independent of j and k in every identifiable model. These authors considered nine models, based on setting α_{jk} and β_{jk} constant in one or more indices:

$$\text{BRD1}:(\alpha,\beta) \quad \text{BRD4}:(\alpha,\beta_k) \quad \text{BRD7}:(\alpha_k,\beta_k)$$
$$\text{BRD2}:(\alpha,\beta_j) \quad \text{BRD5}:(\alpha_j,\beta) \quad \text{BRD8}:(\alpha_j,\beta_k)$$
$$\text{BRD3}:(\alpha_k,\beta) \quad \text{BRD6}:(\alpha_j,\beta_j) \quad \text{BRD9}:(\alpha_k,\beta_j).$$

Interpretation is straightforward. For example, BRD1 is MCAR, and in BRD4 missingness in the first variable is constant, while missingness in the second variable depends on its value. BRD6 through BRD9 saturate the observed-data degrees of freedom, while the lower-numbered ones do not, leaving room for a non-trivial model fit to the observed data.

Rubin, Stern, and Vehovar (1995) conducted several analyses of the data. Their main emphasis was in determining the proportion θ of the population that would attend the plebiscite and vote for independence. The three other combinations of these two binary outcomes would be treated as voting no. Their estimates are reproduced in Table 22.3.

The pessimistic (optimistic) bounds are obtained by setting all incomplete data that can be considered a yes (no) as yes (no). The complete-case estimate for θ is based on the subjects answering all three questions, and the available-case estimate is based on the subjects answering the two questions of interest here. It is noteworthy that both estimates fall outside the pessimistic–optimistic interval and should be disregarded because these seemingly straightforward estimators do not take the decision to treat absences as no's into account and thus discard available information. This confirms that care should be taken with the simple methods and a transition to MAR or more elaborate methods may be required. Rubin, Stern, and Vehovar (1995) considered two MAR models, also reported in

Table 22.3 The Slovenian Public Opinion Survey: Some Estimates of the Proportion θ Attending the Plebiscite and Voting for Independence

Estimation Method	Voting in Favor of Independence $\widehat{\theta}$
Pessimistic–optimistic bounds	$[0.694, 0.905]$
Complete cases	0.928
Available cases	0.929
MAR (2 questions)	0.892
MAR (3 questions)	0.883
MAR	0.782
Plebiscite	0.885

Source: Presented in Rubin, Stern, and Vehovar, 1995; Molenberghs, Kenward, and Goetghebeur, 2001.

Table 22.3, the first one based on the two questions of direct interest only, and the second one using all three. Finally, they considered a single NMAR model, based on the assumption that missingness on a question depends on the answer to that question but not on the other questions. Rubin, Stern, and Vehovar (1995) concluded, owing to the proximity of the MAR analysis to the plebiscite value, that MAR in this and similar cases may be considered a plausible assumption.

Molenberghs, Kenward, and Goetghebeur (2001) and Molenberghs et al. (2007) fitted the BRD models and Table 22.4 summarizes the results. BRD1 produces $\widehat{\theta} = 0.892$, exactly the same as the first MAR estimate obtained by Rubin, Stern, and Vehovar (1995). This comes as no surprise because both models assume MAR and use information from the two main questions. A graphical representation of the original analyses and the BRD models combined is given in Figure 22.2.

22.2.3 Mastitis in dairy cattle

This example, concerning the occurrence of the infectious disease mastitis in dairy cows, was introduced in Diggle and Kenward (1994) and reanalyzed in Kenward (1998). Data were available on the milk yields, in thousands of liters, for 107 dairy cows from a single herd

Table 22.4 The Slovenian Public Opinion Survey: Analysis, Restricted to the Independence and Attendance Questions

Model	Structure	df	Log-Likelihood	$\widehat{\theta}$	C.I.	$\widehat{\theta}_{MAR}$
BRD1	(α, β)	6	-2495.29	0.892	$[0.878, 0.906]$	0.8920
BRD2	(α, β_{j_1})	7	-2467.43	0.884	$[0.869, 0.900]$	0.8915
BRD3	(α_{j_2}, β)	7	-2463.10	0.881	$[0.866, 0.897]$	0.8915
BRD4	(α, β_{j_2})	7	-2467.43	0.765	$[0.674, 0.856]$	0.8915
BRD5	(α_{j_1}, β)	7	-2463.10	0.844	$[0.806, 0.882]$	0.8915
BRD6	$(\alpha_{j_1}, \beta_{j_1})$	8	-2431.06	0.819	$[0.788, 0.849]$	0.8919
BRD7	$(\alpha_{j_2}, \beta_{j_2})$	8	-2431.06	0.764	$[0.697, 0.832]$	0.8919
BRD8	$(\alpha_{j_1}, \beta_{j_2})$	8	-2431.06	0.741	$[0.657, 0.826]$	0.8919
BRD9	$(\alpha_{j_2}, \beta_{j_1})$	8	-2431.06	0.867	$[0.851, 0.884]$	0.8919

Note: Summaries on each of the models BRD1 through BRD9 are presented, with obvious column labels. The column labeled $\widehat{\theta}_{MAR}$ refers to the model corresponding to the given one, with the same fit to the observed data, but with missing-data mechanism of the MAR type.

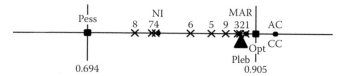

Figure 22.2 The Slovenian Public Opinion Survey. Relative position for the estimates of "proportion of YES votes," based on the models considered in Rubin, Stern, and Vehovar (1995) and on the Baker, Rosenberger, and DerSimonian (1992) models. The vertical lines indicate the nonparametric pessimistic–optimistic bounds. (Pess, pessimistic boundary; Opt, optimistic boundary; MAR, Rubin, Stern, and Vehovar's MAR model; NI, Rubin, Stern, and Vehovar's NMAR model; AC, available cases; CC, complete cases; Pleb, plebiscite outcome. Numbers refer to the BRD models.)

in two consecutive years: Y_{ij} ($i = 1, \ldots, 107; j = 1, 2$). In the first year, all animals were supposedly free of mastitis; in the second year, 27 became infected. Mastitis typically reduces milk yield, and the question of scientific interest is whether the probability of occurrence of mastitis is related to the yield that would have been observed had mastitis not occurred. A graphical representation of the complete data is given in Figure 22.3. An analysis is presented in Section 22.7.3.

22.3 Notation and concepts

Let the random variable Y_{ij} denote the response of interest, for the ith study subject, designed to be measured at occasions t_{ij}, $i = 1, \ldots, N$, $j = 1, \ldots, n_i$. Independence across subjects is assumed. The outcomes can conveniently be grouped into a vector $\boldsymbol{Y}_i = (Y_{i1}, \ldots, Y_{in_i})'$. In addition, define a vector of missingness indicators $\boldsymbol{R}_i = (R_{i1}, \ldots, R_{in_i})'$ with $R_{ij} = 1$ if Y_{ij} is observed and 0 otherwise. In the specific case of dropout, $\boldsymbol{R}_i$ can usefully be replaced by the dropout indicator $D_i = \sum_{j=1}^{n_i} R_{ij}$. Note that the concept of dropout refers to time-ordered variables, such as in longitudinal studies. For a complete sequence, $\boldsymbol{R}_i = \boldsymbol{1}$, where $\boldsymbol{1}$ is a $n_i \times 1$ vector of 1s; and/or $D_i = n_i$. It is customary to split the vector $\boldsymbol{Y}_i$ into observed ($\boldsymbol{Y}_i^o$) and missing ($\boldsymbol{Y}_i^m$) components, respectively.

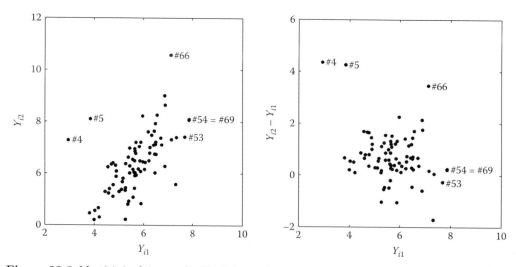

Figure 22.3 Mastitis in dairy cattle. The left panel shows a scatter plot of the second measurement versus the first measurement. The right panel shows a scatter plot of the change versus the baseline measurement.

In principle, one would like to consider the density of the full data $f(\boldsymbol{y}_i, \boldsymbol{r}_i | \boldsymbol{\theta}, \boldsymbol{\psi})$, where the parameter vectors $\boldsymbol{\theta}$ and $\boldsymbol{\psi}$ describe the measurement and missingness processes, respectively. Covariates are assumed to be measured and grouped in a vector $\boldsymbol{x}_i$.

This full density function can be factorized in different ways, each leading to a different framework. The *selection model* (SeM) framework (Chapter 18) is based on the factorization (Rubin, 1976; Little and Rubin, 2002)

$$f(\boldsymbol{y}_i, \boldsymbol{r}_i | \boldsymbol{x}_i, \boldsymbol{\theta}, \boldsymbol{\psi}) = f(\boldsymbol{y}_i | \boldsymbol{x}_i, \boldsymbol{\theta}) f(\boldsymbol{r}_i | \boldsymbol{x}_i, \boldsymbol{y}_i, \boldsymbol{\psi}). \tag{22.1}$$

The first factor is the marginal density of the measurement process, and the second is the density of the missingness process, conditional on the outcomes. As an alternative, one can consider so-called *pattern-mixture models* (PMMs) (Little, 1993, 1994a; Chapter 18 of this volume) using the reversed factorization

$$f(\boldsymbol{y}_i, \boldsymbol{r}_i | \boldsymbol{x}_i, \boldsymbol{\theta}, \boldsymbol{\psi}) = f(\boldsymbol{y}_i | \boldsymbol{x}_i, \boldsymbol{r}_i, \boldsymbol{\theta}) f(\boldsymbol{x}_i, \boldsymbol{r}_i | \boldsymbol{\psi}).$$

This can be seen as a mixture density over different populations, each of which is defined by the observed pattern of missingness.

Instead of using the selection modeling or pattern-mixture modeling frameworks, the measurement and the dropout process can be jointly modeled using a *shared-parameter model* (SPM) (Wu and Carroll, 1988; Wu and Bailey, 1988, 1989; Follman and Wu, 1995; Little, 1995; Chapter 13 and Chapter 19 of this volume). One then assumes there exists a vector of random effects $\boldsymbol{b}_i$, conditional upon which the measurement and dropout processes are independent. This SPM is formulated by way of the factorization

$$f(\boldsymbol{y}_i, \boldsymbol{r}_i | \boldsymbol{x}_i, \boldsymbol{b}_i, \boldsymbol{\theta}, \boldsymbol{\psi}) = f(\boldsymbol{y}_i | \boldsymbol{x}_i, \boldsymbol{b}_i, \boldsymbol{\theta}) f(\boldsymbol{r}_i | \boldsymbol{x}_i, \boldsymbol{b}_i, \boldsymbol{\psi}).$$

Here, $\boldsymbol{b}_i$ are shared parameters, often considered to be random effects and following a specific parametric distribution.

The taxonomy of missing-data mechanisms, introduced by Rubin (1976) and informally described in the Introduction, can easily be formalized using the second factor on the right-hand side of SeM factorization (22.1). A mechanism is MCAR if

$$f(\boldsymbol{r}_i | \boldsymbol{x}_i, \boldsymbol{y}_i, \boldsymbol{\psi}) = f(\boldsymbol{r}_i | \boldsymbol{x}_i, \boldsymbol{\psi}),$$

that is, when the measurement and missingness processes are independent, perhaps conditional on covariates. For a given set of data, MAR holds when

$$f(\boldsymbol{r}_i | \boldsymbol{x}_i, \boldsymbol{y}_i, \boldsymbol{\psi}) = f(\boldsymbol{r}_i | \boldsymbol{x}_i, \boldsymbol{y}_i^o, \boldsymbol{\psi}),$$

strictly weaker than the MCAR condition, but still a simplification of the NMAR case, where missingness depends on the unobserved outcomes $\boldsymbol{y}_i^m$, regardless of the observed outcomes and the covariates.

A final useful concept we need is *ignorability*. Note that the contribution to the likelihood of subject i, based on (22.1), equals

$$L_i = \int f(\boldsymbol{y}_i | \boldsymbol{x}_i, \boldsymbol{\theta}) f(\boldsymbol{r}_i | \boldsymbol{x}_i, \boldsymbol{y}_i^o, \boldsymbol{y}_i^m, \boldsymbol{\psi}) \, d\boldsymbol{y}_i^m. \tag{22.2}$$

In general, (22.2) does not simplify, but under MAR, we obtain

$$L_i = f(\boldsymbol{y}_i^o | \boldsymbol{x}_i, \boldsymbol{\theta}) f(\boldsymbol{r}_i | \boldsymbol{x}_i, \boldsymbol{y}_i^o, \boldsymbol{\psi}).$$

Hence, likelihood and Bayesian inferences for the measurement model parameters $\boldsymbol{\theta}$ can be made without explicitly formulating the missing-data mechanism, provided the parameters $\boldsymbol{\theta}$ and $\boldsymbol{\psi}$ are distinct, meaning that their joint parameter space is the Cartesian product of the two component parameter spaces (Rubin, 1976). In addition, for Bayesian inferences, the priors need to be independent (Little and Rubin, 2002).

It is precisely this result which makes so-called direct-likelihood or Bayesian analyses, valid under MAR, viable candidates for the status of primary analysis.

In spite of direct likelihood's elegance, fundamental model assessment and model selection issues remain, which is the topic of the remainder of this chapter.

22.4 What is different when data are incomplete?

A number of issues arising when analyzing such incomplete data, under MAR as well as under NMAR, are reviewed in Section 22.4.1. Ways of tackling the problems are the subject of Section 22.4.2.

22.4.1 Complexity of model selection and assessment with incomplete data

Model selection and assessment are well-established components of statistical analysis, surrounded by several strands of intuition. Some of these apply generally, while some are particular to certain types of outcomes and/or modeling situations. First, it is researchers' common understanding that "observed $\simeq$ expected" for a well-fitting model, which is usually understood to imply that observed and fitted profiles ought to be sufficiently similar in a longitudinal study, or observed and fitted counts in contingency tables, etc. Second, for the special case of samples from univariate or multivariate normal distributions, the estimators for the mean vector and the variance–covariance matrix are independent, both in a small-sample as well as in an asymptotic sense. Third, in the same normal situation, the least-squares and maximum likelihood estimators are identical as far as mean parameters are concerned, and asymptotically equal for covariance parameters. Fourth, in a likelihood-based context, deviances and related information criteria are considered useful and practical tools for model assessment. Fifth, saturated models are uniquely defined and at the top of the model hierarchy. For contingency tables, such a saturated model is one that exactly reproduces the observed counts.

These five points are based on statisticians' experiences with balanced designs and at the same time complete data, in the sense that the data can be thought of as lining up in a rectangular data matrix. We will now illustrate each of them, grouped into four categories, by means of the running examples and by general considerations.

22.4.1.1   The "observed $\simeq$ expected" relationship

Figure 22.4 shows the observed and fitted mean structures for Models 1, 2, 3, and 7, fitted to the complete version of the growth data set, as reported in Section 22.2.1. Note that observed and fitted means coincide for Model 1. This is in line with general theory, because the model saturates the group by time mean structure *and*, in addition, the data are balanced, in the sense that all subjects are measured at a common set of measurement occasions. While, for the incomplete version of the data, direct likelihood nicely recovers Model 7, observed and expected means do not coincide any longer, not even under Model 1 where the group by age mean structure is saturated (see Figure 22.7). It is important that the discrepancy is seen for the mean at age 10, the only age at which there is missingness.

22.4.1.2   The mean–variance relationship in a normal distribution

Let us consider Table 22.5 to obtain insight into the effect of the variance–covariance structure on the mean structure. We retain an unstructured group by age mean structure and pair it with three covariance structures. Apart from an unstructured residual covariance matrix (Model 1), we also consider a CS structure (Model 7b) and an independence structure (Model 8b).

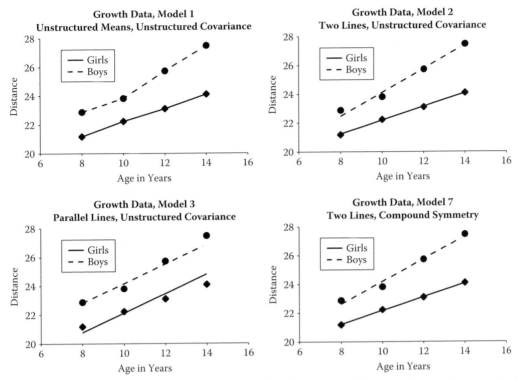

Figure 22.4 The orthodontic growth data: profiles for the complete data, for a selected set of models.

When the data are complete, the choice of covariance structure is immaterial for the point estimates, whereas it is crucial when data are incomplete. Next to an overcorrection of Model 1 at age 10, Model 7b exhibits quite acceptable behavior, but Model 8b coincides with and hence is as bad as CC at age 10, as can be seen in Figure 22.5.

22.4.1.3 The least squares–maximum likelihood difference

Let us now turn to the difference between ordinary least squares and maximum likelihood. While a different issue, it is closely related to the previous two, as will be made clear.

While least-squares regression and normal distribution based regression produce the same point estimator and asymptotically the same precision estimator *for balanced and complete data*, this is no longer true in the incomplete-data setting. The former method is frequentist in nature, the second one likelihood-based. We can easily illustrate this result for a simple,

Table 22.5 The Orthodontic Growth Data: Comparison of Mean Estimates for Boys at Ages 8 and 10, Complete and Incomplete Data, Using Direct Likelihood, an Unstructured Mean Model, and Various Covariance Models

Data	Mean	Covar.	Boys at Age 8	Boys at Age 10
Complete	unstr.	unstr.	22.88	23.81
	unstr.	CS	22.88	23.81
	unstr.	simple	22.88	23.81
Incomplete	unstr.	unstr.	22.88	23.17
	unstr.	CS	22.88	23.52
	unstr.	simple	22.88	24.14

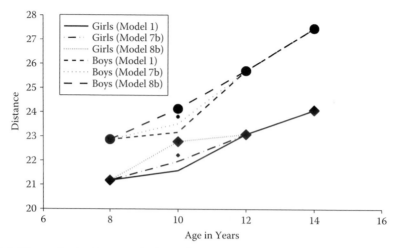

Figure 22.5 The orthodontic growth data: fitted mean profiles to the incomplete data, using maximum likelihood, an unstructured mean model and unstructured covariance matrix (Model 1), CS (Model 7b), and independence (Model 8b) covariance structure.

bivariate normal population with missingness in the second component. Consider a bivariate normal population $(Y_{i1}, Y_{i2})'$ with mean $(\mu_1, \mu_2)'$ and unspecified variance–covariance matrix with variances σ_1^2 and σ_2^2 and covariance σ_{12}. Take a sample of $i = 1, \ldots, N$ subjects. Assume further that d subjects complete the study and $N - d$ drop out after the first measurement occasion.

In a frequentist method, the parameters are estimated using the available information (Verbeke and Molenberghs, 2000; Little and Rubin, 2002), that is, least squares is used. This implies that μ_1 and σ_1^2 would be estimated using all N subjects, whereas only the remaining d contribute to the other three parameters. For the mean parameters, this produces

$$\widehat{\mu_1} = \frac{1}{N} \sum_{i=1}^{N} y_{i1}, \tag{22.3}$$

$$\widetilde{\mu_2} = \frac{1}{d} \sum_{i=1}^{d} y_{i2}. \tag{22.4}$$

Whereas (22.3) is estimated in the same way with likelihood methodology, (22.4) can be shown to change to

$$\widehat{\mu_2} = \frac{1}{N} \left[\sum_{i=1}^{d} y_{i2} + \sum_{i=d+1}^{N} \left\{ \overline{y}_2 + \widehat{\beta_1}(y_{i1} - \overline{y}_1) \right\} \right]. \tag{22.5}$$

Here,

$$\beta_1 = \rho \frac{\sigma_2}{\sigma_1} \tag{22.6}$$

is the regression coefficient of Y_{i2} on $Y_{i1} = y_{i1}$, with ρ following from $\sigma_{12} = \rho \sigma_1 \sigma_2$; and $\overline{y}_1$ and $\overline{y}_2$ are the means of the measurements at the first and second occasion, respectively, among the completers. Several observations can be made. First, under MCAR, the completers and dropouts have equal distributions at the first occasion, and hence the correction term has expectation zero, rendering, again, the frequentist (least squares) and likelihood methods equivalent, *even though they do not produce exactly the same point estimator*. Second, when there is no correlation between the first and second measurements, the regression coefficient $\beta_1 = 0$ and, hence, there is again no correction.

Let us assess the implications for the issues raised above, especially in the context of the orthodontic growth data. The likelihood takes into account the expectation of the missing measurements, given the observed ones. In our data, this only occurs at the age of 10. Comparing the small (all children) with the large (remaining children) bullets and diamonds in Figure 22.5, it is clear that those remaining on study have larger measurements than those removed. The direct-likelihood correction has produced estimates at the age of 10 that are situated below the observed means. Obviously, the likelihood tends to overcorrect in this case. The reason for this is that the estimated correlation between repeated measures at ages 8 and 10 is substantially larger than the correlation between ages 10 and 12. Such variability is not unexpected in relatively small samples. Hence, a careful reflection on the variance–covariance structure is much more important here than when data are complete and balanced. We return to these points in the next section. There are also important consequences for model checking, because the difference between observed and expected quantities can be a function of relatively poor fit *and* the adjustment of the estimates when missingness is not of MCAR type. We return to this in Section 22.4.2.

Additionally, the coefficient β_1 depends on the variance components σ_1^2, σ_{12}, and σ_2^2, as is clear from (22.6), implying that a misspecified variance structure may lead to bias in $\widehat{\mu_2}$. Thus, the well-known independence between the distributions of the estimators for the mean vector and variance–covariance matrix in multivariate normal populations holds, once again, only when the data are balanced and complete.

The adequate performance of Model 7b is due to the fact that the expected mean of a missing age 10 measurement gives equal weight to all surrounding measurements, rather than overweighting the age 8 measurement due to an accidentally high correlation. The zero correlations in Model 8b do not allow for such a correction, and, hence, the information that the age 8, 12, and 14 measurements for the incomplete profiles are relatively low is wasted.

Similar calculations can be performed in the contingency table case as well as in any other likelihood setting (Molenberghs and Kenward, 2007, Part IV).

22.4.1.4 *Deviances and saturated models*

Revisiting Table 22.4, we observe that a deviance comparison between BRD1 and any of BRD2–5 and of the latter with BRD6–9 shows the earlier models suffer from a poor fit. Thus, effectively, we are left with BRD6–9 as candidate models for the SPO data. However, all four models produce exactly the same likelihood at maximum. This is not surprising because the models contain eight parameters, equal to the number of degrees of freedom in the *observed* data. Nevertheless, the estimates of θ differ among these four models. The reason is that θ is a function, not only of the model fit to the observed data, but of the model's *prediction* for the unobserved data, given what has been observed. Thus, model fit and the concept of saturation can be seen either as relative to the observed data, or relative to the complete data, observed and unobserved simultaneously. This, again, poses specific challenges for model selection and assessment of model fit.

All models BRD6–9 being of the NMAR type, it is tempting to conclude that all evidence points to NMAR as the most plausible missing-data mechanism. Notwithstanding this observation, one cannot even so much as formally exclude MAR. Indeed, Molenberghs et al. (2007) have shown that for every NMAR model considered, there is an associated MAR "bodyguard," a model reproducing the same fit as the original NMAR model, but predicting the unobserved data given the observed ones consistent with MAR. Formal derivations are given in Molenberghs et al. (2007) and are summarized in Section 22.5. The corresponding estimates for the proportion θ in favor of independence are presented in the last column of Table 22.4. Let us informally study the relationship and its implications by means of models BRD1, BRD2, BRD7, and BRD9, fitted to the SPO data. BRD1 assumes MCAR, all others NMAR. Only BRD7 and BRD9 saturate the observed-data degrees of freedom.

Table 22.6 The Slovenian Public Opinion Survey: Analysis Restricted to the Independence and Attendance Questions

Fit of BRD7, BRD7(MAR), BRD9, and BRD9(MAR) to Incomplete Data

| 1439 | 78 | 159 | 144 | 54 | 136 |
| 16 | 16 | 32 | | | |

Fit of BRD1 and BRD1(MAR) to Incomplete Data

| 1381.6 | 101.7 | 182.9 | 179.7 | 18.3 | 136.0 |
| 24.2 | 41.4 | 8.1 | | | |

Fit of BRD2 and BRD2(MAR) to Incomplete Data

| 1402.2 | 108.9 | 159.0 | 181.2 | 16.8 | 136.0 |
| 15.6 | 22.3 | 32.0 | | | |

Note: The fit of models BRD1, BRD2, BRD7, and BRD9, and their MAR counterparts, to the observed data is shown.

The incomplete data as observed (see Table 22.2), as predicted by each of the four models, and as predicted by these four models' MAR counterparts, are displayed in Table 22.6. The corresponding predictions of the hypothetical, complete data are presented in Table 22.7.

The fits of models BRD7, BRD9, and their MAR counterparts coincide with the observed data. As follows from Molenberghs et al. (2007), every model produces exactly the same fit as does its MAR counterpart; hence, this is seen for all four models. Because BRD1 is MCAR and hence MAR to begin with, it is the only one coinciding with its MAR counterpart, as indeed BRD1≡BRD1(MAR). Further, while BRD7 and BRD9 produce a different fit to the complete data, BRD7(MAR) and BRD9(MAR) coincide. This is because the fits of BRD7 and BRD9 coincide with respect to their fit to the observed data; because they are saturated, they coincide as such with the incomplete, observed data.

An important observation for model assessment and selection is that the five models BRD6, BRD7, BRD8, BRD9, and BRD6(MAR)≡BRD7(MAR)≡BRD8(MAR)≡BRD9 at the same time saturate the observed-data degrees of freedom and exhibit a dramatically different prediction of the full data. Thus, five perfectly fitting models produce five different estimates for the proportion in favor of independence.

22.4.2 Model selection and assessment with incomplete data

The five issues of Section 22.4.1 originate from the need, when fitting models to incomplete data, to manage two aspects rather than a single one, as schematically represented in Figure 22.6: the contrast between data and model is supplemented with a second contrast between their complete and incomplete versions.

Ideally, we should like to consider Figure 22.6(b), where the comparison is made entirely at the complete level. Because the complete data are, by definition, beyond reach, it is tempting but dangerous to settle for Figure 22.6(c). This would happen if we concluded that Model 1 fit poorly to the orthodontic growth data, as elucidated in Figure 22.7. Such a conclusion would ignore that the model fit is at the complete-data level, accounting for 16 boys and 11 girls at the age of 10, whereas the data only represent the residual 11 boys and 7 girls at the age of 10. Thus, a fair model assessment should be confined to the situations laid out in Figure 22.6(b) and (d). We will start out with the simpler (d) and then return to (b).

Table 22.7 The Slovenian Public Opinion Survey: Analysis Restricted to the Independence and Attendance Questions

	Fit of BRD1 and BRD1(MAR) to Complete Data					
1381.6	101.7	170.4	12.5	176.6	13.0	121.3	9.0
24.2	41.4	3.0	5.1	3.1	5.3	2.1	3.6

	Fit of BRD2 to Complete Data					
1402.2	108.9	147.5	11.5	179.2	13.9	105.0	8.2
15.6	22.3	13.2	18.8	2.0	2.9	9.4	13.4

	Fit of BRD2(MAR) to Complete Data					
1402.2	108.9	147.7	11.3	177.9	12.5	121.2	9.3
15.6	22.3	13.3	18.7	3.3	4.3	2.3	3.2

	Fit of BRD7 to Complete Data					
1439	78	3.2	155.8	142.4	44.8	0.4	112.5
16	16	0.0	32.0	1.6	9.2	0.0	23.1

	Fit of BRD9 to Complete Data					
1439	78	150.8	8.2	142.4	44.8	66.8	21.0
16	16	16.0	16.0	1.6	9.2	7.1	41.1

	Fit of BRD7(MAR) and BRD9(MAR) to Complete Data					
1439	78	148.1	10.9	141.5	38.4	121.3	9.0
16	18	11.8	20.2	2.5	15.6	2.1	3.6

Note: The fit of models BRD1, BRD2, BRD7, and BRD9, and their MAR counterparts, to the hypothetical complete data is shown.

Assessing whether Model 1 fits the incomplete version of the growth data set well can be achieved by comparing the observed means at the age of 10 to its prediction by the model. This implies we have to confine model fit to those children actually observed at the age of 10.

Turning to the analysis of the SPO survey, the principle behind Figure 22.6(d) would lead to the conclusion that the five models BRD6, BRD7, BRD8, BRD9, and BRD6(MAR)≡BRD7(MAR)≡BRD8(MAR)≡BRD9 perfectly fit the observed data, as can be seen in Table 22.6 (first panel). Notwithstanding this, the models drastically differ in their complete-data level fit (Table 22.7) as well as in the corresponding estimates of the proportion in favor of independence, which ranges over $[0.74, 0.89]$. This points to the need for supplementing model assessment, even when carried out in the preferable situation of Figure 22.6(d), with a form of sensitivity analysis.

We will study the two important tasks in turn: (i) model fit to the *observed* data, and (ii) sensitivity analysis. The first of these is discussed next; the latter throughout the remainder of this chapter.

As stated before, model fit to the observed data can be checked by means of either what we will label scenario I, as laid out in Figure 22.6(b), or by means of scenario II of Figure 22.6(d).

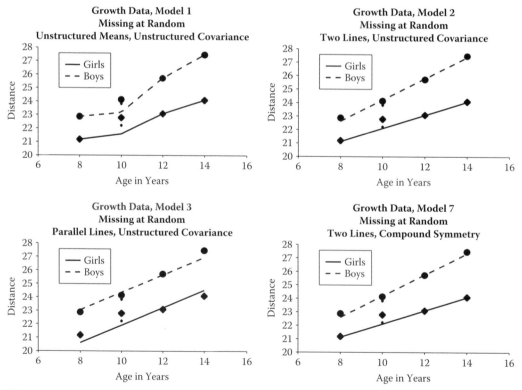

(a)   **Model**
                              Complete   Incomplete
      Raw   Complete   _____
      Data Incomplete _____

(b)   **Model**
                              Complete   Incomplete
      Raw   Complete   ✳   _____
      Data Incomplete _____

(c)   **Model**
                              Complete   Incomplete
      Raw   Complete   _____
      Data Incomplete   ✳   _____

(d)   **Model**
                              Complete   Incomplete
      Raw   Complete   _____
      Data Incomplete _____✳_____

Figure 22.6 Model assessment when data are incomplete. (a) Two dimensions in model assessment exercise when data are incomplete. (b) Ideal situation. (c) Dangerous situation, bound to happen in practice. (d) Comparison of data and model at coarsened, observable level.

Figure 22.7 The orthodontic growth data. Profiles for the growth data set, from a selected set of models. MAR analysis. (The small symbols at age 10 are the observed group means for the complete data set.)

Under scenario I, we conclude BRD6–9 or their MAR counterpart fit perfectly. There is nothing wrong with such a conclusion, as long as we realize *there is more than one model* with this very same property, while at the same time they lead to different substantive conclusions.

Turning to the orthodontic growth data, considering the fit of Model 1 to the data has some interesting ramifications. When the ordinary least-squares fit is considered, only valid under MCAR, one would conclude that there is a perfect fit to the observed means, also at the age of 10. The fit using maximum likelihood would apparently show a discrepancy because the observed mean refers to a reduced sample size while the fitted mean, similar to (22.5), is based on the entire design.

These considerations suggest that caution is required when we consider the fit of a model to an incomplete set of data; moreover, extension and/or modification of the classical model assessment paradigms may be needed. In particular, it is of interest to consider assessment under scenario II.

Gelman et al. (2005) proposed a scenario II method. The essence of their approach is as follows. First, a model, saturated or non-saturated, is fitted to the observed data. Under the fitted model, and assuming ignorable missingness, data sets simulated from the fitted model should "look similar" to the actual data. Therefore, multiple sets of data are sampled from the fitted model, and compared to the data set at hand. Because what one actually observes consists of not only the actually observed outcome data, but also realizations of the missingness process, comparison with the simulated data would also require simulation from, and hence full specification of, the missingness process. This added complexity is avoided by augmenting the observed outcomes with imputations drawn from the fitted model, conditional on the observed responses, and by comparing the so-obtained completed data set with the multiple versions of simulated complete data sets. Such a comparison will usually be based on relevant summary characteristics, such as time-specific averages or standard deviations. As suggested by Gelman et al. (2005), this so-called data-augmentation step could be done multiple times, in line with multiple-imputation ideas from Rubin (1987). However, in cases with a limited amount of missing observations, the between-imputation variability will be far less important than the variability observed between multiple simulated data sets. This is in contrast to other contexts to which the technique of Gelman et al. (2005) has been applied, for example, situations where latent unobservable variables are treated as "missing."

22.4.3 Model assessment for the orthodontic growth data

Let us first apply the method to the orthodontic growth data. The first model considered assumes a saturated mean structure (as in Model 1), with a compound-symmetric covariance structure. Twenty data sets are simulated from the fitted model, and time-specific sample averages are compared to the averages obtained from augmenting the observed data based on the fitted model. The results are shown in Figure 22.8(a)(a). The sample average at age 10, for the girls, is relatively low compared to what would be expected under the fitted model. Because the mean structure is saturated, this may indicate lack of fit of the covariance structure. We therefore extend the model by allowing for gender-specific covariance structures. The results under this new model are presented in Figure 22.8(a)(b). The observed data are now less extreme compared to what is expected under the fitted model. Formal comparison of the two models, based on a likelihood ratio test, indeed rejects the first model in favor of the second one ($p = 0.0003$), with much more between-subject variability for the girls than for the boys, while the opposite is true for the within-subject variability.

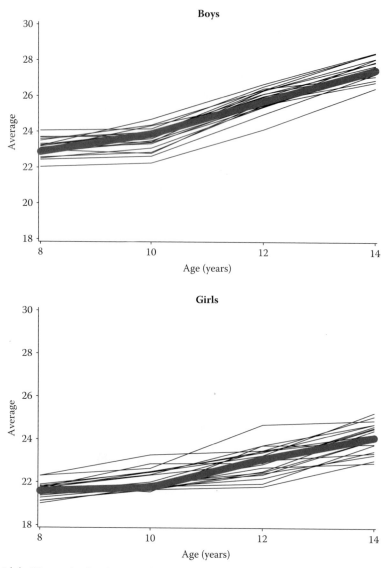

Figure 22.8(a) The orthodontic growth data. Sample averages for the augmented data (bold line), compared to sample averages from 20 simulated data sets, based on the method of Gelman et al. (2005). Both models assume a saturated mean structure and compound-symmetric covariance. Model 1a assumes the same covariance structure for boys and girls, while Model 1b allows gender-specific covariances.

22.4.4 Model assessment for the Slovenian Public Opinion Survey

Let us now turn to the Slovenian Public Opinion Survey data. In such a contingency table case, the above approach can be simplified to comparing the model fit to the complete data, such as presented in Table 22.6, with its counterpart obtained from extending the observed, incomplete data to their complete counterpart by means of the fitted model. Here, we have to distinguish between saturated and non-saturated models. For saturated models, such as BRD6–9 and their MAR counterparts, this is simply the same table as the model fit and, again, all models are seen to fit perfectly. Of course, this statement needs further qualification. It still merely means that these models fit the *incomplete* data perfectly, while each one of them tells a different, unverifiable story about the unobserved data given the

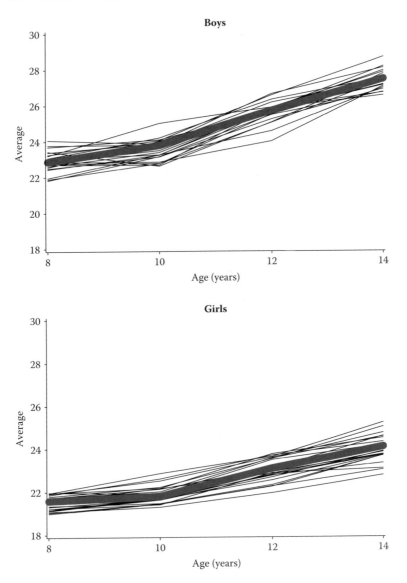

Figure 22.8(b) (*Continued*)

observed ones. In contrast, for the non-saturated models, such as BRD1–5 and their MAR counterparts, a so-completed table is different from the fitted one. To illustrate this, the completed tables are presented in Table 22.8 for the same set of models as in Table 22.6 and Table 22.7.

A number of noteworthy observations can be made. First, BRD1≡BRD1(MAR) exhibit the poorest fit (i.e., the largest discrepancies between this completed table and the model fit as presented in Table 22.7), with an intermediate-quality fit for a model with 7 degrees of freedom, such as BRD2, and a perfect fit for BRD7, BRD9, and their MAR counterparts. Second, compare the data completed using BRD1 (Table 22.8) to the fit of BRD1 (Table 22.7). The data for the group of completers are evidently equal to the original data (Table 22.6) because here no completion takes place; the complete data for the subjects without observations are entirely equal to the model fit (Table 22.7) as, here, there are no data from which to start. The complete data for the two partially classified tables take a

Table 22.8 The Slovenian Public Opinion Survey: Analysis Restricted to the Independence and Attendance Questions

Completed Data Using BRD1≡BRD1(MAR) Fit							
1439	78		148.1	10.9		141.5	38.4	121.3	9.0
16	16		11.9	20.1		2.5	15.6	2.1	3.6

Completed Data Using BRD2 Fit							
1439	78		147.5	11.5		142.4	44.7	105.0	8.2
16	16		13.2	18.8		1.6	9.3	9.4	13.4

Completed Data Using BRD2(MAR) Fit							
1439	78		147.7	11.3		141.4	40.2	121.2	9.3
16	16		13.3	18.7		2.6	13.8	2.3	3.2

Completed Data Using BRD7 Fit							
1439	78		3.2	155.8		142.4	44.8	0.4	112.5
16	16		0.0	32.0		1.6	9.2	0.0	23.1

Completed Data Using BRD9 Fit							
1439	78		150.8	8.2		142.4	44.8	66.8	21.0
16	16		16.0	16.0		1.6	9.2	7.1	41.1

Completed Data Using BRD7(MAR)≡BRD9(MAR) Fit							
1439	78		148.1	10.9		141.5	38.4	121.3	9.0
16	18		11.8	20.2		2.5	15.6	2.1	3.6

Note: Completed versions of the observed data, using the fit of the models BRD1, BRD2, BRD7, and BRD9, and their MAR counterparts.

position in between and hence are not exactly equal to the model fit. Third, note that the above statement needs amending for BRD2 and BRD2(MAR). Now, the first subtable of partially classified subjects exhibits an exact match between completed data and model fit, while this is not true for the second subtable. The reason is that BRD2 allows missingness on the second question to depend on the first one, leading to saturation of the first subtable, whereas missingness on the first question is independent of one's opinion on either question.

While the method is elegant and gives us a handle on the quality of the model fit to the incomplete data while contemplating the completed data and the full model fit, the method is unable to distinguish between the saturated models BRD6–9 and the MAR counterpart, as any method would. This phenomenon points to the need for sensitivity analysis, a topic to which the remainder of the chapter is devoted.

22.5 Every NMAR model has an MAR bodyguard

Over the last decade a variety of models to analyze incomplete multivariate and longitudinal data has been proposed, many of which allow for the missingness mechanism to be NMAR (Diggle and Kenward, 1994; for a review, see Chapter 18 in this volume).

The fundamental problem implied by such models, to which we refer as sensitivity to unverifiable modeling assumptions, has, in turn, sparked off various strands of research in what is now termed *sensitivity analysis*. The nature of sensitivity originates from the fact that an NMAR model is not fully verifiable from the data, rendering the formal distinction between NMAR and random missingness (MAR), where only covariates and observed outcomes influence missingness, hard or even impossible, unless one is prepared to accept the posited NMAR model in an unquestioning way.

We will formally show, in line with the Slovenian Public Opinion Survey analysis of the previous section, that a definitive, data-based distinction between MAR and NMAR is not possible, in the sense that each NMAR model fit to a set of observed data can be reproduced exactly by an MAR counterpart. Of course, such a pair of models will produce different predictions of the unobserved outcomes, given the observed ones.

Such a position is in contrast to the view that one can test for an NMAR mechanism using the data under analysis. Such tests, comparing MAR and NMAR mechanisms, can of course be constructed using conventional statistical methodology as done, for example, by Diggle and Kenward (1994).

Specifically, the following steps are involved: (1) an NMAR model is fitted to the data; (2) the fitted model is reformulated in a PMM form; (3) the density or distribution of the unobserved measurements given the observed ones, and given a particular response pattern, is replaced by its MAR counterpart; and (4) it is established that such an MAR counterpart uniquely exists. We will suppress covariates $\boldsymbol{x}_i$ from the notation, while allowing them to be present.

In the first step, an NMAR model is fitted to the observed set of data. Suppose, for example, that the model is written in SeM format. Then the observed-data likelihood is

$$L = \prod_i \int f(\boldsymbol{y}_i^o, \boldsymbol{y}_i^m | \boldsymbol{\theta}) f(\boldsymbol{r}_i | \boldsymbol{y}_i^o, \boldsymbol{y}_i^m, \boldsymbol{\psi}) d\boldsymbol{y}_i^m. \tag{22.7}$$

Denoting the resulting parameter estimates by $\widehat{\boldsymbol{\theta}}$ and $\widehat{\boldsymbol{\psi}}$, respectively, the fit to the hypothetical full data is

$$f(\boldsymbol{y}_i^o, \boldsymbol{y}_i^m, \boldsymbol{r}_i | \widehat{\boldsymbol{\theta}}, \widehat{\boldsymbol{\psi}}) = f(\boldsymbol{y}_i^o, \boldsymbol{y}_i^m | \widehat{\boldsymbol{\theta}}) f(\boldsymbol{r}_i | \boldsymbol{y}_i^o, \boldsymbol{y}_i^m, \widehat{\boldsymbol{\psi}}). \tag{22.8}$$

For the second step, the full density (22.8) can be re-expressed in PMM form as

$$f(\boldsymbol{y}_i^o, \boldsymbol{y}_i^m | \boldsymbol{r}_i, \widehat{\boldsymbol{\theta}}, \widehat{\boldsymbol{\psi}}) f(\boldsymbol{r}_i | \widehat{\boldsymbol{\theta}}, \widehat{\boldsymbol{\psi}})$$

$$= f(\boldsymbol{y}_i^o | \boldsymbol{r}_i, \widehat{\boldsymbol{\theta}}, \widehat{\boldsymbol{\psi}}) f(\boldsymbol{r}_i | \widehat{\boldsymbol{\theta}}, \widehat{\boldsymbol{\psi}}) f(\boldsymbol{y}_i^m | \boldsymbol{y}_i^o, \boldsymbol{r}_i, \widehat{\boldsymbol{\theta}}, \widehat{\boldsymbol{\psi}}). \tag{22.9}$$

A similar reformulation can be considered for an SPM. In a PMM setting, the model will have been expressed in this form to begin with.

Note that, in line with PMM theory, the final term on the right-hand side of (22.9), $f(\boldsymbol{y}_i^m | \boldsymbol{y}_i^o, \boldsymbol{r}_i, \widehat{\boldsymbol{\theta}}, \widehat{\boldsymbol{\psi}})$, is not identified from the observed data. In this case, it is determined solely from modeling assumptions. Within the PMM framework, identifying restrictions have to be considered (Little, 1994a; Molenberghs et al., 1998; Kenward, Molenberghs, and Thijs, 2003).

In the third step, this factor is replaced by the appropriate MAR counterpart. For this, we need the following lemma, which defines MAR within the PMM framework.

Lemma. In the PMM framework, the missing-data mechanism is MAR if and only if

$$f(\boldsymbol{y}_i^m | \boldsymbol{y}_i^o, \boldsymbol{r}_i, \boldsymbol{\theta}) = f(\boldsymbol{y}_i^m | \boldsymbol{y}_i^o, \boldsymbol{\theta}).$$

This means that, in a given pattern, the conditional distribution of the unobserved components given the observed ones equals the corresponding distribution marginalized over the patterns. The proof is given in Molenberghs et al. (1998).

Using the lemma, it is clear that $f(\boldsymbol{y}_i^m|\boldsymbol{y}_i^o, \boldsymbol{r}_i, \widehat{\boldsymbol{\theta}}, \widehat{\boldsymbol{\psi}})$ needs to be replaced by

$$h(\boldsymbol{y}_i^m|\boldsymbol{y}_i^o, \boldsymbol{r}_i) = h(\boldsymbol{y}_i^m|\boldsymbol{y}_i^o) = f(\boldsymbol{y}_i^m|\boldsymbol{y}_i^o, \widehat{\boldsymbol{\theta}}, \widehat{\boldsymbol{\psi}}), \qquad (22.10)$$

where the $h(\cdot)$ notation is used for brevity. Note that the density in (22.10) follows from the SeM-type marginal density of the complete-data vector. Sometimes, therefore, it may be more convenient to replace the notation $\boldsymbol{y}_i^o$ and $\boldsymbol{y}_i^m$ by one that explicitly indicates which components under consideration are observed and missing in pattern $\boldsymbol{r}_i$:

$$h(\boldsymbol{y}_i^m|\boldsymbol{y}_i^o, \boldsymbol{r}_i) = h(\boldsymbol{y}_i^m|\boldsymbol{y}_i^o) = f\{(y_{ij})_{r_j=0}|(y_{ij})_{r_j=1}, \widehat{\boldsymbol{\theta}}, \widehat{\boldsymbol{\psi}}\}. \qquad (22.11)$$

Thus, (22.11) provides a unique way of extending the model fit to the observed data, within the MAR family. To show formally that the fit remains the same, we consider the observed-data likelihood based on (22.7) and (22.9):

$$
\begin{aligned}
L &= \prod_i \int f(\boldsymbol{y}_i^o, \boldsymbol{y}_i^m|\boldsymbol{\theta}) f(\boldsymbol{r}_i|\boldsymbol{y}_i^o, \boldsymbol{y}_i^m, \boldsymbol{\psi}) d\boldsymbol{y}_i^m \\
&= \prod_i \int f(\boldsymbol{y}_i^o|\boldsymbol{r}_i, \boldsymbol{\theta}, \boldsymbol{\psi}) f(\boldsymbol{r}_i|\boldsymbol{\theta}, \boldsymbol{\psi}) f(\boldsymbol{y}_i^m|\boldsymbol{y}_i^o, \boldsymbol{r}_i, \boldsymbol{\theta}, \boldsymbol{\psi}) d\boldsymbol{y}_i^m \\
&= \prod_i f(\boldsymbol{y}_i^o|\boldsymbol{r}_i, \boldsymbol{\theta}, \boldsymbol{\psi}) f(\boldsymbol{r}_i|\boldsymbol{\theta}, \boldsymbol{\psi}) \\
&= \prod_i \int f(\boldsymbol{y}_i^o|\boldsymbol{r}_i, \boldsymbol{\theta}, \boldsymbol{\psi}) f(\boldsymbol{r}_i|\boldsymbol{\theta}, \boldsymbol{\psi}) h(\boldsymbol{y}_i^m|\boldsymbol{y}_i^o) d\boldsymbol{y}_i^m.
\end{aligned}
$$

The above results justify the following theorem:

Theorem. Every fit to the observed data, obtained from fitting an NMAR model to a set of incomplete data, is exactly reproducible from an MAR model.

The key computational consequence is the need to obtain $h(\boldsymbol{y}_i^m|\boldsymbol{y}_i^o)$ in (22.10) or (22.11). This means, for each pattern, the conditional density of the unobserved measurements given the observed ones needs to be extracted from the marginal distribution of the complete set of measurements. Suggestions for implementation can be found in Molenberghs et al. (2007). Again, the main consequence is that one cannot test NMAR against MAR, without making unverifiable assumptions about the alternative model.

22.6 Regions of ignorance and uncertainty

Acknowledging one's inability to choose definitively between competing models is equivalent to agreeing that there is both *imprecision*, due to (finite) random sampling, and *ignorance*, due to incompleteness. Molenberghs, Kenward, and Goetghebeur (2001) and Kenward, Goetghebeur, and Molenberghs (2001) combined both concepts into *uncertainty*. We will present the method, based on a simple but illustrative running example, the prevalence of HIV in Kenya data. Subsequently, the Slovenian Public Opinion Survey data will be analyzed. The theoretical underpinnings of the approach can be found in Vansteelandt et al. (2006).

22.6.1 Prevalence of HIV in Kenya

In the context of disease monitoring, HIV prevalence is to be estimated in a population of pregnant women in Kenya (Molenberghs, Kenward, and Goetghebeur, 2001). To this end, $N = 787$ samples from Kenyan women were obtained, with the following results: known HIV+, $r = 52$; known HIV−, $n - r = 699$; and (owing to test failure) unknown HIV status, $N - n = 36$. We deduce immediately that the number of HIV+ women lies within

[52, 88] out of 787, producing a *best–worst case interval* of $[6.6, 11.2]\%$. This example, on the one hand, is extremely important from a public health point of view and, on the other hand, provides the simplest setting for sensitivity analysis. Furthermore, the best–worst case interval produces the simplest answer to the sensitivity analysis question. This idea has been used, for example, in the survey literature. While Cochran (1977) considered it and subsequently rejected it on the grounds of overly wide intervals, we believe that in this particular example it does provide useful information and, more importantly, that it leads to a versatile starting point for more elaborate modeling, based on which the width of the intervals can be reduced further.

22.6.2 Uncertainty and sensitivity

To fix ideas, we focus on monotone patterns first. Formally, these can be regarded as coming from two 2×2 tables, of which the probabilities are displayed in Table 22.9. In contrast, we only observe data corresponding to the probabilities as displayed in Table 22.10, where + indicates summation over the corresponding subscript.

A sample from Table 22.10 produces empirical proportions representing the corresponding πs with error. This results in so-called *imprecision*, which is usually captured by way of such quantities as standard errors and confidence intervals. In sufficiently regular problems, imprecision disappears as the sample size tends to infinity and the estimators are consistent. What remains is *ignorance* regarding the redistribution of $\pi_{0,1+}$ and $\pi_{0,2+}$ over the second outcome value. In Table 22.10, there is constrained ignorance regarding the values of $\pi_{0,jk}$ ($j, k = 1, 2$) and hence regarding any derived parameter of scientific interest. For such a parameter, ψ say, a region of possible values that is consistent with Table 22.10 is called the region of ignorance. Analogously, an observed incomplete table leaves ignorance regarding the would-be observed complete table, which in turn leaves imprecision regarding the true complete probabilities. The region of estimators for ψ that is consistent with the observed data provides an estimated region of ignorance. We then conceive a $(1 - \alpha)100\%$ *region of uncertainty* as a larger region in the spirit of a confidence region, designed to capture the combined effects of imprecision and ignorance. It can be seen as a confidence region around the ignorance region.

In standard statistical practice, ignorance is hidden in the consideration of a single identified model. In Section 22.6.3 and Section 22.6.4, we first contrast several identified models for both monotone and non-monotone patterns of missingness, producing a conventional sensitivity analysis. When introducing more formal instruments, it is shown that a conventional assessment can be inadequate and even misleading.

Table 22.9 Theoretical Distribution over Completed Cells for Monotone Patterns

$\pi_{1,11}$	$\pi_{1,12}$	$\pi_{0,11}$	$\pi_{0,12}$
$\pi_{1,21}$	$\pi_{1,22}$	$\pi_{0,21}$	$\pi_{0,22}$

Table 22.10 Theoretical Distribution over Observed Cells for Monotone Patterns

$\pi_{1,11}$	$\pi_{1,12}$	$\pi_{0,1+}$
$\pi_{1,21}$	$\pi_{1,22}$	$\pi_{0,2+}$

Table 22.11 Dropout Models Corresponding to the Setting of Table 22.9

Model	$q_{r\mid jk}$	# par.
1. MCAR	q_r	4
2. MAR	$q_{r\mid j}$	5
3. Protective	$q_{r\mid k}$	5
4. NMAR I	$g(q_{r\mid jk}) = \alpha + \beta_j + \gamma_k$	6
5. M_{sat}	$g(q_{r\mid jk}) = \alpha + \beta_j + \gamma_k + \delta_{jk}$	7

Model	Observed df	Complete df
1. MCAR	Non-Saturated	Non-Saturated
2. MAR	Saturated	Non-Saturated
3. NMAR(0) (Protective)	Saturated	Non-Saturated
4. NMAR(I)	Overspecified	Non-Saturated
5. NMAR(II) (M_{sat})	Overspecified	Saturated

Note: The function $g(\cdot)$ refers to a link function, e.g., the logit link $\text{logit}(p) = \log\{p/(1-p)\}$.

22.6.3 Models for monotone patterns

Applying a selection model factorization to Table 22.9, we get

$$\pi_{r,jk} = p_{jk} q_{r\mid jk},$$

where p_{jk} parameterizes the measurement process and $q_{r\mid jk}$ determines the non-response (or dropout) mechanism. In what follows, we will leave p_{jk} unconstrained and consider various forms for $q_{r\mid jk}$, as listed in Table 22.11.

As a consequence of incompleteness, we can consider degrees of freedom, both at the level of the observed data (Table 22.10), and in terms of the hypothetical complete data (Table 22.9). We refer to these as df(obs) and df(comp), respectively. It is important to realize that identification or saturation of a model is relative to the type of df considered. Model NMAR(0) is also termed "protective" because its missing-data mechanism satisfies the so-called protective assumption, meaning that missingness depends on the unobserved but not further on the observed outcomes. Model NMAR(II) is referred to as "M_{sat}" because it has three measurement parameters and four dropout parameters and therefore saturates df(comp). However, there are only five observable degrees of freedom, rendering this model overspecified when fitted to the observed data.

To ensure a model is identifiable, one has to impose restrictions. Conventional restrictions result from assuming an MCAR or MAR model (models 1 and 2, respectively). Models 2 and 3 both saturate df(obs) and hence are indistinguishable in terms of their fit to the observed data, although they will produce different complete-data tables. In Model 4, dropout is allowed to depend on both measurements but not on their interaction. As a consequence, it overspecifies df(obs) and underspecifies df(comp).

We are dealing here with the simplest possible longitudinal setting, for which there is only a very small number of identifiable models. It is important to see how these observations carry over to settings where a much larger number of identifiable models can be constructed. A natural generalization that allows such an extended class of models is the non-monotone setting.

22.6.4 Models for non-monotone patterns

We now consider the problem of modeling all patterns, monotone and non-monotone alike. The discrepancy between df(comp)= 15 and df(obs)= 8 is larger. A natural class of models

Table 22.12 Theoretical Distribution over Completed Cells for Non-Monotone Patterns

$\pi_{11,11}$ $\pi_{11,12}$	$\pi_{10,11}$ $\pi_{10,12}$	$\pi_{01,11}$ $\pi_{01,12}$	$\pi_{00,11}$ $\pi_{00,12}$
$\pi_{11,21}$ $\pi_{11,22}$	$\pi_{10,21}$ $\pi_{10,22}$	$\pi_{01,21}$ $\pi_{01,22}$	$\pi_{00,21}$ $\pi_{00,22}$

Table 22.13 Theoretical Distribution over Observed Cells for Non-Monotone Patterns

$\pi_{11,11}$ $\pi_{11,12}$	$\pi_{10,1+}$	$\pi_{01,+1}$ $\pi_{01,+2}$	$\pi_{00,++}$
$\pi_{11,21}$ $\pi_{11,22}$	$\pi_{10,2+}$		

for this situation is the one proposed by Baker, Rosenberger, and DerSimonian (1992), presented in Section 22.2.2. The complete-data and observed-data cell probabilities for this setting are presented in Table 22.12 and Table 22.13, respectively. The notation employed in the previous subsection is extended in an obvious way.

22.6.5 Formalizing ignorance and uncertainty

As we have emphasized in Section 22.6.2, the adoption of a single, identified model masks what we have termed ignorance. Examples are Models 1 through 3 in Table 22.11 and Models BRD1–9 (Section 22.2.2) of Section 22.6.4 for the monotone and non-monotone settings, respectively. Among these, a number of different models such as Models 2 and 3 in the monotone case, and BRD6–9, typically saturate df(obs). Naturally, these models cannot be distinguished in terms of their fit to the observed data alone, yet they can produce substantially different inferences. A straightforward sensitivity analysis for the situations considered in Section 22.6.3 and Section 22.6.4 uses a collection of such models. The resulting parameter ranges implied by the estimates of such models are *ad hoc* in nature and lack formal justification as a measure of ignorance. Therefore, we need to parameterize the set of models considered by means of one or more continuous parameters and then to consider all (or at least a range of) models along such a continuum. Such an idea was suggested by Nordheim (1984). Foster and Smith (1998) expanded on this idea, and, by referring to Baker and Laird (1988) and to Rubin, Stern, and Vehovar (1995), they suggested imposing a prior distribution on a range. While this is an obvious way to account for ignorance and still produce a single inference, these authors also noted that the posterior density is, due to the lack of information, often a direct or indirect reproduction of the prior. Little (1994a) presented confidence intervals for a number of values along a range of a sensitivity parameters. In this way, he combined ignorance and imprecision. Similar ideas are found in Little and Wang (1996).

Molenberghs, Kenward, and Goetghebeur (2001), Kenward, Goetghebeur, and Molenberghs (2001), and Vansteelandt et al. (2006) formalized the idea of such ranges of models. A natural way to achieve this goal is to consider models that would be identified if the data were complete, and then fit them to the observed, incomplete data, thereby producing a range of estimates rather than a point estimate. Where such overspecified models have appeared in the literature (Catchpole and Morgan, 1997), various strategies are used, such as applying constraints, to recover identifiability. In contrast, our goal is to use the non-identifiability to delineate the range of inferences consistent with the observed data, that is, to capture ignorance. Maximization of the likelihood function of the overspecified model is a natural approach.

We consider the simple setting of the data on HIV prevalence in Kenya, where r denotes the number of observed successes, $n - r$ the number of observed failures, and $N - n$ the

Table 22.14 Two Transformations of the Observed-Data Likelihood

Model I (MAR)	Model II (NMAR, M_{sat})
Parameterization:	
$\alpha = pq$	$\alpha = pq_1$
$\beta = (1-p)q$	$\beta = (1-p)q_2$
$\gamma = 1-q$	$\gamma = 1 - pq_1 - (1-p)q_2$
	$q_1 = q$
	$q_2 = q\lambda$
Solution:	
$\widehat{p} = \dfrac{\widehat{\alpha}}{\widehat{\alpha} + \widehat{\beta}} = \dfrac{r}{n}$	$pq_1 = \dfrac{r}{N}$
$\widehat{q} = \widehat{\alpha} + \widehat{\beta} = \dfrac{n}{N}$	$(1-p)q_2 = \dfrac{n-r}{N}$
	$\dfrac{r}{q_1} + \dfrac{n-r}{q_2} = N$
	$p : \left[\dfrac{r}{N}, \dfrac{N-n+r}{N} \right]$

number of unclassified subjects. Independent of the parameterization chosen, the observed-data log-likelihood can be expressed in the form

$$\ell = r \log \alpha + (n-r)\log\beta + (N-n)\log(1 - \alpha - \beta),$$

where α is the probability of an observed success and β is the probability of an observed failure. It is sometimes useful to denote $\gamma = 1 - \alpha - \beta$. We consider two models, of which the parameterization is given in Table 22.14. The first one is identified, while the second one is overparameterized. Here, p is the probability of a success (whether observed or not), q_1 (q_2) is the probability of being observed given a success (failure), and λ is the odds of being observed for failures *versus* successes. For Model I, the latter is assumed to be unity. Denote the corresponding log-likelihoods by ℓ_I and ℓ_{II}, respectively. In both cases,

$$\widehat{\alpha} = \frac{r}{N}, \qquad \widehat{\beta} = \frac{n-r}{N}.$$

Maximum likelihood estimators for p and q follow immediately under Model I, either by observing that the moments (α, β) map one-to-one onto the pair (p, q) or by directly solving ℓ_I. The solutions are given in Table 22.14. The asymptotic variance–covariance matrix for p and q is block-diagonal with well-known elements $p(1-p)/n$ and $q(1-q)/N$. Observe that we now obtain only one solution, a strong argument in favor of the current model. The probability of HIV+ is $\widehat{p} = 0.069$ (95% confidence interval $[0.0511, 0.0874]$) and the response probability is $\widehat{q} = 0.954$.

A similar, standard derivation is not possible for Model II, as the triplet (p, q_1, q_2) or, equivalently, the triplet (p, q, λ), is redundant. This follows directly from Catchpole and Morgan (1997) and Catchpole, Morgan, and Freeman (1998), whose theory shows that Model II is rank-deficient and Model I is of full rank. Because Model I is a submodel of Model II and saturates the observed data, so must every solution to ℓ_{II}, implying the relationships

$$pq_1 = \frac{r}{N}, \qquad (1-p)q_2 = \frac{n-r}{N}. \tag{22.12}$$

Constraints (22.12) imply

$$\widehat{p} = \frac{r}{Nq_1} = 1 - \frac{n-r}{Nq_2},$$

and hence

$$\frac{r}{q_1} + \frac{n-r}{q_2} = N.$$

The requirement that $q_1, q_2 \leq 1$ in (22.12) implies an allowable range for p:

$$p \in \left[\frac{r}{N}, \frac{N-n+r}{N} \right]. \tag{22.13}$$

For the prevalence of HIV in Kenya example, the best–worst case range $[0.066, 0.112]$ is recovered. Within this interval of ignorance, the MAR estimate obtained earlier is rather extreme, which is entirely due to the small proportion of HIV+ subjects. To account for the sampling uncertainty, the left estimate can be replaced by a 95% lower limit and the right estimate can be replaced by a 95% upper limit, yielding an interval of uncertainty $[0.0479, 0.1291]$.

Such overspecification of the likelihood can be managed in a more general fashion by considering a minimal set of parameters, $\boldsymbol{\eta}$, conditional upon which the others, $\boldsymbol{\mu}$, are identified, where $\boldsymbol{\psi} = (\boldsymbol{\eta}, \boldsymbol{\mu})$ upon possible reordering. We term $\boldsymbol{\eta}$ the *sensitivity parameter* and $\boldsymbol{\mu}$ the *estimable parameter*. Clearly, there will almost never be a unique choice for $\boldsymbol{\eta}$ and hence for $\boldsymbol{\mu}$. Each value of $\boldsymbol{\eta}$ will produce an estimate $\widehat{\boldsymbol{\mu}}(\boldsymbol{\eta})$. The union of these produces the estimated region of ignorance. A natural estimate of the region of uncertainty is the union of confidence regions based on each $\widehat{\boldsymbol{\mu}}(\boldsymbol{\eta})$. For the HIV+ example, one could choose $\boldsymbol{\mu} = (p, q_1)$ and $\eta = q_2$ or $\boldsymbol{\mu} = (p, q)$ and $\eta = \lambda$. The latter choice motivates our inclusion of λ in Table 22.14 as a sensitivity parameter.

It is not always the case that the range for $\boldsymbol{\eta}$ will be an entire line or real space, and, hence, specific measures may be needed to ensure that $\boldsymbol{\eta}$ is within its allowable range. As the choice of sensitivity parameter is non-unique, a proper choice can greatly simplify the treatment. It will be seen in what follows that the choice of λ as in Table 22.14 is an efficient one from a computational point of view. In contrast, the choice $\theta = q_2 - q_1$ would lead to cumbersome computations and will not be pursued. Of course, what is understood by a proper choice will depend on the context. For example, the sensitivity parameter can be chosen from the nuisance parameters, rather than from the parameters of direct scientific interest. Whether the parameters of direct scientific interest can overlap with the sensitivity set or not is itself an issue (White and Goetghebeur, 1998). For example, if the scientific question is a sensitivity analysis for treatment effect, then one should consider the implications of including the treatment effect parameters in the sensitivity set. There will be no direct estimate of imprecision available for the sensitivity parameter. Alternatively, if, given a certain choice of sensitivity parameter, the resulting profile likelihood has a simple form (analogous to the Box–Cox transformation, where conditioning on the transformation parameter produces essentially a normal likelihood), then such a parameter is an obvious candidate.

Given our choice of sensitivity parameter λ, simple algebra yields estimates for p and q (subscripted by λ to indicate dependence on the sensitivity parameter):

$$p_\lambda = \frac{\widehat{\alpha}\lambda}{\widehat{\beta} + \widehat{\alpha}\lambda} = \frac{\lambda r}{n - r(1-\lambda)},$$

$$q_\lambda = \frac{\widehat{\beta} + \widehat{\alpha}\lambda}{\lambda} = \frac{n - r(1-\lambda)}{N\lambda}.$$

Using the delta method, the estimated asymptotic variance–covariance matrix of p_λ and q_λ has entries

$$\widehat{\mathrm{Var}}(p_\lambda) = \frac{p_\lambda(1-p_\lambda)}{N\lambda q_\lambda}\left[1 + \frac{1-\lambda}{\lambda}(1-p_\lambda)\{1 - p_\lambda q_\lambda(1-\lambda)\}\right], \qquad (22.14)$$

$$\widehat{\mathrm{Cov}}(p_\lambda, q_\lambda) = -\frac{1}{N}p_\lambda(1-p_\lambda)\frac{1-\lambda}{\lambda}q_\lambda,$$

$$\widehat{\mathrm{Var}}(q_\lambda) = \frac{q_\lambda(1-q_\lambda)}{N}\left\{1 + \frac{1-p_\lambda}{1-q_\lambda}\frac{1-\lambda}{\lambda}\right\}.$$

Note that the parameter estimators are asymptotically correlated, except when $\lambda = 1$, that is, under the MAR assumption, or under boundary values ($p_\lambda = 0, 1; q_\lambda = 0$). This is in line with the ignorable nature of the MAR model.

We need to determine the set of allowable values for λ by requiring $0 \leq p_\lambda, q_\lambda, \lambda q_\lambda \leq 1$. These six inequalities reduce to

$$\lambda \in \left[\frac{n-r}{N-r}, \frac{N-(n-r)}{r}\right].$$

Clearly, $\lambda = 1$ is always valid. For the prevalence of HIV in Kenya example, the range equals $\lambda \in [0.951, 1.692]$.

Table 22.15 presents estimates for limiting cases. In particular, results for the HIV example are presented. The interval of ignorance for the success probability is thus seen to be as in (22.13). It is interesting to observe that the success odds estimator is linear in the sensitivity parameter; the resulting interval of ignorance equals

$$\mathrm{odds}(p): \left[\frac{r}{N-r}, \frac{N-n+r}{n-r}\right].$$

For the HIV case, the odds vary between 0.071 and 0.126.

Table 22.15 Prevalence of HIV in Kenya

Estimator	λ	$\lambda = \frac{n-r}{N-r}$	$\lambda = 1$	$\lambda = \frac{N-(n-r)}{r}$
p_λ	$\frac{\lambda r}{n-r(1-\lambda)}$	$\frac{r}{N}$	$\frac{r}{n}$	$\frac{N-n+r}{N}$
q_λ	$\frac{n-r(1-\lambda)}{N\lambda}$	1	$\frac{n}{N}$	$\frac{r}{N-(n-r)}$
$q_\lambda\lambda$	$\frac{n-r(1-\lambda)}{N}$	$\frac{n-r}{N-r}$	$\frac{n}{N}$	1
$\frac{p_\lambda}{1-p_\lambda}$	$\lambda\frac{r}{n-r}$	$\frac{r}{N-r}$	$\frac{r}{n-r}$	$\frac{N-(n-r)}{n-r}$
Prevalence of HIV in Kenya				
p_λ	$\frac{52\lambda}{699+52\lambda}$	0.066	0.069	0.112
q_λ	$\frac{699+52\lambda}{787\lambda}$	1.000	0.954	0.591
$q_\lambda\lambda$	$\frac{699+52\lambda}{787}$	0.951	0.954	1.000
$\frac{p_\lambda}{1-p_\lambda}$	0.074λ	0.071	0.074	0.126

Note: Limiting cases for the sensitivity parameter analysis.

For the success probability, the variance of p_λ is given by (22.14). For the success odds, we obtain

$$\widehat{\mathrm{Var}}\{\mathrm{odds}(p_\lambda)\} = \frac{1}{N\lambda q_\lambda} \frac{p_\lambda}{1-p_\lambda} \left[1 + \frac{1-\lambda}{\lambda}(1-p_\lambda)\{1-p_\lambda q_\lambda(1-\lambda)\} \right],$$

and, for the success logit,

$$\widehat{\mathrm{Var}}\{\mathrm{logit}(p_\lambda)\} = \frac{1}{N\lambda q_\lambda} \frac{1}{p_\lambda(1-p_\lambda)} \left[1 + \frac{1-\lambda}{\lambda}(1-p_\lambda)\{1-p_\lambda q_\lambda(1-\lambda)\} \right].$$

For each λ, a confidence interval C_λ can be constructed for every point within the allowable range of λ. The union of the C_λ is the *interval of uncertainty*, for either p, its odds, or its logit.

Figure 22.9 represents intervals of ignorance and uncertainty for the prevalence of HIV in the Kenya study. Note that the interval of ignorance (the inner interval, indicated by the projecting horizontal lines) is sufficiently narrow to be of practical use. The outer interval, formed by the lowest point of the lower and the highest point of the higher bold dashed lines, represents the interval of uncertainty. This is also sufficiently narrow from a practical viewpoint.

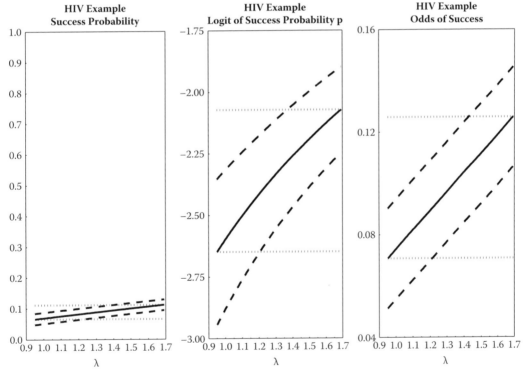

Figure 22.9 The HIV Prevalence Study. Graphical representation of interval of ignorance and interval of uncertainty. The solid bold line represents the point estimates conditional on the sensitivity parameter; its extremes (which are projected on the vertical axis) provide the limits of the interval of ignorance. The dashed bold lines graph the lower and upper confidence limits, conditional on the sensitivity parameter; their extremes form the interval of uncertainty (projecting lines on the vertical axis not shown.)

Table 22.16 The Slovenian Public Opinion Survey: Intervals of Ignorance and Intervals of Uncertainty for the Proportion θ (Confidence Interval) Attending the Plebiscite Resulting from Fitting

Model	df	loglik	$\hat{\theta}$	
			II	IU
Model 10	9	-2431.06	$[0.762, 0.893]$	$[0.744, 0.907]$
Model 11	9	-2431.06	$[0.766, 0.883]$	$[0.715, 0.920]$
Model 12	10	-2431.06	$[0.694, 0.905]$	

22.6.5.1 *Interval of ignorance for the Slovenian public opinion survey*

A sample from Table 22.10 produces empirical proportions, that is, estimates of the πs. This imprecision disappears as the sample size tends to infinity. What remains is ignorance regarding the redistribution of all but the first four πs over the missing outcomes value. This leaves ignorance regarding any probability in which at least one of the first or second indices is equal to 0, and hence regarding any derived parameter of scientific interest. For such a parameter, θ say, a region of possible values that is consistent with Table 22.10 is called a region of ignorance. Analogously, an observed incomplete table leaves ignorance regarding the would-be observed complete table, which in turn leaves imprecision regarding the true complete probabilities. The region of estimates for θ consistent with the observed data provides an estimated region of ignorance. The $(1 - \alpha)100\%$ *region of uncertainty* is a larger region in the spirit of a confidence region, designed to capture the combined effects of imprecision and ignorance. Various ways of constructing regions of ignorance and regions of uncertainty are conceivable. For a single parameter, the regions obviously become intervals.

The estimated intervals of ignorance and intervals of uncertainty are shown in Table 22.16, while a graphical representation of the yes votes is given in Figure 22.10. Model 10 is defined as $(\alpha_{j_2}, \beta_{j_1 j_2})$ with

$$\beta_{j_1 j_2} = \beta_0 + \beta_{j_1} + \beta_{j_2}, \tag{22.15}$$

while Model 11 assumes $(\alpha_{j_1 j_2}, \beta_{j_1})$ and uses

$$\alpha_{j_1 j_2} = \alpha_0 + \alpha_{j_1} + \alpha_{j_2}. \tag{22.16}$$

Finally, Model 12 is defined as $(\alpha_{j_1 j_2}, \beta_{j_1 j_2})$, a combination of both (22.15) and (22.16). Model 10 shows an interval of ignorance which is very close to $[0.741, 0.892]$, the range produced by the models BRD1 through BRD9, while Model 11 is somewhat sharper and

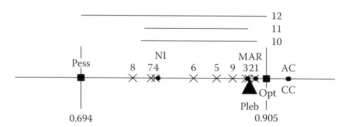

Figure 22.10 The Slovenian Public Opinion survey. Relative position for the estimates of "proportion of yes votes," based on the models considered in Rubin, Stern, and Vehovar (1995) and on the BRD models. The vertical lines indicate the non-parametric pessimistic–optimistic bounds. (Pess, pessimistic boundary; Opt, optimistic boundary; MAR, Rubin, Stern, and Vehovars' MAR model; NI, Rubin, Stern, and Vehovar 's NMAR model; AC, available cases; CC, complete cases; Pleb, plebiscite outcome. Numbers refer to the BRD models. Intervals of ignorance [Models 10 through 12] are represented by horizontal bars.)

just fails to cover the plebiscite value. However, it should be noted that the corresponding intervals of uncertainty contain the true value.

Interestingly, Model 12 virtually coincides with the non-parametric range even though it does not saturate the complete-data degrees of freedom. To do so, not two but in fact seven sensitivity parameters would have to be included. Thus, it appears that a relatively simple sensitivity analysis is sufficient to increase the insight in the information provided by the incomplete data about the proportion of valid yes votes.

22.7 Local and global influence

As is clear from the preceding sections, the conclusions from models for incomplete longitudinal data are sensitive to assumptions that cannot be checked from the data under analysis. Even if the multivariate normal model were the obvious preferred choice for describing the measurement process *if the data were complete*, the analysis of the actually observed, incomplete version would be subject to further, untestable modeling assumptions (Jansen et al., 2006b). The growing body of modeling tools for SeMs (Heckman, 1976; Diggle and Kenward, 1994) requires an understanding of such sensitivities (Glynn, Laird, and Rubin, 1986), as well as tools to deal with it (Draper, 1995; Vach and Blettner, 1995; Copas and Li, 1997).

The sensitivity of NMAR selection models was illustrated by Verbeke and Molenberghs (2000, Chapter 17), who showed, in the context of an onychomycosis study, that excluding a small amount of measurement error drastically changes the likelihood ratio test statistics for the MAR null hypothesis. Kenward (1998) revisited the analysis of the mastitis data performed by Diggle and Kenward (1994). In this study, the milk yields of 107 cows were to be recorded over two consecutive years. While data were complete in the first year, 27 measurements were missing in year 2 because these cows developed mastitis and their milk yield was seriously affected by this, and therefore deemed missing for the purposes of the scientific study. Although in the initial paper there was some evidence for NMAR, Kenward (1998) showed that removing two anomalous profiles from the 107 completely removed this evidence. Kenward also showed that a similar conclusion followed from changing the conditional distribution of the year 2 yield, given the year 1 yield, from a normal to a heavy-tailed t.

Several authors have advocated using local influence tools (Thijs, Molenberghs, and Verbeke, 2000; Molenberghs et al., 2001; Van Steen et al., 2001; Verbeke et al., 2001; Jansen et al., 2006b) for sensitivity analysis purposes. In particular, Molenberghs et al. (2001) revisited the mastitis example. They were able to identify the same two cows also found by Kenward (1998), in addition to another one. However, it is noteworthy that all three are cows with *complete* information, even though local influence methods were originally intended to identify subjects with mechanisms of missingness other than MAR. Thus, an important question concerns the combined nature of the data and model that leads to apparent evidence for a NMAR process. Jansen et al. (2006b) showed that a number of features or aspects, but not necessarily the (outlying) nature of the missingness mechanism in one or a few subjects, may be responsible for an apparent NMAR mechanism. Their work is reviewed in the next section.

We will describe the method for Gaussian outcomes and apply it to the mastitis in dairy cattle data. Details for the non-Gaussian case are discussed in Molenberghs and Kenward (2007); we will apply this method to the Slovenian Public Opinion Survey.

22.7.1 Gaussian outcomes

Thijs, Molenberghs, and Verbeke (2000), Molenberghs et al. (2001), and Verbeke et al. (2001) investigated sensitivity of estimation of quantities of interest, such as treatment effect, growth parameters, or the dropout model parameters, with respect to the dropout model assumptions. They started from the model proposed by Diggle and Kenward (1994). We will first introduce this, and then carry on with local influence analysis.

22.7.1.1   The Diggle–Kenward model for continuous outcomes

We assume that, as before, for subject $i = 1, \ldots, N$, a sequence of measurements Y_{ij} is designed to be measured at time points t_{ij}, $j = 1, \ldots, n_i$, resulting in a vector $\boldsymbol{Y}_i = (Y_{i1}, \ldots, Y_{in_i})'$ of measurements for each participant. If dropout occurs, then $\boldsymbol{Y}_i$ is only partially observed. We denote the occasion at which dropout occurs by $D_i > 1$, and $\boldsymbol{Y}_i$ is split into the (D_i-1)-dimensional observed component $\boldsymbol{Y}_i^o$ and the (n_i-D_i+1)-dimensional missing component $\boldsymbol{Y}_i^m$. In the case of no dropout, we let $D_i = n_i + 1$, and $\boldsymbol{Y}_i$ equals $\boldsymbol{Y}_i^o$. The likelihood contribution of the ith subject, based on the observed data $(\boldsymbol{y}_i^o, d_i)$, is proportional to the marginal density function

$$f(\boldsymbol{y}_i^o, d_i | \boldsymbol{\theta}, \boldsymbol{\psi}) = \int f(\boldsymbol{y}_i, d_i | \boldsymbol{\theta}, \boldsymbol{\psi}) \, d\boldsymbol{y}_i^o$$

$$= \int f(\boldsymbol{y}_i | \boldsymbol{\theta}) f(d_i | \boldsymbol{y}_i, \boldsymbol{\psi}) \, d\boldsymbol{y}_i^o,$$

in which a marginal model for $\boldsymbol{Y}_i$ is combined with a model for the dropout process, conditional on the response; and where $\boldsymbol{\theta}$ and $\boldsymbol{\psi}$ are vectors of unknown parameters in the measurement model and dropout model, respectively.

Let $\boldsymbol{h}_{ij} = (y_{i1}, \ldots, y_{i,j-1})'$ denote the observed history of subject i up to time $t_{i,j-1}$. The Diggle–Kenward model for the dropout process allows the conditional probability for dropout at occasion j, given that the subject was still observed at the previous occasion, to depend on the history $\boldsymbol{h}_{ij}$ and the possibly unobserved current outcome y_{ij}, but not on future outcomes y_{ik}, $k > j$. These conditional probabilities $\Pr(D_i = j | D_i \geq j, \boldsymbol{h}_{ij}, y_{ij}, \boldsymbol{\psi})$ can now be used to calculate the probability of dropout at each occasion:

$$\Pr(D_i = j | \boldsymbol{Y}_i, \boldsymbol{\psi}) \;=\; \Pr(D_i = j | \boldsymbol{h}_{ij}, y_{ij}, \boldsymbol{\psi})$$

$$= \begin{cases} \Pr(D_i = j | D_i \geq j, \boldsymbol{h}_{ij}, y_{ij}, \boldsymbol{\psi}), & j = 2, \\[2mm] \Pr(D_i = j | D_i \geq j, \boldsymbol{h}_{ij}, y_{ij}, \boldsymbol{\psi}) \\[1mm] \qquad \times \displaystyle\prod_{k=2}^{j-1} \left\{ 1 - \Pr(D_i = k | D_i \geq k, \boldsymbol{h}_{ik}, y_{ik}, \boldsymbol{\psi}) \right\}, & j = 3, \ldots, n_i, \\[4mm] \displaystyle\prod_{k=2}^{n_i} \left\{ 1 - \Pr(D_i = k | D_i \geq k, \boldsymbol{h}_{ik}, y_{ik}, \boldsymbol{\psi}) \right\}, & j = n_i + 1. \end{cases}$$

Diggle and Kenward (1994) combine a multivariate normal model for the measurement process with a logistic regression model for the dropout process. More specifically, the measurement model assumes that the vector $\boldsymbol{Y}_i$ of repeated measurements for the ith subject satisfies the linear regression model $\boldsymbol{Y}_i \sim N(X_i\boldsymbol{\beta}, \Sigma_i)$, $i = 1, \ldots, N$. The matrix Σ_i can be left unstructured or assumed to be of a specific form, for example, resulting from a linear mixed model, a factor-analytic structure, or spatial covariance structure (Verbeke and Molenberghs, 2000). A commonly used version of such a logistic dropout model is

$$\text{logit} \left\{ \Pr(D_i = j \mid D_i \geq j, \boldsymbol{h}_{ij}, y_{ij}, \boldsymbol{\psi}) \right\} = \psi_0 \;+\; \psi_1 y_{ij} \;+\; \psi_2 y_{i,j-1}. \tag{22.17}$$

More general models can easily be constructed by including the complete history $\boldsymbol{h}_{ij} = (y_{i1}, \ldots, y_{i,j-1})$, as well as external covariates, in the above conditional dropout model.

22.7.1.2 Local influence for the Diggle–Kenward model

For this, they considered the following perturbed version of dropout model (22.17):

$$\text{logit}\{g(\boldsymbol{h}_{ij}, y_{ij})\} = \text{logit}\left\{\Pr(D_i = j | D_i \geq j, \boldsymbol{y}_i)\right\} = \boldsymbol{h}'_{ij}\boldsymbol{\psi} + \omega_i y_{ij}, \qquad (22.18)$$

where the ω_i are local, individual-specific perturbations around a null model. These should not be confused with subject-specific parameters. Our null model will be the MAR model, corresponding to setting $\omega = 0$ in (22.17). Thus, the ω_i are perturbations that will be used only to derive influence measures (Cook, 1986).

Using this setup, one can study the impact on key model features, induced by small perturbations in the direction (or at least seemingly in the direction) of NMAR. This can be done by constructing local influence measures (Cook, 1986). When small perturbations in a specific ω_i lead to relatively large differences in the model parameters, this suggests that the subject is likely to contribute in a particular way to key conclusions. For example, if such a subject drives the model toward NMAR, then the conditional expectations of the unobserved measurements, given the observed ones, may deviate substantially from the ones under an MAR mechanism (Kenward, 1998). Such an observation is important also for this approach because then the impact on dropout model parameters extends to all functions that include these dropout parameters. One such function is the conditional expectation of the unobserved measurements, given the corresponding dropout pattern, $E(\boldsymbol{y}_i^m | \boldsymbol{y}_i^o, D_i, \boldsymbol{\theta}, \boldsymbol{\psi})$. As a consequence, the corresponding measurement model parameters will be affected as well.

Some caution is needed when interpreting local influence. Even though we may be tempted to conclude that an influential subject drops out non-randomly, this conclusion is misguided because we are not aiming to detect (groups of) subjects who drop out non-randomly but rather subjects who have a considerable impact on the dropout and measurement model parameters. An important observation is that a subject who drives the conclusions toward NMAR may be doing so, not only because his or her true data generating mechanism is of an NMAR type, but also for a wide variety of other reasons, such as an unusual mean profile or autocorrelation structure (Jansen et al., 2006b). Similarly, it is possible that subjects deviating from the bulk of the data because they are generated under NMAR go undetected by this technique. Thus, subjects identified in a local influence analysis should be assessed carefully for their precise impact on the conclusions.

We start by reviewing the key concepts of local influence (Cook, 1986). We denote the log-likelihood function corresponding to (22.18) by $\ell(\boldsymbol{\gamma}|\boldsymbol{\omega}) = \sum_{i=1}^{N} \ell_i(\boldsymbol{\gamma}|\omega_i)$, in which $\ell_i(\boldsymbol{\gamma}|\omega_i)$ is the contribution of the ith individual to the log-likelihood; and $\boldsymbol{\gamma} = (\boldsymbol{\theta}, \boldsymbol{\psi})$ is the s-dimensional vector grouping the parameters of the measurement model and the dropout model, not including the $N \times 1$ vector $\boldsymbol{\omega} = (\omega_1, \omega_2, \dots, \omega_N)'$ of weights defining the perturbation of the MAR model. It is assumed that $\boldsymbol{\omega}$ belongs to an open subset Ω of $\mathbb{R}^N$. For $\boldsymbol{\omega}$ equal to $\boldsymbol{\omega}_0 = (0, 0, \dots, 0)'$, $\ell(\boldsymbol{\gamma}|\boldsymbol{\omega}_0)$ is the log-likelihood function that corresponds to an MAR dropout model.

Let $\widehat{\boldsymbol{\gamma}}$ be the maximum likelihood estimator for $\boldsymbol{\gamma}$, obtained by maximizing $\ell(\boldsymbol{\gamma}|\boldsymbol{\omega}_0)$, and let $\widehat{\boldsymbol{\gamma}}_\omega$ denote the maximum likelihood estimator for $\boldsymbol{\gamma}$ under $\ell(\boldsymbol{\gamma}|\boldsymbol{\omega})$. In the local influence approach, $\widehat{\boldsymbol{\gamma}}_\omega$ is compared to $\widehat{\boldsymbol{\gamma}}$. Sufficiently different estimates suggest that the estimation procedure is sensitive to such perturbations. Cook (1986) proposed measuring the distance between $\widehat{\boldsymbol{\gamma}}_\omega$ and $\widehat{\boldsymbol{\gamma}}$ by the so-called likelihood displacement, defined by $LD(\boldsymbol{\omega}) = 2\{\ell(\widehat{\boldsymbol{\gamma}}|\boldsymbol{\omega}_0) - \ell(\widehat{\boldsymbol{\gamma}}_\omega|\boldsymbol{\omega})\}$. This takes into account the variability of $\widehat{\boldsymbol{\gamma}}$. Indeed, $LD(\boldsymbol{\omega})$ will be large if $\ell(\boldsymbol{\gamma}|\boldsymbol{\omega}_0)$ is strongly curved at $\widehat{\boldsymbol{\gamma}}$, which means that $\boldsymbol{\gamma}$ is estimated with high precision, and small otherwise. Therefore, a graph of $LD(\boldsymbol{\omega})$ versus $\boldsymbol{\omega}$ contains essential information on the influence of perturbations. It is useful to view this graph as the geometric surface formed by the values of the $(N+1)$-dimensional vector $\boldsymbol{\xi}(\boldsymbol{\omega}) = \{\boldsymbol{\omega}', LD(\boldsymbol{\omega})'\}'$, as $\boldsymbol{\omega}$ varies throughout Ω. Because this *influence graph* can only be constructed when $N = 2$, Cook (1986) proposed looking at local influence, that is, at the normal curvatures C_h of $\boldsymbol{\xi}(\boldsymbol{\omega})$ in $\boldsymbol{\omega}_0$, in the direction of some N-dimensional vector $\boldsymbol{h}$ of unit length. Let $\boldsymbol{\Delta}_i$ be the s-dimensional vector defined by

$$\boldsymbol{\Delta}_i = \left.\frac{\partial^2 \ell_i(\boldsymbol{\gamma}|\omega_i)}{\partial \omega_i \partial \boldsymbol{\gamma}}\right|_{\boldsymbol{\gamma}=\widehat{\boldsymbol{\gamma}}, \omega_i=0},$$

and define $\boldsymbol{\Delta}$ as the $(s \times N)$ matrix with $\boldsymbol{\Delta}_i$ as its ith column. Further, let $\ddot{L}$ denote the $(s \times s)$ matrix of second-order derivatives of $\ell(\boldsymbol{\gamma}|\boldsymbol{\omega}_0)$ with respect to $\boldsymbol{\gamma}$, also evaluated at $\boldsymbol{\gamma} = \widehat{\boldsymbol{\gamma}}$. Cook (1986) has then shown that C_h can be calculated as $C_h = 2|\boldsymbol{h}'\boldsymbol{\Delta}'\ddot{L}^{-1}\boldsymbol{\Delta}\boldsymbol{h}|$. Obviously, C_h can be calculated for any direction $\boldsymbol{h}$. One obvious choice is the vector $\boldsymbol{h}_i$, containing one in the ith position and zero elsewhere, corresponding to the perturbation of the ith weight only. This reflects the impact of allowing the ith subject to drop out non-randomly, while the others can only drop out at random. The corresponding local influence measure, denoted by C_i, then becomes $C_i = 2|\boldsymbol{\Delta}_i'\ddot{L}^{-1}\boldsymbol{\Delta}_i|$. Another important direction is $\boldsymbol{h}_{\max}$, that of maximal normal curvature $C_{\max}$. It shows how the MAR model should be perturbed to obtain the largest local changes in the likelihood displacement. $C_{\max}$ is the largest eigenvalue of $-2\,\boldsymbol{\Delta}'\,\ddot{L}^{-1}\,\boldsymbol{\Delta}$, and $\boldsymbol{h}_{\max}$ is the corresponding eigenvector. When a subset $\boldsymbol{\gamma}_1$ of $\boldsymbol{\gamma} = (\boldsymbol{\gamma}_1', \boldsymbol{\gamma}_2')'$ is of special interest, a similar approach can be used, replacing the log-likelihood by the profile log-likelihood for $\boldsymbol{\gamma}_1$, and the methods discussed above for the full parameter vector carry over directly (Lesaffre and Verbeke, 1998).

22.7.2 Applied to the Diggle–Kenward model

Here, we focus on the linear mixed model, combined with (22.17), as in Section 22.7.1. Using the $g(\cdot)$ factor notation, the dropout mechanism is described by

$$f(d_i|\boldsymbol{y}_i, \boldsymbol{\psi}) = \begin{cases} \displaystyle\prod_{j=2}^{n_i}\{1 - g(\boldsymbol{h}_{ij}, y_{ij})\}, & \text{if completer } (d_i = n_i + 1), \\ \displaystyle\prod_{j=2}^{d-1}\{1 - g(\boldsymbol{h}_{ij}, y_{ij})\}g(\boldsymbol{h}_{id}, y_{id}), & \text{if dropout } (d_i = d \le n_i), \end{cases}$$

The log-likelihood contribution for a complete sequence is then

$$\ell_{i\omega} = \log\{f(\boldsymbol{y}_i)\} + \log\{f(d_i|\boldsymbol{y}_i, \boldsymbol{\psi})\},$$

where the parameter dependencies are suppressed for notational ease. The density $f(\boldsymbol{y}_i)$ is multivariate normal, following from the linear mixed model. The contribution from an incomplete sequence is more complicated. Its log-likelihood term is

$$\ell_{i\omega} = \log\{f(y_{i1}, \dots, y_{i,d-1})\} + \sum_{j=2}^{d-1}\log\{1 - g(\boldsymbol{h}_{ij}, y_{ij})\}$$
$$+ \log\left\{\int f(y_{id}|y_{i1}, \dots, y_{i,d-1})g(\boldsymbol{h}_{id}, y_{id})dy_{id}\right\}.$$

Further details can be found in Verbeke et al. (2001). We need expressions for $\boldsymbol{\Delta}$ and $\ddot{L}$. Straightforward derivation shows that the columns $\boldsymbol{\Delta}_i$ of $\boldsymbol{\Delta}$ are given by

$$\left.\frac{\partial^2 \ell_{i\omega}}{\partial \boldsymbol{\theta} \partial \omega_i}\right|_{\omega_i=0} = \boldsymbol{0}, \tag{22.19}$$

$$\left.\frac{\partial^2 \ell_{i\omega}}{\partial \boldsymbol{\psi} \partial \omega_i}\right|_{\omega_i=0} = -\sum_{j=2}^{n_i} \boldsymbol{h}_{ij} y_{ij} g(\boldsymbol{h}_{ij})\{1 - g(\boldsymbol{h}_{ij})\},$$

for complete sequences (no dropout) and by

$$\frac{\partial^2 \ell_{i\omega}}{\partial\boldsymbol{\theta}\partial\omega_i}\bigg|_{\omega_i=0} = \{1 - g(\boldsymbol{h}_{id})\}\frac{\partial\lambda(y_{id}|\boldsymbol{h}_{id})}{\partial\boldsymbol{\theta}}, \tag{22.20}$$

$$\frac{\partial^2 \ell_{i\omega}}{\partial\boldsymbol{\psi}\partial\omega_i}\bigg|_{\omega_i=0} = -\sum_{j=2}^{d-1} \boldsymbol{h}_{ij}y_{ij}g(\boldsymbol{h}_{ij})\{1 - g(\boldsymbol{h}_{ij})\}$$
$$- \boldsymbol{h}_{id}\lambda(y_{id}|\boldsymbol{h}_{id})g(\boldsymbol{h}_{id})\{1 - g(\boldsymbol{h}_{id})\},$$

for incomplete sequences. All expressions are evaluated at $\widehat{\boldsymbol{\gamma}}$, and $g(\boldsymbol{h}_{ij}) = g(\boldsymbol{h}_{ij}, y_{ij})|_{\omega_i=0}$ is the MAR version of the dropout model. In (22.20), we make use of the conditional mean

$$\lambda(y_{id}|\boldsymbol{h}_{id}) = \lambda(y_{id}) + \Sigma_{i,21}\Sigma_{i,11}^{-1}\{\boldsymbol{h}_{id} - \lambda(\boldsymbol{h}_{id})\}. \tag{22.21}$$

The variance matrices follow from partitioning the responses as $(y_{i1}, \ldots, y_{i,d-1}|y_{id})'$.
  The derivatives of (22.21) with respect to the measurement model parameters are

$$\frac{\partial\lambda(y_{id}|\boldsymbol{h}_{id})}{\partial\boldsymbol{\beta}} = \boldsymbol{x}_{id} - \Sigma_{i,21}\Sigma_{i,11}^{-1}X_{i,(d-1)},$$

$$\frac{\partial\lambda(y_{id}|\boldsymbol{h}_{id})}{\partial\boldsymbol{\alpha}} = \left\{\frac{\partial\Sigma_{i,21}}{\partial\boldsymbol{\alpha}} - \Sigma_{i,21}\Sigma_{i,11}^{-1}\frac{\partial\Sigma_{i,11}}{\partial\boldsymbol{\alpha}}\right\}\Sigma_{i,11}^{-1}\{\boldsymbol{h}_{id} - \lambda(\boldsymbol{h}_{id})\},$$

where $\boldsymbol{x}'_{id}$ is the dth row of X_i, and $X_{i,(d-1)}$ indicates the first $(d-1)$ rows X_i. Further, $\boldsymbol{\alpha}$ indicates the subvector of covariance parameters within the vector $\boldsymbol{\theta}$.
  In practice, the parameter $\boldsymbol{\theta}$ in the measurement model is often of primary interest. As $\ddot{L}$ is block-diagonal with blocks $\ddot{L}(\boldsymbol{\theta})$ and $\ddot{L}(\boldsymbol{\psi})$, we have that, for any unit vector $\boldsymbol{h}$, $C_{\boldsymbol{h}}$ equals $C_{\boldsymbol{h}}(\boldsymbol{\theta}) + C_{\boldsymbol{h}}(\boldsymbol{\psi})$, with

$$C_{\boldsymbol{h}}(\boldsymbol{\theta}) = -2\boldsymbol{h}'\left[\frac{\partial^2 \ell_{i\omega}}{\partial\boldsymbol{\theta}\partial\omega_i}\bigg|_{\omega_i=0}\right]' \ddot{L}^{-1}(\boldsymbol{\theta})\left[\frac{\partial^2 \ell_{i\omega}}{\partial\boldsymbol{\theta}\partial\omega_i}\bigg|_{\omega_i=0}\right]\boldsymbol{h},$$

$$C_{\boldsymbol{h}}(\boldsymbol{\psi}) = -2\boldsymbol{h}'\left[\frac{\partial^2 \ell_{i\omega}}{\partial\boldsymbol{\psi}\partial\omega_i}\bigg|_{\omega_i=0}\right]' \ddot{L}^{-1}(\boldsymbol{\psi})\left[\frac{\partial^2 \ell_{i\omega}}{\partial\boldsymbol{\psi}\partial\omega_i}\bigg|_{\omega_i=0}\right]\boldsymbol{h},$$

evaluated at $\boldsymbol{\gamma} = \widehat{\boldsymbol{\gamma}}$. It now immediately follows from (22.19) and (22.20) that *direct* influence on $\boldsymbol{\theta}$ only arises from those measurement occasions at which dropout occurs. In particular, from (22.20) it is clear that the corresponding contribution is large only if (1) the dropout probability was small, but the subject disappeared nevertheless; and (2) the conditional mean "strongly depends" on the parameter of interest. This implies that complete sequences cannot be influential in the strict sense ($C_i(\boldsymbol{\theta}) = 0$) and that incomplete sequences only contribute, in a direct fashion, at the actual dropout time. However, we make an important distinction between direct and indirect influence. It was shown that complete sequences can have an impact by changing the conditional expectation of the unobserved measurements given the observed ones *and given the dropout mechanism*. Thus, a complete observation that has a strong impact on the *dropout model parameters* can still drastically change the measurement model parameters and functions thereof.

22.7.3 Informal sensitivity analysis for the mastitis in dairy cattle data

The data were introduced in Section 22.2.3. Diggle and Kenward (1994) and Kenward (1998) performed several analyses of these data, and we now review these. Local influence ideas are applied in Section 22.7.4.
  In Diggle and Kenward (1994), a separate mean for each group defined by the year of first lactation and a common time effect was considered, together with an unstructured 2×2

Table 22.17 Mastitis in Dairy Cattle: Maximum Likelihood Estimates (Standard Errors) of Random and Non-Random Dropout Models, under Several Deletion Schemes

Parameter	All	(53,54,66,69)	(4,5)	(66)	(4,5,66)
			Random Dropout		
Measurement Model:					
β_0	5.77(0.09)	5.69(0.09)	5.81(0.08)	5.75(0.09)	5.80(0.09)
β_d	0.72(0.11)	0.70(0.11)	0.64(0.09)	0.68(0.10)	0.60(0.08)
σ_1^2	0.87(0.12)	0.76(0.11)	0.77(0.11)	0.86(0.12)	0.76(0.11)
σ_2^2	1.30(0.20)	1.08(0.17)	1.30(0.20)	1.10(0.17)	1.09(0.17)
ρ	0.58(0.07)	0.45(0.08)	0.72(0.05)	0.57(0.07)	0.73(0.05)
Dropout Model:					
ψ_0	−2.65(1.45)	−3.69(1.63)	−2.34(1.51)	−2.77(1.47)	−2.48(1.54)
ψ_1	0.27(0.25)	0.46(0.28)	0.22(0.25)	0.29(0.24)	0.24(0.26)
$\omega = \psi_2$	0	0	0	0	0
−2 Log-Likelihood	280.02	246.64	237.94	264.73	220.23
			Non-Random Dropout		
Measurement Model:					
β_0	5.77(0.09)	5.69(0.09)	5.81(0.08)	5.75(0.09)	5.80(0.09)
β_d	0.33(0.14)	0.35(0.14)	0.40(0.18)	0.34(0.14)	0.63(0.29)
σ_1^2	0.87(0.12)	0.76(0.11)	0.77(0.11)	0.86(0.12)	0.76(0.11)
σ_2^2	1.61(0.29)	1.29(0.25)	1.39(0.25)	1.34(0.25)	1.10(0.20)
ρ	0.48(0.09)	0.42(0.10)	0.67(0.06)	0.48(0.09)	0.73(0.05)
Dropout Model:					
ψ_0	0.37(2.33)	−0.37(2.65)	−0.77(2.04)	0.45(2.35)	−2.77(3.52)
ψ_1	2.25(0.77)	2.11(0.76)	1.61(1.13)	2.06(0.76)	0.07(1.82)
$\omega = \psi_2$	−2.54(0.83)	−2.22(0.86)	−1.66(1.29)	−2.33(0.86)	0.20(2.09)
−2 Log-Likelihood	274.91	243.21	237.86	261.15	220.23
G^2 for NMAR	5.11	3.43	0.08	3.57	0.005

covariance matrix. The dropout model included both Y_{i1} and Y_{i2} and was reparameterized in terms of the size variable $(Y_{i1} + Y_{i2})/2$ and the increment $Y_{i2} - Y_{i1}$. It turned out that the increment was important, in contrast to a relatively small contribution of the size. If this model was deemed to be plausible, MAR would be rejected on the basis of a likelihood ratio test statistic of $G^2 = 5.11$ on 1 degree of freedom.

Kenward (1998) carried out what we term a data-driven sensitivity analysis. He started from the original model in Diggle and Kenward (1994), albeit with a common intercept, because there was no evidence for a dependence on first lactation year. The right-hand panel of Figure 22.3 shows that there are two cows, #4 and #5, with unusually large increments. He conjectured that this might mean that these animals were ill during the first lactation year, producing an unusually low yield, whereas a normal yield was obtained during the second year. He then fitted t-distributions to Y_{i2} given $Y_{i1} = y_{i1}$. Not surprisingly, his finding was that the heavier the tails of the t-distribution, the better the outliers were accommodated. As a result, the difference in fit between the random and non-random dropout models vanished ($G^2 = 1.08$ for a t_2-distribution). Alternatively, removing these two cows and refitting the normal model shows complete lack of evidence for non-random dropout ($G^2 = 0.08$). This latter procedure is similar to a global influence analysis by means of deleting two observations. Parameter estimates and standard errors for random and non-random dropout, under several deletion schemes, are reproduced in Table 22.17. It is clear that the influence on the measurement model parameters is small in the random dropout case, although the gap on the time effect β_d between the random and non-random dropout models is reduced when #4 and #5 are removed.

We now look in more detail at these informal but insightful forms of sensitivity analysis. A simple bivariate Gaussian linear model is used to represent the marginal milk yield in the years (i.e., the yield that would be, or was, observed in the absence of mastitis),

$$\begin{pmatrix} Y_1 \\ Y_2 \end{pmatrix} = N \left\{ \begin{pmatrix} \mu \\ \mu + \delta \end{pmatrix}, \begin{pmatrix} \sigma_1^2 & \rho\sigma_1\sigma_2 \\ \rho\sigma_1\sigma_2 & \sigma_2^2 \end{pmatrix} \right\}.$$

Note that the parameter δ represents the change in average yield between the 2 years. The probability of mastitis is assumed to follow the logistic regression model

$$\text{Pr(dropout)} = \frac{e^{\psi_0 + \psi_1 y_1 + \psi_2 y_2}}{1 + e^{\psi_0 + \psi_1 y_1 + \psi_2 y_2}}.$$

The combined response/dropout model was fitted to the milk yields by maximum likelihood using a generic function maximization routine. In addition, the MAR model ($\psi_2 = 0$) was fitted. This latter is equivalent to fitting separately the Gaussian linear model for the milk yields and logistic regression model for the occurrence of mastitis. These fits produced the parameter estimates, standard errors, and minimized value of twice the negative log-likelihood as displayed in the "All" column of Table 22.17.

Using the likelihoods to compare the fit of the two models, we obtain a difference $G^2 = 5.11$. The corresponding tail probability from the χ_1^2 is 0.02. This test essentially examines the contribution of ψ_2 to the fit of the model. Using the Wald statistic for the same purpose gives a statistic of $(-2.53)^2/0.83 = 9.35$, with corresponding χ_1^2 probability of 0.002. The discrepancy between the results of the two tests suggests that the asymptotic approximations on which these are based are not very accurate in this setting and that the standard error probably underestimates the true variability of the estimate of ψ_2. Indeed, such tests do not have the usual properties of likelihood ratio and Wald tests in regular problems. Nevertheless, there is a suggestion from the change in likelihood that ψ_2 is making a real contribution to the fit of the model. The dropout model estimated from the NMAR setting is

$$\text{logit}\{\text{Pr(mastitis)}\} = 0.37 + 2.25 y_1 - 2.54 y_2. \tag{22.22}$$

Some insight into this fitted model can be obtained by rewriting it in terms of the milk yield totals ($Y_1 + Y_2$) and increments ($Y_2 - Y_1$) as

$$\text{logit}\{\text{Pr(mastitis)}\} = 0.37 - 0.145(y_1 + y_2) - 2.395(y_2 - y_1).$$

The probability of mastitis increases with larger negative increments; that is, those animals who showed (or would have shown) a greater decrease in yield over the 2 years have a higher probability of getting mastitis. The other differences in parameter estimates between the two models are consistent with this: the NMAR dropout model predicts a smaller average increment in yield (δ), with larger second-year variance and smaller correlation caused by greater negative imputed differences between yields.

To gain further insight into these two fitted models, we now take a closer look at the raw data and the predictive behavior of the Gaussian NMAR model. Under an NMAR model, the predicted, or imputed, value of a missing observation is given by the ratio of expectations

$$\widehat{\boldsymbol{y}}_m = \frac{E_{Y_m|Y_o}\{\boldsymbol{y}_m \text{Pr}(\boldsymbol{r} \mid \boldsymbol{y}_o, \boldsymbol{y}_m)\}}{E_{Y_m|Y_o}\{\text{Pr}(\boldsymbol{r} \mid \boldsymbol{y}_o, \boldsymbol{y}_m)\}}. \tag{22.23}$$

Recall that the fitted dropout model (22.22) implies that the probability of mastitis increases with decreasing values of the increment $Y_2 - Y_1$. We therefore plot the 27 imputed values of this quantity together with the 80 observed increments against the first-year yield Y_1.

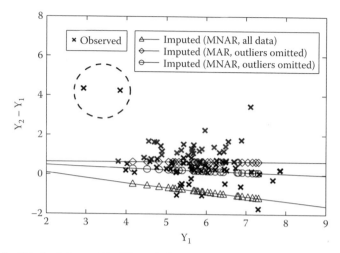

Figure 22.11 Mastitis in dairy cattle. Plot of observed and imputed year 2 — year 1 yield differences against year 1 yield. Two outlying points are circled.

This is presented in Figure 22.11, in which the imputed values are indicated with triangles and the observed values with crosses. Note how the imputed values are almost linear in Y_1; this is a well-known property of the ratio (22.23) within this range of observations. The imputed values are all negative, in contrast to the observed increments, which are nearly all positive. With animals of this age, one would normally expect an increase in yield between the 2 years. The dropout model is imposing very atypical behavior on these animals and this corresponds to the statistical significance of the NMAR component of the model (ψ_2) but, of course, necessitates further scrutiny.

Another feature of this plot is the pair of outlying observed points circled in the top left-hand corner of Figure 22.11. These two animals have the lowest and third lowest yields in the first year, but moderately large yields in the second, leading to the largest positive increments. In a well-husbanded dairy herd, one would expect approximately Gaussian joint milk yields, and these two then represent outliers. It is likely that there is some anomaly, possibly illness, leading to their relatively low yields in the first year. One can conjecture that these two animals are the cause of the structure identified by the Gaussian NMAR model. Under the joint Gaussian assumption, the NMAR model essentially "fills in" the missing data to produce a complete Gaussian distribution. To counterbalance the effect of these two extreme positive increments, the dropout model predicts negative increments for the mastitic cows, leading to the results observed. As a check on this conjecture, we omit these two animals from the data set and refit the MAR and NMAR Gaussian models. The resulting estimates are presented in the "(4, 5)" column of Table 22.17.

The deviance is minimal, and the NMAR model now shows no improvement in fit over MAR. The estimates of the dropout parameters, although still moderately large in an absolute sense, are of the same size as their standard errors which, as mentioned earlier, are probably underestimates. In the absence of the two anomalous animals, the structure identified earlier in terms of the NMAR dropout model no longer exists. The increments imputed by the fitted model are also plotted in Figure 22.11, indicated by circles. Although still lying among the lower region of the observed increments, these are now all positive and lie close to the increments imputed by the MAR model (diamonds). Thus, we have a plausible representation of the data in terms of joint Gaussian milk yields, two pairs of outlying yields and no requirement for an NMAR dropout process.

The two key assumptions underlying the outcome-based NMAR model are, first, the form chosen for the relationship between dropout probability and response and, second, the

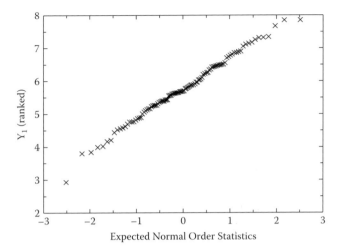

Figure 22.12 Mastitis in dairy cattle. Normal probability plot of the year 1 milk yields.

distribution of the response or, more precisely, the conditional distribution of the possibly unobserved response given the observed response. In the current setting, for the first assumption, if there is dependence of mastitis occurrence on yield, experience with logistic regression tells us that the exact form of the link function in this relationship is unlikely to be critical. In terms of sensitivity, we therefore consider the second assumption, the distribution of the response.

All the data from the first year are available, and a normal probability plot of these, given in Figure 22.12, does not show great departures from the Gaussian assumption. Leaving this distribution unchanged, we therefore examine the effect of changing the conditional distribution of Y_2 given Y_1. One simple and obvious choice is to consider a heavy-tailed distribution, and for this, we use the translated and scaled t_m-distribution with density

$$f(y_2 \mid y_1) = \left\{\sigma\sqrt{m}B(1/2, m/2)\right\}^{-1}\left\{1 + \frac{1}{m}\left(\frac{y_2 - \mu_{2\mid1}}{\sigma}\right)^2\right\}^{-(m+1)/2},$$

where

$$\mu_{2\mid1} = \mu + \delta + \frac{\rho\sigma_2(y_1 - \mu)}{\sigma_1}$$

is the conditional mean of $Y_2 \mid y_1$. The corresponding conditional variance is

$$\frac{m}{m-2}\sigma^2.$$

Relevant parameter estimates from the fits of both MAR and NMAR models are presented in Table 22.18 for three values of m: 2, 10, and 25. Smaller values of m correspond to greater kurtosis and, as m becomes large, the model approaches the Gaussian one used in the previous section. It can be seen from the results for the NMAR model in Table 22.18 that, as the kurtosis increases, the estimate of ψ_2 decreases. Also, the maximized likelihoods of the MAR and NMAR models converge. With 10 and 2 degrees of freedom, there is no evidence at all to support the inclusion of ψ_2 in the model; that is, the MAR model provides as good a description of the observed data as the NMAR, in contrast to the Gaussian-based conclusions. Further, as m decreases, the estimated yearly increment in milk yield δ from the NMAR model increases to the value estimated under the MAR model. In most applications of SeMs, it will be quantities of this type that will be of prime interest, and it is clearly seen

Table 22.18 Mastitis in Dairy Cattle: Details of the Fit of MAR and NMAR Dropout Models, Assuming a t_m-Distribution for the Conditional Distribution of Y_2 Given Y_1

t df	Parameter	MAR	NMAR
25	δ	0.69(0.10)	0.35(0.13)
	ψ_1	0.27(0.24)	2.11(0.78)
	ψ_2		$-2.33(0.88)$
-2 Log-Likelihood		275.54	271.77
10	δ	0.67(0.09)	0.38(0.14)
	ψ_1	0.27(0.24)	1.84(0.82)
	ψ_2		$-1.96(0.95)$
-2 Log-Likelihood		271.22	269.12
2	δ	0.61(0.08)	0.54(0.11)
	ψ_1	0.27(0.24)	0.80(0.66)
	ψ_2		$-0.65(0.73)$
-2 Log-Likelihood		267.87	266.79

Note: Maximum likelihood estimates (standard errors) are shown.

in this example how the dropout model can have a crucial influence on the estimate of this. Comparing the values of the deviance from the t-based model with those from the original Gaussian model, we also see that the former with $m = 10$ or 2 produces a slightly better fit, although no meaning can be attached to the statistical significance of the difference in these likelihood values.

The results observed here are consistent with those from the deletion analysis. The two outlying pairs of measurements identified earlier are not inconsistent with the heavy-tailed t-distribution, so it would require no "filling in" and hence no evidence for non-randomness in the dropout process under the second model. In conclusion, if we consider the data with outliers included, we have two models that effectively fit equally well to the observed data. The first assumes a joint Gaussian distribution for the responses and an NMAR dropout model. The second assumes a Gaussian distribution for the first observation and a conditional t_m-distribution (with small m) for the second given the first, with no requirement for an NMAR dropout component. Each provides a different explanation for what has been observed, with quite a different biological interpretation. In likelihood terms, the second model fits a little better than the first but, as has been stressed repeatedly in this chapter, a key feature of such dropout models is that the distinction between them should not be based on the observed-data likelihood alone. It is always possible to specify models with identical maximized observed-data likelihoods that differ with respect to the unobserved data and dropout mechanism, and such models can have very different implications for the underlying mechanism generating the data. Finally, the most plausible explanation for the observed data is that the pairs of milk yields have joint Gaussian distributions, with no need for an NMAR dropout component, and that two animals are associated with anomalous pairs of yields.

22.7.4 Local influence approach for the mastitis in dairy cattle data

In the previous section, the sensitivity to distributional assumptions of conclusions concerning the randomness of the dropout process was established in the context of the mastitis data. Such sensitivity has led some to conclude that such modeling should be avoided. We have argued that this conclusion is too strong. First, repeated measures tend to be incomplete and therefore the consideration of the dropout process is simply unavoidable. Second, if a non-random dropout component is added to a model and the maximized likelihood changes appreciably, then some real structure in the data has been identified that is not encompassed by the original model. The NMAR analysis may tell us about inadequacies of the original model rather than the adequacy of the NMAR model. It is the interpretation of the identified structure that cannot be made unequivocally from the data under analysis. The mastitis data clearly illustrate this: using external information on the distribution of the response, a plausible explanation of the structure so identified *might* be made in terms of the outlying responses from two animals. However, it should also be noted that absence of structure in the data associated with an NMAR process does not imply that an NMAR process is not operating: different models with similar maximized likelihoods (i.e., with similar plausibility with respect to the observed data) may have completely different implications for the dropout process and the unobserved data. These points together suggest that the appropriate role of such modeling is as a component of a sensitivity analysis.

The analysis of the previous section is characterized by its basis within substantive knowledge about the data. In this section, we will apply the local influence technique to the mastitis data, and see how the results compare with those found in Section 22.7.3. We suggest that, here, a combination of methodology and substantive insight will be the most fruitful approach.

Applying the local influence method to the mastitis data produces Figure 22.13, which suggests that there are four influential animals: #53, #54, #66, and #69. The most striking feature of this analysis is that #4 and #5 are *not* recovered. See also Figure 22.3. It is interesting to consider an analysis with these four cows removed. Details are given in Table 22.17. In contrast to the consequences of removing #4 and #5, the influence on the likelihood ratio test is rather small: $G^2 = 3.43$ instead of the original 5.11. The influence on the measurement model parameters under both random and non-random dropout is small.

It is very important to realize that one should not expect agreement between deletion and our local influence analysis. The latter focuses on the sensitivity of the results with respect to the assumed dropout model; more specifically, on how the results change when the MAR model is extended in the direction of non-random dropout. In particular, all animals singled out so far are complete, and, hence, $C_i(\boldsymbol{\theta}) \equiv 0$, placing all influence on $C_i(\boldsymbol{\psi})$ and $\boldsymbol{h}_{\max,i}$.

Greater insight can also be obtained by studying the approximation for $C_i(\boldsymbol{\psi})$, derived by Verbeke et al. (2001), given by

$$C_i(\boldsymbol{\psi}) \simeq 2(v_i y_{i2}^2) \left\{ v_i \boldsymbol{h}_i' \left(\sum_{i=1}^N v_i \boldsymbol{h}_i \boldsymbol{h}_i' \right)^{-1} \boldsymbol{h}_i \right\}, \tag{22.24}$$

where the factor in the braces equals the hat-matrix diagonal and $v_i = g(y_{i1})[1 - g(y_{i1})]$. For a dropout, y_{i2} has to be replaced by its conditional expectation. The contribution for animal i is made up of three factors. The first factor, v_i, is small for extreme dropout probabilities. The animals that have a very high probability of either remaining or leaving the study are less influential. Cows #4 and #5 have dropout probabilities equal to 0.13 and 0.17, respectively. The 107 cows in the study span the dropout probability interval $[0.13, 0.37]$. Thus, this component rather deflates the influence of animals #4 and #5. Second, (22.24) contains a leverage factor in braces. Third, an animal is relatively more influential when both milk yields are high. We now need to question whether or not this is plausible or relevant. Because both measurements are positively correlated, measurements with both milk yields high or low will not be unusual. In Section 22.7.3, we observed that cows #4 and #5 are unusual on the basis of their *increment*. This is in line with several other applications of similar dropout models (Diggle and Kenward, 1994; Molenberghs, Kenward, and Lesaffre, 1997), where it was found that a strong incremental component pointed to genuine

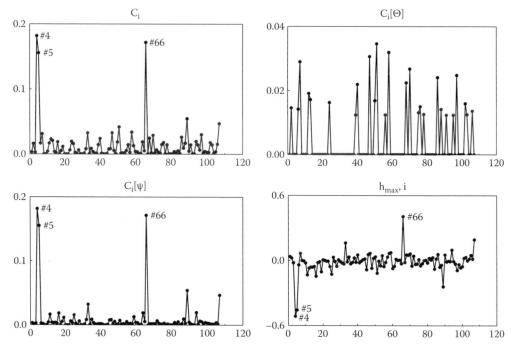

Figure 22.13 Mastitis in dairy cattle. Index plots of C_i, $C_i(\boldsymbol{\theta})$, $C_i(\boldsymbol{\psi})$, and of the components of the direction $\boldsymbol{h}_{\max}$ of maximal curvature, when the dropout model is parameterized as a function of Y_{i1} and Y_{i2}.

non-randomness. In contrast, if correlations among succeeding measurements are not low, the size variable can often be replaced by just the history and, hence, the corresponding model is very close to random dropout.

Even though a dropout model expressed in terms of the outcomes themselves is equivalent to a model in the first variable Y_{i1} and the increment $Y_{i2} - Y_{i1}$, termed an incremental variables representation, we will show that they lead to different perturbation schemes of the form (22.18). At first, this feature can be seen as both an advantage and a disadvantage. The fact that reparameterizations of the linear predictor of the dropout model lead to different perturbation schemes requires careful reflection based on substantive knowledge in order to guide the analysis, such as the considerations on the incremental variable made earlier.

We will present the results of the incremental analysis and then offer further comments on the rationale behind this particular transformation. From the diagnostic plots in Figure 22.14, it is obvious that we recover three influential animals: #4, #5, and #66. Although Kenward (1998) did not consider #66 to be influential, it does appear to be somewhat distant from the bulk of the data (Figure 22.3). The main difference between both types is that the first two were plausibly ill during year 1, and this is not necessarily so for #66. An additional feature is that, in all cases, both $C_i(\boldsymbol{\psi})$ and $\boldsymbol{h}_{\max}$ show the same influential animals. Moreover, $\boldsymbol{h}_{\max}$ suggests that the influence for #66 is different than for the others. It could be conjectured that the latter one pulls the coefficient ω in a different direction. The other values are all relatively small. This could indicate that, for the remaining 104 animals, MAR is plausible, whereas a deviation in the direction of the incremental variable, *with differing signs*, appears to be necessary for the other three animals. At this point, a comparison between $\boldsymbol{h}_{\max}$ for the direct-variable and incremental analyses is useful. Because the contributions h_i sum to 1, these two plots are directly comparable. There is no pronounced influence indication in the direct-variables case and perhaps only random noise is seen. A more formal way to distinguish between signal and noise needs to be developed.

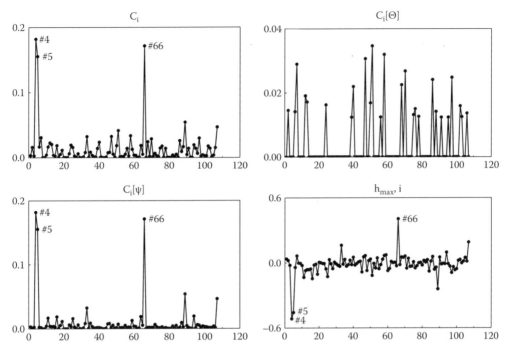

Figure 22.14 Mastitis in dairy cattle. Index plots of C_i, $C_i(\boldsymbol{\theta})$, $C_i(\boldsymbol{\psi})$, and of the components of the direction $\boldsymbol{h}_{\max}$ of maximal curvature, when the dropout model is parameterized in function of Y_{i1} and $Y_{i2} - Y_{i1}$.

In Figure 22.15, we have decomposed (22.24) into its three components: the variance of the dropout probability v_i; the incremental variable $Y_{i2} - Y_{i1}$, which is replaced by its predicted value for a dropout; and the hat-matrix diagonal. In agreement with the preceding discussion, the influence clearly stems from an unusually large increment, which overcomes the fact that v_i actually downplays the influence, because Y_{41} and Y_{51} are comparatively small, and dropout increases with the milk yield in the first year. Furthermore, a better interpretation can be made for the difference in sign of $h_{\max,4}$ and $h_{\max,5}$ versus $h_{\max,66}$.

We have noted already that animals #4 and #5 have relatively small dropout probabilities. In contrast, the dropout probability of #66 is large within the observed range $[0.13, 0.37]$. As the increment is large for these animals, changing the perturbation ω_i can have a large impact on the other dropout parameters ψ_0 and ψ_1. In order to avoid having the effects of the change for #4 and #5 cancel with the effect of #66, the corresponding signs need to be reversed. Such a change implies either that all three dropout probabilities move toward the center of the range or are pulled away from it. (Note that $-\boldsymbol{h}_{\max}$ is another normalized eigenvector corresponding to the largest eigenvalue.)

In the informal approach, extra analyses were carried out with #4 and #5 removed. The resulting likelihood ratio statistic reduces to $G^2 = 0.08$. When only #66 is removed, the likelihood ratio for non-random dropout is $G^2 = 3.57$, very similar to the one when #53, #54, #66, and #69 were removed. Removing all three (#4, #5, and #66) results in $G^2 = 0.005$ (i.e., indicating complete removal of all evidence for non-random dropout). Details are given in Table 22.17.

We now explore the reasons why the transformation of direct outcomes to increments is useful. We have noted already that the associated perturbation schemes (22.18) are different.

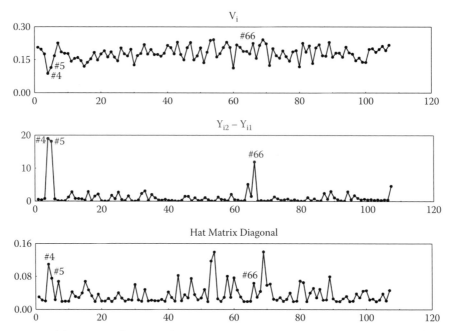

Figure 22.15 Mastitis in dairy cattle. Index plots of the three components of $C_i(\psi)$ when the dropout model is parameterized as a function of Y_{i1} and $Y_{i2} - Y_{i1}$.

An important device in this respect is the equality

$$\psi_0 + \psi_1 y_{i1} + \psi_2 y_{i2} = \psi_0 + (\psi_1 + \psi_2)y_{i1} + \psi_2(y_{i2} - y_{i1}). \qquad (22.25)$$

Equation (22.25) shows that the direct-variables model checks the influence on the random dropout parameter ψ_1, whereas the random dropout parameter in the incremental model is $\psi_1 + \psi_2$. Not only is this a different parameter, it is also estimated with higher precision. One often observes that $\widehat{\psi}_1$ and $\widehat{\psi}_2$ exhibit a similar variance and (strongly) negative correlation, in which case, the linear combination with smallest variance is approximately in the direction of the sum $\psi_1 + \psi_2$. When the correlation is positive, the difference direction $\psi_1 - \psi_2$ is obtained instead. We assess this in the case where all 107 observations are included. The estimated covariance matrix is

$$\begin{pmatrix} 0.59 & -0.54 \\ & 0.70 \end{pmatrix},$$

with correlation -0.84. The variance of $\widehat{\psi}_1 + \widehat{\psi}_2$, on the other hand, is estimated to be 0.21. In this case, the direction of minimal variance is along $(0.74, 0.67)$, which is indeed close to the sum direction. When all three influential subjects are removed, the estimated covariance matrix becomes

$$\begin{pmatrix} 3.31 & -3.77 \\ & 4.37 \end{pmatrix},$$

with correlation -0.9897. Removing only #4 and #5 yields an intermediate situation of which the results are not shown. The variance of the sum is 0.15, which is a further reduction and still close to the direction of minimal variance. These considerations reinforce the claim that an incremental analysis is strongly advantageous. It might therefore be interesting to routinely construct a plot such as in Figure 22.3 or Figure 22.11, even with longer measurement sequences. On the other hand, transforming the dropout model to a size variable $(Y_{i1} + Y_{i2})/2$ will worsen the problem because a poorly estimated parameter associated with Y_{i1} will result.

Finally, observe that a transformation of the dropout model to a size and incremental variable at the same time for the model with all three influential subjects removed gives variance of the size and increment variables of 0.15 and 15.22, respectively. In other words, there is no evidence for an incremental effect, confirming that MAR is plausible.

Although local and global influence are, strictly speaking, not equivalent, it is useful to see how the global influence on $\boldsymbol{\theta}$ can be linked to the behavior of $C_i(\boldsymbol{\psi})$. We observed earlier that all locally influential animals are completers and hence $C_i(\boldsymbol{\theta}) \equiv 0$. Yet, removing #4, #5, and #66 shows some effect on the discrepancy between the random dropout (MAR) and non-random dropout (NMAR) estimates of the time effect β_d. In particular, MAR and NMAR estimates with all three subjects removed are virtually identical (0.60 and 0.63, respectively). Let us do a small thought experiment. Because these animals are influential in $C_i(\boldsymbol{\psi})$, the MAR model could be improved by including incremental terms for these three. Such a model would still imply random dropout. In contrast, allowing a dependence on the increment in *all* subjects will influence $E(Y_{i2}|y_{i1}, \text{dropout})$ for all incomplete observations; hence, the measurement model parameters under NMAR will change. In conclusion, this provides a way to assess the *indirect* influence of the dropout mechanism on the measurement model parameters through local influence methods. In the milk yield data set, this influence is likely to be due to the fact that an exceptional increment that is caused by a different mechanism, perhaps a diseased animal during the first year, is nevertheless treated on an equal footing with the other observations within the dropout model. Such an investigation cannot be done with the case-deletion method because it is not possible to disentangle the various sources of influence.

In conclusion, it is found that an incremental variable representation of the dropout mechanism has advantages over a direct-variable representation. Contrasting our local influence approach with a case-deletion scheme as applied in Kenward (1998), it is argued that the former approach is advantageous, as it allows one to assess direct and indirect influences on the dropout and measurement model parameters, stemming from perturbing the random dropout model in the direction of non-random dropout. In contrast, a case-deletion scheme does not allow one to disentangle the various sources of influence.

22.8 Further methods

In this section, we present a brief overview of some further approaches to sensitivity analysis classes deserving of attention.

Pattern-mixture models can be considered for their own sake, typically because they are appropriate, in given circumstances, to answer a particular scientific question. Furthermore, a range of authors have considered PMMs as a useful contrast to SeMs either (1) to answer the same scientific question, such as marginal treatment effect or time evolution, based on these two rather different modeling strategies; or (2) to gain additional insight by supplementing the SeM results with those from a pattern-mixture approach. PMMs also have a special role in some multiple imputation based sensitivity analyses.

Pattern-mixture models have the advantage that, strictly speaking, only the observed data need to be modeled. However, whenever aspects of the marginal distribution of the outcome are of interest (e.g., the marginal evolution over time), the unobserved measurements must be modeled jointly with the observed ones. Various methods have been proposed in the statistical literature (Molenberghs and Kenward, 2007), all based on very specific assumptions about the conditional distribution of the missing outcomes, given the observed ones. This points to the so-called underidentification of the PMM: within a pattern, there is by definition no information on the unobserved outcomes given the observed ones. There are several ways forward (Molenberghs and Kenward, 2007). First, parametric extrapolation can be used, where the model fitted to the observed data is believed to also hold for the missing data in the same pattern; that is, the parametric model is extrapolated. Second, so-called identifying restrictions (Little, 1993, 1994b) can be used: inestimable parameters of the incomplete patterns are set equal to (functions of) the parameters describing the distribution of the completers. Although some authors perceive this underidentification as a drawback, it can be seen as bringing important advantages. First, assumptions need to be made very explicit. Second, PMMs in this way aid our understanding of the precise nature of sensitivity in SeMs. Third, and related to the previous point, PMMs can serve important roles in a sensitivity analysis.

Examples of pattern-mixture applications can be found in Verbeke, Lesaffre, and Spiessens (2001) or Michiels et al. (2002) for continuous outcomes, and Michiels, Molenberghs, and Lipsitz (1999) for categorical outcomes. Further references include Cohen and Cohen (1983), Allison (1987), Muthén, Kaplan, and Hollis (1987), McArdle and Hamagami (1992), Little and Wang (1996), Little and Yau (1996), Hedeker and Gibbons (1997), Hogan and Laird (1997), Ekholm and Skinner (1998), Molenberghs, Michiels, and Kenward (1998), Verbeke and Molenberghs (2000), Thijs et al. (2002), and Molenberghs and Verbeke (2005). An example within the Bayesian framework is given by Rizopoulos, Verbeke, and Lesaffre (2007). Molenberghs et al. (1998) and Kenward, Molenberghs, and Thijs (2003) studied the relationship between SeMs and PMMs within the context of missing-data mechanisms. The earlier paper presents the PMM counterpart of MAR, and the later one states how PMMs can be constructed such that dropout does not depend on future points in time.

Turning to the shared-parameter framework, one of its main advantages is that it can easily handle non-monotone missingness. Nevertheless, these models are based on very strong parametric assumptions, such as normality for the shared random effect(s). Of course, sensitivities abound in the SeM and PMM frameworks as well, but the assumption of unobserved, random, or latent effects further compounds the issue. Various authors have considered model extensions. An overview is given by Tsonaka, Verbeke, and Lesaffre (2007), who consider SPMs without any parametric assumptions for the shared parameters. A theoretical assessment of the sensitivity with respect to these parametric assumptions is presented in Rizopoulos, Verbeke, and Molenberghs (2007).

Beunckens et al. (2007) proposed a so-called *latent-class mixture model*, bringing together features of the SeM, PMM, and SPM frameworks. Information from the location and evolution of the response profiles (a selection-model concept) and from the dropout patterns (a pattern-mixture idea) is used simultaneously to define latent groups and variables (a shared-parameter feature). This brings several appealing features. First, one uses information in a more symmetric, elegant way. Second, apart from providing a more flexible modeling tool, there is room for use as a sensitivity analysis instrument. Third, a strong advantage over existing methods is the ability to classify subjects into latent groups. If done with due caution, it can enhance substantive knowledge and generate hypotheses. Fourth, while computational burden increases, the proposed method is remarkably stable and acceptable in terms of computation time for the settings considered here. Clearly, neither the proposed model nor any other alternative can be seen as a tool to definitively test for MAR versus NMAR, as amply documented earlier in this chapter. This is why the method's use predominantly lies within the sensitivity analysis context. Such a sensitivity analysis is of use both when it modifies the results of a simpler analysis, for further scrutiny, as well as when it confirms these.

A quite separate, extremely important line of research starts from a semi-parametric standpoint, as opposed to the parametric take on the problem that has prevailed throughout this chapter. Within this paradigm, weighted generalized estimating equations (WGEE), proposed by Robins, Rotnitzky, and Zhao (1994) and Robins and Rotnitzky (1995), play a central role. Rather than jointly modeling the outcome and missingness processes, the centerpiece is inverse probability weighting of a subject's contribution, where the weights are specified in terms of factors influencing missingness, such as covariates and observed outcomes. This methodology is amply discussed in Chapter 20; see also Chapter 23. These ideas are developed in Robins, Rotnitzky, and Scharfstein (1998) and Scharfstein, Rotnitzky, and Robins (1999). Robins, Rotnitzky, and Scharfstein (2000) and Rotnitzky et al. (2001) employ this modeling framework to conduct sensitivity analysis. They allow for the dropout mechanism to depend on potentially unobserved outcomes through the specification of a non-identifiable sensitivity parameter. An important special case for such a sensitivity parameter, τ say, is $\tau = 0$, which the authors term explainable censoring and which is essentially a sequential version of MAR. Conditional upon τ, key parameters, such as treatment effect, are identifiable. By varying τ, sensitivity can be assessed. As such, there is similarity between this approach and the interval of ignorance concept (Section 22.6). There is a connection with PMMs, too, in the sense that, for subjects with the same observed history until a given time $t - 1$, the distribution for those who drop out at t for a given cause is related to the distribution of subjects who remain on study at time t.

22.9 Concluding remarks

First, this chapter has underscored the complexities arising when fitting models to incomplete data. Five generic issues, already arising under MAR, have been brought to the forefront: (i) the classical relationship between observed and expected features is convoluted because one observes the data only partially while the model describes all data; (ii) the independence of mean and variance parameters in a (multivariate) normal model is lost, implying increased sensitivity, even under MAR; (iii) the well-known agreement between the (frequentist) ordinary least squares and maximum likelihood estimation methods for normal models is also lost, as soon as the missing-data mechanism is not of the MCAR type, with related results holding in the non-normal case; (iv) in a likelihood-based context, deviances and related information criteria cannot be used in the same vein as with complete data because they provide no information about a model's prediction of the unobserved data; and, in particular, (v) several models may saturate the observed-data degrees of freedom, while providing a different fit to the complete data; that is, they only coincide in as far as they describe the observed data. As a consequence, different inferences may result from different saturated models.

Second, and based on these considerations, it has been argued that model assessment should always proceed in two steps. In the first step, the fit of a model to the *observed* data should be assessed carefully, while in the second step the sensitivity of the conclusions to the *unobserved data given the observed data* should be addressed. In the first step, one should ensure that the required assessment be carried out under one of two allowable scenarios, as represented by Figures 22.6(b) and (d), thereby carefully avoiding the scenario of Figure 22.6(c), where the model at the complete-data level is compared to the incomplete data; apples and oranges, as it were. The method proposed by Gelman et al. (2005) offers a convenient route to model assessment. These phenomena underscore the fact that fitting a model to incomplete data necessarily encompasses a part that cannot be assessed from the observed data. In particular, whether or not a dropout model is acceptable cannot be determined solely by mechanical model building exercises. Arbitrariness can be removed partly by careful consideration of the plausibility of a model. One should use as much context-derived information as possible. Prior knowledge can give an idea of which models are more plausible. Covariate information can be explicitly included in the model to increase the range of plausible models which can be fit. Moreover, covariates can help explain the dependence between response mechanism and outcomes. Good non-response models in a longitudinal setting should make use of the temporal and/or association structure among the repeated measures. It is worth reiterating that *no* single analysis provides conclusions that are free of dependence in some way or other on untestable assumptions.

Third, and directly following from the previous point, a case has been made for supporting any analysis of incomplete data with carefully conceived and contextually relevant sensitivity analyses, a view that, incidentally, is consistent with the position of the International Conference on Harmonisation (1999) guidelines. Various ways of conducting such a sensitivity analysis have been described and exemplified, without aiming for completeness. Research and development in the field of sensitivity analysis for incomplete data is very active and likely to remain that way for the foreseeable future.

References

Allison, P. D. (1987). Estimation of linear models with incomplete data. *Sociology Methodology* **17**, 71–103.

Baker, S. G. and Laird, N. M. (1988). Regression analysis for categorical variables with outcome subject to non-ignorable non-response. *Journal of the American Statistical Association* **83**, 62–69.

Baker, S. G., Rosenberger, W. F., and DerSimonian, R. (1992). Closed-form estimates for missing counts in two-way contingency tables. *Statistics in Medicine* **11**, 643–657.

Beunckens, C., Molenberghs, G., Verbeke, G., and Mallinckrodt, C. (2007). A latent-class mixture model for incomplete longitudinal Gaussian data. *Biometrics*. To appear.

Catchpole, E. A. and Morgan, B. J. T. (1997). Detecting parameter redundancy. *Biometrika* **84**, 187–196.

Catchpole, E. A., Morgan, B. J. T., and Freeman, S. N. (1998). Estimation in parameter-redundant models. *Biometrika* **85**, 462–468.

Cochran, W. G. (1977). *Sampling Techniques*. New York: Wiley.

Cohen, J. and Cohen, P. (1983). *Applied Multiple Regression/Correlation Analysis for the Behavioral Sciences*, 2nd ed. Hillsdale, NJ: Erlbaum.

Cook, R. D. (1986). Assessment of local influence. *Journal of the Royal Statistical Society, Series B* **48**, 133–169.

Copas, J. B. and Li, H. G. (1997). Inference from non-random samples (with discussion). *Journal of the Royal Statistical Society, Series B* **59**, 55–96.

Diggle, P. J. and Kenward, M. G. (1994). Informative drop-out in longitudinal data analysis (with discussion). *Applied Statistics* **43**, 49–93.

Draper, D. (1995). Assessment and propagation of model uncertainty (with discussion). *Journal of the Royal Statistical Society, Series B* **57**, 45–97.

Ekholm, A. and Skinner, C. (1998). The Muscatine children's obesity data reanalysed using pattern mixture models. *Applied Statistics* **47**, 251–263.

Fitzmaurice, G. M., Molenberghs, G., and Lipsitz, S. R. (1995). Regression models for longitudinal binary responses with informative dropouts. *Journal of the Royal Statistical Society, Series B* **57**, 691–704.

Follmann, D. and Wu, M. (1995). An approximate generalized linear model with random effects for informative missing data. *Biometrics* **51**, 151–168.

Foster, J. J. and Smith, P. W. F. (1998). Model-based inference for categorical survey data subject to non-ignorable non-response. *Journal of the Royal Statistical Society, Series B* **60**, 57–70.

Gelman, A., Van Mechelen, I., Verbeke, G., Heitjan, D. F., and Meulders, M. (2005). Multiple imputation for model checking: Completed-data plots with missing and latent data. *Biometrics* **61**, 74–85.

Glynn, R. J., Laird, N. M., and Rubin, D. B. (1986). Selection modelling versus mixture modelling with non-ignorable nonresponse. In H. Wainer (ed.), *Drawing Inferences from Self Selected Samples*, pp. 115–142. New York: Springer-Verlag.

Heckman, J. J. (1976). The common structure of statistical models of truncation, sample selection and limited dependent variables and a simple estimator for such models. *Annals of Economic and Social Measurement* **5**, 475–492.

Hedeker, D. and Gibbons, R. D. (1997). Application of random-effects pattern-mixture models for missing data in longitudinal studies. *Psychological Methods* **2**, 64–78.

Hogan, J. W. and Laird, N. M. (1997). Mixture models for the joint distribution of repeated measures and event times. *Statistics in Medicine* **16**, 239–258.

International Conference on Harmonisation E9 Expert Working Group (1999). Statistical principles for clinical trials: ICH Harmonised Tripartite Guideline. *Statistics in Medicine* **18**, 1905–1942.

Jansen, I., Beunckens, C., Molenberghs, G., Verbeke, G., and Mallinckrodt, C. (2006a). Analyzing incomplete discrete longitudinal clinical trial data. *Statistical Science* **21**, 52–69.

Jansen, I., Hens, N., Molenberghs, G., Aerts, M., Verbeke, G., and Kenward, M. G. (2006b). The nature of sensitivity in missing not at random models. *Computational Statistics and Data Analysis* **50**, 830–858.

Jennrich, R. I. and Schluchter, M. D. (1986). Unbalanced repeated measures models with structured covariance matrices. *Biometrics* **42**, 805–820.

Kenward, M. G. (1998). Selection models for repeated measurements with nonrandom dropout: An illustration of sensitivity. *Statistics in Medicine* **17**, 2723–2732.

Kenward, M. G., Goetghebeur, E. J. T., and Molenberghs, G. (2001). Sensitivity analysis of incomplete categorical data. *Statistical Modelling* **1**, 31–48.

Kenward, M. G. and Molenberghs, G. (1999) Parametric models for incomplete continuous and categorical longitudinal studies data. *Statistical Methods in Medical Research* **8**, 51–83.

Kenward, M.G., Molenberghs, G., and Thijs, H. (2003). Pattern-mixture models with proper time dependence. *Biometrika* **90**, 53–71.

Laird, N. M. (1994). Discussion of "Informative dropout in longitudinal data analysis," by P. J. Diggle and M. G. Kenward. *Applied Statistics* **43**, 84.

Lesaffre, E. and Verbeke, G. (1998). Local influence in linear mixed models. *Biometrics* **54**, 570–582.

Little, R. J. A. (1993). Pattern-mixture models for multivariate incomplete data. *Journal of the American Statistical Association* **88**, 125–134.

Little, R. J. A. (1994a). A class of pattern-mixture models for normal incomplete data. *Biometrika* **81**, 471–483.

Little, R. J. A. (1994b). Discussion of "Informative dropout in longitudinal data analysis," by P. J. Diggle and M. G. Kenward. *Applied Statistics* **43**, 78.

Little, R. J. A. (1995). Modeling the drop-out mechanism in repeated-measures studies. *Journal of the American Statistical Association* **90**, 1112–1121.

Little, R. J. A. and Rubin, D. B. (2002). *Statistical Analysis with Missing Data*, 2nd ed. New York: Wiley.

Little, R. J. A. and Wang, Y. (1996). Pattern-mixture models for multivariate incomplete data with covariates. *Biometrics* **52**, 98–111.

Little, R. J. A. and Yau, L. (1996). Intent-to-treat analysis for longitudinal studies with drop-outs. *Biometrics* **52**, 1324–1333.

McArdle, J. J. and Hamagami, F. (1992). Modeling incomplete longitudinal and cross-sectional data using latent growth structural models. *Experimental Aging Research* **18**, 145–166.

Michiels, B., Molenberghs, G., and Lipsitz, S. R. (1999). Selection models and pattern-mixture models for incomplete categorical data with covariates. *Biometrics* **55**, 978–983.

Michiels, B., Molenberghs, G., Bijnens, L., and Vangeneugden, T. (2002). Selection models and pattern-mixture models to analyze longitudinal quality of life data subject to dropout. *Statistics in Medicine* **21**, 1023–1041.

Molenberghs, G. and Kenward, M. G. (2007). *Missing Data in Clinical Studies*. Chichester: Wiley.

Molenberghs, G., Kenward, M. G., and Goetghebeur, E. (2001). Sensitivity analysis for incomplete contingency tables: The Slovenian plebiscite case. *Applied Statistics* **50**, 15–29.

Molenberghs, G., Kenward, M. G., and Lesaffre, E. (1997). The analysis of longitudinal ordinal data with non-random dropout. *Biometrika* **84**, 33–44.

Molenberghs, G., Michiels, B., and Kenward, M. G. (1998). Pseudo-likelihood for combined selection and pattern-mixture models for missing data problems. *Biometrical Journal* **40**, 557–572.

Molenberghs, G. and Verbeke, G. (2005). *Models for Discrete Longitudinal Data*. New York: Springer.

Molenberghs, G., Michiels, B., Kenward, M. G., and Diggle, P. J. (1998). Missing data mechanisms and pattern-mixture models. *Statistica Neerlandica* **52**, 153–161.

Molenberghs, G., Goetghebeur, E. J. T., Lipsitz, S. R., and Kenward, M. G. (1999). Non-random missingness in categorical data: strengths and limitations. *American Statistician* **53**, 110–118.

Molenberghs, G., Verbeke, G., Thijs, H., Lesaffre, E., and Kenward, M. G. (2001). Mastitis in dairy cattle: Influence analysis to assess sensitivity of the dropout process. *Computational Statistics and Data Analysis* **37**, 93–113.

Molenberghs, G., Thijs, H., Jansen, I., Beunckens, C., Kenward, M. G., Mallinckrodt, C., and Carroll, R. J. (2004). Analyzing incomplete longitudinal clinical trial data. *Biostatistics* **5**, 445–464.

Molenberghs, G., Beunckens, C., Sotto, C., and Kenward, M. G. (2007). Every missing not at random model has got a missing at random counterpart with equal fit. *Journal of the Royal Statistical Society, Series B*. To appear.

Muthén, B., Kaplan, D., and Hollis, M. (1987). On structural equation modeling with data that are not missing completely at random. *Psychometrika* **52**, 431–462.

Nordheim, E. V. (1984). Inference from nonrandomly missing categorical data: An example from a genetic study on Turner's syndrome. *Journal of the American Statistical Association* **79**, 772–780.

Potthoff, R. F. and Roy, S. N. (1964). A generalized multivariate analysis of variance model useful especially for growth curve problems. *Biometrika* **51**, 313–326.

Rizopoulos D., Verbeke G., and Lesaffre E. (2007). Sensitivity analysis in pattern mixture models using the extrapolation method. Submitted for publication.

Rizopoulos D., Verbeke G., and Molenberghs G. (2008). Shared parameter models under random-effects misspecification. *Biometrika* **94**, 94–95.

Robins, J. M. and Rotnitzky, A. (1995). Semiparametric efficiency in multivariate regression models with missing data. *Journal of the American Statistical Association* **90**, 122–129.

Robins, J. M., Rotnitzky, A., and Scharfstein, D. O. (1998). Semiparametric regression for repeated outcomes with non-ignorable non-response. *Journal of the American Statistical Association* **93**, 1321–1339.

Robins, J. M., Rotnitzky, A., and Scharfstein, D. O. (2000). Sensitivity analysis for selection bias and unmeasured confounding in missing data and causal inference models. In M. E. Halloran and D. A. Berry (eds.), *Statistical Models in Epidemiology, the Environment, and Clinical Trials*, pp. 1–94. New York: Springer.

Robins, J. M., Rotnitzky, A., and Zhao, L. P. (1994). Estimation of regression coefficients when some regressors are not always observed. *Journal of the American Statistical Association* **89**, 846–866.

Rotnitzky, A., Scharfstein, D., Su, T. L., and Robins, J. M. (2001). Methods for conducting sensitivity analysis of trials with potentially nonignorable competing causes of censoring. *Biometrics* **57**, 103–113.

Rubin, D. B. (1976). Inference and missing data. *Biometrika* **63**, 581–592.

Rubin, D. B. (1987). *Multiple Imputation for Nonresponse in Surveys*. New York: Wiley.

Rubin, D. B. (1994). Discussion of "Informative dropout in longitudinal data analysis," by P. J. Diggle and M. G. Kenward. *Applied Statistics* **43**, 80–82.

Rubin, D. B., Stern H. S., and Vehovar V. (1995). Handling "don't know" survey responses: The case of the Slovenian plebiscite. *Journal of the American Statistical Association* **90**, 822–828.

Scharfstein, D. O., Rotnitzky, A., and Robins, J. M. (1999). Adjusting for nonignorable drop-out using semiparametric nonresponse models (with discussion). *Journal of the American Statistical Association* **94**, 1096–1146.

Thijs, H., Molenberghs, G., and Verbeke, G. (2000). The milk protein trial: Influence analysis of the dropout process. *Biometrical Journal* **42**, 617–646.

Thijs, H., Molenberghs, G., Michiels, B., Verbeke, G., and Curran, D. (2002). Strategies to fit pattern-mixture models. *Biostatistics* **3**, 245–265.

Troxel, A. B., Harrington, D. P., and Lipsitz, S. R. (1998). Analysis of longitudinal data with non-ignorable non-monotone missing values. *Applied Statistics* **47**, 425–438.

Tsiatis, A. A. (2006). *Semiparametric Theory and Missing Data*. New York: Springer.

Tsonaka R., Verbeke G., and Lesaffre E. (2007). A semi-parametric shared parameter model to handle non-monotone non-ignorable missingness. Submitted for publication.

Vach, W. and Blettner, M. (1995). Logistic regression with incompletely observed categorical covariates — Investigating the sensitivity against violation of the missing at random assumption. *Statistics in Medicine* **12**, 1315–1330.

Vansteelandt, S., Goetghebeur, E., Kenward, M. G., and Molenberghs, G. (2006). Ignorance and uncertainty regions as inferential tools in a sensitivity analysis. *Statistica Sinica* **16**, 953–979.

Van Steen, K., Molenberghs, G., Verbeke, G., and Thijs, H. (2001). A local influence approach to sensitivity analysis of incomplete longitudinal ordinal data. *Statistical Modelling* **1**, 125–142.

Verbeke, G., Lesaffre, E., and Spiessens, B. (2001). The practical use of different strategies to handle dropout in longitudinal studies. *Drug Information Journal* **35**, 419–434.

Verbeke, G. and Molenberghs, G. (1997). *Linear Mixed Models in Practice: A SAS-Oriented Approach*, Lecture Notes in Statistics 126. New York: Springer.

Verbeke, G. and Molenberghs, G. (2000). *Linear Mixed Models for Longitudinal Data*. New York: Springer.

Verbeke, G., Molenberghs, G., Thijs, H., Lesaffre, E., and Kenward, M. G. (2001). Sensitivity analysis for non-random dropout: A local influence approach. *Biometrics* **57**, 7–14.

White, I. R. and Goetghebeur, E. J. T. (1998). Clinical trials comparing two treatment arm policies: Which aspects of the treatment policies make a difference? *Statistics in Medicine* **17**, 319–340.

Wu, M. C. and Bailey, K. R. (1988). Analysing changes in the presence of informative right censoring caused by death and withdrawal. *Statistics in Medicine* **7**, 337–346.

Wu, M. C. and Bailey, K. R. (1989). Estimation and comparison of changes in the presence of informative right censoring: Conditional linear model. *Biometrics* **45**, 939–955.

Wu, M. C. and Carroll, R. J. (1988). Estimation and comparison of changes in the presence of informative right censoring by modeling the censoring process. *Biometrics* **44**, 175–188.

CHAPTER 23

Estimation of the causal effects of time-varying exposures

James M. Robins and Miguel A. Hernán

Contents

23.1 Introduction

In this chapter we describe methods for the estimation of the causal effect of a time-varying exposure on an outcome of interest from longitudinal data collected in an observational study. The terms "exposure" and "treatment" will be used synonymously and interchangeably. We assume a fixed study population, that is, a closed cohort with a well-defined, known start of follow-up date for each subject. Time will refer to time since start of follow-up, which we also refer to as time since baseline. We only consider estimation of the effect of exposures occurring at or after the start of follow-up because the estimation of the effects of pre-baseline exposures is not possible without making strong, untestable assumptions. We refer to the exposure received at start of follow-up as the baseline exposure. Baseline covariates refer to covariates, including pre-baseline exposure, that occur prior to the baseline exposure. We classify exposures as either fixed or time-varying.

We define an exposure to be fixed if every subject's baseline exposure level determines the subject's exposure level at all later times. Exposures can be fixed because they only occur at the start of follow-up (e.g., a bomb explosion, a one-dose vaccine, a surgical intervention), because they do not change over time (e.g., genotype), or because they evolve over time in a deterministic way (e.g., time since baseline exposure).

Any exposure that is not fixed is said to be time-varying. Some examples of time-varying exposures are a subject's smoking status, a drug whose dose is readjusted according to the patient's clinical response, a surgical intervention that is administered to different study patients at different times from start of follow-up, and the phenotypic expression (say, mRNA level) of a genotype that responds to changing environmental factors.

We shall need to consider time-dependent confounders as well as time-varying exposures. For present purposes, one may consider a time-varying covariate to be a time-dependent confounder if a post-baseline value of the covariate is an independent predictor of (i.e., a risk factor for) both subsequent exposure and the outcome within strata jointly determined by baseline covariates and prior exposure. A more precise definition is given in Section 23.3. For a fixed exposure, time-dependent confounding is absent because baseline exposure fully determines later exposure. As a consequence, in the absence of confounding by unmeasured baseline covariates or model misspecification, conventional methods to adjust for confounding by baseline covariates (e.g., stratification, matching, and/or regression) deliver consistent estimators of the causal effect of a fixed exposure. In contrast, when interest focuses on the causal effect of a time-varying exposure on an outcome, even when confounding by unmeasured factors and model misspecification are both absent, conventional analytic methods may be biased and result in estimates of effect that may fail to have a causal interpretation, regardless of whether or not one adjusts for the time-dependent confounders in the analysis (Robins 1986; Hernán, Hernández-Díaz, and Robins, 2004). In fact, if (i) time-dependent confounding is present, and (ii) within strata of the baseline covariates, baseline exposure predicts the subsequent evolution of the time-dependent confounders, then conventional analytic methods can be biased and falsely find an exposure effect even under the sharp null hypothesis of no net, direct, or indirect effect of exposure on the outcome of any subject (see Section 23.4).

Nearly all exposures of epidemiologic interest are time-varying. However, because of the greater complexity of analytic methods that appropriately control for time-dependent confounding, introductory treatments of causal inference often consider only the case of fixed exposures.

This chapter provides an introduction to causal inference for time-varying exposures in the presence of time-dependent confounding. We will discuss three different methods to estimate the effect of time-varying exposures: the g-computation algorithm formula (the "g-formula"), inverse probability of treatment weighting (IPTW) of marginal structural models (MSMs), and g-estimation of structural nested models (SNMs). We refer to the collection of these methods as "g-methods." If we used only completely saturated (i.e., non-parametric) models, all three methods would give identical estimates of the effect of treatment. However, in realistic longitudinal studies, the data are sparse and high-dimensional. Therefore, possibly misspecified, non-saturated models must be used. As a consequence, the three methods can provide different estimates. The method of choice will then depend both on the causal contrast of primary substantive interest and on the method's robustness to model misspecification (see Section 23.5).

The chapter is organized as follows. First, we review causal inference with fixed exposures. Second, we generalize to time-varying exposures. Third, we analyze a simple hypothetical study of a time-varying exposure using saturated models to illustrate both the bias of conventional analytic methods and the validity of and agreement between the three g-methods. Fourth, we introduce general MSMs and SNMs in order to estimate optimal dynamic treatment regimes. Finally, we examine the strengths and weaknesses of each of our three methods in the analysis of realistic study data. In the interest of brevity, we limit ourselves to the case where (i) covariate and exposure data are collected at fixed equal-spaced intervals (e.g., at weekly clinic visits); (ii) censoring, missed visits, and measurement error are absent; (iii) the outcome is a univariate continuous random variable Y measured at end of follow-up; and (iv) there is no unmeasured confounding. Extensions to settings in which (i) through (iv) are violated can be found in prior work by Robins and collaborators. Violations of (i) are discussed in Robins (1997a, 1998); of (ii) in Robins, Rotnitzky, and Zhao (1995), Robins (2003), and van der Laan and Robins (2003); of (iii) in Robins (1994, 1997a) and Hernán, Brumback, and Robins (2001, 2002); and of (iv) in Robins, Rotnitzky, and Scharfstein (1999) and Brumback et al. (2004).

For consistency with the literature and ease of notation, in this chapter vectors are *not* denoted by boldface type.

23.2 Fixed exposures

We observe on each of N study subjects a fixed, dichotomous exposure A that can take values 0 (unexposed) or 1 (exposed); an outcome Y measured at the end of follow-up; and a vector L of baseline covariates. Capital letters such as Y or A will refer to random variables, that is, variables that can take on different values for different study subjects. Small letters such as y and a refer to the possible values of Y and A. Thus, the random variable A can take on the two values $a = 1$ or $a = 0$. Let Y_a denote the counterfactual or potential outcome for a given subject under exposure level a. For a dichotomous A we have two counterfactual variables $Y_{a=1}$ and $Y_{a=0}$. For example, for a subject whose outcome would be 3 under exposure and 1 under non-exposure, we would write $Y_{a=1} = 3$ and $Y_{a=0} = 1$ and $Y_{a=1} - Y_{a=0} = 2$. If, in the actual study, this subject were exposed, then his observed Y would be 3. That is, Y_a is the random variable representing the outcome Y that would be observed for a given subject were he or she to experience exposure level a. Furthermore, a subject's observed outcome Y is the counterfactual outcome Y_a corresponding to the treatment $A = a$ that the subject actually received. Implicit in our definition of a potential outcome is the assumption that a given subject's response is not affected by other subjects' treatment. This assumption cannot always be taken for granted. For example, it often fails in vaccine efficacy trials conducted within a single city because the vaccine exposure of other subjects can affect the outcome (infection status) of an unvaccinated subject through the mechanism of herd immunity. Standard statistical summaries of uncertainty due to sampling variability (e.g., a confidence interval for a proportion) only have meaning if we assume the N study subjects have been randomly sampled from a large source population of size M, such that N/M is very small. Because we plan to discuss sampling variability, we make this assumption, although we recognize that the assumed source population is ill defined, even hypothetical. Probability statements and expected values will refer to proportions and averages in the source population.

The contrast $Y_{a=1} - Y_{a=0}$ is said to be the individual causal effect of exposure on a subject. The average or mean causal effect in the population is then $E(Y_{a=1} - Y_{a=0}) =$

$E(Y_{a=1}) - E(Y_{a=0})$. We say that the exposure A has a causal effect (protective or harmful) on the mean of the outcome Y if $E(Y_{a=1}) - E(Y_{a=0}) \neq 0$. When Y is a dichotomous outcome variable, then the mean of Y_a equals the risk of Y_a, that is, $E(Y_a) = \Pr(Y_a = 1)$, and we refer to $E(Y_{a=1}) - E(Y_{a=0})$, $E(Y_{a=1})/E(Y_{a=0})$, and

$$\frac{E(Y_{a=1})/\{1 - E(Y_{a=1})\}}{E(Y_{a=0})/\{1 - E(Y_{a=0})\}}$$

as the causal risk difference, causal risk ratio, and causal odds ratio, respectively. Some equivalent statements that denote an average causal effect are: the causal risk difference differs from 0, the causal risk ratio differs from 1, and the causal odds ratio differs from 1.

We now provide conditions, which we refer to as identifiability conditions, under which it is possible to obtain, from observational data, consistent estimators of counterfactual quantities such as $E(Y_a)$ and thus the causal risk difference and the causal risk and odds ratio for binary Y. First, we define some notation. For any random variables, $B \amalg C | L = l$ means B and C are statistically independent within the stratum of subjects in the source population with $L = l$. Thus, if B and C are dichotomous, $B \amalg C | L = l$ says the B–C odds ratio is 1 in the l-stratum-specific 2×2 table of B versus C. $B \amalg C | L$ means B and C are statistically independent in every stratum of L. Thus, if L takes on four possible values, $B \amalg C | L$ implies that the four l-stratum-specific odds ratios are all 1. The three identifiability conditions are (Rosenbaum and Rubin, 1983):

1. Consistency. If $A = a$ for a given subject, then $Y_a = Y$ for that subject.

2. Conditional exchangeability or, equivalently, no unmeasured confounding given data on baseline covariates L, that is,

$$Y_a \amalg A | L = l \text{ for each possible value } a \text{ of } A \text{ and } l \text{ of } L.$$

3. Positivity. If $f_L(l) \neq 0$, then $f_{A|L}(a|l) > 0$ for all a, where $f_L(l) = \Pr(L = l)$ is the population marginal probability that L takes the value l, and $f_{A|L}(a|l) = \Pr(A = a|L = l)$ is the conditional probability that A takes the value a among subjects in the population with L equal to l. (The above assumes L and A are discrete variables. If L and/or A were continuous variables, we would interpret $f_L(l)$ and/or $f_{A|L}(a|l)$ as the marginal density of L and/or the conditional density of A given L, and drop $\Pr(L = l)$ and/or $\Pr(A = a|L = l)$ from the definition of positivity.)

These three conditions generally hold in an ideal two-armed randomized experiment with full compliance. Consistency states that, for a subject who was exposed (i.e., $A = 1$), her potential outcome $Y_{a=1}$ is equal to her observed outcome Y and thus is known (although her outcome $Y_{a=0}$ remains unknown). Analogously, for an unexposed subject (i.e., $A = 0$), her potential outcome $Y_{a=0}$ would equal her observed outcome Y, but $Y_{a=1}$ would remain unknown. Positivity means that the exposure was not deterministically allocated within any level l of the covariates L. That is, not all source population subjects with a given value l of L were assigned to be exposed or unexposed. Note that, even under positivity, all study subjects with $L = l$ could, by chance, be exposed because the study population is a small sample of the source population.

Before explaining conditional exchangeability, we discuss unconditional exchangeability. Unconditional or marginal exchangeability of the exposed and unexposed subgroups of the source population, written as $Y_{a=1} \amalg A$ and $Y_{a=0} \amalg A$, implies that the exposed, had they been unexposed, would have experienced the same distribution of outcomes as the unexposed did. Exchangeability also implies that the previous sentence holds true if one swaps the words "exposed" and "unexposed". Unconditional randomization ensures unconditional exchangeability because the distributions of risk factors in the exposed and unexposed groups are guaranteed to be the same. Conditional exchangeability only requires that exchangeability

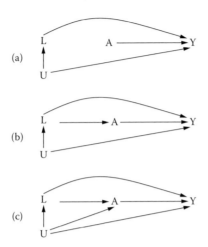

Figure 23.1 DAGs for fixed exposure scenarios.

is achieved within levels of the measured variables in L. For example, conditional — but not unconditional — exchangeability would hold in a randomized experiment in which (i) exposure was randomly assigned within levels l of a baseline covariate L that is an independent risk factor for Y, and (ii) the randomization probabilities $\Pr(A = 1|L = l)$ vary with l.

Unconditional exchangeability and conditional exchangeability can be translated into the language of causal directed acyclic graphs or DAGs (Spirtes, Glymour, and Scheines, 1993; Pearl, 1995). Section 23.7 contains the requisite background material on the representation of counterfactual causal models by causal DAGs. Consider the three causal DAGs of Figure 23.1 (Robins, Hernán, and Brumback, 2000), in which L and U represent vectors of measured and unmeasured baseline causes of Y, respectively. The causal DAG in Figure 23.1a can represent a randomized experiment in which each subject is randomized to exposure with the same probability $\Pr(A = 1)$. Therefore, the conditional probability of exposure does not depend on L or U, that is, $\Pr(A = 1|L = l, U = u) = \Pr(A = 1)$. We then say that there is no confounding by measured variables L or unmeasured variables U. Equivalently, the exposed and the unexposed are unconditionally exchangeable (i.e., $Y_a \amalg A$) because the exposure A and the outcome Y do not share any common causes. When unconditional exchangeability holds, association is causation. That is, the mean outcome had, contrary to fact, all study subjects been exposed to level a (i.e., $E(Y_a)$), equals the mean $E(Y|A = a)$ among the subset of the study population actually treated with a. Hence, for binary Y, the crude risk difference $E(Y|A = 1) - E(Y|A = 0)$ is the causal risk difference $E(Y_{a=1}) - E(Y_{a=0})$, so consistent estimation of the average causal effect is possible, even without data on L.

The causal DAG in Figure 23.1b can represent a randomized experiment in which each subject is randomized to exposure with probability $\Pr(A = 1|L = l)$ that depends on the subject's value of L but not on U, that is, $\Pr(A = 1|L = l, U = u) = \Pr(A = 1|L = l)$. We then say that there is confounding but the measured covariates are sufficient to adjust for it, so there is no unmeasured confounding. Equivalently, the exposed and the unexposed are conditionally exchangeable given L because, even though the exposure A and the outcome Y share some common causes U, the non-causal association between exposure and outcome can be blocked by conditioning on the measured covariates L. In this setting, marginal association is not causation, that is, $E(Y_a) \neq E(Y|A = a)$. However, within a stratum l of L, association is causation, $E(Y_a|L = l) = E(Y|A = a, L = l)$. Furthermore, by using data on L, $E(Y_a)$ can still be consistently estimated.

The causal DAG in Figure 23.1c represents a study in which the conditional probability of exposure $\Pr(A = 1|L = l, U = u)$ depends on the unmeasured variables U as well as the measured variables L and thus cannot possibly represent a randomized experiment. We say that there is unmeasured confounding. Equivalently, the exposed and the unexposed are not conditionally exchangeable given L because we cannot block all non-causal associations between exposure and outcome by conditioning on the measured covariates L. In this setting, neither $E(Y_a|L = l)$ nor $E(Y_a)$ can be consistently estimated, at least without further strong assumptions.

When the (three) identifiability conditions hold, one can use any of the three analytic methods discussed below — g-formula, inverse probability weighting (see Chapter 20), or g-estimation — to consistently estimate $E(Y_a)$. We first describe the g-formula and IPTW. A description of SNMs will be deferred to Section 23.4.

For a given value a of a fixed exposure A and vector L of baseline covariates, the g-formula (based on covariates L) for $E(Y_a)$ is defined to be the weighted sum of the l-stratum-specific means of Y among those exposed to level a in the population with weights equal to the frequency of the L strata. That is,

$$\sum_l E(Y|A = a, L = l)\,\Pr(L = l),$$

where the sum is over all values l of L in the population. Epidemiologists refer to the g-formula for $E(Y_{a=1})$ as the standardized mean of Y in the exposed $(A = 1)$. Note that the g-formula depends on the distribution in the population of the observed variables (A, L, Y). In practice, this distribution will be estimated from the study data.

When L takes values on a continuous scale, then the sum is replaced by an integral, and the g-formula becomes

$$\int E(Y|A = a, L = l)\,dF_L(l).$$

The IPTW formulae for $E(Y_{a=1})$ and $E(Y_{a=0})$ based on L are the mean of Y among the exposed $(A = 1)$ and unexposed $(A = 0)$, respectively, in a pseudo-population constructed by weighting each subject in the population by their subject-specific inverse probability of treatment weight

$$SW = \frac{f(A)}{f(A|L)},$$

where $f(A)$ and $f(A|L)$ are the probability densities $f_A(a)$ and $f_{A|L}(a|l)$ evaluated at the subject's data A, and A and L, respectively. In a randomized experiment, $f(A|L)$ is known by design. In an observational study, it must be estimated from the study data. Consider a subject with $A = 0$ and L equal to a particular value l^*, say. Suppose that two thirds of the population with $L = l^*$ but one third of the total population is exposed. Then, although in the true population each subject counts equally, in the pseudo-population our subject has weight 2 and thus counts as two subjects since $f(A) = \Pr(A = 0) = 1 - 1/3 = 2/3$, $f(A|L) = 1 - 2/3 = 1/3$, and $SW = 2$. In contrast, a second subject with $A = 1$ and $L = l^*$ has $SW = (1/3)/(2/3) = 1/2$ and so only counts as one half of a person.

We refer to the subject-specific SW as stabilized weights and to the pseudo-population created by these weights as a stabilized pseudo-population. In fact, as shown in the next paragraph, the IPTW formula for $E(Y_a)$ does not actually depend on the numerator of SW. Thus, we could alternatively create an unstabilized pseudo-population by weighting each subject by their unstabilized weight

$$W = \frac{1}{f(A|L)}.$$

However, as discussed in later sections, there are other settings in which stabilized or unstabilized weights cannot be used interchangeably.

Mathematically, the respective IPTW formulae for $E(Y_a)$ in the stabilized and unstabilized populations are

$$E\left\{\frac{I(A=a)f(A)}{f(A|L)}Y\right\}\Big/E\left\{\frac{I(A=a)F(A)}{f(A|L)}\right\}$$

and

$$E\left\{\frac{I(A=a)}{f(A|L)}Y\right\}\Big/E\left\{\frac{I(A=a)}{f(A|L)}\right\},$$

as

$$E\left\{N\frac{I(A=a)f(A)}{f(A|L)}\right\}\quad\text{and}\quad E\left\{N\frac{I(A=a)}{f(A|L)}\right\}$$

are the numbers of subjects in the stabilized and unstabilized pseudo-populations with $A=a$, and

$$E\left\{N\frac{I(A=a)f(A)}{f(A|L)}Y\right\}\quad\text{and}\quad E\left\{N\frac{I(A=a)}{f(A|L)}Y\right\}$$

are the sums of their Y values. Here, $I(\cdot)$ is the indicator function such that $I(B)=1$ if B is true, and $I(B)=0$ otherwise. Hernán and Robins (2006a) discuss the mathematical equivalence between the g-formula/standardization and IPTW (based on either stabilized or unstabilized weights) for fixed exposures under positivity. This equivalence extends to time-varying exposures as discussed in Section 23.3. The equivalence for fixed exposures is based on the mathematical identities

$$E\left\{\frac{I(A=a)f(A)}{f(A|L)}Y\right\}\Big/E\left\{\frac{I(A=a)f(A)}{f(A|L)}\right\}=E\left\{\frac{I(A=a)}{f(A|L)}Y\right\}\Big/E\left\{\frac{I(A=a)}{f(A|L)}\right\}$$

$$=\int E(Y|A=a,L=l)\,dF_L(l).$$

When exposure is unconditionally randomized (Figure 23.1a), both the g-formula and the IPTW formula for $E(Y_a)$ are equal to the unadjusted (i.e., crude) mean $E(Y|A=a)$ of Y among those with exposure level a in the population because the exposure A and the covariate L are independent, which implies $F(l)=F(l|a)$ in the g-formula, and $f(A|L)=f(A)$ for IPTW.

On the other hand, when the randomization is conditional on L (Figure 23.1b), then the average causal effect differs from the crude risk difference $E(Y|A=1)-E(Y|A=0)$, and data on L are needed to consistently estimate $E(Y_a)$. The g-formula estimates $E(Y_a)$ by effectively simulating the joint distribution of the variables L, A, and Y that would have been observed in a hypothetical study in which every subject received exposure a. The IPTW method effectively simulates the data that would have been observed had, contrary to fact, exposure been unconditionally randomized. Specifically, both stabilized and unstabilized IPTW create pseudo-populations in which (i) the mean of Y_a is identical to that in the actual study population but (ii) the exposure A is independent of L so that, if the causal graph in Figure 23.1b holds in the actual population, the causal graph in Figure 23.1a with no arrow from L to A will hold in the pseudo-population. The only difference between stabilized and unstabilized IPTW is that, in the unstabilized pseudo-population, $\Pr(A=1)=1/2$, while in the stabilized pseudo-population, $\Pr(A=1)$ is as in the actual population. Thus, $E(Y_a)$ in the actual population is $E_{ps}(Y|A=a)$, where the subscript ps is to remind us that we are taking the average of Y among subjects with $A=a$ in either pseudo-population. For example, suppose in the actual population there are three exposed subjects with SW equal to $1/3$, 2, $1/2$ and Y equal to 3, 6, 4, respectively. Then

$$E(Y_a)=E_{ps}(Y|A=1)=\frac{3\times1/3+6\times2+4\times1/2}{1/3+2+1/2}=5.3,$$

while

$$E(Y|A = 1) = \frac{3 \times 1 + 6 \times 1 + 4 \times 1}{3} = 4.$$

In summary, when the three identifiability conditions hold, the average causal effect $E(Y_{a=1}) - E(Y_{a=0})$ in the population is the crude risk difference $E_{ps}(Y|A = 1) - E_{ps}(Y|A = 0)$ in the pseudo-population.

What about observational studies? Imagine for a moment that the three identifiability conditions — consistency, conditional exchangeability, positivity — are met in a particular observational study. Then there is no conceptual difference between such an observational study and a randomized experiment. Taken together, the three conditions imply that the observational study can be conceptualized as a randomized experiment and hence that the g-formula, IPTW, or g-estimation can also be used to estimate counterfactual quantities like $E(Y_a)$ from the observational data. A difference between randomized experiments and observational studies is that the conditional probability of exposure is not known in the latter and thus needs to be estimated from the data. We discuss this issue in detail in the next section.

The major weakness of observational studies is that, unlike in randomized experiments with full compliance, the three identifiability conditions are not guaranteed by design. Positivity may not hold if subjects with certain baseline characteristics are always exposed (or unexposed) because of prevailing treatment practices in the community. In that case, subjects with those baseline characteristics are often excluded from the study population for purposes of causal inference. Conditional exchangeability will not hold if the exposed and the unexposed differ with respect to unmeasured risk factors as in Figure 23.1(c), that is, if there is unmeasured confounding. Unfortunately, the presence of conditional exchangeability cannot be empirically tested. Even consistency cannot always be taken for granted in observational studies because the counterfactual outcomes themselves are sometimes not well defined, which renders causal inferences ambiguous (Robins and Greenland, 2000; Hernán, 2005). Thus, in observational studies, an investigator who assumes that these conditions hold may be mistaken; hence, causal inference from observational data is a risky business. When the consistency and conditional exchangeability conditions fail to hold, the IPTW and g-formula for $E(Y_a)$ based on L are still well defined and can be estimated from the observed data; however, the formulae no longer equal $E(Y_a)$ and thus do not have the causal interpretation as the mean of Y had all subjects received treatment a. When positivity fails to hold for treatment level a, the IPTW formula remains well defined but fails to equal $E(Y_a)$, while the g-formula is undefined (Hernán and Robins, 2006a).

In summary, causal inference from observational data relies on the strong assumption that the observational study can be likened to a randomized experiment with randomization probabilities that depend on the measured covariates. Often this assumption is not explicit. Although investigators cannot prove that the observational–randomized analogy is 100% correct for any particular study, they can use their subject-matter knowledge to collect data on many relevant covariates and hope to increase the likelihood that the analogy is approximately correct. We next describe how to conceptualize observational studies as randomized experiments when the exposure changes over time.

23.3 Time-varying exposures

To develop methods for the estimation of the causal effects of time-varying exposures, we need to generalize the definition of causal effect and the three identifiability conditions of the previous section. For simplicity, we consider a study of the effect of a time-dependent dichotomous exposure $A(t)$ on a continuous outcome Y measured at end of follow-up at time $K + 1$ from study entry. ($A(t)$ is identical to A_t used in Chapter 20 and Chapter 22.) Subjects change exposure only at weekly clinic visits, so $A(t)$ is recorded at fixed times $t = 0, 1, \ldots, K$, in weeks from baseline. We use overbars to denote history; thus, the exposure history through time (i.e., week) t is $\overline{A}(t) = \{A(0), A(1), \ldots, A(t)\}$. The possible change in exposure at week t occurs after data are available on the history $\overline{L}(t) = \{L(0), L(1), \ldots, L(t)\}$ of a vector of possibly time-dependent covariates. We denote a subject's total exposure and covariate history by $\overline{A} = \overline{A}(K)$ and $\overline{L} = \overline{L}(K)$.

23.3.1 Non-dynamic regimes

To describe causal contrasts for time-varying exposures, we first need to define exposure regimes or plans. For simplicity, we temporarily restrict our description to static (non-dynamic) treatment regimes $\overline{a} = \{a(0), a(1), \ldots, a(K)\}$, where $a(t)$ is 1 if the regime specifies that the subject is to be exposed at time t, and 0 otherwise; and $\overline{a}(t)$ represents exposure history under regime $\overline{a}$ through week t. Note that $\overline{a}(K) = \overline{a}$. Associated with each of the 2^K regimes $\overline{a}$ is the subject's counterfactual outcome $Y_{\overline{a}}$ under exposure regime $\overline{a}$. Some examples of regimes $\overline{a}$ are continuous exposure $\{1, 1, \ldots, 1\}$, no exposure $\{0, 0, \ldots, 0\}$, exposure during the first two periods only $\{1, 1, 0, 0 \ldots, 0\}$, and exposure every other period $\{1, 0, 1, 0, \ldots\}$.

We say that the time-varying exposure $A(t)$ has a causal effect on the average value of Y if $E(Y_{\overline{a}}) - E(Y_{\overline{a}'}) \neq 0$ for at least two regimes $\overline{a}$ and $\overline{a}'$. The g-formula, IPTW, and g-estimation can provide consistent estimators of counterfactual quantities like $E(Y_{\overline{a}})$ under generalizations of our previous definitions of consistency, conditional exchangeability, and positivity. Specifically, the generalized identifiability conditions are:

1. Consistency. If $\overline{A} = \overline{a}$ for a given subject, then $Y_{\overline{a}} = Y$ for that subject.

2. Conditional exchangeability. $Y_{\overline{a}} \perp\!\!\!\perp A(t) | \overline{A}(t-1) = \overline{a}(t-1), \overline{L}(t) = \overline{l}(t)$ for all regimes $\overline{a}$ and all $\overline{l}(t)$.

3. Positivity. If $f_{\overline{A}(t-1), \overline{L}(t)}\{\overline{a}(t-1), \overline{l}(t)\} \neq 0$, then $f_{A(t)|\overline{A}(t-1), \overline{L}(t)}\{a(t)|\overline{a}(t-1), \overline{l}(t)\} > 0$ for all $a(t)$.

The three conditions generally hold in ideal sequentially randomized experiments with full compliance. A sequentially randomized experiment is a randomized experiment in which the exposure value at each successive visit t is randomly assigned with known randomization probabilities (bounded away from 0 and 1) that, by design, may depend on a subject's past exposure $\overline{A}(t-1)$ and covariate history $\overline{L}(t)$ through t. In the setting of time-varying exposures, the assumption of conditional exchangeability is sometimes referred to as the assumption of sequential randomization or the assumption of no unmeasured confounders.

As for fixed exposures, exchangeability and conditional exchangeability can be represented by causal DAGs. The DAGs in Figures 23.2a,b,c are the time-varying analogs of those in Figures 23.1a,b,c, respectively (Robins, Hernán, and Brumback, 2000). Figure 23.2a represents a sequentially randomized experiment in which the randomization probabilities at each time t depend at most on a subject's past exposure history, which is the proper generalization of "no confounding by measured or unmeasured variables" to a sequentially randomized experiment. In particular, the causal DAG in Figure 23.2a implies unconditional or marginal exchangeability, which we write in two different but mathematically equivalent ways: for all t and $\overline{a}$, $Y_{\overline{a}} \perp\!\!\!\perp A(t) | \overline{A}(t-1) = \overline{a}(t-1)$ or $Y_{\overline{a}} \perp\!\!\!\perp \overline{A}$. As with fixed exposures, unconditional exchangeability means that association is causation, $E(Y_{\overline{a}}) = E(Y | \overline{A} = \overline{a})$ and $E(Y_{\overline{a}}) - E(Y_{\overline{a}'}) = E(Y | \overline{A} = \overline{a}) - E(Y | \overline{A} = \overline{a}')$, so data on the measured covariates $\overline{L}$ need not be used to estimate average causal effects.

Figure 23.2b represents a sequentially randomized experiment in which the randomization probabilities at each time t depend on past exposure and measured covariate history but not further on unmeasured covariates, that is, there is confounding by measured covariates but no unmeasured confounding. Thus, the three identifiability conditions hold. In this setting, there is time-dependent confounding by $\overline{L}$, and association is not causation;

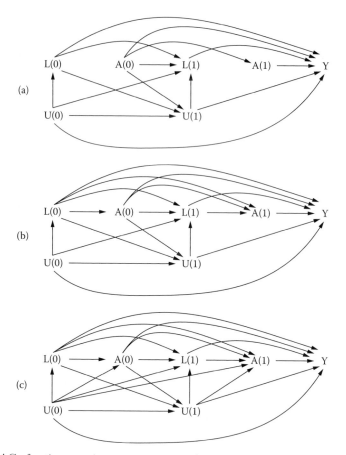

Figure 23.2 DAGs for time-varying exposure scenarios.

however, by using data on $\overline{L}$, $E(Y_{\bar{a}})$ can still be consistently estimated by using g-methods as described below.

Figure 23.2b also motivates the following precise definition of time-dependent confounding due to measured covariates. We say there is confounding for $E(Y_{\bar{a}})$ if $E(Y_{\bar{a}})$ (the mean outcome had, contrary to fact, all study subjects followed regime $\bar{a}$) differs from the mean $E(Y|\overline{A} = \bar{a})$ of Y (equivalently, $Y_{\bar{a}}$) among the subset of subjects who followed regime $\bar{a}$ in the actual study. We say the confounding is solely time-independent (i.e., wholly attributable to baseline covariates) if $E\{Y_{\bar{a}}|L(0)\} = E\{Y|\overline{A} = \bar{a}, L(0)\}$, as would be the case if the only arrows pointing into $A(1)$ in Figure 23.2(b) were from $A(0)$ and $L(0)$. In contrast, if the identifiability conditions hold, but $E\{Y_{\bar{a}}|L(0)\} \neq E\{Y|\overline{A} = \bar{a}, L(0)\}$, we say that time-dependent confounding is present.

Figure 23.2c represents a study in which the probability of exposure depends on variables U that cause Y and are unmeasured and thus cannot possibly represent a sequentially randomized experiment. In Figure 23.2c there is unmeasured confounding, and thus causal effects cannot be consistently estimated.

The expressions for the g-formula and IPTW presented above for fixed exposures need to be generalized for time-varying exposures. For example, with $L(t)$ discrete, the g-formula based on $\overline{L}$ for the counterfactual mean $E(Y_{\bar{a}^*})$ is

$$\sum_{\bar{l}} E(Y|\overline{A} = \bar{a}^*, \overline{L} = \bar{l}) \prod_{k=0}^{K} f\{l(k)|\overline{A}(k-1) = \bar{a}^*(k-1), \overline{L}(k-1) = \bar{l}(k-1)\}, \quad (23.1)$$

where the sum is over all possible $\bar{l}$-histories and $\bar{l}(k-1)$ is the history $\bar{l} = \bar{l}(K)$ through time $k-1$. Experience with this formula will be developed in Section 23.4 by working through an example.

Note that the g-formula for $E(Y_{\bar{a}^*})$ is simply the mean of Y under a joint density $f_{g=\bar{a}^*}(o)$ that differs from the observed density

$$f_{\text{obs}}(o) = f\{y|\overline{A}(k) = \bar{a}(k), \overline{L}(k) = \bar{l}(k)\}$$
$$\times \prod_{k=0}^{K} f\{l(k)|\overline{A}(k-1) = \bar{a}(k-1), \overline{L}(k-1) = \bar{l}(k-1)\}$$
$$\times \prod_{k=0}^{K} f\{a(k)|\overline{A}(k-1) = \bar{a}(k-1), \overline{L}(k) = \bar{l}(k)\}$$

for $O = (\overline{A}, \overline{L}, Y)$ only in that each $f\{a(k)|\overline{A}(k-1) = \bar{a}(k-1), \overline{L}(k) = \bar{l}(k)\}$ is replaced by a degenerate distribution that takes value $a^*(k)$ specified by the regime $\bar{a}^*$ with probability 1.

When applied to data from a sequentially randomized experiment like the one represented in the causal DAG of Figure 23.2b, the g-formula estimates $E(Y_{\bar{a}})$ by effectively simulating the joint distribution of the variables $\overline{L}$, $\overline{A}$, and Y that would have been observed in a hypothetical study where every subject received exposure $\bar{a}$. However, even in a sequentially randomized experiment, $E(Y|\overline{A} = \bar{a}, \overline{L} = \bar{l})$ and $f\{l(k)|\bar{a}(k-1), \bar{l}(k-1)\}$ will not be known, so estimates $\widehat{E}(Y|\overline{A} = \bar{a}, \overline{L} = \bar{l})$ and $\widehat{f}\{l(k)|\bar{a}(k-1), \bar{l}(k-1)\}$ have to be used in the g-formula (23.1). In realistic experiments, these estimates must come from fitting parsimonious non-saturated models. Model misspecification will result in biased estimators of $E(Y_{\bar{a}})$, even though the identifiability conditions hold. Robins and Wasserman (1997) showed that, if the sharp null hypothesis of no effect of exposure on Y is true, that is,

$$Y_{\bar{a}} - Y_{\bar{a}'} = 0 \text{ with probability 1 for all } \bar{a}' \text{ and } \bar{a},$$

then standard non-saturated models $E(Y|\overline{A} = \bar{a}, \overline{L} = \bar{l}; \upsilon)$ and $f\{l(k)|\bar{a}(k-1), \bar{l}(k-1); \omega\}$ based on distinct (i.e., variation-independent) parameters υ and ω cannot all be correct whenever $L(k)$ has any discrete components. As a consequence, in large studies, inference based on the estimated g-formula will result in the sharp null hypothesis being falsely rejected, whenever it is true, even in a sequentially randomized experiment. This phenomenon is referred to as the null paradox of the estimated g-formula. Fortunately, neither IPTW estimation nor g-estimation suffers from the null paradox, and thus both are more robust methodologies. Furthermore, even in a fixed exposure randomized trial, $E(Y|A = a, L = l)$ is unknown and must be estimated by fitting a non-saturated model when L is high-dimensional. The estimated g-formula will generally be biased when the model for $E(Y|A = a, L = l)$ is misspecified and the known randomization probabilities depend on L. However, the null paradox exists only with time-varying exposures.

The IPTW formula based on $\overline{L}$ for the counterfactual mean $E(Y_{\bar{a}})$ is the average of Y among subjects with $\overline{A} = \bar{a}$ in a stabilized or unstabilized pseudo-population constructed by weighting each subject by their subject-specific stabilized IPTW

$$SW = \prod_{k=0}^{K} \frac{f\{A(k)|\overline{A}(k-1)\}}{f\{A(k)|\overline{A}(k-1), \overline{L}(k)\}},$$

or their unstabilized IPTW

$$W = \prod_{k=0}^{K} \frac{1}{f\{A(k)|\overline{A}(k-1), \overline{L}(k)\}},$$

each a product over time-specific weights. When the three identifiability conditions hold, each IPTW method creates a pseudo-population in which (i) the mean of $Y_{\bar{a}}$ is identical to that in the actual population but (ii) like on the DAG in Figure 23.2(a), the randomization probabilities at each time t depend at most on past exposure history. The only difference is that in the unstabilized pseudo-population $\text{Pr}_{ps}\{A(k) = 1|\overline{A}(k-1), \overline{L}(k)\} = 1/2$, while in the stabilized pseudo-population $\text{Pr}_{ps}\{A(k) = 1|\overline{A}(k-1), \overline{L}(k)\}$ is $\text{Pr}\{A(k) = 1|\overline{A}(k-1)\}$ from the actual population. Thus, $E(Y_{\bar{a}})$ in the actual population is $E_{ps}(Y|\overline{A} = \bar{a})$, where the subscript ps refers to either pseudo-population. Hence, the average causal effect $E(Y_{\bar{a}}) - E(Y_{\bar{a}'})$ is $E_{ps}(Y|\overline{A} = \bar{a}) - E_{ps}(Y|\overline{A} = \bar{a}')$.

One can estimate $E_{ps}(Y|\overline{A} = \bar{a})$ from the observed study data by the average of Y among subjects with $\overline{A} = \bar{a}$ in a stabilized or unstabilized pseudo-study population constructed by weighting each study subject by SW or W.

If $f\{A(k)|\overline{A}(k-1), \overline{L}(k)\}$ in SW or W is replaced by an estimator $\widehat{f}\{A(k)|\overline{A}(k-1), \overline{L}(k)\}$ based on a misspecified logistic model for the $\text{Pr}\{A(k) = 1|\overline{A}(k-1), \overline{L}(k\}$, the resulting estimators of $E(Y_{\bar{a}})$ and $E(Y_{\bar{a}}) - E(Y_{\bar{a}'})$ will be biased. In contrast, replacing the numerator of SW with an estimator $\widehat{f}\{A(k)|\overline{A}(k-1)\}$ based on a misspecified model does not result in bias. These remarks apply also to the IPTW estimation of marginal structural models considered in the following subsection. Now, in a sequentially randomized experiment, the denominators of the weights are known by design and so need not be estimated. As a consequence, in contrast to the estimated g-formula, in a sequentially randomized experiment, IPTW estimation unbiasedly estimates $E_{ps}(Y|\overline{A} = \bar{a}) - E_{ps}(Y|\overline{A} = \bar{a}')$ and so is never misleading.

When the three identifiability conditions hold in an observational study with a time-varying exposure, the observational study can be conceptualized as a sequentially randomized experiment, except that the probabilities $f\{A(k)|\overline{A}(k-1), \overline{L}(k)\}$ are unknown and must be estimated. However, the validity of these conditions is not guaranteed by design and is not subject to empirical verification. The best one can do is to use subject-matter knowledge to collect data on many potential time-dependent confounders. Furthermore, even if the identifiability conditions hold, bias in estimation of $E(Y_{\bar{a}})$ and $E(Y_{\bar{a}}) - E(Y_{\bar{a}'})$ can occur (i) when using IPTW estimation due to misspecification of models for $\text{Pr}\{A(k) = 1|\overline{A}(k-1), \overline{L}(k)\}$ and (ii) when using the estimated g-formula due to misspecification of models for $E\{Y|\overline{A} = \bar{a}, \overline{L} = \bar{l}\}$ and $f\{l(k)|\overline{a}(k-1), \overline{l}(k-1)\}$. However, the robustness of IPTW methods to model misspecification can be increased by using doubly robust estimators as described in Bang and Robins (2005); see Chapter 20 for a discussion of doubly robust estimators.

23.3.2 Marginal structural models

If, as is not infrequent in practice, K is of the order of 100 and the number of study subjects is of order 1000, then the 2^{100} unknown quantities $E(Y_{\bar{a}})$ far exceeds the sample size. Thus, very few subjects in the observed study population follow any given regime, so we need to specify a non-saturated model for the $E(Y_{\bar{a}})$ that combines information from many regimes to help estimate a given $E(Y_{\bar{a}})$. The price paid for modeling $E(Y_{\bar{a}})$ is yet another threat to the validity of the estimators due to possible model misspecification.

Suppose for a continuous response Y it is hypothesized that the effect of treatment history $\bar{a}$ on the mean outcome increases linearly as a function of the cumulative exposure $\text{cum}(\bar{a}) = \sum_{t=0}^{K} a(t)$ under regime $\bar{a}$. This hypothesis is encoded in the marginal structural mean model

$$E(Y_{\bar{a}}) = \eta_0 + \eta_1 \text{cum}(\bar{a}) \tag{23.2}$$

for all $\bar{a}$. The model is referred to as a marginal structural model because it models the marginal mean of the counterfactuals $Y_{\bar{a}}$, and models for counterfactuals are often referred to as structural models. There are 2^K different unknown quantities on the left-hand side of model (23.2), one for each of the 2^K different regimes $\bar{a}$, but only two unknown parameters η_0 and η_1 on the right-hand side. It follows that the MSM (23.2) is not a saturated (i.e., non-parametric) model because saturated models must have an equal number of unknowns on both sides of their defining equation. Any unsaturated model may be misspecified. For example, MSM (23.2) would be misspecified if $E(Y_{\bar{a}})$ either depended on some function of the regime $\bar{a}$ other than cumulative exposure (say, cumulative exposure only in the final 5 weeks $\mathrm{cum}_{-5}(\bar{a}) = \sum_{K-5}^{K} a(t)$) or depended non-linearly (say, quadratically) on cumulative exposure. It follows that we need methods both to test whether MSM (23.2) is correctly specified and to estimate the parameters η_0 and η_1. It is important to note that, under the null hypothesis, the MSM is correctly specified with $\eta_1 = 0$. Thus, MSMs are not subject to the null paradox.

MSMs are fit by IPTW as described, for example, by Robins, Hernán, and Brumback (2000). Specifically, Robins (1998) has shown that if we fit the ordinary linear regression model

$$E(Y|\overline{A}) = \gamma_0 + \gamma_1 \mathrm{cum}(\overline{A}) \qquad (23.3)$$

to the observed data by weighted least squares with weights SW or W, then, under the three identifiability conditions, the weighted least-squares estimators of γ_0 and γ_1 are consistent for the causal parameters η_0 and η_1 of the MSM (23.2) (but are inconsistent for the association parameters γ_0 and γ_1 of model [23.3]) because weighted least squares with weights SW or W is equivalent to ordinary least squares (OLS) in the stabilized or unstabilized unconfounded pseudo-population, respectively. In these populations, the association being estimated with OLS is causation.

A robust variance estimator (e.g., as used for GEE models; see Chapter 3) can be used to set confidence intervals for η_0 and η_1 and thus for any $E(Y_{\bar{a}})$ of interest. These intervals remain valid (i.e., are conservative) even when estimates are substituted for the numerator and/or denominator weights, provided the model for the denominator weights is correctly specified. For a non-saturated model like MSM (23.2) the length of the intervals will typically be much narrower when the model is fit with the weights SW than with the weights W, so the SW weights are preferred.

Further, if we fit the model

$$E(Y|\overline{A}) = \gamma_0 + \gamma_1 \mathrm{cum}(\overline{A}) + \gamma_2 \mathrm{cum}_{-5}(\overline{A}) + \gamma_3 \{\mathrm{cum}(\overline{A})\}^2$$

by weighted least squares with weights SW or W, a Wald test on two degrees of freedom of the joint hypothesis $\gamma_2 = \gamma_3 = 0$ is a test of the null hypothesis that MSM (23.2) is correctly specified with high power against the particular directions of misspecification mentioned above, especially if the weights SW are used.

Suppose it is further hypothesized that, for a particular dichotomous component V of the vector of baseline covariates $L(0)$, there might exist a qualitative V-exposure effect modification, with the result that exposure might be harmful to subjects with $V = 0$ and beneficial to those with $V = 1$, or vice versa. To examine this hypothesis, we would elaborate MSM (23.2) as

$$E(Y_{\bar{a}}|V) = \eta_0 + \eta_1 \mathrm{cum}(\bar{a}) + \eta_2 V + \eta_3 \mathrm{cum}(\bar{a})V,$$

an MSM conditional on the baseline covariate V. Qualitative effect modification is present if η_1 and $\eta_1 + \eta_3$ are of opposite signs. We can estimate the model parameters by fitting the ordinary linear regression model $E(Y|\overline{A}, V) = \gamma_0 + \gamma_1 \mathrm{cum}(\overline{A}) + \gamma_2 V + \gamma_3 V \mathrm{cum}(\overline{A})$ by weighted least squares with model weights SW or W. However, Robins (1998) showed that for an MSM defined conditional on a baseline covariate V, confidence intervals will still be valid but narrower if, rather than using weights SW or W, we use the weights

$$SW(V) = \prod_{k=0}^{K} \frac{f\{A(k)|\overline{A}(k-1), V\}}{f\{A(k)|\overline{A}(k-1), \overline{L}(k)\}},$$

which differ from SW by adding V to the conditioning event in the numerator.

23.3.3 Dynamic regimes

So far we have only considered estimation of the mean outcome $E(Y_{\bar{a}})$ under the 2^K static or non-dynamic regimes $\bar{a}$. However, to characterize the optimal treatment strategy, it is usually necessary to consider dynamic regimes as well.

A non-random dynamic regime is a treatment strategy or rule in which the treatment $a(t)$ at time t depends in a deterministic manner on the evolution of a subject's measured time-dependent covariates $\bar{L}(t)$ and, possibly, treatments $\overline{A}(t-1)$ up to t. An example would be the dynamic regime "take the treatment methotrexate at week t if and only if the neutrophil count has been greater than 1000 for three consecutive weeks and the patient was not on treatment at week $t-1$." Mathematically, when $A(t)$ is a binary treatment, a non-random dynamic regime g is a collection of functions $\{g_k\{\bar{a}(k-1), \bar{l}(k)\}; k = 0, \ldots, K\}$, each with range the two-point set $\{0, 1\}$, where $g_k\{\bar{a}(k-1), \bar{l}(k)\}$ specifies the treatment to be taken at k for a subject with past history $\{\bar{a}(k-1), \bar{l}(k)\}$. In our methotrexate example, $g_k\{\bar{a}(k-1), \bar{l}(k)\}$ is 1 if a subject's $a(k-1)$ is zero and his $\bar{l}(k)$ implies that his neutrophil count has been greater than 1000 at weeks $k, k-1$, and $k-2$ (so k must be at least 2); otherwise $g_k\{\bar{a}(k-1), \bar{l}(k)\}$ is 0.

A random dynamic regime is a treatment strategy where the treatment $a(t)$ at time t depends in a probabilistic way on $\bar{l}(t)$ and possibly $\bar{a}(t-1)$. An example would be "if the neutrophil count has been greater than 1000 for three consecutive weeks, randomize the subject to take methotrexate at week t with randomization probability 0.80, otherwise use randomization probability 0.10." Thus, a random dynamic regime is precisely a sequentially randomized experiment.

Now let g represent a regime — dynamic or non-dynamic, deterministic or random — and let Y_g denote the counterfactual outcome had regime g been followed. If high values of the outcome are considered beneficial, then the optimal regime g_{opt} maximizes the average outcome $E(Y_g)$ over all regimes. In fact, we need only try to find the optimal regime among the deterministic regimes as no random strategy can ever be preferred to the optimal deterministic strategy. Furthermore, the above example indicates that this optimal deterministic treatment strategy must be a dynamic regime whenever the treatment is a potentially toxic prescription drug such as methotrexate, as it is essential to temporarily discontinue the drug when a severe toxicity such as neutropenia develops. Random regimes (i.e., ordinary randomized trials and sequentially randomized trials) remain scientifically necessary because, before the trial, it is unknown which deterministic regime is optimal.

Under a slight strengthening of the identifiability conditions, $E(Y_g)$ for a deterministic dynamic regime g can be estimated from the data collected in a sequentially randomized trial by the average of Y among subjects in the unstabilized (but not in the stabilized) pseudo-study population who followed the regime g, that is, subjects whose observed covariate and treatment history is consistent with following regime g. Note that this is our first example of a result that is true for the unstabilized but not the stabilized IPTW estimation. The required strengthening is that we need the "strengthened" identifiability conditions:

1. "Strengthened" consistency. For any regime g, if, for a given subject, we have $A(k) = g_k\{\overline{A}(k-1), \overline{L}(k)\}$ at each time k, then $Y_g = Y$ and $\overline{L}_g(K) = \overline{L}(K)$ for that subject, where $\overline{L}_g(k)$ is the counterfactual L-history through time k under regime g.

Remark: For any regime g for which the treatment at each k does not depend on past treatment history so $g_k\{\overline{a}(k-1), \overline{l}(k)\} = g_k\{\overline{l}(k)\}$, we can write the "strengthened" consistency condition as follows: If $\overline{A} = \overline{g}_K\{\overline{L}(K)\}$ for a given subject, then $Y_g = Y$ and $\overline{L}_g(K) = \overline{L}(K)$ for that subject, where $\overline{g}_k\{\overline{L}(K)\}$ is the treatment through time k of a subject following regime g with covariate history $\overline{L}(k)$.

2. "Strengthened" conditional exchangeability. For any t, $\overline{l}(t)$ and regime g,

$$Y_g \amalg A(t)|\overline{L}(t) = \overline{l}(t), A(k) = g_k\{\overline{A}(k-1), \overline{L}(k)\} \quad \text{for} \quad k = 0, \dots, t-1.$$

Remark: For any regime g for which the treatment at each k does not depend on past treatment history, so $g_k\{\overline{a}(k-1), \overline{l}(k)\} = g_k\{\overline{l}(k)\}$, we can write "strengthened" conditional exchangeability as follows: For all t, $\overline{l}(t)$, and regimes g,

$$Y_g \amalg A(t)|\overline{L}(t) = \overline{l}(t), \quad \overline{A}(t-1) = \overline{g}_{t-1}\{\overline{l}(t-1)\}.$$

3. Positivity. This assumption remains unchanged.

Strengthened conditions 1 and 2 will hold on any causal DAG, such as that corresponding to a sequentially randomized trial, in which all parents of treatment variables $A(m)$ are measured variables. This implication follows from two facts. First, any such causal DAG satisfies both of the following conditions:

1. "Full" consistency: $Y_{\overline{a}} = Y_{\overline{a}^*}$ if $\overline{a}^* = \overline{a}$; $Y = Y_{\overline{a}}$ if $\overline{A} = \overline{a}$; $\overline{L}_{\overline{a}}(m) = \overline{L}_{\overline{a}^*}(m)$ if $\overline{a}^*(m-1) = \overline{a}(m-1)$; $\overline{L}_{\overline{a}}(m) = \overline{L}(m)$ if $\overline{A}(m-1) = \overline{a}(m-1)$, where $\overline{L}_{\overline{a}}(m)$ is the counterfactual L-history through time m under regime $\overline{a}$.

2. "Full" conditional exchangeability

$$(Y_{\overline{\mathcal{A}}}, \overline{L}_{\overline{\mathcal{A}}}) \amalg A(t)|\overline{A}(t-1), \overline{L}(t),$$

where $\overline{\mathcal{A}}$ denotes the set of all 2^K regimes $\overline{a}$, $Y_{\overline{\mathcal{A}}}$ denotes the set of all 2^K counterfactuals $Y_{\overline{a}}$, and $\overline{L}_{\overline{\mathcal{A}}}$ denotes the set of all 2^K counterfactual covariate histories $\overline{L}_{\overline{a}}$ through the end of the study.

Second, the "full" consistency and "full" conditional exchangeability conditions imply both the strengthened conditions, even though the "full" conditions only refer to non-dynamic regimes (Robins, 1986).

Remark: Associated with each regime g with treatment $g_k\{\overline{a}(k-1), \overline{l}(k)\}$ depending on past treatment and covariate history is another regime $g^\triangle$ with treatment $g_k^\triangle\{\overline{l}(k)\}$ depending only on past covariate history such that if "full" consistency holds, any subject following regime g from time zero will have the same treatment, covariate, and outcome history as when following regime $g^\triangle$ from time zero. In particular, $Y_g = Y_{g^\triangle}$ and $\overline{L}_g(K) = \overline{L}_{g^\triangle}(K)$. Specifically, $g^\triangle$ is defined in terms of g recursively by $g_0^\triangle\{l(0)\} = g_0\{\overline{a}(-1) = 0, l(0)\}$ (by convention, $\overline{a}(-1)$ can only take the value zero) and $g_k^\triangle\{\overline{l}(k)\} = g_k[g_{k-1}^\triangle\{\overline{l}(k-1)\}, \overline{l}(k)]$. For the dynamic methotrexate regime g described earlier, $g^\triangle$ is the regime "take methotrexate at k if and only if your $\overline{l}(k)$ implies your neutrophil count has been greater than 1000 for m consecutive weeks and m is an odd number greater than or equal to 3." Requiring m to be odd guarantees that no subject will ever take methotrexate for two consecutive weeks, as specified by regime g. For any regime g for which treatment at each k already does not depend on past treatment history, g and $g^\triangle$ are the identical set of functions. The above definition of $g^\triangle$ in terms of g guarantees that a subject has followed regime g through time t in the observed data (i.e., $A(k) = g_k\{\overline{A}(k-1), \overline{L}(k)\}$ for $k \le t$) if and only if the subject has followed regime $g^\triangle$ through t (i.e., $A(k) = g_k^\triangle\{\overline{L}(k)\}$ for $k \le t$).

"Full" consistency is a natural assumption that we will always make. Therefore, in view of the last remark, unless stated otherwise, we will henceforth use the term "dynamic regime" to refer to dynamic regimes for which the treatment at each k depends on past covariates but not on past treatment history.

The above discussion raises the question of whether or not it is substantively plausible that $E(Y_g)$ is identifiable by the g-formula for non-dynamic g but not for dynamic g, because conditional exchangeability, but neither "full" nor "strengthened" conditional exchangeability holds. Robins (1986) showed that this state of affairs is indeed substantively plausible. For example, it can occur when there exist unmeasured common causes U of treatment $A(k)$ and a covariate $L(t)$, $k < t$, but there do not exist unmeasured common causes of the $A(k)$ and Y. In Section 23.7 we provide a general graphical criterion due to Robins (1997b) that can be used to determine the subset of all regimes g (dynamic and non-dynamic) for which either of the above exchangeability conditions continue to hold. For g in the subset, $E(Y_g)$ remains unidentifiable by the unstabilized IPTW formula.

Of course, very few subjects in the observed study population follow any given regime, so, in practice, we need to combine information from many different regimes to estimate a given $E(Y_g)$. In Section 23.4 and Section 23.5 we show that this combination can be accomplished through g-estimation of nested structural models or IPTW estimation of dynamic MSMs (as defined in Section 23.5). Finally, we note that we can also estimate $E(Y_g)$ under the strengthened identifiability conditions using the g-formula (23.1), modified by replacing $\bar{a}(k-1)$ by $\bar{g}\{\bar{l}(k-1)\}$ and $\bar{a}$ by $\bar{g}\{\bar{l}(K)\}$. However, as discussed above, the estimated g-formula, in contrast to g-estimation of nested structural models, suffers from the null paradox, and thus is less robust.

Under the strengthened identifiability conditions, we will see in Section 23.5 that g-methods can be used to estimate not only $E(Y_g)$ but also the optimal (deterministic) treatment regime from observational data, even though, in most observational studies, subjects are not following any particular deterministic regime. The reason why this strategy succeeds is that we may conceptualize the subjects in such an observational study as following a random dynamic regime, with unknown randomization probabilities that must be estimated from the data.

23.4 Analysis of a hypothetical study

23.4.1 The study

Table 23.1 contains data from a hypothetical study of the effect of antiretroviral therapy on a global health score Y measured at the end of follow-up in 32,000 HIV-infected subjects. Y is a function of CD4 cell count, serum HIV RNA, and certain biochemical measures of possible drug toxicity, with higher values of Y signifying better health. The variables $A(0)$ and $A(1)$ are 1 if a subject received antiretroviral therapy at times $t = 0$ and $t = 1$, respectively, and 0 otherwise. The binary variable $L(1)$ is temporally prior to $A(1)$ and takes on the value 1 if the subject's CD4 cell count was greater than 200 cells/μL at time $t = 1$, and is 0 otherwise. To save space, the table displays one row per combination of values of $A(0)$, $L(1)$, and $A(1)$, rather than one row per subject. For each of the eight combinations, the table provides the number of subjects and the average value of the outcome $E\{Y|A(0), L(1), A(1)\}$. Thus, in row 1 of Table 23.1, $E\{Y|A(0), L(1), A(1)\} = 200$ means $E\{Y|A(0) = 0, L(1) = 1, A(1) = 0\} = 200$. We suppose that sampling variability is absent, and we assume consistency. Further, by inspection of Table 23.1, we can conclude that the positivity condition is satisfied, because otherwise one or more of the eight rows would have had zero subjects.

For the present, we suppose that the data arose from a sequentially randomized trial in which treatment at time 1 is randomly assigned with probability that depends on prior

Table 23.1 The Hypothetical Study Data

Row	$A(0)$	$L(1)$	$A(1)$	N	$E\{Y|A(0), L(1), A(1)\}$
1	0	1	0	2000	200
2	0	1	1	6000	220
3	0	0	0	6000	50
4	0	0	1	2000	70
5	1	1	0	3000	130
6	1	1	1	9000	110
7	1	0	0	3000	230
8	1	0	1	1000	250

covariate history but not on prior exposure. Because our interest is in the implications of time-dependent confounding by $L(1)$, we did not bother to include a measured baseline covariate $L(0)$. Alternatively, one can assume that a measured baseline covariate $L(0)$ exists but that the data in Table 23.1 are from a single stratum $l(0)$ of $L(0)$.

23.4.2 A priori *causal assumptions*

We assume Figure 23.3 is the causal DAG corresponding to this study. In Figure 23.3, U denotes a subject's baseline immunological function, an unmeasured variable that therefore does not appear in Table 23.1. The dotted arrows from $A(0)$ to Y, L to Y, and $A(1)$ to Y emphasize that we do not know, based on prior subject-matter knowledge, whether or not these causal arrows are present; in fact, our goal will be to use the data in Table 23.1 to determine, as far as possible, which of these arrows are present. We will later see that the data from Table 23.1 imply that (i) the arrow from $A(1)$ to Y is present and that (ii) the arrow from $A(0)$ to Y, or the arrow from L to Y, or both are present.

We now describe how, before observing the data, we used our subject-matter knowledge to decide that Figure 23.3 was an appropriate causal DAG. First, note the causal DAG in Figure 23.3, like that in Figure 23.2(b), is not a complete DAG because there do not exist direct arrows from U into either treatment. This is justified by our assumption that the study was a sequentially randomized trial. The absence of these arrows implies strengthened conditional exchangeability, that is, $Y_g \amalg A(0)$ and $Y_g \amalg A(1)|A(0) = a(0), L(1) = l(1)$ for all regimes g, whether static or dynamic. The arrows from U to Y and from U to CD4 cell count $L(1)$ are justified on subject-matter grounds by the well-known effects of immunosuppression on viral load and CD4 cell count. The arrow from $A(0)$ to $L(1)$ is justified by prior knowledge of the effect of antiretroviral therapy on CD4 cell count. The presence of the arrow from $L(1)$ into $A(1)$ and the absence of an arrow from $A(0)$ into $A(1)$ are justified by our knowledge that treatment at time 1 was randomly assigned with probability that depended only on prior covariate history.

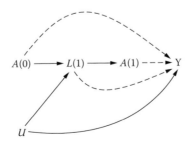

Figure 23.3 Causal DAG in the hypothetical study population.

23.4.3 Testing our causal assumptions

Assumptions concerning causal relations based on subject-matter knowledge can sometimes be mistaken. However, under the sole assumption that the study satisfies conditional exchangeability (implied by the assumption of no arrows from U into either treatment), we can use the data in Table 23.1 to confirm or refute empirically whether or not the arrows argued for on subject-matter grounds are actually present. To carry this out, we assume that the dashed arrows are actually present until we can prove otherwise. If the causal arrow from $A(0)$ to $L(1)$ were not present in Figure 23.3, $A(0)$ and $L(1)$ would be d-separated and thus independent by the causal Markov assumption (see the definitions in Section 23.7). But the data refute independence because $\Pr\{L(1) = 1|A(0) = 1\} = 0.75$ differs from $\Pr\{L(1) = 1|A(0) = 0\} = 0.50$.

Here is an alternative but closely related argument that results in the same conclusion. A causal arrow from $A(0)$ to $L(1)$ exists if the average causal effect $E\{L_{a(0)=1}(1)\} - E\{L_{a(0)=0}(1)\}$ of the fixed exposure $A(0)$ on the outcome $L(1)$ is non-zero. Because there is no confounding for the effect of $A(0)$ on $L(1)$, association is causation, and thus

$$E\{L_{a(0)=1}(1)\} - E\{L_{a(0)=0}(1)\} = E\{L(1)|A(0) = 1\} - E\{L(1)|A(0) = 0\},$$

which is non-zero.

Next, the absence of a causal arrow from $A(0)$ to $A(1)$ in Figure 23.3 implies $A(1)$ and $A(0)$ d-separated, and thus independent, given $L(1)$, which is confirmed by the data in Table 23.1. If the causal arrow from $L(1)$ to $A(1)$ was not present in Figure 23.3, $A(1)$ and $L(1)$ would be d-separated, and thus independent, given $A(0)$, which is refuted by the data in Table 23.1.

23.4.4 Determining which dotted arrows are present

23.4.4.1 A fixed exposure analysis at time 1

We can use the data in Table 23.1 to try to determine which of the dotted arrows in Figure 23.3 are present. In Figure 23.3, if the causal arrow from $A(1)$ to Y was not present, $A(1)$ and Y would be d-separated, and thus independent, given $L(1)$ and $A(0)$, which is refuted by the data in Table 23.1 because, for example, $E\{Y|A(0) = 0, L(1) = 0, A(1) = 1\} = 70$ and $E\{Y|A(0) = 0, L(1) = 0, A(1) = 0\} = 50$. Thus, we conclude that $A(1)$ has a causal effect on Y.

Here is another way to think about this. View the effect of $A(1)$ as that of a fixed baseline exposure in a study beginning at time 1 with baseline covariates $\{A(0), L(1)\}$. Then a causal arrow from $A(1)$ to Y exists if the average causal effect $E\{Y_{a(1)=1}|A(0), L(1)\} - E\{Y_{a(1)=0}|A(0), L(1)\}$ is non-zero in any of the four strata determined by joint levels of $\{A(0), L(1)\}$. But because, by sequential randomization, there is no confounding for the effect of $A(1)$ on Y within levels of $\{A(0), L(1)\}$ (equivalently, all non-causal paths from $A(1)$ to Y are blocked when we condition on $\{A(0), L(1)\}$), conditional association is causation and $E\{Y_{a(1)=1}|A(0), L(1)\} - E\{Y_{a(1)=0}|A(0), L(1)\} = E\{Y|A(1) = 1, A(0), L(1)\} - E\{Y(1)|A(1) = 0, A(0), L(1)\}$, which, for example, is non-zero in the stratum where $A(0) = 0, L(1) = 0$.

We were able to use standard analytic methods (e.g., stratification) to prove the existence of the arrows from $A(1)$ to Y or from $A(0)$ to $L(1)$ because these causal questions were reducible to questions about the effects of fixed treatments.

Our analysis of the effect of $A(1)$ on Y raises alternative interesting points about confounding. Suppose the arrow from $L(1)$ to Y does not exist. Then $L(1)$ would not be a direct cause of Y and thus the only source of confounding for the effect of $A(1)$ on Y (i.e., the causal confounder) would be the unmeasured common cause U; nonetheless data on $L(1)$ still suffice to block backdoor (i.e., non-causal; see Section 23.7) paths from $A(1)$ to Y and thus to control confounding. Further, even were the data in Table 23.1 not available, we would expect (i) $L(1)$ to be associated with exposure $A(1)$ within strata of $A(0)$ (i.e., $L(1) \amalg A(1)|A(0)$ is false) because the path $L(1) \longrightarrow A(1)$ is not blocked by conditioning on $A(0)$; and (ii) $L(1)$ to be an independent risk factor for Y within one or more of the four joint strata of $\{A(1), A(0)\}$ (i.e., $L(1) \amalg Y|A(1), A(0)$ is false or, equivalently, $E\{Y|L(1) = 1, A(1), A(0)\} \neq E\{Y|L(1) = 0, A(1), A(0)\}$), because the path $L(1) \longleftarrow U \longrightarrow Y$ is not blocked by conditioning on $A(1)$ and $A(0)$. In this setting, we follow common practice and refer to $L(1)$ as a confounder for the effect of $A(1)$ on Y (although not as a causal confounder) given data on $A(0)$, because, within levels of $A(0)$, $L(1)$ is the measured risk factor for Y that is used to control confounding. We can empirically confirm that $L(1) \amalg Y|A(1), A(0)$ and $L(1) \amalg A(1)|A(0)$ are both false using the data in Table 23.1.

23.4.4.2 Joint effects, direct effects, and g-methods

Conventional methods, however, may fail to identify the presence (or absence) of causal arrows that correspond to a joint effect of the time-varying exposure $\{A(0), A(1)\}$. A class of joint effects often of interest in epidemiology are the (controlled) direct effects of $A(0)$ on Y not mediated through $A(1)$. With dichotomous exposures, there exist two such direct effects. First, the direct effect of the baseline exposure $A(0)$ when the later exposure $A(1)$ is set (i.e., forced) to be 0 is, by definition, the counterfactual contrast

$$E(Y_{\bar{a}=\{1,0\}}) - E(Y_{\bar{a}=\{0,0\}}) = E(Y_{\bar{a}=\{1,0\}} - Y_{\bar{a}=\{0,0\}}),$$

which is the average of the individual causal effects $Y_{\bar{a}=\{1,0\}} - Y_{\bar{a}=\{0,0\}}$ that quantify the effect of baseline exposure when later exposure is withheld. Note this formal definition for the direct effect of $A(0)$ with $A(1)$ set to zero makes clear that the question of whether $A(0)$ directly affects Y not through $A(1)$ is a question about the effect of joint intervention on $A(0)$ and $A(1)$. The second direct effect is the direct effect of $A(0)$ when the exposure $A(1)$ is set to 1, which, by definition, is the counterfactual contrast

$$E(Y_{\bar{a}=\{1,1\}}) - E(Y_{\bar{a}=\{0,1\}}) = E(Y_{\bar{a}=\{1,1\}} - Y_{\bar{a}=\{0,1\}})$$

that quantifies the effect of the baseline exposure when exposure at time 1 is always given.

When, on the causal DAG in Figure 23.3, the dotted arrows from $A(0)$ to Y and $L(1)$ to Y are both absent, the direct effects $E(Y_{\bar{a}=\{1,0\}}) - E(Y_{\bar{a}=\{0,0\}})$ and $E(Y_{\bar{a}=\{1,1\}}) - E(Y_{\bar{a}=\{0,1\}})$ will both be zero, as then the only sequence of directed arrows from $A(0)$ to Y would go through $A(1)$. If one or both of the direct effects are non-zero, then a sequence of directed arrows from $A(0)$ to Y that avoids $A(1)$ must exist and, thus, one or both of the dotted arrows from $A(0)$ to Y and $L(1)$ to Y must be present. However, to determine from the data in Table 23.1 whether either or both direct effects are non-zero requires appropriate use of methods for causal inference with time-varying exposures like the g-formula, IPTW, or g-estimation, because, as we demonstrate below, conventional methods fail, even when the three identifiability conditions hold.

23.4.4.3 G-formula

If we can estimate the counterfactual means $E(Y_{\bar{a}=\{0,0\}})$, $E(Y_{\bar{a}=\{1,0\}})$, $E(Y_{\bar{a}=\{0,1\}})$, and $E(Y_{\bar{a}=\{1,1\}})$ under the four possible static regimes, we can estimate both direct effects. All four means can be consistently estimated by the g-formula because the three identifiability conditions hold in a sequentially randomized trial. Because the confounder $L(1)$ is a binary variable, the g-formula can be written explicitly as

$$E(Y_{\bar{a}}) = E\{Y|A(0) = a(0), A(1) = a(1), L(1) = 0\} \Pr\{L(1) = 0|A(0) = a(0)\}$$
$$+ E\{Y|A(0) = a(0), A(1) = a(1), L(1) = 1\} \Pr\{L(1) = 1|A(0) = a(0)\}$$

for $\bar{a} = \{a(0), a(1)\}$. Using this formula, the four means under each of the regimes are

$$E(Y_{\bar{a}=\{0,0\}}) = 200 \times \frac{8000}{16,000} + 50 \times \frac{8000}{16,000} = 125,$$

$E(Y_{\bar{a}=\{0,1\}}) = 145$, $E(Y_{\bar{a}=\{1,0\}}) = 155$, $E(Y_{\bar{a}=\{1,1\}}) = 145$.

We conclude that there is a direct effect of $A(0)$ on the mean of Y when $A(1)$ is set to 0 but not when $A(1)$ is set to 1. As a consequence, we know that one or both of the $A(0)$ to Y and $L(1)$ to Y arrows must be present. However, we cannot determine whether or not the causal arrow from $L(1)$ to Y is present, as the causal effect of $L(1)$ on Y cannot be consistently estimated because of the unblockable backdoor path $L(1) \longleftarrow U \longrightarrow Y$. As a consequence of our inability to determine the causal effect of $L(1)$ on Y, we also cannot determine in general whether none (corresponding to no arrow from $A(0)$ to Y), some, or all of the non-zero direct effect of $A(0)$ on Y when $A(1)$ is withheld is due to a direct causal effect of $A(0)$ on Y not through $L(1)$.

23.4.5 Why standard methods fail

We will show that when there exists, as in our study, a post-baseline covariate $L(1)$ that (i) is caused by (or shares a common cause with) baseline exposure $A(0)$ and (ii) is a confounder for the effect of a subsequent exposure $A(1)$ on a response Y, standard analytic methods that use stratification, regression, or matching for covariate adjustment cannot be used to estimate causal contrasts that depend on the joint effects of both the baseline and subsequent exposures. We will see that the difficulty with standard methods is that, to estimate the joint effects of $A(0)$ and $A(1)$, we must adjust for the confounding effect of $L(1)$ to estimate consistently the effect of $A(1)$ on Y; however, if we adjust for the confounding by stratification, regression, or matching on $L(1)$, we cannot consistently estimate the effect of $A(0)$ because the association of $L(1)$ with $A(0)$ results in selection bias, even under the null hypothesis of no causal effect (direct, indirect, or net) of $A(0)$ on Y.

As a specific example, we consider the causal contrast $E(Y_{\bar{a}=\{1,1\}}) - E(Y_{\bar{a}=\{0,1\}})$ representing the direct effect of $A(0)$ on Y when treated with $A(1)$, which we have shown to take the value 0 in our study. If one did not know about g-methods, a natural, but naive, attempt to estimate $E(Y_{\bar{a}=\{1,1\}}) - E(Y_{\bar{a}=\{0,1\}})$ from the data in Table 23.1 would be to calculate the associational contrast $E\{Y|A(0) = 1, A(1) = 1\} - E\{Y|A(0) = 0, A(1) = 1\}$. From Table 23.1, we obtain

$$E\{Y|A(0) = 1, A(1) = 1\} = \frac{1}{10,000}(110 \times 9000 + 250 \times 1000) = 124,$$

$$E\{Y|A(0) = 0, A(1) = 1\} = \frac{1}{8000}(220 \times 6000 + 70 \times 2000) = 182.5.$$

Because this analysis fails to adjust for the confounder $L(1)$ of $A(1)$'s effect on Y, the associational contrast $E\{Y|A(0) = 1, A(1) = 1\} - E\{Y|A(0) = 0, A(1) = 1\} = -58.5$ is non-causal and biased as an estimator of the causal contrast $E(Y_{\bar{a}=\{1,1\}}) - E(Y_{\bar{a}=\{0,1\}}) = 0$. Had the causal DAG in Figure 23.3 not had the arrow from $L(1)$ to $A(1)$, there would have then been no confounding by either the measured factors $L(1)$ or the unmeasured factors U for either of the exposures, and we would have found that association was causation, that is, that $E\{Y|A(0) = 1, A(1) = 1\} - E\{Y|A(0) = 0, A(1) = 1\}$ was equal to $E(Y_{\bar{a}=\{1,1\}}) - E(Y_{\bar{a}=\{0,1\}})$.

It will prove useful later for us to consider the saturated conditional association model

$$E\{Y|A(0) = a(0), A(1) = a(1)\} = \gamma_0 + \gamma_1 a(0) + \gamma_2 a(1) + \gamma_3 a(0)a(1). \qquad (23.4)$$

Table 23.2 The Hypothetical Study Population Collapsed over $L(1)$

$A(0)$	$A(1)$	N		$E\{Y\lvert A(0), A(1)\}$
0	0	8000	87.5	$= \gamma_0$
0	1	8000	182.5	$= \gamma_0 + \gamma_2$
1	0	6000	180	$= \gamma_0 + \gamma_1$
1	1	10000	124	$= \gamma_0 + \gamma_1 + \gamma_2 + \gamma_3$

We can estimate the model parameters by collapsing the population data over $L(1)$, as shown in Table 23.2. We can then calculate the parameter values from the equations

$$E\{Y\lvert A(0) = 0, A(1) = 0\} = \gamma_0,$$
$$E\{Y\lvert A(0) = 0, A(1) = 1\} = \gamma_0 + \gamma_2,$$
$$E\{Y\lvert A(0) = 1, A(1) = 0\} = \gamma_0 + \gamma_1,$$
$$E\{Y\lvert A(0) = 1, A(1) = 1\} = \gamma_0 + \gamma_1 + \gamma_2 + \gamma_3,$$

and the values of $E\{Y\lvert A(0) = a(0), A(1) = a(1)\}$ in Table 23.2. We find $\gamma_0 = 87.5, \gamma_1 = 92.5, \gamma_2 = 95$, and $\gamma_3 = -151$. These parameter estimates are precisely those that result from fitting, by ordinary least squares, a linear model for the outcome Y that contains, as regressors, an intercept, $A(0)$, $A(1)$, and the product $A(0) \times A(1)$. Note that, if we were given the values of the γ-parameters, we could use the above equations in the other direction to calculate the conditional means $E\{Y\lvert A(0) = a(0), A(1) = a(1)\}$.

Upon recognizing that the above associational contrast is biased for $E\{Y_{\bar{a}=\{1,1\}}\} - E\{Y_{\bar{a}=\{0,1\}}\}$ due to uncontrolled confounding by $L(1)$, it is natural to try to adjust for confounding by computing the two l-stratum-specific associations

$$E\{Y\lvert A(0) = 1, L(1) = 0, A(1) = 1\} - E\{Y\lvert A(0) = 0, L(1) = 0, A(1) = 1\}$$
$$= 250 - 70 = 180,$$
$$E\{Y\lvert A(0) = 1, L(1) = 1, A(1) = 1\} - E\{Y\lvert A(0) = 0, L(1) = 1, A(1) = 1\}$$
$$= 110 - 220 = -110,$$

or their weighted average

$$-110 \times \frac{20}{32} + 180 \times \frac{12}{32} = 69.75,$$

with weights determined by the distribution of L in the study population of 32,000. Note that $\Pr(L = 1) = 20/32$. Neither of the l-stratum-specific associations nor their population weighted average is a valid, unbiased estimator of the actual causal contrast $E(Y_{\bar{a}=\{1,1\}}) - E(Y_{\bar{a}=\{0,1\}}) = 0$. The biases in the l-stratum-specific associations reflect the selection bias that is induced when one conditions on a covariate $L(1)$ that is a predictor of Y given $A(0)$ and $A(1)$ and is caused by treatment $A(0)$ (Rosenbaum, 1984; Robins, 1986).

This selection bias can be understood with the help of causal graphs (Hernán, Hernández-Díaz, and Robins, 2004). To do so, consider another study whose causal graph is also given by Figure 23.3, but modified so that all three dotted arrows are absent. The modified graph implies that neither $A(0)$ nor $A(1)$ has a direct, indirect, or net effect on Y. Yet, even in this setting, we would expect that the two l-stratum-specific associations would still remain non-zero and therefore biased for $E(Y_{\bar{a}=\{1,1\}}) - E(Y_{\bar{a}=\{0,1\}}) = 0$. To see why, note the associational l-stratum-specific associations are zero only when Y and $A(0)$ are conditionally independent given $A(1) = 1$ and L. But we would not expect such conditional independence because, on the modified graph, the path $A(0) \longrightarrow L(1) \longleftarrow U \longrightarrow Y$ connecting Y

and $A(0)$ is opened when we condition (i.e., stratify) on the collider $L(1)$ and/or $L(1)$'s descendant $A(1)$ (see definitions in Section 23.7). In our study, conditioning on $A(1) = 1$ and $L(1)$ similarly results in selection bias; however, the presence of the arrow from $A(1)$ to Y and of one or both of the arrows from $A(0)$ to Y and $L(1)$ to Y makes a purely graphical demonstration of the bias less clear.

23.4.6 IPTW and marginal structural models

We now describe how to use IPTW for estimating the counterfactual means $E(Y_{\bar{a}})$ under the four static regimes $\bar{a} = \{a(0), a(1)\}$. The first step is to create a stabilized pseudo-population by weighting the subjects in each row in Table 23.1 by the stabilized weights

$$SW = \frac{f\{A(0)\}f\{A(1)|A(0)\}}{f\{A(0)\}f\{A(1)|A(0), L(1)\}} = \frac{f\{A(1)|A(0)\}}{f\{A(1)|A(0), L(1)\}}.$$

Note that the factor $f\{A(0)\}$ cancels because in our study the potential confounder $L(0)$ is absent. Table 23.3 records the values of $f\{A(1)|A(0)\}$, $f\{A(1)|A(0), L(1)\}$, SW, and the number of subjects in the pseudo-population.

For example, for the first row,

$$f\{A(1)|A(0)\} = \Pr\{A(1) = 0|A(0) = 0\} = 8000/16{,}000 = 0.5,$$

$$f\{A(1)|A(0), L(1)\} = \Pr\{A(1) = 0|A(0) = 0, L(1) = 1\} = 2000/8000 = 0.25.$$

Each of the 2000 subjects in the first row therefore receives the weight $SW = 0.5/0.25 = 2$. Hence, the row contributes $4000 = 2 \times 2000$ subjects to the pseudo-population. The IPTW weights eliminate the arrow between $L(1)$ and $A(1)$ in the pseudo-population as shown in Figure 23.4. The absence of the arrow can be easily confirmed by checking whether or not $A(1) \amalg_{ps} L(1)|A(0)$, where $\amalg_{ps}$ represents independence in the pseudo-population. This conditional independence holds in the pseudo-population of our example because

$$\Pr_{ps}\{A(1) = 1|A(0) = 1, L(1) = 0\} = \Pr_{ps}\{A(1) = 1|A(0) = 1, L(1) = 1\} = 3/8,$$

$$\Pr_{ps}\{A(1) = 1|A(0) = 0, L(1) = 0\} = \Pr_{ps}\{A(1) = 1|A(0) = 0, L(1) = 1\} = 1/2.$$

Therefore, the causal DAG corresponding to the pseudo-population lacks the arrow $L(1)$ to $A(1)$. The absence of this arrow signifies that there is no confounding by $L(1)$ in the pseudo-population and hence that adjustment by stratification is not necessary. That is, $E_{ps}\{Y|A(0) = a(0), A(1) = a(1)\} = E_{ps}(Y_{\bar{a}=\{a(0),a(1)\}})$ in the pseudo-population. Thus, as shown in Table 23.4, we can collapse the pseudo-population data over $L(1)$, obtain $E_{ps}\{Y|A(0), A(1)\}$ for each of the four combinations of values of $A(0)$ and $A(1)$, and conduct an unadjusted analysis.

Table 23.3 The Stabilized Pseudo-Population

$A(0)$	$L(1)$	$A(1)$	N	Exp.	$f\{A(1)|A(0)\}$	$f\{A(1)|A(0), L(1)\}$	SW	N Pseudo-Pop.
0	1	0	2000	200	0.50	0.25	2	4000
0	1	1	6000	220	0.50	0.75	2/3	4000
0	0	0	6000	50	0.50	0.75	2/3	4000
0	0	1	2000	70	0.50	0.25	2	4000
1	1	0	3000	130	0.375	0.25	3/2	4500
1	1	1	9000	110	0.625	0.75	5/6	7500
1	0	0	3000	230	0.375	0.75	1/2	1500
1	0	1	1000	250	0.625	0.25	5/2	2500

Note: The column heading Exp. refers to $E\{Y|A(0), L(1), A(1)\}$.

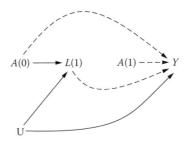

Figure 23.4 Causal DAG in the pseudo-population simulated by IPTW.

For example, the direct effect of $A(0)$ on the mean of Y when $A(1)$ is set to 0 is $E_{ps}\{Y|A(0) = 1, A(1) = 0\} - E\{Y|A(0) = 0, A(1) = 0\} = 155 - 125 = 30$. As expected, the values of $E(Y_{\bar{a}})$ obtained by IPTW (i.e., the values of $E_{ps}\{Y|A(0), A(1)\}$ in the pseudo-population) are equal to those obtained by the g-formula.

In this oversimplified example, we do not need to use models to estimate the inverse probability weights because they can be calculated easily by hand from the data. Also, we do not need models for the counterfactual means $E\{Y_{\bar{a}}\}$ because these means can be calculated by hand. However, for pedagogic purposes, let us consider the saturated marginal structural mean model

$$E(Y_{\bar{a}}) = \eta_0 + \eta_1 a(0) + \eta_2 a(1) + \eta_3 a(0)a(1).$$

We can use the pseudo-population data to calculate the parameters $\eta_0, \eta_1, \eta_2,$ and η_3 because

$$E\{Y_{\bar{a}=\{0,0\}}\} = \eta_0,$$
$$E\{Y_{\bar{a}=\{0,1\}}\} = \eta_0 + \eta_2,$$
$$E\{Y_{\bar{a}=\{1,0\}}\} = \eta_0 + \eta_1,$$
$$E\{Y_{\bar{a}=\{1,1\}}\} = \eta_0 + \eta_1 + \eta_2 + \eta_3,$$

and, therefore, using the estimates for $E(Y_{\bar{a}=\{a(0),a(1)\}})$ in Table 23.4, $\eta_0 = 125$, $\eta_1 = 30$, $\eta_2 = 20$, and $\eta_3 = -30$. This estimation procedure is equivalent to fitting linear model (23.4) to the observed study data by weighted least squares with each subject weighted by SW (e.g., `proc reg` with a `weight` statement in SAS). Because of confounding, the parameters η of the marginal structural mean model differ from the parameters γ of the associational mean model (23.4).

The parameters η can be used to test hypotheses about the joint effect of exposures $A(0), A(1)$. For example, the hypothesis that $A(0)$ has no direct effect on the mean of Y when $A(1)$ is set to 1, that is, $E(Y_{\bar{a}=\{1,1\}}) = E(Y_{\bar{a}=\{0,1\}})$, implies $\eta_0 + \eta_1 + \eta_2 + \eta_3 = \eta_0 + \eta_2$. This would be true only if $\eta_1 + \eta_3 = 0$, which is the case. Similarly, the hypothesis that $A(0)$ has no direct effect on the mean of Y when $A(1)$ is set to 0, $E(Y_{\bar{a}=\{1,0\}}) = E(Y_{\bar{a}=\{0,0\}})$, implies $\eta_0 + \eta_1 = \eta_0$. This would be true only if $\eta_1 = 0$, which is not the case. Suppose that

Table 23.4 The Stabilized Pseudo-Population Collapsed over $L(1)$

$A(0)$	$A(1)$	N	$E\{Y|A(0), A(1)\}$	
0	0	8000	125	$= \theta_0$
0	1	8000	145	$= \theta_0 + \theta_2$
1	0	6000	155	$= \theta_0 + \theta_1$
1	1	10000	145	$= \theta_0 + \theta_1 + \theta_2 + \theta_3$

we had fit the non-saturated misspecified marginal structural model

$$E(Y_{\bar{a}}) = \eta_0 + \eta_1 a(0) + \eta_2 a(1)$$

by weighted least squares with weights SW and then used the parameter estimates to estimate the counterfactual means $E(Y_{\bar{a}})$. Because of misspecification bias, these estimated means would have differed from those obtained from the saturated model and from the g-formula.

23.4.7 Methods for dynamic regimes

23.4.7.1 G-formula

The four regimes $g = \{a(0), a(1)\}$ that we have compared constitute all possible combinations of fixed values of $a(0)$ and $a(1)$ and thus all possible static regimes in our example. We next consider dynamic regimes such as $g = \{1, L(1)\}$, which is the regime "always treat at time 0, treat at time 1 only if $L(1) = 1$." Note the same dynamic regime applied to different people may result in different exposure values. For example, the regime $g = \{1, L(1)\}$ will be the regime $\{1, 1\}$ for those subjects with $L(1) = 1$ under $a(0) = 1$, and $g = \{1, 0\}$ for those with $L(1) = 0$ under $a(0) = 1$. The g-formula for the dynamic regime $g = \{1, L(1)\}$ is the generalization of the g-formula for static regimes in which the exposure $A(1)$ is set to 1 when $L(1) = 1$ and $A(1)$ is set to 0 when $L(1) = 0$. In our example,

$$
\begin{aligned}
E(Y_{g=\{1,L(1)\}}) =& E\{Y|A(0) = 1, A(1) = 1, L(1) = 1\} \Pr\{L(1) = 1|A(0) = 1\} \\
& + E\{Y|A(0) = 1, A(1) = 0, L(1) = 0\} \Pr\{L(1) = 0|A(0) = 1\} \\
=& 110 \times \frac{12}{16} + 230 \times \frac{4}{16} = 140.
\end{aligned}
$$

The above formula can be written more succinctly as

$$E(Y_{g=\{1,L(1)\}}) = \sum_{l(1)} E\{Y|A(0) = 1, A(1) = l(1), L(1) = l(1)\} \Pr\{L(1) = l(1)|A(0) = 1\}.$$

23.4.7.2 IPTW

We now describe how to use IPTW for estimating the counterfactual means $E(Y_{g=\{1,L(1)\}})$. The first step is to create an unstabilized pseudo-population by weighting the subjects in each row in Table 23.1 by the unstabilized weights $W = 1/[f\{A(0)\}f\{A(1)|A(0), L(1)\}]$. Note we use $f\{A(0)\}$ rather than $f\{A(0)|L(0)\}$ because in our study no potential confounder $L(0)$ is present. Table 23.5 records the values of $f\{A(0)\}$, $f\{A(1)|A(0), L(1)\}$, W, and the number of subjects in the unstabilized pseudo-population.

Table 23.5 The Unstabilized Pseudo-Population

$A(0)$	$L(1)$	$A(1)$	N	Exp.	$f\{A(0)\}$	$f\{A(1)|A(0), L(1)\}$	W	N Pseudo-Pop.
0	1	0	2000	200	0.50	0.25	8	16000
0	1	1	6000	220	0.50	0.75	8/3	16000
0	0	0	6000	50	0.50	0.75	8/3	16000
0	0	1	2000	70	0.50	0.25	8	16000
1	1	0	3000	130	0.50	0.25	8	24000
1	1	1	9000	110	0.50	0.75	8/3	24000
1	0	0	3000	230	0.50	0.75	8/3	8000
1	0	1	1000	250	0.50	0.25	8/3	8000

Note: Shorthand column headings are as in Table 23.3.

The IPTW estimate of $E(Y_{g=\{1,L(1)\}})$ is the average of Y among the subjects in the unstabilized pseudo-population who followed regime $g = \{1, L(1)\}$. Only the subjects with $A(0) = 1, L(1) = 1, A(1) = 1$ and $A(0) = 1, L(1) = 0, A(1) = 0$ followed $g = \{1, L(1)\}$. Thus,

$$E(Y_{g=\{1,L(1)\}}) = \frac{24{,}000 \times 110 + 8000 \times 230}{32{,}000} = 140,$$

as was also obtained with the g-formula.

23.4.7.3 G-estimation

G-estimation of structural nested models is a third method for estimation of counterfactual means. The "g-" indicates that g-estimation, like the g-formula, is a general method that can be further used to estimate counterfactual means $E(Y_g)$ under any static or dynamic regime g. We begin with a saturated locally rank-preserving SNM for our example. The model has one equation for each treatment time with one unknown parameter β_0^* in the time 0 equation and a vector β_1^* of four unknown parameters in the time 1 equation:

$$Y_{g=\{a(0),0\}} = Y_{g=\{0,0\}} + \beta_0^* a(0) \qquad (23.5)$$

$$Y_{g=\{a(0),a(1)\}} = Y_{g=\{a(0),0\}} + \beta_{1,1}^* a(1) + \beta_{1,2}^* a(1) L_{g=\{a(0)\}}(1) + \beta_{1,3}^* a(1) a(0)$$
$$+ \beta_{1,4}^* a(1) a(0) L_{g=\{a(0)\}}(1) \qquad (23.6)$$

By evaluating Equation (23.5) at $a(0) = 1$, we see the parameter $\beta_0^* = Y_{g=\{1,0\}} - Y_{g=\{0,0\}}$ represents the subject-specific direct effect of treatment $a(0)$ on the outcome when treatment $a(1)$ is withheld (i.e., set to zero). Under our model, this direct effect β_0^* is exactly the same for every subject. Thus, if $Y_{g=\{0,0\}}$ for subject i exceeds $Y_{g=\{0,0\}}$ for subject j, the same ranking of i and j will hold for $Y_{g=\{1,0\}}$; the model preserves ranks across regimes, and we therefore refer to Equation (23.5) as a rank-preserving model.

The four parameters β_1^* in (23.6) parameterize the effect of an $a(1)$ on Y within the four possible levels of past treatment and covariate history. For example, $\beta_{1,1}^*$ and $\beta_{1,1}^* + \beta_{1,2}^*$ are, respectively, the effect of $a(1)$ on Y when $a(0)$ is withheld among the subset of subjects with $L_{g=\{0\}}(1) = 0$ and the subset with $L_{g=\{0\}}(1) = 1$. Here, $L_{g=\{0\}}(1)$ is the counterfactual value of $L(1)$ when $a(0)$ is withheld. If $\beta_{1,1}^*$ and $\beta_{1,1}^* + \beta_{1,2}^*$ are of opposite sign, then there is a qualitative modification by $L(1)$ of the effect $a(1)$ on Y when $a(0)$ is withheld. Similarly $\beta_{1,1}^* + \beta_{1,3}^*$ and $\beta_{1,1}^* + \beta_{1,3}^* + \beta_{1,4}^*$ are the effect of $a(1)$ when $a(0)$ is taken among the subset of subjects with $L_{g=\{1\}}(1) = 0$ and the subset with $L_{g=\{1\}}(1) = 1$, respectively. If they are of different sign, there is a qualitative modification by $L(1)$ of the effect of $a(1)$ on Y when $a(0)$ is taken. Thus, an SNM models the degree to which the effect of current treatment is modified by past treatment and past time-dependent covariate history. In contrast, non-dynamic MSMs can only model effect modification by baseline covariates V, a subset of $L(0)$.

Finally, we note that if $Y_{g=\{1,0\}}$ for subject i exceeds $Y_{g=\{1,0\}}$ for subject j, we can only be certain that $Y_{g=\{1,1\}}$ for subject i also exceeds $Y_{g=\{1,1\}}$ for subject j, if both have the same values of $L_{g=(1)}$. Because the preservation of the ranking on the counterfactual Y depends on local factors (i.e., the value $L_{g=(1)}$), we refer to Equation (23.6) as a locally rank-preserving model.

We next describe how we can estimate the parameters under the assumption of conditional exchangeability. We then show how to use our parameter estimates to estimate $E(Y_{g=\{1,L(1)\}})$.

G-estimation of the parameters under conditional exchangeability.   We begin by estimating the parameter vector β_1^*. To do so, in Table 23.6 we first use the SNM to calculate the mean of $Y_{\{A(0),0\}}$ in terms of the unknown parameter vector β_1^*. To help understand these

Table 23.6 The g-estimation procedure

$A(0)$	$L(1)$	$A(1)$	N	Y	$Y_{\{A(0),0\}}$	$Y_{\{0,0\}}$
0	1	0	2000	200	200	200
0	1	1	6000	220	$220 - \beta_{1,1}^* - \beta_{1,2}^*$	$220 - \beta_{1,1}^* - \beta_{1,2}^*$
0	0	0	6000	50	50	50
0	0	1	2000	70	$70 - \beta_{1,1}^*$	$70 - \beta_{1,1}^*$
1	1	0	3000	130	130	$130 - \beta_0^*$
1	1	1	9000	110	$110 - \beta_{1,1}^* - \beta_{1,2}^*$ $- \beta_{1,3}^* - \beta_{1,4}^*$	$110 - \beta_0^* - \beta_{1,1}^* - \beta_{1,2}^*$ $- \beta_{1,3}^* - \beta_{1,4}^*$
1	0	0	3000	230	230	$230 - \beta_0^*$
1	0	1	1000	250	$250 - \beta_{1,1}^* - \beta_{1,3}^*$	$250 - \beta_0^* - \beta_{1,1}^* - \beta_{1,3}^*$

calculations, consider the expression $220 - \beta_{1,1}^* - \beta_{1,2}^*$ for the mean of $Y_{\{A(0),0\}} = Y_{\{0,0\}}$ among subjects with $A(0) = 0, L(1) = 1, A(1) = 1$ in the second data row of Table 23.6. By consistency, the observed $L(1)$ of 1 equals $L_{g=\{0\}}(1)$ and the observed mean 220 of Y is the mean of $Y_{g=\{0,1\}}$. By solving (23.6) for $Y_{g=\{0,0\}}$ after substituting $\{0,1\}$ for $\{a(0), a(1)\}$ and 1 for $L_{g=\{a(0)\}}(1)$, we obtain $220 - \beta_{1,1}^* - \beta_{1,2}^*$ upon taking means.

We now estimate β_1^* under the assumption of conditional exchangeability. Conditional exchangeability implies that (i) $Y_{g=\{0,0\}} \amalg A(0)$ and (ii) $Y_{\{a(0),0\}} \amalg A(1)|A(0) = a(0), L(1) = l(1)$. Now condition (ii) implies that, within any of the four joint strata of $\{A(0), L(1)\}$, the mean of $Y_{\{A(0),0\}}$ among subjects with $A(1) = 1$ is equal to the mean among subjects with $A(1) = 0$. Consider first the stratum $\{A(0), L(1)\} = (0,0)$. From data rows 3 and 4 in Table 23.6, we find that the mean when $A(1) = 0$ is 50 and when $A(1) = 1$ is $70 - \beta_{1,1}^*$. Hence, $\beta_{1,1}^* = 20$. Next we equate the means of $Y_{\{A(0),0\}}$ in data rows 1 and 2 corresponding to stratum $\{A(0), L(1)\} = (0,1)$ to obtain $200 = 220 - \beta_{1,1}^* - \beta_{1,2}^*$. Since $\beta_{1,1}^* = 20$, we conclude $\beta_{1,2}^* = 0$.

Continuing, we equate the means of $Y_{\{A(0),0\}}$ in data rows 7 and 8 to obtain $230 = 250 - \beta_{1,1}^* - \beta_{1,3}^*$. Because $\beta_{1,1}^* = 20$, we conclude $\beta_{1,3}^* = 0$. Finally, equating the means of $Y_{\{A(0),0\}}$ in data rows 5 and 6, we obtain $130 = 110 - \beta_{1,1}^* - \beta_{1,2}^* - \beta_{1,3}^* - \beta_{1,4}^*$, so $130 = 110 - 20 - \beta_{1,4}^*$. Thus, $\beta_{1,4}^* = -40$.

To estimate β_0^*, we first substitute $\beta_{1,1}^* = 20$, $\beta_{1,2}^* = \beta_{1,3}^* = 0$, and $\beta_{1,4}^* = -40$ into the expressions for the mean of $Y_{\{A(0),0\}}$ in Table 23.6. We then use (23.5) to obtain the mean of $Y_{\{0,0\}}$ for each data row in Table 23.6 by subtracting $\beta_0^* A(0)$ from the mean of $Y_{\{A(0),0\}}$. Now our assumption $Y_{\{0,0\}} \amalg A(0)$ implies that the means of $Y_{\{0,0\}}$ among the 16,000 subjects with $A(0) = 1$ and the 16,000 subjects with $A(0) = 0$ are identical. The means among subjects with $A(0) = 0$ and those with $A(0) = 1$ are

$$200 \times \frac{8000}{16,000} + 50 \times \frac{8000}{16,000} = 125 \quad \text{and} \quad 130 \times \frac{12,000}{16,000} + 230 \times \frac{4000}{16,000} - \beta_0^* = 155 - \beta_0^*,$$

respectively. Hence, $\beta_0^* = 30$. This method of estimation is referred to as g-estimation.

Estimation of $E(Y_g)$ using locally rank-preserving nested structural models. We now use the above results to estimate various counterfactual population means. Because 125 is the mean of $Y_{g=\{0,0\}}$ both in subjects with $A(0) = 0$ and $A(0) = 1$, we conclude that the population mean $E(Y_{g=\{0,0\}})$ is 125. Further, by (23.5), we have $E(Y_{g=\{1,0\}} - Y_{g=\{0,0\}}) = \beta_0^* = 30$. Thus, $E(Y_{g=\{1,0\}}) = E(Y_{g=\{0,0\}}) + \beta_0^* = 125 + 30 = 155$.

Next, by (23.6), we obtain $E(Y_{g=\{0,1\}} - Y_{g=\{0,0\}}) = E\{\beta_{1,1}^* + \beta_{1,2}^* L_{g=\{0\}}(1)\} = \beta_{1,1}^* + \beta_{1,2}^* E\{L_{g=\{0\}}(1)\} = \beta_{1,1}^* = 20$, as $\beta_{1,2}^* = 0$. Hence, $E(Y_{g=\{0,1\}}) = 145$. By (23.6), we have $E(Y_{g=\{1,1\}} - Y_{g=\{0,0\}}) = E\{\beta_0^* + \beta_{1,1}^* + \beta_{1,3}^* + (\beta_{1,2}^* + \beta_{1,4}^*)L_{g=\{1\}}(1)\} = 30 + 20 + (-40)E\{L_{g=\{1\}}(1)\}$. We conclude that knowledge of the parameters of our SNM is not sufficient to estimate $E(Y_{g=\{1,1\}} - Y_{g=\{0,0\}})$. We also need to know $E\{L_{g=\{1\}}(1)\}$. But, as noted previously, $E\{L_{g=\{1\}}(1)\} = E\{L(1)|A(0) = 1\} = 3/4$, as association is causation for the effect of $A(0)$ on $L(1)$. Thus, $E(Y_{g=\{1,1\}} - Y_{g=\{0,0\}}) = 30 + 20 - 40 \times 3/4 = 20$, so $E(Y_{g=\{1,1\}}) = 145$.

Finally, to obtain $E(Y_{g=\{1,L(1)\}})$, we note that $Y_{g=\{1,L(1)\}} = Y_{g=\{1,1\}}$ if $L_{g=\{1\}}(1) = 1$ and $Y_{g=\{1,L(1)\}} = Y_{g=\{1,0\}}$ if $L_{g=\{1\}}(1) = 0$. Thus, in the three quarters of subjects with $L_{g=\{1\}}(1) = 1$, $Y_{g=\{1,L(1)\}} - Y_{g=\{0,0\}} = Y_{g=\{1,1\}} - Y_{g=\{0,0\}} = \beta_0^* + \beta_{1,1}^* + \beta_{1,3}^* + \beta_{1,2}^* + \beta_{1,4}^* = 30 + 20 - 40 = 10$. In the one quarter of subjects with $L_{g=\{1\}}(1) = 0$, $Y_{g=\{1,L(1)\}} - Y_{g=\{0,0\}} = Y_{g=\{1,0\}} - Y_{g=\{0,0\}} = \beta_0^* = 30$. Thus, the mean of $Y_{g=\{1,L(1)\}} - Y_{g=\{0,0\}}$ is $10 \times 3/4 + 30 \times 1/4 = 15$. Hence, $E(Y_{g=\{1,L(1)\}}) = 125 + 15 = 140$.

All of these results agree with those obtained by the g-formula and by IPTW.

Estimation of $E(Y_g)$ without local rank preservation. We noted above that local rank-preservation implies that the direct effect of treatment $A(0)$ on the outcome when treatment $A(1)$ is withheld is the same for each subject. This assumption is clearly biologically implausible in view of between-subject heterogeneity in unmeasured genetic and environmental background risks. To overcome this limitation, we consider a saturated structural nested mean model (SNMM) that assumes

$$E(Y_{g=\{a(0),0\}}) = E(Y_{g=\{0,0\}}) + \beta_0^* a(0) \quad \text{and}$$

$$E\{Y_{g=\{a_0,a_1\}}|L(1) = l(1), A(0) = a(0)\} = E\{Y_{g=\{a_0,0\}}|L(1) = l(1), A(0) = a(0)\}$$
$$+ \beta_{1,1}^* a(1) + \beta_{1,2}^* a(1)l(1) + \beta_{1,3}^* a(1)a(0) + \beta_{1,4}^* a(1)a(0)l(1).$$

This is a model for unconditional and conditional average treatment effects and thus is totally agnostic as to the question of whether or not there is between-subject heterogeneity in the effect of treatment. Nonetheless, Robins (1994, 1997b) has proved that the previous estimates of the parameters β_0^* and β_1^* and the means $E(Y_g)$ obtained with g-estimation remain valid under the SNMM, provided the strengthened identifiability conditions hold.

23.5 G-estimation in practice and the choice among g-methods

23.5.1 G-estimation of unsaturated structural nested mean models

In practice, we need to combine information from many different regimes to estimate a given $E(Y_g)$. To accomplish this goal, we shall fit an unsaturated additive SNMM by g-estimation.

The general form of an additive SNMM is as follows. Let $\underline{0}(m)$ indicate giving treatment 0 at each treatment time $m, m + 1 \ldots, K$ for any $m = 0, \ldots, K$. For each $m = 0, \ldots, K$,

$$E\{Y_{g=\{\bar{a}(m-1),a(m),\underline{0}(m+1)\}}|\bar{L}_{\bar{a}(m-1)}(m) = \bar{l}(m), \bar{A}(m-1) = \bar{a}(m-1)\}$$
$$= E\{Y_{g=\{\bar{a}(m-1),\underline{0}(m)\}}|\bar{L}_{\bar{a}(m-1)}(m) = \bar{l}(m), \bar{A}(m-1) = \bar{a}(m-1)\} \qquad (23.7)$$
$$+ a(m)\gamma_m\{\bar{a}(m-1), \bar{l}(m), \beta^*\},$$

where (i) $g = \{\bar{a}(m-1), a(m), \underline{0}(m+1)\}$ and $g = \{\bar{a}(m-1), \underline{0}(m)\}$ are non-dynamic regimes that differ only in that the former has treatment $a(m)$ at m while the latter has treatment 0 at time m, while both have treatment $\bar{a}(m-1)$ through $m-1$ and treatment 0 from $m + 1$ to the end of follow-up K; (ii) β^* is an unknown parameter vector; and (iii) $\gamma_m\{\bar{a}(m-1), \bar{l}(m), \beta\}$ is a known function satisfying $\gamma_m\{\bar{a}(m-1), \bar{l}(m), 0\} = 0$ so $\beta^* = 0$ under the null hypothesis of no effect of treatment.

Thus, the SNMM $\gamma_m\{\bar{a}(m-1), \bar{l}(m), \beta^*\}$ models the effect on the mean of Y of a last blip of treatment of magnitude $a(m)$ at m, as a function of (i.e., as modified by) past treatment and covariate history $\{\bar{a}(m-1), \bar{l}(m)\}$. Further, it follows from the consistency condition that we could have replaced $\bar{L}_{g=\{\bar{a}(m-1)\}}(m) = l(m)$ by $\bar{L}(m) = \bar{l}(m)$ in the conditioning event.

In the example of the previous section, we have $K = 1$, $\gamma_1\{\overline{a}(0), \overline{l}(1), \beta^*\} = \beta^*_{1,1} + \beta^*_{1,2}l(1) + \beta^*_{1,3}a(0) + \beta^*_{1,4}a(0)l(1)$ and $\gamma_0\{\overline{a}(-1), \overline{l}(0), \beta^*\} = \beta^*_0$, because $\overline{l}(0)$ and $\overline{a}(-1)$ can both be taken to be identically 0. Other possible choices of $\gamma_m\{\overline{a}(m-1), \overline{l}(m), \beta\}$ include (i) β, (ii) $\beta_0 + \beta_1 m$, and (iii) $\beta_0 + \beta_1 m + \beta_2 a(m-1) + \beta^T_3 l(m) + \beta^T_4 l(m)a(m-1)$.

In model (i), the effect of a last blip of treatment $a(m)$ is the same for all m. Under model (ii), the effect varies linearly with the time m of treatment. Under model (iii), the effect of a last blip of treatment at m is modified by past treatment and covariate history.

We next describe the g-estimation algorithm for estimating the unknown parameter β^* in an observational study under the assumptions of conditional exchangeability and consistency. We note that, to fit an unsaturated SNMM by g-estimation, we do not require positivity to hold.

Fit a pooled logistic regression model

$$\text{logit}[\Pr\{A(m) = 1|\overline{L}(m), \overline{A}(m-1)\}] = \alpha^T W(m) \qquad (23.8)$$

for the probability of treatment at time (i.e., week) m for $m = 0, \dots, K$. Here, $W(m) = w_m\{\overline{L}(m), \overline{A}(m-1)\}$ is a vector of covariates calculated from a subject's covariate and treatment data $\{\overline{L}(m), \overline{A}(m-1)\}$, α^T is a row vector of unknown parameters, and each person-week is treated as an independent observation so each person contributes $K + 1$ observations. An example of $W(m) = w_m\{\overline{L}(m), \overline{A}(m-1)\}$ would be the transpose of the row vector $m, A(m-1), L^T(m), A(m-1)L^T(m), A(m-2), L^T(m-1), L^T(m)A(m-1)A(m-2)$, where $L(m)$ is the vector of covariates measured at time m. Let $\widehat{\alpha}$ be the maximum likelihood estimator of α. (In a sequentially randomized experiment, the preceding step is not required because $\Pr\{A(m) = 1|\overline{L}(m), \overline{A}(m-1)\}$ is known and would not need to be estimated.)

Next, define

$$Y_m(\beta) = Y - \sum_{j=m}^{K} A(j)\gamma_j\{\overline{A}(j-1), \overline{L}(j), \beta\}.$$

Note that, for each β, $Y_m(\beta)$ can be computed from the observed data. For the moment, assume, as in model (i) above, that β is one-dimensional. Let β_{low} and β_{up} be much smaller and larger, respectively, than any substantively plausible value of β^*.

Then, separately, for each β on a grid from β_{low} to β_{up}, say $\beta_{\text{low}}, \beta_{\text{low}} + 0.1, \beta_{\text{low}} + 0.2, \dots, \beta_{\text{up}}$, perform the score test of the hypothesis $\theta = 0$ in the extended logistic model

$$\text{logit}[\Pr\{A(m) = 1|\overline{L}(m), \overline{A}(m-1), Y_m(\beta)\}] = \alpha^T W(m) + \theta Y_m(\beta) \qquad (23.9)$$

that adds the covariate $Y_m(\beta)$ at each time m to the above pooled logistic model. A 95% confidence interval for β^* is the set of β for which an $\alpha = 0.05$ two-sided score test of the hypothesis $\theta = 0$ does not reject. The g-estimate of β^* is the value of β for which the score test takes the value 0 (i.e., the p-value is 1).

A heuristic argument for the validity of the g-estimation algorithm is as follows. Under a locally rank-preserving model, if β was equal to β^*, $Y_m(\beta)$ would equal the counterfactual $Y_{g=\{\overline{A}(m-1), \underline{0}(m)\}}$ in which a subject takes his actual treatment prior to m but no treatment from time m onwards, as shown in the previous section. But, under conditional exchangeability, $Y_{g=\{\overline{A}(m-1), \underline{0}(m)\}}$ and $A(m)$ are conditionally independent given past covariate and treatment history $\overline{L}(m), \overline{A}(m-1)$. That is, $Y_{g=\{\overline{A}(m-1), \underline{0}(m)\}}$ is not a predictor of $A(m)$ given $\{\overline{L}(m), \overline{A}(m-1)\}$, which implies that the coefficient θ of $Y_m(\beta)$ must be zero in the model (23.9) when $\beta = \beta^*$, provided the model $\text{logit}[\Pr\{A(m) = 1|\overline{L}(m), \overline{A}(m-1)\}] = \alpha^T W(m)$ is correctly specified.

Now, we do not know the true value of β. Therefore, any value β for which the data are consistent with the parameter θ of the term $\theta Y_m(\beta)$ being zero might be the true β^*, and thus belongs in our confidence interval. If consistency with the data is defined at the 0.05

level, then our confidence interval will have coverage of 95%. Furthermore, the g-estimate $\widehat{\beta}$ of β^* is that β for which adding the term $\theta Y_m(\beta)$ does nothing to help to predict $A(m)$ whatsoever, which is the β for which the score test of $\theta = 0$ is precisely zero. The g-estimate $\widehat{\beta}$ is also the value of β for which the maximum likelihood estimate of θ in model (23.9) is precisely zero.

It may appear peculiar that a function $Y_m(\beta)$ of the response Y measured at the end of follow-up is being used to predict treatment $A(m)$ at earlier times. However, this peculiarity evaporates when one recalls that, for each β on our grid, we are testing the null hypothesis that $\beta = \beta^*$ and, under this null and a rank-preserving model, $Y_m(\beta)$ is the counterfactual $Y_{g=\{\overline{A}(m-1),\underline{0}(m)\}}$, which we can view as already existing at time m (although we cannot observe its value until time K and then only if treatment in the actual study is withheld from m onwards).

The above arguments are heuristic in the sense that their validity relies on the assumption of local rank-preservation, which is biologically implausible. Nevertheless, Robins (1994, 1997b) proves that the g-estimation algorithm is valid even in the absence of local rank preservation, provided that conditional exchangeability and consistency hold.

Suppose now that the parameter β is a vector. To be concrete, suppose we consider the model with $\gamma_m\{\overline{a}(m-1), \overline{l}(m), \beta\} = \beta_0 + \beta_1 m + \beta_2 a(m-1) + \beta_3 l(m) + \beta_4 l(m)a(m-1)$, so β is five-dimensional, $l(m)$ is one-dimensional, and we would use a five-dimensional grid, one dimension for each component of β. So if we had 20 grid points for each component, we would have 20^5 different values of β on our five-dimensional grid. Now, to estimate five parameters, one requires five additional covariates. Specifically, let $Q_m = q_m\{\overline{L}(m), \overline{A}(m-1)\}$ be a five-dimensional vector of functions of $\overline{L}(m), \overline{A}(m-1)$, such as $q_m^T\{\overline{L}(m), \overline{A}(m-1)\} = \{1, m, A(m-1), L(m), L(m)A(m-1)\}$. We use an extended model that includes five linear functions $Q_m Y_m(\beta)$ of $Y_m(\beta)$ as covariates, such as

$$\text{logit}[\Pr\{A(m) = 1 | \overline{L}(m), \overline{A}(m-1), Y_m(\beta)\}] = \alpha^T W(m) + \theta^T Q_m Y_m(\beta).$$

The particular choice of the functions $Q_m = q_m\{\overline{L}(m), \overline{A}(m-1)\}$ does not affect the consistency of the point estimator, but it determines the width of its confidence interval. See Robins (1994) for the optimal choice of Q_m.

Our g-estimate $\widehat{\beta}$ is the β for which the five-degrees-of-freedom score test that all five components of θ equal zero is precisely zero. A 95% joint confidence interval for β is the set of β on our five-dimensional grid for which the five-degrees-of-freedom score test does not reject at the 5% level. Such an interval is computationally demanding. A less demanding approach is to use a Wald interval $\widehat{\beta}_j \pm 1.96 \, \text{SE}(\widehat{\beta}_j)$ for each component β_j of β centered at its g-estimate $\widehat{\beta}_j$. This gives a univariate 95% large-sample confidence interval for each β_j. (A simultaneous, i.e., joint, 95% large-sample confidence interval for all β_j requires a constant greater than 1.96 in the Wald interval.)

When the dimension of β is greater than 2, finding $\widehat{\beta}$ by search over a grid is generally computationally prohibitive. However, when, as in all the examples we have discussed, $\gamma_m\{\overline{a}(m-1), \overline{l}(m), \beta\} = \beta^T R_m$ is linear in β with $R_m = r_m\{\overline{L}(m), \overline{A}(m-1)\}$ a vector of known functions, then, given the estimator of $\Pr\{A(m) = 1 | \overline{L}(m), \overline{A}(m-1)\}$ expit$\{\widehat{\alpha}^T W(m)\}$, where expit$(u) = \exp(u)/\{1 + \exp(u)\}$, there is an explicit closed-form expression for $\widehat{\beta}$ given by

$$\widehat{\beta} = \left\{ \sum_{i=1,m=0}^{i=N,m=K} X_{im}(\widehat{\alpha}) Q_{im} S_{im}^T \right\}^{-1} \left\{ \sum_{i=1,m=0}^{i=N,m=K} Y_i X_{im}(\widehat{\alpha}) Q_{im} \right\},$$

where $X_{im}(\widehat{\alpha}) = \{A_i(m) - \text{expit}\{\widehat{\alpha}^T W_i(m)\}\}$, $S_{im} = \sum_{j=m}^{j=K} A_i(j) R_{ij}$, and the choice of $Q_{im} = q_m\{\overline{L}_i(m), \overline{A}_i(m-1)\}$ affects efficiency but not consistency. In fact, in the case where $\gamma_m\{\overline{a}(m-1),\overline{l}(m),\beta\,\} = \beta^T R_m$ is linear in β, we can obtain a closed-form doubly robust estimator $\widetilde{\beta}$ of β^* by specifying a working model $\varsigma^T D_m = \varsigma^T d_m\{\overline{L}(m),\overline{A}(m-1)\}$ for $E\{Y_m(\beta^*)|\overline{L}(m),\overline{A}(m-1)\} = E\{Y_{g=\{\overline{A}(m-1),\underline{0}(m)\}}|\overline{L}(m),\overline{A}(m-1)\}$ and defining

$$\begin{pmatrix}\widetilde{\beta} \\ \widetilde{\varsigma}\end{pmatrix} = \left\{\sum_{i=1,m=0}^{i=N,m=K}\begin{pmatrix}X_{im}(\widehat{\alpha})Q_{im} \\ D_{im}\end{pmatrix}(S_{im}^T, D_{im}^T)\right\}^{-1}$$

$$\times \left\{\sum_{i=1,m=0}^{i=N,m=K} Y_i\begin{pmatrix}X_{im}(\widehat{\alpha})Q_{im} \\ D_{im}\end{pmatrix}\right\}.$$

Specifically, $\widetilde{\beta}$ will be a consistent and asymptotically normal (CAN) estimator of β^* if either the model $\varsigma^T D_m$ for $E\{Y_{g=\{\overline{A}(m-1),\underline{0}(m)\}}|\overline{L}(m),\overline{A}(m-1)\}$ is correct or the model $\alpha^T W(m)$ for $\text{logit}[\Pr\{A(m)=1|\overline{L}(m),\overline{A}(m-1)\}]$ is correct.

23.5.2 Monte Carlo estimation of $E(Y_g)$ after g-estimation of an SNMM

Suppose the strengthened identifiability assumptions hold, one has obtained a doubly robust g-estimator $\widetilde{\beta}$ of an SNMM $\gamma_m\{\overline{a}(m-1),\overline{l}(m),\beta\}$, and one wishes to estimate $E(Y_g)$ for a given static or dynamic regime g. To do so, one can use the following steps of a Monte Carlo algorithm:

1. Estimate the mean response $E(Y_{g=\overline{0}_K})$ had treatment always been withheld by the sample average of $Y_0(\widetilde{\beta})$ over the N study subjects. Call the estimate $\widehat{E}(Y_{g=\overline{0}_K})$.

2. Fit a parametric model for $f\{l(k)|\overline{a}(k-1),\overline{l}(k-1)\}$ to the data, pooled over persons and times, and denote the estimate of $f\{l(k)|\overline{a}(k-1),\overline{l}(k-1)\}$ under the model by $\widehat{f}\{l(k)|\overline{a}(k-1),\overline{l}(k-1)\}$.

3. For $v = 1,\dots,V$, do the following:

   (a) Draw $l_v(0)$ from $\widehat{f}\{l(0)\}$.

   (b) Recursively for $k = 1,\dots,K$, draw $l_v(k)$ from $\widehat{f}\{l(k)|\overline{a}_v(k-1),\overline{l}_v(k-1)\}$ with $\overline{a}_v(k-1) = \overline{g}_{k-1}\{\overline{l}_v(k-1)\}$, the treatment history corresponding to the regime g.

   (c) Let $\widehat{\Delta}_{g,v} = \sum_{j=0}^K a_v(j)\gamma_j\{\overline{a}_v(j-1),\overline{l}_v(j),\widetilde{\beta}\}$ be the vth Monte Carlo estimate of $Y_g - Y_{g=\overline{0}_K}$, where $a_v(j) = g_j\{\overline{l}_v(j-1)\}$.

4. Let $\widehat{E}(Y_g) = \widehat{E}(Y_{g=\overline{0}_K}) + \sum_{v=1}^V \widehat{\Delta}_{g,v}/V$ be the estimate of $\widehat{E}(Y_g)$.

If the model for $f\{l(k)|\overline{a}(k-1),\overline{l}(k-1)\}$, the SNMM $\gamma_m\{\overline{a}(m-1),\overline{l}(m),\beta\}$, and either the treatment model $\text{logit}[\Pr\{A(m)=1|\overline{L}(m),\overline{A}(m-1)\}] = \alpha^{*T} W_m$ or the model $E\{Y_{g=\{\overline{A}(m-1),\underline{0}(m)\}}|\overline{L}(m),\overline{A}(m-1)\} = \varsigma^{*T} D_m$ is correctly specified, then $\widehat{E}(Y_g)$ is consistent for $E(Y_g)$. Confidence intervals can be obtained using the non-parametric bootstrap.

Our approach based on g-estimation does not suffer from the null paradox. In fact, under the null hypothesis of no treatment effect, it is the case that misspecification of the model for $f\{l(k)|\overline{a}(k-1),\overline{l}(k-1)\}$ does not result in bias. To understand why, note that, under the null, any SNMM $\gamma_m\{\overline{a}(m-1),\overline{l}(m),\beta\}$ is correctly specified with $\beta^* = 0$ being the true parameter and $\gamma_m\{\overline{a}(m-1),\overline{l}(m),\beta^*\} = 0$. Thus, $\gamma_m\{\overline{a}(m-1),\overline{l}(m),\widetilde{\beta}\}$ will converge to 0 if $\widetilde{\beta}$ is consistent for $\beta^* = 0$, that is, if either the treatment model or the model for $E\{Y_{g=\{\overline{A}(m-1),\underline{0}(m)\}}|\overline{L}(m),\overline{A}(m-1)\}$ is correct. Thus, $\widehat{\Delta}_{g,v}$ will converge to zero and $\widehat{E}(Y_g)$ to $\widehat{E}(Y_{g=\overline{0}_K})$, even if the model for $f\{l(k)|\overline{a}(k-1),\overline{l}(k-1)\}$ is incorrect. We conclude that, under the null, all we require for valid inference is that the conditional exchangeability and consistency assumptions hold and we either know (as in a sequentially randomized experiment) $\Pr\{A(m) = 1|\overline{L}(m), \overline{A}(m-1)\}$ or have a correct model for either $\Pr\{A(m) = 1|\overline{L}(m), \overline{A}(m-1)\}$ or $E\{Y_{g=\{\overline{A}(m-1),\underline{0}(m)\}}|\overline{L}(m), \overline{A}(m-1)\}$.

Suppose that there is no effect modification by past covariate history, as with the SNMM

$$\gamma_m\{\overline{a}(m-1), \overline{l}(m), \beta\} = \beta_0 + \beta_1 m + \beta_2 a(m-1) + \beta_3 a(m-2) + \beta_4 a(m-1)a(m-2). \tag{23.10}$$

Then we can write $\gamma_m\{\overline{a}(m-1), \overline{l}(m), \beta\}$ as $\gamma_m\{\overline{a}(m-1), \beta\}$. In that case, to estimate $E(Y_{g=(\overline{a})})$ for any non-dynamic regime $\overline{a}$, we do not need to use the above Monte Carlo algorithm to simulate the $L(k)$. Rather

$$\widehat{E}(Y_{g=\overline{a}}) = \widehat{E}(Y_{g=\overline{0}_K}) + \sum_{k=0}^{K} a(k)\gamma_k\{\overline{a}(k-1), \widetilde{\beta}\}.$$

However, if one wants to estimate $E(Y_g)$ for a dynamic regime, the previous Monte Carlo algorithm is required.

In fact, an SNMM is an MSM if and only if for all $\overline{a}(m-1)$, $\overline{l}(m)$, and β,

$$\gamma_m\{\overline{a}(m-1), \overline{l}(m), \beta\} = \gamma_m[\overline{a}(m-1), \beta]. \tag{23.11}$$

Specifically, it is a non-dynamic MSM with the functional form

$$E(Y_{g=\overline{a}}) = \eta_0^* + \sum_{k=0}^{K} a(k)\gamma_k\{\overline{a}(k-1), \beta^*\} \text{ does not depend on } \overline{l}(m), \tag{23.12}$$

where $E(Y_{g=\overline{0}_K}) = \eta_0^*$. However, such an SNMM model is not simply an MSM because, in addition to (23.12), it also imposes the additional strong assumption (23.11) that effect modification by past covariate history is absent. In contrast, an MSM such as (23.12) is agnostic as to whether or not there is effect modification by time-varying covariates.

If we specify an SNMM that assumes (23.11), then we can estimate β^* either by g-estimation or IPTW. However, the most efficient g-estimator will be more efficient than the most efficient IPTW estimator when the SNMM (and thus the MSM) is correctly specified because g-estimation uses the additional assumption of no effect modification by past covariates to increase efficiency.

In contrast, suppose the MSM (23.12) is correct but the SNMM (23.10) is incorrect because assumption (23.11) does not hold. Then the g-estimators of β^* and $E(Y_{g=\overline{a}})$ will be biased, while the IPTW estimates remain unbiased. Thus, we have a classic variance–bias trade-off. Given the MSM (23.12), g-estimation can increase efficiency if (23.11) is correct, but introduces bias if (23.11) is incorrect.

23.5.3 Time-varying instrumental variables and g-estimation

Suppose that, at each week m, we obtain data both on whether a drug treatment of interest was prescribed by a subject's physician and on whether or not the subject actually took the drug, based on a blood test, say. We assume that both variables are binary and let $A_p(m)$ and $A_d(m)$, respectively, denote the treatment prescribed and taken in week m. We define $A(m) = \{A_p(m), A_d(m)\}$. Now, in many settings it might be reasonable to assume that we had conditional exchangeability with respect to the prescribed treatment but not with respect to the actual treatment because the covariates influencing a physician's prescriptions have been recorded in the medical record, while the reasons why a given patient does or does not comply with his physician's advice may depend on unmeasured patient characteristics that also directly affect the outcome Y. Thus, we only assume that, for all $\bar{a}, \bar{l}(t)$,

$$Y_{\bar{a}} \amalg A_p(t) | \bar{A}(t-1) = \bar{a}(t-1), \bar{L}(t) = \bar{l}(t). \tag{23.13}$$

If we had (joint) conditional exchangeability for both $A_p(m)$ and $A_d(m)$, the SNMM (23.7) would be precisely equivalent to the SNMM

$$
\begin{aligned}
E\{Y_{g=\{\bar{a}(m-1),a(m),\underline{0}(m+1)\}} | \bar{L}_{g=\{\bar{a}(m-1)\}}(m) &= \bar{l}(m), \bar{A}(m) = \bar{a}(m)\} \\
= E\{Y_{g=\{\bar{a}(m-1),\underline{0}(m)\}} | \bar{L}_{g=\{\bar{a}(m-1)\}}(m) &= \bar{l}(m), \bar{A}(m) = \bar{a}(m)\} \\
+ a(m)\gamma_m\{\bar{a}(m-1), \bar{l}(m), \beta^*\}
\end{aligned} \tag{23.14}
$$

that adds $A(m) = a(m)$ to the conditioning event on both sides of (23.7). However, when only conditional exchangeability for $A_p(m)$ holds as in (23.13), the two models differ. It is only model (23.14) whose parameters can remain identified under the sole restriction given in (23.13). Specifically, given the SNMM (23.14), we can estimate the parameter β^* by g-estimation as above, except that, now, we replace the models for both logit $[\Pr\{A(m) = 1 | \bar{L}(m), \bar{A}(m-1)\}]$ and logit $[\Pr\{A(m) = 1 | \bar{L}(m), \bar{A}(m-1), Y_m(\beta)\}]$ by, respectively, models for logit $[\Pr\{A_p(m) = 1 | \bar{L}(m), \bar{A}(m-1)\}]$ and logit $[\Pr\{A_p(m) = 1 | \bar{L}(m), \bar{A}(m-1), Y_m(\beta)\}]$. This choice reflects the fact that it is only $A_p(m)$ that is conditionally independent of the counterfactuals given $\{\bar{L}(m), \bar{A}(m-1)\}$, and thus only $A_p(m)$ that can be used as the outcome variable in g-estimation.

Suppose we assume that the prescribed dose has no direct effect on the response Y except through the actual dose, that is, the counterfactual outcome $Y_{\bar{a}} = Y_{\bar{a}_p, \bar{a}_d}$ only depends on $\bar{a}_d$. In this case, we can simply write $Y_{\bar{a}}$ as $Y_{\bar{a}_d}$. This assumption is referred to as the exclusion restriction for $\bar{a}_p$ relative to the effect of $\bar{a}_d$ on Y. A variable such as $\bar{a}_p$ that satisfies (23.13) and the exclusion restriction relative to the effect of $\bar{a}_d$ on Y is said to be a time-dependent instrumental variable for the effect of treatment $\bar{a}_d$ on Y. Under this assumption, we can replace $a(m)\gamma_m\{\bar{a}(m-1), \bar{l}(m), \beta^*\}$ by $a_d(m)\gamma_m\{\bar{a}(m-1), \bar{l}(m), \beta^*\}$ in (23.14). However, when conditional exchangeability with respect to the actual treatment $A_d(m)$ does not hold, $A_p(m)$ can still be a non-causal effect modifier; as a consequence, we cannot replace $a(m)\gamma_m\{\bar{a}(m-1), \bar{l}(m), \beta^*\}$ by $a_d(m)\gamma_m\{\bar{a}_d(m-1), \bar{l}(m), \beta^*\}$ in (23.14). In fact, $A_p(m)$ can still be a non-causal effect modifier even if $\bar{a}_p$ satisfies the stronger exclusion restriction of no direct effect of $\bar{a}_p$ on either Y or $\bar{L}$ (except through actual dose $\bar{a}_d$), so both $Y_{\bar{a}}$ and $\bar{L}_{\bar{a}}$ only depend on $\bar{a}_d$. For example, in the SNMM

$$a(m)\gamma_m\{\bar{a}(m-1), \bar{l}(m), \beta^*\} = a_d(m)\{\beta_0^* + \beta_1^* a_p(m-1) + \beta_2^{*T} a_p(m-1)l(m)\},$$

it may still be the case that neither β_1^* nor β_2^{*T} is zero (Hernán and Robins, 2006b).

Furthermore, when conditional exchangeability with respect to the actual treatment $A_d(m)$ does not hold, although the sample average of $Y_0(\tilde{\beta})$ still is consistent for $E(Y_{g=\bar{0}_K})$, it is not possible to consistently estimate $E(Y_g)$ for any other regime g, static or dynamic, without further untestable assumptions such as those in Theorems 8.8 and 8.10 of Robins, Rotnitzky, and Scharfstein (1999) and in Section 7 of Robins (2004).

23.5.4 Dynamic and general SNMMs and MSMs

Both MSMs and SNMMs are models for non-dynamic regimes. Henceforth, we refer to them as non-dynamic MSMs and SNMMs. With the aid of a model for $f\{l(k) | \bar{a}(k-1), \bar{l}(k-1)\}$ and Monte Carlo simulation, we have seen that non-dynamic SNMMs can be used to estimate $E(Y_g)$ for any regime, static or dynamic. Analogously, Robins (1999) shows that with the aid of a model for particular aspects of $f\{y | \bar{a}(K), \bar{l}(K)\} \prod_{k=0}^{K} f\{l(k) | \bar{a}(k-1), \bar{l}(k-1)\}$ and simulation, non-dynamic MSMs can also be used to estimate the mean $E(Y_g)$

of any dynamic regime. However, for non-dynamic MSMs, this calculation is exceedingly difficult, requiring, as an intermediate step, that one solve many integral equations.

Modified versions of both SNMMs and MSMs, which we shall refer to as dynamic regime SNMMs and MSMs, can be used to directly estimate the means $E(Y_g)$ of specific dynamic regimes without requiring the aid of alternative models or simulation. A special case of a dynamic SNMM was considered by Murphy (2003). Robins (2004) built upon her work and introduced a comprehensive class of dynamic SNMMs. A simple dynamic MSM comparing two regimes was considered by Hernán et al. (2006). Dynamic MSMs in full generality were first introduced by Orellana, Rotnitzky, and Robins (2006) and, shortly thereafter, independently by van der Laan and Petersen (2007). Both built on earlier work by Robins (1993) and Murphy, van der Laan, and Robins (2001).

For pedagogic purposes, we shall be ahistorical and first discuss dynamic MSMs.

23.5.4.1 Dynamic and general MSMs

We begin with a simplified version of an example considered by Orellana, Rotnitzky, and Robins (2006) and Robins et al. (2008). We consider an observational study of treatment-naive subjects recently infected with HIV. Subjects return to clinic weekly to have various clinical and laboratory measurements made. Let $L(t)$ be the vector of measurements made at week t, including CD4 cell count. We let $A(t)$ denote the indicator of whether antiretroviral therapy is taken during week t. For simplicity, we assume that, once antiretroviral therapy is begun, it is never stopped. Let Y be a composite health outcome measured at the end of the study at time $K + 1$, higher values of which are preferable. Let $g = x$ denote the dynamic regime "begin antiretroviral therapy the first time t the measured CD4 count falls below x," where x is measured in whole numbers less than 1000. Let $\mathcal{X} = \{0, 1, \dots, 999\}$. Then $\{g = x; x \in \mathcal{X}\}$ denotes the set of all such regimes. Consider the dynamic regime MSM

$$E(Y_{g=x}|V) = h(x, V, \beta^*) \tag{23.15}$$

for the conditional mean of the counterfactual $Y_{g=x}$ given a subset V of the baseline covariates $L(0)$, where

$$h(x, V, \beta) = h_1(x, V, \beta_1) + h_2(V, \beta_2), \tag{23.16}$$

$$h_1(x, V, 0) = 0, \tag{23.17}$$

so that $\beta_1^* = 0$ is the null hypothesis that all regimes in $\{g = x; x \in \mathcal{X}\}$ have the same mean given V.

As an example, for V binary, we might choose $h_1(x, V, \beta_1) = \beta_1^T r(x, v)$, where $\beta_1^T r(x, v) = \beta_{1,1}(x - 350) + \beta_{1,2}(x - 350)^2 + \beta_{1,3}(x - 350)^3 + \beta_{1,4}(x - 350)V$ and $h_2(V, \beta_2) = \beta_{2,1} + \beta_{2,2}V$. Before describing how β^* can be estimated using standard weighted least-squares software, we require the following observations.

Consider a subject who started antiretroviral therapy at a CD4 cell count of 250 in week t whose lowest prior CD4 count was 300. Then this subject's observed data is consistent with having followed regime $g = x$ for $x = 251, 252, \dots, 300$. In fact the subject followed all of these regimes. Consider a subject who never started therapy and whose lowest CD4 count was 225. Then this subject followed regimes $g = x$ for $x = 0, 1, \dots, 225$. Finally, consider a subject who started antiretroviral therapy at a CD4 cell count of 250 in week t whose lowest previous CD4 count was less than 250. Then this subject failed to follow any regime in the set $\{g = x; x \in \mathcal{X}\}$. In contrast, for the non-dynamic MSM $E(Y_{\bar{a}}|V) = h(\bar{a}, V, \beta^*)$, each subject follows one and only one of the regimes whose means are being modeled — the regime $\bar{a}$ corresponding to the subject's actual treatment history $\overline{A}$. It is this difference that makes estimation of a dynamic MSM a bit more involved than estimation of a non-dynamic MSM.

We are now ready to describe our fitting procedure. Let Γ_i be the number of regimes followed by subject i. We create an artificial data set of size $\Gamma = \sum_{i=1}^{N} \Gamma_i$, with each subject $i = 1, \ldots, N$ contributing Γ_i artificial observations $(Y_i, V_i, x_{i1}), (Y_i, V_i, x_{i2}), \ldots, (Y_i, V_i, x_{i\Gamma_i})$, where the x_{ik} denote the regimes followed by subject i. We then fit by weighted least squares with an independence working covariance matrix the regression model

$$E(Y|x, V) = h(x, V, \gamma)$$

to the artificial data set using estimates of the weights

$$SW(x, V) = \prod_{k=0}^{K} \frac{f^*(x|V)}{f\{A(k)|\overline{A}(k-1), \overline{L}(k)\}},$$

where $f^*(x|V)$ is any user-supplied conditional density given V. For example, it could be an estimate of the density of x given V based on the artificial data set. Orellana, Rotnitzky, and Robins (2006) show that this IPTW estimator, $\widehat{\beta}$, say, of γ based on the artificial data set converges to the parameter β^* of our dynamic MSM under the strengthened identifiability conditions. Orellana, Rotnitzky, and Robins (2006) also discuss how to construct more efficient estimators and doubly robust estimators. The optimal treatment regime in the class $\{g = x; x \in \mathcal{X}\}$ for a subject with $V = v$ is estimated as the value of x that maximizes $h(x, v, \widehat{\beta})$ or, equivalently, $h_1(x, v, \widehat{\beta}_1)$ over $x \in \mathcal{X}$.

The general case can be treated using the same notation. Specifically, given any set of regimes $\{g = x; x \in \mathcal{X}\}$ (whether static, dynamic, or both) indexed by x taking values in a (possibly infinite) set X and an MSM (23.15) satisfying (23.16) and (23.17), we can proceed as above, except that, now, the index x need not be a real number, and a different calculation will be required to determine which regimes in $\{g = x; x \in \mathcal{X}\}$ each study subject followed. For example, consider another HIV study where now subjects may repeatedly start and stop therapy, and consider the regimes "take therapy at t if and only if the current white blood count exceeds w and a certain liver function test has value less than b." Here, b and w are non-negative integers in the range 0 to 10,000. Then $x = (w, b)$. An example of a choice for $h_1(x, V, \beta_1)$ is $\beta_{1,1}(b-100) + \beta_{1,2}(b-100)^2 + \beta_{1,3}(b-100)^3 + \beta_{1,4}(w-1000) + \beta_{1,5}(w-1000)^2 + \beta_{1,6}(b-100)V + \beta_{1,7}(w-1000)V + \beta_{1,8}(w-1000)(b-100)$. Note that, from the perspective presented in this paragraph, the general case includes non-dynamic MSMs as a special case. Thus, we henceforth refer to model (23.15) as a general MSM, subsuming all previous MSM categories. However, we have only discussed examples where the number of regimes followed by any subject is finite. Orellana, Rotnitzky, and Robins (2006) extend these dynamic MSM methods to the case where that number is uncountable rather than finite.

MSMs and the positivity condition. Specifying a general MSM can also allow us to weaken positivity requirements. We say that the positivity assumption holds for a regime g if, for all t,

$$f_{\overline{A}(t-1), \overline{L}(t)}[\overline{g}_t\{\overline{l}(t-1)\}, \overline{l}(t)] \neq 0 \text{ implies } f_{A(t)|\overline{A}(t-1), \overline{L}(t)}[g_t\{\overline{l}(t)\}|\overline{g}_t\{\overline{l}(t-1)\}, \overline{l}(t)] > 0.$$

Then, for any regime $g = x^*$ for which the positivity assumption fails, we simply remove any observation (Y, V, x^*) from the artificial data. We can then either interpret our MSM model as a model for $E(Y_{g=x}|V)$, $x \in \mathcal{X}_{\text{pos}} \subset \mathcal{X}$, where $x \in \mathcal{X}_{\text{pos}}$ if $g = x$ satisfies positivity, or as a model for $E(Y_{g=x}|V)$, $x \in \mathcal{X}$. In the latter case, one is identifying $E(Y_{g=x^*}|V)$ for non-positive regimes $g = x^*$ by model-based extrapolation.

Semi-linear MSMs. Recall that an SNMM was guaranteed to be correctly specified under the sharp null hypothesis of no treatment effect with true parameter value $\beta^* = 0$; as a consequence, g-estimation of an SNMM always provides valid inferences in a sequentially randomized experiment when the sharp null holds. In contrast, the MSM (23.15) satisfying

(23.16) and (23.17) is not guaranteed to be correctly specified under the sharp null whenever V is high-dimensional; under the sharp null, $Y_{g=x} = Y$, $\beta_1^* = 0$ so $h_1(x, V, \beta_1^*) = 0$, and thus the MSM reduces to $E(Y|V) = h_2(V, \beta_2)$. But the assumed functional form $h_2(V, \beta_2)$ may be incorrect. Furthermore, the IPTW estimates of β_2 and β_1 are generally correlated. Thus, mispecification of the functional form $h_2(V, \beta_2)$ can result in invalid inferences in a sequentially randomized experiment, even under the sharp null. To overcome this difficulty, we follow Robins (1999) and consider the semi-linear general MSM

$$E(Y_{g=x}|V) = h_1(x, V, \beta_1^*) + h_2^*(V),$$

with $h_2^*(V)$ allowed to be an arbitrary unknown function, so as to prevent bias in the estimation of $h_1(x, V, \beta_1^*)$ from misspecification of a parametric model for $h_2^*(V)$. The estimator $\widetilde{\beta}_1$ that sets to zero the sample average of

$$SW(x, V)\{Y - h_1(x, V, \beta_1^*)\}\{q(x, V) - \int q(x, V)dF^*(x|V)\}$$

over the Γ artificial data vectors (Y, V, x) can be shown to be a CAN estimator of β_1^* when the model $h_1(x, V, \beta_1^*)$ is correct, guaranteeing valid inferences in a sequentially randomized experiment. Robins (1998, 1999) proved this result for non-dynamic MSMs, and Orellana, Rotnitzky, and Robins (2006) showed it for general MSMs. Orellana, Rotnitzky, and Robins (2006) also constructed locally efficient, doubly robust estimators of β_1^* in semilinear general MSMs.

In fact, when $h_1(x, V, \beta_1) = \beta_1^T r(x, V)$ is linear in β_1, it is simple to trick standard weighted least-squares software into computing a CAN estimator of β_1^*. Specifically, we consider the model $h(x, V, \beta) = \beta_1^T r(x, V) + h_2(V, \beta_2)$, with $h_2(V, \beta_2) = \beta_2^T R$, where $R = \sum_x r(x, V)f^*(x|V)$. The first component $\widehat{\beta}_1$ of the aforementioned weighted least-squares estimator $\widehat{\beta} = (\widehat{\beta}_1, \widehat{\beta}_2)$ with estimated weights

$$\widehat{SW}(x, V) = \prod_{k=0}^{K} \frac{f^*(x|V)}{f\{A(k)|\bar{A}(k-1), \bar{L}(k); \widehat{\alpha}\}}$$

applied to the artificial data is a CAN estimator of β_1^* when the model $\alpha^T W(k)$ for $\text{logit}[\Pr\{A(k) = 1|\bar{A}(k-1), \bar{L}(k)\}]$ is correct even when the model $h_2(V, \beta_2) = \beta_2^T R$ for $h_2^*(V)$ is incorrect.

In summary, following Robins (1999), we suggest, when possible, semi-linear general MSMs be substituted for general MSMs.

23.5.4.2 General SNMMs and optimal regime SNMMs

In this section, for reasons discussed below, we need to explicitly consider regimes g in which the treatment $g_m\{\bar{a}(m-1), \bar{l}(m)\}$ specified by the regime g at time m is allowed to depend on both past treatment history and past covariate history. Suppose we are interested in a particular such regime g^*. Then we define a g^*-SNMM to be a model for the effect of treatment $a(m)$ versus treatment 0 at each time m (as a function of treatment and covariate history up to m) when treatment g^* is followed beginning at time $m+1$. Let $Y_{g=\{\bar{a}(m-1), a(m), g^*(m+1)\}}$ be the outcome Y under the regime that follows the non-dynamic regime $\{\bar{a}(m-1), a(m)\}$ through week (time) m and then the regime g^* from $m+1$. Then a g^*-SNMM is defined exactly like the SNMM (23.7), except that $Y_{g=\{\bar{a}(m-1), a(m), g^*(m+1)\}}$ replaces $Y_{g=\{\bar{a}(m-1), a(m), \underline{0}(m+1)\}}$ and $Y_{g=\{\bar{a}(m-1), 0, g^*(m+1)\}}$ replaces $Y_{g=\{\bar{a}(m-1), \underline{0}(m)\}}$. Also, we write the known function $\gamma_m\{\bar{a}(m-1), \bar{l}(m), \beta\}$ as $\gamma_m^{g^*}\{\bar{a}(m-1), \bar{l}(m), \beta\}$ to remind us we are now estimating a g^*-SNMM for a given regime g^*. Note that a g^*-SNMM with g^* the regime where treatment is always withheld is precisely the SNMM (23.7).

To estimate the parameter β^* of $\gamma_m^{g^*}\{\bar{a}(m-1), \bar{l}(m), \beta^*\}$, we use g-estimation as described previously except that we redefine $Y_m(\beta)$ to be

$$Y_m(\beta) = Y + \sum_{j=m}^{K}[g_j^*\{\overline{A}(j-1), \overline{L}(j)\} - A(j)]\gamma_j^{g^*}\{\overline{A}(j-1), \overline{L}(j), \beta\}. \tag{23.18}$$

Again we can motivate this modification by considering a locally rank-preserving version of the model. We say a g^*-SNMM model is locally rank-preserving if $Y_{g=\{\bar{a}(m-1),0,g^*(m+1)\}} = Y_{g=\{\bar{a}(m),g^*(m+1)\}} - a(m)\gamma_m^{g^*}\{\bar{a}(m-1), \bar{l}(m), \beta\}$ with probability 1 for each m. In that case, $Y_m(\beta^*) = Y_{g=\{\overline{A}(m-1),g^*(m)\}}$ and, in particular, $Y_0(\beta^*) = Y_{g*}$. This reflects the fact that, at each time $j \geq m$, $Y_m(\beta^*)$ subtracts from the subject's observed Y the effect $A(j)\gamma_j^{g^*}\{\overline{A}(j-1), \overline{L}(j), \beta^*\}$ of the subject's observed treatment $A(j)$ and replaces it with the effect $g_j^*\{\overline{A}(j-1), \overline{L}(j)\}\gamma_j^{g^*}\{\overline{A}(j-1), \overline{L}(j), \beta^*\}$ of the treatment $g_j^*\{\overline{A}(j-1), \overline{L}(j)\}$ that the subject would have had at j had she, possibly contrary to fact, begun to follow regime g^* at time j.

Even without local rank preservation, it can be proved that, in the absence of model misspecification, under the strengthened identifiability conditions, (i) the g-estimator $\widehat{\beta}$, now based on (23.18), is consistent for the parameter β^* of $\gamma_m^{g^*}\{\bar{a}(m-1), \bar{l}(m), \beta^*\}$; (ii) the sample average $\widehat{E}(Y_{g*}) = N^{-1}\sum_{i=1}^{N}Y_{0,i}(\widehat{\beta})$ of $Y_0(\widehat{\beta})$ is consistent for $E(Y_{g*})$; and (iii) for any other regime g, $\widehat{E}(Y_g) = \widehat{E}(Y_{g*}) + \sum_{v=1}^{V}\widehat{\Delta}_{g,v}^{g^*}/V$ is consistent for $\widehat{E}(Y_g)$ as $V \to \infty$, where the quantity $\widehat{\Delta}_{g,v}^{g^*}$ is defined exactly like $\widehat{\Delta}_{g,v}$ given above except that $a_v(j)\gamma_j\{\bar{a}_v(j-1), \bar{l}_v(j), \widehat{\beta}\}a_v(j)\gamma_j\{\bar{a}_v(j-1), \bar{l}_v(j), \widehat{\beta}\}$ is replaced by the quantity $[a_v(j) - g_j^*\{\bar{a}_v(j-1), \bar{l}_v(j)\}]\gamma_j^{g^*}\{\bar{a}_v(j-1), \bar{l}_v(j), \widehat{\beta}\}$, with $\widehat{\beta}$ based on (23.18). When $\gamma_j^{g^*}\{\overline{A}(j-1), \overline{L}(j), \beta\} = \beta^T R_j$, we have the closed-form expression

$$\widehat{\beta} = \left(\sum_{i=1,m=0}^{i=N,m=K} X_{im}(\widehat{\alpha})Q_{im}S_{im}^T\right)^{-1}$$

$$\times \left\{\sum_{i=1,m=0}^{i=N,m=K} Y_i X_{im}(\widehat{\alpha})Q_{im}\right\} \text{ with } S_{im}^T \text{ redefined as } \sum_{j=m}^{j=K}\{A_i(j) - g_j^*(\overline{A}_i(j-1), \overline{L}_i(j))\}R_{ij}^T.$$

Optimal-regime SNMMs. A primary use of g^*-SNMMs is in attempting to estimate the optimal treatment strategy g_{opt} that maximizes $E(Y_g)$ over all treatment regimes g, including non-dynamic and dynamic regimes in which treatment depends on past covariate history alone, and dynamic regimes in which treatment depends on past covariate and treatment history. To do so, we specify an optimal treatment SNMM, g_{opt}-SNMM, based on a function $\gamma_m^{g_{\text{opt}}}\{\bar{a}(m-1), \bar{l}(m), \beta\}$. As an example, we might specify that

$$\gamma_m^{g_{\text{opt}}}\{\bar{a}(m-1), \bar{l}(m), \beta\} = \beta_0 + \beta_1 m + \beta_2 a(m-1) \tag{23.19}$$

$$+ \beta_3^T l(m) + \beta_4^T l(m)a(m-1) + \beta_5^T l(m-1) + \beta_6^T l(m-1)a(m-1).$$

If the g_{opt}-SNMM were correctly specified, and we knew the true β^*, then we would know the optimal treatment regime. Specifically, the optimal treatment $g_{\text{opt},m}\{\bar{a}(m-1), \bar{l}(m)\}$ at time m given past treatment and covariate history $\{\bar{a}(m-1), \bar{l}(m)\}$ is to take treatment if and only if $\gamma_m^{g_{\text{opt}}}\{\bar{a}(m-1), \bar{l}(m), \beta^*\}$ exceeds zero. That is, $g_{\text{opt},m}\{\bar{a}(m-1), \bar{l}(m)\} = I[\gamma_m^{g_{\text{opt}}}\{\bar{a}(m-1), \bar{l}(m), \beta^*\} > 0]$.

To understand heuristically why this is the case, assume a locally rank-preserving model, and suppose that, at the last treatment time K, a subject has past history $\bar{a}(K-1), \bar{l}(K)$.

If the subject does not take treatment at K, her outcome will be $Y_{g=\{\bar{a}(K-1),0_K\}}$, while if she takes treatment it will be $Y_{g=\{\bar{a}(K-1),1_K\}}$. Now, according to a locally rank-preserving g_{opt}-SNMM, $Y_{g=\{\bar{a}(K-1),0_K\}} = Y_{g=\{\bar{a}(K-1),1_K\}} - \gamma_K^{g_{\text{opt}}}\{\bar{a}(K-1),\bar{l}(K),\beta^*\}$. Because high values of Y are desirable, the optimal treatment choice is to take treatment if and only if $\gamma_K^{g_{\text{opt}}}\{\bar{a}(K-1),\bar{l}(K),\beta^*\}$ exceeds zero. (If $\gamma_K^{g_{\text{opt}}}\{\bar{a}(K-1),\bar{l}(K),\beta^*\}$ is precisely zero, then it does not matter whether treatment is taken; in such cases, we choose not to treat simply to break the "tie.") We continue by backward induction. Specifically, suppose we know the optimal regime from $m+1$ onwards. Consider a subject at time m with past history $\bar{a}(m-1),\bar{l}(m)$. Such a subject will follow the known optimal regime from $m+1$ onwards. But she must decide what treatment to take at m. If she does not take treatment at m, her outcome will be $Y_{g=\{\bar{a}(m-1),0,\underline{g}_{\text{opt}}(m+1)\}}$, while if she takes treatment at m, her outcome will be $Y_{g=\{\bar{a}(m-1),1,\underline{g}_{\text{opt}}(m+1)\}}$. But according to the model,

$$Y_{g=\{\bar{a}(m-1),0,\underline{g}_{\text{opt}}(m+1)\}} = Y_{g=\{\bar{a}(m-1),1,\underline{g}_{\text{opt}}(m+1)\}} - \gamma_m^{g_{\text{opt}}}\{\bar{a}(m-1),\bar{l}(m),\beta^*\},$$ so she should

take treatment if and only if $\gamma_m^{g_{\text{opt}}}\{\bar{a}(m-1),\bar{l}(m),\beta^*\}$ exceeds zero. Even in the absence of local rank preservation, it can be proved that, under the strengthened exchangeability conditions, the optimal decision is to treat if and only if $\gamma_m^{g_{\text{opt}}}\{\bar{a}(m-1),\bar{l}(m),\beta^*\}$ exceeds zero.

Now, if we knew $\gamma_m^{g_{\text{opt}}}\{\bar{a}(m-1),\bar{l}(m),\beta^*\}$ and thus we knew the optimal regime, we would simply have each subject in the population follow the optimal regime beginning at time 0, where at each time m the covariates $L(m)$ must be measured and recorded, so the evolving covariate data necessary to follow the optimal regime will be available.

Thus, it only remains to estimate β^* by g-estimation based on (23.18) in order to obtain an estimate $\hat{g}_{\text{opt},m}\{\bar{a}(m-1),\bar{l}(m)\} = I[\gamma_m^{g_{\text{opt}}}\{\bar{a}(m-1),\bar{l}(m),\hat{\beta}\} > 0]$ of the optimal regime and an estimate $\hat{E}(Y_{g_{\text{opt}}}) = N^{-1}\sum_{i=1}^{N} Y_{0,i}(\hat{\beta})$ of the mean $E(Y_{g_{\text{opt}}})$ of Y when the population is treated optimally. When specialized to the regime g_{opt}, (23.18) becomes

$$Y_m(\beta) = Y + \sum_{j=m}^{K}[g_{\text{opt},j}\{\overline{A}(j-1),\overline{L}(j)\} - A(j)]\gamma_j^{g_{\text{opt}}}\{\overline{A}(j-1),\overline{L}(j),\beta\} \tag{23.20}$$

$$= Y + \sum_{j=m}^{K}(I[\gamma_j^{g_{\text{opt}}}\{\overline{A}(j-1),\overline{L}(j),\beta\} > 0] - A(j))\gamma_j^{g_{\text{opt}}}\{\overline{A}(j-1),\overline{L}(j),\beta\};$$

this differs from the earlier expression in that the regime g_{opt} itself is a function of the parameter β and thus unknown. Nonetheless, one can use g-estimation based on $Y_m(\beta)$ to estimate β and set confidence intervals by searching over a grid of β values. However, when the dimension of β is moderate, so that finding $\hat{\beta}$ by search is computationally prohibitive, there is no longer an explicit closed-form expression for the g-estimate $\hat{\beta}$ based on (23.20), even when $\gamma_m^{g_{\text{opt}}}\{\bar{a}(m-1),\bar{l}(m),\beta\}$ is linear in β, because β now also occurs within an indicator function. In fact, the g-estimate $\hat{\beta}$ is exceedingly difficult to compute. However, the following different, computationally tractable, approach can be used when $\gamma_m^{g_{\text{opt}}}\{\overline{A}(m-1),\overline{L}(m),\beta\}$ is linear in β, that is, $\gamma_m^{g_{\text{opt}}}\{\overline{A}(m-1),\overline{L}(m),\beta\} = R_m^T\beta$, where R_m is a known vector function of $\{\overline{A}(m-1),\overline{L}(m)\}$.

A closed-form estimator of the optimal regime. Suppose for the moment that we have $\gamma_m^{g_{\text{opt}}}\{\bar{a}(m-1),\bar{l}(m),\beta\} = R_m^T\beta_m$ linear in β with a separate, variation-independent parameter vector at each time m, so that $\beta^T = (\beta_0^T,\ldots,\beta_K^T)$. Define $P_m(\hat{\alpha}) = \text{expit}\{\hat{\alpha}^T W(m)\}$ to be the estimate of $\Pr\{A(m) = 1|\overline{L}(m),\overline{A}(m-1)\}$ based on the fit of the model (23.8). Specify a working model $\varsigma_m^T D_m = \varsigma_m^T d_m\{\overline{L}(m),\overline{A}(m-1)\}$ for $E\{Y_m(\beta^*)|\overline{L}(m),\overline{A}(m-1)\}$. Now, beginning with β_K, we recursively obtain the closed-form, doubly robust estimates $\hat{\beta}_m$ of the β_m, with $(\hat{\beta}_m,\tilde{\eta}_m,\hat{\varsigma}_m)$ the OLS estimator of $(\beta_m,\eta_m,\varsigma_m)$ in the regression model

$Y_{m+1}(\widetilde{\underline{\beta}}_{m+1}) = A(m)R_m^T\beta_m + P_m(\widehat{\alpha})R_m^T\eta_m + D_m^T\varsigma_m + \epsilon$. Here, $Y_{K+1}(\widetilde{\underline{\beta}}_{K+1}) = Y$ and

$$Y_{m+1}(\widetilde{\underline{\beta}}_{m+1}) = Y + \sum_{j=m+1}^{K} \{I(R_j^T\widetilde{\beta}_j > 0) - A(j)\}R_j^T\widetilde{\beta}_j.$$

The $\widetilde{\beta}_m$ are CAN for the β_m^* if either all the models $\varsigma_m^T D_m$ are correct or the model (23.8) is correct.

Remark: The $\widetilde{\beta}_m$ are possibly inefficient members of the following general class of estimators. Beginning with $m = K$, we recursively obtain consistent closed-form estimators $\widetilde{\beta}_m(s,q)$ of the β_m, indexed by vectors of functions s, q

$$\widetilde{\beta}_m(s,q) = \left\{\sum_{i=1}^{N} A_i(m)X_{im}(\widehat{\alpha})Q_{im}R_{im}^T\right\}^{-1}\left[\sum_{i=1}^{N}\{Y_{m+1}(\widetilde{\underline{\beta}}_{m+1})_j - S_{im}\}X_{im}(\widehat{\alpha})Q_{im}\right],$$

where $X_m(\widehat{\alpha}) = \{A(m) - \widehat{P}_m(\widehat{\alpha})\}$, and the choice of the $S_m = s_m\{\overline{L}(m), \overline{A}(m-1)\}$ and $Q_m = q_m\{\overline{L}(m), \overline{A}(m-1)\}$ affects efficiency but not consistency when the model (23.8) is correct.

Now suppose that in our g_{opt}-SNMM the same parameter vector β applies to each time m. To be concrete, consider the g_{opt}-SNMM (23.19). In that case, we first estimate a larger model that has a separate, variation-independent parameter vector β_m at each time m; model (23.19) is then the submodel that imposes $\beta_m = \beta$ for all m. Let $\widetilde{\Omega}^{-1}$ be a nonparametric bootstrap estimate of the covariance matrix of $(\widetilde{\beta}_0, \ldots, \widetilde{\beta}_K)$. We then estimate β by an inverse covariance weighted average $\widehat{\beta} = 1_{K+1}^T\widetilde{\Omega}^{-1}(\widetilde{\beta}_0, \ldots, \widetilde{\beta}_K)^T/(1_{K+1}^T\widetilde{\Omega}^{-1}1_{K+1})$ of the $\widetilde{\beta}_m$, where 1_{K+1} is a $(K+1)$-vector with all components equal to 1.

Note that the g_{opt}-SNMM (23.19) is a non-saturated model. For example, it assumes that the optimal regime does not depend on covariate values two weeks in the past or treatment values three weeks in the past, which may be incorrect. If the g_{opt}-SNMM is badly misspecified, then the estimated optimal regime $\widehat{g}_{\mathrm{opt},m}\{\overline{a}(m-1), \overline{l}(m)\}$ may be a poor estimate of the actual optimal regime. Because in realistic studies highly non-saturated g_{opt}-SNMMs must be employed, misspecification can be a serious problem.

We note that, in using a g_{opt}-SNMM to find the optimal regime, it was necessary for us to estimate the treatment strategy

$$g_{\mathrm{opt}} = [g_{\mathrm{opt},0}\{\overline{l}(0)\}, g_{\mathrm{opt},1}\{\overline{a}(0), \overline{l}(1)\}, \ldots, g_{\mathrm{opt},K}\{\overline{a}(K-1), \overline{l}(K)\}]$$

that maximized $E(Y_g)$ over all treatment regimes g, including regimes in which treatment depends on past treatment as well as covariate history. However, as discussed earlier, one can always construct a regime $g_{\mathrm{opt}}^{\Delta} = [g_{\mathrm{opt},1}^{\Delta}\{\overline{l}(0)\}, g_{\mathrm{opt},1}^{\Delta}\{\overline{l}(1)\}, \ldots, g_{\mathrm{opt},K}^{\Delta}\{\overline{l}(K)\}]$ in which treatment depends only on past covariate history such that following regime $g_{\mathrm{opt}}^{\Delta}$ from time 0 onwards is precisely equivalent to following g_{opt} from time 0. Nonetheless, it can be important to know g_{opt} rather than only $g_{\mathrm{opt}}^{\Delta}$, as the following example shows. Suppose that a (random) member of the source population has observed history $\{\overline{A}(m-1), \overline{L}(m)\} = \{\overline{a}(m-1), \overline{l}(m)\}$ (under standard care) that is not consistent with following the optimal regime g_{opt} and comes to our attention only at time m. We wish to intervene beginning at m and give the subject the optimal treatment strategy from time m onwards. Under the strengthened exchangeability conditions, the optimal treatment strategy for such a subject is $[g_{\mathrm{opt},m}\{\overline{a}(m-1), \overline{l}(m)\}, \ldots, g_{\mathrm{opt},K}\{\overline{a}(K-1), \overline{l}(K)\}]$. This strategy can be implemented only if we know (or have a good estimate of) g_{opt}; knowledge of $g_{\mathrm{opt}}^{\Delta}$ does not suffice.

Finally, we return to our hypothetical study of Section 23.4 in order to provide a worked example of optimal regime estimation.

Table 23.7 An Intuitive Approach to Estimate the Optimal Regime

$A(0)$	$L(1)$	$A(1)$	N	Y	$Y_1(\beta_1^*)$	$Y_0(\beta_1^*)$
0	1	0	2000	200	220	235
0	1	1	6000	220	220	235
0	0	0	6000	50	70	85
0	0	1	2000	70	70	85
1	1	0	3000	130	130	130
1	1	1	9000	110	130	130
1	0	0	3000	230	250	250
1	0	1	1000	250	250	250

Estimation of the optimal regime in our hypothetical study.   We estimate the optimal regime twice. First, we use a very intuitive approach that unfortunately does not generalize beyond our simple toy example. Second, we use our closed-form optimal regime estimator to estimate a saturated optimal regime SNMM.

An intuitive approach.   Assume local rank-preservation. Then, by defination, $Y_1(\beta_1^*)$ is the value of Y had subjects followed their observed $A(0)$ and then followed the optimal regime at time 1. Because the subjects in rows 1 and 2 of Table 23.7 are exchangeable, and their treatments only differ at time 1, it is immediate that the mean of $Y_1(\beta_1^*)$ for the subjects in rows 1 and 2 is the greater of 200 and 220. Arguing similarly for rows 3 and 4, 5 and 6, and 7 and 8, we can immediately fill in the $Y_1(\beta_1^*)$ column in Table 23.7 without even explicitly estimating β_1^*. By comparing the $Y_1(\beta_1^*)$ column to the Y column in rows 5 through 8, we discover that, if treatment was taken at time 0, it is optimal to take treatment at time 1 if and only if $L(1) = 0$. Comparing the $Y_1(\beta_1^*)$ and the Y columns in rows 1 through 4, we discover that, if treatment was not taken at time 0, it should be taken at time 1, regardless of $L(1)$. We conclude that the only remaining possibilities for g_{opt} are $g_1 = $ "take treatment at time 0 and then take treatment at time 1 only if $L(1)$ is 0" and $g_2 = $ "do not take treatment at time 0 but take treatment at time 1." Because subjects in rows 1 through 4 are exchangeable with those in rows 5 through 8 and $Y_1(\beta_1^*)$ equals Y_{g_2} for subjects in rows 1 through 4 and equals Y_{g_1} for subjects in rows 5 through 8, we can determine the optimal regime by comparing the mean of $Y_1(\beta_1^*)$ in rows 1 through 4 to that in rows 5 through 8. Now the mean in rows 1 through 4 is

$$220 \times \frac{8000}{16{,}000} + 70 \times \frac{8000}{16{,}000} = 145$$

while that in rows 5 through 8 is

$$130 \times \frac{12{,}000}{16{,}000} + 250 \times \frac{4000}{16{,}000} = 160.$$

We conclude that the regime g_1 is optimal and $E(Y_{g_1}) = 160$. That 160 is the mean of Y_{g_1} can be confirmed using the g-computation algorithm, unstabilized IPTW, or using calculations based on the ordinary SNMM as in Section 23.4.

Closed-form optimal regime estimator.   We now repeat the analysis, but this time using the closed-form optimal regime estimator of a saturated optimal regime SNMM. With the model saturated and no baseline $L(0)$ in our example, we have, with $K = 1$, $\gamma_1^{g_{\mathrm{opt}}}\{\bar{a}(0), \bar{l}(1), \beta^*\} = \beta_{1,1}^* + \beta_{1,2}^* l(1) + \beta_{1,3}^* a(0) + \beta_{1,4}^* a(0) l(1)$ and $\gamma_0^{g_{\mathrm{opt}}}\{\bar{a}(-1), \bar{l}(0), \beta^*\} = \beta_0^*$. We note that, at the last time K, $\gamma_K^{g^*}\{\bar{a}(0), \bar{l}(1), \beta^*\}$ is the same for all g^*-SNMM so, with $K = 1$, $\beta_{1,1}^* = 20$, $\beta_{1,2}^* = \beta_{1,3}^* = 0$, $\beta_{1,4}^* = -40$ as in Section 23.4. (The reader can verify that the OLS estimator of β_1^* described above also returns these values.) Thus, $\gamma_1^{g_{\mathrm{opt}}}\{\bar{a}(0), \bar{l}(1), \beta^*\}$ takes the four values $\gamma_1^{g_{opt}}(0,0) = \beta_{1,1}^* = 20$, $\gamma_1^{g_{opt}}(1,0) = \beta_{1,1}^* = 20$, $\gamma_1^{g_{opt}}(0,1) = \beta_{1,1}^* + \beta_{1,3}^* = 20$, $\gamma_1^{g_{opt}}(1,1) = \beta_{1,1}^* + \beta_{1,2}^* + \beta_{1,3}^* + \beta_{1,4}^* = -20$. Further, $g_{opt,1}\{\overline{a}(0),\overline{l}(1)\} = I[\gamma_1^{g_{opt}}\{\overline{a}(0),\overline{l}(1),\beta^*\} > 0]$ takes the values

$$g_{opt,1}(0,0) = g_{opt,1}(0,1) = g_{opt,1}(1,0) = 1, g_{opt,1}(1,1) = 0. \qquad (23.21)$$

Now, by (23.20) $Y_1(\beta_1^*) = Y + \sum_{j=1}^1 [g_{opt,j}\{\overline{A}(j-1),\overline{L}(j)\} - A(j)]\gamma_j^{g_{opt}}\{\overline{A}(j-1),\overline{L}(j),\beta\}$, so we have that $Y_1(\beta_1^*) = Y$ if $g_{opt,1}\{\overline{A}(0),\overline{L}(1)\} = A(1)$. Consequently, $Y_1(\beta_1^*) = Y$ if $\{A(1) = 1$ and $(\overline{A}(0),\overline{L}(1)) \neq (1,1)\}$ or if $\{A(1) = 0$ and $(\overline{A}(0),\overline{L}(1)) = (1,1)\}$. If $\{\overline{A}(0),\overline{L}(1)\} = (1,1)$ and $A(1) = 1$, $Y_1(\beta_1^*) = Y - (-20) = Y + 20$. If $\{\overline{A}(0),\overline{L}(1)\} \neq (1,1)$, $A(1) = 0$, $Y_1(\beta_1^*) = Y + 20$.

Using the row-specific means of Y given in Table 23.7, we obtain again the same results for the $Y_1(\beta_1^*)$ column as above. Next, in order to estimate the parameter β_0^* of $\gamma_0^{g_{opt}}\{\overline{a}(-1),\overline{l}(0),\beta_0^*\}$, we fit by OLS the regression model $Y_1(\beta_1^*) = \beta_0 A(0)R_0 + \eta_0 P_0 R_0 + \epsilon$, with $R_0 = 1$ and $P_0 = \Pr\{A(0) = 1\} = 1/2$. That is, we fit the model $Y_1(\beta_1^*) = \beta_0 A(0) + \nu_0 + \epsilon$ by OLS, where $\nu_0 = \eta_0/2$. The OLS estimate of β_0 is just the contrast $E\{Y_1(\beta_1^*)|A(0) = 1\} - E\{Y_1(\beta_1^*)|A(0) = 0\}$. Above, we calculated the mean of $Y_1(\beta_1^*)$ to be 145 among subjects with $A(0) = 1$ (rows 1 through 4 of Table 23.7) and 160 among subjects with $A(0) = 0$ (rows 5 through 8). Thus, $\beta_0^* = 15$. Hence, $g_{opt,0} = I[\gamma_0^{g_{opt}}\{\overline{a}(-1),\overline{l}(0),\beta_0^*\} > 0] = I(\beta_0^* > 0) = 1$. We conclude that the optimal treatment $g_{opt,0} = 1$ at time 0 is to "take treatment." The optimal treatment at time 1 given that one followed the optimal treatment at time 0 is, by (23.21), $g_{opt,1}(1,0) = 1$ if $l(1) = 0$ and $g_{opt,1}(1,1) = 0$ if $l(1) = 1$. Thus, we again conclude that the optimal regime is $g_1 = $ "take treatment at time 0 and then take treatment at time 1 only if $L(1)$ is 0." Finally, we compute $Y_0(\beta^*)$ for each subject by adding $\{g_{opt,0} - A(0)\}\gamma_0^{g_{opt}}\{\overline{a}(-1),\overline{l}(0),\beta_0^*\} = \{1 - A(0)\}15$ to $Y_1(\beta_1^*)$. As expected, the mean of $Y_0(\beta^*)$ is the mean of $E(Y_{g_1}) = 160$ both in rows 1 through 4 and rows 5 through 8.

23.6 Strengths and weaknesses

As mentioned previously, owing to the null paradox, methods based on the estimated g-formula should be avoided whenever the null hypothesis of no treatment effect has not yet been excluded. MSMs have the advantage that they are easy to understand and easy to fit with standard, off-the-shelf software that allows for weights. These two points explain their rapid adoption compared to SNMs. The usefulness of MSMs has been extended by the introduction of dynamic MSMs.

However, IPTW estimation of MSMs has four drawbacks not shared by g-estimation of SNMs. First, if the number of time periods is great, the product in the denominator of the weights can become very small for some subjects who then receive inordinately large weights, leading both to bias when the weights must be estimated (and to so-called pseudo-bias [Scharfstein, Rotnitzky, and Robins, 1999] even when they are known) and to imprecision. Problems with large or even truly infinite weights (when positivity does not hold) can be somewhat ameliorated but not cured by using bounded doubly robust estimators (Robins et al., 2008b), adjusting for baseline covariates and then using the covariates in the numerator of the weights, downweighting regimes $g = x$ associated with very small weights, using locally semi-parametric efficient estimators or bounded influence estimators for non-saturated MSMs (as these estimators downweight regimes $g = x$ that result in excessively large weights in a near optimal fashion), and using diagnostics for the undue influence of large weights and for the consequences of truncating large weights (Wang et al., 2006). Second, MSMs cannot be used to estimate causal effects when treatment is confounded but an instrumental variable is available. Third, although not discussed in this chapter, sensitivity analysis models for MSMs are much more restrictive and less useful than those for SNMs. Fourth, SNMs, in contrast to MSMs, allow one to directly model interactions between treatment and evolving time-dependent covariates and so to directly look for qualitative effect modification.

Disadvantages of SNMs compared with MSMs include the following:

1. SNMs cannot be easily used to compare non-dynamic regimes when there is effect modification by a time-dependent covariate.

2. Although SNMMs with a log link can be fit by g-estimation, logistic SNMMs cannot, making SNMMs difficult to use for non-rare dichotomous responses.

3. SNM models for failure time data have had to be based on accelerated failure time-like models that are difficult to fit in the presence of censoring because the objective function is non-smooth.

Problems 2 and 3 may soon be resolved. In an unpublished report, Richardson and Robins (2007) have recently developed methods for fitting risk ratio models to non-rare binary responses that may resolve problem 2. An alternative approach is given in van der Laan, Hubbard, and Jewell (2007). Page, Hernán, and Robins (2005) have developed cumulative incidence structural nested failure time models to solve problem 3 under the rare disease assumption.

In terms of estimation of optimal regimes, both MSMs and SNMs have their distinct place. General MSMs are excellent for estimating the optimal regime in any rather small prespecified classes of regimes (such as the optimal CD4 cell count at which to start therapy) that still may include all logistically feasible regimes, particularly in settings with resource constraints that preclude implementing complex regimes.

In contrast, the method of backward induction on which g-estimation of optimal regime SNMs is based requires that the set of potential regimes from which the optimal is to be selected include all functions of an increasing (in time) amount of information (i.e., of an increasing sigma field). Thus, optimal regime SNMs are useful for estimating the optimal regime in the huge class of dynamic regimes in which treatment at each m can depend on any function of the entire measured past $\bar{l}(m), \bar{a}(m-1)$ (the case considered above) or, as described in Robins (2004, Section 7), in the smaller, but still large, class in which treatment at each m can depend on any function of $\overline{w}(m), \bar{a}(m-1)$ where $W(m)$ is a subvector of the covariates in $L(m)$. Even if $W(m)$ is just CD4 cell count at m, it is possible that the optimal treatment decision at time m may be a complex function of CD4 cell counts at all past times. Such a regime, though optimal, may be logistically impossible to implement, in which case it may be necessary to choose among a smaller class of logistically feasible regimes by fitting a general MSM.

23.7 Appendix: Causal directed acyclic graphs

We define a directed acyclic graph (DAG) G to be a graph whose nodes (vertices) are random variables $V = (V_1, \ldots, V_M)$ with directed edges (arrows) and no directed cycles. We use PA_m to denote the parents of V_m, that is, the set of nodes from which there is a direct arrow into V_m. The variable V_j is a descendant of V_m if there is a sequence of nodes connected by edges between V_m and V_j such that, following the direction indicated by the arrows, one can reach V_j by starting at V_m. For example, consider the causal DAG in Figure 23.1(b) that represents the causal structure of an observational study with no unmeasured confounding or the effect of A on Y. In this DAG, $M = 4$ and we can choose $V_1 = U$, $V_2 = L$, $V_3 = A$, $V_4 = Y$; the parents PA_4 of $V_4 = Y$ are (U, L, A) and the non-descendants of A are (U, L).

Following Spirtes et al. (1993), a causal DAG is a DAG in which (i) the lack of an arrow from node V_j to V_m can be interpreted as the absence of a direct causal effect of V_j on V_m (relative to the other variables on the graph), and (ii) all common causes, even if unmeasured, of any pair of variables on the graph are themselves on the graph. In Figure 23.1(b) the lack of a direct arrow between U and A indicates that unmeasured factors U do not have a direct causal effect (causative or preventive) on the patient's treatment. Also, the inclusion of the measured variables (L, A, Y) implies that the causal DAG must also include the unmeasured common causes U. Note that a causal DAG model makes no reference to and is agnostic as to the existence of counterfactuals.

Our causal DAG is of no practical use unless we make some assumption linking the causal structure represented by the DAG to the statistical data obtained in an epidemiologic study, which we do through the causal Markov assumption. First we give some definitions that apply to any DAG, causal or not.

We say a DAG G represents the joint density of its node variables V if and only if $f(v)$ satisfies the Markov factorization

$$f(v) = \prod_{j=1}^{M} f(v_j \mid pa_j). \tag{23.22}$$

That is, the density $f(v)$ can be factorized as the product of the probability of each variable given its parents. This factorization is equivalent to the statement that the non-descendants of a given variable V_j are independent of V_j conditional on the parents of V_j.

The causal Markov assumption (CMA) states that the joint distribution of the variables on a causal graph satisfies the Markov factorization (23.22). Because of the causal meaning of parents and descendants on a causal DAG, the CMA is equivalent to the statement that, conditional on its direct causes (i.e., parents), a variable V is independent of any variable it does not cause (i.e., any non-descendant of V). The Markov factorization (23.22) logically implies additional statistical independencies and, specifically, it implies that a set of variables A is conditionally independent of another set of variables B given a third set of variables Z if A is d-separated from B given Z on the graph G, written $(A \amalg_{\text{d-sep}} B|Z)_G$, where d-separation, described below, is a statement about the topology of the graph (Pearl, 1995). To check for unconditional (i.e., marginal) independence we make Z the empty set. In the following a path between A and B is any sequence of nodes and edges (where the direction of the arrows is ignored) that connects A to B. A variable C is a collider on a path between variables A and B if the edges on the path that meet at C both have arrows pointing at C.

Unconditional d-separation $(A \amalg_{\text{d-sep}} B)_G$. A variable A is d-separated from variable B on a DAG G if and only if all paths between them are blocked. The path is blocked if there is a collider on the path. If a path is not blocked, we say it is unblocked, active, or open, all of which are synonymous. We say that a set of variables A is d-separated from a set of variables B if and only if each variable in A is d-separated from every variable in B. Thus, $(A \amalg_{\text{d-sep}} B)_G$ if and only if every path from A to B is blocked. If even one path is unblocked, we write $(A \not\amalg_{\text{d-sep}} B)_G$.

Conditional d-separation $(A \amalg_{\text{d-sep}} B|Z)_G$. We say that two variables A and B are d-separated given (or by) a set of variables $Z = (Z_1, \ldots, Z_k)$ if all paths between A and B are blocked where, when we condition on Z, a path between A and B is blocked if (i) there is any variable $Z_m \in Z$ on the path that is not a collider, or (ii) there is a collider on the path such that neither the collider itself nor any of its descendants are in Z. A set of variables A is d-separated from a set of variables B given Z if and only if each variable in A is d-separated from every variable in B given Z.

The CMA allows one to deduce that d-separation implies statistical independence, but does not allow one to deduce that d-connection (i.e., the absence of d-separation) implies statistical dependence. However, d-connected variables will generally be independent only if there is an exact balancing of positive and negative causal effects. Because such precise fortuitous balancing of effects is highly unlikely to occur, we shall henceforth assume that d-connected variables are dependent. This is often referred to as the assumption of faithfulness or stability.

A causal DAG model that includes counterfactuals is a non-parametric structural equation model (NPSEM) (Pearl, 1995). First, some notation. For any random variable W, let $\mathcal{W}$ denote the support (i.e., the set of possible values w) of W. For any $w_1, \dots, w_m$, define $\overline{w}_m = (w_1, \dots, w_m)$. Let R denote any subset of variables in V and let r be a value of R. Then $V_m(r)$ denotes the counterfactual value of V_m when R is set to r. We number the variables in V so that for $j < I$, V_j is not a descendent of V_i.

An NPSEM represented by a DAG G with vertex set V assumes the existence of mutually independent unobserved random variables (errors) ϵ_m and deterministic unknown functions $f_m(pa_m, \epsilon_m)$ such that $V_1 = f_1(\epsilon_1)$ and the one-step ahead counterfactual $V_m(\overline{v}_{m-1}) \equiv V_m(pa_m)$ is given by $f_m(pa_m, \epsilon_m)$, and both V_m and the counterfactuals $V_m(r)$ for any $R \subset V$ are obtained recursively from V_1 and the $V_m(\overline{v}_{m-1})$, $m > 1$. For example, $V_3(v_1) = V_3\{v_1, V_2(v_1)\}$ and $V_3 = V_3\{V_1, V_2(V_1)\}$.

In Figure 23.1b, $Y_a = V_4(v_3) = f_4(V_1, V_2, v_3, \epsilon_4) = f_Y(U, L, a, \epsilon_Y)$ where we define $f_Y = f_4$, $\epsilon_Y = \epsilon_4$, since $Y = V_4$. A DAG G representing an NPSEM is a causal DAG for which the CMA holds because the independence of the error terms ϵ_m both implies the CMA holds and is essentially equivalent to the requirement that all common causes of any variables on the graph are themselves on the causal DAG. Although an NPSEM is a causal DAG, not all causal DAG models are NPSEMs. Indeed, as mentioned above, a causal DAG model makes no reference to and is agnostic about the existence of counterfactuals. In the main body of this chapter, we use the term "causal DAG" to mean a causal DAG representing an NPSEM. All the results for NPSEMs described in this chapter actually hold under the slightly weaker assumptions encoded in the fully randomized, causally interpreted structured tree graph (FRCISTG) models of Robins (1986; see Robins, 2003b, for an additional discussion). All NPSEMs are FRCISTGs, but not all FRCISTGs are NPSEMs.

A graphical condition for g-identifiability. We have the following theorem and corollary given by Robins (1997b) generalizing Pearl and Robins (1995). For consistency with the original source, these are presented using the alternative notation that represents, for example, $A(t)$ as A_t, as described at the beginning of Section 23.3 and used in Chapter 20 and Chapter 22.

Theorem: Given a DAG G whose vertex set consists of the random variables that are elements of the random vectors $Y, X_m, A_m, m = 0, \dots, K$, whose edges are consistent with the (partial) ordering $X_0 A_0 X_1 A_1 \dots . X_K A_K Y$, in the sense that the earlier variables in the ordering are non-descendants of later variables, suppose $X_m = (L_m, U_m)$. We observe Y and, for each m, (A_m, L_m). The U_m are unobserved. Let L_m^* denote an arbitrary element (vertex) in the set of vertices L_m. Consider a set of functions $g = \{g_m; m = 0, \dots, K\}$ where g_m has domain the support $\overline{\mathcal{L}}_m$ of $\overline{L}_m$ and range the support $\mathcal{A}_m$ of A_m. We say that A_m is g-unconnected to a node $L_k^*, k \le m$, if $g_m(\overline{l}_m^{(1)}) = g_m(\overline{l}_m^{(2)})$ whenever $\overline{l}_m^{(1)}$ and $\overline{l}_m^{(2)}$ differ only in l_k^*. Otherwise, A_m is g-connected to $L_k^*, k \le m$. Define the g-formula $b_{g,l}$ and $b_{g,x}$ for g based on l and x, respectively, to be

$$b_{g,l} = \sum_{\overline{l}} E\{Y | \overline{A} = \overline{g}(\overline{l}), \overline{L} = \overline{l}\} \prod_{k=0}^{K} f\{l_k | \overline{A}_{k-1} = \overline{g}_{k-1}(\overline{l}_{k-1}), \overline{L}_{k-1} = \overline{l}_{k-1}\}$$

$$b_{g,x} = \sum_{\overline{x} = (\overline{l}, \overline{u})} E\{Y | \overline{A} = \overline{g}(\overline{l}), \overline{X} = \overline{x}\} \prod_{k=0}^{K} f\{x_k | \overline{A}_{k-1} = \overline{g}_{k-1}(\overline{l}_{k-1}), \overline{X}_{k-1} = \overline{x}_{k-1}\},$$

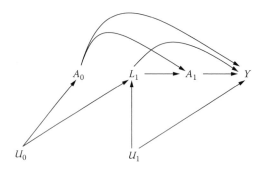

Figure 23.5 Example of a causal DAG.

and note $b_{g,l}$ is a function of the joint distribution of the observables and thus always identified assuming positivity.

(i) A sufficient condition for $b_{g,x} = b_{g,l}$ is that for $m = 0, \dots, K$, A_m and Y are d-separated given $(\overline{L}_m, \overline{A}_{m-1})$ on the DAG G^g_m built from DAG G from the following three rules:

    1. Remove all arrows out of A_m on DAG G.

    2. Remove all arrows into $A_{m+1}, \dots, A_K$ on DAG G.

    3. For $s = m + 1, \dots, K$, add arrows from L^*_j to A_s if A_s is g-connected to $L^*_j, j \leq s$.

(ii) If for $m = 0, \dots, K$, A_m and Y are d-separated given $(\overline{L}_m, \overline{A}_{m-1})$ on the DAG G^g_m and $Y_g \amalg A_m | \overline{X}_m, \overline{A}_{m-1}$, then $Y_g \amalg A_m | \overline{L}_m, \overline{A}_{m-1}$.

Corollary: If the DAG G in the above theorem is a causal DAG representing an NPSEM or FRCISTG, and A_m and Y are d-separated given $(\overline{L}_m, \overline{A}_{m-1})$ on the DAG G^g_m for all m, then $Y_g \amalg A_m | \overline{L}_m, \overline{A}_{m-1}$, and $E(Y_g)$ equals $b_{g,l}$ and so is identified by the g-formula based on the observed data.

Proof: Robins (1986) proved that $E(Y_g) = b_{g,x}$ and $Y_g \amalg A_m | \overline{X}_m, \overline{A}_{m-1}$ if G represents an FRCISTG. Furthermore, Robins (1995) proved an NPSEM is an FRCISTG. The corollary now follows from the preceding theorem.

As an example, consider the DAG G in Figure 23.5. We shall consider the non-dynamic regime $g = \overline{a} = (a_0, a_1)$ and the dynamic regimes $g = \{a_0, g_1(l_1) = l_1)$ and $g = \{a_0, g_1(l_1) = 1 - l_1\}$.

Note that, for the regime $g = \overline{a}$, we have $g_1(l_1) = a_1$. Hence, A_1 is g-unconnected to a node L_1 for $g = \overline{a}$, but A_1 is g-connected to node L_1 for the two dynamic g.

Now the graph G^g_1 is G with the arrow out of A_1 removed for all three g. Further, A_1 and Y are d-separated given (A_0, L_1) on G^g_1. For g non-dynamic, G^g_0 is G with all arrows out of A_0 removed and all arrows into A_1 removed; moreover, A_0 and Y are d-separated on G^g_0. For g dynamic, G^g_0 is G with all arrows out of A_0 removed and a single arrow into A_1 originating at L_1; therefore, A_0 and Y are not d-separated on G^g_0 because of the unblocked path $A_0 \longleftarrow U_0 \longrightarrow L_1 \longleftarrow A_1 \longrightarrow Y$. We conclude that if DAG G in Figure 23.5 represents an NPSEM, then, in the absence of data on U_0 and U_1, $E(Y_{g=\overline{a}})$ is identified and equals $b_{g=\overline{a},l}$. However, for g dynamic, $E(Y_g)$ does not equal $b_{g,l}$, and in fact $E(Y_g)$ is not identified.

Acknowledgments

We thank Dr. Sally Picciotto for her helpful comments. This work was supported by U.S. National Institutes of Health grants R01-AI32475 and R01-HL080644.

References

Bang, H. and Robins, J. M. (2005). Doubly robust estimation in missing data and causal inference models. *Biometrics* **61**, 692–972.

Brumback, B. A., Hernán, M. A., Haneuse, S. J. P. A., and Robins, J. M. (2004). Sensitivity analyses for unmeasured confounding assuming a marginal structural model for repeated measures. *Statistics in Medicine* **23**, 749–767.

Hernán, M. A. (2005). Invited commentary: Hypothetical interventions to define causal effects: afterthought or prerequisite? *American Journal of Epidemiology* **162**, 618–620.

Hernán, M. A., Brumback, B., and Robins, J. M. (2001). Marginal structural models to estimate the joint causal effect of nonrandomized treatments. *Journal of the American Statistical Association* **96**, 440–448.

Hernán, M. A., Brumback B., and Robins, J. M. (2002). Estimating the causal effect of zidovudine on CD4 count with a marginal structural model for repeated measures. *Statistics in Medicine* **21**, 1689–1709.

Hernán, M. A., Hernández-Díaz, S., and Robins, J. M. (2004). A structural approach to selection bias. *Epidemiology* **15**, 615–625.

Hernán, M. A., and Robins, J. M. (2006a). Estimating causal effects from epidemiological data. *Journal of Epidemiology and Community Health* **60**, 578–586.

Hernán, M. A., and Robins, J. M. (2006b). Instruments for causal inference: An epidemiologist's dream? *Epidemiology* **17**, 360–372.

Hernán, M. A., Lanoy, E., Costagliola, D., and Robins, J. M. (2006). Comparison of dynamic treatment regimes via inverse probability weighting. *Basic & Clinical Pharmacology & Toxicology* **98**, 237–242.

Murphy, S. A. (2003). Optimal dynamic treatment regimes. *Journal of the Royal Statistical Society, Series B* **65**, 331–366.

Murphy, S., van der Laan, M. J., and Robins, J. M. (2001). Marginal mean models for dynamic regimes. *Journal of the American Statistical Association* **96**, 1410–1423.

Orellana, L., Rotnitzky, A., and Robins, J. M. (2006). Generalized marginal structural models for estimating optimal treatment regimes. Technical Report, Department of Biostatistics, Harvard School of Public Health.

Page J., Hernán, M. A., and Robins, J. M. (2005). Doubly robust estimation: Structural nested cumulative failure time models, correction of the diagnostic likelihood ratio for verification. Technical Report, Harvard School of Public Health.

Pearl, J. (1995). Causal diagrams for empirical research. *Biometrika* **82**, 669–710.

Pearl, J. and Robins, J. M. (1995). Probabilistic evaluation of sequential plans from causal models with hidden variables. *Uncertainty in Artificial Intelligence, Proceedings of the 11th Conference*, pp. 444–453.

Robins, J. M. (1986). A new approach to causal inference in mortality studies with sustained exposure periods — Application to control of the healthy worker survivor effect. *Mathematical Modelling* **7**, 1393–1512. [Errata (1987) in *Computers and Mathematics with Applications* **14**, 917–921. Addendum (1987) in *Computers and Mathematics with Applications* **14**, 923–945. Errata (1987) to addendum in *Computers and Mathematics with Applications* **18**, 477.]

Robins, J. M. (1993). Analytic methods for estimating HIV treatment and cofactor effects. In D. G. Ostrow and R. Kessler (eds.), *Methodological Issues of AIDS Mental Health Research*, pp. 213–290. New York: Plenum.

Robins, J. M. (1994). Correcting for non-compliance in randomized trials using structural nested mean models. *Communications in Statistics* **23**, 2379–2412.

Robins, J. M. (1997a). Structural nested failure time models. In P. Armitage and T. Colton (eds.), *Encyclopedia of Biostatistics*, pp. 4372–4389. Chichester: Wiley.

Robins, J. M. (1997b). Causal inference from complex longitudinal data. In M. Berkane (ed.), *Latent Variable Modeling and Applications to Causality*, Lecture Notes in Statistics 120, pp. 69–117. New York: Springer.

Robins, J. M. (1998). Marginal structural models. In *1997 Proceedings of the Section on Bayesian Statistical Science*, pp. 1–10. Alexandria, VA: American Statistical Association.

Robins, J. M. (1999). Marginal structural models versus structural nested models as tools for causal inference. In M. E. Halloran and D. Berry (eds.), *Statistical Models in Epidemiology: The Environment and Clinical Trials*, pp. 95–134. New York: Springer-Verlag.

Robins, J. M. (2003a). General methodological considerations. *Journal of Econometrics* **112**, 89–106.

Robins, J. M. (2003b). Semantics of causal DAG models and the identification of direct and indirect effects. In P. Green, N. Hjort, and S. Richardson (eds.), *Highly Structured Stochastic Systems*. Oxford University Press.

Robins, J. M. (2004). Optimal structural nested models for optimal sequential decisions. In D. Y. Lin and P. Heagerty (eds.), *Proceedings of the Second Seattle Symposium on Biostatistics*. New York: Springer.

Robins, J. M. and Greenland, S. (2000). Comment on "Causal inference without counterfactuals" by A. P. Dawid. *Journal of the American Statistical Association* **95**, 477–482.

Robins, J. M., Hernán, M. A., and Brumback, B. (2000). Marginal structural models and causal inference in epidemiology. *Epidemiology* **11**, 550–560.

Robins, J. M., Orellana, L., Hernán, M. A., and Rotnitzky, A. (2008a). Estimation and extrapolation of optimal treatment and testing strategies. *Statistics in Medicine*. To appear.

Robins, J. M., Rotnitzky, A., and Scharfstein, D. (1999). Sensitivity analysis for selection bias and unmeasured confounding in missing data and causal inference models. In M. E. Halloran and D. Berry (eds.), *Statistical Models in Epidemiology: The Environment and Clinical Trials*, pp. 1–92. New York: Springer.

Robins, J. M., Sued, M., Lei-Gomez, Q., and Rotnitzky, A. (2008b). Comment: Performance of double-robust estimators when "inverse probability" weight are height variable. *Statistical Science* (in press).

Robins, J. M., Rotnitzky, A., and Zhao, L.-P. (1995). Analysis of semiparametric regression models for repeated outcomes in the presence of missing data. *Journal of the American Statistical Association* **90**, 106–121.

Robins, J. M. and Wasserman, L. (1997). Estimation of effects of sequential treatments by reparameterizing directed acyclic graphs. In D. Geiger and P. Shenoy (eds.), *Proceedings of the Thirteenth Conference on Uncertainty in Artificial Intelligence*, pp. 409–420. San Francisco: Morgan Kaufmann.

Rosenbaum, P. R. (1984). The consequences of adjustment for a concomitant variable that has been affected by the treatment. *Journal of the Royal Statistical Society, Series A* **147**, 656–666.

Rosenbaum, P. R. and Rubin, D. B. (1983). The central role of the propensity score in observational studies for causal effects. *Biometrika* **70**, 41–55.

Scharfstein, D. O., Rotnitzky, A., and Robins, J. M. (1999). Adjusting for non-ignorable drop-out using semiparametric non-response models. *Journal of the American Statistical Association* **94**, 1096–1120.

Spirtes, P., Glymour, C., and Scheines, R. (1993). *Causation, Prediction, and Search*, Lecture Notes in Statistics 81. New York: Springer.

van der Laan, M. J. (2006). Causal effect models for intention to treat and realistic individualized treatment rules. Working Paper 203, UC Berkeley Division of Biostatistics Working Paper Series. http://www.bepress.com/ucbbiostat/paper203.

van der Laan, M. J. and Robins, J. M. (2003). *Unified Methods for Censored Longitudinal Data and Causality*. New York: Springer.

van der Laan, M. J., Hubbard, A., and Jewell, N. P. (2007). Estimation of treatment effects in randomized trials with non-compliance and a dichotomous outcome. *Journal of the Royal Statistical Society, Series B* **69**, 463–482.

van der Laan, M. J., and Petersen, M. L. (2007). Causal effect models for realistic individualized treatment and intention to treat rules. *International Journal of Biostatistics* **3**(1), Article 3.

Wang, Y., Petersen, M. L., Bangsberg, D., and van der Laan, M. J. (2006). Diagnosing bias in the inverse probability of treatment weighted estimator resulting from violation of experimental treatment assignment. Working Paper 211, UC Berkeley Division of Biostatistics Working Paper Series. http://www.bepress.com/ucbbiostat/paper211.

Author Index

A

Abramowitz, M., 447
Abrams, K. R., 363
Abrams, M., 144
Acock, A., 378
Adams, B. M., 115
Aerts, M., 17, 328, 333, 337, 347, 503, 531, 533
Agresti, A., 13, 45, 74, 377
Airy, G. B., 4, 8
Aitkin, M., 18, 80, 152, 158, 437
Albert, P. S., 168, 183, 223, 424, 437, 439, 442, 443, 444, 445, 448, 449, 450, 495
Alfo, M., 437
Ali, A., 351, 355
Allen, D. M., 5, 7, 275
Allison, P. D., 546
Alonso, A., 39
Altham, P. M. E., 12, 363
Amara, I. A., 97
Andersen, P. K., 351, 353
Anderson, D. A., 18, 80, 152
Anderson, G. L., 10, 53, 200
Anderson, J. A., 492
Anderson, R. D., 80
Anderson, T. W., 20, 413
Andres, R., 367
Angrist, J. D., 148
Ansley, C. F., 258
Anthony, J. C., 359
Apanasovich, T. V., 203, 207, 208
Arenberg, D., 367
Arnold, B. C., 337, 338
Ash, R. B., 234
Ashford, J. R., 12, 17, 18
Ashrafian, H., 351, 355
Asparouhov, T., 143, 148, 151, 152
Ayuso-Mateos, J. L., 148
Azzalini, A., 74, 226

B

Bahadur, R. R., 12, 45, 67, 74
Bailey, K. R., 402, 403, 427, 439, 509
Baker, S. G., 401, 505, 506, 508, 525
Balagtas, C. C., 13, 45, 74

Baltagi, B. H., 100
Bandeen-Roche, K., 153
Bang, H., 61, 471, 472, 564
Bangsberg, D., 592
Banks, H. T., 115
Barnhart, H. X., 62
Baron, A., 377, 378, 379
Barreto, M. L., 359
Bates, D. M., 126, 127, 128, 130, 135, 168, 210, 292, 415
Bauer, D. J., 153
Beal, S. L., 125, 126–127, 131, 132
Bebchuk, J. D., 62, 63
Beck, G. J., 51, 437
Becker, M. P., 13, 45, 74
Beckett, L. A., 376
Belin, T. R., 486
Bellman, R. E., 14
Belsley, D. A., 62
Bennett, J. E., 129
Bergsma, W. P., 45, 74
Berhane, K., 291
Berzuini, C., 357
Best, N. G., 95, 129
Beunckens, C., 502, 513, 514, 546
Bickel, P. J., 454
Bijnens, L., 546
Billingham, L. J., 363
Billingsley, P., 20
Birmingham, J., 416, 471
Bishop, Y. M. M., 50
Blettner, M., 400, 503, 531
Bock, R. D., 151
Boeckmann, A. J., 126–127
Bois, F. Y., 114, 129
Boisvieux, J. F., 125
Bollen, K. A., 379
Bonetti, M., 402
Boneva, L. I., 255
Boos, D. D., 55
Booth, J. G., 19
Borgan, Ø., 351, 353, 359
Borowitz, D., 293, 312
Boscardin, K., 152
Boshuizen, H., 487
Bosker, R. J., 84
Bouvier, A., 415

Subject Index